W9-ARI-668

5.3 Rules for Radicals

Suppose $\sqrt[n]{a}$ and $\sqrt[n]{b}$ are real numbers. Then $\sqrt[n]{ab} = \sqrt[n]{a} \cdot \sqrt[n]{b}$ and $\sqrt[n]{\dfrac{a}{b}} = \dfrac{\sqrt[n]{a}}{\sqrt[n]{b}}$ $(b \neq 0)$.

5.9 Complex numbers $= \{a + bi \mid a$ and b are real numbers$\}$

Imaginary numbers $= \{a + bi \mid a$ and b are real numbers and $b \neq 0\}$

Pure imaginary numbers $= \{bi \mid b$ is a nonzero real number$\}$

Powers of i: $i^2 = -1$, $i^3 = -i$, $i^4 = 1$

6.2 Quadratic Formula

The solutions of the quadratic equation $ax^2 + bx + c = 0$ are $x = \dfrac{-b \pm \sqrt{b^2 - 4ac}}{2a}$.

6.3 The Discriminant

The type of solutions of the quadratic equation $ax^2 + bx + c = 0$, where a, b, and c are integers, can be determined from the discriminant, $b^2 - 4ac$, as follows:

1. $b^2 - 4ac$ is positive and a perfect square (two rational solutions).
2. $b^2 - 4ac$ is positive but not a perfect square (two irrational solutions).
3. $b^2 - 4ac$ is 0 (one rational solution).
4. $b^2 - 4ac$ is negative (two imaginary solutions).

Sum and Product of the Solutions of $ax^2 + bx + c = 0$ $(a \neq 0)$

$x_1 + x_2 = -\dfrac{b}{a}$ and $x_1 \cdot x_2 = \dfrac{c}{a}$

6.5 The Pythagorean Theorem

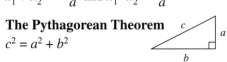

$c^2 = a^2 + b^2$

7.1 Distance Formula

The distance between (x_1, y_1) and (x_2, y_2) is $d = \sqrt{(x_2 - x_1)^2 + (y_2 - y_1)^2}$.

Midpoint Formula

The midpoint of the line segment joining (x_1, y_1) and (x_2, y_2) is $M\left(\dfrac{x_1 + x_2}{2}, \dfrac{y_1 + y_2}{2}\right)$.

7.3 Slope

The slope of the nonvertical line through (x_1, y_1) and (x_2, y_2) is $m = \dfrac{y_2 - y_1}{x_2 - x_1}$.

7.4 Forms of a Linear Equation

Point-slope form: $y - y_1 = m(x - x_1)$ *Slope-intercept form:* $y = mx + b$

Standard form: $ax + by = c$ *Horizontal line:* $y = k$ *Vertical line:* $x = k$

Parallel and Perpendicular Lines

The slopes of parallel lines are equal. The slopes of perpendicular lines are negative reciprocals.

7.6 Variation (k is the constant of variation)

Direct: $y = kx^n$ *Inverse:* $y = \dfrac{k}{x^n}$ *Joint:* $y = kxz$

8.3 Vertex of a Parabola

The vertex of the parabola $y = ax^2 + bx + c$ occurs at $x = \dfrac{-b}{2a}$.

8.4 Standard Form of a Vertical Parabola

The graph of $y = a(x - h)^2 + k$ $(a \neq 0)$ is a vertical parabola with vertex at (h, k). The parabola opens up if $a > 0$ and down if $a < 0$.

Standard Form of a Horizontal Parabola

The graph of $x = a(y - k)^2 + h$ $(a \neq 0)$ is a horizontal parabola with vertex at (h, k). The parabola opens right if $a > 0$ and left if $a < 0$.

8.5 Standard Form of a Circle

An equation of the circle with center at (h, k) and radius r is $(x - h)^2 + (y - k)^2 = r^2$.

ALGEBRA

FOR COLLEGE
STUDENTS

Technical Mathematics
Tensor Calculus
Trigonometry
Vector Analysis

SCHAUM'S SOLVED PROBLEMS SERIES

Each title in this series is a complete and expert source of solved problems with solutions worked out in step-by-step detail.

Titles on the current list include:

3000 Solved Problems in Calculus
2500 Solved Problems in Differential Equations
2000 Solved Problems in Discrete Mathematics
3000 Solved Problems in Linear Algebra
2000 Solved Problems in Numerical Analysis
3000 Solved Problems in Precalculus

BOB MILLER'S MATH HELPERS

Bob Miller's Calc I Helper
Bob Miller's Calc II Helper
Bob Miller's Precalc Helper

McGRAW-HILL PAPERBACKS

Arithmetic and Algebra . . . Again
How to Solve Word Problems in Algebra
Mind Over Math

Available at most college bookstores, or for a complete list of titles and prices, write to:

Schaum Division
The McGraw-Hill Companies, Inc.
11 West 19th Street
New York, NY 10011

ALGEBRA

FOR COLLEGE STUDENTS

Daniel L. Auvil

Kent State University

The McGraw-Hill Companies, Inc.
New York St. Louis San Francisco Auckland Bogotá Caracas Lisbon
London Madrid Mexico City Milan Montreal New Delhi
San Juan Singapore Sydney Tokyo Toronto

McGraw-Hill

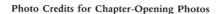

A Division of The **McGraw·Hill** *Companies*

Photo Credits for Chapter-Opening Photos

Benn Mitchell/The Image Bank
Nita Winter/The Image Works
George Goodwin/Monkmeyer
Lionel J-M Delevigne/Stock, Boston
Fredrick D. Bodin/Stock, Boston
Mitch Wojnarowicz/Image Works

R. Matusow/Monkmeyer
Ursula Markus/Photo Researchers
W. Marc Bernsau/The Image Works
AP/Wide World
Gary S. Weber/Photo Researchers
Richard Megna/Fundamental Photographs

Algebra for College Students

This book is printed on acid-free paper.

4   5   6   7   8   9   0   DOW   DOW   9   0   9

ISBN 0-07-003106-1

This book was set in Times Roman by GTS Graphics.
The editors were Michael Johnson, Karen M. Minette, Margery Luhrs, and Ty McConnell;
the design was done by Silvers Design;
the production supervisor was Richard A. Ausburn.
The photo editor was Kathy Bendo;
the photo researcher was Mia Galison.
R. R. Donnelley & Sons Company was printer and binder.

Library of Congress Cataloging-in-Publication Data

Auvil, Daniel L.
    Algebra for college students / Daniel L. Auvil.
        p.   cm.
    Includes indexes.
    ISBN 0-07-003106-1
    1. Algebra.   I. Title.
    QA154.2.A945   1996
    512.9—dc20   95-24774

CONTENTS

Preface   xi

1 THE REAL NUMBER SYSTEM   1

1.1   Sets of Numbers   2

1.2   The Number Line; Absolute Value;
      Inequalities   11

1.3   Properties of Real Numbers   19

1.4   Operations on Real Numbers   26

1.5   Exponents and Order of Operations   35
      Chapter 1 Key Terms   40
      Chapter 1 Key Rules/Steps   40
      Chapter 1 Review Exercises   43
      Chapter 1 Test   44

**2 LINEAR EQUATIONS AND
   INEQUALITIES   45**

2.1   Linear Equations in One Variable   46

2.2   Problem Solving with Formulas   54

2.3   Number, Percent, and Geometry Problems   60

2.4   Money, Mixture, and Motion Problems   68

2.5   Linear Inequalities in One Variable   73

2.6   Compound Inequalities   82

2.7   Absolute-Value Equations and Inequalities   87
      Chapter 2 Key Terms   93
      Chapter 2 Key Rules/Steps   94
      Chapter 2 Review Exercises   96
      Chapter 2 Test   97

3 EXPONENTS AND POLYNOMIALS   99

3.1   Integer Exponents   100

3.2   Rules for Exponents   107

3.3   Scientific Notation: An Application of
      Exponents   112

3.4   Adding and Subtracting Polynomials   116

3.5   Multiplying Polynomials   123

3.6   The Greatest Common Factor; Factoring
      by Grouping   130

3.7   Factoring Trinomials   136

3.8   Special Factorizations   144

3.9   Factoring Strategy   149

3.10 Solving Equations by Factoring   151
      Chapter 3 Key Terms   158
      Chapter 3 Key Rules/Steps   159
      Chapter 3 Review Exercises   161
      Chapter 3 Test   162
      Cumulative Test for Chapters 1, 2, and 3   162

4 RATIONAL EXPRESSIONS   165

4.1   Simplifying Rational Expressions   166

4.2   Multiplying and Dividing Rational
      Expressions   171

4.3   Adding and Subtracting Rational
      Expressions   176

4.4   Complex Fractions   184

4.5   Equations with Rational Expressions   189

4.6   Problem Solving with Rational
      Expressions   194

4.7   Dividing Polynomials   200
      Chapter 4 Key Terms   206
      Chapter 4 Key Rules/Steps   207
      Chapter 4 Review Exercises   209
      Chapter 4 Test   210

5 RADICALS, RATIONAL EXPONENTS, AND COMPLEX NUMBERS   213

5.1 Finding Roots   214
5.2 Rational Exponents   219
5.3 Simplifying Radical Expressions   224
5.4 Multiplying and Dividing Radical Expressions   230
5.5 Adding and Subtracting Radical Expressions   234
5.6 Combinations of Operations   237
5.7 Rationalizing the Denominator   239
5.8 Radical Equations   244
5.9 Complex Numbers   249
     Chapter 5 Key Terms   256
     Chapter 5 Key Rules/Steps   257
     Chapter 5 Review Exercises   258
     Chapter 5 Test   260

6 QUADRATIC EQUATIONS AND INEQUALITIES   261

6.1 Solving Quadratic Equations by Completing the Square   262
6.2 The Quadratic Formula   268
6.3 More about Quadratic Equations   274
6.4 Equations That Lead to Quadratic Equations   280
6.5 Problem Solving with Quadratic Equations   285
6.6 Nonlinear Inequalities   292
     Chapter 6 Key Terms   300
     Chapter 6 Key Rules/Steps   300
     Chapter 6 Review Exercises   302
     Chapter 6 Test   303
     Cumulative Test for Chapters 4, 5, and 6   304

7 LINEAR EQUATIONS AND INEQUALITIES IN TWO VARIABLES   305

7.1 The Rectangular Coordinate System   306
7.2 Graphing Linear Equations   313
7.3 The Slope of a Line   320
7.4 Forms of a Linear Equation   326
7.5 Graphing Linear Inequalities   335
7.6 Variation   341
     Chapter 7 Key Terms   347
     Chapter 7 Key Rules/Steps   347
     Chapter 7 Review Exercises   349
     Chapter 7 Test   350

8 RELATIONS AND FUNCTIONS   353

8.1 Defining Relations and Functions   354
8.2 Function Notation and Operations on Functions   360
8.3 Constant, Linear, and Quadratic Functions   369
8.4 More about Parabolas   379
8.5 The Circle and the Ellipse   389
8.6 The Hyperbola   397
8.7 Special Functions   406
8.8 Increasing/Decreasing Functions; Symmetry   415
     Chapter 8 Key Terms   419
     Chapter 8 Key Rules/Steps   420
     Chapter 8 Review Exercises   424
     Chapter 8 Test   426

9 POLYNOMIAL AND RATIONAL FUNCTIONS   427

9.1 Polynomial Functions and Their Graphs   428
9.2 Synthetic Division; Remainder and Factor Theorems   437
9.3 The Rational Root Test   443
9.4 Zeros of a Function   449
9.5 Rational Functions and Their Graphs   456
     Chapter 9 Key Terms   467
     Chapter 9 Key Rules/Steps   468
     Chapter 9 Review Exercises   471
     Chapter 9 Test   472
     Cumulative Test for Chapters 7, 8, and 9   472

10 EXPONENTIAL AND LOGARITHMIC FUNCTIONS   475

10.1 Exponential Functions   476
10.2 Inverse of a Function   482
10.3 Logarithmic Functions   489
10.4 Properties of Logarithms   495

10.5   Common Logarithms; Natural Logarithms;
         Change of Base   500

10.6   Exponential and Logarithmic Equations   507
         Chapter 10 Key Terms   514
         Chapter 10 Key Rules/Steps   514
         Chapter 10 Review Exercises   516
         Chapter 10 Test   518

11 SYSTEMS OF EQUATIONS AND INEQUALITIES   519

11.1   The Graphing Method   520
11.2   The Substitution Method and the Addition
         Method   525
11.3   Problem Solving with Linear Systems   530
11.4   Linear Systems of Three Equations   535
11.5   Determinants   542
11.6   Cramer's Rule   546
11.7   Solving Linear Systems by Matrices   551
11.8   Nonlinear Systems of Equations   559
11.9   Second-Degree Inequalities; Systems
         of Inequalities   564
11.10   Linear Programming   568
         Chapter 11 Key Terms   574
         Chapter 11 Key Rules/Steps   575
         Chapter 11 Review Exercises   579
         Chapter 11 Test   581

12 SEQUENCES, SERIES, AND PROBABILITY   583

12.1   Defining Sequences and Series   584
12.2   Arithmetic Sequences and Series   589
12.3   Geometric Sequences and Series   593
12.4   Infinite Geometric Series   598
12.5   The Binomial Theorem   601
12.6   Mathematical Induction   607
12.7   Permutations and Combinations   612
12.8   Probability   620
12.9   Compound Probabilities   625
         Chapter 12 Key Terms   633
         Chapter 12 Key Rules/Steps   633
         Chapter 12 Review Exercises   636
         Chapter 12 Test   638
         Cumulative Test for Chapters 10, 11, and
         12   638

ANSWERS   641

APPENDICES   685

Appendix 1:   Abbreviations and Symbols   685
Appendix 2:   Geometric Formulas   689
Appendix 3:   Powers of e 691
Appendix 4:   Common Logarithms   695
Appendix 5:   Natural Logarithms   699

Index   703

PREFACE

Early humans sometimes lived in fear of their environment because they could not explain it. Science provided us with explanations. Today, an understanding of science is even more critical as our society becomes more dependent on technology.

Mathematics is often called the "Queen of the Sciences," because an understanding of mathematics is essential to an understanding of all the other sciences. Basic to the study of mathematics is the study of *algebra,* a generalized form of arithmetic, where variables such as x are used when unknown quantities are involved.

This text presents the topics of *college algebra* at an *intermediate algebra* pace. It is written in a format inspired by the *National Council of Teachers of Mathematics (NCTM) Standards,* AMATYC, and MAA guidelines.

EXERCISE SETS

Mathematics is not a spectator sport—you learn mathematics by doing it. Therefore, the strength of any mathematics text lies in its exercises, and that is particularly true of this text. There are approximately 7,300 exercises in the text, graded in difficulty from simple to challenging, to allow students to gain confidence as they progress through each exercise set. These exercises appear in a wide variety of formats to increase both interest and understanding by encouraging students to think about concepts in a variety of different ways.

Completion Exercises

Completion exercises direct students to fill in a missing word or phrase or a missing symbol or algebraic expression to make a true statement. They encourage understanding of definitions and rules before students actually proceed with solving problems.

Matching Exercises

Matching exercises build association skills by requiring students to recognize patterns and make connections between ideas. For example, a set of matching exercises might ask a student to match a list of equations that appear strikingly similar to a list of graphs that are very dissimilar.

True-False Exercises

True-false exercises point out common errors to avoid when solving problems. More importantly, they discipline students to pay attention to detail and to read mathematics carefully and critically.

Discussion Exercises

Discussion exercises often require deeper thought and insight than the other exercise types. Their goal is to provoke thinking and encourage students to write or talk about mathematics, either individually or in groups, and to think about the reasons behind what they are doing.

Problem-Solving Exercises

Problem-solving exercises help students make connections between mathematics and the real world and learn to use algebra in daily life. They move students beyond skills and toward an understanding of how algebra can be used to perform worthwhile mathematical tasks. They include problems from a wide range of topics, including the physical, social, and life sciences, business and economics, architecture and agriculture, recreation and entertainment, and sports and consumer mathematics.

Getting Acquainted with Your Scientific Calculator

Woven into the exercise sets and complete with keystroke instructions, scientific calculator exercises give students practice in using their calculators and illustrate how technology can be useful in solving problems. These exercises place less emphasis on performing computations and more emphasis on understanding concepts.

A Note about Scientific Calculators

The keystroke instructions that are given in this text apply to many calculators, but they may not apply to your specific calculator. For example, here is how four different calculators may find $\sqrt[3]{8}$ (the cube root of 8).

Calculator 1:   Clear   8   INV y^x 3   =   2
Calculator 2:   Clear   8   2nd y^x 3   =   2
Calculator 3:   Clear   8   Shift x^y 3   =   2
Calculator 4:   Clear   8 $\sqrt[x]{y}$ 3   =   2

You should refer to your owner's manual to see how your particular calculator performs an operation.

Getting Acquainted with Your Graphing Calculator

When appropriate, graphing calculator exercises are included at the end of an exercise set. The intent of these exercises is to enhance learning by enabling students to visualize concepts and to arouse the desire in students to use technology to further explore topics on their own. While the graphing calculator exercises in early chapters (before graphing is discussed in the text) do not involve graphing, they encourage students to become familiar with the keys and menus on their own calculators.

PEDAGOGICAL FEATURES

The style of the text is organized and logical, the writing is clear and direct, and the approach is intuitive and visual. Illustrations are used generously to clarify concepts and stimulate interest. There are more than 1,200 examples, many of the problem-solving variety, carefully sequenced to correspond to the exercise sets. Because spatial reasoning and the ability to visualize a problem is such an important part of mathematics, geometric topics are introduced and reviewed throughout the text. For easy reference, a table of geometric formulas appears in Appendix 2.

Chapter Openers

Each chapter begins with a motivational photo and a caption that describes an applied problem that will be solved later in the chapter. These openers give insight into the usefulness of algebra, emphasize the relevance of mathematics to real life, and let students know that mathematics is an ongoing human activity.

Try Exercise Feature

Placed in the margin near each example is a practice exercise which closely resembles the example. These practice exercises provide immediate reinforcement and allow students to become active participants by attempting similar problems on their own. The practice exercises are taken from the exercise sets and, to aid in identification, are marked with a colored circle when they appear in the exercise sets.

Cautions

These informative statements, most with supporting examples that demonstrate both the correct and the wrong ways to look at a problem, warn students about common mistakes.

Historical Footnotes

Historical material, expressed in brief passages, lend insight into the mathematicians from ancient to modern times who played prominent roles in the development of the mathematics being studied.

Step-by-Step Approach

Whenever a new type of problem is discussed, a box is created listing the steps involved in solving that type of problem. Immediately following this box is an example showing how these steps are used to solve a problem of that type. Students can use these steps until they feel comfortable enough to do that type of problem on their own.

CHAPTER REVIEWS

At the end of each chapter, there is an extensive review containing the following features:

Key Terms

This part of the chapter review redefines all of the terms that were used in the chapter. Each term is listed according to the section in which it was first introduced.

Key Rules/Steps

These pages summarize all of the important rules and steps for solving the various types of exercises found in the chapter. Examples are provided to illustrate the rules and steps being summarized.

Chapter Review Exercises

The chapter review exercises consist of a selection of exercises touching all the major ideas in the chapter. These exercises are keyed to the appropriate section, and they provide students with the opportunity for additional practice, as well as a means to test their understanding of the material in the entire chapter.

Chapter Tests

These sample tests contain questions that are not identified with particular sections. They allow students to measure their readiness to take an actual classroom test over the chapter's material.

ADDITIONAL REVIEWS

Cumulative tests appear after every three chapters. These tests help students judge their mastery of previously covered topics as well as maintain the knowledge base they have developed, prepare for the final exam, and ensure they are prepared for future math and science courses.

Quick Review

The inside covers of the book contain a list of the important definitions and rules covered in the various sections of the book. These pages can be used as a last-minute review before a test or as a handy reference for formulas.

ANSWERS

The answers to the odd-numbered end-of-section exercises (which include the practice exercises), as well as the answers to all of the exercises in the chapter reviews, the chapter tests, and the cumulative tests, are in the back of the book.

SUPPLEMENTS

Student Supplements

A *Student's Solutions Manual* is available through the college bookstore. Prepared by Professor Relja Vulanovic of Kent State University-Stark County Campus, it contains detailed solutions to the odd end-of-section exercises, all of the chapter review exercises, all of the chapter test exercises, and all of the cumulative test exercises. Keystrokes for the TI-81, TI-82, and TI-85 are included in the solutions for all of the graphing calculator exercises.

Mathworks is a self-paced interactive tutorial specifically linked to the text. The Mathworks logo ▣ appears next to each text section for which the tutorial can be used. This tutorial reinforces selected topics and provides unlimited opportunities for students to review concepts and to practice problem solving. It requires virtually *no* computer training and is available for IBM, IBM compatible, and Macintosh computers.

Course *Videotapes* are available for use from instructors. The videotape logo ▭ and reference number appear next to each section for which they can be used.

Instructor Supplements

An *Instructor's Solutions Manual* contains detailed solutions to all of the exercises in the text.

An *Instructor's Resource Manual* contains two forms of multiple-choice and open-ended chapter tests. There are two forms of multiple-choice and open-ended cumulative tests that cover Chapters 1–3, 4–6, 7–9, and 10–12. There are two forms of multiple-choice and open-ended final tests. All of the answers to the tests are provided.

The *Professor's Assistant* is a unique computerized test-generator available to instructors. It is a system that allows the instructor to create tests using both algorithmically generated test questions and those from a standard testbank. This testing system enables the instructor to choose questions either manually, or randomly by section, question type, difficulty level, and other criteria. It is available for IBM, IBM compatible, and Macintosh computers.

A *Printed and Bound Testbank* is also available. It is a hard-copy listing of the questions found in the standard testbank.

For further information about these supplements, please contact your local McGraw-Hill college division sales representative.

ACKNOWLEDGMENTS

The author would like to thank the following people for their contributions to the development of *Algebra for College Students*.

Jean Airington, Navarro College
Carolyn Bernath, Tallahassee Community College
Brian Hayes, Triton College
Bruce Hoelter, Raritan Valley Community College
Daniel Hogan, Hinds Community College
Libby Holt, Florida Community College
Louise Hoover, Clark College
Steven Kahn, Anne Arundel Community College
Jeff Koleno, Lorain County Community College
JoAnn Lewin, Edison Community College
Janice McFatter, Gulf Coast Community College
Lawrence Maher, University of North Texas
Carol Jean Martin, Dodge City Community College
Sunny Norfleet, St. Petersburg Junior College
Calvin Owens, Chowan College
Gwen Parker, Tallahassee Community College
Diane Porter, Troy State University
Bessie Tucker, Jackson State University
Joan Van Glabek, Edison Community College
George Witt, Glendale Community College
Jean Woody, Tulsa Junior College, SE Campus

Thank you to William M. Mays of Gloucester County College for checking the accuracy of the examples, exercises, and answers. Special thanks also go to the staff of McGraw-Hill, including Michael Johnson, Karen Minette, Ty McConnell, and Margery Luhrs.

Daniel L. Auvil

ALGEBRA

FOR COLLEGE
STUDENTS

CHAPTER

1

The Real Number System

Depending on the type of soil in a particular region, it may cost anywhere from $9 to $17 per ft to drill a water well. To hit water, the well must be 130 to 150 ft deep. In Sec. 1.2 you will see how we use a double inequality to express the cost of drilling such a well.

1.1 Sets of Numbers

Numbers are the building blocks of algebra. In this section we provide a brief overview of the various types of numbers used in algebra.

Sets and Elements

A **set** is a collection of objects.* The objects in a set are called **elements,** or **members,** of the set. We use braces, { }, to enclose the elements of a set, and we use commas to separate the elements. For example, the set whose elements are the numbers 1, 2, and 3 is written

$$\{1, 2, 3\}$$

We often use a capital letter to name a set. Therefore we might write

$$A = \{1, 2, 3\} \qquad \text{Read = as "is equal to."}$$

To indicate that 2 is an element of set A, we write

$$2 \in A \qquad \text{Read} \in \text{as "is an element of."}$$

We can also write $2 \in \{1, 2, 3\}$. To indicate that 4 is not an element of A, we write

$$4 \notin A \qquad \text{Read} \notin \text{as "is not an element of."}$$

Two sets are **equal** if they contain exactly the same elements. Therefore,

$$\{1, 2, 3\} = \{3, 1, 2\}$$

Note that **the order in which the elements of a set are listed is unimportant.**

The symbol $\neq$ is read "is not equal to." For example

$$\{1, 2\} \neq \{1, 2, 3\}$$

The set $\{1, 2, 3\}$ is a **finite set,** while the set of **whole numbers,**

$$\{0, 1, 2, 3, 4, 5, \ldots\}$$

is an **infinite set.** The three dots show that the elements continue in the indicated pattern.

The set which contains no elements is called the **empty set,** or **null set.** For example the set of whole numbers between 4 and 5 is the empty set. We use the symbol $\varnothing$ or the symbol { } to represent the empty set.

CAUTION

Here are two correct ways and one incorrect way to write the empty set:

Empty set	*Empty set*	*Not the empty set*
$\varnothing$	{ }	$\{\varnothing\}$ ■

We can identify the elements of a set in two ways. We can *list* the elements of the set, or we can *describe* the elements of the set using a rule. For example, we can write the set of whole numbers between 2 and 7 as $\{3, 4, 5, 6\}$ or as

*HISTORICAL NOTE: Set theory was introduced by Georg Cantor, who was born in Russia in 1845. Cantor moved with his parents to Germany in 1856, studied at the University of Berlin from 1863 to 1869, taught at the University of Halle from 1869 to 1905, and died in a mental hospital in Halle in 1918 (reminding us that sometimes there is a fine line between genius and madness). Cantor was the first to demonstrate that some infinite sets are "larger" than others.

$\{x | x$ is a whole number between 2 and 7$\}$

This last expression is read "the set of all x such that x is a whole number between 2 and 7." This expression is an example of **set-builder notation.** The general form of set-builder notation is shown below:

$$\{x | x \text{ has property } P\}$$

"The set of"   "All elements x"   "Such that"   Rule that describes the elements is written here

The letter x is called a **variable** since it may represent any of the elements in the set.* The value of x is not fixed, but rather, varies. The value of a **constant,** such as 5, remains fixed throughout a particular discussion.

Here are some other examples of sets defined using both the listing method and set-builder notation:

Listing method

$\{0, 2, 4, 6, 8, 10, \ldots\}$
$\{9, 11, 13, 15, 17, \ldots\}$
$\{$red, yellow, blue$\}$

Set-builder notation

$\{x | x$ is an even whole number$\}$
$\{y | y$ is an odd whole number greater than 8$\}$
$\{t | t$ is a primary color$\}$

Try Exercise 45

Write $\{x | x$ is a letter in the word banana$\}$ using the listing method.

Example 1   Write $\{x | x$ is a letter in the word Mississippi$\}$ using the listing method.

Solution: $\{$M, i, s, p$\}$
Note that we do not repeat the elements of a set.   ❑

Example 2   Write $\{0, 1, 2, 3, \ldots, 99\}$ using set-builder notation.

Try Exercise 51

Write $\{0, 1, 2, 3, \ldots, 39\}$ using set-builder notation.

Solution: $\{x | x$ is a whole number less than 100$\}$ ❑

Subsets

Suppose that S is the set of students in your class and W is the set of women students in your class. Then every element of W is also an element of S. We say that W is a *subset* of S.

> **DEFINITION OF SUBSET**
>
> Set A is a **subset** of set B if every element of A is also an element of B.

To denote that A is a subset of B, we write $A \subseteq B$. To denote that A is not a subset of B, we write $A \not\subseteq B$.

Example 3   Insert $\subseteq$ or $\not\subseteq$ in the blank space to make a true statement.

(a) $\{2, 3, 8\}$_____$\{1, 2, 3, 7, 8\}$
(b) $\{0, 5\}$_____$\{3, 4, 5\}$

---

*HISTORICAL NOTE: Our present custom of using letters of the alphabet as variables was introduced by the French mathematician François Viète (1540–1603). Viète was a lawyer and a member of the French Parliament who devoted almost all of his leisure time to mathematics, sometimes isolating himself in his study for days over a mathematical problem. His problem-solving ability enabled him to break the Spanish code, thereby giving France an advantage in its war with Spain.

Solution: (a) Since every element of {2, 3, 8} is also an element of {1, 2, 3, 7, 8}, write

$$\{2, 3, 8\} \subseteq \{1, 2, 3, 7, 8\}$$

(b) Since 0 is an element of {0, 5} and 0 is not an element of {3, 4, 5}, write

$$\{0, 5\} \nsubseteq \{3, 4, 5\}$$ ❑

Try Exercise 57

Insert ⊆ or ⊄ in the blank space to make a true statement:
{4, 6}_____{4, 6, 8}

CAUTION

Do not confuse the symbols ∈ and ⊆. The symbol ∈ is used between an element and a set. The symbol ⊆ is used between two sets.

Correct   *Wrong*

$2 \in \{1, 2, 3\}$ $2 \subseteq \{1, 2, 3\}$

$\{2\} \subseteq \{1, 2, 3\}$ $\{2\} \in \{1, 2, 3\}$ ∎

Note that {0, 5} is *not* a subset of {3, 4, 5}, because {0, 5} contains an element that is not an element of {3, 4, 5}. Since the empty set has no elements, it has no elements that are not in a given set *A*. Therefore **the empty set is a subset of every set.** Also, since every element of a set *A* is also an element of the same set *A*, **every set is a subset of itself.**

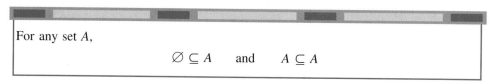

For any set *A*,

$$\varnothing \subseteq A \qquad \text{and} \qquad A \subseteq A$$

Diagrams that depict sets as regions are called **Venn diagrams.*** For example, if *A* = {1, 2, 3} and *B* = {1, 2, 3, 4}, a Venn diagram of *A* and *B* would appear as shown in Fig. 1.1.

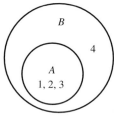

Figure 1.1 A ⊆ B

Set Operations

The process of forming a new set from given sets is called a **set operation.** For example, suppose *F* = {Bob, Sue, Mary} is the set of students who received an A on the first test, and *S* = {Kris, Jim, Mary, Bob} is the set of students who received an A on the second test. Then {Bob, Sue, Mary, Kris, Jim} is the set of students who received an A on either the first test or the second test (or both tests). This set is called the **union** of sets *F* and *S*, written *F* ∪ *S*.

DEFINITION OF SET UNION

For any two sets A and B,

$$A \cup B = \{x \mid x \in A \text{ or } x \in B\}$$

Example 4 Given $A = \{1, 3, 5\}$ and $B = \{2, 3, 4\}$, find $A \cup B$.

<u>**Try Exercise 69**</u>

Given $A = \{1, 2, 4\}$ and $B = \{2, 3, 5, 7\}$, find $A \cup B$.

Solution: $A \cup B = \{1, 3, 5\} \cup \{2, 3, 4\} = \{1, 2, 3, 4, 5\}$ ⃞

In Example 4 the element 3 is listed only once in $A \cup B$ even though 3 is an element of both A and B. A Venn diagram of $A \cup B$ is shown as the shaded portion of Fig. 1.2.

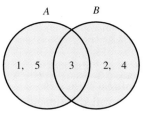

Figure 1.2 $A \cup B$

Returning to our previous example, we note that {Mary, Bob} is the set of students who received an A on both tests. This set is called the **intersection** of sets F and S, written $F \cap S$.

DEFINITION OF SET INTERSECTION

For any two sets A and B,

$$A \cap B = \{x \mid x \in A \text{ and } x \in B\}$$

Example 5 Given $A = \{1, 3, 5\}$ and $B = \{2, 3, 4\}$, find $A \cap B$.

<u>**Try Exercise 71**</u>

Given $B = \{2, 3, 5, 7\}$ and $D = \{1, 2\}$, find $B \cap D$.

Solution: $A \cap B = \{1, 3, 5\} \cap \{2, 3, 4\} = \{3\}$ ⃞

A Venn diagram of $A \cap B$ for Example 5 is shown as the shaded portion of Fig. 1.3.

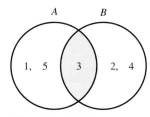

Figure 1.3 $A \cap B$

Example 6 Given $A = \{1, 3, 5\}$, $B = \{2, 3, 4\}$, and $C = \{1, 2, 3\}$, find:

(a) $(A \cap B) \cup C$
(b) $A \cap (B \cup C)$

Solution: (a) Perform the operation in parentheses first.

$$(A \cap B) \cup C = \{3\} \cup C = \{1, 2, 3\}$$

(b) Perform the operation in parentheses first.

$$A \cap (B \cup C) = A \cap \{1, 2, 3, 4\} = \{1, 3\}$$

Note that $(A \cap B) \cup C \neq A \cap (B \cup C)$. ❏

Try Exercise 77

Given $A = \{1, 2, 4\}$, $B = \{2, 3, 5, 7\}$, and $C = \{6, 3, 5, 8\}$, find $A \cup (B \cap C)$.

Real Numbers

Here is a summary of the various sets of numbers we will use throughout the text. You should be able to identify them since many rules of algebra are true for certain kinds of numbers and false for others. We begin by defining the set of **real numbers** as the set of all numbers that can be expressed as decimals. All of the other sets of numbers are subsets of the set of real numbers.*

Sets of Numbers	
Natural numbers (also called **counting numbers** or **positive integers**)	$\{1, 2, 3, 4, 5, 6, \ldots\}$
Whole numbers	$\{0, 1, 2, 3, 4, 5, \ldots\}$
Integers	$\{\ldots, -3, -2, -1, 0, 1, 2, 3, \ldots\}$
Rational numbers	$\left\{\dfrac{a}{b} \;\middle|\; a \text{ and } b \text{ are integers and } b \neq 0\right\}$
Irrational numbers	$\{x \mid x \text{ is a real number that is not rational}\}$
Real numbers	$\{x \mid x \text{ can be expressed as a decimal}\}$

We specify that $b \neq 0$ in the rational number $\frac{a}{b}$ because **division by 0 is undefined.** (Note that if $\frac{5}{0} = n$, then $0 \cdot n = 5$. Since $0 \cdot n = 0$ for all real numbers n, there is no solution to $0 \cdot n = 5$, and $\frac{5}{0}$ is undefined.)

Note that **every natural number is also a whole number,** and **every whole number is also an integer.** Moreover, **every integer is also a rational number** since every integer can be written in the form $\frac{a}{b}$ with $b = 1$. For example $6 = \frac{6}{1}$, $-2 = \frac{-2}{1}$, and $0 = \frac{0}{1}$. Examples of rational numbers that are not integers are $\frac{1}{4}$, $\frac{2}{3}$, and $\frac{31}{11}$.

It can be shown that the decimal form of a rational number either terminates or repeats.

6	Terminating decimal
$\frac{1}{4} = 0.25$	Terminating decimal (divide 1 by 4)
$\frac{2}{3} = 0.666 \cdots$	Repeating decimal (also written $0.\overline{6}$)
$\frac{31}{11} = 2.818181 \cdots$	Repeating decimal (also written $2.\overline{81}$)

This gives us another way of defining the set of rational numbers:

---

*HISTORICAL NOTE: The German mathematician Richard Dedekind (1831–1916) was one of the first to give a precise definition of real numbers. Dedekind utilized what are now known as *Dedekind cuts* to define real numbers.

Rational numbers = $\{x \mid x$ can be expressed as a terminating or a repeating decimal$\}$

Since irrational numbers are real numbers that are not rational, we can define the set of irrational numbers as follows:

Irrational numbers = $\{x \mid x$ can be expressed as a decimal that neither terminates nor repeats$\}$

An example of a decimal that neither terminates nor repeats is

$$5.717117111711117 \cdots$$

Many square roots result in decimals that neither terminate nor repeat. For example,

$$\sqrt{2} = 1.414213562 \cdots \quad \text{Neither terminates nor repeats}$$

Therefore $\sqrt{2}$ is an irrational number, and we can only approximate its value as a decimal.*

$$\sqrt{2} \approx 1.414 \quad \text{Read} \approx \text{as ``is approximately equal to.''}$$

Of course not all square roots are irrational numbers. The square roots $\sqrt{4}$, $\sqrt{9}$, and $\sqrt{36}$ are rational numbers, since $\sqrt{4} = 2$, $\sqrt{9} = 3$, and $\sqrt{36} = 6$. The numbers 4, 9, and 36 are called **perfect squares** since they are natural numbers whose square roots are natural numbers. The first 12 perfect square numbers are listed below:

Perfect squares: 1, 4, 9, 16, 25, 36, 49, 64, 81, 100, 121, 144

The square roots of natural numbers that are not perfect squares are irrational numbers. For example $\sqrt{3}$, $\sqrt{5}$, and $\sqrt{37}$ are irrational numbers. Also, the number π, which is the ratio of the circumference of a circle to its diameter, is an irrational number:

$$\pi \approx 3.1416$$

$$\pi = \frac{C}{d}$$

The relationships between the set of real numbers and its various subsets are illustrated in Fig. 1.4.

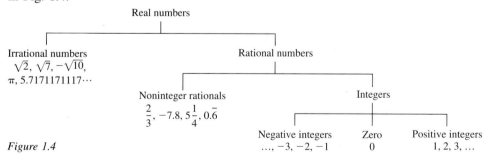

Figure 1.4

---

*HISTORICAL NOTE: The Greek mathematician Pythagoras (ca. 572–495 B.C.) and his followers at first believed that every number was rational. They were so disturbed to find that $\sqrt{2}$ was not rational that, for a while, efforts were made to keep the matter secret.

Example 7   Consider the following set of numbers:

$$\left\{ \frac{1}{8},\ -5,\ 12,\ 10\frac{3}{4},\ \sqrt{100},\ 0.\overline{3},\ 0,\ -9.76,\ \sqrt{12},\ 0.9494494449 \cdots \right\}$$

Identify the elements of this set that are

(a) Natural numbers   (b) Whole numbers   (c) Integers
(d) Rational numbers   (e) Irrational numbers   (f) Real numbers

Solution: (a) The natural numbers are 12 and $\sqrt{100}$ (since $\sqrt{100} = 10$).
(b) The whole numbers are 12, $\sqrt{100}$, and 0.
(c) The integers are -5, 12, $\sqrt{100}$, and 0.
(d) The rational numbers are $\frac{1}{8}$, -5, 12, $10\frac{3}{4}$, $\sqrt{100}$, $0.\overline{3}$, 0, and -9.76.
(e) The irrational numbers are $\sqrt{12}$ and $0.9494494449 \cdots$.
(f) All of the numbers are real numbers.   ❐

Try Exercise 81

Repeat Example 7 with the set
$$\left\{ 6,\ 4\frac{2}{3},\ \sqrt{81},\ -\sqrt{15},\ -24,\ 0.\overline{72},\ 0,\ 9.83,\ \frac{18}{3},\ 0.7575575557 \cdots \right\}.$$

An example of a number that is <u>not</u> a real number is the square root of a negative number, such as $\sqrt{-4}$. The number $\sqrt{-4}$ is called an *imaginary number*. We will discuss imaginary numbers in Chap. 5.

Exercises 1.1

◆◆

Completion Exercises

1. A collection of objects is called a(n) _____.
   The objects in a set are called _____ or _____.

2. The set which contains no elements is called the _____ set or the _____ set.

3. To denote that 6 is an element of set A, we write 6 _____ A.

4. Diagrams that depict sets as regions are called _____ diagrams.

5. Set A is a subset of set B, written A _____ B, if _____.

6. The set of all those elements that belong to either set A or set B (or both) is called the _____ of A and B, written _____.

7. The set of all those elements that are in common to sets A and B is called the _____ of A and B, written _____.

8. The numbers 1, 4, 9, 16, 25, ... are called _____.

True-False Exercises

9. $\{1, 3, 5\} = \{5, 3, 1\}$

10. $\{1, 3, 5\} = \{1, 3, 3, 5\}$

11. Every set is a subset of itself.

12. The empty set is a subset of every set.

Matching Exercises

Match each set of numbers in Exercises 13 through 20 with its description in letters A through H.

13. Natural numbers

14. Whole numbers

15. Positive integers

16. Negative integers

17. Integers

18. Rational numbers

A. $\{0, 1, 2, 3, \ldots\}$

B. $\{x \,|\, x \text{ can be expressed as a decimal}\}$

C. $\{\ldots, -2, -1, 0, 1, 2, \ldots\}$

D. $\{x \,|\, x \text{ can be expressed as a nonterminating, nonrepeating decimal}\}$

E. $\{1, 2, 3, \ldots\}$

F. $\{\ldots, -3, -2, -1\}$

19. Irrational numbers   G. $\{\ldots, -3, -2, -1, 0\}$

20. Real numbers   H. $\{^a/_b \mid a$ and b are integers and $b \neq 0\}$

State whether each set is finite or infinite.

21. $\{3, 5, 6\}$

22. $\{0, 2, 4, 6, 8, \ldots\}$

23. $\{5, 10, 15, \ldots, 50\}$

24. $\varnothing$

25. $\{x \mid x$ is a rational number$\}$

26. $\{y \mid y$ is a whole number less than 1,000,000$\}$

Completion Exercises

Insert $\in$ or $\notin$ in the blank space to make a true statement.

27. 0 _____ $\{-1, 0, 1, 2\}$

28. 94 _____ $\{1, 2, 3, 4, \ldots, 87\}$

29. 24 _____ $\{1, 3, 5, 7, \ldots\}$

30. 1 _____ $\varnothing$

31. $\{6\}$ _____ $\{5, 6, 7\}$

32. $\{0\}$ _____ $\{-1, 0, 1, 2\}$

33. -73 _____ $\{x \mid x$ is an integer$\}$

34. 0 _____ $\{y \mid y$ is a positive integer$\}$

35. $\frac{5}{11}$ _____ $\{z \mid z$ is a rational number$\}$

36. 6.4 _____ $\{z \mid z$ is a rational number$\}$

37. $\sqrt{13}$ _____ $\{t \mid t$ is an irrational number$\}$

38. $\sqrt{49}$ _____ $\{t \mid t$ is an irrational number$\}$

Getting Acquainted with Your Scientific Calculator

To find $\sqrt{3}$ on your calculator, press

Answer
↓

Clear 3 $\sqrt{x}$ 1.732050808

The number 1.732050808 is an approximate value of $\sqrt{3}$. To the nearest thousandth $\sqrt{3} = 1.732$. Find each square root to the nearest thousandth.

39. $\sqrt{5}$

40. $\sqrt{13}$

41. $\sqrt{57}$

42. Try to find $\frac{5}{0}$ using your calculator. Explain your answer.

Write each set using the listing method.

43. $\{y \mid y$ is an even natural number less than 10$\}$

44. $\{m \mid m$ is an odd integer greater than 17$\}$

45. $\{x \mid x$ is a letter in the word banana$\}$

46. $\{x \mid x$ is a letter in the word tooth$\}$

47. $\{t \mid t$ is a whole number between 3 and 8$\}$

48. $\{t \mid t$ is a whole number between 4 and 7$\}$

49. $\{r \mid r$ is an integer between 9 and 10$\}$

50. $\{r \mid r$ is a natural number less than 1$\}$

Write each set using set-builder notation.

51. $\{0, 1, 2, 3, \ldots, 39\}$

52. $\{14, 15, 16, 17, 18\}$

53. $\{5, 6, 7, 8, 9, \ldots\}$

54. $\{\ldots, -4, -2, 0, 2, 4, \ldots\}$

55. $\{$January, June, July$\}$

56. $\{$Tuesday, Thursday$\}$

Completion Exercises

Insert $\subseteq$ or $\nsubseteq$ in the blank space to make a true statement.

57. $\{4, 6\}$ _____ $\{4, 6, 8\}$

58. $\{2, 3, 4, 5\}$ _____ $\{2, 3, 5\}$

59. $\{0\}$ _____ $\{-2, -1, 1, 2\}$

60. $\{7, 3, 1, 5\}$ _____ $\{5, 7, 1, 3\}$

61. $\varnothing$ _____ $\{0, 1, 2\}$

62. $\varnothing$ _____ $\{ \}$

63. 5 _____ $\{3, 4, 5, 9\}$

64. 10 _____ $\{5, 10, 15\}$

65. $\{1, 3, 5, \ldots\}$ _____ $\{1, 2, 3, \ldots, 200\}$

66. $\{x \mid x$ is a female lawyer$\}$ _____ $\{x \mid x$ is a lawyer$\}$

67. List all subsets of $\{a, b\}$.

68. List all subsets of $\{a, b, c\}$.

Suppose $A = \{1, 2, 4\}$, $B = \{2, 3, 5, 7\}$, $C = \{6, 3, 5, 8\}$, and $D = \{1, 2\}$. List the elements in each of the following sets.

69. $A \cup B$

70. $B \cup C$

71. $B \cap D$

72. $A \cap D$

73. $C \cup C$

74. $A \cap A$

75. $A \cap \varnothing$

76. $\varnothing \cup C$

77. $A \cup (B \cap C)$

78. $(A \cap B) \cup C$

79. $D \cup (A \cap C)$

80. $B \cap (C \cap D)$

81. Consider the following set of numbers: $\{6, 4\frac{2}{3}, \sqrt{81}, -\sqrt{15}, -24, 0.\overline{72}, 0, 9.83, \frac{18}{3}, 0.7575575557 \cdots\}$

Identify the elements of this set that are

(a) Natural numbers   (b) Whole numbers

(c) Integers   (d) Rational numbers

(e) Irrational numbers   (f) Real numbers

82. Suppose we name the various sets of numbers as follows:

$$N = \{x \mid x \text{ is a natural number}\}$$
$$W = \{x \mid x \text{ is a whole number}\}$$
$$I = \{x \mid x \text{ is an integer}\}$$
$$Q = \{x \mid x \text{ is a rational number}\}$$
$$H = \{x \mid x \text{ is an irrational number}\}$$
$$R = \{x \mid x \text{ is a real number}\}$$

Insert $\subseteq$ or $\not\subseteq$ in the blank space to make a true statement.

(a) N _____ W (b) W _____ I

(c) W _____ N (d) I _____ Q

(e) Q _____ R (f) H _____ R

True-False Exercises

Assume A, B, and C are sets.

83. $A \cup B = B \cup A$

84. $A \cap B = B \cap A$

85. $A \cap B \subseteq B$

86. $A \subseteq A \cup B$

87. If $A \subseteq B$, then $A \cup B = B$.

88. If $A \subseteq B$, then $A \cap B = A$.

89. If $A \subseteq B$ and $B \subseteq A$, then $A = B$.

90. If $A \subseteq B$ and $B \subseteq C$, then $A \subseteq C$.

91. If $A \subseteq B$ and $4 \in B$, then $4 \in A$.

92. If $7 \in A$, then $7 \in A \cap B$.

Discussion Exercises

93. Explain the difference between $\{\ \}$, $\{0\}$, $\varnothing$, and $\{\varnothing\}$.

94. If $F = \{1\}$ and $G = \{1, \{1\}\}$, then $F \in G$ and $F \subseteq G$. Explain why.

95. Discuss the difference between a finite set and an infinite set.

96. State the two ways of identifying the elements of a set and discuss the advantages and disadvantages of each.

97. Explain the difference between the symbols $\in$ and $\subseteq$.

98. What is the difference between a constant and a variable? Give examples.

99. Explain why $\frac{8}{0}$ is undefined. Is $\frac{0}{8}$ undefined? Why or why not?

100. Discuss the difference between the decimal form of a rational number and the decimal form of an irrational number.

Problem Solving

101. John, Bob, and Sue all became ill with the same symptoms Saturday evening. John had visited Sally that morning, had to walk 2 mi in the snow when his car broke down, and had eaten dinner at Al's Greasy Spoon restaurant. Bob had also visited Sally and eaten at Al's. Sue had eaten at Al's and visited Dan and Dave. Show how a Venn diagram could be used to determine the probable cause of their illness.

Getting Acquainted with Your Graphing Calculator

ROUNDING A NUMBER   Some calculators have a command that allows you to round a number to a given accuracy. Use your calculator to round each number to the indicated accuracy.

102. 43.15 (nearest tenth)

103. 43.15 (nearest whole number)

104. 7.2968 (nearest hundredth)

105. 7.2968 (nearest thousandth)

TESTING A STATEMENT   Some calculators allow you to test a statement to determine whether that statement is true or false. One brand of calculator registers a 1 if the statement is true and a 0 if the statement is false. Use your calculator to test whether each statement is true or false.

106. $3 = 3$

107. $3 \neq 3$

108. $\sqrt{16} = 4$

109. $\sqrt{2} = 1.414$ (Why is this false?)

1.2 The Number Line; Absolute Value; Inequalities

TAPE AU1

In Sec. 1.1 we discussed the meaning of several mathematical symbols. For example the symbol = means "is equal to" and is called an **equality symbol,** or an **equal sign.** Symbols allow us to write mathematical statements in a more abbreviated form. In this section we introduce several other mathematical symbols. These new symbols are defined using a *number line.*

The Real Number Line

A good way to illustrate the set of real numbers is to draw a **number line,** like the one shown in Fig. 1.5. To construct a number line, choose any point on a horizontal line to correspond to the number 0. Draw a *tick mark* to indicate the location of this point. Then choose a point to the right of 0 to correspond to the number 1. Using the distance between 0 and 1 as a unit of measure, locate the remaining integers.

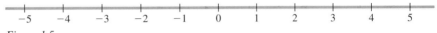

Figure 1.5

Numbers to the right of 0 are called **positive numbers** (note that $2 = +2$). Numbers to the left of 0 are called **negative numbers.** The number **0 is neither positive nor negative.** Positive and negative numbers are collectively called **signed numbers,** or **directed numbers.***

The **graph** of a number is the point on the number line that corresponds to the number. The number is called the **coordinate** of its corresponding point. We indicate the location of a point by drawing a solid dot. Figure 1.6 illustrates the graphs of the real numbers 4, $\sqrt{2}$, and $-10/3$.

Figure 1.6

Opposite of a Number

Two numbers that are the same distance from 0 on the number line but are on opposite sides of 0 are called **opposites,** or **additive inverses,** of each other. For example -3 is the opposite of 3, and 3 is the opposite of -3 (see Fig. 1.7).

Figure 1.7

DEFINITION OF THE OPPOSITE OF A NUMBER

The **opposite,** or **additive inverse,** of the real number a is $-a$.

---

*HISTORICAL NOTE: Positive and negative numbers were used by the Chinese more than 2000 yr ago for bookkeeping purposes. Positive numbers were written in black and negative numbers in red, which gave rise to the expressions "in the black" and "in the red." The symbols $+$ and $-$ first appeared in print in an arithmetic book published in 1489.

Note that the symbol − has three different, though related, meanings:

1. $10 - 8$ means "10 subtract 8."
2. -8 means "negative 8."
3. -8 means "the opposite of 8."

We rely on content to tell us which meaning is intended.

Since − denotes the opposite of a number, then $-(-4)$ denotes the opposite of -4. But the opposite of -4 is 4. That is

$$-(-4) = \text{the opposite of } -4 = 4$$

This suggests the following rule.

Double Negative Rule

For any real number a,

$$-(-a) = a$$

Here are some examples of numbers and their opposites.

Number	*Opposite*
11	-11
-9	$-(-9)$, or 9
$\dfrac{3}{4}$	$-\dfrac{3}{4}$, or $\dfrac{-3}{4}$, or $\dfrac{3}{-4}$
0	-0, or 0

Example 1   Simplify: (a) $-(-2)$, (b) $-(-(-6))$.

Solution: (a) $-(-2) = 2$ Double negative rule
          (b) $-(-(-6)) = -(6)$ Since $-(-6) = 6$
                    $= -6$ ❐

Try Exercise 23

Simplify: $-(-c)$.

Absolute Value

Every nonzero number on the number line has two important properties: a *direction* from 0 and a *distance* from 0. The direction from 0 is indicated by the + or − sign in front of the number. The distance from 0 is called the **magnitude,** or the **absolute value,** of the number.* The absolute value of a number a is written $|a|$.

Figure 1.8 illustrates that the numbers 3 and -3 are not equal, but they are the same distance from 0. We say that 3 and -3 are "equal in absolute value but opposite in sign."

Figure 1.8

---

*HISTORICAL NOTE: The term "absolute value" is derived from the Latin word "absolvere," meaning "to free from," as in "to free from its sign." The absolute value symbol | | was first used by the German mathematician Karl Weierstrass (1815–1897).

Here are some other examples of absolute value:

Number	*Absolute value of the number*
5	$\|5\| = 5$
-8	$\|-8\| = 8$
0	$\|0\| = 0$
$-\dfrac{2}{3}$	$\left\|-\dfrac{2}{3}\right\| = \dfrac{2}{3}$

Here is a formal definition of absolute value.

DEFINITION OF ABSOLUTE VALUE

If a is a real number, then

$$|a| = \begin{cases} a & \text{if } a \text{ is a positive number or } 0 \\ -a & \text{if } a \text{ is a negative number} \end{cases}$$

Since absolute value measures distance, **the absolute value of a number is never negative.** For this reason students are often confused in the definition of absolute value by the statement "$|a| = -a$ if a is a negative number." To clarify this statement, note that if a is a negative number, then $-a$ is a positive number. For example, if $a = -7$, then $-a = -(-7) = 7$.

Example 2   Simplify: (a) $|12|$, (b) $|-15|$, (c) $-|4|$, (d) $-|-9|$, (e) $|10 - 2|$.

Solution:

(a) $|12| = 12$ — Since 12 is a positive number

(b) $|-15| = -(-15) = 15$ — Since -15 is a negative number

(c) $-|4| = -4$ — Find the absolute value of 4. Then take the opposite.

(d) $-|-9| = -9$ — Find the absolute value of -9. Then take the opposite.

(e) $|10 - 2| = |8| = 8$ — Subtract. Then find the absolute value.   ❏

Try Exercise 37

Simplify: $-|-6|$.

Inequalities

A statement that two numbers are not equal is called an **inequality.** For example, $2 \neq 5$ is an inequality. If two numbers are not equal, then one number must be less than the other. The symbol $<$ means "is less than." For example,

$$2 < 5 \text{ means "2 is less than 5"}$$

Instead of saying "2 is less than 5" you could say that "5 is greater than 2," written $5 > 2$.

$$2 < 5 \text{ is equivalent to } 5 > 2$$

One way to keep the symbols $<$ and $>$ straight is to remember that **the inequality symbol always points to the smaller number.***

---

*HISTORICAL NOTE: The symbols $<$ and $>$ were first used by the English mathematician Thomas Harriot (1560–1621).

The graphs of the numbers 2 and 5 are shown in Fig. 1.9.

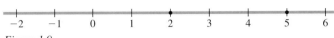

Figure 1.9

Note that the graph of 2 is to the left of the graph of 5. This suggests the following definition.

> ### DEFINITION OF "LESS THAN"
>
> The real number a **is less than** the real number b, written $a < b$, if the graph of a lies to the left of the graph of b on the number line. If $a < b$, then $b > a$.

Example 3   Insert $<$ or $>$ in the blank space to make a true statement.
   (a) -6 _____ -5
   (b) 1 _____ -3
   (c) -2 _____ 0
   (d) $-\dfrac{1}{4}$ _____ $-\dfrac{5}{4}$

Solution: Each of the numbers is graphed in Fig. 1.10.

Figure 1.10

   (a) $-6 < -5$ Since -6 is to the left of -5
   (b) $1 > -3$ Since 1 is to the right of -3
   (c) $-2 < 0$ Since -2 is to the left of 0
   (d) $-\dfrac{1}{4} > -\dfrac{5}{4}$ Since $-\dfrac{1}{4}$ is to the right of $-\dfrac{5}{4}$

Try Exercise 53

Insert $<$ or $>$ in the blank space: -5 _____ -10.

If either $a < b$ or $a = b$, we can write $a \le b$. For example,

$$4 \le 9 \text{ means "4 is less than or equal to 9"}$$

In this case $4 < 9$. We could also write $4 \le 4$, since $4 = 4$. We cannot write $5 \le 4$ since neither $5 < 4$ nor $5 = 4$ is true. In a similar fashion, the symbol $\ge$ means "is greater than or equal to."

Example 4   To run for President, your age, a, must be at least 35 yr. Write this statement as an inequality.

Solution: To be at least 35 your age a must either be 35 (written $a = 35$), or your age a must be greater than 35 (written $a > 35$). Therefore

$$a \ge 35 \text{ yr}$$

Try Exercise 75

Write as an inequality: To vote, your age, a, must be at least 18 yr.

Both $a \ge 35$ yr and 35 yr $\le a$ are correct answers.

Here is a summary of the symbols of inequality.

Symbols of Inequality

Symbol	Meaning	Example
$\neq$	is not equal to	$-3 \neq 3$
$<$	is less than	$1 < 8$
$>$	is greater than	$\frac{1}{2} > \frac{1}{3}$
$\leq$	is less than or equal to	$-100 \leq -99$
$\geq$	is greater than or equal to	$0 \geq 0$

A slash mark negates any of the symbols of inequality. For example $5 \not< 3$ means "5 is not less than 3." Note that $5 \not< 3$ means $5 \geq 3$.

Double Inequalities

To denote that a number is between two other numbers, we combine two separate inequalities to form a *double inequality.* For example since 5 is between 2 and 7, we can combine the two inequalities

$$2 < 5 \qquad \text{and} \qquad 5 < 7$$

to form the double inequality

$$2 < 5 < 7$$

Read $2 < 5 < 7$ as "2 is less than 5 and 5 is less than 7."

You can also denote that 5 is between 2 and 7 by writing

$$7 > 5 > 2$$

which is read "7 is greater than 5 and 5 is greater than 2."

DEFINITION OF A DOUBLE INEQUALITY

If a and b are real numbers and $a < b$, then

$$a < x < b$$

is a **double inequality** that means $a < x$ *and* $x < b$.

Example 5   Write the statement "t is less than 14 but no less than 5" using inequality symbols.

Solution: Since t is less than 14, write $t < 14$. Since t is no less than 5, write $t \geq 5$, or $5 \leq t$. Then combine these two inequalities to form the double inequality

$$5 \leq t < 14 \qquad \qquad \square$$

Try Exercise 73

Write using inequality symbols: t is less than 13 but no less than -5.

Problem Solving

Example 6   In a particular region it costs anywhere from \$9 to \$17 per ft to drill a well. To hit water the well must be anywhere from 130 to 150 ft deep. If C is the cost of drilling the well, write a double inequality that describes the range of values for C.

Solution: The least the well would cost is

$$9 \times 130 = \$1170$$

The most the well would cost is

$$17 \times 150 = \$2550$$

Therefore

$$\$1170 \leq C \leq \$2550$$

Try Exercise 79

A car with a fuel capacity of 13 gal gets 23 mi/gal in the city and 38 mi/gal on the highway. If d is the distance the car can travel on a full tank, write a double inequality that describes the range of values for d.

This means the well will cost at least \$1170 but no more than \$2550.   ☐

Graphing Inequalities

The **graph** of an inequality is the graph of all those values of the variable that make the inequality a true statement.

Example 7   Graph: $\{x | x < 3\}$.

Solution: The graph consists of all those points that lie to the left of 3 on the number line, as shown in Fig. 1.11. Since the inequality symbol is $<$ and not $\leq$, draw the symbol) at 3 to indicate that the point whose coordinate is 3 is not part of the graph. Note that we can also write the inequality $x < 3$ as $3 > x$.   ☐

Try Exercise 81

Graph: $\{x | x < 4\}$.

Figure 1.11

Example 8   Graph: $\{x | x \geq -1\}$.

Solution: The graph is shown in Fig. 1.12. The symbol [at -1 indicates that the point corresponding to -1 is part of the graph. The inequality $x \geq -1$ can also be written as $-1 \leq x$.   ☐

Try Exercise 83

Graph: $\{x | x \geq -3\}$.

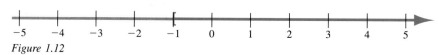

Figure 1.12

Example 9   Graph: $\{x | 0 < x < 4\}$.

Solution: This set consists of all those real numbers between 0 and 4. The graph is shown in Fig. 1.13. The double inequality $0 < x < 4$ can also be written as $4 > x > 0$.   ☐

Try Exercise 85

Graph: $\{x | 5 < x < 6\}$.

Figure 1.13

Example 10   Graph: $\{x | -2 \leq x < 2\}$.

Solution: The graph is shown in Fig. 1.14. Note that the point whose coordinate is -2 is included in the graph, but the point whose coordinate is 2 is not. The double inequality $-2 \leq x < 2$ can also be written as $2 > x \geq -2$.   ☐

Try Exercise 87

Graph: $\{x | -4 \leq x < 1\}$.

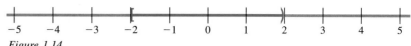

Figure 1.14

Here is a summary of how to graph an inequality.

Set Notation	*Graph*

$\{x\,|\,x < 3\}$

$\{x\,|\,x \geq -2\}$

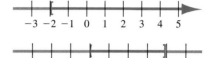

$\{x\,|\,0 < x \leq 4\}$

Exercises 1.2

True-False Exercises

1. The number 0 is neither positive nor negative.
2. The absolute value of a number is never negative.

Completion Exercises

3. Numbers to the right of 0 on the number line are called _____ numbers; numbers to the left of 0 are called _____ numbers.
4. The point on a number line that corresponds to a given number is called the _____ of the number.
5. The distance between 0 and a number a is called the _____ of a and is denoted _____.
6. Two numbers that are the same distance from 0 on the number line but are on opposite sides of 0 are called _____ or _____ of each other.

Graph each set of numbers on a number line.

7. $\{4, 5.5, \frac{3}{4}, -4, -\frac{7}{3}, 0, \sqrt{3}\}$

8. $\left\{-5, 0, \sqrt{2}, -\frac{9}{4}, 3.5, \frac{2}{3}, 5\right\}$

State the opposite of each number.

9. 5

10. -13

11. $-\dfrac{3}{8}$

12. $1\dfrac{4}{9}$

13. 6.99

14. $\sqrt{10}$

Getting Acquainted with Your Scientific Calculator

To find the opposite of 5.4 on your calculator, press

Answer
↓

$\boxed{\text{Clear}}\ 5\ \boxed{\cdot}\ 4\ \boxed{+/-}\ \boxed{\qquad -5.4\qquad}$

Find the opposite of each number on your calculator.

15. 3.9

16. -8

17. 0

18. Try to find $\sqrt{-4}$ on your calculator. Explain your answer.

Discussion Exercises

19. If x is a negative number, is $-x$ a negative number or a positive number? Explain why.
20. If x is a negative number, is $-(-x)$ a negative number or a positive number? Explain why.

Simplify.

21. $-(-3)$

22. $-\left(-4\dfrac{2}{5}\right)$

23. $-(-c)$

24. $-(-(-d))$

25. $-(-(-25))$

26. $-(-(-(-18)))$

Find the absolute value of each number.

27. 4

28. 9

29. -1

30. 0

31. -87.6

32. $-\dfrac{33}{4}$

Simplify.

33. $|8|$

34. $|-13|$

35. $|-16|$

36. $-|27|$

37. $-|-6|$

38. $-|-4|$

39. $|12 - 3|$

40. $-|5 + 1|$

41. $|-19| - |-10|$

42. $|4| + |-9|$

43. Name two numbers whose absolute value is 7.

44. Name two numbers whose absolute value is $\dfrac{1}{2}$.

True-False Exercises

45. For all real numbers a, $|a| = a$.

46. If $a = b$, then $|a| = |b|$.

47. If $a < b$, then $|a| < |b|$.

48. If $|a| = |b|$, then $a = b$.

Completion Exercises

Insert $<$ or $>$ in the blank space to make a true statement.

49. 2 _____ 6

50. 5.08 _____ 5.8

51. 1.1 _____ 1.01

52. 0 _____ -7

53. -5 _____ -10

54. -19 _____ 4

55. $\dfrac{3}{4}$ _____ $\dfrac{5}{7}$

56. $-\dfrac{3}{4}$ _____ $-\dfrac{5}{7}$

57. $|3|$ _____ $|-5|$

58. $|-8|$ _____ $|2|$

59. $|-34|$ _____ $|-14|$

60. $|-18|$ _____ $|-29|$

Write the following word statements using the symbols $<$, $>$, $\leq$, and $\geq$.

61. x is less than 5.

62. x is greater than 8.

63. y is greater than or equal to -2.

64. y is less than or equal to 6.

65. z is at most 15.

66. z is at least 12.

67. m is a positive number.

68. m is a negative number.

69. p is a nonnegative number.

70. p is a nonpositive number.

71. r is between 0 and 7.

72. r is between -3 and 4.

73. t is less than 13 but no less than -5.

74. t is more than 4 but no more than 20.

Problem Solving

Write each word statement as an inequality.

75. To vote, your age, a, must be at least 18 yr.

76. To fight as a middleweight, a boxer's weight, w, can be no more than 160 lb.

77. To enlist in the U. S. Army, your height, h, must fall in the range from 60 in. to 80 in.

78. The temperature, t, on New Year's Day ranged from a morning low of $-4°$ to an afternoon high of $21°$.

79. A car with a fuel capacity of 13 gal gets 23 mi/gal in the city and 38 mi/gal on the highway. If d is the distance the car can travel on a full tank, write a double inequality that describes the range of values for d.

80. A turnpike has a minimum speed of 45 mi/hr and a maximum speed of 65 mi/hr. If d represents the distance you can legally travel on this turnpike in 3 hr, write a double inequality that describes the range of values for d.

Graph each set on a number line. Then rewrite each set so that the inequality symbol(s) points (point) in the opposite direction.

81. $\{x\,|\,x < 4\}$

82. $\{x\,|\,x < -2\}$

83. $\{x\,|\,x \geq -3\}$

84. $\{x\,|\,x \geq 5\}$

85. $\{x\,|\,5 < x < 6\}$

86. $\{x\,|\,-3 \leq x \leq 0\}$

87. $\{x\,|\,-4 \leq x < 1\}$

88. $\{x\,|\,4 < x \leq 6\}$

Discussion Exercises

89. Discuss the meanings of the symbols $=$, $\neq$, $<$, $>$, $\leq$, and $\geq$. Give examples.

90. Suppose $a < b$. Discuss the meaning of the double inequality $a < x < b$. Give examples.

91. Suppose you are given two different real numbers. How do you ultimately decide which number is less?

92. Explain why the following definition of absolute value always gives a nonnegative value for a:

$$|a| = \begin{cases} a & \text{if } a \geq 0 \\ -a & \text{if } a < 0 \end{cases}$$

93. Compare the meanings of the symbols $<$ and $\not>$.

Getting Acquainted with Your Graphing Calculator

ABSOLUTE VALUE   Some calculators have a command that allows you to find the absolute value of a number. One brand uses the symbol "abs" for absolute value. Use your calculator to simplify each expression.

94. $|-15|$

95. $-|9|$

96. $-|-8|$

97. $|23| - |-6|$

Use your calculator to test each statement to determine whether it is true or false.

98. $3 < 5$

99. $0 > -1$

100. $-9 \leq -10$

101. $|-9| \geq |-10|$

1.3 Properties of Real Numbers

TAPE AU1

The rules of algebra depend upon several basic properties of real numbers. In this section we state and give examples of these properties.

In the first group of properties we use the term *algebraic expression*. An **algebraic expression** is a collection of constants, variables, and/or operations (addition, subtraction, multiplication, division, and roots). Here are three examples of algebraic expressions:

$$2x - 5 \qquad 4x - 8y + 3 \qquad \frac{p}{p + 1}$$

Note that $2x$, or $2 \cdot x$, means "2 times x." In algebra we use the symbol $\cdot$ to indicate multiplication rather than $\times$ to avoid confusion with the variable x. Actually there are several ways to indicate the product of the real numbers a and b:

$$a \times b \qquad a \cdot b \qquad ab \qquad a(b) \qquad (a)b \qquad (a)(b)$$

Properties of Equality

If two variables a and b represent the same number, we say that a is **equal** to b and we write $a = b$.*

---

*HISTORICAL NOTE: The symbol $=$ was first used by the English mathematician Robert Recorde (1510–1558), who said of the two parallel line segments that form the symbol, "no two things can be more equal." Recorde was also the personal physician of King Edward VI and Queen Mary.

Properties of Equality

For all real numbers a, b, and c:

$a = a$ **Reflexive property**
If $a = b$, then $b = a$.   **Symmetric property**
If $a = b$ and $b = c$, then $a = c$.   **Transitive property**
If $a = b$, then b can be substituted for a **Substitution property**
in any algebraic expression and the result
is an equivalent expression.

The reflexive property states that a number is equal to itself. The symmetric property states that if one number is equal to a second number, then the second number is equal to the first. The transitive property states that if one number is equal to a second number and the second number is equal to a third number, then the first number is equal to the third. The substitution property states that if two numbers are equal, then one number can be replaced by the other number in an algebraic expression without changing the value of the expression.

Here is an example that illustrates each property.

Example 1   Name the property of equality that justifies each statement. Assume x, k, and y are real numbers. (a) $x + 1 = x + 1$. (b) If $x = 3$, then $3 = x$. (c) If $x = k$ and $k = 15$, then $x = 15$. (d) If $x = 4$ and $x + y = 10$, then $4 + y = 10$.

Solution: (a) Reflexive property, (b) symmetric property, (c) transitive property, (d) substitution property.   ❑

Try Exercise 15

Name the property: If $4 = x$, then $x = 4$.

Properties of Inequality

We now state two basic properties of inequality.

Properties of Inequality

If a and b are real numbers, exactly one   **Trichotomy property**
of the following is true: $a < b$, $a = b$,
or $a > b$.

Suppose a, b, and c are real numbers. If   **Transitive property**
$a < b$ and $b < c$, then $a < c$.

Note that the transitive property of inequality is similar to the transitive property of equality.

Example 2   Name the property of inequality that justifies each statement. Assume x and y are real numbers. (a) $x = 0$, $x < 0$, or $x > 0$. (b) If $x < y$ and $y < 7$, then $x < 7$.

Try Exercise 17

Name the property: If $2 < x$ and $x < y$, then $2 < y$.

Solution: (a) Trichotomy property, (b) transitive property.   ❑

Properties of Addition and Multiplication

The operations of addition and multiplication of real numbers obey several basic properties.

Properties of Addition and Multiplication*

For all real numbers a, b, and c:

Property	Addition	Multiplication
Commutative property	$a + b = b + a$	$ab = ba$
Associative property	$a + (b + c) =$ $(a + b) + c$	$a(bc) = (ab)c$
Identity property	There is a unique number 0, called the **additive identity,** such that $a + 0 = a$ and $0 + a = a$.	There is a unique number 1, called the **multiplicative identity,** such that $a \cdot 1 = a$ and $1 \cdot a = a$.
Inverse property	For each real number a there is a unique real number $-a$, called the **additive inverse** or **opposite** of a, such that $a + (-a) = 0$ and $-a + a = 0$.	For each nonzero real number a there is a unique real number $1/a$, called the **multiplicative inverse** or **reciprocal** of a, such that $a \cdot \dfrac{1}{a} = 1$ and $\dfrac{1}{a} \cdot a = 1$.
Distributive property (of multiplication over addition)	$a(b + c) = ab + ac$	

Using the commutative property of multiplication, we can also write the distributive property as

$$(b + c)a = ba + ca$$

The distributive property is also true when there are more than two numbers inside the parentheses. For example:

$$a(b + c + d + \cdots + z) = ab + ac + ad + \cdots + az$$

This is called the *extended distributive property.*

Here are some examples to give you practice in using the properties of addition and multiplication.

Example 3   Use a commutative property to complete each statement:

(a) $3 + 5 = $ <u> ? </u>   (b) $x \cdot 2 = $ <u> ? </u>

---

*HISTORICAL NOTE: The British mathematicians Augustus DeMorgan (1806–1871), Duncan Gregory (1813–1844), and George Peacock (1791–1858) were the first to notice the presence of a structure to algebra such as the commutative, associative, and distributive properties.

Try Exercise 29

Use a commutative property to complete: $-3 \cdot 1 =$ _____.

Solution: (a) $3 + 5 = 5 + 3$ Change the order.
(b) $x \cdot 2 = 2x$ Change the order.   □

CAUTION

Neither subtraction nor division is a commutative operation.

Subtraction   *Division*

$5 - 3 = 3 - 5$ False $6 \div 2 = 2 \div 6$ False   ■

Example 4   Use an associative property to complete each statement.

(a) $4 + (8 + y) =$ __?__   (b) $\frac{1}{6} \cdot (6m) =$ __?__

Solution: (a) $4 + (8 + y) = (4 + 8) + y$ Change the grouping.
(b) $\frac{1}{6} \cdot (6m) = \left(\frac{1}{6} \cdot 6\right)m$ Change the grouping.   □

Try Exercise 31

Use an associative property to complete: $(y + 1) + 8 =$ _____.

CAUTION

Neither subtraction nor division is an associative operation.

Subtraction   *Division*

$7 - (4 - 1) = (7 - 4) - 1$ $12 \div (6 \div 3) = (12 \div 6) \div 3$
$7 - 3 = 3 - 1$ $12 \div 2 = 2 \div 3$
$4 = 2$ False $6 = \frac{2}{3}$ False   ■

Note that the commutative properties state that we can change the order in an addition or a multiplication problem; the associative properties state that we can change the grouping, or association. Taken together, the commutative and associative properties allow us to rearrange sums in several ways. For example:

$$(2 + 4) + 6 = 6 + 6 = 12$$
$$2 + (4 + 6) = 2 + 10 = 12$$
$$2 + (6 + 4) = 2 + 10 = 12$$

For this reason, we do not have to write parentheses in the expression $2 + 4 + 6$. Neither do we have to write parentheses when computing products like $2 \cdot 4 \cdot 6$. However, the commutative and associative properties do not apply to subtraction or division, so parentheses are needed to clarify expressions like $8 - 5 - 3$ and $12 \div 6 \div 2$.

Example 5   Use an identity property to complete each statement:

(a) $-7 + 0 =$ __?__   (b) $1 \cdot p =$ __?__

Solution: (a) $-7 + 0 = -7$ 0 preserves identities under addition.
(b) $1 \cdot p = p$ 1 preserves identities under multiplication.   □

Try Exercise 33

Use an identity property to complete: $t + 0 =$ _____.

Example 6 Use an inverse property to complete each statement:

$$(a) \; -8 + 8 = \underline{\;\;?\;\;} \qquad (b) \; 5 \cdot \frac{1}{5} = \underline{\;\;?\;\;}$$

Solution: (a) $-8 + 8 = 0$ Opposites add to 0.

(b) $5 \cdot \dfrac{1}{5} = 1$ Reciprocals multiply to 1.   ❑

Try Exercise 35

Use an inverse property to complete: $k \cdot \dfrac{1}{k} = \underline{\;\;\;\;\;}$.

Note that 5 and $\dfrac{1}{5}$ are reciprocals. We find the reciprocal of a number by inverting the number.

Number	*Reciprocal*
$\dfrac{1}{10}$	10
$\dfrac{3}{4}$	$\dfrac{4}{3}$
-2	$-\dfrac{1}{2}$
0	0 does not have a reciprocal since $\dfrac{1}{0}$ is undefined

Example 7 Find the reciprocal of 5.9.

Solution: Since $5.9 = 5\dfrac{9}{10} = \dfrac{59}{10}$, the reciprocal of 5.9 is $\dfrac{10}{59}$.   ❑

Try Exercise 63

Find the reciprocal of 4.7.

The distributive property relates the operations of addition and multiplication. We say that "multiplication distributes over addition." Mathematically this reads as

$$a(b + c) = ab + ac$$

Example 8 Use the distributive property to complete each statement:

(a) $4(m + 2) = \underline{\;\;?\;\;}$ (b) $(6 + 1)7 = \underline{\;\;?\;\;}$
(c) $8(x + y + 5) = \underline{\;\;?\;\;}$

Solution: (a) $4(m + 2) = 4 \cdot m + 4 \cdot 2$ Distribute multiplication over addition.

(b) $(6 + 1)7 = 6 \cdot 7 + 1 \cdot 7$ Distribute from the right.

(c) $8(x + y + 5) = 8 \cdot x + 8 \cdot y + 8 \cdot 5$ Distribute over three numbers.   ❑

Try Exercise 69

Use the distributive property to complete: $4(m + 7) = \underline{\;\;\;\;\;}$.

Simplifying Algebraic Expressions

We can use the properties of real numbers to simplify algebraic expressions.

Example 9 Use the properties of real numbers to simplify each expression: (a) $2(7x)$, (b) $3(5k + 4)$, (c) $-4 + (a + 4)$.

Solution: (a) $2(7x) = (2 \cdot 7)x$ Associative property
                          $= 14x$ Multiply.
            (b) $3(5k + 4) = 3(5k) + 3(4)$ Distributive property
                          $= (3 \cdot 5)k + 3(4)$ Associative property
                          $= 15k + 12$ Multiply.
            (c) $-4 + (a + 4) = -4 + (4 + a)$ Commutative property
                          $= (-4 + 4) + a$ Associative property
                          $= 0 + a$ Inverse property
                          $= a$ Identity property   ❐

Try Exercise 87

Simplify: $5(2k + 3)$.

We can also use the properties of real numbers to prove other properties. A property that is proved using other properties is called a **theorem.** An important theorem in algebra is the **multiplication property of 0.**

Multiplication Property of 0

For any real number a,

$$a \cdot 0 = 0 \qquad \text{and} \qquad 0 \cdot a = 0$$

Proving theorems is an important part of algebra. We want to be certain that the rules we are using are true. However, the primary purpose of this text is to teach you how to use theorems to solve problems, rather than how to prove theorems.

Exercises 1.3

Completion Exercises

Assume all variables represent real numbers.

1. The reflexive property of equality states that $a =$ _____.

2. The symmetric property of equality states that if $a = b$, then _____.

3. The transitive property of equality states that if $a = b$ and $b = c$, then _____.

4. The substitution property of equality states that if $a = b$, then _____.

5. The transitive property of inequality states that if $a < b$ and $b < c$, then _____.

6. The trichotomy property of inequality states that if a and b are real numbers, then exactly one of the following is true: _____.

7. The commutative properties state that _____ (addition) and that _____ (multiplication).

8. The associative properties state that _____ (addition) and that _____ (multiplication).

9. The identity properties state that _____ (addition) and that _____ (multiplication).

10. The inverse properties state that _____ (addition) and that _____ (multiplication).

11. The distributive property (of multiplication over addition) states that _____.

12. The multiplication property of 0 states that _____.

13. The number 0 is called the additive _____, and the number 1 is called the multiplicative _____.

14. The number $-a$ is called the _____ or the _____ of a. If $a \neq 0$, the number $\frac{1}{a}$ is called the _____ or the _____ of a.

Matching Exercises

Match each statement in Exercises 15 through 20 with the property of equality or inequality in letters A through F that justifies the statement. Assume all variables represent real numbers.

15. If $4 = x$, then $x = 4$.

A. Reflexive property of equality

16. $x + 5 = x + 5$.

B. Symmetric property of equality

17. If $2 < x$ and $x < y$, then $2 < y$.

C. Transitive property of equality

18. $x = 3, x < 3,$ or $x > 3$.

D. Substitution property of equality

19. If $x = y + z$ and $y = 9$, then $x = 9 + z$.

E. Trichotomy property

20. If $6x = 18$ and $18 = y$, then $6x = y$.

F. Transitive property of inequality

Completion Exercises

Use the given property to complete each statement. Assume all variables represent real numbers.

21. Reflexive property: $5k + 25 =$ _____.

22. Transitive property: If $c < d$ and $d < 10$, then _____.

23. Transitive property: If $2k = m$ and $m = 7n$, then _____.

24. Symmetric property: If $3c + d = 14$, then _____.

25. Trichotomy property: If x and y are real numbers, then _____.

26. Substitution property: If $x = y$ and $8x = 16$, then _____.

Discussion Exercises

27. Is the relation $<$ reflexive? Explain why or why not.

28. Is the relation $<$ symmetric? Explain why or why not.

Completion Exercises

Use the given property to complete each statement. Assume all variables represent real numbers.

29. $-3 \cdot 1 =$ _____   Commutative property

30. $x + 0 =$ _____   Commutative property

31. $(y + 1) + 8 =$ _____   Associative property

32. $\frac{1}{9} \cdot (9m) =$ _____   Associative property

33. $t + 0 =$ _____   Identity property

34. $1 \cdot z =$ _____   Identity property

35. $k \cdot \dfrac{1}{k} =$ _____   Inverse property

36. $-r + r =$ _____   Inverse property

37. $y \cdot 0 =$ _____   Multiplication property of 0

38. $0 \cdot x =$ _____   Multiplication property of 0

Name the property of addition and/or multiplication that justifies each statement. Assume all variables represent real numbers.

39. $4 + x = x + 4$ 40. $6 \cdot x = x \cdot 6$

41. $2(7y) = (2 \cdot 7)y$

42. $(y + 7) + 1 = y + (7 + 1)$

43. $19 \cdot 1 = 19$ 44. $0 + (-15) = -15$

45. $37 + (-37) = 0$ 46. $\dfrac{1}{8} \cdot 8 = 1$

47. $3 + (p + 5) = 3 + (5 + p)$

48. $4(p \cdot 9) = 4(9p)$ 49. $0 \cdot 20 = 0$

50. $-18 \cdot 0 = 0$ 51. $(6m)2 = 2(6m)$

52. $(m + 9) + 2 = 2 + (m + 9)$

53. $3(z + 7) = 3 \cdot z + 3 \cdot 7$

54. $(z + 5)4 = z \cdot 4 + 5 \cdot 4$

55. $5(k + 1) = (k + 1)5$ 56. $3(k + 7) = 3(7 + k)$

Find the reciprocal of each number. Write your answer in simplest form.

57. 6 58. $\dfrac{1}{7}$

59. $\dfrac{3}{2}$ 60. $-\dfrac{7}{5}$

61. -1 62. 1

63. 4.7 64. 0.35

Getting Acquainted with Your Scientific Calculator

To find the reciprocal of 8 on your calculator, press

Answer
↓

(Clear) 8 (1/x) (0.125)

Use your calculator to find the reciprocal of each number.

65. 2

66. 3

67. −0.03125

68. 0

Completion Exercises

Use the distributive property to complete each statement. Assume all variables represent real numbers.

69. $4(m + 7) = $ _____

70. $(2 + 5)8 = $ _____

71. $(x + y + 45)2 = $ _____

72. $5(x + y + 22) = $ _____

73. $3z + 7z = $ _____

74. $8z + 4z = $ _____

75. $10p + 1p = $ _____

76. $1p + 14p = $ _____

77. $6r + 6s + 6t = $ _____

78. $7r + 7s + 7t = $ _____

Use the properties of real numbers to simplify each expression. Assume all variables represent real numbers.

79. $4(5x)$

80. $(y + 8) + 2$

81. $3(z + 9)$

82. $(z + 5)2$

83. $-6 + (6 + p)$

84. $\dfrac{1}{4}(4m)$

85. $(-t + 47) + t$

86. $-t + (39 + t)$

87. $5(2k + 3)$

88. $6(3k + 2)$

89. $\left(\dfrac{2}{7}r\right)\dfrac{7}{2}$

90. $\left(\dfrac{4}{9}r\right)\dfrac{9}{4}$

Discussion Exercises

91. Is subtraction a commutative operation? Is subtraction an associative operation? Give examples to support your answers.

92. Is division a commutative operation? Is division an associative operation? Give examples to support your answers.

93. What is an algebraic expression? Give examples.

Getting Acquainted with Your Graphing Calculator

STORING A VALUE   If you try to simplify $2(3x + 5)$ on your calculator, you will probably get a number for an answer, rather than $6x + 10$. Your calculator is using the value that has been stored in x to calculate the value of $2(3x + 5)$. The value that was stored in x at the factory was probably 0.

94. Store 4 in x and calculate the value of $2(3x + 5)$. Then calculate the value of $6x + 10$ and observe that the answer is the same.

95. Store 3 in x and 5 in y and calculate the value of $4(7x + 2y)$. Then calculate the value of $28x + 8y$ and observe that the answer is the same.

1.4 Operations on Real Numbers

TAPE AU1

Before you can use algebra to solve problems, you must know how to perform operations on real numbers. In this section we state the rules for addition, subtraction, multiplication, and division of real numbers.*

Addition

If a and b are real numbers, then the **sum** of a and b is written $a + b$.

---

*HISTORICAL NOTE: Although mathematicians had been aware of negative numbers for centuries, and some were even familiar with operations on them, it was not until about 350 years ago that negative numbers were granted widespread acceptance.

Adding Real Numbers

1. To add two numbers having the same sign, add their absolute values. The answer has the same sign as the given numbers.
2. To add two numbers having opposite signs, subtract the smaller absolute value from the larger. The answer has the same sign as the number with the larger absolute value.

Example 1   Add: (a) $-8 + (-2)$, (b) $3 + (-9)$.

Solution: (a) Since the numbers have the same sign, add the absolute values.

$$|-8| + |-2| = 8 + 2 = 10$$

The answer has the same sign as the given numbers:

$$-8 + (-2) = -10$$

(b) Since the numbers have opposite signs, subtract the smaller absolute value from the larger:

$$|-9| - |3| = 9 - 3 = 6$$

The answer has the same sign as -9, since -9 has a larger absolute value than 3:

Try Exercise 9

Add: $-15 + (-2)$.

$$3 + (-9) = -6 \qquad \square$$

CAUTION

To avoid confusion, always use parentheses to separate the symbols $+$, $-$, $\cdot$, and $\div$.

Preferred   *Not preferred*

$-8 + (-2)$ ~~$-8 + -2$~~   ■

When adding real numbers, you should perform the operations with absolute values mentally whenever possible.

Example 2   Find each sum: (a) $-30 + (-6)$, (b) $18 + (-5)$, (c) $4 + (-23)$, (d) $-\dfrac{3}{4} + \dfrac{1}{6}$.

Solution: (a) $-30 + (-6) = -36$ Both numbers are negative, so the sum is negative.

(b) $18 + (-5) = 13$ 18 has the larger absolute value.

(c) $4 + (-23) = -19$ -23 has the larger absolute value.

(d) $-\dfrac{3}{4} + \dfrac{1}{6} = -\dfrac{9}{12} + \dfrac{2}{12}$ Write with the least common denominator (LCD) 12.

$$= -\dfrac{7}{12}$$ $-\dfrac{9}{12}$ has the larger absolute value. $\qquad \square$

Try Exercise 11

Add: $11 + (-18)$.

Subtraction

If a and b are real numbers, then the **difference** of a and b is written $a - b$.

Consider the two problems below:

$$7 - 4 = 3 \quad \text{and} \quad 7 + (-4) = 3$$

Note that subtracting 4 is equivalent to adding the opposite of 4. This suggests the following definition.

> ### DEFINITION OF SUBTRACTION
>
> For any real numbers a and b,
>
> $$a - b = a + (-b)$$

This definition states that to subtract b from a, add the opposite of b to a. In other words, change the sign of the second number and add.

Example 3   Subtract: (a) $3 - 8$, (b) $11 - (-5)$.

Solution:

No change
Change subtraction to addition.
Opposite of 8

(a) $3 - 8 = 3 + (-8)$
$= -5$

No change
Change subtraction to addition.
Opposite of -5

(b) $11 - (-5) = 11 + 5$
$= 16$ ❒

Try Exercise 15

Subtract: $4 - 9$.

Example 4   Find each difference: (a) $-12 - 9$, (b) $-7 - (-13)$.

Solution: (a) $-12 - 9 = -12 + (-9) = -21$
subtract   positive 9   add   negative 9

(b) $-7 - (-13) = -7 + 13 = 6$ ❒
subtract   negative 13   add   positive 13

Try Exercise 17

Subtract: $-8 - (-23)$.

Sometimes several operations of addition and subtraction must be performed in the same problem. In that case we must perform the operation in parentheses (or brackets) first. If there are no parentheses, perform the additions and subtractions from left to right.

Example 5   Simplify: $15 - 9 - 5$.

Solution: Perform the subtraction on the left first.

$$15 - 9 - 5 = 6 - 5 = 1$$ ❒

Try Exercise 25

Simplify: $17 - 8 - 3$.

Example 6   Simplify: $38 - [4 - (-13)] + 20$.

Solution:
$$38 - [4 - (-13)] + 20 \qquad \text{Original problem}$$
$$= 38 - [4 + 13] + 20 \qquad \text{To subtract } -13, \text{ add } 13.$$
$$= 38 - 17 + 20 \qquad \text{Perform the addition in brackets.}$$
$$= 21 + 20 \qquad \text{Subtract, since } - \text{ is on the left.}$$
$$= 41$$

Try Exercise 27

Simplify:
$43 - [7 - (-14)] + 15$.

Multiplication

Recall that if a and b are real numbers, then the **product** of a and b can be written in any of the following ways:

$$a \times b \qquad a \cdot b \qquad ab \qquad a(b) \qquad (a)b \qquad (a)(b)$$

Multiplying Real Numbers

To multiply two numbers, multiply their absolute values.

1. If the numbers have the same sign, the product is positive.
2. If the numbers have opposite signs, the product is negative.

Example 7   Multiply: (a) $-5(-7)$, (b) $(-4)(8)$, (c) $2(-12)$, (d) $-\dfrac{5}{12}\left(\dfrac{8}{9}\right)$.

Solution:
(a) $-5(-7) = 35$ Same sign, product is positive.

(b) $(-4)8 = -32$ Opposite signs, product is negative.

(c) $2(-12) = -24$ Opposite signs, product is negative.

(d) $-\dfrac{5}{12} \cdot \dfrac{8}{9} = -\dfrac{5}{\overset{}{\underset{3}{12}}} \cdot \dfrac{\overset{2}{8}}{9}$ Divide both 8 and 12 by 4.

$$= -\dfrac{10}{27} \qquad \text{Opposite signs, product is negative.}$$

Try Exercise 37

Multiply: $-2(-9)$.

Since multiplication is commutative and associative, we can rearrange products of several numbers in any way we choose.

Example 8   Find each product: (a) $(-2)(-3)(-4)$, (b) $(-5)(-2)(-1)(-6)$.

Solution:
(a) $(-2)(-3)(-4) = 6(-4) = -24$
(b) $(-5)(-2)(-1)(-6) = 10 \cdot 6 = 60$

Try Exercise 43

Multiply: $(-3)(-2)(-11)$.

Note that **the product of an odd number of negative numbers is a negative number,** and **the product of an even number of negative numbers is a positive number.**

Division

If a and b are real numbers, then the **quotient** of a and b ($b \neq 0$) can be written in any of the following ways:

$$\dfrac{a}{b} \qquad a \div b \qquad b\overline{)a}$$

Consider the two problems below:

$$\frac{6}{2} = 3 \quad \text{and} \quad 6 \cdot \frac{1}{2} = 3$$

Note that dividing by 2 is equivalent to multiplying by the reciprocal of 2. This suggests the following definition.

> ### DEFINITION OF DIVISION
> For any real numbers a and b ($b \neq 0$),
> $$\frac{a}{b} = a \cdot \frac{1}{b}$$

Example 9 Divide: (a) $\dfrac{-20}{4}$, (b) $\dfrac{18}{-3}$, (c) $\dfrac{-16}{-8}$.

Solution: (a) $\dfrac{-20}{4} = -20 \cdot \dfrac{1}{4}$ To divide by 4, multiply by $\frac{1}{4}$.

$\qquad\qquad\qquad = -5$ Opposite signs, product is negative.

(b) $\dfrac{18}{-3} = 18\left(-\dfrac{1}{3}\right)$ To divide by -3, multiply by $-\frac{1}{3}$.

$\qquad\qquad = -6$ Opposite signs, product is negative.

(c) $\dfrac{-16}{-8} = -16\left(-\dfrac{1}{8}\right)$ To divide by -8, multiply by $-\frac{1}{8}$.

$\qquad\qquad = 2$ Same sign, product is positive. ❑

Try Exercise 49

Divide: $\dfrac{-45}{9}$.

Since dividing by a number is equivalent to multiplying by the reciprocal of the number, and since a number and its reciprocal have the same sign, the sign rules for division are the same as the sign rules for multiplication.

> **Dividing Real Numbers**
> To divide two numbers, divide their absolute values.
> 1. If the numbers have the same sign, the quotient is positive.
> 2. If the numbers have opposite signs, the quotient is negative.

Example 10 Find each quotient: (a) $\dfrac{-36}{-9}$, (b) $\dfrac{-7}{7}$, (c) $\dfrac{13}{-1}$.

Solution: (a) $\dfrac{-36}{-9} = 4$ Same sign, quotient is positive.

(b) $\dfrac{-7}{7} = -1$ Opposite signs, quotient is negative.

Try Exercise 51

Divide: $\dfrac{-18}{-6}$.

(c) $\dfrac{13}{-1} = -13$ Opposite signs, quotient is negative. ❑

Note that the three fractions

$$-\frac{6}{2} \qquad \frac{-6}{2} \qquad \text{and} \qquad \frac{6}{-2}$$

are all equal to -3, so they must be equal to each other. Also the two fractions

$$\frac{-6}{-2} \qquad \text{and} \qquad \frac{6}{2}$$

are both equal to 3, so they must be equal to each other. These examples suggest the following rules.

Signs of a Fraction

If $b \neq 0$,

$$-\frac{a}{b} = \frac{-a}{b} = \frac{a}{-b} \qquad \text{and} \qquad \frac{-a}{-b} = \frac{a}{b}$$

The forms $-\dfrac{a}{b}$ and $\dfrac{-a}{b}$ are preferred over $\dfrac{a}{-b}$. That is, we usually write $-\dfrac{2}{3}$ or $\dfrac{-2}{3}$, rather than $\dfrac{2}{-3}$.

Example 11   Simplify each fraction: (a) $\dfrac{-6}{-15}$, (b) $-\dfrac{-3}{7}$, (c) $\dfrac{8}{-24}$.

Solution: (a) $\dfrac{-6}{-15} = \dfrac{6}{15} = \dfrac{2}{5}$

(b) $-\dfrac{-3}{7} = \dfrac{-(-3)}{7} = \dfrac{3}{7}$

(c) $\dfrac{8}{-24} = -\dfrac{8}{24} = -\dfrac{1}{3}$ ❏

Try Exercise 65

Simplify: $-\dfrac{7}{-9}$.

Example 12   Find the quotient: $\dfrac{-10}{9} \div \left(\dfrac{15}{-7}\right)$

Solution: $\dfrac{-10}{9} \div \left(\dfrac{15}{-7}\right) = -\dfrac{10}{9} \div \left(-\dfrac{15}{7}\right)$ Reposition the $-$ signs.

$= -\dfrac{10}{9} \cdot \left(-\dfrac{7}{15}\right)$ Invert and multiply.

$= -\dfrac{\overset{2}{\cancel{10}}}{9} \cdot \left(-\dfrac{7}{\underset{3}{\cancel{15}}}\right)$ Divide both 10 and 15 by 5.

$= \dfrac{14}{27}$ Same sign, product is positive.   ❏

Try Exercise 75

Divide: $\dfrac{-5}{8} \div \left(\dfrac{25}{-24}\right)$.

Simplifying Algebraic Expressions

Since subtraction is defined in terms of addition, we can simplify the expression $2(x - 3)$ as follows:

$$
\begin{aligned}
2(x - 3) &= 2[x + (-3)] & &\text{Definition of subtraction} \\
&= 2 \cdot x + 2 \cdot (-3) & &\text{Distributive property} \\
&= 2x + (-6) & &\text{Since } 2 \cdot (-3) = -6 \\
&= 2x - 6 & &\text{Definition of subtraction}
\end{aligned}
$$

In actual practice, we would probably omit some of these steps and simply write

$$
\begin{aligned}
2(x - 3) &= 2x + 2(-3) \\
&= 2x - 6
\end{aligned}
$$

Example 13 Simplify each expression: (a) $x - 6 - 8$, (b) $\dfrac{-18r}{2}$, (c) $-(3m - 7)$.

Solution:

$$
\begin{aligned}
\text{(a) } x - 6 - 8 &= x + (-6) + (-8) & &\text{Change subtraction to addition.} \\
&= x + (-14) & &\text{Add.} \\
&= x - 14 & &\text{Change addition to subtraction.}
\end{aligned}
$$

$$
\begin{aligned}
\text{(b) } \frac{-18r}{2} &= \frac{-18}{2} \cdot \frac{r}{1} & &\text{Write as a product of two fractions.} \\
&= -9r & &\text{Divide.}
\end{aligned}
$$

$$
\begin{aligned}
\text{(c) } -(3m - 7) &= -1 \cdot (3m - 7) & &\text{Since } -a = -1 \cdot a \\
&= -1 \cdot 3m + (-1)(-7) & &\text{Distributive property} \\
&= -3m + 7 & &\text{Multiply.} \qquad \square
\end{aligned}
$$

Try Exercise 97

Simplify: $-(4m - 9)$.

Problem Solving

Example 14 For the past five years, the low temperatures on Christmas Day in Snowtown, U.S.A., were $-6°$, $2°$, $-11°$, $0°$, and $-5°$. Find the average of these temperatures.

Solution: Add the numbers and divide by 5:

$$
\text{Average} = \frac{-6 + 2 + (-11) + 0 + (-5)}{5} = \frac{-20}{5} = -4
$$

Try Exercise 105

A gambler bet on the first four horse races at a local track with the following results: lost $6, won $7.20, lost $10, won $4.80. Find the gambler's average winnings per race.

The average low temperature was $-4°$. $\qquad \square$

Exercises 1.4

◆◇◆

Completion Exercises

Assume a and b are real numbers.

1. The expression $a + b$ is called the _____ of a and b. The expression $a - b$ is called the _____ of a and b.

2. The expression $a \cdot b$ is called the _____ of a and b. The expression $\frac{a}{b}$ is called the _____ of a and b.

3. The definition of subtraction states that $a - b =$ _____.

4. The definition of division states that if $b \neq 0$, then $\frac{a}{b} =$ _____.

Discussion Exercises

5. Explain the sign rules for adding two real numbers.

6. Explain the sign rules for multiplying or dividing two real numbers.

7. What is the difference between $4 - 3$ and $4(-3)$?

8. Are $-\frac{2}{7}$ and $\frac{-2}{-7}$ equal? Why or why not? Are $-\frac{2}{7}$ and $\frac{-2}{7}$ equal? Why or why not?

Find each sum or difference.

9. $-15 + (-2)$ 10. $-3 + (-13)$

11. $11 + (-18)$ 12. $-17 + 12$

13. $-6 + 17$ 14. $15 + (-8)$

15. $4 - 9$ 16. $15 - (-7)$

17. $-8 - (-23)$ 18. $-22 - 5$

19. $0 - 10$ 20. $0 - (-6)$

21. $-3 - (-16)$ 22. $-14 - (-4)$

23. $-\frac{7}{8} + \frac{5}{12}$ 24. $\frac{3}{4} - \left(-\frac{1}{10}\right)$

Add or subtract as indicated.

25. $17 - 8 - 3$ 26. $2 - (-24) - 6$

27. $43 - [7 - (-14)] + 15$ 28. $58 - [3 - (-9)] + 10$

29. $-11 - (6 - 4) + (-18 + 7)$

30. $-14 - (-2 + 8) + (13 - 21)$

31. $-|12| - |-4| - (|7| - |-2|)$

32. $-|-16| - |5| - (|-8| - |3|)$

33. $-\frac{5}{6} - \left(\frac{1}{4} - \frac{2}{9}\right)$ 34. $-\frac{2}{3} - \left(\frac{1}{2} - \frac{7}{9}\right)$

35. $70 - 36 - (14 - 48) - [81 - (-4 - 7 + 3)]$

36. $54 - 28 - (16 - 30) - [63 - (-5 - 8 + 4)]$

Find each product.

37. $-2(-9)$ 38. $(-5)(-8)$

39. $(-7)3$ 40. $-4(6)$

41. $\left(-\frac{18}{25}\right)\left(-\frac{7}{12}\right)$ 42. $\frac{8}{12}\left(-\frac{10}{27}\right)$

43. $(-3)(-2)(-11)$ 44. $(-4)(5)(-10)$

45. $(-4)(-5)(-2)(-1)$ 46. $(-7)(-8)15(-4)2$

Find each quotient.

47. $-12 \div (-4)$ 48. $-15 \div 5$

49. $\frac{-45}{9}$ 50. $\frac{16}{-4}$

51. $\frac{-18}{-6}$ 52. $\frac{-24}{-8}$

53. $\frac{8}{-8}$ 54. $\frac{-12}{12}$

55. $\frac{-19}{1}$ 56. $\frac{7}{-1}$

57. $\frac{-32}{0}$ 58. $\frac{0}{-15}$

 ### *Getting Acquainted with Your Scientific Calculator*

To find $-8 + 3$ on your calculator, press

Perform each operation on your calculator.

59. $-6.34 + (-5.9)$

60. $-3.28 - (-7.5)$

61. $84.2(-7.35)$

62. $\dfrac{-46.72}{6.4}$

63. $\dfrac{0}{-8.17}$

64. $\dfrac{-24.3}{0}$

Simplify each fraction.

65. $-\dfrac{7}{-9}$

66. $-\dfrac{-3}{10}$

67. $\dfrac{-5}{20}$

68. $\dfrac{8}{-32}$

69. $-\dfrac{-48}{-60}$

70. $-\dfrac{-36}{-54}$

Find each quotient.

71. $-\dfrac{2}{3} \div \dfrac{3}{8}$

72. $\dfrac{3}{4} \div \left(-\dfrac{2}{5}\right)$

73. $\dfrac{6}{-5} \div (-4)$

74. $\dfrac{-10}{9} \div (-6)$

75. $\dfrac{-5}{8} \div \left(\dfrac{25}{-24}\right)$

76. $\dfrac{5}{-12} \div \left(\dfrac{-15}{8}\right)$

Simplify each expression.

77. $x - 9 + 3$

78. $x - 4 - 7$

79. $-8 + y - 2$

80. $-6 + y + 1$

81. $z - 14 - z$

82. $z + 19 - z$

83. $-4(5p)$

84. $2(-8p)$

85. $-\dfrac{3}{7}\left(-\dfrac{7}{3}h\right)$

86. $-\dfrac{9}{4}\left(-\dfrac{4}{9}h\right)$

87. $\dfrac{-18r}{3}$

88. $\dfrac{30r}{-6}$

89. $\dfrac{5t}{-5}$

90. $\dfrac{-7t}{7}$

91. $\dfrac{-k}{-1}$

92. $\dfrac{k}{-1}$

93. $9(3x - 4)$

94. $-3(9x - 7)$

95. $-12\left(\dfrac{5y}{6} + \dfrac{1}{2}\right)$

96. $16\left(\dfrac{3y}{4} + \dfrac{1}{8}\right)$

97. $-(4m - 9)$

98. $-(1 - 6m)$

99. $-(-a + 5b - 2c)$

100. $-(-4a - 2b + c)$

Problem Solving

Use operations on real numbers to solve each problem.

101. The temperature is $-17°$. If it rises $4°$, find the new temperature.

102. A checking account has a balance of $\$-8$. If $\$25$ is deposited into the account, find the new balance.

103. Determine the difference in elevation between Mount Everest at 29,028 ft above sea level and the Dead Sea at 1286 ft below sea level.

104. Determine the difference in temperature between an afternoon high of $-3°$ and a morning low of $-15°$.

105. A gambler bet on the first four horse races at a local track with the following results: lost $\$6$, won $\$7.20$, lost $\$10$, won $\$4.80$. Find the gambler's average winnings per race.

106. During the course of a football game, a quarterback is tackled six times with the following results: 5 yd loss, 3 yd gain, 7 yd loss, no gain, 1 yd loss, 8 yd loss. Find the average gain per carry.

Discussion Exercises

107. Does $|a - b| = |b - a|$ for all real numbers a and b? Why or why not?

108. Does $|a + b| = |a| + |b|$ for all real numbers a and b? Why or why not?

Getting Acquainted with Your Graphing Calculator

SUBTRACT VERSUS OPPOSITE OF   Your calculator probably has two different keys, namely ⊟ and ⊡, to reflect "subtract" and "opposite of," respectively. Try each exercise below with both keys, and discuss which is the correct key for each exercise and why.

109. $5 - 3$

110. $5(-3)$

1.5 Exponents and Order of Operations

TAPE AU1

In the previous section we learned how to add, subtract, multiply, and divide real numbers. In this section we learn a fifth operation—raising a number to a power. We shall also learn which operation to perform first when we are solving a problem that contains several different operations.

Exponents

If a and b are integers and $a \cdot b = c$, we say that a and b are **factors** of c. For example since $2 \cdot 3 = 6$, we say that 2 and 3 are factors of 6.

$$2 \cdot 3 = 6$$

Factors   Product

Other factors of 6 are 1, 6, -1, -2, -3, and -6.

In algebra we write repeated factors such as

$$5 \cdot 5 \cdot 5 \qquad \text{and} \qquad x \cdot x \cdot x \cdot x$$

using *exponents*. For example, we write $5 \cdot 5 \cdot 5$ as 5^3. The number 5 is called the **base** and the number 3 is called the **exponent.** The expression 5^3 is called a **power of 5.** The product $x \cdot x \cdot x \cdot x$ is written x^4 and is called a power of x.

DEFINITION OF EXPONENTIAL NOTATION*

If a is a real number and n is a natural number, then

$$a^n = \underbrace{a \cdot a \cdot a \cdot \cdots \cdot a}_{\mathbf{n} \ factors \ of \ \mathbf{a}}$$

Example 1 Evaluate: (a) 7^2, (b) 4^3, (c) 2^4, (d) 9^1, (e) $(-3)^2$, (f) -3^2.

Solution:

(a) $7^2 = 7 \cdot 7 = 49$	Read 7^2 as "7 squared."
(b) $4^3 = 4 \cdot 4 \cdot 4 = 64$	Read 4^3 as "4 cubed."
(c) $2^4 = 2 \cdot 2 \cdot 2 \cdot 2 = 16$	Read 2^4 as "2 to the fourth power."
(d) $9^1 = 9$	Read 9^1 as "9 to the first power."
(e) $(-3)^2 = (-3)(-3) = 9$	The base is -3.
(f) $-3^2 = -(3 \cdot 3) = -9$	The base is 3.

Try Exercise 7

Evaluate: 3^5.

CAUTION

The base in -6^2 is 6, *not* -6. Therefore to evaluate -6^2, square the 6 and *then* apply the negative sign.

---

*HISTORICAL NOTE: Exponential notation (in the form used today) was introduced by the French mathematician and philosopher René Descartes (1596–1650) in his book *La géométrie.* In this book, Descartes used the symbol ∞ rather than = for equality, but, other than that, it was the first mathematical text whose notation was essentially the same as the notation in use today.

Correct   *Wrong*

$-6^2 = -36$ ~~$-6^2 = 36$~~

That is, $(-6)^2 = (-6)(-6) = 36$, but $-6^2 = -(6 \cdot 6) = -36$.   ■

Order of Operations

Consider the problem

$$3 + 2 \cdot 5$$

Do we add first and get 25, or do we multiply first and get 13? To avoid this confusion, we have agreed on the following **order of operations.**

Order of Operations

Perform operations in the following order.

1. Do all operations above and below a fraction bar separately.
2. Do all operations in parentheses ( ) and brackets [ ]. Apply the innermost grouping symbols first and work outward.
3. Apply all exponents.
4. Do all multiplications and divisions, working from left to right.
5. Do all additions and subtractions, working from left to right.

Since we multiply in step 4 and add in step 5, we multiply before adding. Therefore we multiply first in the expression $3 + 2 \cdot 5$ and get

$$3 + 2 \cdot 5 = 3 + 10 = 13$$

The following examples further illustrate how to apply the order of operations rules.

Example 2   Evaluate: $16 - 2 \cdot 3^2 + 8$.

Solution: $16 - 2 \cdot 3^2 + 8 = 16 - 2 \cdot 9 + 8$ Apply the exponent.
                                  $= 16 - 18 + 8$ Multiply before adding or subtracting.

                                  $= -2 + 8$ Subtract since $-$ is on the left.

Try Exercise 27 $= 6$ Add.   ❏

Evaluate: $13 - 4 \cdot 3^2 + 9$.

Example 3   Evaluate: $(9 - 27) \div 2 + 2$.

Solution: $(9 - 27) \div 2 + 2 = -18 \div 2 + 2$ Subtract, since $-$ is in parentheses.

                              $= -9 + 2$ Divide before adding.

Try Exercise 31 $= -7$ Add.   ❏

Evaluate: $(4 + 18) \div 2 + 4$.

Example 4   Evaluate: $7 + 3(2[17 - 4(8 - 5)])$.

Solution:

$7 + 3(2[17 - 4(8 - 5)])$	Original problem
$= 7 + 3(2[17 - 4(3)])$	Apply innermost grouping symbols.
$= 7 + 3(2[17 - 12])$	Multiply before subtracting.
$= 7 + 3(2[5])$	Apply innermost grouping symbols.
$= 7 + 30$	Multiply in any order.
$= 37$	Add. ❑

Try Exercise 45

Evaluate:
$5 + 5(2[12 - 3(9 - 7)])$.

Example 5   Evaluate: $\dfrac{3(-7) + (-9)(-5)}{2(-4 - 6)}$.

Solution: Do all operations above and below the fraction bar separately:

$$\frac{3(-7) + (-9)(-5)}{2(-4 - 6)} = \frac{-21 + 45}{2(-10)} = \frac{24}{-20} = -\frac{6}{5}$$ ❑

Evaluating Algebraic Expressions

An algebraic expression takes on different values as its variable(s) takes on different values.

Try Exercise 49

Evaluate:
$\dfrac{4(-8) + (-6)(-10)}{2(-3 - 3)}$.

Example 6   Evaluate: $3x - 5y - 9$ at $x = 4$ and $y = -2$.

Solution: Replace x with 4 and y with -2 and follow the order of operations:

$$3x - 5y - 9 = 3(4) - 5(-2) - 9$$
$$= 12 + 10 - 9$$
$$= 13$$ ❑

Try Exercise 65

Evaluate: $2x - 4y + 7$ at $x = 3$ and $y = -5$.

Example 7   Evaluate: $\dfrac{-a^2 - b^3}{4c}$ at $a = 7$, $b = -1$, and $c = 3$.

Solution: Replace a with 7, b with -1, and c with 3 and follow the order of operations:

$$\frac{-a^2 - b^3}{4c} = \frac{-7^2 - (-1)^3}{4(3)} = \frac{-49 - (-1)}{12} = \frac{-49 + 1}{12}$$
$$= \frac{-48}{12}$$
$$= -4$$ ❑

Try Exercise 73

Evaluate: $\dfrac{-a^2 - b^3}{4c}$ at $a = -2$, $b = 4$, and $c = -1$.

Problem Solving

Example 8   A woman's boutique sold six hats at a profit of $15 per hat, nine purses at a profit of $12 per purse, and four scarves at a loss of $3 per scarf. Determine the net profit on the 19 sales.

Solution: Treat the profits as positive numbers and the losses as negative numbers:

$6(15) + 9(12) + 4(-3) = 90 + 108 + (-12)$	Multiply before adding.
$= 186$	Add in any order.

Try Exercise 79

A furniture store sold seven chairs at a profit of $90 per chair, four rugs at a loss of $23 per rug, and 11 lamps at a profit of $35 per lamp. Determine the net profit on the 22 sales.

The net profit was $186.   ❑

Exercises 1.5

Completion Exercises

1. If a and b are integers and $a \cdot b = c$, we say that a and b are _____ of c.

2. In 3^4, the number 3 is called the _____, and the number 4 is called the _____.

3. The expression 3^4 is called a _____ of 3.

4. If a is a real number and n is a natural number, then $a^n =$ _____.

True-False Exercises

5. The base in $(-4)^2$ is -4.

6. The base in -4^2 is -4.

Evaluate.

7. 3^5
8. 6^2
9. 10^1
10. 8^1
11. 1^4
12. 0^3
13. $(-5)^2$
14. $(-4)^4$
15. -5^2
16. -4^4
17. $\left(-\dfrac{3}{4}\right)^3$
18. $\left(-\dfrac{2}{5}\right)^5$

Getting Acquainted with Your Scientific Calculator

To find 5^2 on your calculator, press

Answer
↓

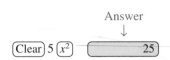

To find $(-4)^3$ on your calculator, press

Answer
↓

Note:

Some scientific calculators will not raise a negative number to a power. When calculating $(-4)^3$, your calculator may give the incorrect answer 64.

Find each power on your calculator.

19. 9^2
20. 6.6^3
21. $(-7)^5$
22. 4^{10}

Discussion Exercises

23. Explain the difference between $(-3)^2$ and -3^2.

24. Both $(-2)^3$ and -2^3 are equal to -8, but for different reasons. Discuss why.

Evaluate using the order of operations.

25. $7 + 8 \cdot 9 - 3$
26. $25 - 5 \cdot 2 + 6$
27. $13 - 4 \cdot 3^2 + 9$
28. $17 - 2 \cdot 5^2 + 3$
29. $28 - 4 \div 2$
30. $-24 \div 6 - 3$
31. $(4 + 18) \div 2 + 4$
32. $(8 - 32) \div 6 - 2$
33. $1 - 6^2 - 4^3$
34. $(7 - 2)^2 - 5 - 8 \cdot 4$
35. $3^3 \div 9 - 5(3 - 4)$
36. $2^3 \div 4 - 2(3 - 8)$
37. $50 \div 5 \cdot 5 \div 50$
38. $50 \cdot 5 \div 5 \cdot 5$
39. $8[7 - (4 + 3)]$
40. $4[13 - 3(4 - 1)]$
41. $3(2[12 - 4(8 - 6)])$
42. $6[-19 + 5(6 - 14)]$
43. $32 \div [-6 - 11(3 - 5)]$
44. $21 \div [-3 - 5(7 - 9)]$
45. $5 + 5(2[12 - 3(9 - 7)])$
46. $8 + 2(3[19 - 4(6 - 2)])$
47. $\dfrac{6 - 3(2 + 4)}{27 - 10}$
48. $\dfrac{7 - 2(5 + 1)}{29 - 10}$
49. $\dfrac{4(-8) + (-6)(-10)}{2(-3 - 3)}$
50. $\dfrac{(-5)(-9) + (-7)3}{2(-3 - 2)}$
51. $\dfrac{5(-2) + 3(-1) - 9}{-6 - 7 - 2(-1)}$
52. $\dfrac{4(-1) + 3(-2) - 23}{-1 - 1 - 13(-1)}$
53. $\left[\dfrac{9 - 5(-3)}{8 - 4}\right]\left[\dfrac{5 + (-10)}{-2 - 3}\right]$
54. $\left[\dfrac{-7 + 3}{-1 + 5}\right]\left[\dfrac{28 + 4(-10)}{2 - 8}\right]$

55. $\dfrac{-3^2[1 - 2(5 - 4)^2]^2}{2 \cdot 4 - 6^2 \div 3 + 1}$

56. $\dfrac{-2^2[9 - 3(8 - 6)^2]^2}{2 \cdot 4 - 6^2 \div 2 + 7}$

57. $\dfrac{19 - 3\left[\dfrac{4 + 6}{5}\right] + 9}{34 + 34 \div 2 - 17}$

58. $\dfrac{12 - 3\left[\dfrac{7 + 5}{4 - 8}\right] + 2}{-6 + 4\left[\dfrac{1 - 4}{6 - 3}\right] - 9}$

59. $\dfrac{-1}{6}\left(-\dfrac{9}{5}\right) + \left(\dfrac{1}{-5}\right) \div \left(-\dfrac{2}{7}\right)$

60. $\left(\dfrac{5}{8}\right)^2 \div \left(\dfrac{7}{12} - \dfrac{3}{4}\right)$

Getting Acquainted with Your Scientific Calculator

Your calculator should follow the order of operations. To check, calculate $3 + 2 \cdot 5$ as follows:

Answer
↓

(Clear) 3 (+) 2 (×) 5 (=)   [13]

If your calculator multiplied first (as it should), the answer will be 13. If it added first, the answer will be 25. Try these problems on your calculator.

61. $3 \cdot 4^2 - 7$

62. $-20 \div 5 \cdot 2$

63. $8 + 12 \div 4 + 2$

64. $6 - 9(3 + 2)^2$

Evaluate each algebraic expression at $x = 3$ and $y = -5$.

65. $2x - 4y + 7$

66. $5x - 3y - 1$

67. $\dfrac{x}{y + 5}$

68. $\dfrac{y}{x - 3}$

69. $-y^2 + xy - 9x$

70. $-x^2 - xy + 2y$

Evaluate each algebraic expression at $a = -2$, $b = 4$, and $c = -1$.

71. $6|a| - 3|b| - |c|$

72. $(a + 1)^3 - (b + c)^2$

73. $\dfrac{-a^2 - b^3}{4c}$

74. $\dfrac{2(ab - bc + ac)}{ab}$

Problem Solving

Solve each word problem by setting up the appropriate operations and then applying the order of operations.

75. An elevator contained eight people. At the first stop four people got off and two got on. At the second stop one person got off. How many people are now on the elevator?

76. A bus was loaded with 29 people. At the first stop seven people got off and three got on. At the second stop four people got off and five got on. How many people are now on the bus?

77. The temperature is $-19°$. If it rises at the rate of $4°$ per hour for 3 hr and then falls $5°$, what is the new temperature?

78. A party of spelunkers (cave explorers) is at an elevation of -73 ft. If the party ascends at the rate of 6 ft/hr for 4 hr and then descends 8 ft, what is the new elevation?

79. A furniture store sold seven chairs at a profit of $90 per chair, four rugs at a loss of $23 per rug, and 11 lamps at a profit of $35 per lamp. Determine the net profit on the 22 sales.

80. A used car dealer sold three cars at a loss of $75 per car, six cars at a profit of $95 per car, and four cars at a profit of $150 per car. Determine the net profit on the 13 sales.

Discussion Exercises

81. What is the purpose of exponential notation?

82. Why do we have an order of operations? Why do we multiply before we add and not vice-versa?

Getting Acquainted with Your Graphing Calculator

USING PARENTHESES A scientific calculator may compute -3^2 incorrectly as 9, and may not compute $(-3)^2$ at all. However, your graphing calculator should compute -3^2 correctly as -9, and $(-3)^2$ correctly as 9. Try these problems on your calculator.

83. (a) $(-3)^2$, (b) -3^2

84. (a) $(-15.23)^2$, (b) -15.23^2

Evaluate, inserting grouping symbols as shown. (*Hint:* You will have to put parentheses around the numerator and the denominator in Exercise 86 to make certain the operations above and below the fraction bar are done separately. Otherwise, your calculator will do the problem $-6^2 + 5(-4) \div 2 - 18 \div 3$ and get an incorrect answer of -52.)

85. $6 + 2(4[15 - 8(3 - 7)])$

86. $\dfrac{-6^2 + 5(-4)}{2 - 18 \div 3}$

87. $-|8 + 3| - (|4| - |9|)$

88. Store -5 in x and evaluate $-x^2 + 3x + 7$.

You can calculate x^3 by using an exponential key or by pressing $\boxed{x}$ three times. Similarly, your calculator assumes multiplication between a constant and a variable, so you can calculate $2x^3$ by pressing $2\,\boxed{x}\,\boxed{x}\,\boxed{x}$. Store 5 in x and -2 in y and calculate without using an exponential key or a multiplication key.

89. x^3

90. $2x^3$

91. $x^2 + 3x + 7$

92. $-x^2 - xy + 4y^2$

Chapter 1 Key Terms

1.1 Set A collection of objects

Element or **member** (of a set) An object in a set

Equal sets Sets that contain exactly the same elements

Finite set A set with a limited number of elements

Infinite set A set with an unlimited number of elements

Empty (null) set The set with no elements

Set-builder notation A method for identifying the elements of a set without listing them, written $\{x \mid x$ has property $P\}$

Variable A symbol that can represent any of the elements in a set

Constant A symbol that represents only one value throughout a particular discussion

Subset Set A is a subset of set B, written $A \subseteq B$, if every element of A is also an element of B.

Venn diagram A diagram that depicts sets as regions

Set operation A process that forms a new set from given sets

Perfect squares $1, 4, 9, 16, 25, \ldots$

1.2 Number line A line used to illustrate the set of real numbers

Positive number A number whose graph lies to the right of 0 on the number line

Negative number A number whose graph lies to the left of 0 on the number line

Signed (directed) numbers Positive and negative numbers

Graph (of a number) The point on the number line that corresponds to the number

Coordinate (of a point on the number line) The number that corresponds to the point

Opposite (additive inverse) The opposite of a is $-a$.

Absolute value or **magnitude** (of a real number) The distance between the number and 0

Inequality A statement that two numbers are not equal

Less than The real number a is less than the real number b, written $a < b$, if the graph of a lies to the left of the graph of b on the number line.

Double inequality $a < x < b$

Graph (of an inequality) The graph of all those values of the variable that make the inequality a true statement

1.3 Algebraic expression A collection of constants, variables, and/or operations ($+$, $-$, $\times$, $\div$, and roots)

Additive identity The number 0

Multiplicative identity The number 1

Multiplicative inverse (reciprocal) If $a \neq 0$, the multiplicative inverse of a is $1/a$.

Theorem A property that is proved using other properties

1.4 Sum (of a and b) $a + b$

Difference (of a and b) $a - b$

Product (of a and b) $a \cdot b$

Quotient (of a and b, $b \neq 0$) a/b

1.5 Factor If a and b are integers and $a \cdot b = c$, then a and b are factors of c.

Base In a^n, the base is a.

Exponent In a^n, the exponent is n.

Power (of a) a^n

Order of operations An agreed-upon order for performing arithmetic operations

Chapter 1 Key Rules/Steps

1.1 Sets of Numbers

Let $A = \{3, 5, 7\}$ and $B = \{3, 4, 5, 6, 7\}$.

Symbol	Meaning	Example
$\in$	is an element of	$5 \in A$
$\notin$	is not an element of	$6 \notin A$
$\subseteq$	is a subset of	$A \subseteq B$
$\not\subseteq$	is not a subset of	$B \not\subseteq A$

Set-Builder Notation

$\{x \mid x$ is an odd whole number$\} = \{1, 3, 5, 7, \ldots\}$
$\{y \mid y$ is a letter in the word book$\} = \{k, o, b\}$

For any set A, $A \subseteq A$ and $\varnothing \subseteq A$

Examples: $\{1, 5\} \subseteq \{1, 5\}$ and $\varnothing \subseteq \{1, 5\}$

Set Operations

Union: $A \cup B = \{x \mid x \in A$ or $x \in B\}$
Example: $\{1, 3\} \cup \{2, 3, 4\} = \{1, 2, 3, 4\}$

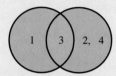

Intersection: $A \cap B = \{x \mid x \in A$ and $x \in B\}$

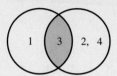

Example: $\{1, 3\} \cap \{2, 3, 4\} = \{3\}$

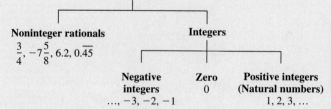

Real numbers
$\{x \mid x$ is a decimal$\}$

Irrational numbers
$\{x \mid x$ is a decimal that neither terminates nor repeats$\}$
or
$\{x \mid x$ is a real number that is not rational$\}$
$\sqrt{3}$, π, $0.8484484448\cdots$

Rational numbers
$\{x \mid x$ is a terminating or repeating decimal$\}$
or
$\{\frac{a}{b} \mid a$ and b are integers and $b \neq 0\}$

Noninteger rationals
$\frac{3}{4}$, $-7\frac{5}{8}$, 6.2, $0.\overline{45}$

Integers

Negative integers
$\ldots, -3, -2, -1$

Zero
0

Positive integers (Natural numbers)
$1, 2, 3, \ldots$

Whole numbers $= \{0, 1, 2, 3, \ldots\}$

1.2 The Number Line; Absolute Value; Inequalities

Double Negative Rule

For any real number a, $-(-a) = a$.
Example: $-(-5) = 5$

Absolute Value

If a is a real number, then

$$|a| = \begin{cases} a & \text{if } a \geq 0 \\ -a & \text{if } a < 0 \end{cases}$$

Examples: $|6| = 6$ and $|-8| = -(-8) = 8$

Symbol	Meaning	Example
$=$	is equal to	$9 = 9$
$\neq$	is not equal to	$-5 \neq 5$
$<$	is less than	$0 < 1$
$>$	is greater than	$-6 > -60$
$\leq$	is less than or equal to	$\frac{1}{5} \leq \frac{1}{4}$
$\geq$	is greater than or equal to	$10 \geq 10$

Graphing Inequalities

$\{x \mid x < 2\}$

$\{x \mid x \geq 0\}$

$\{x \mid -1 < x \leq 1\}$

1.3 Properties of Real Numbers

Properties of Equality

For all real numbers a, b, and c:

Reflexive property: $a = a$
Example: $m + 3 = m + 3$
Symmetric property: If $a = b$, then $b = a$.
Example: If $x = 2$, then $2 = x$.
Transitive property: If $a = b$ and $b = c$, then $a = c$.
Example: If $x = 4$ and $4 = y$, then $x = y$.
Substitution property: If $a = b$, then b can be substituted for a in any algebraic expression and the result is an equivalent expression.
Example: If $x = -5$, then $2x + 3 = 2(-5) + 3$.

Properties of Inequality

Let a, b, and c be real numbers.

Trichotomy property

$a < b$ or $a = b$ or $a > b$
Example: $x < 1$ or $x = 1$ or $x > 1$

Transitive property

If $a < b$ and $b < c$, then $a < c$.
Example: If $x < 6$ and $6 < y$, then $x < y$.

Properties of Addition and Multiplication

For all real numbers a, b, and c:

Commutative properties
$$a + b = b + a \text{ and } ab = ba$$
Examples: $x + 7 = 7 + x$ and $4(-3) = (-3)4$

Associative properties
$$a + (b + c) = (a + b) + c \text{ and } a(bc) = (ab)c$$
Examples: $-5 + (6 + 9) = (-5 + 6) + 9$
$$2(8y) = (2 \cdot 8)y$$

Identity properties
$$a + 0 = 0 + a = a \text{ and } a \cdot 1 = 1 \cdot a = a$$
Examples: $-9 + 0 = 0 + (-9) = -9$ and $3 \cdot 1 = 1 \cdot 3 = 3$

Inverse properties
$$a + (-a) = -a + a = 0 \text{ and } a \cdot \frac{1}{a} = \frac{1}{a} \cdot a = 1 \ (a \neq 0)$$

Examples: $7 + (-7) = -7 + 7 = 0$ and $4 \cdot \frac{1}{4} = \frac{1}{4} \cdot 4 = 1$

Distributive property
$$a(b + c) = ab + ac$$
Example: $5(m + 3) = 5 \cdot m + 5 \cdot 3$

Multiplication Property of 0

For any real number a,

$$a \cdot 0 = 0 \cdot a = 0.$$
Example: $(-37) \cdot 0 = 0 \cdot (-37) = 0$

1.4 Operations on Real Numbers

Adding Real Numbers

Same sign: Add the absolute values. The answer has the same sign as the given numbers.
Examples: $3 + 5 = 8$ and $-4 + (-7) = -11$

Opposite signs: Subtract the smaller absolute value from the larger. The answer has the same sign as the number with the larger absolute value.
Examples: $-15 + 2 = -13$ and $9 + (-6) = 3$

Definition of Subtraction

For any real numbers a and b, $a - b = a + (-b)$.
Examples: $3 - 7 = 3 + (-7) = -4$
$$-10 - 5 = -10 + (-5) = -15$$
$$-1 - (-8) = -1 + 8 = 7$$

Multiplying Real Numbers

Multiply the absolute values.
Same sign: Product is positive.
Examples: $2 \cdot 6 = 12$ and $(-3)(-5) = 15$

Opposite signs: Product is negative.
Examples: $(-1)8 = -8$ and $4(-9) = -36$

Definition of Division

For any real numbers a and b $(b \neq 0)$, $\dfrac{a}{b} = a \cdot \dfrac{1}{b}$.

Examples: $\dfrac{6}{2} = 6 \cdot \dfrac{1}{2}$ and $\dfrac{-15}{5} = -15 \cdot \dfrac{1}{5}$

Dividing Real Numbers

Divide the absolute values.
Same sign: Quotient is positive.

Examples: $\dfrac{12}{4} = 3$ and $\dfrac{-7}{-1} = 7$

Opposite signs: Quotient is negative.

Examples: $\dfrac{-18}{2} = -9$ and $\dfrac{10}{-5} = -2$

Signs of a Fraction

If $b \neq 0$, $-\dfrac{a}{b} = \dfrac{-a}{b} = \dfrac{a}{-b}$ and $\dfrac{-a}{-b} = \dfrac{a}{b}$.

Examples: $-\dfrac{2}{5} = \dfrac{-2}{5} = \dfrac{2}{-5}$ and $\dfrac{-3}{-8} = \dfrac{3}{8}$

1.5 Exponents and Order of Operations

Definition of Exponential Notation

If a is a real number and n is a natural number, then

$$a^n = \underbrace{a \cdot a \cdot a \cdots a}_{n \text{ factors of } a}$$

Example: $a^5 = a \cdot a \cdot a \cdot a \cdot a$

Order of Operations

Perform operations in the following numerical order.
1. Do all operations above and below a fraction bar separately.

   Example: $\dfrac{12 + 6}{4 - 2} = \dfrac{18}{2} = 9$

2. Do all operations in grouping symbols.
3. Apply all exponents
4. Do all multiplications and divisions, from left to right.
5. Do all additions and subtractions, from left to right.
Example:
$$10 - 4(2 \cdot 3^2 - 16) + 5 = 10 - 4(2 \cdot 9 - 16) + 5$$
$$= 10 - 4(2) + 5$$
$$= 10 - 8 + 5 = 2 + 5 = 7$$

Evaluating Algebraic Expressions

To evaluate $6x - 2y + 3$ at $x = -4$ and $y = -5$, write
$6(-4) - 2(-5) + 3 = -24 + 10 + 3 = -11$.

Chapter 1 Review Exercises

Assume all variables in this review represent real numbers.

[1.1] State whether each set is finite or infinite.

1. $\{x \mid x \text{ is an integer}\}$ 2. $\{8, 9, 10, \ldots, 100\}$

Insert $\in$ or $\notin$ in the blank space to make a true statement.

3. -5.71_____$\{x \mid x \text{ is a rational number}\}$
4. $\{33\}$_____$\{1, 3, 5, \ldots\}$

Write each set using the listing method.

5. $\{p \mid p \text{ is a whole number less than 7}\}$
6. $\{y \mid y \text{ is a natural number divisible by 5}\}$

Write each set using set-builder notation.

7. $\{10, 12, 14, 16, \ldots\}$ 8. $\{a, e, i, o, u\}$

Insert $\subseteq$ or $\nsubseteq$ in the blank space to make a true statement.

9. 20_____$\{17, 18, 19, 20\}$
10. $\{x \mid x \text{ is an irrational number}\}$_____$\{x \mid x \text{ is a real number}\}$

Suppose $A = \{3, 7, 9\}$, $B = \{2, 5, 7\}$, and $C = \{1, 3, 5\}$. List the elements in each set.

11. $(A \cap B) \cup C$ 12. $(A \cup B) \cup (C \cap \varnothing)$

13. Consider the set $\{12, -4.9, \sqrt{27}, 0, 5\frac{3}{4}, -99, 8.\overline{7}\}$. Identify the elements that are

   (a) Natural numbers   (b) Whole numbers
   (c) Integers   (d) Rational numbers
   (e) Irrational numbers (f) Real numbers

14. List all subsets of $\{a, b, c, d\}$.

[1.2]

15. Graph each number on a number line:

   $$\left\{2, -\frac{7}{4}, \sqrt{10}\right\}.$$

Simplify.

16. $-(-(-6.28))$ 17. $-|-14|$
18. $|3 + 7|$ 19. $|-9| - |1|$

Insert $<$ or $>$ in the blank space to make a true statement.

20. -5.02_____-5.2
21. $|-11|$_____$|4|$

Use inequality symbols to write the following word statements.

22. z is at most 6.

23. m is more than 8 but no more than 18.

24. The temperature, t, ranged from a morning low of $-3°$ to an afternoon high of $24°$.

Graph each set on a number line. Then rewrite each set so that the inequality symbol(s) points (point) in the opposite direction.

25. $\left\{x \mid x > -\dfrac{10}{3}\right\}$ 26. $\{t \mid 0 \leq t < 7\}$

[1.3] Name the property of equality or inequality that justifies each statement.

27. If $x = 1$, then $1 = x$.
28. If $-4 < y$ and $y < z$, then $-4 < z$.
29. If $t = 10$ and $p = 2t$, then $p = 2(10)$.
30. $m = 0$ or $m > 0$ or $m < 0$.

Use the given property to complete each statement.

31. Reflexive property: $z + 9 = $ __?__ .
32. Transitive property: If $x = 11$ and $11 = y$, then __?__ .

Name the property of addition and/or multiplication that justifies each statement.

33. $\frac{3}{7} \cdot 1 = \frac{3}{7}$ 34. $0(-6) = 0$
35. $8(ab) = (8a)b$
36. $4(5y + 9) = 4(5y) + 4(9)$
37. $-z + z = 0$ 38. $8(ab) = (ab)8$

Use the given property to complete each statement.

39. $0 + 4 = $ __?__   Commutative property
40. $3(9x) = $ __?__   Associative property
41. $y + 0 = $ __?__   Identity property
42. $\frac{5}{9} \cdot \frac{9}{5} = $ __?__   Inverse property

Find the reciprocal of each number. Simplify your answer.

43. -8 44. 2.6

Use the distributive property to complete each statement.

45. $2(p + 14) = $ __?__   46. $5y + 7y = $ __?__

Simplify each expression.

47. $-19 + (x + 19)$ 48. $(5y)2$
49. $\frac{1}{7}(7z)$ 50. $4(3p + 8)$

[1.4] Perform the indicated operations.

51. $12 - 15 - (-3)$

52. $62 - |-8 - 2| - (25 - 17)$

53. $(-\frac{7}{8} + \frac{1}{6}) - \frac{5}{12}$

54. $-7(2)(-3)$

55. $12(-4)\frac{11}{18}$

56. $\frac{-2}{15} \div (-\frac{3}{4})$

Simplify each expression.

57. $x - 12 - x$

58. $-8 + y + 19$

59. $-3(-z)$

60. $-4(2r + 7)$

61. $-\dfrac{-54}{-9}$

62. $-\dfrac{-6}{0}$

63. $\dfrac{24t}{-6}$

64. $-10(-\frac{4}{5}m)$

65. $\frac{5}{12}(6p - \frac{3}{10})$

66. $-(-7a - b + 13)$

[1.5] Perform the indicated operations.

67. $(-\frac{5}{3})^3$

68. -8^2

69. $16 - 2 \cdot 4^2 + 9$

70. $-48 \div 4 \cdot 2$

71. $2^5 \div 8 - 3(1 - 7)$

72. $6[25 - 15(5 - 3)]$

73. $12 - 2[18 - 3(9 - 5)]$

74. $\dfrac{-4}{9} \div \left(-\dfrac{1}{18}\right) - \left(\dfrac{3}{2}\right)^2$

75. $\dfrac{-5(7) + 19 - 8(-3)}{-1 - 6 - (-4)(-2)}$

76. $\dfrac{-19 - 5^2 + 9(4)}{20 + 16\left(\dfrac{13 + 7}{3 - 11}\right) + 22}$

Evaluate each expression at $x = -4$ and $y = 2$.

77. $3x^2 - xy + 7y$

78. $\dfrac{5y - |x| - 12}{2(x + y)}$

79. The temperature is 6°. It falls 2° per hour for 5 hr, and then rises 3°. Find the new temperature.

Chapter 1 Test

Assume all variables represent real numbers.

1. Write $\{1, 2, 3, \ldots, 88\}$ using set-builder notation.

2. List all subsets of $\{0, 9\}$.

3. If $A = \{1, 4, 6\}$, $B = \{6, 7, 8\}$, and $C = \{1, 5, 7\}$, find $(A \cup C) \cap (B \cup \varnothing)$.

Graph each set on a number line. Then rewrite each set so that the inequality symbol(s) points (point) in the opposite direction.

4. $\{x | x \le -\dfrac{5}{2}\}$

5. $\{y | 0 < y < 4\}$

In the set $\{-\sqrt{7}, \dfrac{18}{5}, 0.375, -\sqrt{36}, 0.1010010001 \cdots\}$, identify the elements that are

6. Integers

7. Rational numbers

Simplify each expression.

8. $\dfrac{0}{-32}$

9. $20\left(\dfrac{3x}{5} - \dfrac{7}{4}\right)$

10. $-(6y + 9) + 9$

11. $|16| - |-3| + |-7 - 1| - (-8)$

Use inequality symbols to write the following word statements.

12. z is a negative number.

13. p is at least 2 and at most 15.

Name the property that justifies each statement.

14. $x - 4 = x - 4$

15. $6 + (9 + 2) = 6 + (2 + 9)$

16. $3a + 3b = 3(a + b)$

17. If $y = k$ and $k = -7$, then $y = -7$.

True or false. A and B are sets.

18. If $A \subseteq B$ and $6 \in A$, then $6 \in B$.

19. If $9 \in A$, then $9 \in A \cup B$.

20. The reciprocal of $-\dfrac{3}{11}$ is $\dfrac{11}{3}$.

Perform the indicated operations.

21. $14 - 4 \cdot 2 - 3$

22. $(23 - 52) - (-6 - 5 + 8)$

23. $(4 + 1)^2 - 2^2 + 3^2$

24. $6 + 2[-7 - 5(3 - 6)]$

25. $-\dfrac{15}{-12} \div (-10)$

26. $-\dfrac{4}{3} - \left(\dfrac{1}{6} - \dfrac{5}{9}\right)$

27. $\dfrac{(-4)7 - (-30)}{6 - 8} \cdot \dfrac{9 - 5(6)}{-1 - 2}$

28. $\dfrac{5(-8) - (-2)(-7) + (-9)}{(-3)4 + (-1)(-6) - (-3)}$

29. Evaluate $t^2 - 3r^2 - 4$ at $t = -4$ and $r = 2$.

30. Find the average of 26, -11, 3, -19, and -2.

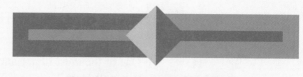

CHAPTER

Linear Equations and Inequalities

2

An earthquake sets off two different types of waves. P-waves travel at 5 mi/sec and S-waves travel at 2.75 mi/sec. In Sec. 2.4 we find the distance to the center of an earthquake by measuring the time interval between the two types of waves.

2.1 Linear Equations in One Variable

One of the most important problem-solving tools in algebra is an *equation*. In this section you will learn how to solve a linear equation in one variable. One of the first steps in solving a linear equation is to simplify each side of the equation.

Simplifying Expressions

In Sec. 1.5 we noted that **factors** are expressions that are related by multiplication. On the other hand, **terms** are expressions that are related by addition.

$$a \cdot b \cdot c \qquad a + b + c$$

three factors   three terms

Here are some examples of algebraic expressions and their terms:

Expression	*Terms*
$2x - 7$	$2x, -7$
$5y^2 - 4y + 8$	$5y^2, -4y, 8$
$a^2 + \dfrac{a - b}{2} + b^2$	$a^2, \dfrac{a - b}{2}, b^2$

The numerical factor in a term is called the **numerical coefficient,** or simply the **coefficient,** of the term.

Term	*Coefficient*
$4x^3$	4
$-y^2$	-1
$\dfrac{m}{3}$	$\dfrac{1}{3}$

Like terms are terms that have the same variables with the same exponents.

Like terms	*Unlike terms*
$5x, 3x$	$5x, 3y$ (different variables)
$8y^2, 2y^2$	$4y^2, 4y$ (different exponents)
$-rs, 6rs$	$8r, 7$

Note that **like terms may differ only in their coefficients.**

We can use the distributive property to add or subtract like terms. This is called **combining like terms.** For example, to add the like terms $2x$ and $7x$ we write

$$2x + 7x = (2 + 7)x \qquad \text{Distributive property}$$
$$= 9x \qquad \text{Add the coefficients.}$$

Here are some more examples of combining like terms:

$$y^2 + 12y^2 = 1y^2 + 12y^2 = (1 + 12)y^2 = 13y^2$$
$$3m - 9m = (3 - 9)m = -6m$$
$$-10p - p = -10p - 1p = (-10 - 1)p = -11p$$
$$-8r + 8r = (-8 + 8)r = 0r = 0$$
$$4a + 5b + 1 \text{ cannot be simplified}$$

Note that like terms are combined simply by combining (adding or subtracting) their coefficients. Unlike terms cannot be combined.

If an algebraic expression contains several terms, you may need to use the commutative and associative properties to rearrange the terms so that the like terms are together. This will make the terms easier to combine.

Example 1 Simplify each expression.

(a) $5x - 6x - 9x + 3$ (b) $10m + 13 - 2m - 24$
(c) $-8p^2 - 4 - 6p + p^2 + 15$

Solution:

(a) $5x - 6x - 9x + 3 = (5 - 6 - 9)x + 3$ Distributive property
$= -10x + 3$ Combine the coefficients.

Since addition is commutative, the answers $3 + (-10x)$ and $3 - 10x$ are also correct answers. However we usually write the variable term first.

(b) $10m + 13 - 2m - 24 = 10m - 2m + 13 - 24$ Rearrange the terms.
$= 8m - 11$ Combine like terms.

(c) $-8p^2 - 4 - 6p + p^2 + 15 = -8p^2 + p^2 - 6p - 4 + 15$
$= -7p^2 - 6p + 11$ ❐

Try Exercise 7

Simplify:
$11m + 18 - 3m - 25$.

Sometimes we must use the distributive property to remove parentheses before we can combine like terms.

Example 2 Simplify: $3 + 2(6k - 7) - (5k - 1)$.

Solution:

$3 + 2(6k - 7) - (5k - 1)$ Original expression
$= 3 + 2(6k - 7) - 1 \cdot (5k - 1)$ Since $-a = -1 \cdot a$
$= 3 + 12k - 14 - 5k + 1$ Distributive property
$= 7k - 10$ Combine like terms.   ❐

Try Exercise 19

Simplify:
$7 + 3(4k - 5) - (9k - 2)$.

Solving Linear Equations

An **equation** is a statement that two algebraic expressions are equal. Here are three examples of equations:

$$2x + 3 = 9 \qquad 3(y - 4) = y + 8 \qquad \frac{m}{2} - 1 = \frac{3m}{4} - 6$$

Since the highest power of the variable in each of these equations is 1, these equations are called *first-degree equations,* or *linear equations.**

DEFINITION OF A LINEAR EQUATION

A **linear equation** in the variable x is an equation that can be written in the form

$$ax + b = 0$$

where a and b are real numbers and $a \neq 0$.

---

*HISTORICAL NOTE: Linear equations are found in the Rhind Papyrus, a scroll of Egyptian origin that is about 1 ft high and 18 ft long and made from the pith of the papyrus plant. It is named after the English historian A. Henry Rhind, who purchased it in Egypt in 1858. It is also called the Ahmes Papyrus in honor of the scribe who copied it in 1650 B.C. from material dated from the period 2000 B.C. to 1800 B.C.

If an equation becomes a true statement when its variable is replaced by a real number, then that number is called a **solution** of the equation. For example, 3 is a solution of the equation $x + 2 = 5$, since the equation becomes a true statement when x is replaced by 3. We also say that 3 *satisfies* the equation $x + 2 = 5$. The set $\{3\}$ is called the **solution set** of the equation.

To **solve** an equation means to find its solution set. We solve a linear equation by writing a sequence of *equivalent equations* until we reach an equation of the form

$$x = \text{a number}$$

Equivalent equations are equations that have the same solution set. For example the three equations below are equivalent, since each has the same solution set $\{7\}$:

$$3x + 8 = 29$$
$$3x = 21 \quad \longleftarrow \text{equivalent equations}$$
$$x = 7$$

In order to write a sequence of equivalent equations, we use the **addition** and **multiplication properties of equality.**

Suppose a, b, and c are real numbers.

Addition Property of Equality

If $a = b$, then

$$a + c = b + c$$

Multiplication Property of Equality

If $a = b$, then

$$ac = bc \quad (c \neq 0)$$

The addition property states that we can add the same number to both sides of an equation and the result is an equivalent equation. Since subtracting a number is the same as adding the opposite of the number, we can use the addition property to subtract the same number from both sides of an equation.

The multiplication property states that we can multiply both sides of an equation by the same nonzero number and the result is an equivalent equation. Since dividing by a number is the same as multiplying by the reciprocal of the number, we can use the multiplication property to divide both sides of an equation by the same nonzero number.

Example 3   Solve: $2x + 9 = -11$.

Solution: Our goal is to isolate x on either side of the equation. To do this, use the addition property to subtract 9 from both sides. This will eliminate 9 from the left side:

$2x + 9 = -11$	Original equation
$2x + 9 - 9 = -11 - 9$	Subtract 9 from both sides.
$2x = -20$	Simplify each side.

To eliminate the coefficient 2, use the multiplication property to divide both sides by 2:

$$\frac{2x}{2} = \frac{-20}{2}$$ Divide both sides by 2.

$$x = -10$$ Simplify each side.

You can check this solution by replacing x in the original equation with -10.

Check: $2x + 9 = -11$ Original equation
$2(-10) + 9 = -11$ Let $x = -10$.
$-20 + 9 = -11$
$-11 = -11$ True

Try Exercise 31

Solve: $3x + 8 = -10$.

The number -10 checks, so the solution set is $\{-10\}$.*   ❐

If an equation contains like terms on the same side of the equation, make sure you combine the like terms before using the addition or multiplication property.

Example 4   Solve: $4y - 8 - 5y = 4 - 13$.

Solution: Combine like terms on each side of the equation:

$4y - 8 - 5y = 4 - 13$ Original equation
$-y - 8 = -9$ Simplify each side.
$-y - 8 + 8 = -9 + 8$ Add 8 to both sides.
$-y = -1$ Simplify each side.
$\dfrac{-y}{-1} = \dfrac{-1}{-1}$ Divide both sides by -1.
$y = 1$ Simplify each side.

Check: $4y - 8 - 5y = 4 - 13$ Original equation
$4(1) - 8 - 5(1) = 4 - 13$ Let $y = 1$.
$4 - 8 - 5 = -9$
$-9 = -9$ True

Try Exercise 39

Solve: $7y - 6 - 8y = 5 - 14$.

The number 1 checks, so the solution set is $\{1\}$.   ❐

Sometimes we must use the distributive property to remove parentheses so that we can combine like terms.

Example 5   Solve: $p - (3p - 19) = 3p + 21$.

Solution:

$p - (3p - 19) = 3p + 21$ Original equation
$p - 3p + 19 = 3p + 21$ Distribute -1.
$-2p + 19 = 3p + 21$ Simplify the left side.

Use the addition property to get the variable terms on one side and the constant terms on the other side. One way to proceed is to subtract $3p$ from both sides. (We could also begin by subtracting 19 from both sides.)

---

*HISTORICAL NOTE: The Dutch mathematician Albert Girard (1590–1633) was one of the first to admit that the solution of an equation could be a negative number. It was Girard's view of positive and negative numbers that spawned the real number line.

$$-2p + 19 - 3p = 3p + 21 - 3p \qquad \text{Subtract } 3p \text{ from both sides.}$$
$$-5p + 19 = 21 \qquad \text{Simplify each side.}$$
$$-5p + 19 - 19 = 21 - 19 \qquad \text{Subtract 19 from both sides.}$$
$$-5p = 2 \qquad \text{Simplify each side.}$$
$$\frac{-5p}{-5} = \frac{2}{-5} \qquad \text{Divide both sides by } -5.$$
$$p = -\frac{2}{5} \qquad \text{Simplify each side.}$$

Try Exercise 45

Solve:
$p - (4p - 17) = 4p + 22.$

The solution set is $\left\{ -\dfrac{2}{5} \right\}$. Check in the original equation. ❐

To solve an equation with fractions, multiply both sides of the equation by the *least common denominator*. The **least common denominator (LCD)** of a set of fractions is the smallest number that is divisible (without remainder) by all of the denominators of the fractions. Multiplying by the LCD will clear the equation of fractions.

Example 6   Solve: $\dfrac{r}{3} + 2 = \dfrac{5r}{12} - \dfrac{3}{4}$.

Solution:

$$12\left(\frac{r}{3} + 2 \right) = 12\left(\frac{5r}{12} - \frac{3}{4} \right) \qquad \begin{array}{l}\text{Multiply both sides by the}\\ \text{LCD 12.}\end{array}$$
$$12\left(\frac{r}{3} \right) + 12(2) = 12\left(\frac{5r}{12} \right) - 12\left(\frac{3}{4} \right) \qquad \text{Distribute 12 over every term.}$$
$$4r + 24 = 5r - 9 \qquad \text{Simplify each side.}$$

In this equation we decide to isolate the variable on the right side.

$$4r + 24 - 4r = 5r - 9 - 4r \qquad \text{Subtract } 4r \text{ from both sides.}$$
$$24 = r - 9 \qquad \text{Simplify each side.}$$
$$24 + 9 = r - 9 + 9 \qquad \text{Add 9 to both sides.}$$
$$33 = r \qquad \text{Simplify each side.}$$

Try Exercise 57

Solve: $\dfrac{r}{3} + 1 = \dfrac{5r}{12} - \dfrac{3}{4}$.

The solution set is $\{33\}$. Check in the original equation. ❐

Here is a summary of the steps in solving a linear equation in one variable. Not all equations require all of these steps.

Solving a Linear Equation in One Variable

1. Clear fractions by multiplying both sides by the LCD. Distribute the LCD over every term, even if the term does not contain a fraction.
2. Use the distributive property to remove parentheses.
3. Simplify each side by combining like terms.
4. Use the addition property to collect all variable terms on one side of the equation and all constant terms on the other side.
5. Use the multiplication property to write the variable term with a coefficient of 1.
6. Check the solution in the original equation.

We illustrate these steps in the next example.

Example 7   Solve: $\dfrac{3m-4}{6} + 2 = 1 - \dfrac{m-3}{9}$.

Solution: Step 1:   Clear fractions by multiplying both sides by the LCD 18. Distribute 18 over every term.

$$18\left(\frac{3m-4}{6}\right) + 18(2) = 18(1) - 18\left(\frac{m-3}{9}\right)$$

$$3(3m-4) + 36 = 18 - 2(m-3)$$

Careful! Write parentheses here.

Step 2:   Use the distributive property to remove parentheses.

$$9m - 12 + 36 = 18 - 2m + 6$$

Step 3:   Simplify each side by combining like terms.

$$9m + 24 = -2m + 24$$

Step 4:   Use the addition property to collect all variable terms on one side of the equation and all constant terms on the other side.

$$9m + 24 + 2m = -2m + 24 + 2m \qquad \text{Add } 2m.$$
$$11m + 24 = 24$$
$$11m + 24 - 24 = 24 \ \ - 24 \qquad \text{Subtract 24.}$$
$$11m = 0$$

Step 5:   Use the multiplication property to write the variable term with a coefficient of 1.

$$\frac{11m}{11} = \frac{0}{11} \qquad \text{Divide by 11.}$$
$$m = 0$$

Step 6:   Check the solution in the original equation.

$$\text{\textit{Check:}} \quad \frac{3(0)-4}{6} + 2 = 1 - \frac{0-3}{9} \qquad \text{Let } m = 0.$$
$$-\frac{2}{3} + 2 = 1 + \frac{1}{3}$$
$$1\frac{1}{3} = 1\frac{1}{3} \qquad \text{True}$$

The number 0 checks, so the solution set is {0}.   ❏

Try Exercise 61

Solve:
$\dfrac{3m-2}{9} + 1 = 2 - \dfrac{m-1}{6}$.

Conditional Equations, Identities, and Contradictions

All of the equations we have solved so far in this section were *conditional equations*. A **conditional equation** is an equation that is true for some values of the variable and false for others. For example, the equation

$$x + 4 = 7$$

is a conditional equation because it is true when $x = 3$ and false for any other value of x.

An **identity** is an equation that is true for all values of the variable, so long as both sides of the equation are defined. For example, the equation

$$x + 1 = x + 1$$

is an identity since it is true for all values of x.

A **contradiction** is an equation that is false for all values of the variable. For example, the equation

$$x + 1 = x$$

is a contradiction since it is false for all values of x.

Example 8   Classify each as a conditional equation, an identity, or a contradiction.

(a) $2 + 3 - 3x = x + 5 - 4x$ (b) $4x + 6 = x + 3(x - 2)$

Solution: (a)

$2 + 3 - 3x = x + 5 - 4x$	Original equation
$5 - 3x = -3x + 5$	Simplify each side.
$5 - 3x + 3x = -3x + 5 + 3x$	Add $3x$.
$5 = 5$	True

Since $5 = 5$ is a true statement, the original equation is true for all values of x. Therefore the original equation is an identity and the solution set is all real numbers.

(b)

$4x + 6 = x + 3(x - 2)$	Original equation
$4x + 6 = x + 3x - 6$	Distributive property
$4x + 6 = 4x - 6$	Simplify the right side.
$4x + 6 - 4x = 4x - 6 - 4x$	Subtract $4x$.
$6 = -6$	False

Since $6 = -6$ is a false statement, the original equation is false for all values of x. Therefore the original equation is a contradiction and the solution set is $\varnothing$.   ❏

Try Exercise 67

Classify as a conditional equation, an identity, or a contradiction:
$9 + 2 - 2x = x + 11 - 3x$.

Exercises 2.1

◆◇◆

True-False Exercises

1. $6x^2$ and $6x$ are like terms.

2. $-5a^2b$ and $8ab^2$ are like terms.

3. $3(y - 1) = 15 + 7y$ is a linear equation.

4. $4x^2 + 9 = 0$ is a linear equation.

Simplify each expression.

5. $6y + 5y + 3 + 7$

6. $7y + 2y + 8 + 5$

7. $11m + 18 - 3m - 25$

8. $15m - 13 - 5m + 7$

9. $9y - 3 - 5y - y$

10. $-10y + 2y - 9 - 3y$

11. $-6r - 7r - 4 + 3r - 2$

12. $4r - 3 - 7r - 5 - 6r$

13. $-6p^2 - 3p + 8 + 4p^2 - 19$

14. $2p - 8p^2 - 5p + 18 - p^2$

15. $2(x + 6) + 7x - 10 - 2$

16. $3(x - 5) + 20 + 4x - 5$

17. $-(8y - 1) + 4(7 + 2y)$

18. $-(6y + 5) - 2(9 - 3y)$

19. $7 + 3(4k - 5) - (9k - 2)$

20. $8 + 2(4k + 3) - (7k - 6)$

21. $x^2 + 3(-x^2 + 8) - 9(2x + 1) + x^2$

22. $3x^2 - 3(-x - 7) - 8(2x + 1) + x^2$

Completion Exercises

23. A statement that two algebraic expressions are equal is called a(n) _____.

24. Since the number 3 makes $2x + 5 = 11$ a true statement when it is substituted for x, 3 is called a(n) _____ of $2x + 5 = 11$.

25. The set of all solutions of an equation is called the _____ of the equation.

26. To _____ an equation means to find its solution set.

27. The number -4 is called the _____ of the term $-4x^3$.

28. Since the equations $5x + 4 = 14$ and $5x = 10$ have the same solution set, they are called _____ equations.

29. The addition property of equality states that if $a = b$, then _____.

30. The multiplication property of equality states that if $a = b$ and $c \neq 0$, then _____.

Solve each equation.

31. $3x + 8 = -10$

32. $5x - 10 = -15$

33. $2n + 13 = 13$

34. $3n + 19 = 19$

35. $-p - 9 = 6$

36. $-p + 8 = 12$

37. $10 = -12 - 4z$

38. $16 = -17 - 9z$

39. $7y - 6 - 8y = 5 - 14$

40. $-4 - 6y + 5y = -16 + 7$

41. $6t - 2t + 4 - 3t = t + 14 - 2t$

42. $8 + 3t + 1 - 2t = 5t + 3 - 6t$

43. $3(x + 2) - 5 = x + 3$

44. $5(x + 2) - 9 = x + 5$

45. $p - (4p - 17) = 4p + 22$

46. $p - (5p - 21) = 5p + 28$

47. $8 - 3(4x - 2) + x = -10 - 4x + 45$

48. $9 - 7(2x - 3) + 4x = 70 - 2x - 8$

49. $7 - 2[r - 5(r - 1)] = 1 + 3r - 6[9 - (1 - r)]$

50. $11 - 3[r - 3(r - 1)] = 3 + 5r - 6[7 - (1 - r)]$

Solve each equation involving fractions.

51. $\frac{4}{9}y = 36$

52. $\frac{2}{3}y = -24$

53. $8 = -\frac{7}{2}p$

54. $-10 = -\frac{6}{5}p$

55. $d - \frac{13}{3} = \frac{7d}{3} - \frac{9}{2}$

56. $d - \frac{27}{8} = \frac{7d}{4} - \frac{7}{2}$

57. $\frac{r}{3} + 1 = \frac{5r}{12} - \frac{3}{4}$

58. $\frac{r}{2} + 2 = \frac{7r}{12} - \frac{2}{3}$

59. $\frac{k + 6}{5} - \frac{13}{2} = \frac{7}{10} - \frac{k - 3}{4}$

60. $\frac{k + 7}{2} - \frac{1}{6} = \frac{2}{3} - \frac{k + 9}{9}$

61. $\frac{3m - 2}{9} + 1 = 2 - \frac{m - 1}{6}$

62. $\frac{2m - 3}{6} + 2 = 3 - \frac{m - 1}{9}$

63. $3.1y - 6 = 4.5y + 2.4$

64. $6.2y - 1.1 = 8.9y + 7$

65. $0.05h + 0.25(8 - h) = 1.2(4)$

66. $0.35h + 0.05(6 - h) = 2.4(2)$

Classify each as a conditional equation, an identity, or a contradiction.

67. $9 + 2 - 2x = x + 11 - 3x$

68. $5 - 8 + 5x = x - 3 + 4x$

69. $5(x + 2) - x = 3x + 25$

70. $4(x - 2) + 2x = 5x + 4$

71. $6x - 10 = 4x + 2(x + 5)$

72. $7x + 8 = 3x + 4(x - 2)$

73. $3y - (5y - 3) = 2(4 - y) + 1$

74. $6y - (9y - 3) = 4 + 3(1 - y)$

75. $\frac{m + 1}{6} - \frac{m}{4} = \frac{1}{6} - \frac{m}{12}$

76. $\frac{m + 3}{9} - \frac{m}{6} = \frac{1}{3} - \frac{m}{18}$

Discussion Exercises

77. Discuss the difference between terms and factors. Give examples.

78. Discuss the difference between a conditional equation, an identity, and a contradiction. Give examples.

79. What are like terms, and how are they combined?

80. Explain what is meant by the least common denominator of a set of fractions.

81. What is a linear equation in one variable?

82. Discuss the steps in solving a linear equation in one variable.

Getting Acquainted with Your Graphing Calculator

83. Store -7 in x and calculate the value of each expression. Then store another value in x and calculate again. (a) $2x^2 + 3x^2$; (b) $5x^2$. Do $2x^2 + 3x^2$ and $5x^2$ have the same value for every value of x?

84. Check that 3 is the solution of $2(x - 5) + 6 = 5 - x$ by storing 3 in x and calculating the value of each side.

85. $3(x - 2) + 7 = 3x + 1$ is an identity. Store several different values in x and show the statement is true for each value.

86. $3(x - 2) + 8 = 3x + 1$ is a contradiction. Store several different values in x and show the statement is false for each value.

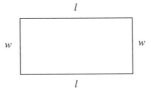

2.2 Problem Solving with Formulas

Many problems in algebra require the use of a *formula*. In this section you will learn what a formula is and how to use a formula to solve a problem. Appendix 2 contains a list of the geometric formulas that we will use in this text.

Formulas

A **formula** is an equation that relates two or more real-world quantities. For example,

$$P = 2l + 2w$$

is a formula because the variables P, l, and w represent the perimeter (distance around), length, and width of a rectangle, respectively (see Fig. 2.1).

Figure 2.1

Solving for a Specified Variable

The formula $P = 2l + 2w$ is said to be **solved for P** because P is isolated on one side of the equation. Sometimes this formula is easier to use if it is solved for one of the other variables. We solve a formula for one of its variables in much the same way that we solve a linear equation.

Example 1   Solve $P = 2l + 2w$ for l. Then find the length of a rectangle with perimeter 58 ft and width 12 ft.

Solution:

$$P = 2l + 2w \qquad \text{Original formula}$$
$$P - 2w = 2l + 2w - 2w \qquad \text{Subtract } 2w \text{ from both sides.}$$
$$P - 2w = 2l \qquad \text{Simplify the right side.}$$
$$\frac{P - 2w}{2} = \frac{2l}{2} \qquad \text{Divide both sides by 2.}$$
$$\frac{P - 2w}{2} = l \qquad \text{Simplify the right side.}$$

Using the symmetric property of equality, we can also write this formula as $l = \dfrac{P - 2w}{2}$. Another way to write this formula is $l = \dfrac{P}{2} - w$.

To find l when $P = 58$ and $w = 12$, substitute these values into the formula:

$$l = \frac{P - 2w}{2}$$
$$l = \frac{58 - 2(12)}{2} = \frac{58 - 24}{2} = \frac{34}{2} = 17$$

Try Exercise 5

Do Example 1 with $P = 72$ ft and $w = 14$ ft.

The length is 17 ft. ❏

The formula

$$d = rt$$

relates distance d, rate r, and time t.

Example 2 Solve $d = rt$ for t. Then determine how long it will take to drive 132 mi at 55 mi/hr.

Solution:

$$d = rt \qquad \text{Original formula}$$
$$\frac{d}{r} = \frac{rt}{r} \qquad \text{Divide by } r.$$
$$\frac{d}{r} = t \qquad \text{Simplify the right side.}$$

Now substitute $d = 132$ and $r = 55$:

$$t = \frac{d}{r}$$
$$t = \frac{132}{55} = 2.4$$

Try Exercise 9

Do Example 2 with $d = 45$ mi and $r = 18$ mi/hr.

It will take 2.4 hr. ❏

The amount of money A in an account that pays simple interest is given by the formula

$$A = P + Prt$$

where P is the principal (original amount), r is the yearly interest rate *in decimal form,* and t is the time in years.

Example 3 Solve $A = P + Prt$ for P. Then determine the principal that must be invested at 5 percent simple interest to amount to $870 in 4 yr.

Solution: Use the distributive property to write the right side as a product.

$$A = P + Prt \qquad \text{Original formula}$$
$$A = P(1 + rt) \qquad \text{Distributive property}$$
$$\frac{A}{1 + rt} = \frac{P(1 + rt)}{1 + rt} \qquad \text{Divide by } 1 + rt.$$
$$\frac{A}{1 + rt} = P \qquad \text{Simplify the right side.}$$

Now substitute $A = 870$, $r = 0.05$, and $t = 4$:

$$P = \frac{A}{1 + rt}$$

$$P = \frac{870}{1 + (0.05)(4)} = \frac{870}{1 + 0.2} = \frac{870}{1.2} = 725$$

Try Exercise 11

Do Example 3 with $A = \$1096$, $r = 12\%$, and $t = 5$ yr.

The required principal is $725. ❑

Many geometric formulas contain the irrational number π (Greek letter *pi*). The number π is the ratio of the circumference of a circle (distance around) to the diameter (distance across), as shown in Fig. 2.2.

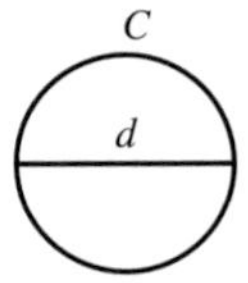

$$\pi = \frac{C}{d}$$

Figure 2.2

Since π is an irrational number, it is a decimal that neither terminates nor repeats. An approximate value of π is

$$\pi \approx 3.1415926536*$$

One formula that contains π is

$$V = \pi r^2 h$$

where V is the volume of a right circular cylinder, r is the radius of the base, and h is the height (see Fig. 2.3).

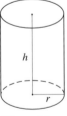

Figure 2.3

---

*HISTORICAL NOTE: The Rhind Papyrus (ca. 2000 B.C.) uses a value of $3\frac{1}{6}$ for π. The Greek mathematician Archimedes (ca. 287–212 B.C.) estimated π as 22/7. In 1767, the Swiss mathematician Johann Lambert proved that π is an irrational number and therefore is a nonterminating, nonrepeating decimal. In 1989, David and Gregory Chudnovsky of Columbia University used a computer to calculate π to a billion decimal places.

Example 4 Solve $V = \pi r^2 h$ for h. Then find the height of a tin can with volume 942 cm³ and radius 5 cm. Use $\pi = 3.14$.

Solution:

$$V = \pi r^2 h \qquad \text{Original formula}$$

$$\frac{V}{\pi r^2} = \frac{\pi r^2 h}{\pi r^2} \qquad \text{Divide by } \pi r^2.$$

$$\frac{V}{\pi r^2} = h \qquad \text{Simplify the right side.}$$

Substitute $V = 942$, $r = 5$, and $\pi = 3.14$:

$$h = \frac{V}{\pi r^2}$$

$$h = \frac{942}{(3.14)(5)^2} = \frac{942}{(3.14)(25)} = \frac{942}{78.5} = 12$$

The height is 12 cm.   ❑

Try Exercise 17

If $V = 254{,}340$ cm³ and $r = 30$ cm in Example 4, find h.

Here are the steps in solving a formula for one of its variables. Some formulas do not require all of these steps.

Solving for a Specified Variable

1. Clear fractions by multiplying both sides by the LCD.
2. Use the distributive property to remove parentheses.
3. Use the addition property to collect all terms containing the specified variable on one side and all other terms on the other side.
4. Use the distributive property to write the side that contains the specified variable as a product of that variable and a sum of other terms.
5. Use the multiplication property to write the specified variable with a coefficient of 1.

The area A of a trapezoid (see Fig. 2.4) is given by the formula

$$A = \frac{h}{2}(b_1 + b_2)$$

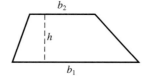

Figure 2.4

where h is the height, and b_1 and b_2 are the lengths of the two bases. The numbers 1 and 2 are called **subscripts,** and b_1 and b_2 are read "b sub 1" and "b sub 2," respectively. Subscripts allow us to use the same letter (in this case b) to represent different numbers.

Example 5 Solve $A = \dfrac{h}{2}(b_1 + b_2)$ for b_1.

Solution: Step 1: Clear fractions by multiplying both sides by the LCD, 2.

$$2(A) = 2\left[\frac{h}{2}(b_1 + b_2)\right]$$

$$2A = \left(2 \cdot \frac{h}{2}\right)(b_1 + b_2) \quad \text{Associative property}$$

$$2A = h(b_1 + b_2)$$

Step 2: Use the distributive property to remove parentheses.

$$2A = hb_1 + hb_2$$

Step 3: Collect all terms containing b_1 on one side and all other terms on the other side.

$$2A - hb_2 = hb_1 + hb_2 - hb_2 \quad \text{Subtract } hb_2.$$

$$2A - hb_2 = hb_1$$

Step 4: (Not needed)

Step 5: Write b_1 with a coefficient of 1.

$$\frac{2A - hb_2}{h} = \frac{hb_1}{h} \quad \text{Divide by } h.$$

$$\frac{2A - hb_2}{h} = b_1$$ ❐

Try Exercise 33

Solve $A = \frac{h}{2}(b_1 + b_2)$ for b_2.

Here is another approach to Example 5:

$$A = \frac{h}{2}(b_1 + b_2) \qquad \text{Original formula}$$

$$2A = h(b_1 + b_2) \qquad \text{Multiply both sides by 2.}$$

$$\frac{2A}{h} = b_1 + b_2 \qquad \text{Divide both sides by } h.$$

$$\frac{2A}{h} - b_2 = b_1 \qquad \text{Subtract } b_2 \text{ from both sides.}$$

This answer is equivalent to the answer obtained in Example 5.

Exercises 2.2

◆◆

Completion Exercises

1. An equation that relates two or more real-world quantities is called a(n) _____.

2. If the variable T is isolated on one side of a formula, the formula is said to be _____.

3. The ratio of the circumference of a circle to its diameter equals the irrational number _____.

4. Numbers that are written to the right and slightly below a letter, allowing us to use that letter to represent different quantities, are called _____.

Problem Solving

For each exercise, select the appropriate formula either from the examples in the section or from Appendix 2. Then solve the formula for the unknown variable. Finally, substitute the known variables to find the value of the unknown variable. Use $\pi = 3.14$.

5. A rectangle has perimeter 72 ft and width 14 ft. Find the length.

6. The area and length of a rectangular room are 221 ft² and 17 ft, respectively. Find the width.

7. How high is a rectangular walk-in freezer that is 6 ft wide and 9 ft long if the storage capacity of the freezer is 378 ft³?

8. The base of a flower garden in the shape of a triangle is 26 m. If the area of the garden is 195 m², find the height.

9. How long will it take to cycle 45 mi at 18 mi/hr?

10. At what speed must you drive to travel 84 mi in $2\frac{2}{5}$ hr?

11. Determine the principal that must be invested at 12 percent simple interest to amount to $1096 in 5 yr.

12. Find the simple interest rate required if an investment of $1475 is to grow to $2360 in 6 yr.

13. The length of a circular race track is 455.3 yd. Find the diameter of the track.

14. What is the radius of a circular Ferris wheel whose circumference is 113.04 yd?

15. If the Celsius temperature is $-5°$, what is the corresponding Fahrenheit temperature?*

16. Determine the Fahrenheit temperature that corresponds to a Celsius temperature of $-15°$.*

17. An oil drum has the shape of a right circular cylinder. Find the height of the drum if it holds 254,340 cm³ of oil and the radius is 30 cm.

18. A pile of sand has the shape of a right circular cone. Find the height of the pile if it contains 508,680 cm³ of sand and the diameter is 180 cm.

19. How long will it take $100 invested at 16 percent simple interest to triple?

20. The speed of sound is 1100 ft/sec. If you see some lightning 3.5 sec before you hear the thunder, how far away is the lightning?

Getting Acquainted with Your Scientific Calculator

To find π on your calculator, press

Answer
↓

[Clear] [π] [3.141592654]

*HISTORICAL NOTE: The Celsius and Fahrenheit temperature scales were named after their inventors, the Swedish astronomer Anders Celsius (1701–1744) and the German physicist Gabriel Fahrenheit (1686–1736).

Since π is irrational, its decimal form neither terminates nor repeats, so this is an approximate value of π.

21. To the nearest thousandth, find the surface area of a sphere whose radius is 23.7 cm.

22. To the nearest thousandth, find the volume of a sphere whose diameter is 14.9 m.

Discussion Exercises

23. Why is it wrong to say that $A = P + Prt$ solved for P is $P = A - Prt$?

24. Why is it wrong to write $l = \dfrac{P - 2w}{2}$ as $l = P - w$?

Solve each formula for the specified variable.

25. $A = \frac{1}{3}Bh$ for B 26. $s = \frac{1}{2}(a + b + c)$ for a

27. $A = P + Pr$ for r 28. $s = vt + \frac{1}{2}at^2$ for v

29. $c^2 = a^2 + b^2$ for b^2 30. $E = mc^2$ for c^2

31. $T = \pi r(l + r)$ for l 32. $T = 2\pi r(h + r)$ for h

33. $A = \dfrac{h}{2}(b_1 + b_2)$ for b_2 34. $A = \dfrac{h}{2}(b_1 + b_2)$ for h

35. $l = a + (n - 1)d$ for n

36. $S = \dfrac{n}{2}[2a + (n - 1)d]$ for d

37. $\dfrac{P_1V_1}{T_1} = \dfrac{P_2V_2}{T_2}$ for T_2 38. $\dfrac{P_1V_1}{T_1} = \dfrac{P_2V_2}{T_2}$ for V_1

39. $S = \dfrac{a}{1 - r}$ for r 40. $S = \dfrac{a - lr}{1 - r}$ for r

Problem Solving

41. Find the area of the shaded region in Fig. 2.5.

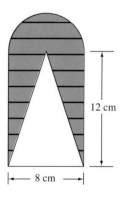

12 cm

8 cm

Figure 2.5

42. Find the area of the shaded region in Fig. 2.6.

14 cm

8 cm

Figure 2.6

43. Determine the grazing area of a horse that is tied with a 40-ft rope to one corner of a shed that is 20 ft by 20 ft.

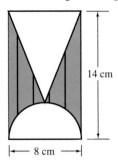

44. A car averages 30 mi/hr over the first mile of a 2-mi track. What must it average over the second mile in order to have an average speed of 60 mi/hr for the entire 2-mi lap?

45. Two bicyclists 50 mi apart start toward each other, one traveling at 10 mi/hr and the other at 15 mi/hr. At the same instant, a fly leaves the handlebars of the first bike and travels toward the second bike. Upon reaching the second bike, the fly immediately turns and heads back toward the first bike. If the fly continues in this back-and-forth fashion at 20 mi/hr until the two bikes meet, what is the total distance traveled by the fly?

46. Find the radius x of the circle inscribed in the triangle in Fig. 2.7.

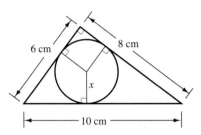

6 cm   8 cm

x

10 cm

Figure 2.7

 Getting Acquainted with Your Graphing Calculator

47. Use your calculator to show that the statement $\pi = 3.14$ is false. Why is this statement false?

48. The expression $1.8x + 32$ represents the Fahrenheit temperature when the Celsius temperature is x. Find the Fahrenheit temperature that corresponds to each Celsius temperature by storing the given value in x and evaluating $1.8x + 32$. (a) 100°C (b) 0°C (c) −40°C

(*Hint:* You may be able to do this without retyping $1.8x + 32$. For example, on one brand of calculator you can enter $y_1 = 1.8x + 32$ and then find the value of y_1 for each value of x.)

2.3 Number, Percent, and Geometry Problems

 TAPE AU3

In this section we illustrate how to use algebraic equations to solve real-life problems. To do this, we must first be able to translate the problems from words to algebra.

*Translating from English to Math**

Here are some examples of how to translate a word phrase to an algebraic expression:

---

*HISTORICAL NOTE: Symbolic algebra (that is, "$x + 6$" instead of "the sum of a number and six") did not reach its full maturity until the publication in 1637 of *La géométrie* by René Descartes.

Word phrase	**Algebraic expression**
Addition	
The sum of a number and 5	$x + 5$
Six more than a number	$y + 6$
A number increased by 2	$n + 2$
A number plus 9	$t + 9$
Twelve added to a number	$p + 12$
Subtraction	
The difference of a number and 1	$x - 1$
Three less than a number	$y - 3$
A number decreased by 7	$z - 7$
Ten minus a number	$10 - m$
Four subtracted from a number	$r - 4$
Multiplication	
The product of a number and 8	$y \cdot 8$ or $8y$
Twice a number	$2k$
A number multiplied by 14	$14n$
Five times a number	$5p$
Three-fourths of a number	$\frac{3}{4}x$ or $\frac{3x}{4}$
Six percent of a number	$0.06z$
Division	
The quotient of a number and 6	$\frac{x}{6}$
Nine divided by a number	$\frac{9}{y}$
One-third of a number	$\frac{1}{3}m$ or $\frac{m}{3}$

We can put combinations of these word phrases together to create more complicated algebraic expressions.

Word phrase	**Algebraic expression**
The sum of twice a number and 3	$2x + 3$
Twice the sum of a number and 3	$2(x + 3)$
Four less than 6 times a number	$6x - 4$
The quotient of a number and 10 is decreased by 8	$\frac{x}{10} - 8$
The distance traveled in t hours at 55 mi/hr	$55t$ miles
The value of x nickels and y dimes	$5x + 10y$ cents
The simple interest on $n + 75$ dollars invested at 12 percent for 1 yr	$0.12(n + 75)$ dollars
The amount of pure acid in $80 - p$ kilograms of a solution that is 5 percent acid	$0.05(80 - p)$ kilograms

When two or more quantities are related, we usually represent the most basic quantity by a variable. Then we represent the other quantities as algebraic expressions in terms of that variable. For example, to represent two *consecutive integers* (such as 5 and 6, or 71 and 72) we might represent the smaller integer by x. The larger integer is then

$x + 1$. (You could also represent the larger integer by x and the smaller integer by $x - 1$.)

Word phrase	Algebraic expression
Two consecutive even integers (such as 12 and 14)	x and $x + 2$
Three consecutive odd integers (such as 7, 9, and 11)	x, $x + 2$, and $x + 4$
The speed of two trains, if one is traveling 3 times as fast as the other	r and $3r$
Sue's age now and Sue's age 5 yr ago	a and $a - 5$
The number of males and the number of females in a class of 30 students	m and $30 - m$

Solving Word Problems

We will now illustrate how to solve a word problem.* Before you begin to solve the problem, you should read the problem carefully. You may need to read it several times to determine the information that is given and the information you are asked to find. Having read the problem, you can use the following steps to solve the problem:

To Solve a Word Problem

1. Write down the unknown quantities and represent one of them by a variable, say x. Then write the other unknown quantities in terms of x.
2. Use the information given in the problem to write an equation involving x. Sometimes a chart or a diagram is helpful in keeping track of the unknown quantities and in writing the equation.
3. Solve the equation to find the value of x. Then use this value to find the values of the other unknown quantities.
4. Check your solution in the words of the original problem.

We shall now look at several types of word problems. Some of these problems may seem artificial, and some of the solutions may seem easy to obtain using trial and error. Keep in mind, however, that the purpose of these examples is to teach you how to translate a word problem into the language of algebra, and then use the rules of algebra to solve the problem. In doing this, you will be improving your problem-solving skills.

Number Problems

In a number problem, the unknown quantities are numbers.

Example 1   One number is 4 less than twice another number. The sum of the numbers is 119. Find the numbers.

Solution: Step 1:   Write down the unknown quantities and represent one of them by x. Then write the other unknown quantity in terms of x.

---

*HISTORICAL NOTE: Word problems have long been of interest to mathematicians. The following problem is from the Rhind Papyrus (ca. 2000 B.C.): If a certain number, two-thirds of it, half of it, and one-seventh of it are added together, the result is 97. What is the number?

$$x = \text{first number}$$
$$2x - 4 = \text{second number}$$

Step 2:   Write an equation involving x.

First number	plus	second number	is	119
↓	↓	↓	↓	↓
x	$+$	$(2x - 4)$	$=$	119

Step 3:   Solve the equation to find x. Then use this value to find the value of $2x - 4$.

$$3x - 4 = 119 \qquad \text{Combine like terms.}$$
$$3x = 123 \qquad \text{Add 4.}$$
$$x = 41 \qquad \text{Divide by 3.}$$

The first number is 41. The second number is $2x - 4 = 2(41) - 4 = 78$.

Step 4:   Check in the words of the original problem.
78 is 4 less than twice 41, and the sum of 78 and 41 is 119.   ❐

Try Exercise 33

One number is 3 less than twice another number. The sum of the numbers is 111. Find the numbers.

CAUTION

Always check the solution to a word problem in the words of the original problem, *not* in the original equation. You may have made a mistake when you translated the words into an equation.   ■

We will solve Examples 2 through 6 using the same four steps.

Example 2   Find three consecutive integers such that 3 times the first minus the third is 12 more than the second.

Solution:
$$x = \text{first integer}$$
$$x + 1 = \text{second integer}$$
$$x + 2 = \text{third integer}$$

3 times the first	minus	the third	is	12	more than	the second
↓	↓	↓	↓	↓	↓	↓
$3x$	$-$	$(x + 2)$	$=$	12	$+$	$(x + 1)$

$$3x - x - 2 = 12 + x + 1 \qquad \text{Remove parentheses.}$$
$$2x - 2 = x + 13 \qquad \text{Simplify each side.}$$
$$x - 2 = 13 \qquad \text{Subtract } x.$$
$$x = 15 \qquad \text{Add 2.}$$

The first integer is 15, the second integer is $x + 1 = 15 + 1 = 16$, and the third integer is $x + 2 = 15 + 2 = 17$. Check in the words of the original problem.   ❐

Try Exercise 37

Find three consecutive integers such that 3 times the first minus the third is 8 more than the second.

Percent Problems

When you solve a percent problem, remember to write the percent in decimal form.

Example 3   A videocassette recorder is marked down 15 percent to $416.50. What was the original price of the recorder?

Solution: $x = $ original price of the recorder

Original price	minus	markdown	is	sale price
↓	↓	↓	↓	↓
x	−	15% of x	=	416.50

$$x - 0.15x = 416.5 \qquad \text{Since } 15\% = 0.15$$

Note that $x - 0.15x = 1x - 0.15x = (1 - 0.15)x = 0.85x$.

$$0.85x = 416.5 \qquad \text{Combine like terms.}$$
$$x = 490 \qquad \text{Divide by 0.85.}$$

The original price was $490. Check in the words of the original problem.   ❐

Try Exercise 39

A necklace is marked down 35 percent to $568.75. What was the original price of the necklace?

Example 4   How can $6300 be invested, part at a low-risk 7 percent and the rest at a higher-risk 11 percent, so that the interest will be the same on each investment?

Solution: $x = $ amount invested at 7%
$$6300 - x = \text{amount invested at 11\%}$$

A chart is helpful in organizing the information.

	Amount invested	Interest earned	
7% investment	x	$0.07x$	⎤ Since $I = Prt$
11% investment	$6300 - x$	$0.11(6300 - x)$	⎦ and $t = 1$ yr

Interest on 7% investment	equals	interest on 11% investment
↓	↓	↓
$0.07x$	=	$0.11(6300 - x)$

$$0.07x = 693 - 0.11x \qquad \text{Distribute 0.11.}$$
$$0.18x = 693 \qquad \text{Add 0.11}x.$$
$$x = 3850 \qquad \text{Divide by 0.18.}$$

A sum of $3850 should be invested at 7 percent, and $6300 - x = 6300 - 3850 = 2450 should be invested at 11 percent. Check in the words of the original problem.   ❐

Try Exercise 43

How can $7500 be invested, part at a low-risk 8 percent and the rest at a higher-risk 12 percent, so that the interest will be the same on each investment?

Geometry Problems

In a geometry problem, we are trying to find the unknown parts of a geometric figure.

Example 5 A tennis court for singles play is 24 ft longer than twice the width. If the perimeter is 210 ft, find the width and length.

Solution:
$$x = \text{width of the court}$$
$$2x + 24 = \text{length of the court}$$

A tennis court is a rectangle, as shown in Fig. 2.8.

Figure 2.8

Twice the width	plus	twice the length	equals	the perimeter
↓	↓	↓	↓	↓
$2x$	$+$	$2(2x + 24)$	$=$	210

$$
\begin{array}{ll}
2x + 4x + 48 = 210 & \text{Distribute 2.} \\
6x + 48 = 210 & \text{Combine like terms.} \\
6x = 162 & \text{Subtract 48.} \\
x = 27 & \text{Divide by 6.}
\end{array}
$$

The width is 27 ft, and the length is $2x + 24 = 2(27) + 24 = 78$ ft. Check in the words of the original problem. ❒

Try Exercise 47

The length of a soccer field is 30 yd less than twice the width. If the perimeter is 390 yd, find the width and length.

Example 6 The second angle of a triangle is 7° more than the first angle, and the third angle is 41° less than 4 times the second angle. (Fig. 2.9). Find the measures of the three angles.

Solution:
$$x = \text{first angle}$$
$$x + 7 = \text{second angle}$$
$$4(x + 7) - 41 = \text{third angle}$$

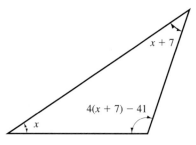

Figure 2.9

To write the equation, we use the fact that the sum of the interior angles of any triangle is 180°.

First angle	plus	second angle	plus	third angle	equals	180
↓	↓	↓	↓	↓	↓	↓
x	$+$	$(x + 7)$	$+$	$4(x + 7) - 41$	$=$	180

$$x + x + 7 + 4x + 28 - 41 = 180 \qquad \text{Remove parentheses.}$$
$$6x - 6 = 180 \qquad \text{Combine like terms.}$$
$$6x = 186 \qquad \text{Add 6.}$$
$$x = 31 \qquad \text{Divide by 6.}$$

Try Exercise 51

Determine the measures of the three interior angles of a triangle if the second angle is 5° more than twice the first and the third angle is 17° less than twice the second.

The first angle is 31°, the second angle is $x + 7 = 31 + 7 = 38°$, and the third angle is $4(x + 7) - 41 = 4(31 + 7) - 41 = 111°$. Check in the words of the original problem.   ❒

Exercises 2.3

Write each word phrase as an algebraic expression.

1. Twelve less than the product of a number and 5

2. Ten more than the quotient of a number and 6

3. Two-thirds of a number decreased by 9

4. Three-fifths of a number increased by 1

5. The sum of three times a number and 7

6. The difference of twice a number and 4

7. Three times the sum of a number and 7

8. Twice the difference of a number and 4

Discussion Exercises

9. Why is it correct to write the sum of a number and 8 as either $x + 8$ or $8 + x$, but the difference of a number and 8 must be written as $x - 8$ and not $8 - x$?

10. Why is it correct to write the product of a number and 3 as either $x \cdot 3$ or $3x$, but the quotient of a number and 3 must be written as $\dfrac{x}{3}$ and not $\dfrac{3}{x}$?

Completion Exercises

11. The number of hours in p days is _____.

12. The distance traveled in t hours at 48 mi/hr is _____.

13. The number of words typed in m minutes at 35 words/min is _____.

14. The earnings of a person who works h hours at $11 per hour is _____.

15. The cost of x records and y tapes at $9.50 per record and $14.50 per tape is _____.

16. The number of calories in x eggs and y muffins at 75 calories per egg and 100 calories per muffin is _____.

17. The simple interest on $x + 280$ dollars invested at 9 percent for 1 year is _____.

18. The amount of salt in $55 - x$ kilograms of a solution that is 25 percent salt is _____.

19. The average distance traveled each day if 1200 mi are driven in k days is _____.

20. The amount each winner of a $1,000,000 lottery receives if there are n winning tickets is _____.

For each exercise, choose a variable to represent one of the quantities. Then represent the other quantities in terms of that variable.

21. Three consecutive integers

22. Three consecutive even integers

23. Four consecutive odd integers

24. The speed of two runners if one is traveling twice as fast as the other

25. Bob's age now and Bob's age 7 yr ago

26. Pam's age now and Pam's age in 4 yr

27. The amount of antifreeze and the amount of water in a radiator that contains 15 qt of solution

28. The amount given to each of two sisters if $1500 is divided (not necessarily equally) between them

Problem Solving

For each exercise, represent the unknown quantities in terms of a single variable. Then write an equation involving that variable. Finally, solve the equation to find the unknown quantities.

29. Six times a number plus 13 is 61. Determine the number.

30. Four times a number plus 19 is 51. Determine the number.

31. If 3 times a number is 7 more than 5 times the number, find the number.

32. If twice a number is 5 more than 6 times the number, find the number.

33. One number is 3 less than twice another number. The sum of the numbers is 111. Find the numbers.

34. One number is 6 less than 3 times another number. The sum of the numbers is 126. Find the numbers.

35. Two consecutive even integers have a sum of 234. Find the integers.

36. One fourth the sum of three consecutive integers is 18. Find the integers.

37. Find three consecutive integers such that 3 times the first minus the third is 8 more than the second.

38. Find three consecutive odd integers such that the smallest subtracted from twice the largest is 26.

39. A necklace is marked down 35 percent to $568.75. What was the original price of the necklace?

40. A furniture store has a standard markup of 55 percent. How much did the store pay for a cocktail table that it has priced at $224.75?

41. The price of a television set plus a 6 percent tax is $630.70. Find the price of the television set itself.

42. The enrollment at a small liberal arts college decreases 8 percent to 2438 students. Find the original enrollment.

43. How can $7500 be invested, part at a low-risk 8 percent and the rest at a higher-risk 12 percent, so that the interest will be the same on each investment?

44. How can $8800 be invested, part at a low-risk 9 percent and the rest at a higher-risk 13 percent, so that the interest will be the same on each investment?

45. If $5000 is invested at 6 percent, how much additional money must be invested at 7 percent so that the total yearly interest from both investments is $510?

46. If $4000 is invested at 8 percent, how much additional money must be invested at 9 percent so that the total yearly interest from both investments is $500?

47. The length of a soccer field is 30 yd less than twice the width. If the perimeter is 390 yd, find the width and length.

48. The length of a basketball court is 18 ft less than twice the width. If the perimeter is 300 ft, find the width and length.

49. Two angles are complementary angles if the sum of their measures is 90°. Find two complementary angles if the measure of one angle is 14° less than the other.

50. Two angles are supplementary angles if the sum of their measures is 180°. Find two supplementary angles if the measure of one angle is 26° more than the other.

51. Determine the measures of the three interior angles of a triangle if the second angle is 5° more than twice the first and the third angle is 17° less than twice the second.

52. An isosceles triangle has two equal sides. Find the length of all three sides if the base is two-thirds of the length of each of the equal sides, and the perimeter is 48 ft.

53. A rancher has 13.5 mi of fencing to enclose a rectangular pasture. The rancher wants the length l to be 3 times the width w, and the rancher wants to divide the pasture in half with a fence parallel to the width (see Fig. 2.10). Find the dimensions of the pasture.

Figure 2.10

54. A farmer has 6.4 mi of fencing to enclose a rectangular grazing area. One side of the grazing area lies at the base of a cliff and needs no additional fencing. If the length l is

4 times the width w and the grazing area is divided into three equal areas as shown in Fig. 2.11, find the dimensions of the grazing area.

Discussion Exercises

55. Why is it important to check in the words of the original problem, rather than in the original equation?

56. Discuss the steps in solving a word problem.

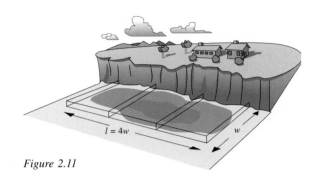

Figure 2.11

TAPE AU3

2.4 Money, Mixture, and Motion Problems

In this section we will solve some other types of word problems. However, we use the same four steps to solve the problems in this section that we used to solve the problems in Sec. 2.3.

Money Problems

To solve a money problem, we keep track of the *value* of the various quantities.

Example 1   A collection of nickels, dimes, and quarters has a total value of $3.80. If there are two more dimes than nickels, and 3 times as many quarters as nickels, how many of each kind of coin is in the collection?

Solution:

$$x = \text{number of nickels}$$
$$x + 2 = \text{number of dimes}$$
$$3x = \text{number of quarters}$$

A chart is helpful in organizing the information.

	Number of coins	*Value of coins, ¢*
Nickels	x	$5x$
Dimes	$x + 2$	$10(x + 2)$
Quarters	$3x$	$25(3x)$

When you write the equation, make sure you write both sides in the same units. In this case we write both sides in cents.

Value of nickels	plus	value of dimes	plus	value of quarters	is	total value
↓	↓	↓	↓	↓	↓	↓
$5x$	$+$	$10(x + 2)$	$+$	$25(3x)$	$=$	380

$$5x + 10x + 20 + 75x = 380 \quad \text{Remove parentheses.}$$
$$90x + 20 = 380 \quad \text{Combine like terms.}$$
$$90x = 360 \quad \text{Subtract 20.}$$
$$x = 4 \quad \text{Divide by 90.}$$

Try Exercise 1

A collection of nickels, dimes, and quarters has a total value of $4.35. If there are three more quarters than dimes, and twice as many nickels as dimes, how many of each kind of coin is in the collection?

The collection contains 4 nickels, $x + 2 = 4 + 2 = 6$ dimes, and $3x = 3(4) = 12$ quarters. Check in the words of the original problem.   ❏

CAUTION

Make sure the units are consistent when you write your equation. Here are two correct ways and two wrong ways to write the equation for Example 1.

Correct

$$5x + 10(x + 2) + 25(3x) = 380$$

or

$$0.05x + 0.10(x + 2) + 0.25(3x) = 3.80$$

Wrong

$$5x + 10(x + 2) + 25(3x) = 3.80$$

or

$$0.05x + 0.10(x + 2) + 0.25(3x) = 380 \quad ■$$

Mixture Problems

Suppose we mix a solution that is 12% acid with a solution that is 30% acid. To determine the amount of each solution needed to produce 72 kg of a 20% acid solution, we simply keep track of the acid during the mixing process.

Example 2   One solution is 12% acid and another solution is 30% acid. How many kilograms of each should be used to make 72 kg of a solution that is 20% acid?

Solution:

$$x = \text{number of kg of 12\% solution}$$
$$72 - x = \text{number of kg of 30\% solution}$$

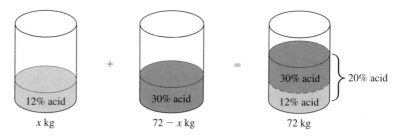

$$x \text{ kg} \qquad 72 - x \text{ kg} \qquad 72 \text{ kg}$$

We use a chart to organize the information.

	Kilograms of solution	*Kilograms of acid*
12% Solution	x	$0.12x$
30% Solution	$72 - x$	$0.30(72 - x)$
20% Solution	72	$0.20(72)$

Since no acid is gained or lost when the solutions are mixed, we can write the following equation.

Kg of acid in 12% solution	plus	kg of acid in 30% solution	equals	kg of acid in 20% solution
↓	↓	↓	↓	↓
$0.12x$	$+$	$0.30(72 - x)$	$=$	$0.20(72)$

$$\begin{aligned}
0.12x + 21.6 - 0.30x &= 14.4 && \text{Remove parentheses.} \\
21.6 - 0.18x &= 14.4 && \text{Combine like terms.} \\
-0.18x &= -7.2 && \text{Subtract 21.6.} \\
x &= 40 && \text{Divide by } -0.18.
\end{aligned}$$

Forty kilograms of 12% solution and $72 - x = 72 - 40 = 32$ kg of 30% solution should be used. Check in the words of the original problem.   ❑

Try Exercise 9

One alloy contains 20% gold, and another contains 28% gold. How many grams of each should be melted together to produce 72 g of an alloy that is 22% gold?

Motion Problems

To solve a motion problem, we use the formula

$$d = rt$$

where d is distance, r is rate, and t is time.

Our first motion problem involves finding the distance from a seismic station (a laboratory which measures the shock waves from an earthquake) to the epicenter (the center of the earthquake). An earthquake sets off two major types of vibrations, or waves, in the earth: *P-waves* (primary waves) and *S-waves* (secondary waves). P-waves travel at approximately 5 mi/sec and cause the earth to move in the same direction as the wave. S-waves travel at approximately 2.75 mi/sec and cause the earth to move at right angles to the direction of the wave.

Example 3 A seismograph measures a time interval of 36 sec between the P-wave and the S-wave of an earthquake. How long did it take each wave to arrive at the seismic station? How far away is the epicenter?

Solution:

$$t = \text{number of seconds P-wave travels}$$
$$t + 36 = \text{number of seconds S-wave travels}$$

Collect the information in a chart.

	Rate	Time	Distance
P-wave	5	t	$5t$
S-wave	2.75	$t + 36$	$2.75(t + 36)$

Since $d = rt$

Since both waves travel the same distance, we can write the following equation:

Distance traveled by P-wave	equals	distance traveled by S-wave
↓	↓	↓
$5t$	$=$	$2.75(t + 36)$

$$5t = 2.75t + 99 \qquad \text{Distribute 2.75.}$$
$$2.25t = 99 \qquad \text{Subtract 2.75}t.$$
$$t = 44 \qquad \text{Divide by 2.25.}$$

The P-wave takes 44 sec to arrive, and the S-wave takes $t + 36 = 44 + 36 = 80$ sec. The epicenter is (5 mi/sec)(44 sec) = 220 mi away. Check in the words of the original problem. ❐

Try Exercise 15

A seismograph measures a time interval of 27 sec between the P-wave and the S-wave of an earthquake. How long did it take each wave to arrive at the seismic station? How far away is the epicenter?

Example 4   How far upstream can a motorboat travel at 3 mi/hr and still return down-stream at 5 mi/hr in a total time of 2 hr?

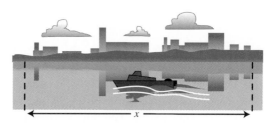

Solution: $x =$ distance in miles boat travels upstream

	Distance	Rate	Time
Upstream	x	3	$x/3$
Downstream	x	5	$x/5$

Since $t = \dfrac{d}{r}$

Since the total time is 2 hr, we can write the following equation:

Time upstream	plus	time downstream	is	total time
↓	↓	↓	↓	↓
$\dfrac{x}{3}$	$+$	$\dfrac{x}{5}$	$=$	2

$$15\left(\frac{x}{3}\right) + 15\left(\frac{x}{5}\right) = 15(2) \qquad \text{Multiply by the LCD 15.}$$

$$5x + 3x = 30 \qquad \text{Simplify each side.}$$

$$8x = 30 \qquad \text{Combine like terms.}$$

$$x = 3.75 \qquad \text{Divide by 8.}$$

The boat can travel 3.75 miles upstream (and back) in 2 hr. Check in the words of the original problem.   ❐

Try Exercise 19

How far upstream can a motorboat travel at 8 mi/hr and still return downstream at 12 mi/hr in a total time of 4 hr?

There are usually several different ways to set up a word problem. In Example 4 we could have avoided fractions by setting up the problem as follows:

$x =$ number of hours boat travels upstream

	Rate	Time	Distance
Upstream	3	x	$3x$
Downstream	5	$2 - x$	$5(2 - x)$

Since $d = rt$

Distance upstream	equals	distance downstream
↓	↓	↓
$3x$	$=$	$5(2 - x)$

Solving this equation gives a time upstream of $x = 1.25$ hr. Therefore the distance upstream is $3x = 3(1.25) = 3.75$ mi.

Exercises 2.4

For each exercise, represent the unknown quantities in terms of a single variable. Then write an equation involving the variable. Finally, solve the equation to find the unknown quantities.

1. A collection of nickels, dimes, and quarters has a total value of $4.35. If there are three more quarters than dimes, and twice as many nickels as dimes, how many of each kind of coin is in the collection?

2. A coin changer contains $3.35 in nickels, dimes, and quarters. If the changer contains two more dimes than quarters, and twice as many nickels as quarters, how many of each kind of coin is in the changer?

3. A piggy bank contains 40 nickels and dimes worth $3.05. How many nickels and how many dimes are there?

4. The total receipts for a football game were $1030 for 500 tickets sold. If adult tickets were $3.50 and student tickets were $1.50, how many of each were sold?

5. Linda's day job pays $12 per hour, and her night job pays $9.50 per hour. During one week Linda worked a total of 46 hr and earned a total of $527. How many hours did she work at each job?

6. Tony has two part-time jobs, one paying $7.50 per hour and the other paying $5 per hour. If Tony worked a total of 28 hr one week and earned a total of $190, how many hours did he work at each job?

7. A chemist wants to mix a 70% acid solution with 30 kg of a 25% acid solution to make a 45% acid solution. How much 70% solution should be used?

8. A winemaker wants to mix a 10% alcohol wine with 20 kg of a 55% alcohol wine to make a 35% wine cooler. How much 10% alcohol wine should be used?

9. One alloy contains 20% gold, and another contains 28% gold. How many grams of each should be melted together to produce 72 g of an alloy that is 22% gold?

10. One alloy contains 15% silver, and another contains 40% silver. How many grams of each should be melted together to produce 25 g of an alloy that is 20% silver?

11. How many pounds of pure alcohol should be added to 5 lb of a 40% alcohol solution to obtain a 50% alcohol solution?

12. How many pounds of pure formaldehyde should be added to 15 lb of a 40% formaldehyde solution to obtain a 50% formaldehyde solution?

13. Two cars 315 mi apart start toward each other on the same road. One car travels 5 mi/hr faster than the other. If they meet in 3 hr, find the speed of each car.

14. Two ships 270 mi apart start toward each other. One ship travels 5 mi/hr faster than the other. If they meet in 6 hr, find the speed of each ship.

15. A seismograph measures a time interval of 27 sec between the P-wave and the S-wave of an earthquake. How long did it take each wave to arrive at the seismic station? How far away is the epicenter?

16. A seismograph measures a time interval of 45 sec between the P-wave and the S-wave of an earthquake. How long did it take each wave to arrive at the seismic station? How far away is the epicenter?

17. A motorist starts out on a trip traveling 45 mi/hr. A half hour later a second motorist leaves from the same point and travels the same route at 50 mi/hr. How long does it take the second motorist to overtake the first?

18. A jogger leaves the starting point of a trail traveling 10 mi/hr. One-quarter hour later, another jogger begins the same trail traveling 12 mi/hr. How long does it take the second jogger to overtake the first?

19. How far upstream can a motorboat travel at 8 mi/hr and still return downstream at 12 mi/hr in a total time of 4 hr?

20. How far can a plane fly with the wind at 280 mi/hr and still return against the wind at 220 mi/hr if it has enough fuel to last 10 hr?

21. How long will it take a runner who travels at 12 mi/hr to lap a runner who travels at 8 mi/hr on a 1-mi oval track?

Start

22. One runner finishes a race in 1 hr. A second runner finishes 20 min later. If the rate of the faster runner is 2 mi/hr more than the rate of the slower runner, find the rate of each runner.

23. The cost of renting a car is $110 per week plus 30¢ per mile. How many miles can be traveled in 1 week for $305?

24. A salesperson earns $175 per week plus a 15 percent commission on all sales. What must the total sales in a particular week be for the salesperson to earn $415 that week?

25. The grades on Amy's first three accounting tests were 85, 89, and 97. What score must she get on the final exam, which is counted as two tests, to have a final average of 93?

26. Stacey is half as old as her sister. In 7 years she will be the same age her sister was 5 years ago. Find the present ages of Stacey and her sister.

27. A theater normally seats 4000 with tickets selling for $8.50. How many seats would have to be added so that tickets could be sold for $8 without changing the total revenue realized from a sold-out theater?

28. Each multiple-choice question on an exam has five choices. To discourage guessing, a student's final score is determined by subtracting one-fourth of a point for each incorrect answer from the number of correct answers. If a student who answered all 75 questions on a particular exam received a final score of 45, how many answers were correct?

Discussion Exercise

29. Algebra is a generalized form of arithmetic. To illustrate what this means, consider the following puzzle:

Think of a number.	7, for example
Add 8.	$7 + 8 = 15$
Multiply by 3.	$3(15) = 45$
Subtract 9.	$45 - 9 = 36$
Multiply by 2.	$2(36) = 72$
Divide by 6.	$\dfrac{72}{6} = 12$
Subtract the original number	$12 - 7 = 5$

How can you use algebra to show that the result is always 5, no matter what number you start with?

2.5 Linear Inequalities in One Variable

TAPE AU2

In Secs. 2.1 through 2.4, we learned to solve equations and word problems that led to equations. In this section, we learn to solve inequalities and word problems that lead to inequalities.

Solving Linear Inequalities

An equation is a statement that two algebraic expressions are equal. An **inequality** is a statement that two algebraic expressions are *not equal.* Here are some examples of inequalities:

$$x + 4 < 10 \qquad 3y - 12 \geq 0 \qquad 1 \leq 2(p + 7)$$

Since the highest power of the variable in each of these inequalities is 1, these inequalities are called *first-degree inequalities,* or *linear inequalities.*

DEFINITION OF A LINEAR INEQUALITY

A **linear inequality** in the variable x is an inequality that can be written in the form

$$ax + b < 0$$

where a and b are real numbers and $a \neq 0$.

In general, the definitions and rules we state for $<$ are also valid for $>$, $\leq$, and $\geq$. Therefore $ax + b > 0$, $ax + b \leq 0$, and $ax + b \geq 0$ are also linear inequalities (assuming $a \neq 0$). Note that a linear inequality is a linear equation whose equal sign has been replaced with one of the symbols $<$, $>$, $\leq$, or $\geq$.

A **solution** of a linear inequality is a number that makes the inequality a true statement when it is substituted for the variable. While a linear equation has just one solution, a linear inequality has an infinite number of solutions. The collection of all solutions is called the **solution set.** For example, the solution set of the inequality $x + 1 > 6$ consists of all those numbers greater than 5, written $\{x \mid x > 5\}$. To **solve** an inequality means to find its solution set.

Since linear inequalities are similar in form to linear equations, we solve a linear inequality in much the same way that we solve a linear equation. That is, we write a sequence of **equivalent inequalities** (inequalities with the same solution set) until we isolate the variable on one side of the inequality. We write this sequence of equivalent inequalities using the **addition** and **multiplication properties of inequality.**

Addition Property of Inequality

Suppose a, b, and c are real numbers. If $a < b$, then

$$a + c < b + c.$$

The addition property states that we can add the same number to both sides of an inequality and the result is an equivalent inequality. Since subtracting a number is the same as adding the opposite of the number, we can use the addition property to subtract the same number from both sides of an inequality.

Multiplication Property of Inequality

Suppose a, b, and c are real numbers.

1. Suppose $c > 0$ (c is a positive number). If $a < b$, then

$$ac < bc.$$

2. Suppose $c < 0$ (c is a negative number). If $a < b$, then

$$ac > bc.$$

The multiplication property states that if we multiply both sides of an inequality by a *positive* number, the result is an equivalent inequality; however **if we multiply both sides by a *negative* number, we must reverse the direction of the inequality symbol.** Since dividing by a number is the same as multiplying by the reciprocal of the number, these same rules apply when dividing both sides of an inequality by a number.

Multiplying by a negative		*Dividing by a negative*	
$4 < 5$	True	$9 \geq -6$	True
$-2(4) > -2(5)$	Multiply by -2, reverse $<$ to $>$.	$\dfrac{9}{-3} \leq \dfrac{-6}{-3}$	Divide by -3, reverse $\geq$ to $\leq$.
$-8 > -10$	True	$-3 \leq 2$	True

Example 1   Solve $3x + 5 < 17$. Then graph the solution set.

Solution:

$$3x + 5 < 17 \qquad \text{Original inequality}$$
$$3x + 5 - 5 < 17 - 5 \qquad \text{Subtract 5 from both sides.}$$
$$3x < 12 \qquad \text{Simplify each side.}$$
$$\frac{3x}{3} < \frac{12}{3} \qquad \text{Divide both sides by 3.}$$
$$x < 4 \qquad \text{Simplify each side.}$$

Try Exercise 17

Solve $3x + 7 < 13$ and graph the solution set.

The solution set is $\{x \mid x < 4\}$. The graph is shown in Fig. 2.12.   ❑

Figure 2.12

The set of numbers $\{x \mid x < 4\}$, whose graph is shown in Fig. 2.12, is called an **interval.** We can write this set more simply as follows:

$$\{x \mid x < 4\} = (-\infty, 4)$$

The symbol $-\infty$ is read "negative infinity," and the expression $(-\infty, 4)$ is called **interval notation.**

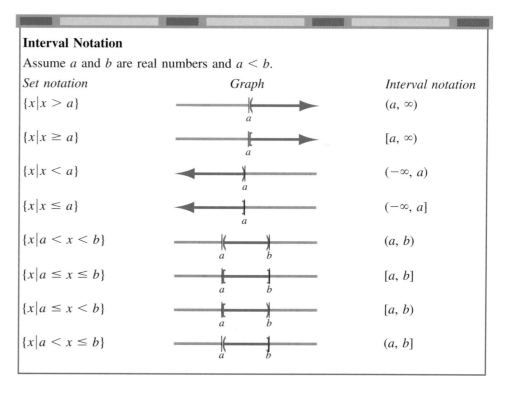

Interval Notation

Assume a and b are real numbers and $a < b$.

Set notation	*Graph*	*Interval notation*
$\{x \mid x > a\}$		(a, ∞)
$\{x \mid x \geq a\}$		$[a, \infty)$
$\{x \mid x < a\}$		$(-\infty, a)$
$\{x \mid x \leq a\}$		$(-\infty, a]$
$\{x \mid a < x < b\}$		(a, b)
$\{x \mid a \leq x \leq b\}$		$[a, b]$
$\{x \mid a \leq x < b\}$		$[a, b)$
$\{x \mid a < x \leq b\}$		$(a, b]$

Note that a square bracket, [or], indicates that the endpoint is included in the interval. A parenthesis, (or), indicates that the endpoint is not included. The symbols ∞ and $-\infty$ do not represent real numbers. They simply indicate that the interval is unbounded on that end. We always use a parenthesis with ∞ and $-\infty$.

An interval that includes both of its endpoints is a **closed interval.** Therefore [2, 7] is the closed interval from 2 to 7. An interval that includes neither of its endpoints is an **open interval.** Therefore (−5, 1) is the open interval from −5 to 1, and (−∞, −3) is the open interval from $-\infty$ to −3. An interval that includes one endpoint but not the other is called a **half-open interval.** Therefore (4, 6] and [−8, ∞) are half-open intervals.

From now on we will write the solution set of an inequality in interval notation.

Example 2   Solve $-y - 3y - 7 \le 15 - 14$. Then graph the solution set.

Solution:

$$
\begin{array}{ll}
-y - 3y - 7 \le 15 - 14 & \text{Original inequality} \\
-4y - 7 \le 1 & \text{Simplify each side.} \\
-4y - 7 + 7 \le 1 + 7 & \text{Add 7 to both sides.} \\
-4y \le 8 & \text{Simplify each side.}
\end{array}
$$

To solve for y, divide both sides by −4. Since −4 is a negative number, reverse the direction of the inequality symbol.

$$
\begin{array}{ll}
\dfrac{-4y}{-4} \ge \dfrac{8}{-4} & \begin{array}{l}\text{Divide by } -4, \\ \text{reverse} \le \text{to} \ge.\end{array} \\
y \ge -2 & \text{Simplify each side.}
\end{array}
$$

Try Exercise 23

Solve $-y - 5y - 2 \le 16 - 12$ and graph the solution set.

The solution set is $[-2, \infty)$. The graph is shown in Fig. 2.13.   ❐

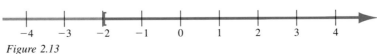

Figure 2.13

Example 3   Solve $8 - 4(r + 1) > 2r + 14$. Then graph the solution set.

Solution:

$$
\begin{array}{ll}
8 - 4(r + 1) > 2r + 14 & \text{Original inequality} \\
8 - 4r - 4 > 2r + 14 & \text{Distribute } -4. \\
-4r + 4 > 2r + 14 & \text{Simplify the left side.} \\
-4r + 4 - 2r > 2r + 14 - 2r & \text{Subtract } 2r \text{ from both sides.} \\
-6r + 4 > 14 & \text{Simplify each side.} \\
-6r + 4 - 4 > 14 - 4 & \text{Subtract 4 from both sides.} \\
-6r > 10 & \text{Simplify each side.} \\
\dfrac{-6r}{-6} < \dfrac{10}{-6} & \begin{array}{l}\text{Divide by } -6, \\ \text{reverse} > \text{to} <.\end{array} \\
r < -\dfrac{5}{3} & \text{Simplify each side.}
\end{array}
$$

Try Exercise 29

Solve $7 - 3(r + 1) > 3r + 18$ and graph the solution set.

The solution set, $\left(-\infty, -\frac{5}{3}\right)$, is graphed in Fig. 2.14.   ❐

Figure 2.14

Here is a summary of the steps in solving a linear inequality in one variable. Note the similarity between these steps and the steps for solving a linear equation.

Solving a Linear Inequality in One Variable

1. Clear fractions by multiplying both sides by the LCD. Distribute the LCD over every term.
2. Use the distributive property to remove parentheses.
3. Simplify each side by combining like terms.
4. Use the addition property to collect all variable terms on one side of the inequality and all constant terms on the other side.
5. Use the multiplication property to write the variable term with a coefficient of 1. Reverse the inequality symbol if you multiply or divide both sides by a negative number.

We illustrate these steps in Example 4.

Example 4   Solve $\dfrac{m}{3} - \dfrac{m - 6}{4} \geq \dfrac{m}{6} + 1$. Then graph the solution set.

Solution: Step 1:   Clear fractions by multiplying both sides by the LCD 12. Distribute 12 over every term.

$$12\left(\frac{m}{3}\right) - 12\left(\frac{m - 6}{4}\right) \geq 12\left(\frac{m}{6}\right) + 12(1)$$

Careful! Write parentheses here.

$$4m - 3(m - 6) \geq 2m + 12$$

Step 2:   Use the distributive property to remove parentheses.

$$4m - 3m + 18 \geq 2m + 12$$

Step 3:   Simplify each side by combining like terms.

$$m + 18 \geq 2m + 12$$

Step 4:   Use the addition property to collect all variable terms on one side of the inequality and all constant terms on the other side.

$$m + 18 - 2m \geq 2m + 12 - 2m \qquad \text{Subtract } 2m.$$
$$-m + 18 \geq 12$$
$$-m + 18 - 18 \geq 12 - 18 \qquad \text{Subtract } 18.$$
$$-m \geq -6$$

Step 5:   Use the multiplication property to write the variable term with a coefficient of 1. Reverse the inequality symbol since we are dividing by a negative number.

$$\frac{-m}{-1} \leq \frac{-6}{-1} \qquad \text{Divide by } -1.$$
$$m \leq 6$$

The solution set, $(-\infty, 6]$, is graphed in Fig. 2.15.   ❐

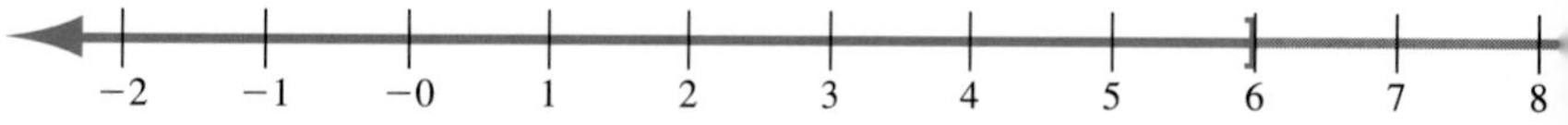

Try Exercise 45

Solve: $\dfrac{m}{3} - \dfrac{m - 2}{2} \geq \dfrac{m}{4} - 4$.

Figure 2.15

Just as some equations (namely, identities) are true for all values of the variable, so are some inequalities. And just as some equations (namely, contradictions) are false for all values of the variable, so are some inequalities.

Example 5   Solve each inequality and graph the solution set.
(a) $3p - 10p - 8 < 8 - 7p$ (b) $12h \geq 25 - 3(8 - 4h)$

Solution: (a) $\begin{aligned} 3p - 10p - 8 &< 8 - 7p \qquad &&\text{Original inequality} \\ -7p - 8 &< 8 - 7p \qquad &&\text{Simplify the left side.} \\ -8 &< 8 \qquad &&\text{Add } 7p. \end{aligned}$

Since $-8 < 8$ is a true statement, the solution set is all real numbers, written $(-\infty, \infty)$. The graph is shown in Fig. 2.16.

Figure 2.16

(b) $\begin{aligned} 12h &\geq 25 - 3(8 - 4h) \qquad &&\text{Original inequality} \\ 12h &\geq 25 - 24 + 12h \qquad &&\text{Distributive property} \\ 12h &\geq 1 + 12h \qquad &&\text{Simplify the right side.} \\ 0 &\geq 1 \qquad &&\text{Subtract } 12h. \end{aligned}$

Since $0 \geq 1$ is a false statement, the solution set is $\varnothing$. There is no graph.   ❐

Try Exercise 47

Solve $4p - 8p - 1 < 6 - 4p$ and graph the solution set.

Neither inequality in Example 5 is actually a linear inequality. Since the variable drops out, neither inequality can be written in the form $ax + b < 0$ with $a \neq 0$.

Double Inequalities

Recall that the double inequality

$$1 < x < 5$$

means that x is between 1 and 5, as shown in Fig. 2.17.

Figure 2.17

We solve the double inequality in Example 6 by isolating the variable in the middle.

Example 6   Solve $-4 \leq 5x - 4 \leq 11$. Then graph the solution set.

Solution:
$$\begin{aligned} -4 &\leq 5x - 4 \leq 11 \qquad &&\text{Original inequality} \\ -4 + 4 &\leq 5x - 4 + 4 \leq 11 + 4 \qquad &&\text{Add 4 to each part.} \\ 0 &\leq 5x \leq 15 \qquad &&\text{Simplify each part.} \\ \frac{0}{5} &\leq \frac{5x}{5} \leq \frac{15}{5} \qquad &&\text{Divide each part by 5.} \\ 0 &\leq x \leq 3 \qquad &&\text{Simplify each part.} \end{aligned}$$

Try Exercise 53

Solve $-5 \leq 4x - 5 \leq 3$ and graph the solution set.

The solution set is $[0, 3]$. The graph is shown in Fig. 2.18.   ❐

Figure 2.18

If we multiply or divide each part of a double inequality by a negative number, we must reverse the direction of *both* inequality symbols.

Example 7 Solve $3 < 1 - 2y \le 10$. Then graph the solution set.

Solution:

$$3 < 1 - 2y \le 10 \qquad \text{Original inequality}$$
$$3 - 1 < 1 - 2y - 1 \le 10 - 1 \qquad \text{Subtract 1 from each part.}$$
$$2 < -2y \le 9 \qquad \text{Simplify each part.}$$
$$\frac{2}{-2} > \frac{-2y}{-2} \ge \frac{9}{-2} \qquad \text{Divide each part by } -2, \text{ reverse both inequality symbols.}$$
$$-1 > y \ge -\frac{9}{2} \qquad \text{Simplify each part.}$$

Double inequalities may be easier to read when the lesser numbers are to the left, since then the order of the inequality follows the order of the number line. Therefore, we often write this last double inequality as

$$-\tfrac{9}{2} \le y < -1$$

Try Exercise 57

Solve $4 < 1 - 3y \le 8$ and graph the solution set.

The solution set, $[-\tfrac{9}{2}, -1)$, is graphed in Fig. 2.19.   ❑

Figure 2.19

Here is an example of an applied problem that can be solved by solving an inequality.

Problem Solving

Example 8 A graphics company can print posters at a cost of $4 per poster plus a daily overhead of $600. How many posters must the company print and sell each day at $5.20 per poster for the daily profit to be greater than 10 percent of the daily cost?

Solution: $x =$ the number of posters printed and sold each day

The daily cost C is

cost per poster	times	number of posters	plus	daily overhead
↓	↓	↓	↓	↓
$C = 4$	·	x	+	600

$$C = 4x + 600$$

The daily revenue (income) R is

price per poster	times	number of posters
↓	↓	↓
$R = 5.20$	·	x

$$R = 5.20x$$

The daily profit P is revenue minus cost. Therefore

revenue   minus   cost

↓   ↓   ↓

$$P = 5.20x \qquad - \qquad (4x + 600)$$
$$P = 1.2x - 600$$

Since profit must be greater than 10 percent of cost, we can write the following inequality:

Daily   is greater   10% of

profit   than   daily cost

↓   ↓   ↓

$$1.2x - 600 \qquad > \qquad 0.10(4x + 600)$$

Now solve this inequality:

$1.2x - 600 > 0.4x + 60$	Distributive property
$0.8x > 660$	Subtract $0.4x$, add 600.
$x > 825$	Divide by 0.8.

Therefore, more than 825 posters must be printed and sold each day for daily profit to be greater than 10 percent of daily cost.   ❏

Try Exercise 69

A textile plant can produce T-shirts at a cost of $4 per T-shirt plus a daily overhead of $1600. How many T-shirts must the plant produce and sell each day at $7.20 per T-shirt for the daily profit to be greater than 20 percent of the daily cost?

Exercises 2.5

Matching Exercises

Match each interval in Exercises 1 through 4 with its graph in letters A through D.

1. $(2, \infty)$

2. $(-\infty, 2)$

3. $[-2, 2)$

4. $(-2, 2]$

A.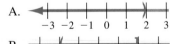

B.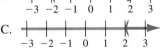

C.
```
  +  +  +  +  +  K——▶
 -3 -2 -1  0  1  2  3
```

D.

True-False Exercises

5. $x^2 > 9$ is a linear inequality.

6. $\dfrac{y}{3} - 7 \geq -4$ is a linear inequality.

7. If you add a negative number to both sides of an inequality, you must reverse the direction of the inequality symbol.

8. If you divide both sides of an inequality by a negative number, you must reverse the direction of the inequality symbol.

Completion Exercises

9. A number that makes an inequality a true statement when it is substituted for the variable is called a(n) _____ of the inequality.

10. A statement that two algebraic expressions are not equal is called a(n) _____.

11. To _____ an inequality means to find its solution set.

12. Inequalities that have the same solution set are called _____ inequalities.

Solve each inequality. Then graph the solution set.

13. $4x > -12$

14. $6x > -18$

15. $-5x > 25$

16. $-8x > 32$

17. $3x + 7 < 13$

18. $2x + 5 < 11$

19. $-t + 10 \geq 9$

20. $-t + 17 \geq 15$

21. $4 \leq -1 - 5r$

22. $2 \leq -4 - 3r$

23. $-y - 5y - 2 \leq 16 - 12$

24. $-y - 3y + 14 \leq 21 - 11$

25. $15 - 6y \leq 9y - 15$

26. $10 - 3y \leq 7y - 10$

27. $7z + 9 \geq -3z + 9$

28. $5z + 1 \geq -2z + 1$

29. $7 - 3(r + 1) > 3r + 18$

30. $5 - 2(r + 1) > 2r + 13$

31. $2(2q - 1) \leq 6 - (q + 8)$

32. $2(2q - 1) \leq 2 - (q + 4)$

33. $-(5 - h) + 3 - 6h \geq 7 - h - 18$

34. $-(1 + h) - 4 - 5h \geq -8 - h - 16$

Solve each inequality involving fractions.

35. $\frac{2}{3}x \leq -2$

36. $\frac{3}{4}x \leq -3$

37. $20 > -\frac{5}{4}p$

38. $35 > -\frac{7}{5}p$

39. $\frac{3m - 5}{4} < 1$

40. $\frac{2m + 3}{7} < 1$

41. $\frac{1}{3}t + \frac{7}{12} > \frac{1}{2}t + \frac{3}{4}$

42. $\frac{1}{9}t + \frac{5}{18} > \frac{1}{6}t + \frac{2}{3}$

43. $1 < 11 - \frac{3u + 2}{5}$

44. $1 < 10 - \frac{4u + 5}{5}$

45. $\frac{m}{3} - \frac{m - 2}{2} \geq \frac{m}{4} - 4$

46. $\frac{3m}{7} - \frac{m - 4}{3} \geq 4 + \frac{2m}{7}$

Solve each inequality and graph the solution set.

47. $4p - 8p - 1 < 6 - 4p$

48. $-3p + p + 4 < 5 - 2p$

49. $15h \geq 19 - 3(3 - 5h)$

50. $18h \geq 15 - 2(6 - 9h)$

Solve each double inequality. Then graph the solution set.

51. $1 < t + 2 < 4$

52. $3 < t - 1 < 5$

53. $-5 \leq 4x - 5 \leq 3$

54. $-9 \leq 5x + 6 \leq 6$

55. $6 \leq -3r < 9$

56. $8 \leq -2r < 10$

57. $4 < 1 - 3y \leq 8$

58. $5 < 3 - 2y \leq 10$

59. $2 < 5 - \frac{1}{2}x < 1$

60. $3 < 4 - \frac{1}{3}x < 2$

61. $x - 10 < 5x - 6 < x + 10$

62. $x - 6 < 4x - 3 < x + 9$

Problem Solving

63. Five more than 6 times a number x is between -7 and 17. What are the possible values of x?

64. If two-thirds of a number is subtracted from 8, the result is no more than twice the number. Find all such numbers.

65. The perimeter of a square must be between 48 in. and 64 in. Find all possible values for the length of one side of the square.

66. The perimeter of an equilateral triangle (three equal sides) must be between 39 in. and 51 in. Find all possible values for the length of one side of the triangle.

67. An indoor tennis club has two plans. With Plan A you pay a membership fee of $100 and an additional $12 an hour for court time. With Plan B you simply pay $14.50 an hour for court time. How many hours of court time must be rented to make Plan A the better deal?

68. To earn a B in chemistry, your average on five tests must be at least 80 and less than 90. Your scores on the first four tests are 96, 83, 79, and 94. What score on the fifth test will give you a B for the course?

69. A textile plant can produce T-shirts at a cost of $4 per T-shirt plus a daily overhead of $1600. How many T-shirts must the plant produce and sell each day at $7.20 per T-shirt for the daily profit to be greater than 20 percent of the daily cost?

70. A small electronics firm can produce calculators at a cost of $5 per calculator plus a daily overhead of $1400. How many calculators must the firm produce and sell each day at $13.50 per calculator for the firm's daily profit to be greater than 30 percent of the daily cost?

Discussion Exercises

71. How could you check the solution of a linear inequality? Give examples.

72. What is a linear inequality in one variable?

73. Explain the addition property of inequality.

74. How does the multiplication property of inequality differ from the multiplication property of equality?

75. Discuss the steps in solving a linear inequality in one variable.

76. Explain the differences between open, closed, and half-open intervals. Give examples.

77. What is the purpose of interval notation?

Getting Acquainted with Your Graphing Calculator

78. The solution set of $3x - (x + 5) < x + 4$ is $(-\infty, 9)$. Store a value from $(-\infty, 9)$ in x and show the statement is true. Then store a value from $[9, \infty)$ in x and show the statement is false.

79. The solution set of $1 \leq 7 - 2x \leq 11$ is $[-2, 3]$. Store a value from $[-2, 3]$ in x and show that both of the statements $1 \leq 7 - 2x$ and $7 - 2x \leq 11$ are true. Then store a value from $(-\infty, -2)$ or $(3, \infty)$ in x and show that at least one of the statements $1 \leq 7 - 2x$ or $7 - 2x \leq 11$ is false.

2.6 Compound Inequalities

TAPE AU2

In Sec. 2.5 we learned to solve a double inequality by isolating x in the middle. A double inequality is actually a special kind of *compound inequality*. Note that the double inequality $2 < x - 3 < 7$ can be written as two separate inequalities joined by the connective word "and":

$$2 < x - 3 \qquad \text{and} \qquad x - 3 < 7$$

A **compound inequality** is two inequalities joined by the connective word "and" or the connective word "or."

Here are two examples of compound inequalities:

1. $x - 1 \leq 4$ and $3x \geq 6$
2. $r + 10 \leq 12$ or $r - 5 > 1$

The Connective Word "And"

The solution set of a compound inequality with the connective word "and" is the *intersection* of the solution sets of the two inequalities that make up the compound inequality.

Example 1   Solve the compound inequality $x - 1 \leq 4$ and $3x \geq 6$.

Solution: Solve each part separately:

$$x - 1 \leq 4 \qquad \text{and} \qquad 3x \geq 6$$
$$x \leq 5 \qquad\qquad\qquad x \geq 2$$

The top graph in Fig. 2.20 illustrates the solution set of the first part, and the middle graph illustrates the solution set of the second part. The bottom graph illustrates the intersection of the two solution sets (the numbers in common to the first two graphs). Therefore the solution set of the compound inequality is

$$(-\infty, 5] \cap [2, \infty) = [2, 5] \qquad\qquad\qquad \square$$

Try Exercise 29

Solve the compound inequality $x - 3 \leq 5$ and $2x \geq 8$, and graph the solution set.

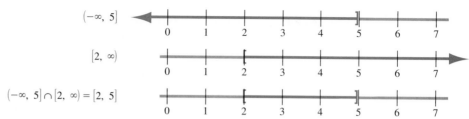

Figure 2.20

Example 2   Solve the compound inequality $2y + 1 < -7$ and $-5y + 4 \geq 4$.

Solution: Solve each part separately.

$$2y + 1 < -7 \qquad \text{and} \qquad -5y + 4 \geq 4$$
$$2y < -8 \qquad\qquad\qquad -5y \geq 0$$
$$y < -4 \qquad\qquad\qquad\quad y \leq 0$$

From Fig. 2.21 we see that the intersection of the two intervals $(-\infty, -4)$ and $(-\infty, 0]$ is $(-\infty, -4)$. ❑

Try Exercise 31

Solve the compound inequality $3y + 1 < -8$ and $-2y + 9 \geq 9$, and graph the solution set.

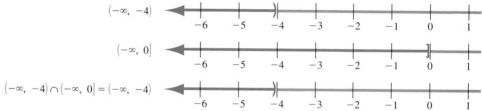

Figure 2.21

Example 3   Solve the compound inequality $-2h > -2$ and $h - 4 > -1$.

Solution: Solve each part separately:

$$-2h > -2 \qquad \text{and} \qquad h - 4 > -1$$
$$h < 1 \qquad\qquad\qquad\qquad h > 3$$

From Fig. 2.22 we see that there are no numbers in common to the two intervals $(-\infty, 1)$ and $(3, \infty)$. Therefore the solution set is $\varnothing$. ❑

Try Exercise 33

Solve the compound inequality $-3h > -6$ and $h - 2 > 3$, and graph the solution set.

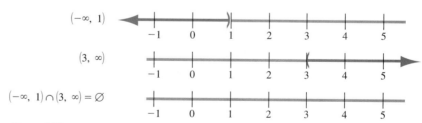

Figure 2.22

The Connective Word "Or"

The solution set of a compound inequality with the connective word "or" is the *union* of the solution sets of the two inequalities that make up the compound inequality.

Example 4   Solve the compound inequality $r + 10 \leq 12$ or $r - 5 > 1$.

Solution: Solve each part separately:

$$r + 10 \leq 12 \qquad \text{or} \qquad r - 5 > 1$$
$$r \leq 2 \qquad\qquad\qquad r > 6$$

The top graph in Fig. 2.23 illustrates the solution set of the first part, and the middle graph illustrates the solution set of the second part. The bottom graph illustrates the union of the two solution sets (the numbers in either the first graph, or the second graph, or both graphs). Therefore the solution set of the compound inequality is

$$(-\infty, 2] \cup (6, \infty)$$

Since the two intervals do not overlap, the answer cannot be simplified further.   ❑

Try Exercise 35

Solve the compound inequality $r + 11 \leq 14$ or $r - 4 > 0$, and graph the solution set.

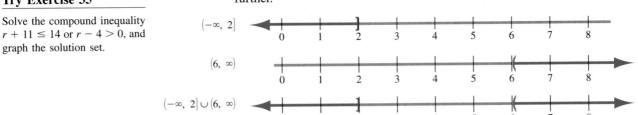

Figure 2.23

<u>CAUTION</u>

Since the solution in Example 4 consists of the two disjoint (no numbers in common) intervals $(-\infty, 2]$ and $(6, \infty)$, you cannot write the answer as a double inequality.

Two correct ways to write the answer   *Two wrong ways to write the answer*

1. $(-\infty, 2] \cup (6, \infty)$ 1. $2 \geq r > 6$
2. $r \leq 2$ or $r > 6$ 2. $r \leq 2$ and $r > 6$

The double inequality $2 \geq r > 6$ incorrectly implies that $2 > 6$. The compound inequality $r \leq 2$ and $r > 6$ has no solution, since no value of r is both less than or equal to 2 and at the same time greater than 6.   ∎

Example 5   Solve the compound inequality $k + 1 > 0$ or $8k > 16$.

Solution: Solve each part separately:

$$k + 1 > 0 \qquad \text{or} \qquad 8k > 16$$
$$k > -1 \qquad\qquad\qquad k > 2$$

From Fig. 2.24 we see that the union of the two intervals $(-1, \infty)$ and $(2, \infty)$ is $(-1, \infty)$.   ❑

Try Exercise 37

Solve the compound inequality $k + 4 > -1$ or $6k > 6$, and graph the solution set.

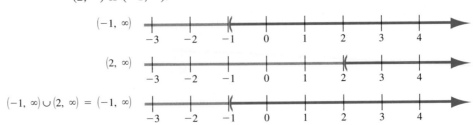

Figure 2.24

Example 6   Solve the compound inequality $\frac{4}{3}t \leq 14$ or $4t - 17 \geq 19$.

Solution: Solve each part separately:

$$\frac{4}{3}t \leq 14 \qquad \text{or} \qquad 4t - 17 \geq 19$$

$$\frac{3}{4}(\frac{4}{3}t) \leq \frac{3}{4}(14) \qquad\qquad 4t \geq 36$$

$$t \leq \frac{21}{2} \qquad\qquad\qquad t \geq 9$$

Try Exercise 39

Solve the compound inequality $\frac{4}{5}t \leq 10$ or $5t - 29 \geq 21$, and graph the solution set.

From Fig. 2.25 we see that the union of the two intervals $(-\infty, \frac{21}{2}]$ and $[9, \infty)$ is the set of all real numbers, $(-\infty, \infty)$.   ❏

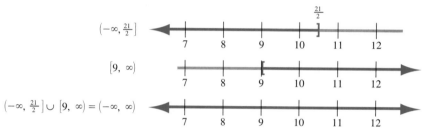

Figure 2.25

Here is a summary of the steps in solving a compound inequality.

To Solve a Compound Inequality

1. Solve each part separately.
2. If the connective word is "and," the solution set is the *intersection* of the two intervals obtained in step 1.
3. If the connective word is "or," the solution set is the *union* of the two intervals obtained in step 1.

Exercises 2.6

Matching Exercises

Match each compound inequality in Exercises 1 through 6 with its graph in letters *A* through *F*.

1. $x > 0$ and $x < 3$ A.

2. $x > 0$ or $x < 3$ B.

3. $x > 0$ and $x > 3$ C.

4. $x < 0$ and $x > 3$ D.

5. $x < 0$ or $x > 3$ E.

6. $x < 0$ or $x < 3$ F.

Completion Exercises

7. A compound inequality is two inequalities joined by either the word "_____" or the word "_____."

8. The solution set of a compound inequality with the

connecting word "and" is the _____ of the solution sets of the two inequalities that make up the compound inequality.

9. The solution set of a compound inequality with the connecting word "or" is the _____ of the solution sets of the two inequalities that make up the compound inequality.

10. A double inequality is actually two inequalities joined by the word "_____."

True-False Exercises

11. The compound inequality $x > 6$ or $x < -1$ can be written $-1 > x > 6$.

12. The compound inequality $x > -4$ or $x < 7$ can be written $-4 < x < 7$.

Perform the indicated union or intersection.

13. $(-4, 3) \cap (0, 6)$

14. $(-\infty, 5] \cap [2, \infty)$

15. $(-1, 4) \cup [3, 8]$

16. $(-\infty, -2] \cap (10, \infty)$

17. $(-\infty, -6) \cup (-\infty, -4)$

18. $[5, \infty) \cap (9, \infty)$

19. $(-\infty, 1) \cup (2, \infty)$

20. $(0, 4) \cup (5, 8)$

21. $(-\infty, 2] \cap (2, \infty)$

22. $(-3, \infty) \cup [4, 5]$

23. $[1, 2] \cup (0, \infty)$

24. $(-\infty, 11) \cup [11, \infty)$

Solve each compound inequality and graph the solution set.

25. $x > 1$ and $x < 6$

26. $y \le 3$ or $y \le -2$

27. $p > -1$ and $p \ge -4$

28. $m < 7$ or $m \ge 0$

29. $x - 3 \le 5$ and $2x \ge 8$

30. $4x \le 20$ and $x + 1 \ge 2$

31. $3y + 1 < -8$ and $-2y + 9 \ge 9$

32. $-6y - 7 \le -7$ and $2y - 3 > 5$

33. $-3h > -6$ and $h - 2 > 3$

34. $h - 8 > 1$ and $-5h > -5$

35. $r + 11 \le 14$ or $r - 4 > 0$

36. $r + 15 \le 17$ or $r + 4 > 7$

37. $k + 4 > -1$ or $6k > 6$

38. $3k < 9$ or $k - 2 < -6$

39. $\frac{4}{5}t \le 10$ or $5t - 29 \ge 21$

40. $5t + 23 \le -12$ or $\frac{8}{9}t \ge -12$

41. $-\frac{3}{8}p > \frac{5}{4}$ or $3p + 10 > 0$

42. $\dfrac{6x - 7}{5} < -1$ or $\dfrac{1 - 3x}{4} < 0$

Write each double inequality as a compound inequality and solve. Then graph the solution set.

43. $-7 < 2x - 5 \le 9$

44. $-5 \le 3x + 4 < 19$

45. $2y - 1 < y + 2 < 6y + 12$

46. $5m \le 12m - 7 \le 8m$

47. $15 - t \le t + 15 < 9t - 9$

48. $h + 1 \le \frac{2}{3}h \le h - 2$

Problem Solving

Find all numbers that satisfy the given conditions.

49. The sum of a number and 9 is less than -6 or greater than 14.

50. The difference of a number and 7 is greater than 18 or less than -2.

51. Twice the difference of a number and 3 is more than -11 but no more than 0.

52. Three times the sum of a number and 5 is less than 8 but no less than 0.

53. A number increased by 12 is at least twice the number, and the difference of 5 and the number is at most one-half of the number.

54. A number decreased by 4 is at most twice the number, and the difference of 10 and the number is at least one-third of the number.

Discussion Exercises

55. What type of graph on the number line can be described by a double inequality and what type cannot?

56. Discuss the steps in solving a compound inequality.

2.7 Absolute-Value Equations and Inequalities

TAPE AU4

In Sec. 1.2 you learned that the absolute value of a real number is the distance between the number and 0 on the number line. In this section you will learn to solve equations and inequalities that involve absolute value.

Absolute-Value Equations

An **absolute-value equation** is an equation that involves the absolute value of a variable quantity. For example

$$|x| = 3$$

is an absolute-value equation. Since $|x|$ represents the distance between x and 0, the solutions of this equation are those numbers whose distance to 0 is 3 units, namely, $x = 3$ or $x = -3$ (see Fig. 2.26).

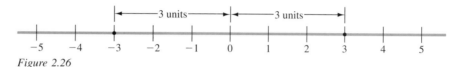

Figure 2.26

This suggests the following rule.

To Solve $|x| = a$

If a is a positive number, then

$$|x| = a \text{ is equivalent to } x = a \text{ or } x = -a$$

Example 1   Solve each equation: (a) $|y| = 10$, (b) $|p| = 0$, (c) $|r| = -4$.

Solution: (a) Since 10 is a positive number, $|y| = 10$ is equivalent to $y = 10$ or $y = -10$. Therefore the solution set is $\{10, -10\}$.

(b) Since the right side of the equation $|p| = 0$ is not a positive number, we cannot use the rule for solving $|x| = a$. Instead, we simply note that the only number whose absolute value is 0 is 0 itself. Therefore the solution set is $\{0\}$.

(c) Again, the right side of $|r| = -4$ is not a positive number, so we cannot use the rule for solving $|x| = a$. Instead, we note that there is no number whose absolute value is -4. Therefore the solution set is $\varnothing$.   ❑

Try Exercise 15

Solve: $|y| = 8$.

Example 2   Solve: $|2m - 3| = 7$.

Solution: The equation

$$|2m - 3| = 7$$

is equivalent to

$$2m - 3 = 7 \qquad \text{or} \qquad 2m - 3 = -7$$

Solve each part separately:

$$2m = 10 \qquad \text{or} \qquad 2m = -4$$
$$m = 5 \qquad\qquad\qquad m = -2$$

Check each number in the original equation.

Check: m = 5	*Check: m* = −2
$\|2m - 3\| = 7$	$\|2m - 3\| = 7$
$\|2(5) - 3\| = 7$	$\|2(-2) - 3\| = 7$
$\|7\| = 7$ True	$\|-7\| = 7$ True

Try Exercise 17

Solve: $\|2m - 5\| = 9$.

Since both numbers check, the solution set is {5, −2}.   ❏

We now consider equations of the form

$$|x| = |y|$$

This equation is true if x and y are equal or if x and y are opposites.

To Solve $|x| = |y|$

$|x| = |y|$ is equivalent to $x = y$ or $x = -y$

It is not necessary to consider the equations $-x = -y$ and $-x = y$ in the rule above, since $-x = -y$ is equivalent to $x = y$, and $-x = y$ is equivalent to $x = -y$.

Example 3   Solve: $\|2z - 1\| = \|z + 4\|$.

Solution: The equation

$$|2z - 1| = |z + 4|$$

is equivalent to

$$2z - 1 = z + 4 \qquad \text{or} \qquad 2z - 1 = -(z + 4)$$

Solve each part separately:

$$z = 5 \qquad \text{or} \qquad 2z - 1 = -z - 4$$
$$3z = -3$$
$$z = -1$$

Try Exercise 25

Solve: $\|2z + 3\| = \|z + 9\|$.

The solution set is {5, −1}. Check in the original equation.   ❏

Example 4   Solve: $\|4x + 5\| = \|8 - 4x\|$.

Solution: The original equation is equivalent to

$$4x + 5 = 8 - 4x \qquad \text{or} \qquad 4x + 5 = -(8 - 4x)$$

Solve each part separately:

$$8x = 3 \qquad \text{or} \qquad 4x + 5 = -8 + 4x$$
$$x = \tfrac{3}{8} \qquad\qquad\qquad 5 = -8$$

Try Exercise 29

Solve: $\|5x - 9\| = \|6 - 5x\|$.

Since $5 = -8$ is a false statement, the only solution is $x = \tfrac{3}{8}$. Therefore the solution set is $\{\tfrac{3}{8}\}$. Check in the original equation.   ❏

Absolute-Value Inequalities

An **absolute-value inequality** is an inequality that involves the absolute value of a variable quantity. We will study two basic types of absolute-value inequalities: $|x| < a$ and $|x| > a$.

Recall that the solutions of the equation $|x| = 3$ are those numbers whose distance to 0 is 3 units. Therefore the solutions of the inequality $|x| < 3$ are those numbers whose distance to 0 is *less* than 3 units, namely, those numbers that satisfy the double inequality $-3 < x < 3$ (see Fig. 2.27).

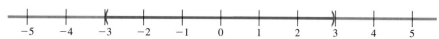

Figure 2.27

This suggests the following rule.

To Solve $|x| < a$

If a is a positive number, then

$$|x| < a \text{ is equivalent to } -a < x < a$$

Note that if a is a positive number, then $|x| \le a$ is equivalent to $-a \le x \le a$.

Example 5   Solve each inequality: (a) $|x| \le 8$, (b) $|m| < -9$, (c) $|t| \le 0$.

Solution: (a) Since 8 is a positive number, $|x| \le 8$ is equivalent to $-8 \le x \le 8$. Therefore the solution set is the interval $[-8, 8]$.
(b) The absolute value of a number cannot be less than a negative number. Therefore the solution set is $\varnothing$.
(c) The absolute value of a number cannot be less than 0. However, $|t| = 0$ when $t = 0$. Therefore the solution set is $\{0\}$.   ◻

Try Exercise 33

Solve: $|x| \le 4$.

Example 6   Solve: $\left| \dfrac{8r + 2}{3} \right| < 6$.

Solution: The original inequality is equivalent to

$$-6 < \frac{8r + 2}{3} < 6$$

Solve by isolating r in the middle:

$$-18 < 8r + 2 < 18 \qquad \text{Multiply each part by 3.}$$
$$-20 < 8r < 16 \qquad \text{Subtract 2 from each part.}$$
$$-\tfrac{5}{2} < r < 2 \qquad \text{Divide each part by 8 and reduce } -\tfrac{20}{8}.$$

Try Exercise 43

Solve: $\left| \dfrac{4r + 1}{3} \right| < 5$.

The solution set is the interval $(-\tfrac{5}{2}, 2)$, as shown in Fig. 2.28.   ◻

Figure 2.28

Example 7   Solve: $2|3p + 6| - 35 \leq -5$.

Solution: First isolate the absolute value:

$$2|3p + 6| - 35 \leq -5 \qquad \text{Original inequality}$$
$$2|3p + 6| \leq 30 \qquad \text{Add 35 to both sides.}$$
$$|3p + 6| \leq 15 \qquad \text{Divide both sides by 2.}$$
$$-15 \leq 3p + 6 \leq 15 \qquad \text{Write an equivalent double inequality.}$$
$$-21 \leq 3p \leq 9 \qquad \text{Subtract 6 from each part.}$$
$$-7 \leq p \leq 3 \qquad \text{Divide each part by 3.}$$

Try Exercise 47

Solve: $3|2p + 1| - 25 \leq -4$.

The solution set is the interval $[-7, 3]$, as shown in Fig. 2.29.   ❑

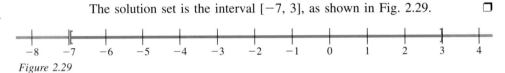

Figure 2.29

As we have just seen, the solutions of $|x| < 3$ are those numbers whose distance to 0 is less than 3 units. The solutions of $|x| > 3$ are those numbers whose distance to 0 is *greater* than 3 units, namely, those numbers that satisfy either the inequality $x < -3$ or the inequality $x > 3$ (see Fig. 2.30).

Figure 2.30

This suggests the following rule.

To Solve $|x| > a$

If a is a positive number, then

$$|x| > a \text{ is equivalent to } x < -a \text{ or } x > a$$

Note that if a is a positive number, then $|x| \geq a$ is equivalent to $x \leq -a$ or $x \geq a$.

Example 8   Solve: $|t - 2| > 4$.

Solution: Since 4 is a positive number, this inequality is equivalent to

$$t - 2 < -4 \qquad \text{or} \qquad t - 2 > 4$$

Solve each part separately:

$$t < -2 \qquad \text{or} \qquad t > 6$$

Try Exercise 51

Solve: $|t - 3| > 5$.

The solution set is $(-\infty, -2) \cup (6, \infty)$, as shown in Fig. 2.31.   ❑

Figure 2.31

Example 9   Solve: $11 \leq |5 - 4x|$.

Solution: First write the inequality with the absolute value on the left side:

$$|5 - 4x| \geq 11$$

Since 11 is a positive number, this inequality is equivalent to

$$5 - 4x \leq -11 \quad \text{or} \quad 5 - 4x \geq 11$$

Solve each part separately:

$$-4x \leq -16 \quad \text{or} \quad -4x \geq 6$$

$$x \geq 4 \qquad\qquad x \leq -\tfrac{3}{2}$$

Try Exercise 59

Solve: $13 \leq |5 - 4x|$.

The solution set is $(-\infty, -\tfrac{3}{2}] \cup [4, \infty)$, as shown in Fig. 2.32.   ❑

Figure 2.32

Example 10   Solve: $|3w + 10| > -2$.

Solution: Since $|3w + 10| \geq 0$ for any value of w, $|3w + 10| > -2$ for any value of w. Therefore the solution set is the set of all real numbers, $(-\infty, \infty)$.   ❑

Try Exercise 63

Solve: $|3w - 14| \geq -3$.

Here is a summary of the rules we discussed in this section.

Solving Absolute-Value Equations and Inequalities

If a is a positive number, then the following hold true:

1. $|x| = a$ is equivalent to $x = a$ or $x = -a$.
2. $|x| = |y|$ is equivalent to $x = y$ or $x = -y$.
3. $|x| < a$ is equivalent to $-a < x < a$.
4. $|x| > a$ is equivalent to $x < -a$ or $x > a$.

If a is a negative number, then the following hold true:

1. $|x| = a$ has solution set $\varnothing$.
2. $|x| < a$ has solution set $\varnothing$.
3. $|x| > a$ has solution set $(-\infty, \infty)$.

If $a = 0$, use the following properties:

1. $|x| = 0$ is equivalent to $x = 0$.
2. $|x| < 0$ has solution set $\varnothing$.
3. $|x| > 0$ has solution set $(-\infty, 0) \cup (0, \infty)$.

Exercises 2.7

Completion Exercises

1. An equation that involves the absolute value of a variable quantity is called a(n) _____ equation.

2. An inequality that involves the absolute value of a variable quantity is called a(n) _____ inequality.

3. The equation $|x| = 5$ is equivalent to _____ or _____.

4. The equation $|x| = |y|$ is equivalent to _____ or _____.

5. The inequality $|x| < 7$ is equivalent to _____.

6. The inequality $|x| > 10$ is equivalent to _____ or _____.

Matching Exercises

Match each equation or inequality in Exercises 7 through 14 with its graph in letters *A* through *H*.

7. $|x| = 2$ A.

8. $|y - 2| = 0$ B.

9. $|p| < 2$ C.

10. $|t| \leq 2$ D.

11. $|z| \geq 2$ E.

12. $|k| > 2$ F.

13. $|r + 2| = -2$ G.

14. $|h - 2| > 0$ H.

Solve each equation.

15. $|y| = 8$

16. $|y| = 12$

17. $|2m - 5| = 9$

18. $|2m - 7| = 3$

19. $|5x| + 3 = 18$

20. $|4x| + 5 = 17$

21. $|7t + 2| + 3 = 0$

22. $|8t + 3| + 5 = 0$

23. $|\frac{3}{4}h + 6| - 21 = 0$

24. $|\frac{3}{5}h - 9| - 30 = 0$

25. $|2z + 3| = |z + 9|$

26. $|3z + 4| = |2z + 11|$

27. $|6y - 1| = |3y + 2|$

28. $|5y - 4| = |2y + 5|$

29. $|5x - 9| = |6 - 5x|$

30. $|4x - 9| = |3 - 4x|$

31. $|3b - 5| = |b - 5|$

32. $|3b + 9| = |b + 9|$

Solve each inequality.

33. $|x| \leq 4$

34. $|x| < 2$

35. $|p - 4| \leq 1$

36. $|p + 2| < 5$

37. $|h - 15| < 0$

38. $|h + 11| < 0$

39. $|t + 1| \leq 0$

40. $|t - 6| \leq 0$

41. $|3z + 9| \leq 6$

42. $|2z + 5| \leq 3$

43. $\left| \dfrac{4r + 1}{3} \right| < 5$

44. $\left| \dfrac{8r + 1}{3} \right| < 5$

45. $\frac{1}{2}|8k| - 12 < 0$

46. $\frac{1}{3}|6k| - 4 < 0$

47. $3|2p + 1| - 25 \leq -4$

48. $4|2p + 5| - 33 \leq -5$

Solve each inequality.

49. $|x| > 1$

50. $|x| \geq 5$

51. $|t - 3| > 5$

52. $|t + 1| > 8$

53. $|6h - 1| \geq 11$

54. $|5h - 3| \geq 7$

55. $|7r + 12| \geq 0$

56. $|9r - 16| \geq 0$

57. $|p + 4| > 0$

58. $|p - 1| > 0$

59. $13 \leq |5 - 4x|$

60. $11 \leq |3 - 4x|$

61. $|\frac{3}{4}m| - 6 > 0$

62. $|\frac{4}{5}m| - 8 > 0$

63. $|3w - 14| \geq -3$

64. $|2w + 17| \geq -1$

Solve each equation or inequality.

65. $|3x - 5| = 7$

66. $|3x - 4| = 8$

67. $|2y - 3| < 5$

68. $|2y - 7| < 3$

69. $|7 + 2r| > 1$

70. $|1 + 2r| > 5$

71. $|3t - 4| = |2t - 1|$

72. $|2t - 3| = |t - 5|$

73. $|3 - 4x| \leq 5$

74. $|1 - 4x| \geq 9$

Problem Solving

Find all numbers that satisfy the given conditions.

75. If 4 is added to a number, the absolute value of the sum is 9.

76. If 3 is subtracted from a number, the absolute value of the difference is 8.

77. The absolute value of one-third of a number is no greater than 6.

78. The absolute value of one-half of a number is no less than 5.

79. If 5 is added to twice a number, the absolute value of the result is greater than 7.

80. If 6 is added to 3 times a number, the absolute value of the result is less than 9.

Discussion Exercises

81. Why do we not say that $|x| = |y|$ is equivalent to the *four* equations $x = y$, $x = -y$, $-x = y$, and $-x = -y$?

82. The solution set of $|\text{any algebraic expression}| \geq -1$ is $(-\infty, \infty)$. Why?

83. The solution set of $|\text{any algebraic expression}| \leq -1$ is $\varnothing$. Why?

84. The solution set of $|x - 4| = |4 - x|$ is $(-\infty, \infty)$. Why?

85. Discuss the procedure for solving an absolute-value equation.

86. Discuss the procedure for solving an absolute-value inequality.

Getting Acquainted with Your Graphing Calculator

87. Store each number in x and test whether $|2x - 3| = 5$ is true or false: (a) 4, (b) -1, (c) 0. Which of the numbers are solutions?

88. Store each number in x and test whether $|2x - 3| < 5$ is true or false: (a) 4, (b) -1, (c) 0. Which of the numbers are solutions?

Chapter 2 Key Terms

2.1 **Factors** Expressions that are related by multiplication
Terms Expressions that are related by addition
(Numerical) coefficient The numerical factor in a term
Like terms Terms that have the same variables with the same exponents
Combining like terms Adding and/or subtracting like terms
Equation A statement that two algebraic expressions are equal
Linear (first-degree) equation (in x) An equation that can be written in the form $ax + b = 0$ $(a \neq 0)$
Solution (of an equation) A value of the variable that makes the equation a true statement
Solution set (of an equation) The set of all solutions of the equation
Solve (an equation) To find the solution set of the equation
Equivalent equations Equations that have the same solution set
Least common denominator (LCD) The smallest number that is divisible by all of the given denominators
Conditional equation An equation that is true for some values of the variable and false for others
Identity An equation that is true for all values of the variable, so long as both sides are defined

Contradiction An equation that is false for all values of the variable

2.2 **Formula** An equation that relates two or more real-world quantities
Solving for a variable Isolating that variable on one side of an equation
Subscripts Numbers written to the right and slightly below a letter to allow us to use that letter to represent different numbers

2.5 **Inequality** A statement that two algebraic expressions are not equal
Linear (first-degree) inequality (in x) An inequality that can be written in one of the forms $ax + b < 0$ or $ax + b \leq 0$ $(a \neq 0)$
Solution (of an inequality) A value of the variable that makes the inequality a true statement
Solution set (of an inequality) The set of all solutions of the inequality
Solve (an inequality) To find the solution set of the inequality
Equivalent inequalities Inequalities that have the same solution set
Interval notation An abbreviated way of writing intervals of numbers
Closed interval An interval that includes both of its endpoints

Open interval An interval that includes neither of its endpoints

Half-open interval An interval that includes one endpoint but not the other

2.6 **Compound inequality** Two inequalities joined by the word "and" or the word "or"

2.7 **Absolute-value equation** An equation that involves the absolute value of a variable quantity

Absolute-value inequality An inequality that involves the absolute value of a variable quantity

Chapter 2 Key Rules/Steps

2.1 Linear Equations in One Variable

Combining Like Terms

$3x + 5x = (3 + 5)x = 8x$
$-9m - m = -10m$
$2y^2 - 6y^2 + y = -4y^2 + y$
$4 + 2(5p - 3) - (p + 8) = 4 + 10p - 6 - p - 8$
$\qquad\qquad\qquad\qquad = 9p - 10$

Addition Property of Equality

If $a = b$, then $a + c = b + c$.
Example: If $x + 2 = 5$, then $x + 2 + (-2) = 5 + (-2)$.

Multiplication Property of Equality

If $a = b$, then $ac = bc$ $(c \neq 0)$.
Example: If $2x = 6$, then $\frac{1}{2} \cdot 2x = \frac{1}{2} \cdot 6$.

To Solve the Linear Equation $\dfrac{x - 8}{2} + 1 = \dfrac{x}{6} + 2$

1. Clear fractions by multiplying both sides by the LCD.

$$6\left(\frac{x - 8}{2}\right) + 6(1) = 6\left(\frac{x}{6}\right) + 6(2)$$
$$3(x - 8) + 6 = x + 12$$

2. Use the distributive property to remove parentheses.

$$3x - 24 + 6 = x + 12$$

3. Simplify each side by combining like terms.

$$3x - 18 = x + 12$$

4. Collect all variable terms on one side and all constant terms on the other side.

$$2x = 30$$

5. Write the variable term with a coefficient of 1.

$$x = 15$$

6. Check in the original equation.

$$\frac{15 - 8}{2} + 1 = \frac{15}{6} + 2$$
$$\frac{9}{2} = \frac{9}{2} \quad \text{True}$$

2.2 Problem Solving with Formulas

To Solve the Formula $C = \frac{5}{9}(F - 32)$ for F

1. Clear fractions by multiplying both sides by the LCD.

$$9 \cdot C = 9 \cdot \tfrac{5}{9}(F - 32)$$
$$9C = 5(F - 32)$$

2. Use the distributive property to remove parentheses.

$$9C = 5F - 160$$

3. Collect all terms with F on one side and all other terms on the other side.

$$9C + 160 = 5F$$

4. Write the side with F as a product of F and the sum of other terms.

There is only one term involving F, so this step is done.

5. Write F with a coefficient of 1.

$$\frac{9C + 160}{5} = F$$

2.3 Number, Percent, and Geometry Problems

To Solve a Word Problem

Two consecutive even integers have a sum of 34. Find the integers.

1. Call one unknown x and write the other in terms of x.

$$x = \text{smaller integer}$$
$$x + 2 = \text{larger integer}$$

2. Write an equation involving x.

$$x + x + 2 = 34$$

3. Solve the equation.

$$2x = 32$$
$$x = 16$$
$$x + 2 = 18$$

4. Check in the words of the original problem.

16 and 18 are consecutive even integers whose sum is 34.

2.4 Money, Mixture, and Motion Problems

A collection of 20 nickels and dimes is worth $1.65. How many nickels and how many dimes are there?

	Number	Value
Nickels	x	$5x$
Dimes	$20 - x$	$10(20 - x)$

$$5x + 10(20 - x) = 165$$
$$5x + 200 - 10x = 165$$
$$-5x = -35$$
$$x = 7 \text{ nickels}$$
$$20 - x = 13 \text{ dimes}$$

2.5 Linear Inequalities in One Variable

Addition Property of Inequality

If $a < b$, then $a + c < b + c$.

Example: If $x - 3 < 7$, then $x - 3 + 3 < 7 + 3$.

Multiplication Property of Inequality

If $a < b$ and $c > 0$, then $ac < bc$.

If $a < b$ and $c < 0$, then $ac > bc$.

Examples: If $\frac{1}{4}x < 2$, then $4 \cdot \frac{1}{4}x < 4 \cdot 2$.

If $-x < 6$, then $-1(-x) > -1(6)$.

Interval Notation

Open interval: $(2, \infty) = \{x \mid x > 2\}$

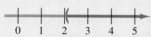

Closed interval: $[2, 4] = \{x \mid 2 \leq x \leq 4\}$

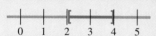

Half-open interval: $[2, 4) = \{x \mid 2 \leq x < 4\}$

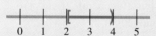

To Solve the Linear Inequality $x - 6 \geq 7 + 3(2x - 1)$

1. Clear fractions by multiplying both sides by the LCD.

   There are no fractions, so this step is done.

2. Use the distributive property to remove parentheses.

   $$x - 6 \geq 7 + 6x - 3$$

3. Simplify each side by combining like terms.

   $$x - 6 \geq 6x + 4$$

4. Collect all variable terms on one side and all constant terms on the other side.

   $$-5x \geq 10$$

5. Write the variable term with a coefficient of 1.
   Divide by -5, reverse $\geq$ to $\leq$.

   $$\frac{-5x}{-5} \leq \frac{10}{-5}$$
   $$x \leq -2$$

To Solve the Double Inequality $-3 < 2x + 7 < 5$

Isolate x in the middle.

$$-3 - 7 < 2x + 7 - 7 < 5 - 7$$
$$-10 < 2x < -2$$
$$-5 < x < -1$$

2.6 Compound Inequalities

To Solve $x - 1 > 0$ and $3x < 12$

$$x > 1 \text{ and } x < 4$$
$$1 < x < 4$$

To Solve $-x > 0$ or $x - 2 \geq 0$

$$x < 0 \text{ or } x \geq 2$$

2.7 Absolute-Value Equations and Inequalities

Let a be a positive number.

1. $|x| = a$ is equivalent to $x = a$ or $x = -a$.

2. $|x| = |y|$ is equivalent to $x = y$ or $x = -y$.

3. $|x| < a$ is equivalent to $-a < x < a$.

4. $|x| > a$ is equivalent to $x < -a$ or $x > a$.

   Note: $|x| \geq 0$ is true for all real numbers x.
   $|x| = 0$ is true only when $x = 0$.
   $|x| < 0$ is false for all real numbers x.

To Solve $|x - 2| = 1$

$$x - 2 = 1 \quad \text{or} \quad x - 2 = -1$$
$$x = 3 \qquad\qquad x = 1$$

To Solve $|x - 3| = |x - 5|$

$$x - 3 = x - 5 \quad \text{or} \quad x - 3 = -(x - 5)$$
$$-3 = -5 \qquad\qquad x - 3 = -x + 5$$
$$\varnothing \qquad\qquad\qquad 2x = 8$$
$$x = 4$$

To Solve $|3x| \leq 6$

$$-6 \leq 3x \leq 6$$
$$-2 \leq x \leq 2$$

To Solve $|x + 1| > 4$

$$x + 1 < -4 \quad \text{or} \quad x + 1 > 4$$
$$x < -5 \qquad\qquad x > 3$$

Chapter 2 Review Exercises

[2.1] Simplify each expression.

1. $3x - 7x - 5 - x$

2. $4(6y - 9) - (10y + 8)$

3. $-p^2 - 2(5p - 1) + 6(p + 3) - p^2$

Solve each equation.

4. $-x + 8 = 17$

5. $7y - 5 = 3y + 23$

6. $4(r - 1) - 3(2r - 5) = 7 + 4r$

7. $-(3z - 4) + 2 - z = -(-3 + 2z) - (z - 5)$

8. $\dfrac{x}{2} + \dfrac{x}{6} = 2$

9. $\dfrac{2p}{3} - \dfrac{1}{12} = \dfrac{3p}{4} + \dfrac{1}{6}$

10. $\dfrac{h - 3}{3} - \dfrac{1}{5} = \dfrac{2}{3} - \dfrac{h - 2}{15}$

11. $\dfrac{3m - 2}{4} = 2 - \dfrac{m - 1}{6}$

Classify as a conditional equation, an identity, or a contradiction.

12. $5(x - 2) = 5x - 2$

13. $2y - (3y + 4) = 1 - (y + 5)$

[2.2] Solve each formula for the specified variable.

14. $A = P + Pr$ for P

15. $V = \pi r^2 h$ for h

16. $s = \frac{1}{2}(a + b + c)$ for c

17. $\dfrac{1}{R} = \dfrac{1}{R_1} + \dfrac{1}{R_2}$ for R_1

Solve each word problem by using the appropriate formula. Use $\pi = 3.14$.

18. At what rate must you travel to cover 270 mi in 5 hr?

19. A rectangular billboard that is 16 ft wide has a perimeter of 106 ft. Find the length of the billboard.

20. You see an auto accident 1.5 sec before you hear the crash. How far away is the accident? The speed of sound is 1100 ft/sec.

21. How much oil would it take to fill a cylindrical drum of height 4 ft and radius $1\frac{1}{4}$ ft?

22. How much money must be invested at 9 percent simple interest to amount to $812.70 in 8 yr?

[2.3] Write each word phrase as an algebraic expression.

23. Twice the difference of a number and 6

24. The sum of 17 and 6 times a number

25. The number of minutes in t days

26. The amount of acid in $40 - x$ grams of a solution that is 35% acid

Represent one of the quantities by a variable. Then represent the other quantities in terms of that variable.

27. Three consecutive odd integers

28. Sally's age now, Sally's age in 2 years, and Sally's age 6 years ago

Solve each word problem.

29. Three times a number is 8 less than 5 times the number. Find the number.

30. The sum of two numbers is 85, and the larger number is 4 times the smaller. Find the numbers.

31. Find three consecutive integers such that 4 times the first minus twice the third is 14 more than the second.

32. A television set is marked down 25 percent to $330. What was the original price of the set?

33. How can $9600 be invested, part at 6 percent and the rest at 10 percent, so that the interest will be the same on each investment?

34. The length of a badminton court for singles play is 10 ft more than twice the width. If the perimeter is 122 ft, find the width and length.

[2.4] Solve each word problem.

35. A collection of dimes and quarters is worth $1.80. If there are twice as many dimes as quarters, how many dimes and how many quarters are there?

36. The total receipts for a school play were $1378 for 400 seats sold. If adult tickets were $4.50 and student tickets were $2.50, how many of each were sold?

37. One alloy contains 22% gold and another contains 30% gold. How many grams of each should be melted together to produce 52 g of an alloy that is 28% gold?

38. A ship leaves port traveling 15 mi/hr. Two and one-half hours later, a second ship leaves the same port traveling the same course at 20 mi/hr. How long does it take the second ship to overtake the first?

39. How far can a plane fly with the wind at 325 mi/hr and still return against the wind at 275 mi/hr if it has enough fuel to last 6 hr?

40. Adam is twice as old as his brother. Three years ago he was 3 times as old as his brother. Find the present ages of Adam and his brother.

[2.5] Solve each inequality. Then graph the solution set.

41. $\frac{3}{4}x > 24$

42. $-3x \le 6$

43. $2t - 4 < 6$

44. $-y + 11 \ge 8$

45. $-2 \le x + 2 \le 5$

46. $-8 < 4t - 8 \le 0$

47. $-\frac{2}{5} \le \frac{1 - 2q}{10} < \frac{1}{2}$

48. $t - 8 < 3t - 4 < t + 8$

Solve each inequality.

49. $7m - 3 > 3m + 5$

50. $5 - (p + 8) < 3(p - 1)$

51. $3(2x - 3) \ge 9 - (x + 4)$

52. $6(3 - 5h) - 1 - h > 7 - 14h - 11h$

53. $4 \le 5 - \frac{2z + 9}{-3}$

54. $\frac{3}{4}y - 1 \le \frac{1}{2}y + \frac{1}{4}$

55. $\frac{m - 4}{3} \le -1 - \frac{m + 1}{3}$

56. $\frac{2t}{3} - \frac{t - 4}{12} \ge \frac{t + 3}{6}$

Solve each word problem.

57. Eight more than 4 times a number x is between -4 and 28. What are the possible values for x?

58. To earn an A in biology your average on six tests must be at least 93. Your scores on the first five tests are 87, 96, 91, 100, and 85. What score on the sixth test will give you an A for the course?

[2.6] Perform the indicated union or intersection.

59. $(-6, 4) \cap (-1, 2)$

60. $[0, 8) \cap (5, 9]$

61. $(-3, 5) \cap [7, \infty)$

62. $(-\infty, 0] \cup (0, \infty)$

63. $(-4, -2) \cup (2, 4)$

64. $(-\infty, -11) \cup (-\infty, -8]$

Solve each compound inequality and graph the solution set.

65. $5y - 7 < 3$ or $-y < -4$

66. $11 - 2x \le 7$ and $3x \le 18$

67. $\frac{2}{3}p > 12$ and $-\frac{p}{5} \le 0$

68. $m + 5 \ge 2$ or $-7 < m - 6$

69. $-\frac{5}{3}t > -\frac{35}{9}$ or $\frac{t}{-4} > -\frac{1}{2}$

70. $\frac{z + 1}{5} \ge 1$ and $\frac{1 - z}{6} \ge 1$

Write each double inequality as a compound inequality and solve. Then graph the solution set.

71. $k + 8 < 5k - 4 < 3k + 10$

72. $9x \le 15x + 12 \le 11x - 5$

[2.7] Solve each equation or inequality.

73. $|4x| = 8$

74. $|2z - 5| < 3$

75. $|4t - 3| > 7$

76. $|2p - 3| = 9$

77. $|3y - 2| = |4y - 5|$

78. $|4 - m| \ge 1$

79. $\left|\frac{2x - 1}{3}\right| \le 5$

80. $|y - 3| < 0$

81. $|r - 6| \ge 0$

82. $|p - 5| + 11 = 8$

Find all numbers that satisfy the given conditions.

83. If 7 is added to a number, the absolute value of the sum is 4.

84. If 5 is added to twice a number, the absolute value of the result is less than 7.

Chapter 2 Test

Perform the indicated union or intersection.

1. $(-3, 0] \cap [-1, 5)$

2. $(2, \infty) \cup [4, \infty)$

Classify as a conditional equation, an identity, or a contradiction.

3. $2(x + 3) = 3x - (x - 6)$

4. $8y - 5y - 7 = -7 - 3y$

Solve each equation.

5. $t - (2t - 1) = 3$

6. $5p - (p + 6) = 2 - (3p - 8)$

7. $\frac{m}{3} - \frac{m}{4} = 2$

8. $\frac{h + 2}{6} - \frac{h}{2} + \frac{1}{3} = 0$

Solve each formula for the specified variable.

9. $A = \frac{1}{2}bh$ for b

10. $P = 2l + 2w$ for w

Solve each inequality. Then graph the solution set.

11. $-\frac{2}{3}x \geq 6$

12. $7 - 3r < 2(r + 1)$

13. $\frac{3}{4}y - 2 > \frac{1}{2}y + \frac{1}{4}$

14. $-1 < \frac{1 - m}{2} \leq 3$

Solve each compound inequality and graph the solution set.

15. $x + 13 \geq 9$ and $6x - 10 < 8$

16. $7 - y \geq 6$ or $\frac{9}{5}y < 3$

Solve each equation or inequality.

17. $|5x| = 15$

18. $|2x + 3| = |3x + 7|$

19. $|p + 4| - 2 > 1$

20. $\left|\frac{2m - 3}{5}\right| \leq 1$

Solve each word problem.

21. Determine the Fahrenheit temperature F that corresponds to a Celsius temperature C of $-25°$. Use $F = \frac{9}{5}C + 32$.

22. Find the simple interest rate required if an investment of $1200 is to grow to $1680 in 5 years. Use $A = P + Prt$.

23. Two consecutive even integers have a sum of 34. Find the integers.

24. If $3400 is invested at 7 percent, how much additional money must be invested at 11 percent so that the total yearly interest from both investments is $546?

25. Determine the measures of the three interior angles of a triangle if the second angle is 15° more than the first, and the third angle is 3 times the second. (*Hint:* $\angle A + \angle B + \angle C = 180°$.)

26. Nancy's day job pays $10 per hour, and her night job pays $6.50 per hour. One week Nancy worked a total of 48 hours and earned a total of $438. How many hours did she work at each job?

27. A chemist wants to mix a 60% alcohol solution with 10 kg of a 35% solution to make a 40% alcohol solution. How much 60% solution should be used?

28. Two trains 380 mi apart start toward each other. One train travels 5 mi/hr faster than the other. If they meet in 4 hr, find the speed of each train. (*Hint:* $d = rt$.)

Exponents and Polynomials

When you borrow money to purchase a car or a house, how does the bank determine the size of your monthly payment? In Sec. 3.2 we state a formula that gives the size of your payment in terms of the amount borrowed, the interest rate, and the number of years you take to pay back the loan.

FOR SALE

E. Geoffroy Ward
REALTOR

377-3454 | 540-0660

3.1 Integer Exponents

TAPE AU5

In Sec. 1.5 we defined natural number exponents. In this section we extend the definition of exponents to include the *zero exponent* and *negative exponents*. We will also state rules that allow us to multiply and divide exponential expressions, and we will see how to use exponents to solve problems involving bacterial growth, compound interest, radioactive decay, and caffeine levels in the bloodstream.

Recall that exponents allow us to write repeated factors in a shortened form:

$$4 \cdot 4 = 4^2 \qquad \text{Read } 4^2 \text{ as "4 squared."}$$
$$9 \cdot 9 \cdot 9 = 9^3 \qquad \text{Read } 9^3 \text{ as "9 cubed."}$$
$$\underbrace{a \cdot a \cdot a \cdot \cdots \cdot a}_{n \text{ factors of } a} = a^n \qquad \text{Read } a^n \text{ as "}a \text{ to the } n\text{th power."}$$

The number n in a^n is called the **exponent,** and a is called the **base.** The expression a^n is called an *exponential expression.*

The Product Rule

Consider the product of a^2 and a^3:

$$a^2 \cdot a^3 = \underbrace{a \cdot a}_{\text{two factors}} \cdot \underbrace{a \cdot a \cdot a}_{\text{three factors}} = \underbrace{a \cdot a \cdot a \cdot a \cdot a}_{\text{five factors}} = a^5$$

Note that $a^2 \cdot a^3 = a^{2+3} = a^5$. This suggests the following rule.

Product Rule for Exponents

If a is a real number and m and n are natural numbers, then
$$a^m \cdot a^n = a^{m+n}$$

The product rule states that **to multiply two powers having the same base, keep the base and add the exponents.** We can verify the product rule for *any* natural numbers m and n as follows:

$$a^m \cdot a^n = \underbrace{a \cdot a \cdot \cdots \cdot a}_{m \text{ factors}} \cdot \underbrace{a \cdot a \cdot \cdots \cdot a}_{n \text{ factors}} = \underbrace{a \cdot a \cdot a \cdot \cdots \cdot a}_{m + n \text{ factors}} = a^{m+n}$$

Example 1   Simplify each expression.

    (a) $x^4 \cdot x^5$ (b) $2^4 \cdot 2^5$ (c) $7 \cdot 7^3 \cdot 7^9$

Solution: (a) $x^4 \cdot x^5 = x^{4+5} = x^9$ Keep the base, add the exponents.
                 (b) $2^4 \cdot 2^5 = 2^{4+5} = 2^9$ **Do *not*** multiply the bases.
                 (c) $7 \cdot 7^3 \cdot 7^9 = 7^{1+3+9} = 7^{13}$ Since $7 = 7^1$.   ❏

Try Exercise 11

Simplify: $x^2 \cdot x^6$.

CAUTION

The product rule for exponents does not apply when the bases are different. Therefore $2^{20} \cdot 3^{15}$ cannot be simplified using the product rule for exponents. ∎

Example 2   Simplify each expression.

    (a) $(-3r^2)(6r^8)$ (b) $(4a^3b^7)(5a^6b)$

Solution: Use the commutative and associative properties to group the coefficients and group the powers of each variable.

$$\text{(a) } (-3r^2)(6r^8) = (-3 \cdot 6)(r^2 r^8)$$
$$= -18r^{10}$$
$$\text{(b) } (4a^3b^7)(5a^6b) = (4 \cdot 5)(a^3a^6)(b^7b)$$
$$= 20a^9b^8$$

Try Exercise 19

Simplify: $(-4r^3)(6r^9)$.

The Zero Exponent

We can use the product rule to define the **zero exponent.** Consider the expression a^0. If the product rule is to apply when one of the exponents is 0, then

$$a^0 \cdot a^2 = a^{0+2} = a^2$$

Since $a^0 \cdot a^2 = a^2$, this suggests that $a^0 = 1$.

DEFINITION OF THE ZERO EXPONENT

If $a \neq 0$, then

$$a^0 = 1$$

The expression 0^0 is undefined.

Example 3 Evaluate each expression. Assume $y \neq 0$.

(a) 10^0 (b) $(-8)^0$ (c) -8^0 (d) $(6y)^0$ (e) $6y^0$

Solution:
(a) $10^0 = 1$
(b) $(-8)^0 = 1$
(c) $-8^0 = -(8^0) = -(1) = -1$
(d) $(6y)^0 = 1$
(e) $6y^0 = 6(y^0) = 6(1) = 6$

Try Exercise 27

Simplify: -7^0.

CAUTION

The base of a power consists only of the symbol that lies immediately to the left of the exponent, unless parentheses indicate otherwise.

Correct
$5x^0 = 5(x^0) = 5(1) = 5$
$-p^0 = -(p^0) = -(1) = -1$
$-8^2 = -(8^2) = -(64) = -64$
$(-8)^2 = (-8)(-8) = 64$

Wrong
$5x^0 = 1$
$-p^0 = 1$
$-8^2 = 64$

Negative Exponents

We can also use the product rule to define **negative exponents.*** Consider the expression a^{-2}. If the product rule is to apply for negative exponents, then

$$a^{-2} \cdot a^2 = a^{-2+2} = a^0 = 1$$

Since $a^{-2} \cdot a^2 = 1$, this suggests that $a^{-2} = \dfrac{1}{a^2}$.

---

*HISTORICAL NOTE: The English mathematician John Wallis (1616–1703) was the first to fully explain the zero exponent and negative exponents. He also helped found the Royal Society, one of the oldest scientific organizations still in existence, and he was the first to use the symbol ∞ for infinity.

DEFINITION OF NEGATIVE EXPONENTS

If $a \neq 0$ and n is any natural number, then

$$a^{-n} = \frac{1}{a^n}$$

Example 4 Simplify each expression. Assume $x \neq 0$.

(a) 4^{-3} (b) 2^{-1}

(c) $(-3)^{-4}$ (d) -3^{-4}

(e) $(5x)^{-2}$ (f) $5x^{-2}$

Solution: (a) $4^{-3} = \dfrac{1}{4^3} = \dfrac{1}{64}$ (b) $2^{-1} = \dfrac{1}{2^1} = \dfrac{1}{2}$

                 (c) $(-3)^{-4} = \dfrac{1}{(-3)^4} = \dfrac{1}{81}$ (d) $-3^{-4} = -\dfrac{1}{3^4} = -\dfrac{1}{81}$

                 (e) $(5x)^{-2} = \dfrac{1}{(5x)^2} = \dfrac{1}{25x^2}$ (f) $5x^{-2} = 5 \cdot \dfrac{1}{x^2} = \dfrac{5}{x^2}$ ❏

Try Exercise 31

Simplify: 9^{-2}.

CAUTION

A negative exponent can produce either a positive answer or a negative answer:

$$(-3)^{-2} = \frac{1}{(-3)^2} = \frac{1}{9} \quad \text{but} \quad (-3)^{-3} = \frac{1}{(-3)^3} = -\frac{1}{27} \quad ■$$

If you carefully study the two equations below, you will see two shortcuts that we sometimes use when working with negative exponents:

$$\left(\frac{3}{5}\right)^{-2} = \frac{1}{\left(\frac{3}{5}\right)^2} = \frac{1}{\frac{9}{25}} = 1 \div \frac{9}{25} = 1 \cdot \frac{25}{9} = \left(\frac{5}{3}\right)^2$$

$$\frac{1}{4^{-2}} = \frac{1}{\frac{1}{4^2}} = \frac{1}{\frac{1}{16}} = 1 \div \frac{1}{16} = 1 \cdot \frac{16}{1} = 4^2$$

These examples suggest the following two shortcuts.

Negative Exponent Shortcuts

If $a \neq 0$ and $b \neq 0$ and n is a natural number, then

$$\left(\frac{a}{b}\right)^{-n} = \left(\frac{b}{a}\right)^n \quad \text{and} \quad \frac{1}{a^{-n}} = a^n$$

Example 5 Evaluate each expression. Assume $p \neq 0$.

(a) $\left(-\dfrac{2}{7}\right)^{-3}$ (b) $\dfrac{6}{p^{-5}}$ (c) $2^{-1} + 5^{-1}$

Solution: (a) $\left(-\dfrac{2}{7}\right)^{-3} = \left(-\dfrac{7}{2}\right)^{3} = \left(-\dfrac{7}{2}\right)\left(-\dfrac{7}{2}\right)\left(-\dfrac{7}{2}\right) = -\dfrac{343}{8}$

(b) $\dfrac{6}{p^{-5}} = 6 \cdot \dfrac{1}{p^{-5}} = 6p^5$

(c) $2^{-1} + 5^{-1} = \dfrac{1}{2} + \dfrac{1}{5} = \dfrac{5}{10} + \dfrac{2}{10} = \dfrac{7}{10}$

Try Exercise 41

Simplify: $(-\frac{2}{5})^{-3}$.

The Quotient Rule

Consider the quotient of a^6 and a^2:

$$\dfrac{a^6}{a^2} = \dfrac{\overbrace{a \cdot a \cdot a \cdot a \cdot a \cdot a}^{\text{six factors}}}{\underbrace{a \cdot a}_{\text{two factors}}} = \overbrace{a \cdot a \cdot a \cdot a}^{\text{four factors}} = a^4$$

Note that $\dfrac{a^6}{a^2} = a^{6-2} = a^4$. This suggests the following rule.

Quotient Rule for Exponents

If $a \neq 0$ and m and n are integers, then

$$\dfrac{a^m}{a^n} = a^{m-n}$$

Example 6 Simplify each expression. Assume no variable is 0.

(a) $\dfrac{y^9}{y^3}$ (b) $\dfrac{5^{18}}{5^4}$ (c) $\dfrac{z^{-1}}{z}$ (d) $\dfrac{2}{2^{-10}}$ (e) $\dfrac{m^{-16}}{m^{-13}}$

Solution: (a) $\dfrac{y^9}{y^3} = y^{9-3} = y^6$ Keep the base, subtract the exponents.

(b) $\dfrac{5^{18}}{5^4} = 5^{18-4} = 5^{14}$ Do *not* divide the bases.

(c) $\dfrac{z^{-1}}{z} = z^{-1-1} = z^{-2} = \dfrac{1}{z^2}$ Since $z = z^1$

(d) $\dfrac{2}{2^{-10}} = 2^{1-(-10)} = 2^{1+10} = 2^{11}$

(e) $\dfrac{m^{-16}}{m^{-13}} = m^{-16-(-13)} = m^{-16+13} = m^{-3} = \dfrac{1}{m^3}$

Try Exercise 61

Simplify: $\dfrac{m^{-3}}{m^{-9}}$.

CAUTION

The quotient rule for exponents does not apply when the bases are different. Therefore $\dfrac{7^{25}}{5^{10}}$ cannot be simplified using the quotient rule for exponents.

Example 7   Simplify each expression. Assume no variable is 0.

$$\text{(a)} \ \frac{-18k^{25}}{6k^5} \qquad \text{(b)} \ \frac{36x^7y^{-2}}{4x^2y^{12}}$$

Solution: (a) $\dfrac{-18k^{25}}{6k^5} = \dfrac{-18}{6} \cdot \dfrac{k^{25}}{k^5}$

$$= -3 \cdot k^{25-5}$$

$$= -3k^{20}$$

(b) $\dfrac{36x^7y^{-2}}{4x^2y^{12}} = \dfrac{36}{4} \cdot \dfrac{x^7}{x^2} \cdot \dfrac{y^{-2}}{y^{12}}$

$$= 9x^{7-2}y^{-2-12}$$

$$= 9x^5y^{-14}$$

$$= 9x^5 \cdot \dfrac{1}{y^{14}}$$

$$= \dfrac{9x^5}{y^{14}} \qquad \square$$

Try Exercise 65

Simplify: $\dfrac{-35k^{50}}{7k^5}$.

Sometimes we must use both the product and the quotient rules to simplify a given expression.

Example 8   Simplify each expression. Assume $z \neq 0$.

$$\text{(a)} \ \frac{3^{-2} \cdot 3^{14}}{3^{-5}} \qquad \text{(b)} \ \frac{(4z^{-8})(-9z^{-5})}{6z^{-3}}$$

Solution: (a) $\dfrac{3^{-2} \cdot 3^{14}}{3^{-5}} = \dfrac{3^{-2+14}}{3^{-5}} = \dfrac{3^{12}}{3^{-5}} = 3^{12-(-5)} = 3^{12+5} = 3^{17}$

(b) $\dfrac{(4z^{-8})(-9z^{-5})}{6z^{-3}} = \dfrac{4(-9)z^{-8}z^{-5}}{6z^{-3}} = \dfrac{-36z^{-13}}{6z^{-3}} = -6z^{-13-(-3)}$

$$= -6z^{-10} = \dfrac{-6}{z^{10}} \qquad \square$$

Try Exercise 75

Simplify: $\dfrac{6^{-4} \cdot 6^{16}}{6^{-9}}$.

We can also use the product and the quotient rules when the exponents are variables.

Example 9   Simplify each expression. Assume no variable is 0.

$$\text{(a)} \ x^n \cdot x^{2n+1} \qquad \text{(b)} \ \frac{y^{3m+5}}{y^{m-2}}$$

Solution: (a) $x^n \cdot x^{2n+1} = x^{n+2n+1} = x^{3n+1}$

(b) $\dfrac{y^{3m+5}}{y^{m-2}} = y^{3m+5-(m-2)} = y^{3m+5-m+2} = y^{2m+7} \qquad \square$

Try Exercise 81

Simplify: $x^n \cdot x^{n+1}$.

Problem Solving

Example 10   A common bacteria known as *Escherichia coli* is capable of doubling its number every hour. If a culture that contains five cells doubles its number every hour, how many cells will it contain in 24 hr?

Solution: After 1 hr, the culture contains $(5) \cdot 2 = 5 \cdot 2^1$ cells.
After 2 hr, the culture contains $(5 \cdot 2^1) \cdot 2 = 5 \cdot 2^2$ cells.
After 3 hr, the culture contains $(5 \cdot 2^2) \cdot 2 = 5 \cdot 2^3$ cells.

.

.

.

Try Exercise 89

A bacteria culture that contains three cells doubles its number every hour. How many cells will the culture contain in 18 hr?

After 24 hr, the culture contains $5 \cdot 2^{24} = 83{,}886{,}080$ cells.   ❑

Exercises 3.1

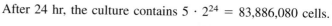

Completion Exercises

1. If $x \neq 0$, then $x^0 =$ _____.

2. The exponent of y in $5y^8$ is _____.

3. The base of the exponent 3 in $6m^3$ is _____.

4. The base of the exponent 2 in $(-4)^2$ is _____.

5. The base of the exponent 2 in -4^2 is _____.

6. To multiply two powers having the same base, _____ the base and _____ the exponents.

7. To divide two powers having the same base, _____ the base and _____ the exponents.

8. If $p \neq 0$, then $p^{-7} =$ _____.

9. If $r \neq 0$, then $\dfrac{1}{r^{-10}} =$ _____.

10. If neither a nor b is 0, then $\left(\dfrac{a}{b}\right)^{-n} =$ _____.

Simplify each expression.

11. $x^2 \cdot x^6$

12. $x^3 \cdot x^7$

13. $3^2 \cdot 3^6$

14. $2^3 \cdot 2^7$

15. $5 \cdot 5^4 \cdot 5^7$

16. $6^5 \cdot 6^3 \cdot 6$

17. $(p + 3)^5 (p + 3)^{10}$

18. $(p - 4)^8 (p - 4)^3$

19. $(-4r^3)(6r^9)$

20. $(3r^5)(-8r^2)$

21. $(5a^4b)(7a^3b^{10})$

22. $(2ab^6)(9a^4b^8)$

Discussion Exercises

23. The expression -5^0 is equal to -1, not 1. Explain why.

24. The expression -9^2 is equal to -81, not 81. Explain why.

Simplify each expression. Assume no variable is 0.

25. 14^0

26. $(-11)^0$

27. -7^0

28. -18^0

29. $(9y)^0$

30. $9y^0$

31. 9^{-2}

32. 6^{-1}

33. -2^{-4}

34. $(-5)^{-3}$

35. $6x^{-2}$

36. $(6x)^{-2}$

37. $4 \cdot 2^{-5}$

38. $9 \cdot 3^{-5}$

39. $\left(\tfrac{1}{10}\right)^{-1}$

40. $\left(-\tfrac{2}{7}\right)^{-2}$

41. $\left(-\tfrac{2}{5}\right)^{-3}$

42. $\left(-\tfrac{3}{4}\right)^{-3}$

43. $\dfrac{1}{5^{-2}}$

44. $\dfrac{1}{8^{-2}}$

45. $\dfrac{9}{p^{-6}}$

46. $\dfrac{4}{p^{-5}}$

47. $\dfrac{3}{(2r)^{-4}}$

48. $\dfrac{2}{3r^{-3}}$

49. $4^{-1} + 5^{-1}$

50. $2^{-1} + 3^{-1}$

51. $2 \cdot 6^0 - 3 \cdot 6^{-1}$

52. $5 \cdot 10^0 - 2 \cdot 10^{-1}$

Getting Acquainted with Your Scientific Calculator

To find 2^{-3} on your calculator, press

$$\boxed{\text{Clear}}\ 2\ \boxed{y^x}\ 3\ \boxed{+/-}\ \boxed{=}\qquad\boxed{0.125}$$

Find each power on your calculator.

53. 101^0

54. 5^{-4}

55. 10^{-3}

56. 1.6^{-2}

Simplify each expression. Assume no denominator is 0.

57. $\dfrac{y^6}{y^3}$

58. $\dfrac{y^8}{y^2}$

59. $\dfrac{z^{-4}}{z}$

60. $\dfrac{z}{z^{-5}}$

61. $\dfrac{m^{-3}}{m^{-9}}$

62. $\dfrac{m^{-10}}{m^{-5}}$

63. $\dfrac{(3p-1)^{100}}{(3p-1)^{99}}$

64. $\dfrac{(6p+1)^{101}}{(6p+1)^{100}}$

65. $\dfrac{-35k^{50}}{7k^5}$

66. $\dfrac{-42k^{36}}{6k^9}$

67. $\dfrac{18x^3y}{9x^{15}y^{-4}}$

68. $\dfrac{60x^6y^{-5}}{12x^4y}$

Simplify each expression. Assume no variable is 0.

69. $7^{15}\cdot 7^{-6}$

70. $5^{-9}\cdot 5^{17}$

71. $x^{-3}\cdot x^{-2}\cdot x$

72. $x^{-4}\cdot x\cdot x^{-5}$

73. $-2m^3(4m^{-5})(-9m^{-7})$

74. $-3m^{-2}(-6m^7)(-5m^{-10})$

75. $\dfrac{6^{-4}\cdot 6^{16}}{6^{-9}}$

76. $\dfrac{10^{-5}}{10^{12}\cdot 10^{-3}}$

77. $\dfrac{y^{-3}y^3}{y^4y^{-6}}$

78. $\dfrac{y^{-4}y}{y^2y^{-2}}$

79. $\dfrac{(9z^{-7})(-12z^{-5})}{18z^{-2}}$

80. $\dfrac{(-6z^{-7})(10z^{-3})}{15z^{-6}}$

Simplify each expression. Assume any variable in an exponent is an integer and no variable is 0.

81. $x^n\cdot x^{n+1}$

82. $x^{3k-1}\cdot x$

83. $x^{-2}\cdot x^{2r}\cdot x^{3r+4}$

84. $\dfrac{y^q}{y^{3q}}$

85. $\dfrac{y^{q+1}}{y^{2q+1}}$

86. $\dfrac{y^{5m+7}}{y^{2m-3}}$

Problem Solving

87. The quantity Q of caffeine that will still be in your bloodstream after t hours is given by the formula

$$Q = A\cdot 2^{-t/6}$$

where A is the amount of caffeine consumed. Suppose you consume 100 mg of caffeine (about 1 cup of coffee). How much caffeine will still be in your bloodstream after (a) 0 hours? (b) 6 hours? (c) 12 hours?

88. The radioactive substance strontium 90 decays according to the formula

$$Q = A\cdot 2^{-t/28}$$

where A represents the original amount and Q represents the quantity that remains after t years. If the original amount is 500 g, find the quantity that remains after (a) 0 yr, (b) 28 yr, (c) 56 yr.

89. A bacteria culture that contains three cells doubles its number every hour. How many cells will the culture contain in 18 hr?

90. A baseball card that was purchased for \$2 triples its value each year. What will the value of the card be in 7 years?

91. Construct a table that enumerates all the possible ways that you can complete a true-false test that contains (a) one question, (b) two questions, (c) three questions. In how many ways can you complete a true-false test containing 10 questions? Containing n questions?

92. Construct a table that enumerates all the possible ways that you can select the winner of (a) one basketball game, (b) two basketball games, (c) three basketball games. In how many ways can you select the winner of 10 basketball games? Of n basketball games?

Getting Acquainted with Your Scientific Calculator

The value V of a savings account that compounds interest annually is given by the formula

$$V = P(1 + r)^t$$

where P is the principal (original amount), r is the interest rate *in decimal form,* and t is the time in years.*

---

*HISTORICAL NOTE: The formula $V = P(1 + r)^t$ was known to the Babylonians, perhaps as early as 2000 B.C.

93. Suppose $1000 was invested for 5 yr in an account that paid 8 percent interest compounded annually.

    (a) Find the value of the account.

    (b) Find the value of the account if the principal is doubled.

    (c) Find the value of the account if the interest rate is doubled.

    (d) Find the value of the account if the time is doubled.

94. A piece of property is purchased for $90,000. If the property appreciates at the rate of 11 percent per year, what is the value of the property after 4 yr?

Discussion Exercises

95. Why don't we multiply the bases when we simplify $2^3 \cdot 2^4$?

96. Why don't we divide the bases when we simplify $\dfrac{5^6}{5^2}$?

97. When you raise a number to a negative exponent, is the answer always, sometimes, or never a negative number? Give examples.

98. Why is 7^0 not equal to 0? Why is 6^{-1} not equal to -6?

3.2 Rules For Exponents

TAPE AU5

In this section we state three more rules for exponents. As is the case with the product rule and the quotient rule, these new rules are true for all integer exponents—positive, negative, and zero. At the end of the section we use exponents (with the aid of a calculator) to calculate the size of a car payment, to calculate the size of a house payment, and to calculate the value of a retirement fund.

The Power Rules

To illustrate the first power rule, consider the expression $(a^2)^3$.

$$(a^2)^3 = (a^2)(a^2)(a^2) = a^{2+2+2} = a^6$$

Note that $(a^2)^3 = a^{2 \cdot 3} = a^6$. This suggests the following rule.

Power-to-a-Power Rule

If a is a real number and m and n are integers, then

$$(a^m)^n = a^{mn}$$

Example 1   Simplify each expression: (a) $(x^3)^4$, (b) $(3^{-2})^{-7}$, (c) $(p^5)^{-4}$.

Solution: Each expression is a power to a power. Therefore keep the base and multiply the exponents.

(a) $(x^3)^4 = x^{3 \cdot 4} = x^{12}$

(b) $(3^{-2})^{-7} = 3^{(-2)(-7)} = 3^{14}$

(c) $(p^5)^{-4} = p^{5(-4)} = p^{-20} = \dfrac{1}{p^{20}}$ $(p \neq 0)$ ☐

Try Exercise 5

Simplify: $(x^2)^4$.

CAUTION

Do not confuse the power-to-a-power rule with the product rule.

$$(a^3)^7 = a^{21} \qquad \text{but} \qquad a^3 \cdot a^7 = a^{10}$$ ∎

The two equations below illustrate what happens when we raise the product ab or the quotient $\frac{a}{b}$ to a power:

$$(ab)^3 = (ab)(ab)(ab) = (a \cdot a \cdot a) \cdot (b \cdot b \cdot b) = a^3 b^3$$

$$\left(\frac{a}{b}\right)^3 = \left(\frac{a}{b}\right)\left(\frac{a}{b}\right)\left(\frac{a}{b}\right) = \frac{a \cdot a \cdot a}{b \cdot b \cdot b} = \frac{a^3}{b^3}$$

Note that $(ab)^3 = a^3 b^3$ and $\left(\frac{a}{b}\right)^3 = \frac{a^3}{b^3}$. These two equations suggest the following rules.

Suppose a and b are real numbers and n is an integer.

Product-to-a-Power Rule

$$(ab)^n = a^n b^n$$

Quotient-to-a-Power Rule

$$\left(\frac{a}{b}\right)^n = \frac{a^n}{b^n} \qquad (b \neq 0)$$

The product-to-a-power rule states that **to raise a product to a power, raise each factor to the power.** The quotient-to-a-power rule states that **to raise a quotient to a power, raise both the numerator and the denominator to the power.**

Example 2   Simplify each expression: (a) $(2y)^4$, (b) $(-9mn)^2$, (c) $5(-3k)^3$, (d) $\left(\frac{3}{4}\right)^2$, (e) $\left(\frac{-2}{z}\right)^5$.

Solution: (a) $(2y)^4 = 2^4 y^4 = 16y^4$
(b) $(-9mn)^2 = (-9)^2 m^2 n^2 = 81m^2 n^2$
(c) $5(-3k)^3 = 5(-3)^3 k^3 = 5(-27)k^3 = -135k^3$

(d) $\left(\frac{3}{4}\right)^2 = \frac{3^2}{4^2} = \frac{9}{16}$

(e) $\left(\frac{-2}{z}\right)^5 = \frac{(-2)^5}{z^5} = \frac{-32}{z^5} = -\frac{32}{z^5} \qquad (z \neq 0)$

Try Exercise 15

Simplify: $4(-2k)^3$.

CAUTION

Correct

$(3x)^2 = 9x^2$

$\left(\frac{a}{b}\right)^4 = \frac{a^4}{b^4}$

Wrong

$(3x)^2 = 3x^2$

$\left(\frac{a}{b}\right)^4 = \frac{a^4}{b}$

All of the rules for exponents are summarized on the next page.

3.2 • Rules for Exponents

Rules for Exponents

Suppose a and b are real numbers and m and n are integers.

Zero exponent	$a^0 = 1$	$(a \neq 0)$
Negative exponent	$a^{-n} = \dfrac{1}{a^n}$	$(a \neq 0)$
Negative exponent shortcuts	$\left(\dfrac{a}{b}\right)^{-n} = \left(\dfrac{b}{a}\right)^n$	$(a \neq 0, b \neq 0)$
	$\dfrac{1}{a^{-n}} = a^n$	$(a \neq 0)$
Product rule	$a^m \cdot a^n = a^{m+n}$	
Quotient rule	$\dfrac{a^m}{a^n} = a^{m-n}$	$(a \neq 0)$
Power-to-a-power rule	$(a^m)^n = a^{mn}$	
Product-to-a-power rule	$(ab)^n = a^n b^n$	
Quotient-to-a-power rule	$\left(\dfrac{a}{b}\right)^n = \dfrac{a^n}{b^n}$	$(b \neq 0)$

In the following examples we use a combination of several rules for exponents to simplify the given expression.

Example 3   Simplify each expression. Assume $q \neq 0$.

(a) $(x^6 y^8)^3$ (b) $\left(\dfrac{p^2}{q^7}\right)^5$

Solution: (a) $(x^6 y^8)^3 = (x^6)^3 (y^8)^3$ Product-to-a-power rule

$= x^{18} y^{24}$ Power-to-a-power rule

(b) $\left(\dfrac{p^2}{q^7}\right)^5 = \dfrac{(p^2)^5}{(q^7)^5}$ Quotient-to-a-power rule

$= \dfrac{p^{10}}{q^{35}}$ Power-to-a-power rule

Try Exercise 23

Simplify: $\left(\dfrac{p^3}{q^5}\right)^6$.

Example 4   Simplify each expression. Assume no variable is 0.

(a) $(r^{-4} r)^{-5}$ (b) $(6^{-2} k^9)^{-1}$ (c) $\left(\dfrac{m^{-3}}{n^6}\right)^{-2}$

Solution: (a) $(r^{-4} r)^{-5} = (r^{-4+1})^{-5} = (r^{-3})^{-5} = r^{(-3)(-5)} = r^{15}$

(b) $(6^{-2} k^9)^{-1} = (6^{-2})^{-1}(k^9)^{-1} = 6^2 k^{-9} = 36 \cdot \dfrac{1}{k^9} = \dfrac{36}{k^9}$

(c) $\left(\dfrac{m^{-3}}{n^6}\right)^{-2} = \dfrac{(m^{-3})^{-2}}{(n^6)^{-2}} = \dfrac{m^6}{n^{-12}} = m^6 \cdot \dfrac{1}{n^{-12}} = m^6 n^{12}$

Try Exercise 27

Simplify: $(3^{-3} k^{10})^{-1}$.

Example 5   Simplify: $(-2x^{-2}y)^3(x^7y^{-4})^{-5}$.

Solution: $(-2x^{-2}y)^3(x^7y^{-4})^{-5} = (-2)^3(x^{-2})^3y^3(x^7)^{-5}(y^{-4})^{-5}$

$$= (-8)x^{-6}y^3x^{-35}y^{20}$$

$$= -8x^{-41}y^{23}$$

$$= -\frac{8y^{23}}{x^{41}}$$ ❑

Try Exercise 37

Simplify:
$(-4x^{-3}y^4)^3(x^9y^{-2})^{-3}$.

Example 6   Simplify: $\left(\dfrac{a^3b^4c^{-6}}{a^{-2}b^{-7}c}\right)^{-2}$.

Solution: $\left(\dfrac{a^3b^4c^{-6}}{a^{-2}b^{-7}c}\right)^{-2} = \left(\dfrac{a^5b^{11}}{c^7}\right)^{-2} = \left(\dfrac{c^7}{a^5b^{11}}\right)^2 = \dfrac{c^{14}}{a^{10}b^{22}}$ ❑

Try Exercise 53

Simplify: $\left(\dfrac{a^6b^5c^{-4}}{a^{-3}b^{-8}c}\right)^{-3}$

The rules for exponents also apply when the exponents are variables.

Example 7   Simplify each expression. Assume m and n are integers and no variable is 0.

(a) $(x^mx^{-2})^5$ (b) $\dfrac{y^n(y^{-2n})^3}{y^{4n}}$

Solution: (a) $(x^mx^{-2})^5 = (x^{m-2})^5 = x^{5(m-2)} = x^{5m-10}$

(b) $\dfrac{y^n(y^{-2n})^3}{y^{4n}} = \dfrac{y^ny^{-6n}}{y^{4n}} = \dfrac{y^{-5n}}{y^{4n}} = y^{-5n-4n} = y^{-9n} = \dfrac{1}{y^{9n}}$ ❑

Try Exercise 67

Simplify: $(x^mx^{-1})^4$.

Exercises 3.2

Completion Exercises

1. The expression $(2p^4)^3$ can be simplified to _____ .

2. To raise a power to a power, _____ the base and _____ the exponents.

3. To raise a quotient to a power, raise both the _____ and the _____ to that power.

4. To raise a product to a power, raise each _____ to that power.

Simplify each expression. Write your answer with positive exponents. Assume no variable is 0.

5. $(x^2)^4$

6. $(x^5)^3$

7. $(2^{-3})^{-7}$

8. $(3^{-7})^2$

9. $(p^6)^{-1}$

10. $(p^{-4})^{-5}$

11. $(3y)^4$

12. $(4y)^2$

13. $(-7mn)^2$

14. $(-5mn)^3$

15. $4(-2k)^3$

16. $3(-2k)^4$

17. $(\frac{3}{7})^2$

18. $(\frac{4}{5})^2$

19. $\left(\dfrac{-1}{z}\right)^5$

20. $\left(\dfrac{-2}{z}\right)^3$

21. $(x^4y^7)^3$

22. $(x^8y^6)^5$

23. $\left(\dfrac{p^3}{q^5}\right)^6$

24. $\left(\dfrac{p^4}{q^2}\right)^7$

25. $(rr^{-5})^{-4}$

26. $(r^{-6}r)^{-5}$

27. $(3^{-3}k^{10})^{-1}$

28. $(2^{-5}k^8)^{-1}$

29. $(-2u^3v^{-2})^{-2}$

30. $(-3u^{-4}v^2)^{-2}$

31. $\left(\dfrac{m^{-6}}{n^5}\right)^{-4}$

32. $\left(\dfrac{m^7}{n^{-1}}\right)^{-3}$

33. $\left(\dfrac{3x}{y^{-2}}\right)^{-2}$

34. $\left(\dfrac{2x}{y^{-3}}\right)^{-2}$

35. $(z^3)^{-2}(z^2z^{-4})^{-7}$

36. $(z^{-5}z^3)^{-9}(z^{-4})^2$

37. $(-4x^{-3}y^4)^3(x^9y^{-2})^{-3}$

38. $(-5x^3y^7)^3(x^{-2}y^8)^{-4}$

39. $9r^2s(-3r^4s^4)^3(r^8s^{10})^4$

40. $(-4r^4s^5)^3(8r^2s^3)(r^6s^6)^5$

41. $(2p^{-5}q^{-1})^{-3}(4p^6q^{-7})^2$

42. $(3p^{-3}q^2)^{-3}(6p^{-6}q^9)^2$

43. $\dfrac{2a^{-1}}{5b^{-1}}$

44. $\dfrac{3a^{-1}}{4b^{-1}}$

45. $\dfrac{ax^{-2}}{by^{-3}}$

46. $\dfrac{ax^{-3}}{by^{-4}}$

47. $\dfrac{-18(a^3b^5)^2}{6a^2b}$

48. $\dfrac{-16(a^2b^{10})^2}{8ab^5}$

49. $\dfrac{4b^3}{5m}\left(\dfrac{2m^2}{b^7}\right)^5$

50. $\dfrac{3b^4}{5m}\left(\dfrac{3m^3}{b^6}\right)^3$

51. $\dfrac{(3t^{-4})^2}{t^{-8}}$

52. $\dfrac{(2t^{-3})^2}{t^{-6}}$

53. $\left(\dfrac{a^6b^5c^{-4}}{a^{-3}b^{-8}c}\right)^{-3}$

54. $\left(\dfrac{a^{-5}b^{-8}c^3}{a^2b^{-6}c^{-1}}\right)^{-5}$

55. $\dfrac{5^9(y^3)^{-2}y^0}{5^{-1}y^{-4}y^4}$

56. $\dfrac{7^{10}(y^{-4})^2y^{-8}}{7^{-1}y^{-3}y^0}$

57. $\left(\dfrac{4x^{-5}z^3}{5y^4}\right)^{-2}$

58. $\left(\dfrac{5x^{-3}z^2}{4y^2}\right)^{-2}$

59. $\dfrac{5a^{-1}}{8b^{-1}}\left(\dfrac{a^{-11}b^4}{c^{-5}}\right)^2$

60. $\dfrac{3a^{-2}}{7b^{-2}}\left(\dfrac{a^6b^{-14}}{c^{-3}}\right)^2$

61. $\left(\dfrac{2^{-1}x^7}{5^{-1}y^{-2}}\right)^4\left(\dfrac{5x^{-3}}{5^{-1}y^4}\right)^{-2}$

62. $\left(\dfrac{3^{-1}x^5}{4^{-2}y^{-4}}\right)^3\left(\dfrac{4^{-2}x^{-6}}{4^{-5}y^7}\right)^{-2}$

63. $\dfrac{(2^{-1}w^{-2})^{-4}(-3w^2)^2}{4w(9^{-1}w^{-3})^{-3}}$

64. $\dfrac{(3^{-1}w^{-4})^{-3}(-5w^3)^2}{9w(25^{-1}w^{-2})^{-3}}$

Simplify each expression. Assume m and n are integers and no variable is 0.

65. $(x^{-7m}x^{2m})^{-2}$

66. $(x^mx^{-4m})^{-5}$

67. $(x^mx^{-1})^4$

68. $(x^{3m}x^{-2})^3$

69. $\dfrac{y^n(y^{-3n})^2}{y^{5n}}$

70. $\dfrac{y^{-n}(y^{6n})^{-1}}{y^{8n}}$

71. $\dfrac{(z^{4+m})^2z^{-m}}{(z^2)^{-m}}$

72. $\dfrac{(z^{3+2m})^2z^{-2m}}{(z^3)^{-m}}$

73. Write 8^{10} as a power of 2.

74. Write 9^{20} as a power of 3.

75. Write 25^{4m} as a power of 5. Assume m is an integer.

76. Write 16^{3m} as a power of 2. Assume m is an integer.

 Getting Acquainted with Your Scientific Calculator

To calculate the size p of a monthly loan payment, you can use the formula

$$p = \dfrac{Ar}{12[1 - (1 + r/12)^{-12t}]}$$

where A is the amount borrowed, r is the interest rate *in decimal form,* and t is the time in years.

77. To purchase a new car you borrow $10,000 at 9 percent interest. If the loan is to be paid back over 5 yr, find your monthly car payment. How much total interest will you pay?

78. To purchase a home you borrow $60,000 at 12 percent interest. Complete the following chart:

Number of years	Monthly payment	Total amount paid
15	$720.10	$129,618
20	?	?
30	?	?

What is the smallest your monthly payment can be no matter how many years you take to pay back the loan?

79. Suppose you make monthly payments, each of size p, into a retirement fund that earns an interest rate r (in decimal form) compounded monthly. The value V of your fund after t years is given by the formula

$$V = p\left[\dfrac{(1 + r/12)^{12t} - 1}{r/12}\right]$$

Find V if $p = \$60$, $r = 10$ percent, and $t = 50$ yr.

Discussion Exercises

80. Why do we not add the exponents when we simplify $(4^3)^2$?

81. Why is $(2x)^3$ not equal to $2x^3$?

82. Why don't we just square the numerator when we simplify $(\frac{5}{7})^2$?

TAPE AU5

3.3 Scientific Notation: An Application of Exponents

Scientists often deal with numbers that are very large or very small. For example, a light year (the distance light travels in 1 yr) is about 5,870,000,000,000 mi, and the diameter of a red corpuscle is 0.000075 cm. In this section we use exponents to write numbers like these in a more compact form, called *scientific notation*.

DEFINITION OF SCIENTIFIC NOTATION

A positive real number is in **scientific notation** when it is written in the form

$$a \times 10^n$$

where $1 \le a < 10$ and n is an integer.

To see how to write a number in scientific notation, consider what happens when we multiply a number by a positive power of 10:

$$5.87 \times 10^1 = 58.7 \qquad \text{Decimal point moves one place to the right.}$$

$$5.87 \times 10^2 = 587 \qquad \text{Decimal point moves two places to the right.}$$

$$5.87 \times 10^3 = 5870 \qquad \text{Decimal point moves three places to the right.}$$

These equations tell us that **to move the decimal point k places to the right, multiply by 10^k.**

Example 1   Write 5,870,000,000,000 in scientific notation.

Solution: We must write 5,870,000,000,000 as the product of a number between 1 and 10 (including 1) and an integer power of 10. Note that 5.87 is between 1 and 10. (Find this number by placing the decimal point to the right of the first nonzero digit.) To get 5,870,000,000,000 from 5.87 we must move the decimal point twelve places to the right:

$$5.870000000000$$

twelve places to the right

Therefore multiply 5.87 by 10^{12}:

$$5.87 \times 10^{12} \qquad \square$$

Try Exercise 13

Write 8,090,000,000 in scientific notation.

When writing a number in scientific notation it is customary to use the symbol $\times$ for multiplication to avoid confusing the dot multiplication symbol with the decimal point.

Example 2   Write each number in scientific notation:
(a) 6,040,000, (b) 100,000, (c) 8.2.

Solution: (a) $6,040,000 = 6.04 \times 10^6$
(b) $100,000 = 1 \times 10^5$, or simply 10^5
(c) $8.2 = 8.2 \times 10^0$, or simply 8.2 $\square$

Try Exercise 15

Write 10,000,000 in scientific notation.

Now let us see what happens when we multiply a number by a *negative* power of 10:

$$7.5 \times 10^{-1} = 7.5 \times \frac{1}{10} = \frac{7.5}{10} = 0.75$$

Decimal point moves one place to the left.

$$7.5 \times 10^{-2} = 7.5 \times \frac{1}{10^2} = \frac{7.5}{100} = 0.075$$

Decimal point moves two places to the left.

$$7.5 \times 10^{-3} = 7.5 \times \frac{1}{10^3} = \frac{7.5}{1000} = 0.0075$$

Decimal point moves three places to the left.

These equations tell us that **to move the decimal point k places to the left, multiply by 10^{-k}.**

Example 3   Write 0.000075 in scientific notation.

Solution: We must write 0.000075 as the product of a number between 1 and 10 (including 1) and an integer power of 10. Note that 7.5 is between 1 and 10. (Find this number by placing the decimal point to the right of the first nonzero digit.) To get 0.000075 from 7.5 we must move the decimal point five places to the left:

$$0.000075$$

five places to the left

Therefore, multiply 7.5 by 10^{-5}:

$$7.5 \times 10^{-5} \qquad \square$$

Try Exercise 17

Write 0.000000384 in scientific notation.

Example 4   Write each number in scientific notation:
(a) 0.00000009, (b) 0.5243, (c) 0.0001.

Solution: (a) $0.00000009 = 9 \times 10^{-8}$
(b) $0.5243 = 5.243 \times 10^{-1}$
(c) $0.0001 = 1 \times 10^{-4}$, or simply 10^{-4} $\qquad \square$

Try Exercise 19

Write 0.000001 in scientific notation.

To convert from scientific notation to standard form we reverse the process.

Example 5   Write each number in standard form:
(a) 5×10^9, (b) 4.3×10^{-6}, (c) 7.162×10^2, (d) 2.08×10^{-3}.

Solution: (a) $5 \times 10^9 = 5,000,000,000$ Nine places to the right
(b) $4.3 \times 10^{-6} = 0.0000043$ Six places to the left
(c) $7.162 \times 10^2 = 716.2$ Two places to the right
(d) $2.08 \times 10^{-3} = 0.00208$ Three places to the left $\qquad \square$

Try Exercise 35

Write 9.907×10^2 in standard form.

To take full advantage of scientific notation, we must be able to perform operations on numbers that are written in scientific notation.

Example 6   Multiply: (2000)(34,000).

Solution: $(2000)(34,000) = (2 \times 10^3)(3.4 \times 10^4)$ Write each number in scientific notation.

$= (2 \times 3.4)(10^3 \times 10^4)$ Group the numbers, group the powers of 10.

$= 6.8 \times 10^7$ Multiply each group. $\qquad \square$

Try Exercise 45

Multiply: (3000)(21,000).

Example 7   Perform the indicated operations on $\dfrac{(8{,}000{,}000{,}000)(0.072)}{(0.002)(240{,}000{,}000)}$

Solution:

$$\dfrac{(8{,}000{,}000{,}000)(0.072)}{(0.002)(240{,}000{,}000)} = \dfrac{(8 \times 10^9)(7.2 \times 10^{-2})}{(2 \times 10^{-3})(2.4 \times 10^8)} \quad \begin{array}{l}\text{Write each number in}\\\text{scientific notation.}\end{array}$$

$$= \dfrac{(8)(7.2)}{(2)(2.4)} \times \dfrac{10^9 10^{-2}}{10^{-3} 10^8} \quad \begin{array}{l}\text{Group the numbers,}\\\text{group the powers of 10.}\end{array}$$

$$= 12 \times 10^{9+(-2)-(-3)-8} \quad \text{Operate on each group.}$$

$$= 12 \times 10^2 \quad \text{Simplify the exponent.}$$

Note that this last expression is not in scientific notation because 12 is not between 1 and 10. Therefore, write 12 in scientific notation and continue:

$$= 1.2 \times 10 \times 10^2 \quad \text{Since } 12 = 1.2 \times 10$$

$$= 1.2 \times 10^3 \quad \text{Since } 10 \times 10^2 = 10^3 \quad \square$$

We can also write negative numbers in scientific notation. Here are two examples:

$$-477{,}000 = -4.77 \times 10^5$$

$$-0.00018 = -1.8 \times 10^{-4}$$

Try Exercise 53

Simplify:
$\dfrac{(9{,}000{,}000{,}000)(0.084)}{(0.003)(210{,}000{,}000)}$.

Problem Solving

Example 8   Radio waves travel at the speed of light (approximately 186,000 mi/sec). How long would it take a radio message from earth to reach a spaceship 4,650,000 mi away?

Solution: We use the formula $d = rt$. Since we know the distance and the rate, the unknown quantity is time.

$$t = \dfrac{d}{r} \quad \text{Solve } d = rt \text{ for } t.$$

$$= \dfrac{4{,}650{,}000}{186{,}000} \quad \text{Substitute the known values.}$$

$$= \dfrac{4.65 \times 10^6}{1.86 \times 10^5} \quad \text{Convert to scientific notation.}$$

$$= 2.5 \times 10 \quad \begin{array}{l}\text{Divide the numbers, divide}\\\text{the powers of 10.}\end{array}$$

The radio message would take 25 sec to reach the spaceship. $\square$

Try Exercise 63

The speed of light is approximately 186,000 mi/sec. How long does it take light rays from the sun 93,000,000 mi away to reach the earth?

Exercises 3.3

Matching Exercises

Match each number in scientific notation in Exercises 1 through 4 with an equivalent number in standard form in letters A through D.

1. 4.6×10^2 A. 0.46

2. 4.6×10^{-3} B. 4600

3. 4.6×10^{-1} C. 0.0046

4. 4.6×10^3 D. 460

True-False Exercises

5. The number 2,001,000 written in scientific notation is 20.01×10^5.

6. Multiplying a number by 10^6 moves the decimal point of the number six places to the right.

Completion Exercises

7. Multiplying a number by 10^{-9} moves the decimal point of the number _____ places to the _____.

8. A positive real number is in scientific notation when it is written as $a \times 10^n$ where _____ $\leq a <$ _____ and n is a(n) _____.

Write each number in scientific notation.

9. 30,000

10. 9000

11. 0.005

12. 0.0003

13. 8,090,000,000

14. 701,000,000

15. 10,000,000

16. 1,000,000

17. 0.000000384

18. 0.0000000568

19. 0.000001

20. 0.000000001

21. 275.1

22. 83.45

23. 5.59

24. 4.41

25. 0.119×10^5

26. 33.1×10^{-6}

27. 74.3×10^{-4}

28. 0.268×10^3

Write each number in standard form.

29. 5.5×10^9

30. 8.8×10^6

31. 4×10^{-3}

32. 7×10^{-2}

33. 1.9×10^{-5}

34. 2.4×10^{-8}

35. 9.907×10^2

36. 5.029×10^2

37. 6.332×10^0

38. 4.118×10^0

39. 1.01×10^{-1}

40. 7.07×10^{-1}

Discussion Exercises

41. What is the purpose of scientific notation?

42. Why do you think scientific notation specifies that the number a in $a \times 10^n$ be a number between 1 and 10?

Arrange the numbers in order from smallest to largest.

43. 2.7×10^5, 8.97, 4.3×10^{-4}, 6.5×10^{-2}

44. 3.09×10^{-1}, 1.81×10^3, 5.44, 9.9×10^{-6}

Write the following exercises in scientific notation and then perform the indicated operations. Write your answer in scientific notation.

45. $(3000)(21,000)$ 46. $(4000)(22,000)$

47. $(40,000,000,000)(0.000000002)$

48. $(20,000,000,000)(0.000000003)$

49. $(0.000008)(5,000,000)$ 50. $(0.000006)(5,000,000)$

51. $\dfrac{0.00000042}{7,000,000,000}$ 52. $\dfrac{0.00000035}{5,000,000,000}$

53. $\dfrac{(9,000,000,000)(0.084)}{(0.003)(210,000,000)}$

54. $\dfrac{(6,000,000,000)(0.065)}{(0.002)(130,000,000)}$

55. $\dfrac{(0.00000004)(65,000,000)}{(50,000,000,000)(0.8)}$

56. $\dfrac{(0.00000005)(44,000,000)}{(80,000,000,000)(0.5)}$

Problem Solving

57. The average lifespan of a human is about 2,400,000,000 sec. Write this number in scientific notation.

58. The internal temperature of the sun is about 20,000,000° C. Write this number in scientific notation.

59. The mass of a water molecule is 3×10^{-23} g. Write this number in standard form.

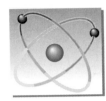

60. The least stable nuclear particles known have a lifespan of 1.6×10^{-24} sec. Write this number in standard form.

61. A *googol* is the number 1 followed by 100 zeros. Write this number in scientific notation.

62. A *picosecond* is one trillionth of a second. Write this number in scientific notation.

63. The speed of light is approximately 186,000 mi/sec. How long does it take light rays from the sun 93,000,000 mi away to reach the earth?

64. Radio waves travel at the speed of light (approximately 186,000 mi/sec). How long would it take a radio message from an astronaut on the moon 235,000 mi away to reach the earth?

65. The mass of the earth is about 6×10^{27} g. A gram corresponds to 1.1×10^{-6} ton. Convert the mass of the earth to tons.

66. A computer can perform 8.5×10^6 calculations/sec. How many calculations can it perform in 10 hr?

 Getting Acquainted with Your Scientific Calculator

To square 5,000,000 on your calculator, press

$$\boxed{\text{Clear}} \ 5{,}000{,}000 \ \boxed{x^2} \ \boxed{\quad 2.5 \quad 13 \quad}$$

This means that $(5{,}000{,}000)^2 = 2.5 \times 10^{13}$. Your calculator displays the answer in scientific notation because the answer has too many digits to be displayed in standard form.

Perform the indicated operations on your calculator. Write your answer in scientific notation. Round approximate answers to three significant figures.

67. $(4{,}000{,}000)(60{,}000)$

68. $(310{,}000)^2$

69. $\dfrac{0.0000578}{90{,}000{,}000}$

70. $(20{,}000)^{-3}$

You can enter a number into your calculator in scientific notation using a key labeled $\boxed{\text{EE}}$ or $\boxed{\text{EXP}}$. For example, since $8{,}400{,}000{,}000{,}000 = 8.4 \times 10^{12}$, you can enter this number into your calculator by pressing

$$\boxed{\text{Clear}} \ 8.4 \ \boxed{\text{EE}} \ 12 \ \boxed{\quad 8.4 \quad 12 \quad}$$

Perform the indicated operations by entering the numbers into your calculator in scientific notation. Write your answer in scientific notation. Round approximate answers to three significant figures.

71. $(8{,}400{,}000{,}000{,}000)(700{,}000{,}000{,}000)$

72. $\dfrac{(43{,}720{,}000{,}000)(0.6517)}{(0.000000000888)(103{,}000{,}000)}$

 Getting Acquainted with Your Graphing Calculator

Set your calculator in scientific notation mode and perform the following operations.

73. $(2.75)(3.49)$

74. 5^{10}

75. $1 \div 80$

76. $(2.5)^{-6}$

 ## 3.4 Adding and Subtracting Polynomials

TAPE AU6

A type of algebraic expression that has particular importance in mathematics is a *polynomial*. In this section we define a polynomial, learn how to add and subtract polynomials, and solve applied problems that involve polynomials.

Monomials

In order to define a polynomial, we first define a *monomial*.

DEFINITION OF A MONOMIAL

A **monomial in** x is an expression of the form

$$ax^n$$

where a is a real number and n is a whole number. The number a is called the **coefficient** of the monomial. If $a \neq 0$, the **degree** of the monomial is n.

Here are some examples of monomials:

Monomial	Coefficient	Degree	
$8x^3$	8	3	
$\frac{2}{5}y$	$\frac{2}{5}$	1	
$-z^4$	-1	4	
6	6	0	(Since $6 = 6 \cdot x^0$)

The following algebraic expressions are *not* monomials:

Not a monomial	Reason
$7x^{-5}$	Negative exponent on x
$\dfrac{3}{y}$	Since $\dfrac{3}{y} = 3y^{-1}$
$9m^{1/2}$	Fractional exponent on m (we will define fractional exponents in Chap. 5)

Polynomials

We are now ready to define a polynomial.

DEFINITION OF A POLYNOMIAL

A **polynomial** is either a monomial or a finite sum of monomials.

The **degree** of a polynomial is the same as the degree of its highest-degree monomial. The coefficient of the highest-degree monomial is called the **leading coefficient** of the polynomial. The term that does not contain a variable is called the **constant term** of the polynomial.

Polynomial	Degree	Leading coefficient	Constant term
$3x^2 + 4x + 10$	2	3	10
$9y - 5$	1	9	-5
$2p + 6p^8 + p^{13}$	13	1	0

Recall that **terms** are related by addition. Therefore the polynomial $x^2 - 7x + 8$, which can be written $x^2 + (-7x) + 8$, has three terms, namely, x^2, $-7x$, and 8. Therefore, a **monomial** is a polynomial with one term. A polynomial with two terms is called a **binomial**, and a polynomial with three terms is called a **trinomial**.

Monomials	Binomials	Trinomials
$5x$	$y - 3$	$z^2 - 6z - 9$
$-4t^9$	$2r^2 + r$	$7p^4 + 3p^2 + 1$
-1	$k^4 + 16$	$4h + 11h^5 - 2h^3$

A polynomial with four or more terms is not given a special name. An example of a polynomial with four terms is

$$x^3 - 8x^2 + 2x - 15$$

Since the exponents on the variable x decrease from left to right, we say that this polynomial is written in **descending order.** We usually write polynomials in descending order to make them easier to read.

Example 1   Write each polynomial in descending order.

(a) $3x + x^2 - 12$ (b) $6 + 5y$ (c) $m - 4m^3 - 9m^5 + 2$

Solution: (a) $x^2 + 3x - 12$
(b) $5y + 6$
(c) $-9m^5 - 4m^3 + m + 2$ ❑

Try Exercise 15

Write in descending order:
$4x + x^2 - 13$.

Monomials may contain more than one variable. The **degree** of a monomial in several variables is the sum of the exponents on all of the variables. For example,

$$8x^2y^3z \text{ is a monomial of degree } 2 + 3 + 1 = 6$$

The **degree** of a polynomial in several variables is the same as the degree of its highest-degree monomial. For example,

$$2p^3m - 9p^2m^5 + 5m^6 \text{ is a polynomial of degree } 7$$

Evaluating Polynomials Using P(x) Notation

We often name polynomials with capital letters. For example, we might use the letter P to name the polynomial $4x^2 - 7x + 9$. To denote that this is a polynomial in x, we write $P(x)$. The symbol $P(x)$ is read "P of x."

$$P(x) = 4x^2 - 7x + 9$$

CAUTION

In $P(x) = 4x^2 - 7x + 9$, $P(x)$ does *not* mean P times x. The variable x represents a number, but the variable P does *not* represent a number. P is the *name* of the polynomial $4x^2 - 7x + 9$.   ∎

In a similar fashion, we can name a polynomial in the variable y or z:

$$Q(y) = y^3 + 3y$$
$$R(z) = 2z^3 - 5z^2 + 7z - 10$$

A polynomial takes on different values as its variable takes on different values. To **evaluate** a polynomial in x at a particular value of x, say $x = 3$, simply replace each x in the polynomial by 3. Then simplify using the order of operations.

Example 2   Suppose $P(x) = 4x^2 - 7x + 9$. Find each of the following:
(a) $P(3)$, (b) $P(-2)$, (c) $P(-k)$.

Solution: (a) $P(x) = 4x^2 - 7x + 9$ Original polynomial
$$P(3) = 4(3)^2 - 7(3) + 9$$ Replace x with 3.
$$\left.\begin{aligned} &= 4(9) - 21 + 9 \\ &= 36 - 21 + 9 \\ &= 24 \end{aligned}\right\} \quad \text{Order of operations}$$

(b) $P(-2) = 4(-2)^2 - 7(-2) + 9$ Replace x with -2.
$$\left.\begin{aligned} &= 4(4) + 14 + 9 \\ &= 39 \end{aligned}\right\} \quad \text{Order of operations}$$

(c) $P(-k) = 4(-k)^2 - 7(-k) + 9$ Replace x with $-k$.
$$= 4k^2 + 7k + 9$$ Simplify.   ❑

Try Exercise 25

If $P(x) = 5x^2 - 8x + 13$, find
(a) $P(3)$, (b) $P(-2)$, (c) $P(-k)$.

Adding and Subtracting Polynomials

Recall from Sec. 2.1 that **like terms** are terms that have the same variables with the same exponents. We combine like terms by combining their coefficients.

Example 3   Simplify each polynomial by combining like terms.
(a) $2p^3 - 5p^3 - p^3$ (b) $3m + 6n + 5m + 8n$
(c) $-4x^2y + 7xy^2 - 9x^2y - 2xy^2$

Solution: (a) $2p^3 - 5p^3 - p^3 = (2 - 5 - 1)p^3$ Distributive property
$$= -4p^3$$ Combine the coefficients.

(b) Use the commutative and associative properties to group the like terms:

$$3m + 6n + 5m + 8n = (3m + 5m) + (6n + 8n)$$
$$= 8m + 14n$$

This expression cannot be simplified further.

(c) $-4x^2y + 7xy^2 - 9x^2y - 2xy^2 = (-4x^2y - 9x^2y) + (7xy^2 - 2xy^2)$
$$= -13x^2y + 5xy^2$$ ❑

Try Exercise 35

Simplify: $5m - 9n + 8m + 3n$.

Once we have mastered combining like terms, it is an easy transition to adding and subtracting polynomials.

Adding Polynomials

To add two polynomials, add their like terms.

We can add polynomials in either horizontal form or vertical form. We illustrate the horizontal form in Example 4 and the vertical form in Example 5.

Example 4   Add $5t^2 - 2t - 7$ and $-11t^2 + 6t - 9$ in horizontal form.

Solution: Use the commutative and associative properties to get the like terms together:

$$(5t^2 - 2t - 7) + (-11t^2 + 6t - 9) = 5t^2 - 11t^2 - 2t + 6t - 7 - 9$$
$$= -6t^2 + 4t - 16$$ ❑

Try Exercise 43(a)

Add in horizontal form:
$2t^2 - 6t - 1, \; -8t^2 + 10t - 3$.

Example 5   Add $5t^2 - 2t - 7$ and $-11t^2 + 6t - 9$ in vertical form.

Solution: Line up the like terms in columns. Then add the columns.

$$5t^2 - 2t - 7$$
$$\underline{(+)\quad -11t^2 + 6t - 9}$$
$$-6t^2 + 4t - 16 \qquad \square$$

Try Exercise 43(b)

Add in vertical form:
$2t^2 - 6t - 1,\ -8t^2 + 10t - 3.$

Example 6   Add $-2r^3 + 9r^5 - 3r^2$ and $6r^4 - 2r^2 - r^5 + 8.$

Solution: Write each polynomial in descending order, arranging the like terms in columns. Then add the columns.

$$9r^5 \qquad\quad - 2r^3 - 3r^2$$
$$\underline{(+)\quad -r^5 + 6r^4 \qquad\quad - 2r^2 + 8}$$
$$8r^5 + 6r^4 - 2r^3 - 5r^2 + 8 \qquad \square$$

Try Exercise 47

Add:
$$8r^5 + 5r^3 - r^2 \qquad - 3$$
$$-2r^5 \qquad\quad -9r^2 + 6r$$

You should be able to add polynomials both in horizontal form and in vertical form. You should also be able to subtract polynomials using either form. Recall that to subtract two numbers, we change the sign of the second number and add. That is, to subtract b from a we write $a - b = a + (-b)$. We subtract two polynomials in the same way.

Subtracting Polynomials

To subtract two polynomials, change the sign of each term in the second polynomial. Then add like terms.

Example 7   Subtract $4p^2 + 7p - 8$ from $9p^2 - 6p + 11$ in horizontal form.

Solution:

$(9p^2 - 6p + 11) - (4p^2 + 7p - 8)$	Write in horizontal form.
$= 9p^2 - 6p + 11 - 4p^2 - 7p + 8$	Change each sign in the second polynomial.
$= 9p^2 - 4p^2 - 6p - 7p + 11 + 8$	Rearrange terms.
$= 5p^2 - 13p + 19$	Add like terms. $\square$

Try Exercise 51(a)

Subtract $3p^2 + 8p - 2$ from $8p^2 - 6p + 14$ in horizontal form.

Example 8   Subtract $4p^2 + 7p - 8$ from $9p^2 - 6p + 11$ in vertical form.

Solution: Line up the like terms in columns:

$$9p^2 - 6p + 11$$
$$\underline{(-)\quad 4p^2 + 7p - \ 8}$$

Change each sign in the second polynomial and add:

$$9p^2 - \ \ 6p + 11$$
$$\underline{(+)\quad -4p^2 - \ \ 7p + \ \ 8} \qquad \text{All signs changed}$$
$$5p^2 - 13p + 19 \qquad \text{Add like terms.} \quad \square$$

Try Exercise 51(b)

Subtract $3p^2 + 8p - 2$ from $8p^2 - 6p + 14$ in vertical form.

Example 9   Subtract $-x^3 + 8x^2y^2 + 7y^3$ from $x^3 + 5x^2y^2 - 7y^3.$

Solution: Line up the like terms in columns.

$$x^3 + 5x^2y^2 - 7y^3$$
$$\underline{(-)\quad -x^3 + 8x^2y^2 + 7y^3}$$

Change each sign in the second polynomial and add:

$$x^3 + 5x^2y^2 - 7y^3$$
$$(+)\quad x^3 - 8x^2y^2 - 7y^3$$
$$\overline{2x^3 - 3x^2y^2 - 14y^3}$$

Try Exercise 53

Subtract:
$$x^3 + 6xy^2 - 5y^3$$
$$-x^3 + 4xy^2 + 5y^3$$

Problem Solving

Example 10 Under ideal conditions, the stopping distance in feet (including reaction time) of a car traveling r miles per hour is given by the polynomial $S(r) = 0.044r^2 + 1.1r$. Determine the stopping distance of a car traveling 25 mi/hr.

Solution:

$$S(r) = 0.044r^2 + 1.1r \qquad \text{Original polynomial}$$
$$S(25) = 0.044(25)^2 + 1.1(25) \qquad \text{Replace } r \text{ with 25.}$$
$$= 0.044(625) + 27.5$$
$$= 27.5 + 27.5 \qquad \text{Order of operations}$$
$$= 55$$

Try Exercise 77

Find the stopping distance of a car traveling
(a) 30 mi/hr, (b) 60 mi/hr.

The stopping distance is 55 ft.

Exercises 3.4

Completion Exercises

1. A monomial in x is an expression of the form ax^n where a is a(n) _____ and n is a(n) _____.

2. The number 5 is called the _____ of the monomial $5x^3$.

3. The exponent 3 is called the _____ of the monomial $5x^3$.

4. A polynomial is either a _____ or a finite sum of _____.

5. The coefficient of the highest-degree monomial in a polynomial is called the _____ coefficient of the polynomial.

6. The term in a polynomial that does not contain a variable is called the _____ term.

7. A polynomial with two terms is called a _____.

8. A polynomial with three terms is called a _____.

True-False Exercises

9. The expression $8x^{-2}$ is a monomial.

10. The expression $4y^{1/2}$ is a monomial.

11. The expression $m^2 + m + \dfrac{1}{m}$ is a polynomial.

12. The polynomial $10 + 5p + p^2$ is written in descending order.

13. The degree of $15x^2y^3$ is 3.

14. Like terms are terms that have the same coefficients.

Write each polynomial in descending order. State the degree, the leading coefficient, and the constant term of the polynomial. Identify any monomials, binomials, or trinomials.

15. $4x + x^2 - 13$

16. $7 + 3y$

17. z^3

18. $5p - p^4$

19. 8

20. $-\frac{1}{4} + \frac{1}{3}r^6 + \frac{5}{9}r^{12}$

21. $m^2 - 2m^5 - 9m + 6$

22. $-14xy^4z^2$

23. $4x^2t^2 + 3x^3t^4 - 6x^5t$

For each polynomial, find (a) $P(3)$, (b) $P(-2)$, and (c) $P(-k)$.

24. $P(x) = 4x + 2$

25. $P(x) = 5x^2 - 8x + 13$

26. $P(x) = 2x^3 - 3x^2 + 5x + 1$

For each polynomial, find (a) $Q(2)$, (b) $Q(-1)$, and (c) $Q(2a)$.

27. $Q(y) = y^3 - y$

28. $Q(y) = -y^2 + 2y - 9$

29. $Q(y) = -y^4 - y^2 + 4y + 6$

For each polynomial, find (a) $R(0)$, (b) $R(\frac{2}{3})$, and (c) $R(z^2)$.

30. $R(z) = 9z^3 - 3z^2 + 5z - 7$

31. $R(z) = 3z^3 - z^2 - 4z + 9$

 Getting Acquainted with Your Scientific Calculator

To find $P(7.5)$ where $P(x) = 8x^3 - 4x^2 + 5.2x - 13.9$, press

$\boxed{\text{Clear}}\ 8\ \boxed{\times}\ 7.5\ \boxed{y^x}\ 3\ \boxed{-}\ 4\ \boxed{\times}\ 7.5$

$\boxed{x^2}\ \boxed{+}\ 5.2\ \boxed{\times}\ 7.5\ \boxed{-}\ 13.9\ \boxed{=}\ \boxed{\quad 3175.1\quad}$

32. Find $P(2.35)$ where $P(x) = 4x^3 - 8x^2 + 7.2x - 11.7$. Round your answer to three decimal places.

33. Find $Q(-0.64)$ where $Q(y) = y^5 - 3y^3 + 7y^2 - y - 10$. Round your answer to three decimal places.

Simplify each polynomial by combining like terms.

34. $3a^2 - 7a - 2a + 5$

35. $5m - 9n + 8m + 3n$

36. $2r^2 - r^3 - 10r^2 + r$

37. $-6x^2y + 9xy^2 - 4xy^2 - 3x^2y$

38. $3 + [(k + 1) - (k - 9)]$

39. $4u - [2u^2 - (3u + 4)]$

40. $t^2 - (3t^3 - t^2 - 4) + 5$

41. $m - (m - [m - (2m + n)] - 2n)$

Add each pair of polynomials (a) in horizontal form, (b) in vertical form.

42. $3p + 7, 5p - 9$

43. $2t^2 - 6t - 1, -8t^2 + 10t - 3$

44. $4x^2 - 3xy + 9y^2, 2x^2 + 3xy - 11y^2$

Add each pair of polynomials.

45. $\begin{array}{l} z^3 - 2z^2 + 8z - 17 \\ z^3 - 13z^2 - z - 6 \end{array}$

46. $\begin{array}{l} z^3 + 16z^2 - 4z - 13 \\ z^3 - 7z^2 - 9z + 5 \end{array}$

47. $\begin{array}{l} 8r^5 + 5r^3 - r^2 - 3 \\ -2r^5 - 9r^2 + 6r \end{array}$

48. $\begin{array}{l} -6r^4 - 2r^3 + 4r - 8 \\ 3r^4 - 3r^2 - r + 19 \end{array}$

Subtract the second polynomial from the first (a) in horizontal form, (b) in vertical form.

49. $5x + 7, 2x + 4$

50. $6x + 7, 4x - 1$

51. $8p^2 - 6p + 14, 3p^2 + 8p - 2$

52. $9p^2 + 3p - 17, 5p^2 - 2p - 7$

Subtract the second polynomial from the first.

53. $\begin{array}{l} x^3 + 6xy^2 - 5y^3 \\ -x^3 + 4xy^2 + 5y^3 \end{array}$

54. $\begin{array}{l} -x^3 + 4x^2y + 8y^3 \\ x^3 - 2x^2y - 8y^3 \end{array}$

55. $\begin{array}{l} a^3 + 3a^2 - 7a - 5 \\ a^3 - 4a^2 - 9a + 12 \end{array}$

56. $\begin{array}{l} a^3 - 3a^2 + 6a - 4 \\ a^3 + 5a^2 - 11a - 7 \end{array}$

57. $\begin{array}{l} -m^5 + m \\ 3m^5 + 2m^4 - 8m^2 - m \end{array}$

58. $\begin{array}{l} 4m^4 - m^2 \\ -m^4 - 3m^3 + 6m^2 - 1 \end{array}$

Perform the indicated operations.

59. $(3x^2 - 2x + 7) + (-5x^2 + 9x + 4)$

60. $(2x^2 + 11x - 1) + (-8x^2 - 3x + 2)$

61. $(6m^3 + 4m^2 - 5) - (2m^3 - m^2 - 9m + 3)$

62. $(8m^3 - 3m^2 + 2) - (5m^3 - 2m^2 + m - 7)$

63. $(9r^3 - r^2s + 6s^3) - (3r^3 - rs^2 - 5s^3)$

64. $(5r^3 + r^3s^2 - 4s^3) - (8r^3 + r^2s^3 + 2s^3)$

65. $(a - 3b + c) - (4a + b - c) + (3a + 4b - c)$

66. $(a - 2b + c) - (2a - b + c) + (a + b - 2c)$

67. $(8p^3 - pz^2 + 9z) - [(4p^3 + pz^2 - 3p) - (-7p + z)]$

68. $(2p^2 + 3p^2z - z^2) - [(5p^3 - 6p^2z + z) - (-9z^2 - z)]$

69. $(3x^{2m} - 5x^m + 7) + (x^{3m} - 4x^{2m} + 8x^m - 10)$

70. $(x^{4m} + 2x^{3m} - 6x^{2m} - x^m) + (7x^{3m} - 4x^{2m} + x^m)$

71. Subtract $5z^3 - 6z^2 + 11$ from the sum of $4z^2 - 9z + 8$ and $z^3 - z^2 - 12z$.

72. Subtract $6z^3 - 4z - 9$ from the sum of $2z^3 - 5z^2 - 7$ and $-3z^2 - 11z + 8$.

Problem Solving

73. Find the sum of three consecutive integers given that the smallest integer is n.

74. Find the sum of three consecutive even integers given that the smallest integer is n.

75. If each of the n teams in a coed volleyball league plays every other team twice, then the total number of league games is given by the polynomial $T(n) = n^2 - n$. Find the total number of league games played if the league consists of (a) six teams, (b) twelve teams.

76. An arrow is projected upward from the top of a building 35 ft high with a bow that gives it an initial velocity of 96 ft/sec. If we ignore air resistance, the height in feet of the arrow after t seconds is given by the polynomial $H(t) = -16t^2 + 96t + 35$. Find the height of the arrow after (a) 2 sec, (b) 4 sec.

77. Under ideal conditions, the stopping distance in feet (including reaction time) of a car traveling r miles per hour is given by the polynomial $S(r) = 0.044r^2 + 1.1r$. Find the stopping distance of a car traveling (a) 30 mi/hr, (b) 60 mi/hr.

78. Suppose the cost of producing x pairs of skis is given by the polynomial $C(x) = 0.01x^3 - 0.05x^2 + 25x + 150$. Find the cost of producing (a) 10 pairs of skis, (b) 20 pairs of skis.

Discussion Exercises

79. How do you determine the degree of a polynomial?

80. What does it mean to evaluate a polynomial at $x = a$?

81. Explain how to add two polynomials.

82. Explain how to subtract two polynomials.

Getting Acquainted with Your Graphing Calculator

Type $2.4(x)^3 - 5.6(x)^2 + 7.2(x) - 12.8$ into your calculator. Then find each of the following by replacing the x in this polynomial by the given value.

83. $P(2)$ 84. $P(-2)$

85. $P(3.5)$

NOTE:

Many calculators allow you another way to do Exercises 83, 84, and 85. For example, on one brand of calculator you can enter $y_1 = 2.4x^3 - 5.6x^2 + 7.2x - 12.8$ and then find the value of y_1 for each value of x.

 ## 3.5 Multiplying Polynomials

TAPE AU6

In Sec. 3.4 we learned how to add and subtract polynomials. In this section we learn how to multiply polynomials.

Multiplying Two Monomials

Recall that we multiply monomials by using the commutative and associative properties to group the coefficients and group the powers of each variable.

Example 1 Find each product.

(a) $(4x^3)(-2x^6)$ (b) $(3y^2p^7)(5y^4p)$

Solution: (a) $(4x^3)(-2x^6) = 4(-2)x^3x^6 = -8x^9$
(b) $(3y^2p^7)(5y^4p) = 3 \cdot 5y^2y^4p^7p = 15y^6p^8$ ◻

Try Exercise 1

Multiply: $(5x^2)(-3x^5)$.

Multiplying a Monomial and a Polynomial

To multiply a monomial and a polynomial with two or more terms, we use the distributive property.

Example 2   Find each product.

(a) $-2m(m^2 + 5m - 8)$ (b) $5x^2y^3(-4x^3 - 6x^2y + 3y^2)$

Solution: (a) $-2m(m^2 + 5m - 8)$ Original expression
$= -2m(m^2) + (-2m)(5m) + (-2m)(-8)$ Distribute $-2m$.
$= -2m^3 - 10m^2 + 16m$ Multiply the monomials.
(b) $5x^2y^3(-4x^3 - 6x^2y + 3y^2)$ Original expression
$= 5x^2y^3(-4x^3) + 5x^2y^3(-6x^2y) + 5x^2y^3(3y^2)$ Distribute $5x^2y^3$.
$= -20x^5y^3 - 30x^4y^4 + 15x^2y^5$ Multiply the monomials.
❒

Try Exercise 7

Multiply: $-3m(m^2 + 9m - 7)$.

Multiplying Two Binomials

Consider the product of $a + b$ and $c + d$. Think of $c + d$ as a single number and distribute $c + d$ over the sum $a + b$:

$$(a + b)(c + d) = a(c + d) + b(c + d)$$

Now distribute a over $c + d$ and distribute b over $c + d$:

$$= ac + ad + bc + bd$$

Note that the first term of the answer is the product of the first terms of the binomials:

First terms: $a \cdot c$

$$(a + b)(c + d)$$

The second term ad is the product of the two outer terms:

$$(a + b)(c + d)$$

Outer terms: $a \cdot d$

The third term bc is the product of the two inner terms:

$$(a + b)(c + d)$$

Inner terms: $b \cdot c$

The fourth term bd is the product of the two last terms:

Last terms: $b \cdot d$

$$(a + b)(c + d)$$

By remembering the key word **FOIL** (**F**irst, **O**uter, **I**nner, **L**ast), you can perform the multiplication in one step.

The FOIL Method

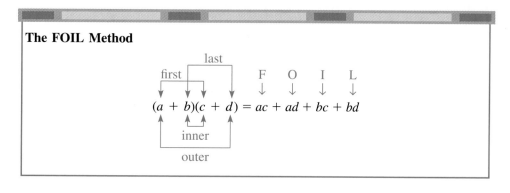

$$(a + b)(c + d) = ac + ad + bc + bd$$

After using the FOIL method to multiply two binomials, try to simplify your answer by combining like terms.

Example 3   Find each product.

(a) $(2x + 5)(x - 4)$ (b) $(3y - 2)(4y + 7)$

Solution:

(a) $(2x + 5)(x - 4) = 2x^2 - 8x + 5x - 20$

$$= 2x^2 - 3x - 20$$

(b) $(3y - 2)(4y + 7) = 12y^2 + 21y - 8y - 14$

$$= 12y^2 + 13y - 14$$ ❏

Try Exercise 15

Multiply: $(3x + 8)(x - 2)$.

Example 4   Find each product.

(a) $(5m + 7n)(m + 2n)$ (b) $(6a^2 - 5b)(3a^2 - b)$

Solution:

(a) $(5m + 7n)(m + 2n) = 5m^2 + 10mn + 7mn + 14n^2$

$$= 5m^2 + 17mn + 14n^2$$

(b) $(6a^2 - 5b)(3a^2 - b) = 18a^4 - 6a^2b - 15a^2b + 5b^2$

$$= 18a^4 - 21a^2b + 5b^2$$ ❏

Try Exercise 19

Multiply: $(3m + 7n)(4m + 5n)$.

Special Binomial Products

There are certain products involving two binomials that occur so often that they merit special attention. First, note what happens when we use the FOIL method to multiply $a + b$ and $a - b$. The binomials $a + b$ and $a - b$ are called **conjugates** of each other.

$$\overset{\text{F \quad O \quad I \quad L}}{\underset{\downarrow \quad \downarrow \quad \downarrow \quad \downarrow}{}}$$
$$(a + b)(a - b) = a^2 - ab + ab - b^2$$
$$= a^2 - b^2$$

Hence we have the following special product.

Product of Two Conjugates
$$(a + b)(a - b) = a^2 - b^2$$

Note that **the product of two conjugates is the square of the first term minus the square of the second term.** The expression $a^2 - b^2$ is called a *difference of two squares.*

Example 5   Find each product.

(a) $(k + 8)(k - 8)$ \qquad (b) $(3z + 2)(3z - 2)$
(c) $(5p + 7q)(5p - 7q)$ \qquad (d) $(4x + y^3)(4x - y^3)$

Solution: \qquad (a) $\underset{\downarrow \ \ \downarrow \ \downarrow \ \ \downarrow \ \ \ \downarrow \ \ \ \downarrow}{(a + b)(a - b) = a^2 - b^2}$
$$(k + 8)(k - 8) = k^2 - 8^2 = k^2 - 64$$

(b) $(3z + 2)(3z - 2) = (3z)^2 - 2^2 = 9z^2 - 4$
(c) $(5p + 7q)(5p - 7q) = (5p)^2 - (7q)^2 = 25p^2 - 49q^2$
(d) $(4x + y^3)(4x - y^3) = (4x)^2 - (y^3)^2 = 16x^2 - y^6$ ❑

Try Exercise 35

Multiply: $(k + 6)(k - 6)$.

Another binomial product that deserves special attention is the square of a binomial. Observe what happens when we square the binomial $a + b$:

$$\overset{\text{F \quad O \quad I \quad L}}{\underset{\downarrow \quad \downarrow \quad \downarrow \quad \downarrow}{}}$$
$$(a + b)^2 = (a + b)(a + b) = a^2 + ab + ab + b^2$$
$$= a^2 + 2ab + b^2$$

A similar rule holds for $(a - b)^2$.

Square of a Binomial
$$(a + b)^2 = a^2 + 2ab + b^2$$
$$(a - b)^2 = a^2 - 2ab + b^2$$

Note that **the square of a binomial is the square of the first term, plus twice the product of the two terms, plus the square of the last term.** The two expressions $a^2 + 2ab + b^2$ and $a^2 - 2ab + b^2$ are called *perfect-square trinomials.*

Example 6 Find each product.

(a) $(h + 5)^2$ (b) $(r - 7)^2$
(c) $(2t + 5)^2$ (d) $(3x^2 - 4y)^2$

Solution:

(a) $(a + b)^2 = a^2 + 2 \ a \ b + b^2$

$(h + 5)^2 = h^2 + 2 \cdot h \cdot 5 + 5^2$
$\quad\quad\quad = h^2 + 10h + 25$

(b) $(r - 7)^2 = r^2 - 2 \cdot r \cdot 7 + 7^2$
$\quad\quad\quad = r^2 - 14r + 49$
(c) $(2t + 5)^2 = (2t)^2 + 2(2t)(5) + 5^2$
$\quad\quad\quad = 4t^2 + 20t + 25$
(d) $(3x^2 - 4y)^2 = (3x^2)^2 - 2(3x^2)(4y) + (4y)^2$
$\quad\quad\quad = 9x^4 - 24x^2y + 16y^2$

Try Exercise 43

Multiply: $(r - 9)^2$.

CAUTION

You cannot square a binomial simply by squaring both terms.

Correct

$(x + 3)^2 = x^2 + 6x + 9$
$(a + b)^2 = a^2 + 2ab + b^2$
$(a - b)^2 = a^2 - 2ab + b^2$

Wrong

$(x + 3)^2 = x^2 + 9$
$(a + b)^2 = a^2 + b^2$
$(a - b)^2 = a^2 - b^2$

We can apply the special product rules to multiply more complicated polynomials.

Example 7 Find each product.

(a) $[(p + m) + 1]^2$ (b) $[2x + (3y + 5)][2x - (3y + 5)]$

Solution:

(a) Treat this as the square of a binomial:

$(a + b)^2 = a^2 + 2 \ a \ b + b^2$

$[(p + m) + 1]^2 = (p + m)^2 + 2(p + m)1 + 1^2$
$\quad\quad\quad = p^2 + 2pm + m^2 + 2p + 2m + 1$

(b) Treat this as a product of two conjugates:

$(a + b)(a - b) = a^2 - b^2$

$[2x + (3y + 5)][2x - (3y + 5)] = (2x)^2 - (3y + 5)^2$
$\quad\quad\quad = 4x^2 - (9y^2 + 30y + 25)$
$\quad\quad\quad = 4x^2 - 9y^2 - 30y - 25$

Try Exercise 57

Multiply: $[(p - m) + 1]^2$.

Multiplying Two Polynomials

Now consider the product of the binomial $a + b$ and the trinomial $c + d + e$. Think of $a + b$ as a single number and distribute $a + b$ over the sum, $c + d + e$:

$(a + b)(c + d + e) = (a + b)c + (a + b)d + (a + b)e$

Now distribute c, d, and e over the sum, $a + b$:

$$= ac + bc + ad + bd + ae + be$$

If you study this answer carefully, you will find that you can obtain the same result simply by multiplying each term of $a + b$ by each term of $c + d + e$. This suggests the following rule.

Multiplying Polynomials

To multiply any two polynomials, multiply each term of one polynomial by each term of the other polynomial.

We can multiply two polynomials in either horizontal form or vertical form. We illustrate the horizontal form in Example 8 and the vertical form in Examples 9 and 10.

Example 8 Multiply: $(3x + 2y)(x^2 + 5xy - 7y)$.

Solution: Multiply each term in the second polynomial first by $3x$, and then by $2y$:

$$(3x + 2y)(x^2 + 5xy - 7y)$$

$$= 3x(x^2) + 3x(5xy) + 3x(-7y) + 2y(x^2) + 2y(5xy) + 2y(-7y)$$
$$= 3x^3 + 15x^2y - 21xy + 2x^2y + 10xy^2 - 14y^2$$

Now simplify by combining like terms:

$$= 3x^3 + 17x^2y - 21xy + 10xy^2 - 14y^2 \qquad \square$$

Try Exercise 63

Multiply:
$(2x + 5y)(x^2 - 3xy + 9y)$.

Example 9 Multiply: $(4p + 7)(5p + 2)$.

Solution: Multiply each term of $5p + 2$ first by 7, and then by $4p$. Make sure you line up like terms in columns as you multiply.

$$
\begin{array}{r}
5p + 2 \\
4p + 7 \\
\hline
35p + 14 \\
20p^2 + 8p \\
\hline
20p^2 + 43p + 14
\end{array}
$$

$\leftarrow 7$ times $5p + 2$
$\leftarrow 4p$ times $5p + 2$, align like terms.
$\leftarrow$ Add like terms. $\square$

Try Exercise 73

Multiply: $\begin{array}{r} 6p + 5 \\ 4p + 3 \end{array}$

Example 10 Multiply: $(2m - 5)(4m^3 - 3m + 6)$.

Solution:

$$
\begin{array}{r}
4m^3 - 3m + 6 \\
2m - 5 \\
\hline
-20m^3 \qquad + 15m - 30 \\
8m^4 \qquad - 6m^2 + 12m \\
\hline
8m^4 - 20m^3 - 6m^2 + 27m - 30
\end{array}
$$

$\leftarrow -5$ times $4m^3 - 3m + 6$
$\leftarrow 2m$ times $4m^3 - 3m + 6$
$\leftarrow$ Add like terms. $\square$

Try Exercise 77

Multiply: $\begin{array}{r} 5m^3 + 8m - 11 \\ 3m + 4 \end{array}$

Exercises 3.5

Find each product.

1. $(5x^2)(-3x^5)$ 2. $(-2x^4)(6x^7)$

3. $(-a^2b^7c^9)(-9a^{11}b^3)(-2b^5c)$

4. $(-10a^{12}b^2c^4)(4b^5c^3)(-ab^8)$

5. $2r(6r + 5)$ 6. $4r(3r - 7)$

7. $-3m(m^2 + 9m - 7)$ 8. $-8t^3(2 - 4t - t^2)$

9. $4x^5y^4(-5x^3 - 3xy^2 + 9y^3)$

10. $-5k^2z(k^4 - 2k^3z^2 + k^2z^3 - 5z^4)$

✎ Discussion Exercises

11. What do the letters in the word FOIL represent? Can you use FOIL to multiply any two polynomials?

12. State the general rule for multiplying any two polynomials.

Find each product using the FOIL method.

13. $(p + 5)(p + 3)$ 14. $(p - 7)(p + 2)$

15. $(3x + 8)(x - 2)$ 16. $(5x - 3)(x - 4)$

17. $(-r - 5)(7r - 6)$ 18. $(-3r + 4)(-r + 8)$

19. $(3m + 7n)(4m + 5n)$ 20. $(9m - 4n)(m - 2n)$

21. $(6p - 5q)(2p + 3q)$ 22. $(5p + 2q)(3p - 4q)$

23. $(t^2 - 4)(t^2 + 9)$ 24. $(t^2 - 1)(t^2 - 16)$

25. $(x - \frac{1}{3}h)(x - \frac{2}{3}h)$ 26. $(x - \frac{3}{2}h)(x + \frac{1}{2}h)$

Matching Exercises

Match each expression in Exercises 27 through 30 with an equivalent expression in letters A through D.

27. $(x + 3)^2$ A. $x^2 + 9$

28. $(x + 3)(x - 3)$ B. $x^2 - 6x + 9$

29. $(x - 3)^2$ C. $x^2 - 9$

30. $x^2 + 3^2$ D. $x^2 + 6x + 9$

✎ Discussion Exercises

31. Explain why $(x + y)^2$ is not equal to $x^2 + y^2$. Substitute numbers for x and y to illustrate your point. For example, let $x = 2$ and $y = 3$.

32. What is the difference between $(x + 2)^3$ and $x + 2^3$?

33. Explain, in words, how to find the product of two conjugates.

34. Explain, in words, how to square a binomial.

Find each special binomial product.

35. $(k + 6)(k - 6)$ 36. $(k + 9)(k - 9)$

37. $(2z + 5)(2z - 5)$ 38. $(4z + 3)(4z - 3)$

39. $(3p + 8q)(3p - 8q)$ 40. $(6p + 7q)(6p - 7q)$

41. $(5x + y^3)(5x - y^3)$ 42. $(3x + y^3)(3x - y^3)$

43. $(r - 9)^2$ 44. $(r + 4)^2$

45. $(3t + 4)^2$ 46. $(6t + 5)^2$

47. $(5a - 2b)^2$ 48. $(4a - 3b)^2$

49. $(6k + \frac{3}{4}m)^2$ 50. $(9k - \frac{5}{6}m)^2$

Problem Solving

51. Write an expression that represents the area of Fig. 3.1.

52. Write an expression that represents the area of Fig. 3.2.

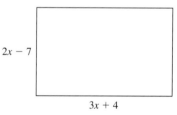

Figure 3.1

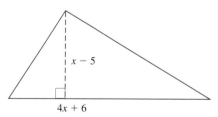

Figure 3.2

Find the following products.

53. $2ay(4ax - 5by)^2$ 54. $5xb(3ax + 2by)^2$

55. $-m^3p(mp + 3q)(mp - 3q)$

56. $-mp^2(mp - 5q)(mp + 5q)$

57. $[(p - m) + 1]^2$ 58. $[(2x - 5) - y]^2$

59. $[(k + h) + 5][(k + h) - 5]$

60. $[10 + (3a + b)][10 - (3a + b)]$

Multiply.

61. $(3m - 6)(m^2 - 5m + 2)$ 62. $(4m + 7)(m^2 + 3m - 5)$

63. $(2x + 5y)(x^2 - 3xy + 9y)$

64. $(3x - 2y)(x^2 + 6xy - 8y)$

65. $(4k - 3)(3k^3 + 5k^2 - 2k - 9)$

66. $(2k + 5)(4k^3 - 6k^2 - 3k + 7)$

67. $(5p^2 + 9p - 3)(2p^2 - p - 6)$

68. $(3p^2 - 6p + 4)(p^2 + 2p - 8)$

69. $(r - 4)^3$ 70. $(r + 5)^3$

71. $(2x + 5y)^3$ 72. $(4x - 3y)^3$

73. $\begin{array}{r} 6p + 5 \\ 4p + 3 \\ \hline \end{array}$ 74. $\begin{array}{r} 8p + 3 \\ 5p - 2 \\ \hline \end{array}$

75. $\begin{array}{r} 7y^2 - 4y + 9 \\ 2y - 3 \\ \hline \end{array}$ 76. $\begin{array}{r} 9y^2 + 4y - 1 \\ 3y + 5 \\ \hline \end{array}$

77. $\begin{array}{r} 5m^3 + 8m - 11 \\ 3m + 4 \\ \hline \end{array}$ 78. $\begin{array}{r} 9m^3 - 5m^2 + 7 \\ 2m - 3 \\ \hline \end{array}$

79. $\begin{array}{r} x^3 - 6x^2 - 2x + 7 \\ 5x^2 - 3 \\ \hline \end{array}$ 80. $\begin{array}{r} x^3 + 4x^2 - 8x - 9 \\ 4x^2 + 2 \\ \hline \end{array}$

Simplify.

81. $5p^2q - 6[p - p(pq - q)] - pq$

82. $6pq^2 - 2[q - q(pq - p)] - pq$

83. $m^2[m - (m - 2)(m + 3)]$

84. $m^2[m - (m + 8)(m - 1)]$

85. $(k + 10)(k - 10) - (k - 6)^2$

86. $(k + 4)^2 - (k + 7)(k - 7)$

Find each product. Assume all exponents are integers.

87. $5x^m(2x^{m+1} - 3x + 7x^{-m})$

88. $(4y^n + 3)(4y^n - 3)$

89. $(9z^k - p^{2k})(2z^k + 5p^{3k})$ 90. $(6r^{a-1} + t^{5a})^2$

3.6 The Greatest Common Factor; Factoring by Grouping

TAPE AU7

In Sec. 3.5 we learned how to multiply two polynomials. In Secs. 3.6, 3.7, and 3.8, we will learn how to reverse the process; that is, we will learn how to write a given polynomial as a product of two other polynomials. This procedure is called **factoring,** and it is one of the most important skills in algebra.

$$\underset{\text{factoring}}{\overset{\text{multiplying}}{2x(3x + 5) = 6x^2 + 10x}}$$

Factors

We can factor 18 by writing 18 as the product of 2 and 9:

$$18 = 2 \cdot 9$$

The integers 2 and 9 are called **factors,** or **divisors,** of 18. We can factor 18 still further as follows:

$$18 = 2 \cdot 9$$
$$= 2 \cdot 3 \cdot 3 = 2 \cdot 3^2$$

The integers 2 and 3 are called *prime numbers* because they cannot be factored into the product of two other positive integers.

A **prime number** is an integer greater than 1 whose only positive factors (divisors) are 1 and itself. The first 10 primes are listed below*:

$$\{2, 3, 5, 7, 11, 13, 17, 19, 23, 29, \ldots\}$$

When a number is written as a product of primes, the number is said to be in **prime factored form.**

Example 1   Write 360 in prime factored form.

Solution: Divide 360 by the smallest prime 2. Then divide that quotient by 2. Continuing to divide each new quotient by its smallest prime divisor, we obtain the sequence of divisions below:

$$\begin{array}{r} 2\,)\overline{360} \\ 2\,)\overline{180} \\ 2\,)\overline{90} \\ 3\,)\overline{45} \\ 3\,)\overline{15} \\ 5 \end{array}$$

Therefore, $360 = 2 \cdot 2 \cdot 2 \cdot 3 \cdot 3 \cdot 5 = 2^3 \cdot 3^2 \cdot 5$.   ❏

Try Exercise 11

Write 675 in prime factored form.

The Greatest Common Factor

The **greatest common factor** (**GCF**) of a set of integers is the largest integer that is a factor of every integer in the set. Example 2 shows how to find the greatest common factor when it is not obvious from a simple inspection.

Example 2   Find the greatest common factor of 72, 120, and 144.

Solution: Write each number in prime factored form:
$$72 = 2^3 \cdot 3^2$$
$$120 = 2^3 \cdot 3 \cdot 5$$
$$144 = 2^4 \cdot 3^2$$

Now take the lowest power of each prime. The lowest power of 2 is 2^3. The lowest power of 3 is 3. The lowest power of 5 is 5^0 (or 1). Multiply these powers to obtain the greatest common factor:

$$\text{GCF} = 2^3 \cdot 3 = 24$$ ❏

Try Exercise 19

Find the GCF of 84, 120, and 180.

Here are the steps in finding the greatest common factor.

To Find the Greatest Common Factor (GCF)

1. Write each number in prime factored form.
2. Take the lowest power of each prime factor. Only those prime factors that appear in *every* factorization will appear in the GCF.
3. Multiply the powers taken in step 2. If there are none, the GCF is 1.

---

*HISTORICAL NOTE: During the third century B.C., the Greek mathematician Euclid proved in his monumental work *Elements* that there are an infinite number of primes. The *Elements* is considered by many to be the most significant mathematics textbook ever written.

We can use these same steps to find the greatest common factor of a set of terms containing variables.

Example 3   Find the greatest common factor of $12x^3y^5$, $30x^4y^2$, and $24x^5y^7$.

Solution: Write each term in prime factored form:

$$12x^3y^5 = 2^2 \cdot 3 \cdot x^3 \cdot y^5$$
$$30x^4y^2 = 2 \cdot 3 \cdot 5 \cdot x^4 \cdot y^2$$
$$24x^5y^7 = 2^3 \cdot 3 \cdot x^5 \cdot y^7$$

Multiply together the lowest power of each prime factor:

$$\text{GCF} = 2 \cdot 3 \cdot x^3 \cdot y^2 = 6x^3y^2 \qquad \square$$

Try Exercise 23

Find the GCF of $60x^5y^7$, $45x^3y^9$, and $75x^6y^4$.

The first step in factoring a polynomial is to find the greatest common factor of all the terms of the polynomial. Then use the distributive property to "factor out" the greatest common factor.

Example 4   Factor: $4x + 12$.

Solution: The greatest common factor of $4x$ and 12 is 4. Use the distributive property to factor out 4:

$$\begin{aligned} 4x + 12 &= 4 \cdot x + 4 \cdot 3 && \text{The GCF is 4.} \\ &= 4(x + 3) && \text{Distributive property} \end{aligned}$$

Therefore the factors of $4x + 12$ are 4 and $x + 3$. To check, multiply 4 and $x + 3$. The result should be $4x + 12$. $\qquad \square$

Try Exercise 29

Factor: $9x + 18$.

When factoring a polynomial of two or more terms, we usually do not factor the monomials. That is why we wrote the answer to Example 4 as $4(x + 3)$ rather than $2 \cdot 2(x + 3)$.

Example 5   Factor each polynomial.

(a) $6x^3 + 15x^2$ (b) $a^3 + a$ (c) $-5p^5 + 10p^4 - 15p^3$

Solution: (a) $\begin{aligned}[t] 6x^3 + 15x^2 &= 3x^2 \cdot 2x + 3x^2 \cdot 5 && \text{The GCF is } 3x^2. \\ &= 3x^2(2x + 5) && \text{Factor out the GCF.} \end{aligned}$

(b) $\begin{aligned}[t] a^3 + a &= a \cdot a^2 + a \cdot 1 && \text{The GCF is } a. \\ &= a(a^2 + 1) && \text{Factor out the GCF.} \end{aligned}$

(c) $\begin{aligned}[t] -5p^5 &+ 10p^4 - 15p^3 \\ &= 5p^3(-p^2) + 5p^3(2p) - 5p^3(3) && \text{The GCF is } 5p^3. \\ &= 5p^3(-p^2 + 2p - 3) && \text{Factor out the GCF.} \qquad \square \end{aligned}$

Try Exercise 31

Factor: $8x^3 + 12x^2$.

We could have used $-5p^3$ as the greatest common factor in Example 5(c). This gives the factorization

$$-5p^3(p^2 - 2p + 3)$$

Both factorizations are correct.

Remember that **you can always check a factorization by multiplication.**

Example 6   Factor each polynomial.

(a) $18r^4t^9 - 24r^8t^3 + 36r^5t^5$ (b) $3x(m + n) - 7y(m + n)$
(c) $(k - 9) + (k - 9)^2$

Solution: (a) The GCF is $6r^4t^3$.

$$18r^4t^9 - 24r^8t^3 + 36r^5t^5 = 6r^4t^3(3t^6) - 6r^4t^3(4r^4) + 6r^4t^3(6rt^2)$$
$$= 6r^4t^3(3t^6 - 4r^4 + 6rt^2)$$

(b) The GCF is $m + n$.

$$3x(m + n) - 7y(m + n) = (m + n)(3x - 7y)$$

(c) The GCF is $k - 9$.

$$(k - 9) + (k - 9)^2 = (k - 9) \cdot 1 + (k - 9) \cdot (k - 9)$$
$$= (k - 9)[1 + (k - 9)]$$
$$= (k - 9)(k - 8)$$ ❐

Try Exercise 39

Factor:
$45r^8t^4 - 30r^2t^8 - 60r^6t^3$.

Just as some numbers are prime, so are some polynomials prime. For now, we shall say that a polynomial is a **prime polynomial** when it cannot be written as the product of two simpler polynomials with integer coefficients. For example,

$$3x - 8 \text{ is a prime polynomial}$$
$$y^2 + 1 \text{ is a prime polynomial}$$

Factoring by Grouping

If a polynomial has four or more terms, sometimes we can find a common factor by grouping the terms. This procedure is called **factoring by grouping.**

Example 7   Factor: $xy + 7x + 2y + 14$.

Solution: Note that there is no factor common to all four terms (except 1). However, we can group the first two terms and factor out x, and group the last two terms and factor out 2:

$$(xy + 7x) + (2y + 14) = x(y + 7) + 2(y + 7)$$

This procedure exposes the common binomial factor $y + 7$, which we factor out as follows:

Try Exercise 55

Factor: $xy + 11x + 3y + 33$.

$$= (y + 7)(x + 2)$$ ❐

In Example 7 we could have grouped the first and third terms and the second and fourth terms:

$$xy + 7x + 2y + 14 = xy + 2y + 7x + 14 \qquad \text{Rearrange terms.}$$
$$= y(x + 2) + 7(x + 2) \qquad \text{Factor out } y, \text{ factor out 7.}$$
$$= (x + 2)(y + 7) \qquad \text{Factor out } x + 2.$$

Since multiplication is commutative, this answer is the same as $(y + 7)(x + 2)$.

Example 8   Factor: $m^3 - 5m^2 + 3m - 15$.

Solution: Factor m^2 from the first two terms and 3 from the last two terms:

$$m^3 - 5m^2 + 3m - 15 = m^2(m - 5) + 3(m - 5)$$

Factor out the common binomial factor $m - 5$:

Try Exercise 59

Factor: $m^3 - 2m^2 + 8m - 16$.

$$= (m - 5)(m^2 + 3)$$ ❐

Example 9   Factor: $z^2a^2 - 6z^2 + a^2 - 6$.

Solution: Factor z^2 from the first two terms and 1 from the last two terms:
$$z^2a^2 - 6z^2 + a^2 - 6 = z^2(a^2 - 6) + 1(a^2 - 6)$$

Factor out the common binomial factor $a^2 - 6$:

Try Exercise 65

Factor: $z^2a^2 - 3z^2 + a^2 - 3$.

$$= (a^2 - 6)(z^2 + 1) \qquad ❐$$

Sometimes we must factor out a negative number so that the signs of the common binomial factors are the same.

Example 10   Factor: $6p^2 + 8p - 15p - 20$.

Solution: Factor $2p$ from the first two terms and -5 from the last two terms:
$$6p^2 + 8p - 15p - 20 = 2p(3p + 4) - 5(3p + 4)$$

Try Exercise 71

Factor: $12p^2 + 16p - 9p - 12$.

$$= (3p + 4)(2p - 5) \qquad ❐$$

It should be pointed out that factoring by grouping works on many polynomials with four terms, but it will not work on *every* polynomial with four terms.

CAUTION

Since the third term in Example 10 has a negative coefficient, you must be careful if you use parentheses to group the terms.

Wrong   *Correct*

~~$(6p^2 + 8p) - (15p - 20)$~~ $(6p^2 + 8p) + (-15p - 20)$ ∎

Exercises 3.6

True-False Exercises

1. The number 1 is a prime number.

2. The number 91 is a prime number.

3. The prime factored form of 12 is $3 \cdot 4$.

4. The prime factored form of $5x^3y + 5x^2y$ is $5x^2(xy + y)$.

Completion Exercises

5. Since $6 = 2 \cdot 3$, we call 2 and 3 _____ of 6.

6. The process of writing a polynomial as a product of two other polynomials is called _____.

7. An integer greater than 1 whose only positive factors are 1 and itself is called a _____ number.

8. The largest integer that is a factor of every integer in a set of integers is called the _____ of the set of integers.

Write each number in prime factored form.

9. 20

10. 50

11. 675

12. 567

13. 59

14. 71

15. 5040

16. 3240

Find the greatest common factor of each set of terms.

17. 8, 20

18. 12, 18

19. 84, 120, 180

20. 72, 90, 108

21. $6p^3, 9p^2$

22. $10p^4, 15p^5$

23. $60x^5y^7, 45x^3y^9, 75x^6y^4$

24. $56x^8y^5, 42x^{10}y^6, 70x^6y^9$

25. $7a(2m + 5), 9b(2m + 5)$

26. $3b(4m - 9), 11a(4m - 9)$

27. $3z - 1, (3z - 1)^2$

28. $(6z + 7)^2, 6z + 7$

Factor out the greatest common factor, including -1 if the leading coefficient is negative.

29. $9x + 18$

30. $4x - 20$

31. $8x^3 + 12x^2$

32. $12x^4 - 18x^3$

33. $a^5 + a$

34. $a^7 + a$

35. $28m^3 - 14m^2 + 35m$

36. $25m^3 + 15m^2 + 45m$

37. $-4p^5 + 12p^3 - 16p^2$

38. $-3p^6 - 12p^5 + 9p^3$

39. $45r^8t^4 - 30r^2t^8 - 60r^6t^3$

40. $24r^6t^3 - 36r^4t^4 - 48r^3t^4$

41. $-16k^{11}m^8h^9 - 24k^7m^{10}h^6$

42. $-27k^{13}m^7h^{11} - 36k^9m^9h^5$

43. $\pi a^2bc + \pi ab^2c - \pi abc^2$

44. $\pi a^2b^3c^2 - \pi a^3b^2c^2 - \pi a^2b^2c^3$

45. $54x^8z^4 - 108x^5z^3 - 36x^3z^2 + 72xz$

46. $36x^9z^5 + 48x^7z^4 - 24x^5z^3 - 144x^2z^2$

47. $5x(m + n) - 9y(m + n)$ 48. $2x(m - n) + 7y(m - n)$

49. $(k - 3) + (k - 3)^2$ 50. $(k + 6) + (k + 6)^2$

51. $(p + 2)(3p + 5) - (p + 2)(p - 4)$

52. $(p - 7)(5p - 3) - (p - 7)(2p + 1)$

53. $(y - 2)^3 6(y + 8)^5 + (y + 8)^6 3(y - 2)^2$

54. $(y + 3)^6 4(y - 5)^3 + (y - 5)^4 6(y + 3)^5$

Factor by grouping.

55. $xy + 11x + 3y + 33$

56. $xy - 13x + 2y - 26$

57. $p^2 - 5p + 4p - 20$

58. $p^2 + 7p + 2p + 14$

59. $m^3 - 2m^2 + 8m - 16$

60. $m^3 + 2m^2 + 5m + 10$

61. $z^3 + z^2 + z + 1$

62. $z^3 - z^2 + z - 1$

63. $5x^2 - 5xy + 2x - 2y$

64. $3x^2 + 3xy + 7x + 7y$

65. $z^2a^2 - 3z^2 + a^2 - 3$

66. $z^2a^2 + 6z^2 + a^2 + 6$

67. $8r^2 + rt^2 + 48rt + 6t^3$

68. $9r^3 + r^2t + 63rt + 7t^2$

69. $2k^3 - k^2 - 10k + 5$

70. $3k^3 + k^2 - 18k - 6$

71. $12p^2 + 16p - 9p - 12$

72. $10p^2 - 15p - 8p + 12$

73. $6m^4 + 30m^3 + 8m + 40$

74. $6m^4 + 9m^3 + 36m + 54$

Factor each polynomial. Assume m and n are integers.

75. $3y^{n+3} + 6y^{n+2} - 9y^2$ 76. $2y^{n+2} - 8y^{n+1} - 4y$

77. $z^{3m} - 6z^{2m} + 2z^m - 12$ 78. $z^{3m} + 5z^{2m} - 3z^m - 15$

Discussion Exercises

79. How can you check any factorization?

80. Is $(x + 2)(x - 3)$ equal to $(x - 3)(x + 2)$? Why or why not?

81. When should you consider factoring by grouping?

82. Outline the steps in finding the greatest common factor.

Problem Solving

83. The expression $\pi rl + \pi r^2$ represents the total surface area of a right circular cone with radius r and lateral height l (see Fig. 3.3). Factor this expression.

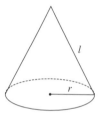

Figure 3.3

84. The expression $2\pi r^2 + 2\pi rh$ represents the total surface area of a right circular cylinder with radius r and height h (see Fig. 3.4). Factor this expression.

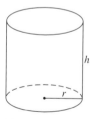

Figure 3.4

85. If P dollars are invested at an interest rate r compounded annually, the value of the account after 2 yr is given by $P(1 + r) + P(1 + r)r$. Factor this expression.

86. If P dollars are invested at an interest rate r compounded annually, the value of the account after 3 yr is given by $P(1 + r)^2 + P(1 + r)^2 r$. Factor this expression.

3.7 Factoring Trinomials

In Sec. 3.5 we used the FOIL method to multiply two binomials:

$$
\begin{array}{cccc}
\text{F} & \text{O} & \text{I} & \text{L} \\
\downarrow & \downarrow & \downarrow & \downarrow
\end{array}
$$

$$
\begin{aligned}
(x + 2)(x + 3) &= x^2 + 3x + 2x + 6 \\
&= x^2 + 5x + 6
\end{aligned}
$$

In this section we learn to reverse the procedure. That is, we learn how to write the trinomial $x^2 + 5x + 6$ as the product of two binomials.

Factoring $x^2 + bx + c$

We begin by factoring trinomials whose leading coefficient is 1.

Example 1   Factor: $x^2 + 5x + 6$.

Solution: Begin by writing

$$x^2 + 5x + 6 = (x \quad)(x \quad)$$

trinomial   two binomials

Since the last term of the trinomial is 6, list all pairs of integers whose product is 6:

$$1 \cdot 6 = 6$$
$$2 \cdot 3 = 6$$
$$(-1)(-6) = 6$$
$$(-2)(-3) = 6$$

This suggests four possible factorizations of $x^2 + 5x + 6$:

$$
\begin{array}{cccc}
(x + 1)(x + 6) & (x + 2)(x + 3) & (x - 1)(x - 6) & (x - 2)(x - 3) \\
x & 2x & -x & -2x \\
6x & 3x & -6x & -3x
\end{array}
$$

Only the second factorization produces the correct middle term of $5x$. Therefore

$$x^2 + 5x + 6 = (x + 2)(x + 3)$$

◻

Try Exercise 13

Factor: $x^2 + 8x + 15$.

We could also factor the trinomial of Example 1 as follows:

$$x^2 + 5x + 6 = (-x - 2)(-x - 3)$$

However, we shall always factor a trinomial so that the leading coefficients of both binomial factors are positive.

Here are the steps in factoring a trinomial whose leading coefficient is 1.

> **To Factor $x^2 + bx + c$**
> 1. List all pairs of integers whose product is c.
> 2. Find the pair of integers from step 1 whose sum is b. Suppose these integers are m and n.
> 3. Factor the trinomial as follows:
> $$x^2 + bx + c = (x + m)(x + n)$$
> 4. Check by multiplication.

We illustrate these steps in Example 2.

Example 2 Factor: $y^2 - 10y + 21$.

Solution: We are looking for a pair of integers whose product is 21 and whose sum is -10. Since the product is positive, the integers must have the same sign. Since the sum is negative, both integers must be negative.

Step 1: List all pairs of (nega-   Step 2: Find the pair whose sum is
tive) integers whose -10.
product is 21.

Product is 21   *Sum is -10*
$(-1)(-21)$ $-1 + (-21) = -22$ No
$(-3)(-7)$ $-3 + (-7) = -10$ Yes

The desired integers are -3 and -7.

Step 3: Factor the trinomial as follows.
$$y^2 - 10y + 21 = [y + (-3)][y + (-7)]$$
$$= (y - 3)(y - 7)$$

Step 4: Check by multiplication.

$$\begin{array}{cccc} F & O & I & L \\ \downarrow & \downarrow & \downarrow & \downarrow \end{array}$$
$$(y - 3)(y - 7) = y^2 - 7y - 3y + 21 = y^2 - 10y + 21 \quad \square$$

Try Exercise 15

Factor: $y^2 - 9y + 14$.

If a trinomial is not already in descending order, write the trinomial in descending order before attempting to factor it.

Example 3 Factor: $-4p + p^2 - 12$.

Solution: Write the trinomial in descending order as $p^2 - 4p - 12$. Then find a pair of integers whose product is -12 and whose sum is -4. Since the product is negative, the integers must have opposite signs. Since the sum is negative, consider only those integers whose sum is negative.

Product is -12   *Sum is -4*
$1(-12)$ $1 + (-12) = -11$ No
$2(-6)$ $2 + (-6) = -4$ Yes
$3(-4)$

The desired integers are 2 and -6. Therefore
$$-4p + p^2 - 12 = p^2 - 4p - 12 = (p + 2)(p - 6)$$

Try Exercise 19

Factor: $-7p + p^2 - 18$.

Check by multiplication. $\square$

In Example 3, note that we stopped as soon as we found a pair of integers whose product was -12 and whose sum was -4. We stopped because **a polynomial has one and only one prime factored form.** (*Note:* Some polynomials are prime, so they are already in prime factored form. Also, we can change the appearance of any prime factored form by factoring out 1 or -1. However, since 1 is not a prime, this is not considered to be a different prime factored form.)

Example 4   Factor: $m^2 - 6m - 9$.

Solution:

Product is -9	*Sum is* -6	
$1(-9)$	$1 + (-9) = -8$	No
$3(-3)$	$3 + (-3) = 0$	No
$(-1)9$	$-1 + 9 = 8$	No

There is no pair of integers whose product is -9 and whose sum is -6. Therefore $m^2 - 6m - 9$ is a prime polynomial.   ❏

Try Exercise 23

Factor: $m^2 + 2m - 4$.

You can use the following test to determine quickly whether or not a trinomial is factorable. The reason this test works will be explained in Sec. 6.3.

Test for Factorability

The trinomial $ax^2 + bx + c$, where a, b, and c are integers, will factor into two binomials with integer coefficients if and only if $b^2 - 4ac$ is a perfect square.

To illustrate the test for factorability, note that, for Example 1,

$$b^2 - 4ac = 5^2 - 4(1)(6) = 25 - 24 = 1$$

Since $1 = 1^2$, 1 is a perfect square, so the trinomial $x^2 + 5x + 6$ is factorable. However, in Example 4,

$$b^2 - 4ac = (-6)^2 - 4(1)(-9) = 36 + 36 = 72$$

Since 72 is *not* a perfect square, the trinomial $m^2 - 6m - 9$ is not factorable.

If the leading coefficient of the trinomial is -1, begin by factoring out -1 from the trinomial.

Example 5   Factor: $24 - 5t - t^2$.

Solution:

$$24 - 5t - t^2 = -t^2 - 5t + 24 \qquad \text{Descending order}$$
$$= -1(t^2 + 5t - 24) \qquad \text{Factor out } -1.$$
$$= -(t - 3)(t + 8) \qquad \text{Factor the trinomial.}$$

We can check this answer as follows:

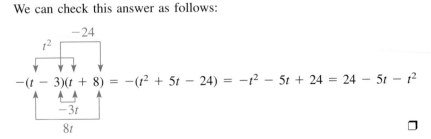

$$-(t - 3)(t + 8) = -(t^2 + 5t - 24) = -t^2 - 5t + 24 = 24 - 5t - t^2$$

❏

Try Exercise 25

Factor: $16 - 6t - t^2$.

We can use the same steps to factor a trinomial in two variables.

Example 6   Factor: $r^2 - 7rt - 8t^2$.

Solution:
$$r^2 - 7rt - 8t^2 = (r + 1t)(r - 8t)$$
$$= (r + t)(r - 8t)$$
❑

Try Exercise 29

Factor: $r^2 - 11rt - 12t^2$.

CAUTION

When factoring a trinomial in two variables, make sure you write both variables in your answer.

Correct   *Wrong*

$x^2 + 8xy + 15y^2 = (x + 3y)(x + 5y)$ ~~$x^2 + 8xy + 15y^2 = (x + 3)(x + 5)$~~   ∎

Always factor out the greatest common factor before attempting any other type of factoring.

Example 7   Factor: $12z^3 + 36z^2 - 48z$.

Solution:
$$12z^3 + 36z^2 - 48z = 12z(z^2 + 3z - 4) \qquad \text{Factor out the GCF.}$$
$$= 12z(z - 1)(z + 4) \qquad \text{Factor the trinomial.} \quad ❑$$

Try Exercise 35

Factor: $12z^3 - 24z^2 - 96z$.

CAUTION

Make sure you write the GCF as part of your final answer.

Correct   *Wrong*

$5x^2 - 15x - 50 = 5(x^2 - 3x - 10)$ ~~$5x^2 - 15x - 50 = 5(x^2 - 3x - 10)$~~
$\qquad\qquad = 5(x + 2)(x - 5)$ ~~$= (x + 2)(x - 5)$~~   ∎

Factoring $ax^2 + bx + c$

We use a similar approach to factor trinomials of the form $ax^2 + bx + c$ with $a \neq 1$.

To Factor $ax^2 + bx + c$ ($a \neq 1$) by Grouping

1. List all pairs of integers whose product is ac.
2. Find the pair of integers from step 1 whose sum is b. Suppose these integers are m and n.
3. Rewrite the trinomial, splitting the middle term as follows:
$$bx = mx + nx$$
4. Factor the resulting polynomial of four terms by grouping.
5. Check by multiplication.

We illustrate these steps in Example 8.

Example 8   Factor: $6x^2 + 7x - 5$.

Solution: First note that $ac = 6(-5) = -30$ and $b = 7$. Therefore we must find a pair of integers whose product is -30 and whose sum is 7. Since the product is negative, the integers must have opposite signs. Since the sum is positive, consider only those integers whose sum is positive.

Step 1: List all pairs of integers whose product is -30.

Step 2: Find the pair whose sum is 7.

Product is -30	*Sum is 7*	
$(-1)30$	$-1 + 30 = 29$	No
$(-2)15$	$-2 + 15 = 13$	No
$(-3)10$	$-3 + 10 = 7$	Yes

The desired integers are -3 and 10.

Step 3: Rewrite the trinomial, splitting the middle term as follows: $7x = -3x + 10x$.

$$6x^2 + 7x - 5 = 6x^2 - 3x + 10x - 5$$

Step 4: Factor by grouping:

$$= 3x(2x - 1) + 5(2x - 1)$$
$$= (2x - 1)(3x + 5)$$

Step 5: Check by multiplication.

$$\begin{array}{cccc} F & O & I & L \\ \downarrow & \downarrow & \downarrow & \downarrow \end{array}$$

Try Exercise 41

Factor: $6x^2 + 7x - 3$.

$$(2x - 1)(3x + 5) = 6x^2 + 10x - 3x - 5 = 6x^2 + 7x - 5 \qquad \square$$

Note that if we had written the middle term in Example 8 as $10x - 3x$, rather than $-3x + 10x$, the result would have been the same:

$$\begin{aligned} 6x^2 + 7x - 5 &= 6x^2 + 10x - 3x - 5 && \text{Split the middle term.} \\ &= 2x(3x + 5) - 1(3x + 5) && \text{Factor out } 2x, \text{ factor out } -1. \\ &= (3x + 5)(2x - 1) && \text{Factor out } 3x + 5. \end{aligned}$$

Only the order of the factors has been changed. Since multiplication is commutative, the two answers are equal.

Example 9 Factor: $4y^2 - 12y + 9$.

Solution: Note that $ac = 4 \cdot 9 = 36$ and $b = -12$. Therefore, find a pair of integers whose product is 36 and whose sum is -12. Since the product is positive, the integers must have the same sign. Since the sum is negative, both integers must be negative.

Product is 36	*Sum is -12*	
$(-1)(-36)$	-37	No
$(-2)(-18)$	-20	No
$(-3)(-12)$	-15	No
$(-4)(-9)$	-13	No
$(-6)(-6)$	-12	Yes

Rewrite the trinomial, splitting the middle term:

$$4y^2 - 12y + 9 = 4y^2 - 6y - 6y + 9$$

Factor by grouping:

Try Exercise 47

Factor: $4y^2 - 20y + 25$.

$$\begin{aligned} &= 2y(2y - 3) - 3(2y - 3) \\ &= (2y - 3)(2y - 3) \\ &= (2y - 3)^2 \qquad \square \end{aligned}$$

Sometimes it is faster to factor a trinomial simply by trying various combinations of binomials. Here is a shortcut that may allow you to eliminate some of the combinations immediately: **If all terms of a trinomial are positive, all terms of both binomial factors will be positive.**

Example 10   Factor: $2p^2 + 9p + 7$.

Solution: Since all terms of the trinomial are positive, all terms of both binomial factors will be positive. Therefore write
$$2p^2 + 9p + 7 = (\underline{\quad}p + \underline{\quad})(\underline{\quad}p + \underline{\quad})$$

The first term of the trinomial is $2p^2$, so the first terms of the binomial factors must be $2p$ and p:

$$2p^2 + 9p + 7 = (2p + \underline{\quad})(p + \underline{\quad})$$

The last term of the trinomial is 7, so the last terms of the binomial factors must be 1 and 7. However, there are two ways to position the 1 and the 7:

$$(2p + 1)(p + 7) \qquad (2p + 7)(p + 1)$$
$$14p \qquad\qquad 2p$$

The second factorization produces the correct middle term of $9p$. Therefore

$$2p^2 + 9p + 7 = (2p + 7)(p + 1). \qquad \square$$

Try Exercise 53

Factor: $3p^2 + 10p + 7$.

Example 11   Factor: $3m^2 - 4m - 4$.

Solution: The first term of the trinomial is $3m^2$, so the first terms of the binomial factors must be $3m$ and m:
$$3m^2 - 4m - 4 = (3m \qquad)(m \qquad)$$

There are three ways to factor -4: $1(-4)$, $2(-2)$, and $(-1)4$. Therefore there are six possible factorizations to consider:

$$(3m + 1)(m - 4) \qquad (3m - 4)(m + 1)$$
$$(3m + 2)(m - 2) \qquad (3m - 2)(m + 2)$$
$$(3m - 1)(m + 4) \qquad (3m + 4)(m - 1)$$

Only the factorization $(3m + 2)(m - 2)$ produces the correct middle term of $-4m$. Therefore

Try Exercise 55

$$3m^2 - 4m - 4 = (3m + 2)(m - 2) \qquad \square$$

Factor: $5m^2 - 12m - 9$.

Here is another shortcut that may help when factoring a trinomial: **If a trinomial has no common factor (except 1), then neither binomial factor will have a common factor (except 1).**

Example 12   Factor: $12x^2 + 17x - 40$.

Solution: There are three ways to factor 12: $1 \cdot 12$, $2 \cdot 6$, and $3 \cdot 4$. There are eight ways to factor -40: $1(-40)$, $2(-20)$, $4(-10)$, $5(-8)$, $(-1)40$, $(-2)20$, $(-4)10$, and $(-5)8$. This means there are 48 possible combinations to consider. However, since the trinomial has no common factor, neither binomial factor will have a common factor. Therefore we can immediately eliminate 40 of these combinations. For example, we can eliminate the combination $(3x + 5)(4x - 8)$ since the second binomial has a common factor of 2. This leaves only the following 8 combinations to consider:

$$(x + 40)(12x - 1) \qquad (x - 40)(12x + 1)$$
$$(x + 8)(12x - 5) \qquad (x - 8)(12x + 5)$$
$$(3x + 40)(4x - 1) \qquad (3x - 40)(4x + 1)$$
$$(3x + 8)(4x - 5) \qquad (3x - 8)(4x + 5)$$

Only the factorization $(3x + 8)(4x - 5)$ produces the correct middle term of $17x$. Therefore

$$12x^2 + 17x - 40 = (3x + 8)(4x - 5) \qquad \square$$

Try Exercise 61

Factor: $36x^2 + 13x - 40$.

Here is a summary of the steps in factoring a trinomial by trial and error.

To Factor the Trinomial $ax^2 + bx + c$ by Trial and Error

1. Write the trinomial in descending order. Factor out the GCF, including -1 if the leading coefficient is negative.
2. Write $ax^2 + bx + c = (\underline{\quad} x + \underline{\quad})(\underline{\quad} x + \underline{\quad})$.
3. Fill in the first blank in each binomial with a pair of (positive) integers whose product is a. Fill in the second blank in each binomial with a pair of integers whose product is c.
4. Try all combinations in step 3 until you obtain the correct middle term. The following shortcuts may help:
   (a) If all terms of the trinomial are positive, all terms of both binomial factors will be positive.
   (b) If the trinomial has no common factor, then neither binomial factor will have a common factor.
5. Check by multiplication.

Example 13   Factor: $9x^4y - 15x^3y^2 - 36x^2y^3$.

Solution: $9x^4y - 15x^3y^2 - 36x^2y^3 = 3x^2y(3x^2 - 5xy - 12y^2)$ Factor out the GCF.
$$= 3x^2y(3x + 4y)(x - 3y) \quad \text{Factor the trinomial.}$$
$\square$

Try Exercise 65

Factor:
$10x^4y - 46x^3y^2 - 20x^2y^3$.

Sometimes we can use the technique of factoring trinomials, along with an appropriate substitution, to factor more complicated polynomials. This is called **factoring by substitution.**

Example 14   Factor: $5z^4 + 9z^2 - 18$.

Solution: Let $u = z^2$. Then $u^2 = (z^2)^2 = z^4$. Using these substitutions the original polynomial becomes

$$5u^2 + 9u - 18$$

Factor this trinomial:

$$5u^2 + 9u - 18 = (5u - 6)(u + 3)$$

Now substitute $u = z^2$ and $u^2 = z^4$:

$$5z^4 + 9z^2 - 18 = (5z^2 - 6)(z^2 + 3)$$ ❑

Try Exercise 77

Factor: $5z^4 + 17z^2 - 12$.

Example 15   Factor: $2(m + k)^2 + 15(m + k) - 50$.

Solution: Let $u = m + k$. Then $u^2 = (m + k)^2$, and the original polynomial becomes
$$2u^2 + 15u - 50$$

Factor this trinomial:

$$2u^2 + 15u - 50 = (2u - 5)(u + 10)$$

Now substitute $u = m + k$ and $u^2 = (m + k)^2$:

$$2(m + k)^2 + 15(m + k) - 50 = [2(m + k) - 5][(m + k) + 10]$$
$$= (2m + 2k - 5)(m + k + 10)$$ ❑

Try Exercise 83

Factor:
$2(m + k)^2 + 17(m + k) - 30$.

The ability to factor trinomials quickly and accurately is acquired only through practice. There is no substitute for doing lots of problems. As you gain experience in factoring, you will find that you can do much of the work in your head. For some problems, you will soon be able to write down the correct factors immediately.

Exercises 3.7

Matching Exercises

Match each trinomial in Exercises 1 through 6 with its factored form in letters A through F.

1. $x^2 + 5x - 6$
2. $x^2 - 5x + 6$
3. $x^2 + x - 6$
4. $x^2 + 5x + 6$
5. $x^2 - x - 6$
6. $x^2 - 5x - 6$

A. $(x + 2)(x + 3)$
B. $(x - 2)(x + 3)$
C. $(x + 1)(x - 6)$
D. $(x - 2)(x - 3)$
E. $(x - 1)(x + 6)$
F. $(x + 2)(x - 3)$

Completion Exercises

7. You should write a trinomial in _____ order before attempting to factor it.

8. A polynomial has _____ (how many) prime factored form(s).

9. If the leading coefficient of a trinomial is negative, begin by factoring out _____.

10. Always factor out the _____ before attempting any other type of factoring.

True-False Exercises

11. If all terms of a trinomial are positive, all terms of both binomial factors will be positive.

12. If a trinomial has no common factor, then neither binomial factor will have a common factor.

Factor each trinomial.

13. $x^2 + 8x + 15$
14. $x^2 + 7x + 10$
15. $y^2 - 9y + 14$
16. $y^2 - 14y + 33$
17. $m^2 - 3m - 18$
18. $m^2 - 2m - 24$
19. $-7p + p^2 - 18$
20. $-5p + p^2 - 24$
21. $35 + 12k + k^2$
22. $30 + 11k + k^2$
23. $m^2 + 2m - 4$
24. $m^2 - 10m - 25$
25. $16 - 6t - t^2$
26. $12 + 4t - t^2$
27. $x^2 + 11xy + 10y^2$
28. $x^2 + 14xy + 13y^2$
29. $r^2 - 11rt - 12t^2$
30. $r^2 - 17rt - 18t^2$
31. $a^2b^2 + 4ab - 21$
32. $a^2b^2 + 2ab - 35$
33. $5k^2 + 5k + 5$
34. $7k^2 - 7k + 7$

35. $12z^3 - 24z^2 - 96z$

36. $18z^3 + 54z^2 - 180z$

37. $2x^2 + 7x + 5$

38. $2x^2 + 7x + 3$

39. $3y^2 - 5y + 2$

40. $3y^2 - 8y + 5$

41. $6x^2 + 7x - 3$

42. $6x^2 + 13x - 5$

43. $12r^2 - 17r - 5$

44. $15r^2 - 14r - 8$

45. $4t^2 + 5t - 12$

46. $6t^2 + 5t - 10$

47. $4y^2 - 20y + 25$

48. $9y^2 - 12y + 4$

49. $6 + 17x + 5x^2$

50. $12 + 11x + 2x^2$

51. $12 - 2k^2 - 5k$

52. $15 - 2k^2 - 7k$

53. $3p^2 + 10p + 7$

54. $5p^2 + 8p + 3$

55. $5m^2 - 12m - 9$

56. $7m^2 - 12m - 4$

57. $9p^2 - 34pq - 8q^2$

58. $9p^2 - 14pq - 8q^2$

59. $3a^2b^2 + 2ab - 56$

60. $2a^2b^2 + 5ab - 42$

61. $36x^2 + 13x - 40$

62. $24x^2 - 5x - 36$

63. $8r^2 + 28r - 288$

64. $12r^2 + 128r - 192$

65. $10x^4y - 46x^3y^2 - 20x^2y^3$

66. $9x^3y^2 - 39x^2y^3 - 30xy^4$

67. What polynomial has $(3x - 5)(3x + 5)$ as its prime factored form?

68. What polynomial has $(x + 2)(x^2 - 2x + 4)$ as its prime factored form?

 Discussion Exercises

69. The prime factored form of $6x^2 + 3x - 30$ is *not* $(3x - 6)(2x + 5)$. Why?

70. Discuss the steps in factoring $x^2 + bx + c$.

71. Discuss the steps in factoring $ax^2 + bx + c$ by grouping.

72. Discuss the steps in factoring $ax^2 + bx + c$ by trial and error.

73. How can you tell whether a trinomial with integer coefficients will factor into two binomials with integer coefficients without actually trying to factor the trinomial?

74. What is factoring by substitution?

Factor each polynomial.

75. $t^4 + 5t^2 + 6$

76. $t^4 + 7t^2 + 12$

77. $5z^4 + 17z^2 - 12$

78. $7z^4 + 5z^2 - 18$

79. $w^6 - 5w^3 - 14$

80. $w^6 - 2w^3 - 15$

81. $(x - y)^2 - 8(x - y) + 12$

82. $(x + y)^2 - 10(x + y) + 16$

83. $2(m + k)^2 + 17(m + k) - 30$

84. $3(m - k)^2 + 11(m - k) - 20$

Factor each polynomial. Assume m and n are positive integers.

85. $x^{2m} + 4x^m + 4$

86. $x^{2m} - 2x^m + 1$

87. $2y^{4n} - 9y^{2n} + 10$

88. $3y^{4n} - 14y^{2n} + 15$

 ## 3.8 Special Factorizations

TAPE AU8

Certain types of polynomials occur so often that we give them special attention when it comes to factoring. These polynomials are the *difference of two squares, perfect-square trinomials,* and the *sum and difference of two cubes.* We can factor each of these polynomials using a special rule. You need to memorize these rules so you can factor these polynomials without using trial and error.

Difference of Two Squares

Recall from Sec. 3.5 that

$$(a + b)(a - b) = a^2 - b^2$$

If we turn this equation around, we have the rule for factoring $a^2 - b^2$, which is called a **difference of two squares.**

Difference of Two Squares

$$a^2 - b^2 = (a + b)(a - b)$$

Example 1 Factor each polynomial: (a) $4p^2 - 25$, (b) $16z^2 - 81k^2$, (c) $r^6 - t^4$.

Solution:

$$a^2 - b^2 = (a + b)(a - b)$$

(a) $4p^2 - 25 = (2p)^2 - 5^2 = (2p + 5)(2p - 5)$
(b) $16z^2 - 81k^2 = (4z)^2 - (9k)^2 = (4z + 9k)(4z - 9k)$
(c) $r^6 - t^4 = (r^3)^2 - (t^2)^2 = (r^3 + t^2)(r^3 - t^2)$ ☐

Try Exercise 17

Factor: $9p^2 - 25$.

CAUTION

The sum of two squares $a^2 + b^2$ is a prime polynomial and does not factor.

Wrong   *Wrong*   *Wrong*
~~$x^2 + 4 = (x + 2)(x - 2)$~~   ~~$x^2 + 4 = (x + 2)(x + 2)$~~   ~~$x^2 + 4 = (x - 2)(x - 2)$~~ ∎

Example 2 Factor: $-6x^5y + 6xy$.

Solution: First factor out the GCF, $-6xy$:
$$-6x^5y + 6xy = -6xy(x^4 - 1)$$

Then factor the difference of two squares, $x^4 - 1$:
$$= -6xy(x^2 + 1)(x^2 - 1)$$

Finally, factor $x^2 - 1$. Remember $x^2 + 1$ is prime.

Try Exercise 31

Factor: $-8x^5y^2 + 8xy^2$.

$$= -6xy(x^2 + 1)(x + 1)(x - 1)$$ ☐

Example 3 Factor: $m^2 - (3n + 7)^2$.

Solution: Treat this polynomial as $a^2 - b^2$ with $a = m$ and $b = 3n + 7$.

$$a^2 - b^2 = (a + b)(a - b)$$

Try Exercise 73

Factor: $m^2 - (2n + 5)^2$.

$$m^2 - (3n + 7)^2 = [m + (3n + 7)][m - (3n + 7)]$$
$$= (m + 3n + 7)(m - 3n - 7)$$ ☐

Perfect-Square Trinomials

In Sec. 3.5 we learned the rules for squaring a binomial. If we turn these rules around, we have two more special factoring formulas.

Perfect-Square Trinomials

$$a^2 + 2ab + b^2 = (a + b)^2$$
$$a^2 - 2ab + b^2 = (a - b)^2$$

Each of the trinomials above is called a **perfect-square trinomial,** since each is the square of a binomial. Note that the first and last terms of a perfect-square trinomial are perfect squares. The middle term is twice the product of the two terms of the squared binomial.

Example 4 Factor: $x^2 + 8x + 16$.

Solution: Note that x^2 and 16 are perfect squares ($16 = 4^2$). Since the middle term is positive, try

$$(x + 4)^2$$

To check, take twice the product of the two terms of the squared binomial:

$$2(x)(4) = 8x$$

Since $8x$ is the middle term of the trinomial, the trinomial is a perfect-square trinomial and

$$x^2 + 8x + 16 = (x + 4)^2. \qquad \square$$

Try Exercise 37

Factor: $x^2 + 6x + 9$.

Example 5   Factor: $121y^2 - 110y + 25$.

Solution: Since $121y^2 = (11y)^2$ and $25 = 5^2$, the terms $121y^2$ and 25 are perfect squares. Try

$$(11y - 5)^2$$

Now check the middle term:

$$2(11y)(-5) = -110y$$

Therefore

Try Exercise 41

$$121y^2 - 110y + 25 = (11y - 5)^2 \qquad \square$$

Factor: $4y^2 - 20y + 25$.

Example 6   Factor: $(m - p)^2 + 16(m - p) + 64$.

Solution: Since $(m - p)^2$ and 64 are perfect squares, try
$$[(m - p) + 8]^2$$

Check the middle term:

$$2(m - p)(8) = 16(m - p)$$

Therefore

Try Exercise 75

$$(m - p)^2 + 16(m - p) + 64 = [(m - p) + 8]^2 \qquad \square$$

Factor:
$(m - p)^2 + 4(m - p) + 4$.

In Example 7 we use three techniques of factoring—grouping, perfect-square trinomial, and the difference of two squares—to factor the given polynomial.

Example 7   Factor: $r^2 - 2r + 1 - t^2$.

Solution:

$$
\begin{aligned}
r^2 - 2r + 1 - t^2 &= (r^2 - 2r + 1) - t^2 && \text{Group.} \\
&= (r - 1)^2 - t^2 && \text{Perfect-square trinomial} \\
&= [(r - 1) + t][(r - 1) - t] && \text{Difference of two squares} \\
&= (r + t - 1)(r - t - 1) && \text{Rearrange the terms.} \quad \square
\end{aligned}
$$

Try Exercise 79

Factor: $r^2 - 12r + 36 - t^2$.

Sum and Difference of Two Cubes

Earlier in this section we noted that we can factor the difference of two squares $a^2 - b^2$, but not the sum of two squares $a^2 + b^2$. We shall now see how to factor both a **sum** and a **difference of two cubes** using the following rules.

Sum and Difference of Two Cubes

$$a^3 + b^3 = (a + b)(a^2 - ab + b^2)$$
$$a^3 - b^3 = (a - b)(a^2 + ab + b^2)$$

It may help you to learn these rules if you remember that each factorization has exactly one negative sign.

We can verify the first rule by computing the product on the right side:

$$
\begin{array}{r}
a^2 - \; ab + b^2 \\
a + b \\
\hline
a^2b - ab^2 + b^3 \\
a^3 - a^2b + ab^2 \\
\hline
a^3 \qquad\qquad\quad + b^3
\end{array}
$$

To verify the second rule, compute the product $(a - b)(a^2 + ab + b^2)$.

Example 8   Factor: $z^3 + 27$.

Solution:

$$a^3 + b^3 = (a + b)(a^2 - ab + b^2)$$
$$\downarrow \;\; \downarrow \qquad \downarrow \;\; \downarrow\downarrow \qquad \downarrow \qquad \downarrow$$
$$z^3 + 27 = z^3 + 3^3 = (z + 3)(z^2 - z \cdot 3 + 3^2)$$
$$= (z + 3)(z^2 - 3z + 9)$$

This answer cannot be factored further. ❏

Try Exercise 53

Factor: $z^3 + 8$.

Example 9   Factor: $8p^3 - 1$.

Solution:

$$a^3 \; - b^3 = (a \; - b) \; (a^2 \; + \; ab \; + b^2)$$
$$\downarrow \qquad \downarrow \qquad \downarrow \quad \downarrow \quad \downarrow \qquad \downarrow \qquad \downarrow$$
$$8p^3 - 1 = (2p)^3 - 1^3 = (2p - 1)[(2p)^2 + (2p)1 + 1^2]$$
$$= (2p - 1)(4p^2 + 2p + 1)$$

This answer cannot be factored further. ❏

Try Exercise 57

Factor: $27p^3 - 1$.

CAUTION

The polynomials $a^2 - ab + b^2$ and $a^2 + ab + b^2$ are prime. Do not confuse these polynomials with the perfect-square trinomials $a^2 - 2ab + b^2$ and $a^2 + 2ab + b^2$. Also, do not confuse $a^3 + b^3$ with $(a + b)^3$.

Correct

$a^3 + b^3 = (a + b)(a^2 - ab + b^2)$

Wrong

$a^3 + b^3 = (a + b)(a^2 - 2ab + b^2)$
$a^3 + b^3 = (a + b)^3$ ∎

Example 10   Factor: $1000m^3 - 27n^3$.

Solution:
$$1000m^3 - 27n^3 = (10m)^3 - (3n)^3$$
$$= (10m - 3n)[(10m)^2 + (10m)(3n) + (3n)^2]$$
$$= (10m - 3n)(100m^2 + 30mn + 9n^2) \qquad ❏$$

Try Exercise 61

Factor: $343m^3 - 1000n^3$.

Example 11   Factor: $128x^6y + 250y^4$.

Solution: First factor out the GCF, $2y$:
$$128x^6y + 250y^4 = 2y(64x^6 + 125y^3)$$

Then factor the sum of two cubes in parentheses:

$$= 2y[(4x^2)^3 + (5y)^3]$$
$$= 2y(4x^2 + 5y)[(4x^2)^2 - (4x^2)(5y) + (5y)^2]$$
$$= 2y(4x^2 + 5y)(16x^4 - 20x^2y + 25y^2) \qquad ❏$$

Try Exercise 67

Factor: $16x^6y + 128y^4$.

Exercises 3.8

Matching Exercises

Match each polynomial in Exercises 1 through 8 with its factored form in letters A through H.

1. $x^2 - 16$
2. $x^2 + 8x + 16$
3. $x^2 + 10x + 16$
4. $x^3 + 64$
5. $x^2 - 6x - 16$
6. $x^2 - 8x + 16$
7. $x^2 + 16$
8. $x^3 - 64$

A. $(x + 4)^2$
B. $(x + 4)(x^2 - 4x + 16)$
C. $(x - 4)^2$
D. $(x + 2)(x - 8)$
E. Prime polynomial
F. $(x - 4)(x^2 + 4x + 16)$
G. $(x + 2)(x + 8)$
H. $(x + 4)(x - 4)$

Completion Exercises

9. The difference of two squares $a^2 - b^2$ factors into (_____)(_____).
10. The perfect-square trinomial $a^2 + 2ab + b^2$ factors into (_____)(_____).
11. The sum of two cubes $a^3 + b^3$ factors into (_____)(_____).
12. The difference of two cubes $a^3 - b^3$ factors into (_____)(_____).

True-False Exercises

13. The sum of two squares $x^2 + y^2$ is a prime polynomial.
14. The second factor in the sum-of-two-cubes formula is a perfect-square trinomial.

Factor each polynomial.

15. $x^2 - 16$
16. $x^2 - 64$
17. $9p^2 - 25$
18. $4p^2 - 81$
19. $1 - 49h^2$
20. $1 - 36h^2$
21. $4z^2 - 121k^2$
22. $9z^2 - 100k^2$
23. $16y^2 + 9m^2$
24. $4y^2 + 25m^2$
25. $100a^2b^2 - 81$
26. $16a^2b^2 - 49$
27. $r^6 - 4t^4$
28. $9r^4 - t^6$
29. $3p^4 - 432$
30. $5p^4 - 605$
31. $-8x^5y^2 + 8xy^2$
32. $-14x^6y + 14x^2y$

Getting Acquainted with Your Scientific Calculator

33. Calculate $85^2 - 15^2$ using your calculator. Then do the same calculation by first factoring the expression.
34. Calculate $75^3 + 25 \cdot 75^2$ using your calculator. Then do the same calculation by first factoring the expression.

Problem Solving

35. The area of the ring-shaped region shown in Fig. 3.5 is given by $\pi R^2 - \pi r^2$. Write this expression in prime factored form.

Figure 3.5

36. The volume of the cylindrical shell shown in Fig. 3.6 is given by $\pi R^2 h - \pi r^2 h$. Write this expression in prime factored form.

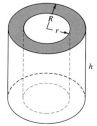

Figure 3.6

Factor each polynomial.

37. $x^2 + 6x + 9$
38. $x^2 - 4x + 4$
39. $x^2 - 2x + 1$
40. $z^2 + 2z + 1$
41. $4y^2 - 20y + 25$
42. $9y^2 + 24y + 16$
43. $64k^2 + 16km + m^2$
44. $100k^2 - 20km + m^2$
45. $14r - 49 - r^2$
46. $18r - r^2 - 81$
47. $36p^2 + 143pq + 121q^2$
48. $144p^2 - 125pq + 25q^2$
49. $12m^3 + 96m^2n + 192mn^2$
50. $8m^2n + 96mn^2 + 288n^3$
51. $16t^4 + 40t^2 + 25$
52. $64t^4 - 48t^2 + 9$

Factor each polynomial.

53. $z^3 + 8$
54. $z^3 - 64$
55. $y^3 - 125$
56. $y^3 + 1000$
57. $27p^3 - 1$
58. $125p^3 + 1$
59. $64r^3 + t^3$
60. $r^3 - 27t^3$
61. $343m^3 - 1000n^3$
62. $512m^3 + 729n^3$

63. $a^3b^3 + c^3$

64. $a^3 - b^3c^3$

65. $-4k^4 - 108k$

66. $-6k^4 - 48k$

67. $16x^6y + 128y^4$

68. $12x^6y^2 - 324y^5$

Discussion Exercises

69. How can you recognize a perfect-square trinomial?

70. Explain why $a^2 + b^2$ is a prime polynomial.

71. You could factor $x^6 - 64$ as a difference of two squares or as a difference of two cubes. Which method is better, and why?

72. What is wrong with the following "proof"?

$a = b$	a and b denote the same number.
$a^2 = ab$	Multiply both sides by a.
$a^2 - b^2 = ab - b^2$	Subtract b^2 from both sides.
$(a + b)(a - b) = b(a - b)$	Factor each side.
$a + b = b$	Divide both sides by $a - b$.
$b + b = b$	Replace a with b.
$2b = b$	Combine like terms.
$2 = 1$!?!	Divide both sides by b.

Factor each polynomial.

73. $m^2 - (2n + 5)^2$

74. $m^2 - (3n - 4)^2$

75. $(m - p)^2 + 4(m - p) + 4$

76. $(m + p)^2 - 6(m + p) + 9$

77. $(x + 4)^2 - 10(x + 4) + 25$

78. $(x - 5)^2 + 8(x - 5) + 16$

79. $r^2 - 12r + 36 - t^2$

80. $r^2 + 20r + 100 - t^2$

81. $9a^2 + 24ab + 16b^2 - c^2$

82. $4a^2 - 20ab + 25b^2 - c^2$

83. $(a + b)^3 - 27$

84. $(a - b)^3 + 8$

Factor each polynomial. Assume m, n, and k are integers.

85. $4x^{2m} - 25$

86. $9x^{2m} - 64$

87. $9p^{2n} + 24p^n + 16$

88. $16p^{2n} - 40p^n + 25$

89. $x^{3k} - 125$

90. $x^{3k} + 27$

3.9 Factoring Strategy

TAPE AU7

In this section we mix up various types of factoring problems from Secs. 3.6, 3.7, and 3.8. This will give you practice in recognizing which factoring technique to use on a given problem. Here is a summary of the strategy we will use to factor a polynomial.

Factoring Strategy

1. Factor out the greatest common factor, including -1 if the leading coefficient is negative.
2. To factor a binomial, try one of the factorizations below:

$a^2 - b^2 = (a + b)(a - b)$	Difference of two squares
$a^2 + b^2$	Prime polynomial
$a^3 + b^3 = (a + b)(a^2 - ab + b^2)$	Sum of two cubes
$a^3 - b^3 = (a - b)(a^2 + ab + b^2)$	Difference of two cubes

3. To factor a trinomial, try one of the factorizations below:

$a^2 + 2ab + b^2 = (a + b)^2$	Perfect-square trinomial
$a^2 - 2ab + b^2 = (a - b)^2$	Perfect-square trinomial
$x^2 + bx + c$	See Sec. 3.7.
$ax^2 + bx + c$	See Sec. 3.7.

4. To factor a polynomial with more than three terms, try grouping.
5. Continue to factor until you reach prime factored form.
6. Check by multiplication.

We illustrate this strategy with several examples.

Example 1   Factor: $6x^5y - 24x^4y + 12x^3y$.

Solution: Factor out the GCF, $6x^3y$:

$$6x^5y - 24x^4y + 12x^3y = 6x^3y(x^2 - 4x + 2)$$

The trinomial $x^2 - 4x + 2$ is prime, so this answer cannot be factored further.   ❑

Try Exercise 15

Factor: $4x^6y - 16x^5y - 24x^4y$.

Example 2   Factor: $32 + 4z - z^2$.

Solution: Write the trinomial in descending order and factor out -1:

$$32 + 4z - z^2 = -z^2 + 4z + 32$$
$$= -1(z^2 - 4z - 32)$$

Find a pair of integers whose product is -32 and whose sum is -4. The desired integers are 4 and -8:

$$= -(z + 4)(z - 8)$$ ❑

Try Exercise 21

Factor: $36 - 9z - z^2$.

Example 3   Factor: $a^2b + a^2 - 4b - 4$.

Solution: There is no common factor (except 1). Since the polynomial has four terms, try grouping.

$$\begin{aligned} a^2b + a^2 - 4b - 4 &= a^2(b + 1) - 4(b + 1) && \text{Factor out } a^2, \text{ factor out } -4. \\ &= (b + 1)(a^2 - 4) && \text{Factor out } b + 1. \\ &= (b + 1)(a + 2)(a - 2) && \text{Factor } a^2 - 4. \end{aligned}$$ ❑

Try Exercise 25

Factor: $a^2b + a^2 - 16b - 16$.

Example 4   Factor: $30r^2 + 35rt - 100t^2$.

Solution: Factor out the GCF, 5:

$$30r^2 + 35rt - 100t^2 = 5(6r^2 + 7rt - 20t^2)$$

Factor the trinomial using the methods of Sec. 3.7:

$$= 5(3r - 4t)(2r + 5t)$$ ❑

Try Exercise 29

Factor: $24r^2 + 2rt - 70t^2$.

Example 5   Factor: $24p^4 + 375pq^3$.

Solution: Factor out the GCF, $3p$:

$$24p^4 + 375pq^3 = 3p(8p^3 + 125q^3)$$

Factor the sum of two cubes:

$$\begin{aligned} &= 3p[(2p)^3 + (5q)^3] \\ &= 3p(2p + 5q)[(2p)^2 - (2p)(5q) + (5q)^2] \\ &= 3p(2p + 5q)(4p^2 - 10pq + 25q^2) \end{aligned}$$ ❑

Try Exercise 33

Factor: $16p^3q^2 - 1458q^5$.

Example 6   Factor: $16x^2 - 8x + 1 - y^2$.

Solution: Since there are four terms, try grouping.

$$\begin{aligned} 16x^2 - 8x + 1 - y^2 &= (16x^2 - 8x + 1) - y^2 && \text{Group.} \\ &= (4x - 1)^2 - y^2 && \text{Perfect-square trinomial} \\ &= [(4x - 1) + y][(4x - 1) - y] && \text{Difference of two squares} \\ &= (4x + y - 1)(4x - y - 1) && \text{Rearrange the terms.} \end{aligned}$$ ❑

Try Exercise 37

Factor: $25x^2 - 10x + 1 - y^2$.

Exercises 3.9

Matching Exercises

Match each polynomial in Exercises 1 through 6 with its factored form in letters A through F.

1. $a^2 + 2ab + b^2$
2. $a^2 - b^2$
3. $a^3 - b^3$
4. $a^2 - 2ab + b^2$
5. $a^2 + b^2$
6. $a^3 + b^3$

A. $(a + b)(a - b)$
B. $(a - b)^2$
C. Prime polynomial
D. $(a + b)(a^2 - ab + b^2)$
E. $(a + b)^2$
F. $(a - b)(a^2 + ab + b^2)$

True-False Exercises

7. You should factor out the greatest common factor before attempting any other type of factorization.

8. To factor a polynomial with more than three terms, you should consider factoring by grouping.

9. To factor $x^2 + bx + c$, you need to find a pair of integers whose product is c and whose sum is b.

10. To factor $ax^2 + bx + c$, you can use trial and error, or you can split the middle term and factor by grouping.

Factor each polynomial.

11. $x^2 + 3x - 40$
12. $x^2 - 7x - 30$
13. $9y^2 - 49z^2$
14. $8p^2 + 162q^2$
15. $4x^6y - 16x^5y - 24x^4y$
16. $ab - 7b + 3a - 21$
17. $k^3 + 5k^2 + k + 5$
18. $4a^2 - 36ar + 81r^2$
19. $8t^3 + 125$
20. $27t^3 - 64$

21. $36 - 9z - z^2$
22. $a^3 + 2a^2 + a$
23. $x^4y^4 - 16$
24. $m^4n^4 - mn$
25. $a^2b + a^2 - 16b - 16$
26. $4x^2 - 13x - 12$
27. $-3p^3 + 300p$
28. $-2p^3 + 288p$
29. $24r^2 + 2rt - 70t^2$
30. $5m^4 - 8m^3 + 5m - 8$
31. $4\pi R^2 - 4\pi r^2$
32. $36y^2k + 120yk^2 + 100k^3$
33. $16p^3q^2 - 1458q^5$
34. $96a^2b - 48ab - 72a + 36$
35. $m^2 - n^2 + 10m + 10n$ 36. $-6k^6 - 48$
37. $25x^2 - 10x + 1 - y^2$ 38. $4a^2 - b^2 + 8b - 16$
39. $5x^4 - 22x^2y - 15y^2$ 40. $12p^6 - 3p^3 - 9$
41. $64z^6 - 1$ 42. $z^6 - 729$
43. $(x - 3)^4 8(x + 6)^7 + (x + 6)^8 4(x - 3)^3$
44. $(x + 2)^6 3(x - 4)^2 + (x - 4)^3 6(x + 2)^5$

Factor each polynomial. Assume m, n, and k are integers.

45. $9x^{4m} - 100y^{2n}$ 46. $4r^{2k} + 12r^k + 9$
47. $p^{6n} - 1$ 48. $14z^{3m} + 25z^{2m} - 25z^m$
49. $1000 - (a + b)^3$
50. $x^2 - 8xy + 16y^2 - 9a^2 + 6ab - b^2$

 Discussion Exercises

51. When is it a good idea to factor out -1?

52. What are some of the ways to factor a binomial?

53. How can you check your factorization?

 # 3.10 Solving Equations by Factoring

TAPE AU9

In Chapter 2 we solved linear equations like $2x + 7 = 0$ by isolating the variable on one side of the equation. In this section we use factoring to solve equations like

$$x^2 + 2x - 15 = 0 \qquad 3y^2 + 11y = 4 \qquad \text{and} \qquad 5z^2 = 10z$$

The three equations above are called *quadratic equations,* or *second-degree* equations.

DEFINITION OF A QUADRATIC EQUATION

A **quadratic equation** in the variable x is an equation that can be written in the form

$$ax^2 + bx + c = 0$$

where a, b, and c are real numbers and $a \neq 0$. This form is called the **standard form** of a quadratic equation.

We can solve many quadratic equations by writing the equation in standard form, factoring the left side, and applying the **zero factor property.**

Zero Factor Property

Suppose p and q are real numbers.

$$\text{If } p \cdot q = 0, \text{ then } p = 0 \text{ or } q = 0$$

The zero factor property states that if the product of two numbers is 0, then one or both of the numbers must be 0. We now show how to use the zero factor property to solve a quadratic equation.

Example 1   Solve: $x^2 + 2x - 15 = 0$.

Solution: The equation is in standard form with $a = 1$, $b = 2$, and $c = -15$. Factor the left side:

$$x^2 + 2x - 15 = 0$$
$$(x - 3)(x + 5) = 0$$

According to the zero factor property, if the product of $x - 3$ and $x + 5$ is 0, then

$$x - 3 = 0 \quad \text{or} \quad x + 5 = 0$$

Solve each of these linear equations and get

$$x = 3 \quad \text{or} \quad x = -5$$

Therefore the solution set is $\{3, -5\}$. You can check each solution by substituting it into the original equation.   ❏

Try Exercise 17

Solve: $x^2 + 3x - 10 = 0$.

We can summarize the steps in solving a quadratic equation by factoring as follows.

To Solve a Quadratic Equation by Factoring

1. Write the equation in the standard form $ax^2 + bx + c = 0$. If a is negative, multiply both sides by -1.
2. Factor the left side.
3. Set each factor equal to 0, and solve the resulting linear equations.
4. Check each solution in the original equation.

We illustrate these steps in Example 2.

Example 2   Solve: $3y^2 + 11y = 4$.

Solution: Step 1:   Write the equation in standard form.

$$3y^2 + 11y = 4 \qquad \text{Original equation}$$
$$3y^2 + 11y - 4 = 0 \qquad \text{Subtract 4 from both sides.}$$

Step 2:   Factor the left side.

$$(3y - 1)(y + 4) = 0$$

Step 3:   Set each factor equal to 0, and solve the resulting linear equations.

$$3y - 1 = 0 \qquad \text{or} \qquad y + 4 = 0$$
$$3y = 1 \qquad\qquad\qquad y = -4$$
$$y = \tfrac{1}{3}$$

Step 4:   Check each solution in the original equation.

Check: $y = \tfrac{1}{3}$ *Check:* $y = -4$
$$3y^2 + 11y = 4 \qquad\qquad\qquad 3y^2 + 11y = 4$$
$$3(\tfrac{1}{3})^2 + 11(\tfrac{1}{3}) = 4 \qquad\qquad 3(-4)^2 + 11(-4) = 4$$
$$3(\tfrac{1}{9}) + \tfrac{11}{3} = 4 \qquad\qquad\qquad 3(16) - 44 = 4$$
$$\tfrac{1}{3} + \tfrac{11}{3} = 4 \qquad\qquad\qquad\qquad 48 - 44 = 4$$
$$\tfrac{12}{3} = 4 \quad \text{True} \qquad\qquad\qquad\qquad 4 = 4 \quad \text{True}$$

Try Exercise 21

Solve: $3y^2 + 23y = 8$.

Both solutions check, so the solution set is $\{\tfrac{1}{3}, -4\}$. ❏

CAUTION

We can apply the zero factor property only when one side of the equation is 0.

Wrong   *Wrong*
~~$(x + 1)(x + 7) = 6$~~   ~~$(x + 1)(x + 7) = 6$~~
~~$x + 1 = 6$ or $x + 7 = 6$~~   ~~$x + 1 = 2$ or $x + 7 = 3$~~   ∎

Example 3   Solve: $5z^2 = 10z$.

Solution:
$$5z^2 = 10z \qquad \text{Original equation}$$
$$5z^2 - 10z = 0 \qquad \text{Write in standard form.}$$
$$5z(z - 2) = 0 \qquad \text{Factor the left side.}$$
$$5z = 0 \qquad \text{or} \qquad z - 2 = 0 \qquad \text{Set each factor equal to 0.}$$
$$z = 0 \qquad\qquad\qquad z = 2 \qquad \text{Solve each linear equation.}$$

Try Exercise 25

Solve: $4z^2 = 20z$.

The solution set is $\{0, 2\}$. Check in the original equation. ❏

CAUTION

Do not divide both sides of the equation in Example 3 by $5z$. In doing so you lose the solution $z = 0$. The danger in dividing both sides by a variable expression, such as $5z$, is that you may be dividing by 0. Remember, division by zero is undefined.

Correct

$$5z^2 = 10z$$
$$5z^2 - 10z = 0$$
$$5z(z - 2) = 0$$
$$z = 0 \quad \text{or} \quad z = 2$$

Wrong

$$\require{cancel}$$

$$\cancel{5}z^2 = 10\cancel{z}$$

$$\frac{5z^2}{5z} \diagdown \frac{10z}{5z}$$

$$z = 2$$ ■

Example 4   Solve: $-r^2 - 20r - 100 = 0$.

Solution: Since the leading coefficient is negative, multiply both sides by -1:

$$-1(-r^2 - 20r - 100) = -1(0) \qquad \text{Multiply by } -1.$$
$$r^2 + 20r + 100 = 0 \qquad \text{Distribute } -1 \text{ over every term.}$$

Factor, set each factor equal to 0, and solve:

$$(r + 10)(r + 10) = 0$$

$$r + 10 = 0 \qquad \text{or} \qquad r + 10 = 0$$
$$r = -10 \qquad\qquad\qquad r = -10$$

Try Exercise 29

Solve: $-r^2 - 24r - 144 = 0$.

The solution set is $\{-10\}$. Check in the original equation.   ❐

Note that the equation of Example 4 had only one solution. On the other hand, the equation $x^2 = -9$ has *no* real solution, since no real number squared is -9. **A quadratic equation will have 0, 1, or 2 real solutions.**

Example 5   Solve: $(5p - 6)(p + 1) = p(p - 1) + 75$.

Solution: Write the equation in standard form:

$$(5p - 6)(p + 1) = p(p - 1) + 75 \qquad \text{Original equation}$$
$$5p^2 - p - 6 = p^2 - p + 75 \qquad \text{Multiply out each side.}$$
$$4p^2 - 81 = 0 \qquad \text{Collect terms on the left side.}$$

Factor, set each factor equal to 0, and solve:

$$(2p + 9)(2p - 9) = 0$$
$$2p + 9 = 0 \qquad \text{or} \qquad 2p - 9 = 0$$
$$2p = -9 \qquad\qquad\qquad 2p = 9$$
$$p = -\tfrac{9}{2} \qquad\qquad\qquad p = \tfrac{9}{2}$$

Try Exercise 41

Solve:
$(5p - 3)(4p + 1) = p(4p - 7) + 46$.

The solution set is $\{-\tfrac{9}{2}, \tfrac{9}{2}\}$. Check in the original equation.   ❐

Higher-Degree Equations

We can use the method of factoring to solve certain types of higher-degree equations.

Example 6   Solve: $x^3 + 7x^2 = 8x$.

Solution: This is a third-degree equation.

$$x^3 + 7x^2 = 8x \qquad \text{Original equation}$$
$$x^3 + 7x^2 - 8x = 0 \qquad \text{Subtract } 8x.$$
$$x(x^2 + 7x - 8) = 0 \qquad \text{Factor out } x.$$
$$x(x - 1)(x + 8) = 0 \qquad \text{Factor the trinomial.}$$

The product on the left side will be 0 if any of the factors is 0. Therefore set each factor equal to 0 and solve:

$$x = 0 \quad \text{or} \quad x - 1 = 0 \quad \text{or} \quad x + 8 = 0$$
$$x = 1 \qquad\qquad x = -8$$

Try Exercise 45

Solve: $x^3 + 6x^2 = 27x$.

The solution set is $\{0, 1, -8\}$. Check in the original equation. ❒

Sometimes we cannot solve a quadratic equation (or a higher-degree equation) by factoring because we cannot factor the left side. In Chap. 6 you will learn methods for solving *any* quadratic equation, including those that cannot be solved by factoring.*

Problem Solving with Quadratic Equations

We can use quadratic equations to solve certain types of word problems. We solve the word problems in this section using the same four steps we used in Sec. 2.3. The only difference in this section is that the equation we write in step 2 is a quadratic equation.

Example 7   The base of a triangle is 3 m less than twice the height. If the area of the triangle is 27 m², find the height and the base.

Solution: Write down the unknown quantities and represent one of them by x. Then write the other unknown in terms of x:

$$x = \text{height of triangle}$$
$$2x - 3 = \text{base of triangle}$$

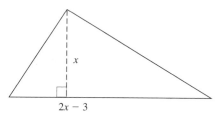

Figure 3.7

Write an equation involving x (see Fig. 3.7):

$$A = \tfrac{1}{2}bh \qquad\qquad \text{Area of a triangle}$$
$$27 = \tfrac{1}{2}(2x - 3)x \qquad \text{Substitute for } A, b, \text{ and } h.$$

To clear the fraction, multiply both sides by the LCD 2:

$$2 \cdot 27 = 2 \cdot \tfrac{1}{2}(2x - 3)x \qquad \text{Multiply by 2.}$$
$$54 = (2x - 3)x \qquad\qquad \text{Simplify each side.}$$

In this case it is easier to collect terms on the right side:

$$54 = 2x^2 - 3x \qquad \text{Multiply out the right side.}$$
$$0 = 2x^2 - 3x - 54 \qquad \text{Subtract 54.}$$
$$0 = (2x + 9)(x - 6) \qquad \text{Factor.}$$

---

*HISTORICAL NOTE: A method for solving any quadratic equation was known by the Babylonians, perhaps as early as 2000 B.C. However, it was not until the 16th century that algebraic methods for solving general third-degree and fourth-degree equations were discovered. Later it was proven that no such algebraic methods could ever be devised for solving fifth-degree or higher-degree equations.

Set each factor equal to 0 and solve:

$$2x + 9 = 0 \qquad \text{or} \qquad x - 6 = 0$$
$$2x = -9 \qquad\qquad\qquad x = 6$$
$$x = -\tfrac{9}{2}$$

We ignore the solution $x = -\tfrac{9}{2}$ since the height of the triangle cannot be a negative number. Therefore the height of the triangle is 6 m, and the base is $2x - 3 = 2(6) - 3 = 9$ m.   ❐

Try Exercise 57

The base of a triangle is 1 m less than 3 times the height. If the area of the triangle is 7 m², find the height and the base.

Example 8   A ball is thrown upward from the top of a building 48 ft high with an initial velocity of 32 ft/sec (see Fig. 3.8). If we ignore air resistance, the height h in feet, above ground level of the ball after t seconds is given by the formula $h = -16t^2 + 32t + 48$.* How long does it take the ball to strike the ground?

Solution: Since the height of the ball when it strikes the ground is 0 feet, substitute $h = 0$ in the formula:

$$h = -16t^2 + 32t + 48 \qquad \text{Original formula}$$
$$0 = -16t^2 + 32t + 48 \qquad \text{Replace } h \text{ with } 0.$$

Write the equation in standard form and solve by factoring:

$$16t^2 - 32t - 48 = 0 \qquad \text{Collect terms on the left side.}$$
$$16(t^2 - 2t - 3) = 0 \qquad \text{Factor out 16.}$$
$$16(t + 1)(t - 3) = 0 \qquad \text{Factor the trinomial.}$$

The factor 16 cannot equal 0. Therefore set the other two factors equal to 0 and solve for t:

$$t + 1 = 0 \qquad \text{or} \qquad t - 3 = 0$$
$$t = -1 \qquad\qquad\qquad t = 3$$

The time cannot be -1 sec. Therefore it takes 3 sec for the ball to strike the ground.   ❐

Try Exercise 63

A slingshot fires a stone vertically upward with an initial velocity of 112 ft/sec. Ignoring air resistance, the height h, in feet, of the stone after t seconds is given by the formula $h = -16t^2 + 112t$. How long will it take the stone to hit the ground?

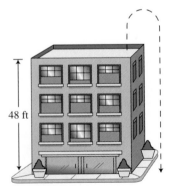

Figure 3.8

48 ft

---

*HISTORICAL NOTE: The formula $h = -16t^2 + 32t + 48$ is due largely to the work of the Italian astronomer and physicist Galileo Galilei (1564–1643) and the English mathematician and physicist Isaac Newton (1642–1727). Newton, along with Archimedes (ca. 287 B.C.–212 B.C.) and Carl Friedrich Gauss (1777–1855), is considered to be one of the three greatest mathematicians who ever lived. In deference to the mathematicians who preceded him, Newton once remarked, "If I have seen farther than others, it is because I have stood on the shoulders of giants."

Exercises 3.10

Completion Exercises

1. An equation of the form $ax^2 + bx + c = 0$ $(a \neq 0)$ is called a(n) _____ equation.

2. The form $ax^2 + bx + c = 0$ is called the _____ form of a quadratic equation.

3. The zero factor property states that if $p \cdot q = 0$, then _____ or _____.

4. A quadratic equation will have _____ (how many) solutions.

Matching Exercises

Match each quadratic equation in Exercises 5 through 8 with its solution set in letters A through D.

5. $(x - 4)(x - 1) = 0$ A. $\{0, 4\}$

6. $y^2 = 16$ B. $\{4\}$

7. $p(p - 4) = 0$ C. $\{4, -4\}$

8. $(m - 4)^2 = 0$ D. $\{4, 1\}$

True-False Exercises

9. If $p \cdot q = 6$, then $p = 2$ or $q = 3$.

10. If $(x + 2)(x + 3) = 1$, then $x + 2 = 1$ or $x + 3 = 1$.

Discussion Exercises

11. Why do we specify that $a \neq 0$ in the quadratic equation $ax^2 + bx + c = 0$?

12. What is the danger in dividing both sides of an equation by a variable expression?

13. Suppose that to solve $5(x - 2)(x + 7) = 0$, you write

    $$5 = 0 \quad \text{or} \quad x - 2 = 0 \quad \text{or} \quad x + 7 = 0$$

    How do you handle the equation $5 = 0$, and what are the solutions of the original equation?

14. Outline the steps in solving a quadratic equation by factoring.

Solve each equation.

15. $x^2 - 5x + 4 = 0$

16. $x^2 - 7x + 6 = 0$

17. $x^2 + 3x - 10 = 0$

18. $p^2 - 2p - 24 = 0$

19. $m^2 + 15m + 36 = 0$

20. $2z^2 - 15z + 7 = 0$

21. $3y^2 + 23y = 8$

22. $5k^2 + 6k + 1 = 0$

23. $12r^2 - 7r - 10 = 0$

24. $x^2 = 18x - 81$

25. $4z^2 = 20z$

26. $y^2 = 121$

27. $0 = 9t^2 - 16$

28. $-6m^2 - 3m = 0$

29. $-r^2 - 24r - 144 = 0$

30. $5x^2 - 65x + 150 = 0$

31. $28 - 22y - 18y^2 = 0$

32. $121m^2 + 25 = 110m$

33. $10p^2 + 19p = 19p$

34. $z^2 - \frac{1}{4} = 0$

35. $\frac{x^2}{2} - \frac{3}{4}x = 0$

36. $2y^2 - \frac{y}{6} - \frac{5}{4} = 0$

37. $(r - 9)(r + 11) = 21$

38. $z(9z + 7) = z - 1$

39. $(2k + 1)^2 + 3k^2 = 6k + 1$

40. $(3k + 1)^2 - k^2 = 9k + 1$

41. $(5p - 3)(4p + 1) = p(4p - 7) + 46$

42. $(7p - 6)(5p + 4) = 2p(5p - 1) + 57$

Solve each higher-order equation.

43. $(y + 1)(4y^2 - 7y - 36) = 0$

44. $(y - 1)(6y^2 - 5y - 14) = 0$

45. $x^3 + 6x^2 = 27x$

46. $x^3 + 8x^2 = 20x$

47. $2z^4 = 35z^2 - 3z^3$

48. $3z^4 = 28z^2 - 5z^3$

49. $p^4 - 5p^2 + 4 = 0$

50. $p^4 - 10p^2 + 9 = 0$

51. $x^3 + 2x^2 - x - 2 = 0$

52. $x^3 - 5x^2 - 4x + 20 = 0$

Problem Solving

53. If 4 times a number is added to the square of the number, the result is 3 more than twice the number. Find the number.

54. The sum of the squares of two consecutive integers is 112 less than the square of their sum. Find the integers.

55. The square of the sum of two consecutive integers is 180 more than the sum of their squares. Find the integers.

56. Determine three consecutive odd integers such that the square of the smallest added to the product of the other two is 268.

57. The base of a triangle is 1 m less than 3 times the height. If the area of the triangle is 7 m², find the height and the base.

58. The height of a triangle is 5 m less than twice the base. If the area of the triangle is 9 m², find the base and the height.

59. Given the revenue equation $R = 125x - \frac{1}{4}x^2$ and the cost equation $C = 15x$, determine the breakeven points (*Hint:* R = revenue, C = cost, x = no. of units, and breakeven occurs when revenue equals cost.)

60. Given the revenue equation $R = 175x - \frac{1}{4}x^2$ and the cost equation $C = 20x$, determine the breakeven points. (*Hint:* R = revenue, C = cost, x = no. of units, and breakeven occurs when revenue equals cost.)

61. If we ignore air resistance, the distance d, in feet, that an object falls in t seconds is given by the formula $d = 16t^2$. A bolt falls from a bridge 196 ft high. How long does it take to hit the water below?

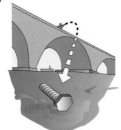

62. An engagement ring is thrown downward from a window 128 ft high with a velocity of 32 ft/sec. If we ignore air resistance, the distance d, in feet, that the ring travels in t

seconds is given by the formula $d = 16t^2 + 32t$. How long does it take the ring to hit the street below?

63. A slingshot fires a stone vertically upward with an initial velocity of 112 ft/sec. Ignoring air resistance, the height h, in feet, of the stone after t seconds is given by the formula $h = -16t^2 + 112t$. How long will it take the stone to hit the ground?

64. An arrow is shot upward from a cliff overlooking a beach. The initial velocity of the arrow is 112 ft/sec and the height of the cliff is 60 ft. Ignoring air resistance, the height h, in feet, of the arrow above the beach after t seconds is given by the formula $h = -16t^2 + 112t + 60$. When will the height of the arrow be 220 ft?

Getting Acquainted with Your Graphing Calculator

Check that the given number is a solution of the given equation by storing the given number in x and then testing the equation to determine whether it is true or false.

65. $6x^2 + 19x + 15 = 0; x = -1.5$

66. $8x^3 + 7x = 15(2x^2 - 3); x = 2.25$

Chapter 3 Key Terms

3.1 **Exponent** In a^n, the exponent is n.
Base In a^n, the base is a.

3.3 **Scientific notation** A positive real number is in scientific notation when it is written in the form $a \times 10^n$, where $1 \le a < 10$ and n is an integer.

3.4 **Monomial** (in x) An expression of the form ax^n, where a is a real number and n is a whole number
Coefficient (of a monomial) The coefficient of ax^n is a.
Degree (of a monomial) If $a \ne 0$, the degree of ax^n is n.
Polynomial A monomial or a finite sum of monomials
Degree (of a polynomial) Equals the degree of the highest-degree monomial in the polynomial
Leading coefficient (of a polynomial) The coefficient of the highest-degree monomial in the polynomial
Constant term (of a polynomial) The term of the polynomial that does not contain a variable
Terms Expressions that are related by addition
Monomial A polynomial with one term
Binomial A polynomial with two terms
Trinomial A polynomial with three terms
Descending order The order of a polynomial whose terms decrease in degree from left to right

Degree (of a monomial in several variables) Equals the sum of the exponents on all of the variables
Degree (of a polynomial in several variables) Equals the degree of its highest-degree monomial
Like terms Terms that have the same variables with the same exponents

3.5 **FOIL** (First, Outer, Inner, Last) A method for multiplying two binomials
Conjugates The binomials $a + b$ and $a - b$ are conjugates of each other.

3.6 **Factoring** Writing a number (or a polynomial) as a product of simpler numbers (polynomials)
Factors (divisors) Since $(x + 2)(x + 3) = x^2 + 5x + 6$, the binomials $x + 2$ and $x + 3$ are factors, or divisors, of $x^2 + 5x + 6$.
Prime number An integer greater than 1 whose only positive factors are 1 and itself
Prime factored form The factored form of a number (or a polynomial) that contains only prime numbers (or prime polynomials)
Greatest common factor (GCF) The GCF of a set of integers is the largest integer that is a factor of every integer in the set.

Prime polynomial A polynomial that cannot be written as a product of simpler polynomials with integer coefficients.

3.10 **Quadratic equation** (in x) An equation that can be written in the form $ax^2 + bx + c = 0$ $(a \neq 0)$
Standard form (of a quadratic equation) The form $ax^2 + bx + c = 0$

Chapter 3 Key Rules/Steps

3.1 Integer Exponents

Let a and b be real numbers and m and n be integers. Assume no denominator is 0.

Product Rule for Exponents

$a^m \cdot a^n = a^{m+n}$ Example: $x^3 \cdot x^5 = x^8$

Zero Exponent

If $a \neq 0$, then $a^0 = 1$.   Example: $(-8)^0 = 1$

Negative Exponents

$a^{-n} = \dfrac{1}{a^n}$ $\dfrac{1}{a^{-n}} = a^n$ $\left(\dfrac{a}{b}\right)^{-n} = \left(\dfrac{b}{a}\right)^{n}$

Examples:

$4^{-2} = \dfrac{1}{4^2}$ $\dfrac{1}{p^{-6}} = p^6$ $\left(\dfrac{2}{5}\right)^{-3} = \left(\dfrac{5}{2}\right)^{3}$

Quotient Rule for Exponents

$\dfrac{a^m}{a^n} = a^{m-n}$ Example: $\dfrac{m^{-3}}{m^{-5}} = m^{-3-(-5)} = m^2$

3.2 Rules for Exponents

Let a and b be real numbers and m and n be integers. Assume no denominator is 0.

Power-to-a-Power Rule

$(a^m)^n = a^{mn}$ Example: $(2^3)^7 = 2^{21}$

Product-to-a-Power Rule

$(ab)^n = a^n b^n$ Example: $(4y)^2 = 4^2 y^2$

Quotient-to-a-Power Rule

$\left(\dfrac{a}{b}\right)^n = \dfrac{a^n}{b^n}$ Example: $\left(\dfrac{t}{10}\right)^3 = \dfrac{t^3}{10^3}$

3.3 Scientific Notation: An Application of Exponents

$3{,}200{,}000 = 3.2 \times 10^6$

$0.000707 = 7.07 \times 10^{-4}$

$23.45 \times 10^8 = 2.345 \times 10^9$

3.4 Adding and Subtracting Polynomials

Evaluate $P(x) = 3x^2 - 5x - 9$ at $x = -2$

$P(-2) = 3(-2)^2 - 5(-2) - 9 = 3(4) + 10 - 9 = 13$

Add: $(x^2 - 9) + (4x^2 - 6x + 7) = 5x^2 - 6x - 2$

Subtract: $(5x^3 - 2x^2 + 9) - (x^3 + 10x^2 - x + 1)$
$= 5x^3 - 2x^2 + 9 - x^3 - 10x^2 + x - 1$
$= 4x^3 - 12x^2 + x + 8$

3.5 Multiplying Polynomials

FOIL

first	outer	inner	last
↓	↓	↓	↓

$(2x + 3y)(5x - 4y) = 10x^2 - 8xy + 15xy - 12y^2$
$= 10x^2 + 7xy - 12y^2$

Product of Two Conjugates

$(a + b)(a - b) = a^2 - b^2$
Example: $(p + 8)(p - 8) = p^2 - 64$

Square of a Binomial

$(a + b)^2 = a^2 + 2ab + b^2$
$(a - b)^2 = a^2 - 2ab + b^2$
Examples: $(y + 5)^2 = y^2 + 10y + 25$
$(3x - 1)^2 = 9x^2 - 6x + 1$

Multiplying Polynomials

Multiply each term of one polynomial by each term of the other polynomial.

$(3x - 5)(2x^2 - x + 4)$
$= 6x^3 - 3x^2 + 12x - 10x^2 + 5x - 20$
$= 6x^3 - 13x^2 + 17x - 20$

3.6 The Greatest Common Factor; Factoring by Grouping

To Find the Greatest Common Factor of $72x^3y^3$ and $90xy^4$

1. Write in prime factored form.

$72x^3y^3 = 2^3 \cdot 3^2 \cdot x^3 \cdot y^3$
$90xy^4 = 2 \cdot 3^2 \cdot 5 \cdot x \cdot y^4$

2 & 3. Multiply together the lowest powers of the prime factors.

$\text{GCF} = 2 \cdot 3^2 \cdot x \cdot y^3 = 18xy^3$

Factoring Out the GCF

$72x^3y^3 - 90xy^4 = 18xy^3(4x^2 - 5y)$

Factoring by Grouping

$$m^3 - 2m^2 + 3m - 6 = m^2(m - 2) + 3(m - 2)$$
$$= (m - 2)(m^2 + 3)$$

3.7 Factoring Trinomials

Factoring $x^2 - 8x + 15$

1. List all pairs of integers whose product is $c = 15$.

$$1 \cdot 15 \quad 3 \cdot 5 \quad (-1)(-15) \quad (-3)(-5)$$

2 & 3. Find the pair whose sum is $b = -8$. Then use these numbers to factor the trinomial.

Since $-3 + (-5) = -8$, write
$x^2 - 8x + 15 = (x - 3)(x - 5)$.

4. Check by multiplication.

Test for Factorability

The trinomial $ax^2 + bx + c$, where a, b, and c are integers, will factor into two binomials with integer coefficients if and only if $b^2 - 4ac$ is a perfect square.

$x^2 - x - 12$ is factorable,

since $b^2 - 4ac = (-1)^2 - 4(1)(-12)$
$= 49$ (a perfect square)

$3x^2 + 5x + 1$ is not factorable,

since $b^2 - 4ac = 5^2 - 4(3)(1)$
$= 13$ (not a perfect square)

Factoring $3x^2 - x - 10$ by Grouping

1. List all pairs of integers whose product is $ac = -30$.

$$1(-30) \quad 2(-15) \quad 3(-10) \quad 5(-6)$$
$$(-1)30 \quad (-2)15 \quad (-3)10 \quad (-5)6$$

2 & 3. Find the pair whose sum is $b = -1$. Then use these numbers to split the middle term.

Since $5 + (-6) = -1$, write
$3x^2 - x - 10 = 3x^2 + 5x - 6x - 10$

4. Factor by grouping.

$$3x^2 - x - 10 = 3x^2 + 5x - 6x - 10$$
$$= x(3x + 5) - 2(3x + 5)$$
$$= (3x + 5)(x - 2)$$

5. Check by multiplication.

To Factor $ax^2 + bx + c$ by Trial and Error

1. Write in descending order. Factor out the GCF, including -1 if the leading coefficient is negative.

2. Write $ax^2 + bx + c = (\underline{\hspace{0.5em}} x + \underline{\hspace{0.5em}})(\underline{\hspace{0.5em}} x + \underline{\hspace{0.5em}})$.

3. Fill in the first blank in each binomial with a pair of (positive) integers whose product is a. Fill in the second blank in each binomial with a pair of integers whose product is c.

4. Try all combinations in Step 3 until you obtain the correct middle term. The following shortcuts may help:

(a) If all terms of the trinomial are positive, all terms of both binomial factors will be positive.

(b) If the trinomial has no common factor, then neither binomial factor will have a common factor.

5. Check by multiplication.

3.8 Special Factorizations

Difference of Two Squares

$a^2 - b^2 = (a + b)(a - b)$
$a^2 + b^2$ is a prime polynomial.
Examples: $9x^2 - 25 = (3x + 5)(3x - 5)$
             $p^2 + 16$ is a prime polynomial.

Perfect-Square Trinomials

$a^2 + 2ab + b^2 = (a + b)^2$
$a^2 - 2ab + b^2 = (a - b)^2$
Examples: $m^2 + 20m + 100 = (m + 10)^2$
             $4x^2 - 12xy + 9y^2 = (2x - 3y)^2$

Sum and Difference of Two Cubes

$a^3 + b^3 = (a + b)(a^2 - ab + b^2)$
$a^3 - b^3 = (a - b)(a^2 + ab + b^2)$
Examples: $p^3 + 64 = (p + 4)(p^2 - 4p + 16)$
             $8z^3 - 1 = (2z - 1)(4z^2 + 2z + 1)$

3.9 Factoring Strategy

1. Factor out the GCF, including -1 if the leading coefficient is negative.

2. To factor a binomial, see if it is a difference of two squares, a sum of two cubes, or a difference of two cubes.

3. To factor a trinomial, see if it is a perfect-square trinomial. If not, see Sec. 3.7.

4. To factor a polynomial with more than three terms, try grouping.

5. Continue to factor until you reach prime factored form.

6. Check by multiplication.

3.10 Solving Equations by Factoring

Zero Factor Property

Suppose p and q are real numbers.

If $p \cdot q = 0$, then $p = 0$ or $q = 0$.

Example: If $x(x - 2) = 0$, then $x = 0$ or $x - 2 = 0$.

To Solve $5x^2 + 18x = 8$ by Factoring

1. Write in standard form.

$$5x^2 + 18x - 8 = 0$$

2. Factor the left side.

$$(5x - 2)(x + 4) = 0$$

3. Set each factor equal to 0, and solve the resulting equations.

$$5x - 2 = 0 \quad \text{or} \quad x + 4 = 0$$
$$x = \tfrac{2}{5} \qquad\qquad x = -4$$

4. Check each solution in the original equation.

Chapter 3 Review Exercises

[3.1] Simplify each expression. Use only positive exponents in your answer. Assume n is an integer and no variable is 0.

1. 8^0

2. $-4x^0$

3. 6^{-2}

4. $5y^{-1}$

5. $\left(-\dfrac{3}{4}\right)^{-3}$

6. $\dfrac{2}{7p^{-5}}$

7. $x \cdot x^3 \cdot x^5$

8. $3^{-18} \cdot 3^4$

9. $(-4a^9b^2)(7ab^6)$

10. $-p(2p^3k^{-10})(-5p^{-1}k)$

11. $\dfrac{t^{10}}{t^2}$

12. $\dfrac{11^{-4}}{11^{-6}}$

13. $\dfrac{-48m^{-8}h}{8m^3h^{-1}}$

14. $\dfrac{y^{2n}}{y^{n+1}}$

[3.2] Simplify. Use only positive exponents in your answer. Assume m is an integer and no variable is 0.

15. $(2p)^5$

16. $(-5a^{-3}b^4)^2$

17. $\left(\dfrac{x^7}{y^3}\right)^4$

18. $\left(\dfrac{-z}{k^{-1}}\right)^{-3}$

19. $m^2(4m^{-7})^{-2}(-3m)$

20. $\dfrac{2^{-3}(4^{-1}t^3t)^{-2}}{t^3(t^{-5}t^4)^3}$

21. $\left(\dfrac{3^{-9}x^{-6}}{x^2y^{-7}}\right)^2\left(\dfrac{xy^{-5}}{3^4x^4y^{-1}}\right)^{-3}$

22. $(x^{-10m}x^m)^{-1}$

[3.3] Write each number in scientific notation.

23. 34,000,000

24. 0.007

Write each number in standard form.

25. 9.02×10^5

26. 8.911×10^{-4}

Write each exercise in scientific notation and then perform the indicated operations. Write your answer in scientific notation.

27. $\dfrac{(6,000,000,000)(32,000)}{500,000}$

28. $\dfrac{(0.000005)(0.0000066)}{20,000}$

29. A *light year* is the distance light travels in a year. Since light travels at 1.86×10^5 mi/sec, and there are approximately 3.15×10^7 sec in a year, how many miles are in a light year?

[3.4] Write each polynomial in descending order. State the degree, the leading coefficient, and the constant term of the polynomial. Identify any monomials, binomials, or trinomials.

30. $3 + 4x - 6x^2$

31. $-x^3yz^2$

32. $-7 + p$

Evaluate each polynomial at the given values.

33. $Q(y) = -y^2 + y - 9$; $Q(3)$, $Q(0)$

34. $R(z) = 4z^3 + 6z^2 - 3z + 1$; $R(\tfrac{1}{2})$, $R(-k)$

35. Add.
$$\begin{array}{r} 4x^2 - 6x + 9 \\ 2x^2 + 3x - 5 \end{array}$$

36. Subtract.
$$\begin{array}{r} 6y^3 - 2y^2 \quad\;\; + 5 \\ 3y^3 + 7y^2 - y - 5 \end{array}$$

Simplify.

37. $(7m^2 - m + 9) + (m^2 + m - 6)$

38. $5x^2 - [10x^2 - (x^2 + 4xy - 7y^2) - xy]$

39. Subtract the largest of three consecutive odd integers from the sum of the other two integers, given that the middle integer is n.

[3.5] Find each product and simplify.

40. $3y^2(4y^2 - 6y + 7)$

41. $(4z + 9)(2z - 7)$

42. $(2x - 5y)^2$

43. $(3a - 8b)(3a + 8b)$

44. $(6t^2 - 5)(t^2 + 4)$

45. $(4r - 5)(2r^3 + r^2 - 3r + 6)$

46. Multiply.
$$\begin{array}{r} 2x^3 - 8x^2 + 1 \\ 3x - 5 \end{array}$$

Simplify. Assume k is an integer and $x \neq 0$.

47. $t[t^2 - (t + 4)(t - 4)]$

48. $-10x^k(x^{k+1} - 2x + x^{-k})$

[3.6] Write each number in prime factored form.

49. 75

50. 5544

Find the greatest common factor.

51. 420, 280, 252

52. $12x^{13}y^2$, $18x^9y^{11}$

Factor out the greatest common factor, including -1 if the leading coefficient is negative. Assume n is a positive integer.

53. $18p^5 - 36p^3 + 27p^2$

54. $-8x^7y^2z^3 + 20x^5y^3z^2$

55. $3x(5a + b) - y(5a + b)$

56. $x^{5n} - 3x^{3n} + x^{2n}$

Factor by grouping.

57. $rs + r + 2s + 2$

58. $x^3 - 7x^2 + 2x - 14$

59. $5a^2 + 4ab - 15ba - 12b^2$

60. $pk - p - k + 1$

[3.7] Factor each trinomial. Assume k is a positive integer.

61. $x^2 - 2x - 8$

62. $36 - 12p + p^2$

63. $y^2 + 3yz - 28z^2$

64. $2m^2 + 5m - 12$

65. $18a^2 - 37ab - 20b^2$

66. $6x^3y^2 - 17x^2y^3 + 12xy^4$

67. $8r^4 - 14r^2 - 15$

68. $21y^{6k} + 5y^{3k} - 4$

[3.8] Factor each polynomial. Assume n is a positive integer.

69. $64x^2 - 9$

70. $4a^2 - 121b^2$

71. $10p^4 - 160k^4$

72. $x^5y^5 - 625xy$

73. $-100 - z^2 - 20z$

74. $45a^2 - 60ab + 20b^2$

75. $p^3 + 27$

76. $8x^3 - 125y^3$

77. $2t^3 - 128$

78. $x^2 - y^2 + 2y - 1$

79. $(m + 6)^3 - 1000$

80. $25x^{2n} - 81$

[3.9] Factor each polynomial.

81. $12r^3t^2 - 24r^2t^3 + 18rt^4$

82. $x^3 + 3x^2 - 4x - 12$

83. $4p^2 - 28p + 49$

84. $3z^6 + 21z^3 - 24$

85. $81 - y^4$

86. $5a^4b^3 + 1080a$

[3.10] Solve each equation.

87. $x^2 + 7x + 10 = 0$

88. $y^2 = 6 - y$

89. $8r^2 = 10r$

90. $(t + 2)(t - 2) = 5$

91. $(3m + 1)^2 - (2m)^2 = 2(m + 5)$

Solve each word problem.

92. The length of a rectangle is 4 m longer than twice the width. If the area of the rectangle is 70 m², find the length and width.

93. A ball is thrown upward from a balcony 64 ft high with an initial velocity of 48 ft/sec. If we ignore air resistance, the height h, in feet, of the ball after t seconds is given by the formula $h = -16t^2 + 48t + 64$. How long will it take the ball to hit the ground?

Chapter 3 Test

Write each number in scientific notation.

1. 160,000

2. 0.000759

Simplify. Use only positive exponents in your answer. Assume no variable is 0.

3. $6 \cdot 2^{-3}$

4. $(-7p)^0$

5. $3m(3^{-2}m^5)^{-1}$

6. $(8x^{-4}y^9)(-5x^{10}y^{-3})$

7. $\left(\dfrac{1}{4}\right)^{-1} + \dfrac{1}{5^{-2}}$

8. $\dfrac{a^2}{6b^2}\left(\dfrac{4b^3}{a^2}\right)^2$

Perform the indicated operations and simplify.

9. $(5x^3 - x^2 + 4) - (x^2 + 2x - 9)$

10. $p^2 - [p - (p + 3)^2]$

11. $(2y - 7)(3y + 4) - 5y(y - 1)$

12. $(3r - 4)(r^3 + 2r^2 - 5r + 7)$

13. If $P(t) = t^3 + 3t^2 - 4t - 13$, find $P(-2)$.

14. Subtract.
$$\begin{array}{r} 8p^3 \qquad\;\; - p + 9 \\ 2p^3 + 6p^2 - 5p - 9 \end{array}$$

15. Write in scientific notation and perform the indicated operations. Write your answer in scientific notation.
$$\dfrac{(45,000,000,000)(0.08)}{0.000000000024}$$

Factor each polynomial.

16. $y^2 - 4y - 12$

17. $a^2 + 6ab + 9b^2$

18. $64p^3 - 27$

19. $rs - 5r - 2s + 10$

20. $24z^4 + 24z^3 - 90z^2$

21. $t^4 - 16$

22. $5m^3 + 40$

23. $x^2 - 8x + 16 - y^2$

Solve each equation.

24. $2x^2 + 5x = 12$

25. $6y^2 - 18y = 0$

26. $8 - 12p = (3p - 2)^2$

27. If 6 times a number is subtracted from the square of the number, the result is 10 more than 3 times the number. Find the number.

Cumulative Test for Chapters 1, 2, and 3

1. List all subsets of $\{1, 2, 3\}$.

2. Write $\{\ldots, -3, -2, -1, 0, 1, 2, 3, \ldots\}$ using set-builder notation.

3. If $A = \{2, 4, 5\}$, $B = \{1, 2, 3\}$, and $C = \{3, 4, 5\}$, find $A \cup (B \cap C)$.

4. Use inequality notation to write the phrase "x is no more than 7."

5. Graph $\{x \mid -1 < x \le 3\}$ on the number line.

6. Name the property that justifies each statement.

    (a) If $x = 4$, then $4 = x$

    (b) $y + 7 = 7 + y$

    (c) $6(a + b) = 6a + 6b$

7. State the reciprocal of $\frac{2}{3}$.

8. Simplify $5 + 5[-3 - 4(1 - 7)]$.

9. Write $[0, \infty) \cap (3, \infty)$ as a single interval.

10. True or false: $4(x - 2) = 5x - (x + 2)$ is an identity.

11. Write 0.0000913 in scientific notation.

12. Multiply $(2x - 5)(x^2 - 3x + 7)$ and simplify your answer.

13. If $P(x) = -x^2 + 7x - 11$, find $P(-3)$.

14. Solve $A = P + Prt$ for P.

15. Solve the compound inequality $-9x > 0$ or $3x - 5 \ge 7$. Then graph the solution set.

Simplify. Do not leave negative exponents in your answer. Assume no variable is 0.

16. $6x^{-2}(x^3)^5x^0$

17. $(3x^5y^{-7})^{-1}$

Factor each polynomial.

18. $25x^2 - 16$

19. $m^3 + 27$

20. $x^3 - 7x^2 + 4x - 28$

21. $8p^2 + 2p - 15$

Solve each equation or inequality.

22. $\dfrac{x + 3}{4} - 1 = \dfrac{x}{3}$

23. $2m - 5(m + 1) > 7 - m$

24. $|2p - 7| = 5$

25. $|3t - 4| < 16$

26. $5x^2 - 10x = 0$

27. $3y^2 + 11y = 4$

Solve each word problem.

28. How long will it take to travel 780 mi if your average speed is 50 mi/hr? Use $d = rt$.

29. A collection of 40 nickels and dimes is worth $2.85. How many nickels and how many dimes are in the collection?

30. An executive leaves home for the airport, driving 40 mi/hr. Fifteen minutes later, her husband discovers the briefcase she left behind and starts after her at 60 mi/hr. How long does it take him to catch her?

31. The sum of the squares of two consecutive odd integers is 202. Find the integers.

CHAPTER

4

Rational
Expressions

Suppose one painter can paint a room
in 4 hr, and another painter can paint
the same room in 3 hr. In Sec. 4.6 we
determine how long it will take the
two painters to paint the room if they
work together.

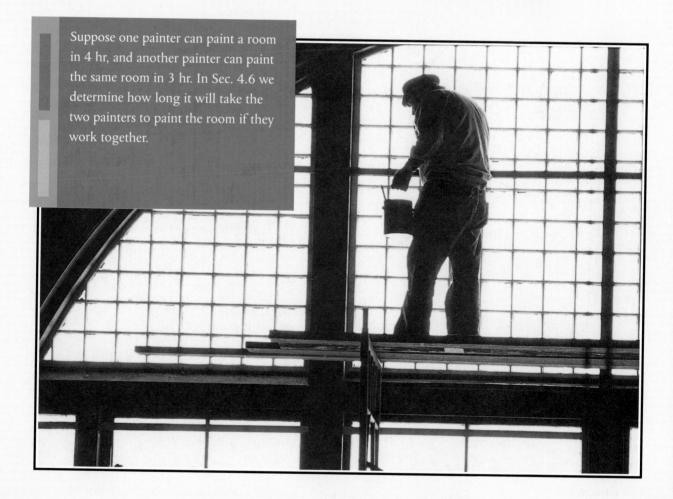

4.1 Simplifying Rational Expressions

In arithmetic, we often use fractions to solve problems. In algebra, we often use fractions that contain variables to solve problems. In this chapter we will study a particular type of fraction that contains variables, called a *rational expression.*

Recall from Sec. 1.1 that a rational number is a quotient of two integers, where the divisor is not 0. In a similar way, we define a rational expression as a quotient of two *polynomials,* where the divisor is not 0.

> **DEFINITION OF A RATIONAL EXPRESSION**
>
> A **rational expression** is an expression of the form
>
> $$\frac{P}{Q}$$
>
> where P and Q are polynomials and $Q \neq 0$.

Each of the expressions

$$\frac{x}{x-1}, \qquad \frac{3y-2}{y^2-5y+6}, \qquad \text{and} \qquad m^3 + 8$$

is a rational expression, since each is the quotient of two polynomials. Note that $m^3 + 8$ is the quotient of the two polynomials $m^3 + 8$ and 1. Since *any* polynomial can be written as the quotient of itself and 1, *every polynomial is a rational expression.*

Example 1 Find all values of the variable that make each rational expression undefined.

(a) $\dfrac{3}{x-2}$ (b) $\dfrac{5y}{y^2+6y-7}$ (c) $\dfrac{p^2-1}{10}$

Solution: (a) A rational expression is undefined when its denominator is 0. Therefore set the denominator equal to 0 and solve.

$$x - 2 = 0$$
$$x = 2$$

The rational expression is undefined when $x = 2$.
(b) Set the denominator equal to 0 and solve.

$$y^2 + 6y - 7 = 0$$
$$(y + 7)(y - 1) = 0$$

$$y + 7 = 0 \qquad \text{or} \qquad y - 1 = 0$$
$$y = -7 \qquad \text{or} \qquad y = 1$$

The rational expression is undefined when $y = -7$ or when $y = 1$.
(c) The denominator is never 0 (it is always 10), so the rational expression is never undefined. ❑

Try Exercise 15

What values make
$\dfrac{-6y}{y^2 + 3y - 10}$ undefined?

From now on, when we write a rational expression we will assume that those values that make the denominator 0 are excluded.

The rules for operating on rational expressions are essentially the same as the rules for operating on ordinary fractions. For example, to simplify a fraction we can factor the numerator and the denominator and then divide out common factors:

$$\frac{10}{15} = \frac{2 \cdot 5}{3 \cdot 5} = \frac{2 \cdot \overset{1}{\cancel{5}}}{3 \cdot \underset{1}{\cancel{5}}} = \frac{2}{3}$$

We can simplify a rational expression in the same way.

Fundamental Principle of Rational Expressions

If P, Q, and K are polynomials and neither Q nor K is 0, then

$$\frac{P \cdot K}{Q \cdot K} = \frac{P}{Q}$$

The fundamental principle of rational expressions states that we can divide the numerator and the denominator of a rational expression by the same nonzero polynomial and the result is an equivalent rational expression.

Example 2   Simplify each rational expression.

(a) $\dfrac{7xy^3}{21x^4y^2}$ (b) $\dfrac{z + 5}{4z + 20}$

Solution: (a) Write the numerator and the denominator in prime factored form. Then divide out common factors.

$$\frac{7xy^3}{21x^4y^2} = \frac{\overset{1}{7} \cdot \overset{1}{\cancel{x}} \cdot \overset{1}{\cancel{y}} \cdot \overset{1}{\cancel{y}} \cdot y}{3 \cdot \underset{1}{7} \cdot \underset{1}{\cancel{x}} \cdot x \cdot x \cdot x \cdot \underset{1}{\cancel{y}} \cdot \underset{1}{\cancel{y}}} = \frac{y}{3x^3}$$

(b) Factor the numerator and the denominator and divide out the common factor, $z + 5$.

$$\frac{z + 5}{4z + 20} = \frac{1(\overset{1}{\cancel{z + 5}})}{4(\underset{1}{\cancel{z + 5}})} = \frac{1}{4}$$

Try Exercise 31

Simplify: $\dfrac{4xy^2}{12x^4y^3}$.

Here is a summary of the steps in simplifying a rational expression.

To Simplify a Rational Expression

1. Write the numerator and the denominator in prime factored form.
2. Divide both the numerator and the denominator by any factors they have in common.

We illustrate these steps in Example 3.

Example 3   Simplify: $\dfrac{r^2 - 9}{r^2 + 6r + 9}$.

Solution: Step 1:   Write the numerator and the denominator in prime factored form.

$$\frac{r^2 - 9}{r^2 + 6r + 9} = \frac{(r + 3)(r - 3)}{(r + 3)(r + 3)}$$

Step 2:   Divide both the numerator and the denominator by the common factor $r + 3$.

$$= \frac{\overset{1}{\cancel{(r + 3)}}(r - 3)}{\underset{1}{\cancel{(r + 3)}}(r + 3)}$$

$$= \frac{r - 3}{r + 3}$$

This answer cannot be simplified further.   ❐

Try Exercise 37

Simplify: $\dfrac{r^2 - 100}{r^2 + 20r + 100}$.

CAUTION

Remember that *factors* are related by multiplication, but *terms* are related by addition. **We can divide out common factors, but we cannot divide out common terms.**

Wrong

Wrong

■

Example 4   Simplify: $\dfrac{4m - 5}{5 - 4m}$.

Solution: Factor out 1 from the numerator and -1 from the denominator. Then divide out the common factor, $4m - 5$.

$$\frac{4m - 5}{5 - 4m} = \frac{1(4m - 5)}{-1(-5 + 4m)} = \frac{1\cancel{(4m - 5)}}{-1\cancel{(4m - 5)}} = \frac{1}{-1} = -1 \quad ❐$$

Try Exercise 43

Simplify: $\dfrac{3m - 7}{7 - 3m}$.

Did you notice in Example 4 that the numerator and the denominator are opposites? Whenever you divide two opposites, the result is always -1.

CAUTION

Note the difference in the following three fractions:

$$\frac{x + y}{y + x} = 1 \qquad \frac{x - y}{y - x} = -1 \qquad \frac{x + y}{x - y} \text{ cannot be simplified}$$

■

Example 5   Simplify: $\dfrac{5t^2 + 10t - 75}{30t - 10t^2}$.

Solution:
$$\frac{5t^2 + 10t - 75}{30t - 10t^2} = \frac{5(t^2 + 2t - 15)}{-10t(t - 3)}$$ Factor numerator and denominator.

$$= \frac{5(t + 5)(t - 3)}{-1 \cdot 5 \cdot 2t(t - 3)}$$

$$= \frac{5(t + 5)(t - 3)}{-1 \cdot 5 \cdot 2t(t - 3)}$$ Divide out common factors.

$$= \frac{t + 5}{-2t}$$

$$= -\frac{t + 5}{2t}$$

We could also write the answer as $\dfrac{-t - 5}{2t}$.

Try Exercise 47

Simplify: $\dfrac{6t^2 - 30t + 36}{8t - 4t^2}$.

Example 6 Simplify: $\dfrac{a^3 - 8b^3}{a^2 - 2ab - 4a + 8b}$.

Solution: The numerator is a difference of two cubes. The denominator can be factored by grouping.

$$\frac{a^3 - 8b^3}{a^2 - 2ab - 4a + 8b} = \frac{a^3 - (2b)^3}{a(a - 2b) - 4(a - 2b)}$$

$$= \frac{(a - 2b)(a^2 + 2ab + 4b^2)}{(a - 2b)(a - 4)}$$

$$= \frac{a^2 + 2ab + 4b^2}{a - 4}$$

Try Exercise 57

Simplify:
$\dfrac{a^3 - 27b^3}{a^2 - 3ab - 4a + 12b}$.

Exercises 4.1

❖❖

Completion Exercises

1. A quotient of two polynomials, where the divisor is not 0, is called a(n) _____.

2. We cannot cancel the common _____ (terms, factors) of a rational expression; we can only cancel the common _____ (terms, factors).

3. The x in $\dfrac{3x}{x^2 + 9}$ is a _____ (term, factor) of the numerator.

4. The x^2 in $\dfrac{3x}{x^2 + 9}$ is a _____ (term, factor) of the denominator.

5. If the denominator is not 0, the expression $\dfrac{p - 5}{5 - p}$ simplifies to _____.

6. The fundamental principle of rational expressions states that if neither Q nor K is 0, then $\dfrac{P \cdot K}{Q \cdot K} =$ _____.

True-False Exercises

7. The rational expression $\dfrac{x}{x - 1}$ is undefined when $x = 0$.

8. We can cancel the y's in $\dfrac{y + 6}{y - 3}$.

9. $\dfrac{\sqrt{x}}{x + 1}$ is a rational expression.

10. Every polynomial is also a rational expression.

Find all values of the variable that make each rational expression undefined.

11. $\dfrac{x + 1}{4x}$

12. $\dfrac{x + 8}{3x - 10}$

13. $\dfrac{9}{y^2 - 36}$

14. $\dfrac{-1}{y^2 - 16}$

15. $\dfrac{-6y}{y^2 + 3y - 10}$

16. $\dfrac{4y}{y^2 - 4y - 21}$

17. $\dfrac{2z^2 + 3z - 5}{z^2 - z}$

18. $\dfrac{3z^2 - z - 4}{z^2 + z}$

19. $\dfrac{p^2 - 25}{25}$

20. $\dfrac{p^2 - 100}{100}$

21. $\dfrac{t^2}{t^2 + 1}$

22. $\dfrac{t^2}{t^2 + 9}$

Matching Exercises

Match each rational expression in Exercises 23 through 26 with its simplified form in letters A through D.

23. $\dfrac{x + 2}{x - 2}$

A. -1

24. $\dfrac{x - 2}{2 - x}$

B. x

25. $\dfrac{x^2 + 2x}{x + 2}$

C. Cannot be simplified

26. $\dfrac{x - 2}{x - 2}$

D. 1

Simplify each rational expression.

27. $\dfrac{5x}{9x}$

28. $\dfrac{7x}{8x}$

29. $\dfrac{8b^2}{14}$

30. $\dfrac{9b^2}{15}$

31. $\dfrac{4xy^2}{12x^4y^3}$

32. $\dfrac{7xy^4}{14x^3y^2}$

33. $\dfrac{z + 3}{6z + 18}$

34. $\dfrac{py - 4pz}{qy - 4qz}$

35. $\dfrac{k^3 + 6k^2}{6k^3 + 36k^2}$

36. $\dfrac{20t + 15}{16t^2 - 9}$

37. $\dfrac{r^2 - 100}{r^2 + 20r + 100}$

38. $\dfrac{r^2 - 64}{r^2 - 16r + 64}$

39. $\dfrac{(w - 8)^2}{(w - 8)^6}$

40. $\dfrac{(w + 6)^3}{(w + 6)^6}$

41. $\dfrac{x^2 - y^2}{(x + y)^2}$

42. $\dfrac{x^2 - y^2}{(x - y)^2}$

43. $\dfrac{3m - 7}{7 - 3m}$

44. $\dfrac{2m - 9}{9 - 2m}$

45. $\dfrac{a + 10}{10 + a}$

46. $\dfrac{a + b}{b + a}$

47. $\dfrac{6t^2 - 30t + 36}{8t - 4t^2}$

48. $\dfrac{10t^2 - 10t - 60}{18t - 6t^2}$

49. $\dfrac{12p^2 + 17pq - 40q^2}{4p^2 + 3pq - 10q^2}$

50. $\dfrac{15p^2 + 2pq - 24q^2}{3p^2 - 5pq - 12q^2}$

51. $\dfrac{2z^3 - 28z^2k + 98zk^2}{21k^2z + 39kz^2 - 6z^3}$

52. $\dfrac{3z^2k - 30zk^2 + 75k^3}{25k^3 + 70k^2z - 15kz^2}$

53. $\dfrac{r^3 - s^3}{s^2 - r^2}$

54. $\dfrac{5 - m}{m^3 - 125}$

55. $\dfrac{xy^2 - y + xyz - z}{xy^2 - y - xyz + z}$

56. $\dfrac{z - xyz + y^2 - xy^3}{z - xyz - y^2 + xy^3}$

57. $\dfrac{a^3 - 27b^3}{a^2 - 3ab - 4a + 12b}$

58. $\dfrac{a^3 - 64b^3}{a^2 - 4ab - 2a + 8b}$

59. $\dfrac{(p - 5)2 - p}{(p - 5)p - 50}$

60. $\dfrac{(q^2 + 2)4 - 8}{(q^2 + 2)q}$

61. $\dfrac{(x + h)^2 - x^2}{h}$

62. $\dfrac{(x + h)^3 - x^3}{h}$

Problem Solving

Rational expressions are used in many formulas. Solve each applied problem by making the appropriate substitution into the formula.

63. The number of degrees, a, in each interior angle of a regular polygon having n sides is given by the formula

$$a = \dfrac{180n - 360}{n}$$

Find the number of degrees in each interior angle of a regular

(a) pentagon (5 sides)   (b) hexagon (6 sides)

64. According to experiments conducted by A. J. Clark, the response r of a frog's heart to the injection of x units of acetylcholine is

$$r = \dfrac{100x}{15 + x}$$

Find r when $x = 22.5$.

Getting Acquainted with Your Scientific Calculator

To evaluate $\dfrac{3x}{8-x}$ when $x=4$, press

Clear 3 $\times$ 4 $\div$ (8 $-$ 4) $=$ [3]

NOTE:

If you don't insert parentheses, your calculator will compute $\dfrac{3\cdot 4}{8}-4$, instead of $\dfrac{3\cdot 4}{8-4}$.

65. The sharp reduction in pitch of an automobile horn as it passes you is called the *Doppler effect.** It occurs because the speed of the car increases the normal frequency of the horn as the car approaches you, and decreases the normal frequency as the car moves away. The formula that describes the Doppler effect is

$$f = \frac{1100F}{1100-v}$$

where f is the observed frequency, F is the actual frequency, and v is the velocity of the car. If a horn produces a sound with a frequency F of 483 vibrations per second, what is the observed frequency f under the following conditions:

(a) As the car approaches at 50 mi/hr? (*Hint:* Let $v=50$.)

(b) As the car moves away at 50 mi/hr? (*Hint:* Let $v=-50$.)

66. According to the cube law of politics, the percent y of seats won by a political party is

$$y = \frac{x^3}{3x^2 - 300x + 10{,}000}$$

where x is the percent of total votes for that party. Find y if $x=60$.

Discussion Exercises

67. Explain why both statements below are false. Substitute numbers for x to illustrate your point.

(a) FALSE: $\dfrac{x+3}{3} = x$

(b) FALSE: $\dfrac{x+3}{3} = x+1$

68. Discuss the steps in simplifying a rational expression.

Simplify each rational expression. Assume m, n, and k, are integers.

69. $\dfrac{3r^4 + 7r^2 - 20}{r^2 + 4}$

70. $\dfrac{x^{3m} - 5x^{2m}}{3x^{2m} - 15x^m}$

71. $\dfrac{5y^n + 45}{y^{2n} - 81}$

72. $\dfrac{z^{2k} + 2z^k - 3}{z^{2k} - 2z^k - 15}$

---

*HISTORICAL NOTE: The Doppler effect is named after the Austrian physicist Christian Johann Doppler (1803–1853), who first explained this phenomenon. The Doppler effect is used in radar to determine the velocity of the object under surveillance.

TAPE AU10

4.2 Multiplying and Dividing Rational Expressions

In Sec. 4.1 we defined rational expressions and learned how to simplify them. In this section we learn how to multiply and divide rational expressions.

Multiplying Rational Expressions

Recall that we multiply fractions by multiplying their numerators and multiplying their denominators:

$$\frac{2}{3} \cdot \frac{5}{7} = \frac{2\cdot 5}{3\cdot 7} = \frac{10}{21}$$

We multiply rational expressions in the same way.

Multiplying Rational Expressions

If $\dfrac{P}{Q}$ and $\dfrac{R}{S}$ are rational expressions, then

$$\frac{P}{Q} \cdot \frac{R}{S} = \frac{P\cdot R}{Q\cdot S}$$

Example 1 Multiply: $\dfrac{3a^3}{8} \cdot \dfrac{2}{a^2}$.

Solution: Multiply the numerators and multiply the denominators.

$$\frac{3a^3}{8} \cdot \frac{2}{a^2} = \frac{3a^3 \cdot 2}{8 \cdot a^2} = \frac{6a^3}{8a^2}$$

Simplify by dividing out common factors.

$$= \frac{\overset{3}{\cancel{6}}\overset{a}{\cancel{a^3}}}{\underset{4}{\cancel{8}}\underset{1}{\cancel{a^2}}}$$

$$= \frac{3a}{4}$$ ❐

Try Exercise 5

Multiply: $\dfrac{7a^3}{12} \cdot \dfrac{2}{a^2}$.

An easier way to do Example 1 is to divide out the common factors *before* multiplying.

$$\frac{3a^3}{8} \cdot \frac{2}{a^2} = \frac{3\overset{a}{\cancel{a^3}}}{\underset{4}{\cancel{8}}} \cdot \frac{\overset{1}{2}}{\cancel{a^2}} = \frac{3a}{4}$$

Here is a summary of the steps in multiplying rational expressions.

To Multiply Rational Expressions

1. Write all numerators and denominators in prime factored form.
2. Divide out common factors.
3. Multiply the remaining factors in the numerators and multiply the remaining factors in the denominators. Leave your answer in factored form.

We illustrate these steps in Example 2.

Example 2 Multiply: $\dfrac{p^2 - 4}{p^2} \cdot \dfrac{8p^2 - 48p}{p^2 - 4p - 12}$

Solution: Step 1: Write all numerators and denominators in prime factored form.

$$\frac{p^2 - 4}{p^2} \cdot \frac{8p^2 - 48p}{p^2 - 4p - 12} = \frac{(p + 2)(p - 2)}{p^2} \cdot \frac{8p(p - 6)}{(p + 2)(p - 6)}$$

Step 2: Divide out common factors.

$$= \frac{\cancel{(p + 2)}(p - 2)}{\underset{p}{\cancel{p^2}}} \cdot \frac{8p\cancel{(p - 6)}}{\cancel{(p + 2)}\cancel{(p - 6)}}$$

Step 3: Multiply the remaining factors in the numerators and multiply the remaining factors in the denominators.

Try Exercise 23

Multiply:

$(p^2 - 9) \cdot \dfrac{9p^2 - 45p}{p^2 - 2p - 15}$.

$$= \frac{8(p - 2)}{p}$$

We could multiply out the numerator, but it is common to leave the answer in factored form. ❐

Example 3   Multiply: $\dfrac{6x^2 + 13xy - 28y^2}{2x^2 - 3xy - 35y^2} \cdot \dfrac{25y^2 - x^2}{3x - 4y}$.

Solution:
$$\dfrac{6x^2 + 13xy - 28y^2}{2x^2 - 3xy - 35y^2} \cdot \dfrac{25y^2 - x^2}{3x - 4y}$$

$$= \dfrac{(3x - 4y)(2x + 7y)}{(2x + 7y)(x - 5y)} \cdot \dfrac{(5y + x)(5y - x)}{3x - 4y}$$
  Factor all numerators and denominators.

$$= \dfrac{(3x - 4y)(2x + 7y)}{(2x + 7y)(x - 5y)} \cdot \dfrac{(5y + x)(5y - x)}{3x - 4y}$$
  Divide out common factors.

  Remember, $\dfrac{5y - x}{x - 5y} = -1$.

$$= -(5y + x) \text{ or } -5y - x$$
  Multiply the remaining factors. ❏

Dividing Rational Expressions

Recall that we divide two fractions by inverting the divisor and multiplying.

$$\dfrac{2}{3} \div \dfrac{5}{7} = \dfrac{2}{3} \cdot \dfrac{7}{5} = \dfrac{14}{15}$$

We divide two rational expressions in the same way.

Dividing Rational Expressions

If $\dfrac{P}{Q}$ and $\dfrac{R}{S}$ are rational expressions and $\dfrac{R}{S} \neq 0$, then

$$\dfrac{P}{Q} \div \dfrac{R}{S} = \dfrac{P}{Q} \cdot \dfrac{S}{R} = \dfrac{P \cdot S}{Q \cdot R}$$

The rational expressions $\dfrac{R}{S}$ and $\dfrac{S}{R}$ are called **reciprocals** of each other because their product is 1. Therefore **to divide two rational expressions, multiply the first rational expression by the reciprocal of the second.**

Example 4   Divide: $\dfrac{5m}{n} \div \dfrac{75}{2m}$.

Solution: Multiply the first rational expression by the reciprocal of the second.

$$\dfrac{5m}{n} \div \dfrac{75}{2m} = \dfrac{5m}{n} \cdot \dfrac{2m}{75}$$
  Invert and change to multiplication.

$$= \dfrac{\overset{1}{5m}}{n} \cdot \dfrac{2m}{\underset{15}{75}}$$
  Divide out common factors.

$$= \dfrac{2m^2}{15n}$$
  Multiply the remaining factors. ❏

Example 5   Divide: $\dfrac{z^3 - z^2}{2z^2 + 12z} \div \dfrac{z^2 - 1}{6z + 6}$.

Try Exercise 29

Multiply: $\dfrac{10x^2 - 7xy - 12y^2}{5x^2 - 26xy - 24y^2}$ $\cdot \dfrac{36y^2 - x^2}{2x - 3y}$.

Try Exercise 9

Divide: $\dfrac{3m}{n} \div \dfrac{60}{7m}$.

Solution:
$$\frac{z^3 - z^2}{2z^2 + 12z} \div \frac{z^2 - 1}{6z + 6} = \frac{z^3 - z^2}{2z^2 + 12z} \cdot \frac{6z + 6}{z^2 - 1}$$

Invert and change to multiplication.

$$= \frac{z^2(z - 1)}{2z(z + 6)} \cdot \frac{6(z + 1)}{(z + 1)(z - 1)}$$

Factor.

$$= \frac{\overset{z}{z^2}(\cancel{z - 1})}{2\cancel{z}(z + 6)} \cdot \frac{\overset{3}{6}(\cancel{z + 1})}{(\cancel{z + 1})(\cancel{z - 1})}$$

Divide out common factors.

$$= \frac{3z}{z + 6}$$

Multiply the remaining factors. ❑

Try Exercise 35

Divide:
$$\frac{z^3 - 4z^2}{3z^2 + 15z} \div \frac{z^2 - 16}{6z + 30}.$$

Example 6 Divide: $\dfrac{q - 3}{2} \div \dfrac{q^2 + 4q + 4}{5q - 15}$.

Solution:
$$\frac{q - 3}{2} \div \frac{q^2 + 4q + 4}{5q - 15} = \frac{q - 3}{2} \cdot \frac{5q - 15}{q^2 + 4q + 4}$$

Invert and change to multiplication.

$$= \frac{q - 3}{2} \cdot \frac{5(q - 3)}{(q + 2)^2}$$

Factor.

$$= \frac{5(q - 3)^2}{2(q + 2)^2}$$

Multiply.

Note that there are no common factors to be divided out. ❑

Try Exercise 41

Divide:
$$\frac{q - 4}{7} \div \frac{q^2 + 6q + 9}{3q - 12}.$$

Example 7 Divide: $\dfrac{ms - mt - ns + nt}{m - n} \div (s^3 - t^3)$.

Solution: The reciprocal of $s^3 - t^3$ is $\dfrac{1}{s^3 - t^3}$.

$$\frac{ms - mt - ns + nt}{m - n} \div (s^3 - t^3)$$

$$= \frac{ms - mt - ns + nt}{m - n} \cdot \frac{1}{s^3 - t^3}$$

$$= \frac{m(s - t) - n(s - t)}{m - n} \cdot \frac{1}{(s - t)(s^2 + st + t^2)}$$

$$= \frac{(s - t)(m - n)}{m - n} \cdot \frac{1}{(s - t)(s^2 + st + t^2)}$$

$$= \frac{1}{s^2 + st + t^2}$$ ❑

Try Exercise 47

Divide:
$$\frac{ms + mt + ns + nt}{m + n} \div (t^3 + s^3).$$

If a problem involves both multiplication and division, perform the operations from left to right.

Example 8 Simplify: $\dfrac{3x^2 + 19x + 20}{3x^2 - 23x - 8} \div \dfrac{x^2 + 4x - 5}{x^2 + 8x - 9} \cdot \dfrac{x^2 - 10x + 16}{x^2 - 11x + 18}$.

Solution: Divide first, since division is on the left.

$$\left(\frac{3x^2 + 19x + 20}{3x^2 - 23x - 8} \div \frac{x^2 + 4x - 5}{x^2 + 8x - 9} \right) \cdot \frac{x^2 - 10x + 16}{x^2 - 11x + 18}$$

$$= \left(\frac{3x^2 + 19x + 20}{3x^2 - 23x - 8} \div \frac{x^2 + 8x - 9}{x^2 + 4x - 5} \right) \cdot \frac{x^2 - 10x + 16}{x^2 - 11x + 18}$$

Since multiplication is associative, we can remove the parentheses. Factor and divide out the common factors.

$$= \frac{(3x+4)(x+5)}{(3x+1)(x-8)} \cdot \frac{(x+9)(x-1)}{(x+5)(x-1)} \cdot \frac{(x-2)(x-8)}{(x-2)(x-9)}$$

$$= \frac{(3x+4)(x+9)}{(3x+1)(x-9)}$$

Try Exercise 49

Simplify: $\dfrac{5x^2+17x+14}{5x^2-29x-6}$
$\div \dfrac{x^2-x-6}{x^2+5x-24}$
$\cdot \dfrac{x^2-10x+24}{x^2+4x-32}.$

Exercises 4.2

Completion Exercises

1. The rational expressions $\dfrac{x}{x+1}$ and $\dfrac{x+1}{x}$ are _____ of each other.

2. The product of two reciprocals is _____.

3. If $\dfrac{P}{Q}$ and $\dfrac{R}{S}$ are rational expressions, then $\dfrac{P}{Q} \cdot \dfrac{R}{S} =$ _____.

4. If $\dfrac{P}{Q}$ and $\dfrac{R}{S}$ are rational expressions, and $\dfrac{R}{S} \neq 0$, then $\dfrac{P}{Q} \div \dfrac{R}{S} =$ _____.

Multiply or divide as indicated. Write your answer in simplest form.

5. $\dfrac{7a^3}{12} \cdot \dfrac{2}{a^2}$

6. $\dfrac{4a^4}{15} \cdot \dfrac{3}{a^3}$

7. $\dfrac{5x^2}{10y^3} \cdot \dfrac{4y^4}{35x^5}$

8. $\dfrac{8x^3}{20y} \cdot \dfrac{15y^2}{18x^6}$

9. $\dfrac{3m}{n} \div \dfrac{60}{7m}$

10. $\dfrac{2m^2}{n} \div \dfrac{54}{5m^2}$

11. $\dfrac{k^2p^4}{14p^2} \div \dfrac{k^8p^2}{24k^3}$

12. $\dfrac{k^3p^2}{16p^3} \div \dfrac{k^7p^3}{18k^4}$

13. $\dfrac{(-6ab^5)^2}{9a(b^2c^3)^3} \div \dfrac{45(a^4c^2)^4}{-30a^7bc^2}$

14. $\dfrac{20ab^6c^4}{-12(a^3b)^5} \div \dfrac{(-4b^2c^5)^2}{24a^2(b^2c)^3}$

Multiply or divide as indicated. Write your answer in simplest form.

15. $\dfrac{6z+18}{5} \cdot \dfrac{25z}{7z+21}$

16. $\dfrac{11z}{3z-9} \cdot \dfrac{4z-12}{22}$

17. $\dfrac{x^2y^2}{xy-1} \div \dfrac{x}{y}$

18. $\dfrac{x^3y^3}{xy+1} \div \dfrac{y}{x}$

19. $\dfrac{k^2+4k}{k^3} \div (k+4)$

20. $\dfrac{k^2-6k}{k^4} \div (k-6)$

21. $(a^2-1) \div \dfrac{1-a}{a}$

22. $(a^2-25) \div \dfrac{5-a}{a}$

23. $(p^2-9) \cdot \dfrac{9p^2-45p}{p^2-2p-15}$

24. $(p^2-16) \cdot \dfrac{8p^2+16p}{p^2-2p-8}$

25. $\dfrac{8m+16n}{m^2+4mn+4n^2} \cdot \dfrac{2n+m}{2}$

26. $\dfrac{m^2+2mn+n^2}{6m+6n} \cdot \dfrac{12}{n+m}$

27. $\dfrac{s^2+s-12}{s^2-2s-3} \cdot \dfrac{s+1}{8s^3+32s^2}$

28. $\dfrac{s^2+3s-4}{s^2-6s+5} \cdot \dfrac{s-5}{7s^3+28s^2}$

29. $\dfrac{10x^2-7xy-12y^2}{5x^2-26xy-24y^2} \cdot \dfrac{36y^2-x^2}{2x-3y}$

30. $\dfrac{15x^2-4xy-4y^2}{5x^2-43xy-18y^2} \cdot \dfrac{81y^2-x^2}{3x-2y}$

31. $\dfrac{m^2+5m+6}{m^2+4m+3} \div \dfrac{m^2-4}{m^2-3m-4}$

32. $\dfrac{m^2+4m-12}{m^2+2m-8} \div \dfrac{m^2+7m+6}{m^2-1}$

33. $\dfrac{c^2 - d^2}{3c^2 + 11cd - 4d^2} \cdot \dfrac{c^2 + 3cd - 4d^2}{(c - d)^2}$

34. $\dfrac{2c^2 - 9cd - 5d^2}{c^2 - d^2} \cdot \dfrac{(c + d)^2}{c^2 - 4cd - 5d^2}$

35. $\dfrac{z^3 - 4z^2}{3z^2 + 15z} \div \dfrac{z^2 - 16}{6z + 30}$

36. $\dfrac{z^4 - 3z^3}{2z^2 + 14z} \div \dfrac{z^2 - 9}{4z + 28}$

37. $\dfrac{8w^2 + 14w - 15}{6w^2 + 7w - 20} \cdot \dfrac{9w^2 - 16}{16w^2 - 9}$

38. $\dfrac{20w^2 - 23w + 6}{8w^2 - 26w + 15} \cdot \dfrac{4w^2 - 25}{25w^2 - 4}$

39. $\dfrac{2k^2 - 21km - 36m^2}{k^2 - 11km - 12m^2} \div \dfrac{10k + 15m}{k^3 - km^2}$

40. $\dfrac{3k^2 + 7km - 20m^2}{k^2 + 6km + 8m^2} \div \dfrac{9k - 15m}{k^2m - 4m^3}$

41. $\dfrac{q - 4}{7} \div \dfrac{q^2 + 6q + 9}{3q - 12}$

42. $\dfrac{q + 3}{2} \div \dfrac{q^2 - 8q + 16}{5q + 15}$

43. $\dfrac{x^3 + 64y^3}{x + 4y} \div \dfrac{x^2 - 4xy + 16y^2}{x - y}$

44. $\dfrac{x + y}{x - 3y} \div \dfrac{x^2 + 3xy + 9y^2}{x^3 - 27y^3}$

45. $\dfrac{ax - ay - bx + by}{x^3} \div \dfrac{y^2 - x^2}{x^2}$

46. $\dfrac{ax + ay - bx - by}{b^5} \div \dfrac{b^2 - a^2}{b^2}$

47. $\dfrac{ms + mt + ns + nt}{m + n} \div (t^3 + s^3)$

48. $\dfrac{ms + mt - ns - nt}{m - n} \div (s^3 + t^3)$

49. $\dfrac{5x^2 + 17x + 14}{5x^2 - 29x - 6} \div \dfrac{x^2 - x - 6}{x^2 + 5x - 24} \cdot \dfrac{x^2 - 10x + 24}{x^2 + 4x - 32}$

50. $\dfrac{4x^2 - 17x - 15}{4x^2 + 13x + 3} \div \dfrac{x^2 + 3x - 10}{x^2 - 9x + 14} \cdot \dfrac{x^2 + 9x + 18}{x^2 - x - 42}$

51. $\dfrac{8y^2 - 18y - 5}{12y^2 - 5y - 2} \div (4y^2 - 20y + 25) \div \dfrac{1}{2y - 3y^2}$

52. $\dfrac{9y^2 + 6y - 8}{12y^2 - 23y + 10} \div (9y^2 + 24y + 16) \div \dfrac{1}{5y - 20y^2}$

Discussion Exercises

53. Outline the steps in multiplying two rational expressions.

54. Explain how to divide two rational expressions.

55. Why is it important to factor all numerators and denominators and cancel common factors before you multiply two rational expressions, rather than multiplying first and then factoring and canceling?

4.3 Adding and Subtracting Rational Expressions

TAPE AU10

Now that we have learned how to multiply and divide rational expressions, we shall discuss how to add and subtract rational expressions.

Adding and Subtracting Rational Expressions with Like Denominators

Recall that we can add or subtract fractions having the same denominator by adding or subtracting their numerators and placing the sum or difference over the common denominator.

$$\frac{3}{7} + \frac{2}{7} = \frac{3 + 2}{7} = \frac{5}{7} \qquad \text{and} \qquad \frac{3}{7} - \frac{2}{7} = \frac{3 - 2}{7} = \frac{1}{7}$$

We add or subtract rational expressions in the same way.

Adding and Subtracting Rational Expressions

If $\dfrac{P}{Q}$ and $\dfrac{R}{Q}$ are rational expressions, then

$$\frac{P}{Q} + \frac{R}{Q} = \frac{P + R}{Q} \qquad \text{and} \qquad \frac{P}{Q} - \frac{R}{Q} = \frac{P - R}{Q}$$

Example 1   Add or subtract as indicated.

$$\text{(a) } \frac{5x}{8} + \frac{y}{8} \qquad \text{(b) } \frac{4p}{z} - \frac{7p}{z} \qquad \text{(c) } \frac{5}{9r^2} - \frac{1}{9r^2} + \frac{2}{9r^2}$$

Solution: The denominators are the same in each case, so operate on the numerators and place the result over the common denominator. Then simplify your answer if possible.

(a) $\dfrac{5x}{8} + \dfrac{y}{8} = \dfrac{5x + y}{8}$ Add the numerators, keep the same denominator.

(b) $\dfrac{4p}{z} - \dfrac{7p}{z} = \dfrac{4p - 7p}{z} = \dfrac{-3p}{z}$ or $-\dfrac{3p}{z}$ Subtract the numerators, keep the same denominator.

(c) $\dfrac{5}{9r^2} - \dfrac{1}{9r^2} + \dfrac{2}{9r^2} = \dfrac{5 - 1 + 2}{9r^2} = \dfrac{6}{9r^2} = \dfrac{2}{3r^2}$ ◻

Try Exercise 13

Simplify: $\dfrac{7}{15r^2} - \dfrac{1}{15r^2} + \dfrac{4}{15r^2}$.

To Add or Subtract Rational Expressions with Like Denominators

1. Add or subtract the numerators and place the sum or difference over the common denominator. Do *not* add or subtract the denominators.
2. If possible, simplify your answer by writing the numerator and the denominator in prime factored form and dividing out common factors.

We illustrate these steps in Example 2.

Example 2   Add: $\dfrac{3x}{x^2 + 2x - 15} + \dfrac{15}{x^2 + 2x - 15}$.

Solution: Step 1:   Add the numerators and place the sum over the common denominator.

$$\frac{3x}{x^2 + 2x - 15} + \frac{15}{x^2 + 2x - 15} = \frac{3x + 15}{x^2 + 2x - 15}$$

Step 2:   Factor and divide out common factors.

$$= \frac{3(x + 5)}{(x - 3)(x + 5)}$$

$$= \frac{3}{x - 3}$$ ◻

Try Exercise 21

Add:

$$\frac{3x}{x^2 + x - 12} + \frac{12}{x^2 + x - 12}.$$

Example 3   Add: $\dfrac{a^2 - 25}{a - 5} + \dfrac{25}{5 - a}$.

Solution: The denominators are not the same, but they are opposites. To make the denominators the same, multiply numerator and denominator of the second rational expression by -1.

$$\frac{a^2 - 25}{a - 5} + \frac{25}{5 - a} = \frac{a^2 - 25}{a - 5} + \frac{25}{5 - a} \cdot \frac{-1}{-1}$$

$$= \frac{a^2 - 25}{a - 5} + \frac{-25}{a - 5}$$

$$= \frac{a^2 - 25 + (-25)}{a - 5}$$

$$= \frac{a^2 - 50}{a - 5}$$

You could also perform the addition by multiplying numerator and denominator of the first rational expression by -1. ❏

Try Exercise 25

Add: $\dfrac{a^2 + 36}{a - 6} + \dfrac{36}{6 - a}$.

The Least Common Denominator

To add or subtract rational expressions having different denominators, first find the least common denominator. The **least common denominator (LCD)** of a set of rational expressions is the simplest polynomial that can be divided by the denominator of each rational expression.

When the denominators are simple, we can determine the LCD by inspection. If the LCD is not obvious from a simple inspection of the denominators, we can determine the LCD as follows:

To Find the Least Common Denominator

1. Write each denominator in prime factored form using exponents.
2. Multiply together the highest power of each factor.

We illustrate these steps in Example 4.

Example 4   Find the LCD of $\dfrac{3}{8x}$ and $\dfrac{5}{2x^2}$.

Solution: You may be able to see immediately that the simplest polynomial that can be divided by both denominators is $8x^2$. If not, you can find the LCD as follows:

Step 1:   Write each denominator in prime factored form using exponents.

$$8x = 2^3 \cdot x$$
$$2x^2 = 2 \cdot x^2$$

Step 2:   Multiply together the highest power of each factor. The highest power of 2 is 2^3, and the highest power of x is x^2. Therefore

$$LCD = 2^3 \cdot x^2 = 8x^2$$

Note that the numerators of the fractions play no role in determining the LCD. ❏

Try Exercise 33(a)

Find the LCD: $\dfrac{4}{9x} + \dfrac{7}{3x^2}$.

Example 5   Find the LCD of $\dfrac{1}{24x^2y}$ and $\dfrac{7}{90xy^3}$.

Solution:
$$\left.\begin{array}{l} 24x^2y = 2^3 \cdot 3 \cdot x^2 \cdot y \\ 90xy^3 = 2 \cdot 3^2 \cdot 5 \cdot x \cdot y^3 \\ \text{LCD} = 2^3 \cdot 3^2 \cdot 5 \cdot x^2 \cdot y^3 \\ \quad\quad\; = 360x^2y^3 \end{array}\right\}$$
Prime factored form
Highest power of each factor

Try Exercise 37(a)

Find the LCD:
$\dfrac{1}{18xy^2} + \dfrac{3}{40x^3y}.$

Example 6   Find the LCD of p and $\dfrac{p^2}{p+4}$.

Solution: Since $p = \dfrac{p}{1}$, both denominators are already in prime factored form. Therefore
$$\text{LCD} = 1 \cdot (p+4) = p+4$$

Note that when the denominators have no common factor, the LCD is simply the product of the denominators.

Try Exercise 41(a)

Find the LCD: $p - \dfrac{p^2}{p+9}.$

Example 7   Find the LCD of $\dfrac{r}{r-2}$ and $\dfrac{r-1}{r+3}$.

Solution: Both denominators are already in prime factored form. Since they have no common factor, the LCD is the product of the two denominators.
$$\text{LCD} = (r-2)(r+3)$$

Note that we generally leave the LCD in factored form.

Try Exercise 47(a)

Find the LCD:
$\dfrac{r}{r-4} - \dfrac{r-3}{r+1}.$

Example 8   Find the LCD of $\dfrac{2}{m^2-m}$ and $\dfrac{2}{m^2-2m+1}$.

Solution:
$$\left.\begin{array}{l} m^2 - m = m(m-1) \\ m^2 - 2m + 1 = (m-1)^2 \\ \text{LCD} = m(m-1)^2 \end{array}\right\}$$
Prime factored form
Highest power of each factor

Try Exercise 53(a)

Find the LCD:
$\dfrac{6}{m^2-2m} - \dfrac{6}{m^2-4m+4}.$

Example 9   Find the LCD of $\dfrac{3x-3}{x^2-x-2}$, $\dfrac{x+5}{x^2+4x+3}$, and $\dfrac{5}{x^2+x-6}$.

Solution:
$$\left.\begin{array}{l} x^2 - x - 2 = (x+1)(x-2) \\ x^2 + 4x + 3 = (x+1)(x+3) \\ x^2 + x - 6 = (x-2)(x+3) \end{array}\right\}$$
Prime factored form
$$\text{LCD} = (x+1)(x-2)(x+3)$$
Highest power of each factor

Try Exercise 69(a)

Find the LCD:
$\dfrac{2x-4}{x^2-4x+3} + \dfrac{3x}{x^2+x-2}$
$- \dfrac{3x-4}{x^2-x-6}.$

Adding and Subtracting Rational Expressions with Unlike Denominators

To add or subtract rational expressions having different denominators, first write each expression with the least common denominator.

Example 10   Add: $\dfrac{3}{8x} + \dfrac{5}{2x^2}$.

Solution: From Example 4, we know that the LCD is $8x^2$. Use the fundamental principle from Sec. 4.1 to write each expression with the LCD.

$$\dfrac{3}{8x} = \dfrac{3 \cdot x}{8x \cdot x} = \dfrac{3x}{8x^2} \qquad \text{Multiply numerator and denominator by } x.$$

$$\dfrac{5}{2x^2} = \dfrac{5 \cdot 4}{2x^2 \cdot 4} = \dfrac{20}{8x^2} \qquad \text{Multiply numerator and denominator by } 4.$$

Now add the numerators and place the sum over the common denominator.

$$\frac{3}{8x} + \frac{5}{2x^2} = \frac{3x}{8x^2} + \frac{20}{8x^2} = \frac{3x + 20}{8x^2}$$ ❏

Try Exercise 33(b)

Add: $\frac{4}{9x} + \frac{7}{3x^2}$.

Example 11 Add: $\frac{1}{24x^2y} + \frac{7}{90xy^3}$.

Solution: From Example 5, we know that the LCD is $360x^2y^3$.

$$\frac{1}{24x^2y} + \frac{7}{90xy^3} = \frac{1 \cdot 15y^2}{24x^2y \cdot 15y^2} + \frac{7 \cdot 4x}{90xy^3 \cdot 4x} \qquad \text{Fundamental principle}$$

$$= \frac{15y^2}{360x^2y^3} + \frac{28x}{360x^2y^3} \qquad \text{Multiply out.}$$

$$= \frac{15y^2 + 28x}{360x^2y^3} \qquad \text{Add the numerators.}$$ ❏

Try Exercise 37(b)

Add: $\frac{1}{18xy^2} + \frac{3}{40x^3y}$.

Example 12 Subtract $p - \frac{p^2}{p + 4}$.

Solution: From Example 6, we know that the LCD is $p + 4$.

$$p - \frac{p^2}{p + 4} = \frac{p(p + 4)}{1(p + 4)} - \frac{p^2}{p + 4} \qquad \text{Fundamental principle}$$

$$= \frac{p^2 + 4p}{p + 4} - \frac{p^2}{p + 4} \qquad \text{Multiply out.}$$

$$= \frac{p^2 + 4p - p^2}{p + 4} \qquad \text{Subtract the numerators.}$$

$$= \frac{4p}{p + 4} \qquad \text{Combine like terms.}$$ ❏

Try Exercise 41(b)

Subtract: $p - \frac{p^2}{p + 9}$.

Here is a summary of the steps in adding or subtracting rational expressions.

To Add or Subtract Rational Expressions with Unlike Denominators

1. Factor the denominators to find the LCD.
2. Use the fundamental principle of rational expressions to write each expression with the LCD.
3. Multiply out the numerators. Leave the denominators in factored form.
4. Add or subtract the numerators and place the sum or difference over the LCD.
5. Factor the numerator of the answer and divide out common factors.

We illustrate these steps in Example 13.

Example 13 Subtract: $\frac{r}{r - 2} - \frac{r - 1}{r + 3}$.

Solution: Step 1: Find the LCD. From Example 7, we know that the LCD is $(r - 2)(r + 3)$.

Step 2:   Write each expression with the LCD.

$$\frac{r}{r-2} - \frac{r-1}{r+3} = \frac{r(r+3)}{(r-2)(r+3)} - \frac{(r-1)(r-2)}{(r+3)(r-2)}$$

Step 3:   Multiply out the numerators. Leave the denominators in factored form.

$$= \frac{r^2 + 3r}{(r-2)(r+3)} - \frac{r^2 - 3r + 2}{(r+3)(r-2)}$$

Step 4:   Subtract the numerators. (Be sure to write parentheses around the second numerator.)

$$= \frac{r^2 + 3r - (r^2 - 3r + 2)}{(r-2)(r+3)}$$

$$= \frac{r^2 + 3r - r^2 + 3r - 2}{(r-2)(r+3)}$$

$$= \frac{6r - 2}{(r-2)(r+3)}$$

Step 5:   Factor the numerator and divide out common factors.

$$= \frac{2(3r-1)}{(r-2)(r+3)}$$ ❑

Try Exercise 47(b)

Subtract: $\dfrac{r}{r-4} - \dfrac{r-3}{r+1}$.

CAUTION

When you subtract two rational expressions, make sure you write parentheses around the second numerator if it contains more than one term.

Correct

$$\frac{5}{x+3} - \frac{x+2}{x+3} = \frac{5 - (x+2)}{x+3}$$

$$= \frac{5 - x - 2}{x+3}$$

$$= \frac{3 - x}{x+3}$$

Wrong

$$\frac{5}{x+3} - \frac{x+2}{x+3} = \frac{5 - x + 2}{x+3}$$

$$= \frac{7 - x}{x+3}$$ ∎

Example 14   Subtract: $\dfrac{2}{m^2 - m} - \dfrac{2}{m^2 - 2m + 1}$.

Solution: From Example 8, we know that the LCD is $m(m-1)^2$.

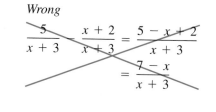

$$\frac{2}{m(m-1)} - \frac{2}{(m-1)^2} = \frac{2(m-1)}{m(m-1)(m-1)} - \frac{2m}{(m-1)^2 m}$$

$$= \frac{2m - 2}{m(m-1)^2} - \frac{2m}{m(m-1)^2}$$

$$= \frac{2m - 2 - 2m}{m(m-1)^2}$$

$$= \frac{-2}{m(m-1)^2}$$ ❑

Try Exercise 53(b)

Subtract:

$\dfrac{6}{m^2 - 2m} - \dfrac{6}{m^2 - 4m + 4}$.

Example 15 Subtract: $\dfrac{3x - 3}{x^2 - x - 2} - \dfrac{x + 5}{x^2 + 4x + 3} - \dfrac{5}{x^2 + x - 6}$.

Solution: From Example 9, we know that the LCD is $(x + 1)(x - 2)(x + 3)$.

$$\dfrac{3x - 3}{(x + 1)(x - 2)} - \dfrac{x + 5}{(x + 1)(x + 3)} - \dfrac{5}{(x - 2)(x + 3)}$$

$$= \dfrac{(3x - 3)(x + 3)}{(x + 1)(x - 2)(x + 3)} - \dfrac{(x + 5)(x - 2)}{(x + 1)(x + 3)(x - 2)} - \dfrac{5(x + 1)}{(x - 2)(x + 3)(x + 1)}$$

$$= \dfrac{3x^2 + 6x - 9}{(x + 1)(x - 2)(x + 3)} - \dfrac{x^2 + 3x - 10}{(x + 1)(x + 3)(x - 2)} - \dfrac{5x + 5}{(x - 2)(x + 3)(x + 1)}$$

Write parentheses!

$$= \dfrac{3x^2 + 6x - 9 - (x^2 + 3x - 10) - (5x + 5)}{(x + 1)(x - 2)(x + 3)}$$

$$= \dfrac{3x^2 + 6x - 9 - x^2 - 3x + 10 - 5x - 5}{(x + 1)(x - 2)(x + 3)}$$

$$= \dfrac{2x^2 - 2x - 4}{(x + 1)(x - 2)(x + 3)}$$

$$= \dfrac{2(x^2 - x - 2)}{(x + 1)(x - 2)(x + 3)}$$

$$= \dfrac{2(x + 1)(x - 2)}{(x + 1)(x - 2)(x + 3)}$$

$$= \dfrac{2}{x + 3} \qquad \qquad \square$$

Try Exercise 69(b)

Simplify:
$$\dfrac{2x - 4}{x^2 - 4x + 3} + \dfrac{3x}{x^2 + x - 2} - \dfrac{3x - 4}{x^2 - x - 6}.$$

Exercises 4.3

True-False Exercises

1. To add two rational expressions, add the numerators and add the denominators.

2. To subtract two rational expressions, subtract the numerators and keep the same denominator.

3. $\dfrac{x}{2} + \dfrac{y}{2} = \dfrac{x + y}{2}$

4. $\dfrac{1}{x} + \dfrac{1}{y} = \dfrac{1}{x + y}$

5. $\dfrac{x}{x + 3} - \dfrac{x + 1}{x + 3} = \dfrac{x - x + 1}{x + 3} = \dfrac{1}{x + 3}$

6. The numerators play no role in determining the least common denominator.

Completion Exercises

7. If $\dfrac{P}{Q}$ and $\dfrac{R}{Q}$ are rational expressions, then $\dfrac{P}{Q} - \dfrac{R}{Q} =$ _____.

8. If $\dfrac{P}{Q}$ and $\dfrac{R}{Q}$ are rational expressions, then $\dfrac{P}{Q} + \dfrac{R}{Q} =$ _____.

9. The simplest polynomial that can be divided by each denominator of a set of rational expressions is called the _____.

10. If the denominators of a set of rational expressions have no common factor (except 1), the LCD is the _____ of the denominators.

Add or subtract as indicated. Write your answer in simplest form.

11. $\dfrac{3x}{5} + \dfrac{y}{5}$

12. $\dfrac{5p}{z} - \dfrac{9p}{z}$

13. $\dfrac{7}{15r^2} - \dfrac{1}{15r^2} + \dfrac{4}{15r^2}$

14. $\dfrac{9}{25r^3} - \dfrac{3}{25r^3} + \dfrac{4}{25r^3}$

15. $\dfrac{k+4}{k+1} - \dfrac{3}{k+1}$

16. $\dfrac{k-6}{k+2} + \dfrac{8}{k+2}$

17. $\dfrac{6}{q} - \dfrac{q^3+6}{q}$

18. $\dfrac{8}{q} - \dfrac{q^2+8}{q}$

19. $\dfrac{t^2}{t-5} - \dfrac{25}{t-5}$

20. $\dfrac{t^2}{t+3} - \dfrac{9}{t+3}$

21. $\dfrac{3x}{x^2+x-12} + \dfrac{12}{x^2+x-12}$

22. $\dfrac{5x}{x^2-3x-10} + \dfrac{10}{x^2-3x-10}$

23. $\dfrac{x(x+3)}{(x-3)(x+3)} - \dfrac{3(x-3)}{(x+3)(x-3)}$

24. $\dfrac{x(x+4)}{(x-4)(x+4)} - \dfrac{4(x-4)}{(x+4)(x-4)}$

25. $\dfrac{a^2+36}{a-6} + \dfrac{36}{6-a}$

26. $\dfrac{a^2-64}{a-8} - \dfrac{64}{8-a}$

27. $\dfrac{x}{x^2-y^2} - \dfrac{y}{y^2-x^2}$

28. $\dfrac{x}{x^2-y^2} + \dfrac{y}{y^2-x^2}$

For each problem (a) find the LCD, (b) use the LCD to perform the indicated operations. Write your answer in simplest form.

29. $\dfrac{1}{3a} + \dfrac{1}{5a}$

30. $\dfrac{1}{2a} + \dfrac{1}{7a}$

31. $\dfrac{2}{b} - \dfrac{1}{b^2}$

32. $\dfrac{3}{b} - \dfrac{1}{b^3}$

33. $\dfrac{4}{9x} + \dfrac{7}{3x^2}$

34. $\dfrac{5}{16x} + \dfrac{3}{4x^2}$

35. $\dfrac{6}{y} - 3y$

36. $4y - \dfrac{8}{y}$

37. $\dfrac{1}{18xy^2} + \dfrac{3}{40x^3y}$

38. $\dfrac{1}{27x^2y^2} + \dfrac{7}{60xy^4}$

39. $\dfrac{4}{z} - \dfrac{4}{z-4}$

40. $\dfrac{6}{z} - \dfrac{6}{z-3}$

41. $p - \dfrac{p^2}{p+9}$

42. $p - \dfrac{p^2}{p+7}$

43. $\dfrac{8}{q(q+2)} + \dfrac{4}{q+2}$

44. $\dfrac{18}{q(q+3)} + \dfrac{6}{q+3}$

45. $\dfrac{10}{h-10} - \dfrac{5}{h-5}$

46. $\dfrac{12}{h-12} - \dfrac{4}{h-4}$

47. $\dfrac{r}{r-4} - \dfrac{r-3}{r+1}$

48. $\dfrac{r}{r-2} - \dfrac{r-5}{r+1}$

49. $\dfrac{-4}{t+10} - \dfrac{2t}{5t+50}$

50. $\dfrac{-4}{t+6} - \dfrac{2t}{3t+18}$

51. $\dfrac{1}{2z+18} - \dfrac{z}{z^2-81}$

52. $\dfrac{1}{2z+16} - \dfrac{z}{z^2-64}$

53. $\dfrac{6}{m^2-2m} - \dfrac{6}{m^2-4m+4}$

54. $\dfrac{5}{m^2+m} - \dfrac{5}{m^2+2m+1}$

55. $\dfrac{a^2}{a^2-16b^2} + \dfrac{b}{a^2+4ab}$

56. $\dfrac{a^2}{ab-6b^2} - \dfrac{b}{a^2-36b^2}$

57. $\dfrac{1}{k^2+k-12} + \dfrac{1}{k^2-k-6}$

58. $\dfrac{1}{k^2+3k-10} + \dfrac{1}{k^2-k-2}$

59. $\dfrac{4m}{m-1} - m - 2$

60. $\dfrac{5m}{m+1} - m - 3$

61. $1 - \dfrac{1}{r+1} - \dfrac{r}{(r+1)^2}$

62. $1 + \dfrac{1}{r-1} + \dfrac{r}{(r-1)^2}$

63. $\dfrac{4y}{6y+3} - \dfrac{1}{6y^2+3y} + \dfrac{1}{y}$

64. $\dfrac{9y}{12y-4} - \dfrac{1}{12y^2-4y} - \dfrac{1}{y}$

65. $\dfrac{x}{4x+8} - \dfrac{x+10}{x^2-4} + \dfrac{3}{x-2}$

66. $\dfrac{x}{2x+6} - \dfrac{x+15}{x^2-9} + \dfrac{3}{x-3}$

67. $\dfrac{m}{m^2+5m+4} - \dfrac{m-1}{m^2+2m+1} - \dfrac{1}{m^2+4m}$

68. $\dfrac{m}{m^2-m-2} - \dfrac{m+2}{m^2-4m+4} - \dfrac{1}{m^2+m}$

69. $\dfrac{2x-4}{x^2-4x+3} + \dfrac{3x}{x^2+x-2} - \dfrac{3x-4}{x^2-x-6}$

70. $\dfrac{4x+4}{x^2+2x-3} - \dfrac{x-2}{x^2+5x+6} - \dfrac{6}{x^2+x-2}$

71. Show that $\dfrac{a}{b} + \dfrac{c}{d} = \dfrac{ad+bc}{bd}$.

72. Show that $\dfrac{a}{b} - \dfrac{c}{d} = \dfrac{ad-bc}{bd}$

Problem Solving

73. Find the sum of the reciprocals of two consecutive odd integers if the smaller integer is *n*.

74. The velocity of a meteor from outside our solar system is given by the formula

$$v^2 = MG\left(\frac{2}{R} + \frac{1}{a}\right)$$

Express the right side of this formula as one fraction.

75. Discuss the steps in adding or subtracting rational expressions with like denominators.

76. Explain the steps in finding the least common denominator.

77. Outline the steps in adding or subtracting rational expressions with unlike denominators.

4.4 Complex Fractions

TAPE AU10

Sometimes an applied problem leads to a fraction that contains other fractions. In this section we will learn how to simplify such a fraction, called a *complex fraction*.

> **DEFINITION OF A COMPLEX FRACTION**
>
> A **complex fraction** is a fraction that contains other fractions in its numerator or denominator or both. A fraction that is not a complex fraction is called a **simple fraction**.

Here are some examples of complex fractions:

$$\frac{\dfrac{3}{4}}{\dfrac{7}{8}} \qquad \frac{1 + \dfrac{1}{x}}{x} \qquad \frac{\dfrac{9b}{a} - \dfrac{1}{ab}}{3 + \dfrac{a}{b}}$$

We identify the parts of a complex fraction as follows:

main fraction bar $\longrightarrow$ $$\frac{\left.\dfrac{1}{x} - \dfrac{1}{y}\right\}}{\left.\dfrac{x - y}{y}\right\}}$$ ← numerator of complex fraction

← denominator of complex fraction

The fractions $\dfrac{1}{x}, \dfrac{1}{y}$, and $\dfrac{x - y}{y}$ are called **secondary fractions** of the complex fraction above.

There are two basic methods for simplifying a complex fraction.

To Simplify a Complex Fraction

Method 1: Simplify the numerator and the denominator separately. Then invert the denominator and multiply.

Method 2: Multiply both numerator and denominator by the LCD of all the secondary fractions.

We illustrate both methods, side by side, in Examples 1 and 2.

Example 1 Simplify: $\dfrac{\dfrac{m+3}{2m}}{\dfrac{m-3}{m}}$.

Solution:

Method 1

Both the numerator and the denominator are already simplified. Therefore, invert the denominator and multiply.

$$\frac{\dfrac{m+3}{2m}}{\dfrac{m-3}{m}} = \frac{m+3}{2m} \cdot \frac{m}{m-3}$$

$$= \frac{m+3}{2(m-3)}$$

Method 2

Multiply numerator and denominator by the LCD of all the secondary fractions. The LCD of $2m$ and m is $2m$.

$$\frac{\dfrac{m+3}{2m}}{\dfrac{m-3}{m}} = \frac{\left(\dfrac{m+3}{2m}\right)2m}{\left(\dfrac{m-3}{m}\right)2m}$$

$$= \frac{m+3}{2(m-3)}$$

Note that the answer is the same no matter which method is used.   ❏

Try Exercise 7

Simplify: $\dfrac{\dfrac{m+1}{3m}}{\dfrac{m-1}{m}}$.

Example 2 Simplify: $\dfrac{\dfrac{1}{xy}+1}{\dfrac{1}{x^2}+\dfrac{y}{x}}$.

Solution:

Method 1

Simplify the numerator and denominator separately.

$$\frac{\dfrac{1}{xy}+1}{\dfrac{1}{x^2}+\dfrac{y}{x}} = \frac{\dfrac{1}{xy}+\dfrac{xy}{xy}}{\dfrac{1}{x^2}+\dfrac{xy}{x^2}}$$

$$= \frac{\dfrac{1+xy}{xy}}{\dfrac{1+xy}{x^2}}$$

Invert the denominator and multiply.

$$= \frac{1+xy}{xy} \cdot \frac{x^2}{1+xy}$$

$$= \frac{x}{y}$$

Method 2

Multiply numerator and denominator by the LCD of all the secondary fractions. The LCD of xy, x^2, and x is x^2y.

$$\frac{\dfrac{1}{xy}+1}{\dfrac{1}{x^2}+\dfrac{y}{x}} = \frac{\left(\dfrac{1}{xy}+1\right)x^2y}{\left(\dfrac{1}{x^2}+\dfrac{y}{x}\right)x^2y}$$

Distribute x^2y over every term.

$$= \frac{\dfrac{1}{xy} \cdot x^2y + 1 \cdot x^2y}{\dfrac{1}{x^2} \cdot x^2y + \dfrac{y}{x} \cdot x^2y}$$

$$= \frac{x+x^2y}{y+xy^2}$$

$$= \frac{x(1+xy)}{y(1+xy)}$$

$$= \frac{x}{y}$$ ❏

Try Exercise 17

Simplify: $\dfrac{\dfrac{1}{xy}-1}{\dfrac{1}{x^2}-\dfrac{y}{x}}$.

Note that in Method 1 you are working with the numerator and the denominator of the complex fraction separately. In Method 2 you are working with the entire complex fraction at once. For some problems Method 1 is easier; for other problems Method 2 is easier.

Example 3   Simplify: $\dfrac{k + 3 - \dfrac{4k}{k + 2}}{7 - \dfrac{9}{k}}$.

Solution: We use Method 1 and simplify the numerator and denominator separately.

$$\dfrac{\dfrac{k+3}{1} - \dfrac{4k}{k+2}}{\dfrac{7}{1} - \dfrac{9}{k}} = \dfrac{\dfrac{(k+3)(k+2)}{1(k+2)} - \dfrac{4k}{k+2}}{\dfrac{7 \cdot k}{1 \cdot k} - \dfrac{9}{k}}$$

$$= \dfrac{\dfrac{k^2 + 5k + 6 - 4k}{k + 2}}{\dfrac{7k - 9}{k}}$$

Simplify the numerator of the top fraction and invert and multiply.

$$= \dfrac{k^2 + k + 6}{k + 2} \cdot \dfrac{k}{7k - 9}$$

$$= \dfrac{k(k^2 + k + 6)}{(k + 2)(7k - 9)} \qquad \square$$

Try Exercise 31

Simplify: $\dfrac{k + 1 - \dfrac{3k}{k + 6}}{3 - \dfrac{2}{k}}$.

Negative exponents often lead to complex fractions.

Example 4   Simplify: $\dfrac{t^{-1}}{t^{-2} + 1}$.

Solution: First write with positive exponents.

$$\dfrac{t^{-1}}{t^{-2} + 1} = \dfrac{\dfrac{1}{t}}{\dfrac{1}{t^2} + 1}$$

Then use Method 2 and multiply each term in the numerator and the denominator by t^2.

$$= \dfrac{\dfrac{1}{t} \cdot t^2}{\dfrac{1}{t^2} \cdot t^2 + 1 \cdot t^2}$$

$$= \dfrac{t}{1 + t^2} \qquad \square$$

Try Exercise 41

Simplify: $\dfrac{t^{-2}}{t^{-1} + 1}$.

Problem Solving

Example 5 Suppose you drive the 5 mi to work at 40 mi/hr and drive home at 60 mi/hr. Find your average speed for the round trip.

Solution: Your average speed is *not* 50 mi/hr. This is because you drive at 40 mi/hr for a longer period of time than you drive at 60 mi/hr. Instead, use the formula $r = d/t$, where r is rate, d is distance, and t is time.

$$\text{Average speed} = \frac{\text{distance of round trip}}{\text{time of round trip}}$$

$$= \frac{5 + 5}{\dfrac{5}{40} + \dfrac{5}{60}} \leftarrow \text{Since } t = \frac{d}{r}$$

$$= \frac{10}{\dfrac{1}{8} + \dfrac{1}{12}}$$

Now, simplify the complex fraction by multiplying every term by 24.

$$\text{Average speed} = \frac{10 \cdot 24}{\dfrac{1}{8} \cdot 24 + \dfrac{1}{12} \cdot 24} = \frac{240}{3 + 2} = \frac{240}{5} = 48$$

Your average speed was 48 mi/hr. ◻

Try Exercise 49

Suppose you walk the 2 mi to the beach at 6 mi/hr and walk home at 4 mi/hr. Find your average speed for the round trip.

Exercises 4.4

Completion Exercises

1. A fraction that contains other fractions in its numerator or denominator or both is called a(n) _____ .

2. A fraction that is not a complex fraction is called a(n) _____ .

3. The fractions that are contained in the numerator or in the denominator of a complex fraction are called _____ .

4. We can simplify a complex fraction using Method 1 by simplifying the numerator and the denominator sepa-rately, and then _____ ; or, using Method 2 by multiplying the numerator and the denominator by _____ .

True-False Exercises

5. $\dfrac{1 + 3 \cdot 4}{15}$ is a simple fraction.

6. $\dfrac{1 + \dfrac{1}{2}}{6}$ is a complex fraction.

Simplify each complex fraction.

7. $\dfrac{\dfrac{m+1}{3m}}{\dfrac{m-1}{m}}$

8. $\dfrac{\dfrac{m+2}{m}}{\dfrac{m-2}{5m}}$

9. $\dfrac{\dfrac{24x^4y^8}{10y}}{\dfrac{18x^7y^5}{15x^2}}$

10. $\dfrac{\dfrac{20x^6}{36x^3y^8}}{\dfrac{30x^4}{27x^2y^5}}$

11. $\dfrac{\dfrac{3k-6z}{k^2+kz}}{\dfrac{2k-4z}{kz+z^2}}$

12. $\dfrac{\dfrac{5k+20z}{2k^2+kz}}{\dfrac{4k+16z}{2kz+z^2}}$

13. $\dfrac{3d}{\dfrac{d}{10}+\dfrac{d}{6}}$

14. $\dfrac{2d}{\dfrac{d}{15}+\dfrac{d}{6}}$

15. $\dfrac{\dfrac{6}{r}-\dfrac{6}{t}}{\dfrac{6}{r}+\dfrac{6}{t}}$

16. $\dfrac{\dfrac{2}{r}+\dfrac{2}{t}}{\dfrac{2}{r}-\dfrac{2}{t}}$

17. $\dfrac{\dfrac{1}{xy}-1}{\dfrac{1}{x^2}-\dfrac{y}{x}}$

18. $\dfrac{1+\dfrac{1}{xy}}{\dfrac{x^2}{y^2}+\dfrac{x}{y^3}}$

19. $\dfrac{16-\dfrac{1}{m^2}}{4+\dfrac{1}{m}}$

20. $\dfrac{5-\dfrac{1}{m}}{25-\dfrac{1}{m^2}}$

21. $\dfrac{\dfrac{1}{x}-\dfrac{1}{y}}{\dfrac{x-y}{y}}$

22. $\dfrac{\dfrac{x-y}{x}}{\dfrac{1}{x}-\dfrac{1}{y}}$

23. $\dfrac{\dfrac{b^2}{a^2}-1}{\dfrac{1}{7a}-\dfrac{b}{7a^2}}$

24. $\dfrac{1-\dfrac{a^2}{b^2}}{\dfrac{a}{8b^2}-\dfrac{1}{8b}}$

25. $\dfrac{\dfrac{c}{9}-\dfrac{1}{c}}{1+\dfrac{c+6}{c}}$

26. $\dfrac{\dfrac{c}{4}-\dfrac{1}{c}}{1+\dfrac{c+4}{c}}$

27. $\dfrac{\dfrac{5}{d+3}+1}{1-\dfrac{2}{3+d}}$

28. $\dfrac{1-\dfrac{7}{d+1}}{\dfrac{4}{1+d}+1}$

29. $\dfrac{p+3+\dfrac{12}{p-5}}{p+5+\dfrac{16}{p-5}}$

30. $\dfrac{p+5+\dfrac{7}{p-3}}{p+3-\dfrac{7}{p-3}}$

31. $\dfrac{k+1-\dfrac{3k}{k+6}}{3-\dfrac{2}{k}}$

32. $\dfrac{k-4+\dfrac{2k}{k-2}}{8-\dfrac{3}{k}}$

33. $x-\dfrac{1}{1+\dfrac{1}{x-1}}$

34. $x+\dfrac{1}{1+\dfrac{1}{x+1}}$

35. $1+\dfrac{2}{2+\dfrac{1}{2+\dfrac{1}{2+\dfrac{1}{x}}}}$

36. $1-\dfrac{1}{3-\dfrac{1}{3-\dfrac{1}{3-\dfrac{1}{x}}}}$

Discussion Exercises

37. Explain how to simplify a complex fraction using (a) Method 1, (b) Method 2.

38. Simplify the two complex fractions below and note that they are not equal. Then explain why they are not equal.

(a) $\dfrac{\dfrac{6}{6}}{6}$ (b) $\dfrac{6}{\dfrac{6}{6}}$

Write with positive exponents and simplify.

39. $\dfrac{1-y^{-2}}{1+y^{-2}}$

40. $\dfrac{y^{-1}+1}{y^{-1}-1}$

41. $\dfrac{t^{-2}}{t^{-1}+1}$

42. $\dfrac{t^{-2}-1}{t^{-1}}$

43. $\dfrac{m+3n}{m^{-2}-(3n)^{-2}}$

44. $\dfrac{m^{-2}-(2n)^{-2}}{m+2n}$

45. $(x^{-1}+y^{-1})^{-1}$

46. $(x^{-1}-y^{-1})^{-1}$

Problem Solving

47. The efficiency E of a jack is given by the formula

$$E=\dfrac{\dfrac{x}{2}}{x+\dfrac{1}{2}}.$$

Determine E if $x=\dfrac{4}{5}$.

48. The efficiency E of a jack is given by the formula

$$E = \frac{\dfrac{x}{2}}{x + \dfrac{1}{2}}.$$

Determine E if $x = \dfrac{3}{8}$.

49. Suppose you walk the 2 mi to the beach at 6 mi/hr and walk home at 4 mi/hr. Find your average speed for the round trip.

50. A lumberjack travels 3 mi/hr going up a 40-ft tree and 12 mi/hr coming down. Find the lumberjack's average speed for the round trip.

51. A skier rides the ski lift up the slope at 2 mi/hr and skis down the slope at 30 mi/hr. Find the skier's average speed for the round trip.

52. A path up a hill is 1 mi long. A hiker walks up the path at 1 mi/hr and runs down the path at 12 mi/hr. Find the hiker's average speed for the round trip.

4.5 Equations with Rational Expressions

TAPE AU11

We learned how to solve linear equations in Chapter 2 and quadratic equations in Chapter 3. In this section we learn how to solve equations that contain rational expressions.

One way to solve an equation with rational expressions is to multiply both sides of the equation by the least common denominator. This will clear the equation of fractions, and we can proceed as before.

Example 1   Solve: $\dfrac{x}{2} + \dfrac{x}{3} = 10$.

Solution: Multiply both sides by the LCD, 6.

$$6\left(\frac{x}{2} + \frac{x}{3}\right) = 6(10)$$

$$6\left(\frac{x}{2}\right) + 6\left(\frac{x}{3}\right) = 6(10) \qquad \text{Distribute 6 over every term.}$$

$$3x + 2x = 60 \qquad \text{Simplify.}$$

$$5x = 60 \qquad \text{Combine like terms.}$$

$$x = 12 \qquad \text{Divide both sides by 5.}$$

The solution set is {12}. Check by substituting 12 for x in the original equation.   ❐

Try Exercise 9

Solve: $\dfrac{x}{2} + \dfrac{x}{5} = 14$.

CAUTION

Students often confuse the two problems below.

Problem 1   Solve: $\dfrac{x}{2} + \dfrac{x}{3} = 10$.

Solution:
$$6\left(\frac{x}{2} + \frac{x}{3}\right) = 6(10)$$
$$3x + 2x = 60$$
$$5x = 60$$
$$x = 12$$

Problem 2   Add: $\dfrac{x}{2} + \dfrac{x}{3}$.

Solution:
$$\frac{x}{2} + \frac{x}{3} = \frac{x \cdot 3}{2 \cdot 3} + \frac{x \cdot 2}{3 \cdot 2}$$
$$= \frac{3x}{6} + \frac{2x}{6}$$
$$= \frac{5x}{6}$$

Note that we *solve* the equation to get a value for x. We *add* the two rational expressions to get a third rational expression.   ∎

Example 2   Solve: $\dfrac{1}{6x} - \dfrac{1}{4x} = \dfrac{1}{8}$.

Solution:

$$24x\left(\dfrac{1}{6x} - \dfrac{1}{4x}\right) = 24x\left(\dfrac{1}{8}\right)$$ Multiply by the LCD, $24x$.

$$24x\left(\dfrac{1}{6x}\right) - 24x\left(\dfrac{1}{4x}\right) = 24x\left(\dfrac{1}{8}\right)$$ Distributive property

$$4 - 6 = 3x$$ Simplify.
$$-2 = 3x$$ Combine like terms.
$$-\dfrac{2}{3} = x$$ Divide by 3.

The number $-\frac{2}{3}$ checks, so the solution set is $\{-\frac{2}{3}\}$.   ❏

Try Exercise 15

Solve: $\dfrac{1}{8x} - \dfrac{1}{4x} = \dfrac{1}{12}$.

CAUTION

Students often make the mistake of inverting each term of an equation. Note how this mistake produces an incorrect solution to Example 2.

Wrong

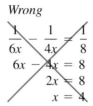

$$6x - 4x = 8$$
$$2x = 8$$
$$x = 4$$

A **proportion** is an equation of the form

$$\dfrac{a}{b} = \dfrac{c}{d}$$

where $b \neq 0$ and $d \neq 0$. If we multiply both sides by bd, we have

$$bd \cdot \dfrac{a}{b} = \dfrac{c}{d} \cdot bd$$
$$ad = bc$$

That is, we can solve a proportion by cross-multiplying.*

Cross-Product Rule

Assume $b \neq 0$ and $d \neq 0$.

$$\text{If } \dfrac{a}{b} = \dfrac{c}{d}, \text{ then } ad = bc$$

We illustrate the cross-product rule in Example 3.

Example 3   Solve: $\dfrac{5}{y + 8} = \dfrac{1}{y - 8}$.

*HISTORICAL NOTE: Problems involving proportions appear in the Rhind papyrus (ca. 2000 B.C.), and a general theory of proportions is found in Euclid's *Elements* (ca. 300 B.C.). Pythagoras (ca. 572–495 B.C.), in what is perhaps the oldest law of physics, noted that vibrating strings whose lengths are proportional will produce the same note one or more octaves apart.

Solution: Since the equation is a proportion, we can use the cross-product rule.

$$\frac{5}{y+8} \nearrow \frac{1}{y-8}$$ Cross-product rule

$$5(y-8) = (y+8)1$$ Multiply.

$$5y - 40 = y + 8$$ Distributive property

$$4y = 48$$ Subtract y, add 40.

$$y = 12$$ Divide by 4.

The number 12 checks, so the solution set is $\{12\}$.   ☐

Try Exercise 25

Solve: $\dfrac{3}{y+3} = \dfrac{1}{y-3}$.

Example 4   Solve: $\dfrac{4}{r} + 2 = \dfrac{2r}{r-4}$.

Solution: We cannot use the cross-product rule because the left side is not a single fraction. Instead, multiply every term by the LCD, $r(r-4)$.

$$r(r-4)\frac{4}{r} + r(r-4)2 = r(r-4)\frac{2r}{r-4}$$

$$(r-4)4 + 2r(r-4) = 2r^2$$ Divide out common factors.

$$4r - 16 + 2r^2 - 8r = 2r^2$$ Distributive property

$$2r^2 - 4r - 16 = 2r^2$$ Simplify.

$$-4r = 16$$ Subtract $2r^2$, add 16.

$$r = -4$$ Divide by -4.

The number -4 checks, so the solution set is $\{-4\}$.   ☐

Try Exercise 35

Solve: $\dfrac{8}{r} + 5 = \dfrac{5r}{r-2}$.

Example 5   Solve: $\dfrac{5}{p-3} - \dfrac{6}{p+3} = \dfrac{30}{p^2-9}$.

Solution: Since $p^2 - 9 = (p+3)(p-3)$, the LCD is $(p+3)(p-3)$. Multiply every term by the LCD.

$$(p+3)(p-3)\frac{5}{p-3} - (p+3)(p-3)\frac{6}{p+3} = (p+3)(p-3)\frac{30}{p^2-9}$$

$$(p+3)5 - (p-3)6 = 30$$

$$5p + 15 - 6p + 18 = 30$$

$$-p + 33 = 30$$

$$-p = -3$$

$$p = 3$$

But the number 3 does not check in the original equation because it makes two of the denominators 0.

Check: $\dfrac{5}{p-3} - \dfrac{6}{p+3} = \dfrac{30}{p^2-9}$ Original equation

$$\frac{5}{3-3} - \frac{6}{3+3} = \frac{30}{3^2-9}$$ Substitute $p = 3$.

$$\frac{5}{0} - \frac{6}{6} = \frac{30}{0}$$ False, since division by 0 is undefined

Since the only proposed solution, 3, does not check, there is no solution. The solution set is $\varnothing$, the empty set.   ☐

Try Exercise 51

Solve:

$\dfrac{6}{p-4} - \dfrac{7}{p+4} = \dfrac{48}{p^2-16}$.

When you solve an equation that contains a variable in a denominator, you must check your answer to make sure it does not make a denominator equal 0. A solution that is obtained during the solving process that does not check in the original equation is called an **extraneous solution** and must be discarded.

We can summarize the steps in solving an equation with rational expressions as follows.

To Solve an Equation with Rational Expressions

1. Factor the denominators to find the LCD.
2. Multiply every term on both sides by the LCD. This will clear the equation of fractions.
3. Solve the equation resulting from Step 2.
4. Check all solutions in the original equation. If any solution makes a denominator equal 0, it is an extraneous solution and must be discarded.

Note:

If the equation is of the form $\dfrac{a}{b} = \dfrac{c}{d}$, it is a proportion, and it may be easier to use the cross-product rule.

We illustrate these steps in Example 6.

Example 6   Solve: $\dfrac{m - 7}{m^2 + m - 20} + \dfrac{3}{2m + 10} = 1$.

Solution: Step 1:   Factor the denominators to find the LCD.

$$\frac{m - 7}{(m + 5)(m - 4)} + \frac{3}{2(m + 5)} = 1$$

The LCD is $2(m + 5)(m - 4)$.

Step 2:   Multiply every term by the LCD.

$$2(m + 5)(m - 4)\frac{m - 7}{(m + 5)(m - 4)} + 2(m + 5)(m - 4)\frac{3}{2(m + 5)} = 2(m + 5)(m - 4)1$$

Divide out common factors.

$$2(m - 7) + 3(m - 4) = 2(m^2 + m - 20)$$

Step 3:   Solve the equation resulting from Step 2.

$$2m - 14 + 3m - 12 = 2m^2 + 2m - 40$$
$$5m - 26 = 2m^2 + 2m - 40$$
$$0 = 2m^2 - 3m - 14$$
$$0 = (2m - 7)(m + 2)$$
$$2m - 7 = 0 \quad \text{or} \quad m + 2 = 0$$
$$2m = 7 \qquad\qquad m = -2$$
$$m = \frac{7}{2}$$

Try Exercise 57

Solve:
$\dfrac{m + 9}{m^2 - 7m - 8} + \dfrac{3}{2m + 2} = 1$.

Step 4:   Check all solutions in the original equation. Both numbers check, so the solution set is $\{\frac{7}{2}, -2\}$.   ◻

Exercises 4.5

True-False Exercises

1. One way to solve an equation with rational expressions is to multiply both sides by the least common denominator.

2. Multiplying both sides of an equation with rational expressions by the least common denominator will clear the equation of fractions.

3. We can only use the cross-product rule when the equation is a proportion.

4. When you solve an equation that contains a variable in a denominator, you must check each answer to make sure it does not make a denominator equal 0.

Completion Exercises

5. An equation of the form $\dfrac{a}{b} = \dfrac{c}{d}$ is called a(n) _____.

6. We can solve a proportion by _____.

7. A solution that is obtained during the solving process that does not check in the original equation is called a(n) _____.

8. Without even solving the equation $\dfrac{x}{x-4} = \dfrac{x}{x-4} + 1$, we know that the number _____ could not be a solution.

Solve each equation.

9. $\dfrac{x}{2} + \dfrac{x}{5} = 14$

10. $\dfrac{x}{4} + \dfrac{x}{5} = 18$

11. $\dfrac{4y - 2}{3} = \dfrac{3y + 11}{6}$

12. $\dfrac{3y - 1}{2} = \dfrac{7y + 6}{8}$

13. $z = \dfrac{1}{z}$

14. $z = \dfrac{4}{z}$

15. $\dfrac{1}{8x} - \dfrac{1}{4x} = \dfrac{1}{12}$

16. $\dfrac{1}{4x} - \dfrac{1}{3x} = \dfrac{1}{6}$

17. $3 - \dfrac{5}{2k} = \dfrac{3}{k} - \dfrac{5}{2}$

18. $2 - \dfrac{4}{3k} = \dfrac{2}{k} - \dfrac{4}{3}$

19. $\dfrac{5x - 3}{x + 4} = 0$

20. $\dfrac{7x - 2}{x + 6} = 0$

21. $\dfrac{4t + 13}{2t - 1} = 1$

22. $\dfrac{5t + 18}{3t - 4} = 1$

23. $1 - \dfrac{3}{2p} = \dfrac{1}{p^2}$

24. $1 + \dfrac{5}{6p} = \dfrac{1}{p^2}$

25. $\dfrac{3}{y + 3} = \dfrac{1}{y - 3}$

26. $\dfrac{5}{y + 2} = \dfrac{3}{y - 2}$

27. $\dfrac{5z + 2}{4z - 2} = \dfrac{z - 3}{2z + 3}$

28. $\dfrac{3z - 1}{5z - 2} = \dfrac{z + 2}{3z + 4}$

29. $\dfrac{w}{w + 9} = \dfrac{w}{w - 9}$

30. $\dfrac{w}{w - 4} = \dfrac{w}{w + 4}$

31. $\dfrac{x}{x + 10} + 1 = \dfrac{x - 8}{x + 10}$

32. $\dfrac{x}{x - 7} + 1 = \dfrac{x + 11}{x - 7}$

33. $\dfrac{t}{t - 4} + \dfrac{2}{t - 4} + t = 0$

34. $\dfrac{t}{t - 5} + \dfrac{3}{t - 5} + t = 0$

35. $\dfrac{8}{r} + 5 = \dfrac{5r}{r - 2}$

36. $\dfrac{6}{r} + 3 = \dfrac{3r}{r - 3}$

37. $\dfrac{1}{p} + \dfrac{1}{p + 1} = \dfrac{5}{6}$

38. $\dfrac{1}{p} + \dfrac{1}{p - 1} = \dfrac{7}{12}$

39. $\dfrac{x}{x - 2} + 4 = \dfrac{2}{x - 2}$

40. $\dfrac{x}{x - 3} + 3 = \dfrac{x}{x - 3}$

41. $\dfrac{t + 1}{3t + 15} + \dfrac{t}{t + 5} = \dfrac{5}{3}$

42. $\dfrac{t + 2}{2t + 8} + \dfrac{t}{t + 4} = \dfrac{1}{2}$

43. $\dfrac{m}{m + 1} = \dfrac{3}{m - 3} + 2$

44. $\dfrac{m}{m + 2} = -\dfrac{4}{m - 1} + 2$

45. $1 = \dfrac{1}{y + 2} + \dfrac{12}{(y + 2)^2}$

46. $1 = \dfrac{1}{y + 3} + \dfrac{20}{(y + 3)^2}$

47. $\dfrac{3}{r^2 - 3r} - \dfrac{1}{r - 3} = 1$

48. $\dfrac{2}{r^2 + r} + \dfrac{2}{r + 1} = 1$

49. $\dfrac{2}{z + 1} - \dfrac{4}{z - 1} = \dfrac{z}{z^2 - 1}$

50. $\dfrac{4}{z + 2} - \dfrac{5}{z - 2} = \dfrac{z}{z^2 - 4}$

51. $\dfrac{6}{p - 4} - \dfrac{7}{p + 4} = \dfrac{48}{p^2 - 16}$

52. $\dfrac{3}{p-6} - \dfrac{4}{p+6} = \dfrac{48}{p^2-36}$

53. $\dfrac{2}{q^2+q-2} = \dfrac{1}{q+2} + \dfrac{1}{q-1}$

54. $\dfrac{2}{q^2+q-12} = \dfrac{1}{q+4} + \dfrac{1}{q-3}$

55. $\dfrac{3}{4r-12} - \dfrac{1}{r-3} = \dfrac{1}{6r+12}$

56. $\dfrac{5}{6r-12} - \dfrac{1}{r-2} = \dfrac{1}{4r+12}$

57. $\dfrac{m+9}{m^2-7m-8} + \dfrac{3}{2m+2} = 1$

58. $\dfrac{m-7}{m^2-3m-10} + \dfrac{3}{5m+10} = 1$

59. $\dfrac{u-4}{u+5} - \dfrac{u+1}{u} = \dfrac{25}{u^2+5u}$

60. $\dfrac{u-3}{u+6} - \dfrac{u+1}{u} = \dfrac{14}{u^2+6u}$

Discussion
Exercises

61. Can you write the equation $\dfrac{1}{2x} - \dfrac{1}{x} = \dfrac{1}{4}$ as $2x - x = 4$? Why or why not?

62. Can you use the cross-product rule on the equation $\dfrac{x}{2} = \dfrac{4}{x} + 1$? Why or why not?

63. Discuss the difference in the two problems below.

Problem 1	**Problem 2**
Solve: $\dfrac{x}{2} + \dfrac{x}{4} = 8$.	Add: $\dfrac{x}{2} + \dfrac{x}{4}$.

64. Outline the steps in solving an equation with rational expressions.

Solve the equation or perform the indicated operation, as appropriate.

65. $\dfrac{x}{2} + \dfrac{x}{5}$

66. $\dfrac{x+1}{3} - \dfrac{x}{6} = 0$

67. $\dfrac{1}{y} - \dfrac{1}{y^2} = 0$

68. $\dfrac{1}{y} + \dfrac{5}{y^2}$

69. $\dfrac{m}{m-1} + \dfrac{4}{m-4}$

70. $\dfrac{m}{m+1} - \dfrac{9}{m+9}$

71. $\dfrac{2p}{p+3} - \dfrac{3}{p-3} = 1$

72. $\dfrac{2p}{p+1} - \dfrac{1}{p-1} = 1$

Getting Acquainted with Your Graphing Calculator

73. Check that 6 is a solution of $\dfrac{x}{x+3} = 1 - \dfrac{2}{x}$ by storing 6 in x and then calculating the value of each side of the equation.

74. Check that 6 is a solution of $\dfrac{x}{x+3} = 1 - \dfrac{2}{x}$ by storing 6 in x and then showing the equation is a true statement.

4.6 Problem Solving with Rational Expressions

In this section we learn how to solve applied problems that involve rational expressions.

TAPE AU11

Formulas

Recall that a **formula** is an equation that relates two or more real-world quantities. For example, suppose two resistors, one having a resistance of R_1 Ω and the other having a resistance of R_2 Ω, are connected in parallel, as shown in Fig. 4.1. The total resistance R of the electrical circuit is given by the formula

$$\frac{1}{R} = \frac{1}{R_1} + \frac{1}{R_2}$$

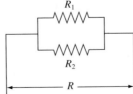

Figure 4.1

Example 1   Find the resistance in the other branch of a parallel-wired circuit if the resistance in one branch is 6 Ω and the total resistance is 2 Ω.

Solution:

$$\frac{1}{R} = \frac{1}{R_1} + \frac{1}{R_2} \qquad \text{Write down the appropriate formula.}$$

$$\frac{1}{2} = \frac{1}{6} + \frac{1}{R_2} \qquad \text{Let } R = 2 \text{ and } R_1 = 6.$$

To solve for R_2, multiply both sides by the LCD, $6R_2$.

$$6R_2 \cdot \frac{1}{2} = 6R_2 \cdot \frac{1}{6} + 6R_2 \cdot \frac{1}{R_2}$$

$$3R_2 = R_2 + 6 \qquad \text{Divide out common factors.}$$

$$2R_2 = 6 \qquad \text{Subtract } R_2.$$

$$R_2 = 3 \qquad \text{Divide by 2.}$$

The resistance in the other branch is 3 Ω.   ❑

Try Exercise 7
_____

If $\dfrac{1}{R} = \dfrac{1}{R_1} + \dfrac{1}{R_2}$, find R_1 if $R = 6$ and $R_2 = 12$.

Solving for a Specified Variable

Recall that we **solve** a formula for a specified variable by isolating that variable on one side of the equation.

Example 2   Solve: $\dfrac{1}{R} = \dfrac{1}{R_1} + \dfrac{1}{R_2}$ for R_1.

Solution:

$$RR_1R_2 \cdot \frac{1}{R} = RR_1R_2 \cdot \frac{1}{R_1} + RR_1R_2 \cdot \frac{1}{R_2} \qquad \text{Multiply by the LCD, } RR_1R_2.$$

$$R_1R_2 = RR_2 + RR_1 \qquad \text{Divide out common factors.}$$

$$R_1R_2 - RR_1 = RR_2 \qquad \text{Collect all terms with } R_1 \text{ on one side.}$$

$$R_1(R_2 - R) = RR_2 \qquad \text{Factor out } R_1.$$

$$R_1 = \frac{RR_2}{R_2 - R} \qquad \text{Divide by } R_2 - R.$$
❑

Try Exercise 25
_____

Solve: $\dfrac{1}{R} = \dfrac{1}{R_1} + \dfrac{1}{R_2}$ for R_2.

Word Problems

We solve the word problems in this section using the same steps we used in Sec. 2.3. The only difference in this section is that the equation we write in Step 2 is an equation with rational expressions.

Example 3   The sum of the reciprocals of two consecutive odd integers is equal to 8 divided by the product of the two integers. Find the integers.

Solution: Step 1:   Write down the unknown quantities and represent one of them by x. Then represent the other unknown quantity in terms of x. Since consecutive odd integers differ by 2, write

$$x = \text{first integer}$$
$$x + 2 = \text{second integer}$$

Step 2:   Write an equation involving x.

$$\underset{\downarrow}{\underset{\text{first integer}}{\text{Reciprocal of}}} + \underset{\downarrow}{\underset{\text{second integer}}{\text{reciprocal of}}} = \underset{\downarrow}{\underset{\text{product of integers}}{\text{8 divided by}}}$$

$$\frac{1}{x} + \frac{1}{x + 2} = \frac{8}{x(x + 2)}$$

Step 3: Solve to find x. Then use this value to find $x + 2$. Multiply by the LCD, $x(x + 2)$.

$$x(x + 2)\frac{1}{x} + x(x + 2)\frac{1}{x + 2} = x(x + 2)\frac{8}{x(x + 2)}$$

$$\begin{aligned}(x + 2) + x &= 8 && \text{Divide out common factors.}\\ 2x + 2 &= 8 && \text{Combine like terms.}\\ 2x &= 6 && \text{Subtract 2.}\\ x &= 3 && \text{Divide by 2.}\end{aligned}$$

The first integer is 3. The second integer is $x + 2 = 3 + 2 = 5$.

Step 4: Check your solution in the words of the original problem. 3 and 5 are consecutive odd integers. The sum of their reciprocals is $\frac{1}{3} + \frac{1}{5} = \frac{8}{15}$, which is 8 divided by their product. ❏

Try Exercise 35

The sum of the reciprocals of two consecutive odd integers is equal to 12 divided by the product of the two integers. Find the integers.

Another type of word problem that leads to an equation with rational expressions is a work problem. The key to solving a work problem lies in the following assumption: If it takes 4 hr to do a job, then the portion of the job done in 1 hr is $\frac{1}{4}$. This suggests the following rule.

Rule for Solving Work Problems

If it takes x hours to do a job, then the portion of the job done in 1 hr is $\frac{1}{x}$.

We use this rule in Examples 4 and 5.

Example 4   Two painters work at different rates. The first painter can paint a room in 3 hr. The second painter takes 4 hr to paint the same room. How long would it take the two painters working together to paint the room?

Solution: Let $x =$ the number of hours it would take the painters to paint the room working together.

$$\underset{\downarrow}{\underset{\substack{\text{painted by first}\\\text{painter in 1 hr}}}{\text{Portion of room}}} + \underset{\downarrow}{\underset{\substack{\text{painted by second}\\\text{painter in 1 hr}}}{\text{portion of room}}} = \underset{\downarrow}{\underset{\substack{\text{painted by both}\\\text{together in 1 hr}}}{\text{portion of room}}}$$

$$\frac{1}{3} + \frac{1}{4} = \frac{1}{x}$$

To clear fractions, multiply every term by the LCD, $12x$.

$$12x \cdot \frac{1}{3} + 12x \cdot \frac{1}{4} = 12x \cdot \frac{1}{x}$$

$$4x + 3x = 12 \qquad \text{Divide out common factors.}$$
$$7x = 12 \qquad \text{Combine like terms.}$$
$$x = \frac{12}{7} \qquad \text{Divide by 7.}$$

It would take $1\frac{5}{12}$ hr, or 1 hr and 25 min, to paint the room working together. Check the solution in the words of the original problem.   ◻

Try Exercise 37

Two paper hangers work at different rates. The first worker can wallpaper a wall in 2 hr. The second worker takes 3 hr to wallpaper the same wall. How long would it take the two workers working together to paper the wall?

Example 5   An old computer takes 5 hr longer to process the company payroll than does a new computer. If the two computers work together, the job takes 6 hr. How long does it take each computer to do the job on its own?

Solution:

$$x = \text{number of hours it takes the new computer}$$
$$x + 5 = \text{number of hours it takes the old computer}$$

Portion of job done by new computer in 1 hr	+	portion of job done by old computer in 1 hr	=	portion of job done by both computers in 1 hr
↓	↓	↓	↓	↓
$\dfrac{1}{x}$	$+$	$\dfrac{1}{x+5}$	$=$	$\dfrac{1}{6}$

Multiply every term by the LCD, $6x(x + 5)$. Then divide out the common factors.

$$6x(x+5)\frac{1}{x} + 6x(x+5)\frac{1}{x+5} = 6x(x+5)\frac{1}{6}$$
$$6(x+5) + 6x = x(x+5)$$

Simplify each side.

$$6x + 30 + 6x = x^2 + 5x$$
$$12x + 30 = x^2 + 5x$$

This is a quadratic equation. Collect terms on the right side and solve by factoring.

$$0 = x^2 - 7x - 30$$
$$0 = (x - 10)(x + 3)$$

$$x - 10 = 0 \qquad \text{or} \qquad x + 3 = 0$$
$$x = 10 \qquad\qquad\qquad x = -3$$

The new computer's time cannot be -3 hr. Therefore the new computer's time is 10 hr. The old computer's time is $x + 5 = 15$ hr. Check the solution in the words of the original problem.   ◻

Try Exercise 41

An old computer takes 6 hr longer to process the company payroll than does a new computer. If the two computers work together, the job takes 4 hr. How long does it take each computer to do the job on its own?

Another type of word problem that can lead to an equation with rational expressions is a motion problem. We can solve motion problems by making a chart and using the formula $d = rt$, which can also be written in the form $r = \dfrac{d}{t}$ or $t = \dfrac{d}{r}$.

Example 6   A motorboat travels 15 mi/hr in still water. If the boat takes the same amount of time to travel 2 mi upstream as it does to travel 3 mi downstream, find the speed of the current.

Solution: Let x = the speed of the current in miles per hour. To find the rate of the boat upstream (against the current), subtract the speed of the current from the speed of the boat. To find the rate of the boat downstream (with the current), add the speed of the current to the speed of the boat. Organize the information as in the chart below:

	d	r	t
Upstream	2	$15 - x$	$\dfrac{2}{15 - x}$
Downstream	3	$15 + x$	$\dfrac{3}{15 + x}$

since $t = \dfrac{d}{r}$

Since the boat spends the same amount of time traveling upstream as it does traveling downstream, we can write the following equation:

$$\text{Time upstream} = \text{time downstream}$$
$$\downarrow \qquad \qquad \downarrow \qquad \qquad \downarrow$$
$$\frac{2}{15 - x} = \frac{3}{15 + x}$$

The equation is a proportion, so we can cross-multiply.

$2(15 + x) = 3(15 - x)$	Cross-product rule
$30 + 2x = 45 - 3x$	Distributive property
$5x = 15$	Add $3x$, subtract 30.
$x = 3$	Divide by 5.

Try Exercise 45

A motorboat travels 10 mi/hr in still water. If the boat takes the same amount of time to travel 2 mi upstream as it does to travel 3 mi downstream, find the speed of the current.

The speed of the current is 3 mi/hr. Check the solution in the words of the original problem.   ❑

Exercises 4.6

◆◆◆

Completion Exercises

1. An equation that relates two or more real-world quantities is a(n) _____.

2. To solve a formula for one of its variables means to _____.

3. If it takes 3 hr to do a job, then the portion of the job done in 1 hr is _____.

4. If it takes x days to do a job, then the portion of the job done in 1 day is _____.

5. The formula $d = rt$, solved for t, is $t =$ _____.

6. The formula $d = rt$, solved for r, is $r =$ _____.

The formula $\dfrac{1}{R} = \dfrac{1}{R_1} + \dfrac{1}{R_2}$ is used to determine the resistance in a parallel-wired electrical circuit.

7. Find R_1 if $R = 6$ and $R_2 = 12$.

8. Find R if $R_1 = 10$ and $R_2 = 30$.

The formula $\dfrac{1}{f} = \dfrac{1}{d_o} + \dfrac{1}{d_i}$ is used to determine the focal length of a lens.

9. Find f if $d_o = 15$ and $d_i = 5$.

10. Find d_o if $f = 4$ and $d_i = 20$.

The relationship between the pressure, volume, and temperature of a gas is given by the formula $\dfrac{P_1V_1}{T_1} = \dfrac{P_2V_2}{T_2}$.

11. Find V_2 if $P_1 = 75$, $V_1 = 8$, $T_1 = 250$, $P_2 = 50$, and $T_2 = 275$.

12. Find T_1 if $P_1 = 135$, $V_1 = 18$, $P_2 = 90$, $V_2 = 12$, and $T_2 = 150$.

Getting Acquainted with Your Scientific Calculator

To find p if $\dfrac{1}{p} = \dfrac{1}{2} - \dfrac{1}{3}$, press

[Clear] 2 [1/x] [−] 3 [1/x] [=] [1/x]   6

The *sidereal* period of a planet is the time required for the planet to make one complete orbit about the sun. The *synodic period* of a planet is the time required for the planet, the sun, and the Earth to repeat the same configuration. A planet that orbits closer to the sun than does the Earth is called an *inferior planet*, while one that orbits farther from the sun is a *superior planet*. Mercury and Venus then are the only inferior planets.

13. For an inferior planet, the relationship between the sidereal period S_i and the synodic period S_y is

$$\frac{1}{S_i} = \frac{1}{E} + \frac{1}{S_y}$$

where E is the sidereal period of the Earth (365.26 days). Find the sidereal period of Venus, which has a synodic period of 583.92 days.

14. For a superior planet, the relationship between the sidereal period S_i and the synodic period S_y is

$$\frac{1}{S_i} = \frac{1}{E} - \frac{1}{S_y}$$

where E is the sidereal period of the Earth (365.26 days). Find the sidereal period of Mars, which has a synodic period of 779.95 days.

Discussion Exercises

Exercises 15 and 16 illustrate the steps in solving particular equations for x. Explain what was done at each step and why it was done.

15. $y = \dfrac{1}{x} + 3$ *Original equation*

Step 1: $xy = 1 + 3x$ _____
Step 2: $xy - 3x = 1$ _____
Step 3: $x(y - 3) = 1$ _____

Step 4: $x = \dfrac{1}{y - 3}$ _____

16. $y = \dfrac{x}{x + 2}$ *Original equation*

Step 1: $y(x + 2) = x$ _____
Step 2: $yx + 2y = x$ _____
Step 3: $2y = x - yx$ _____
Step 4: $2y = x(1 - y)$ _____

Step 5: $\dfrac{2y}{1 - y} = x$ _____

Solve each equation for the specified variable.

17. $C = \dfrac{Q}{V}$ for V

18. $I = \dfrac{V}{R}$ for R

19. $F = \dfrac{gm_1m_2}{d^2}$ for g

20. $F = \dfrac{gm_1m_2}{d^2}$ for m_2

21. $\dfrac{P_1V_1}{T_1} = \dfrac{P_2V_2}{T_2}$ for P_1

22. $\dfrac{P_1V_1}{T_1} = \dfrac{P_2V_2}{T_2}$ for T_1

23. $I = \dfrac{E}{R + r}$ for r

24. $I = \dfrac{E}{R + r}$ for R

25. $\dfrac{1}{R} = \dfrac{1}{R_1} + \dfrac{1}{R_2}$ for R_2

26. $\dfrac{1}{f} = \dfrac{1}{f_1} + \dfrac{1}{f_2}$ for f

27. $x = \dfrac{1}{y} - 2$ for y

28. $x = \dfrac{1}{y} + 5$ for y

29. $d = \dfrac{r}{1 + rt}$ for r

30. $d = \dfrac{r}{1 - rt}$ for r

31. $y = \dfrac{x + 3}{x - 3}$ for x

32. $y = \dfrac{x - 4}{x + 4}$ for x

Problem Solving

33. What number must be added to the numerator and the denominator of the fraction $\frac{10}{13}$ to make the result equal $\frac{4}{5}$?

34. One number is 3 times another. If the sum of their reciprocals is $\frac{2}{9}$, find the numbers.

35. The sum of the reciprocals of two consecutive odd integers is equal to 12 divided by the product of the two integers. Find the integers.

36. If the third of three consecutive even integers is divided by the first, the result is $\frac{1}{10}$ more than the quotient of the second and the first. Find the integers.

37. Two paper hangers work at different rates. The first worker can wallpaper a wall in 2 hr. The second worker takes 3 hr to wallpaper the same wall. How long would it take the two workers working together to paper the wall?

38. John can plow the family's field in 7 hr using the old tractor; his wife can plow the field with the new tractor in just 5 hr. How long will it take them to plow the field if they work together?

39. Working together, two part-time employees can stuff 55,000 envelopes in 10 days. One employee can stuff envelopes twice as fast as the other. How long would it take each employee to stuff the envelopes working alone?

40. Working together, two secretaries can type a 1300-page manuscript in 8 days. One secretary can type twice as fast as the other. How long would it take each secretary to type the manuscript working alone?

41. An old computer takes 6 hr longer to process the company payroll than does a new computer. If the two computers work together, the job takes 4 hr. How long does it take each computer to do the job on its own?

42. An old printing press takes 8 hr longer to put out a daily paper than a new press. If the two presses work together, the job takes 3 hr. How long does it take each press to put out the paper on its own?

43. One inlet pipe can fill an empty tank in 6 min; another takes 18 min. A pump can empty a full tank in 9 min. How long will it take to fill the empty tank if both pipes are open and the pump is on?

44. One inlet pipe can fill an empty swimming pool in 3 days; another takes 6 days. A pump can empty a full pool in 4 days. How long will it take to fill the empty pool if both pipes are open and the pump is on?

45. A motorboat travels 10 mi/hr in still water. If the boat takes the same amount of time to travel 2 mi upstream as it does to travel 3 mi downstream, find the speed of the current.

46. A helicopter can travel 80 mi/hr in still air. If the helicopter takes the same amount of time to travel 2 mi against the wind as it does to travel 3 mi with the wind, find the speed of the wind.

47. An express train travels 160 mi in the same time that a freight train travels 100 mi. If the express travels 30 mi/hr faster than the freight, find the speed of each.

	d	r	t
Freight	100	x	
Express	160	$x + 30$	

48. A plane travels 840 mi in the same time that a car travels 240 mi. If the plane travels 150 mi/hr faster than the car, find the speed of each.

	d	r	t
Car	240	x	
Plane	840	$x + 150$	

49. A motorist drove 110 mi before running out of gas and walking 2 mi to a gas station. The rate at which the motorist drove was 10 times the rate at which the motorist walked. If the total time spent driving and walking was 3 hr, what was the rate at which the motorist walked?

50. A racer cycled 75 mi of a 78-mi course before breaking a wheel and walking the rest of the way. The rate at which the racer cycled was 5 times the rate at which the racer walked. If the total time spent cycling and walking was 4 hr, what was the rate at which the racer walked?

Discussion Exercise

51. Outline the steps in solving a word problem.

4.7 Dividing Polynomials

TAPE AU6

In Chapter 3 we learned how to add, subtract, and multiply polynomials. We also learned how to divide two polynomials when both polynomials were monomials. For example,

$$\frac{x^2}{x} = x \qquad \frac{10y^3}{2y} = 5y^2 \qquad \text{and} \qquad \frac{-24a^2b^5}{4a^2b^2} = -6b^3$$

In this section we learn how to divide *any* two polynomials.

Dividing by a Monomial

We begin by discussing how to divide a polynomial having two or more terms by a monomial. Recall the rule for adding and subtracting fractions.

$$\frac{a}{d} + \frac{b}{d} - \frac{c}{d} = \frac{a + b - c}{d}$$

If we turn this rule around, we have

$$\frac{a + b - c}{d} = \frac{a}{d} + \frac{b}{d} - \frac{c}{d}$$

This suggests the following rule for dividing by a monomial.

Dividing by a Monomial

To divide a polynomial by a monomial, divide each term of the polynomial by the monomial.

Example 1   Divide $9x^2 + 6x - 12$ by 3.

Solution:

$$\frac{9x^2 + 6x - 12}{3} = \frac{9x^2}{3} + \frac{6x}{3} - \frac{12}{3} \qquad \text{Divide each term by 3.}$$

$$= 3x^2 + 2x - 4 \qquad \text{Simplify each fraction.}$$

Check by multiplying $3x^2 + 2x - 4$ by 3. The result should be the polynomial $9x^2 + 6x - 12$. ❑

Try Exercise 5

Divide: $\dfrac{8x^2 + 4x - 6}{2}$.

Example 2   Divide: $\dfrac{6y^2 - 4y + 8}{2y}$.

Solution:

$$\frac{6y^2 - 4y + 8}{2y} = \frac{6y^2}{2y} - \frac{4y}{2y} + \frac{8}{2y} \qquad \text{Divide each term by } 2y.$$

$$= 3y - 2 + \frac{4}{y} \qquad \text{Simplify each fraction.}$$

Try Exercise 7

Divide: $\dfrac{9y^2 - 18y + 12}{3y}$.

Note that the answer is not a polynomial, due to the term $\dfrac{4}{y}$. ❑

Example 3   Divide: $\dfrac{64r^2 - 28s^2 + 10r^3s^3 - 4r^2s^2}{4r^2s^2}$.

Solution:

$$\frac{64r^2 - 28s^2 + 10r^3s^3 - 4r^2s^2}{4r^2s^2} \qquad \text{Original problem}$$

$$= \frac{64r^2}{4r^2s^2} - \frac{28s^2}{4r^2s^2} + \frac{10r^3s^3}{4r^2s^2} - \frac{4r^2s^2}{4r^2s^2} \qquad \text{Divide each term by } 4r^2s^2.$$

$$= \frac{16}{s^2} - \frac{7}{r^2} + \frac{5}{2}rs - 1 \qquad \text{Simplify each fraction.}$$ ❑

Try Exercise 15

Divide:
$\dfrac{64r^2 - 48s^2 + 10r^2s^3 - 8r^2s^2}{8r^2s^2}$.

Long Division of Polynomials

If the divisor contains two or more terms, we use a procedure called **long division of polynomials** to find the quotient. This procedure is illustrated in Example 4. As you read through this example, note the similarity between long division of polynomials and long division of whole numbers.

Example 4   Divide $3x^2 + 17x + 10$ by $x + 5$.

Solution: Step 1:   Write the divisor and the dividend in descending order in long division form.

$$x + 5 \overline{)3x^2 + 17x + 10}$$

$$\underbrace{\quad}_{\uparrow \atop \text{divisor}} \quad \underbrace{\qquad\qquad}_{\uparrow \atop \text{dividend}}$$

Step 2: Divide x into $3x^2$ and get $3x$.

$$\dfrac{3x^2}{x} = 3x$$

$$\begin{array}{r} 3x \\ x + 5 \overline{)3x^2 + 17x + 10} \end{array}$$

Step 3: Multiply $3x$ by $x + 5$ and get $3x^2 + 15x$.

times

$$\begin{array}{r} \boxed{3x} \\ (x + 5)\overline{)3x^2 + 17x + 10} \\ 3x^2 + 15x \end{array}$$

equals

Step 4: Subtract $3x^2 + 15x$ from $3x^2 + 17x$ and get $2x$. Then bring down the next term in the dividend, 10. This gives a new dividend of $2x + 10$.

$$\begin{array}{r} 3x \\ x + 5 \overline{)3x^2 + 17x + 10} \\ \underline{3x^2 + 15x} \downarrow \\ 2x + 10 \end{array}$$

Step 5: Divide x into $2x$ and get 2.

$$\dfrac{2x}{x} = 2$$

$$\begin{array}{r} 3x \;+\; 2 \\ x + 5 \overline{)3x^2 + 17x + 10} \\ \underline{3x^2 + 15x} \\ 2x + 10 \end{array}$$

Step 6: Multiply 2 by $x + 5$ and get $2x + 10$.

times

$$\begin{array}{r} 3x \boxed{+ \; 2} \\ (x + 5)\overline{)3x^2 + 17x + 10} \\ \underline{3x^2 + 15x} \\ 2x + 10 \end{array}$$

equals $\longrightarrow 2x + 10$

Step 7: Subtract $2x + 10$ from $2x + 10$ and get the remainder, 0.

$$\begin{array}{r} 3x \;+\; 2 \\ x + 5 \overline{)3x^2 + 17x + 10} \\ \underline{3x^2 + 15x} \\ 2x + 10 \\ \underline{2x + 10} \\ 0 \leftarrow \text{remainder} \end{array}$$

The quotient is $3x + 2$ and the remainder is 0.

Step 8: Check by multiplying the divisor $x + 5$ by the quotient $3x + 2$, and then adding the remainder 0. The result should be the dividend $3x^2 + 17x + 10$.

Divisor · quotient + remainder = dividend

$$\begin{array}{cccc}
\downarrow & \downarrow & \downarrow & \downarrow \\
(x+5)\cdot(3x+2)+ & 0 & = 3x^2 + 17x + 10 \\
3x^2 + 17x + 10 & = 3x^2 + 17x + 10 & \text{True} \quad \square
\end{array}$$

Try Exercise 17

Divide: $\dfrac{3x^2 + 14x + 8}{x + 4}$.

CAUTION

Remember, to subtract two polynomials, mentally change the sign of each term in the second polynomial and then add.

Correct

$$\begin{array}{r}
5x^2 - 7x \\
(-) \quad 5x^2 - 3x \\
\hline
-4x
\end{array}$$

Wrong

$$\begin{array}{r}
5x^2 - 7x \\
(-) \quad 5x^2 - 3x \\
\hline
-10x
\end{array}$$

∎

Here is a summary of the steps in long division of polynomials.

Long Division of Polynomials

1. Write the divisor and the dividend in descending order. Write any missing power with a coefficient of 0.
2. Divide the first term of the divisor into the first term of the dividend. This gives the first term of the quotient.
3. Multiply the term of the quotient obtained in Step 2 by every term of the divisor.
4. Subtract and bring down the next term to get the new dividend.
5. Repeat Steps 2, 3, and 4 with the new dividend. Continue repeating these steps until the degree of the new dividend is less than the degree of the divisor. This last dividend is the remainder.
6. Check by multiplying the divisor by the quotient and then adding the remainder. The result should equal the original dividend.

Example 5   Divide: $\dfrac{10y^3 - y^2 + 12y - 9}{5y - 3}$.

Solution: Follow the steps and arrange your work as shown below.

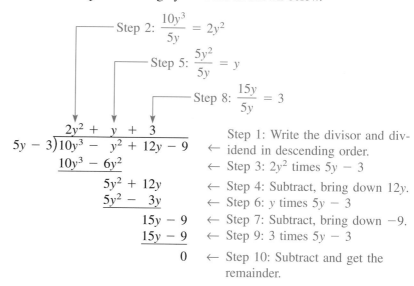

Step 2: $\dfrac{10y^3}{5y} = 2y^2$

Step 5: $\dfrac{5y^2}{5y} = y$

Step 8: $\dfrac{15y}{5y} = 3$

$$\begin{array}{r}
2y^2 + y + 3 \\
5y - 3\overline{)10y^3 - y^2 + 12y - 9} \\
\underline{10y^3 - 6y^2} \\
5y^2 + 12y \\
\underline{5y^2 - 3y} \\
15y - 9 \\
\underline{15y - 9} \\
0
\end{array}$$

← Step 1: Write the divisor and dividend in descending order.
← Step 3: $2y^2$ times $5y - 3$
← Step 4: Subtract, bring down $12y$.
← Step 6: y times $5y - 3$
← Step 7: Subtract, bring down -9.
← Step 9: 3 times $5y - 3$
← Step 10: Subtract and get the remainder.

Try Exercise 23

Divide:
$$\frac{10y^3 - 23y^2 + y - 15}{2y - 5}.$$

The quotient is $2y^2 + y + 3$ and the remainder is 0. Check by multiplication. ❐

Example 6   Divide: $\dfrac{4p + 8p^3 + 11}{1 + 2p}.$

Solution: Write the divisor and the dividend in descending order. Write $0p^2$ for the missing power of p in the dividend so that like terms will line up.

$$
\begin{array}{r}
4p^2 - 2p\ + 3 \\
2p + 1\overline{)8p^3 + 0p^2 + 4p + 11} \\
\underline{8p^3 + 4p^2} \\
-4p^2 + 4p \\
\underline{-4p^2 - 2p} \\
6p + 11 \\
\underline{6p + \ \ 3} \\
8
\end{array}
$$

Try Exercise 27

Divide: $\dfrac{11p + 9p^3 + 10}{1 + 3p}.$

The quotient is $4p^2 - 2p + 3$ and the remainder is 8. ❐

To check the answer to Example 6, show that the following equation is true:

$$(2p + 1)(4p^2 - 2p + 3) + 8 = 8p^3 + 4p + 11$$

Note that if we turn this equation around and divide both sides by $2p + 1$, we get

$$\frac{8p^3 + 4p + 11}{2p + 1} = \frac{(2p + 1)(4p^2 - 2p + 3)}{2p + 1} + \frac{8}{2p + 1}$$

which simplifies to

$$\frac{8p^3 + 4p + 11}{2p + 1} = 4p^2 - 2p + 3 + \frac{8}{2p + 1}$$

The left side of this equation is the original division problem. Therefore the right side is the answer to the original division problem. We shall write our answers in this form from now on.

CAUTION

When you write the answer to a long division problem involving polynomials, make sure you write the addition symbol in front of the fraction containing the remainder; otherwise, multiplication is assumed.

Correct

$$4p^2 - 2p + 3 + \frac{8}{2p + 1}$$

Wrong

$$4p^2 - 2p + 3\,\frac{8}{2p + 1}$$ ■

Example 7   Divide: $\dfrac{6m^4 + m^3 - 5}{3m^2 - 4m}.$

Solution: Arrange your work as follows:

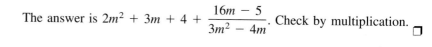

$$3m^2 - 4m + 0 \overline{)\begin{array}{l} 2m^2 + 3m + 4 \\ 6m^4 + m^3 + 0m^2 + 0m - 5 \end{array}}$$

$$\underline{6m^4 - 8m^3 + 0m^2}$$
$$9m^3 + 0m^2 + 0m$$
$$\underline{9m^3 - 12m^2 + 0m}$$
$$12m^2 + 0m - 5$$
$$\underline{12m^2 - 16m + 0}$$

Stop, since degree is less than degree of divisor. $\rightarrow 16m - 5$

Try Exercise 35

Divide: $\dfrac{20m^4 + 13m^3 - 8}{5m^2 - 3m}$.

The answer is $2m^2 + 3m + 4 + \dfrac{16m - 5}{3m^2 - 4m}$. Check by multiplication. ◻

Exercises 4.7

True-False Exercises

1. To divide a polynomial by a monomial, divide each term of the polynomial by the monomial.

2. Use long division to divide two polynomials when the divisor contains two or more terms.

3. To subtract two polynomials, mentally change the sign of each term in the second polynomial, and then add.

4. If we divide $x^2 + 5x + 10$ by $x + 2$, the quotient is $x + 3$ and the remainder is 4. Therefore, we can write the answer as
$$x + 3\frac{4}{x + 2}$$

Divide.

5. $\dfrac{8x^2 + 4x - 6}{2}$

6. $\dfrac{10x^2 - 15x + 20}{5}$

7. $\dfrac{9y^2 - 18y + 12}{3y}$

8. $\dfrac{8y^2 + 16y + 24}{4y}$

9. $(z^3 + z^2 - z) \div z$

10. $(z^5 - z^3 - z^2) \div z^2$

11. $\dfrac{6a^5 - 24a^4 + 10a^3 - 18a}{-6a^3}$

12. $\dfrac{-16a^5 + 8a^4 + 20a^3 - 32a}{-8a^3}$

13. $\dfrac{27m^2n^2 - 18mn - 3m}{9m^2n}$

14. $\dfrac{50m^2n^2 - 75mn + 5n}{25mn^2}$

15. $\dfrac{64r^2 - 48s^2 + 10r^2s^3 - 8r^2s^2}{8r^2s^2}$

16. $\dfrac{48r^3 + 36s^3 - 20r^4s^4 - 12r^3s^3}{12r^3s^3}$

Use long division of polynomials to divide.

17. $\dfrac{3x^2 + 14x + 8}{x + 4}$

18. $\dfrac{3x^2 + 11x + 10}{x + 2}$

19. $\dfrac{6m^2 + 11m - 10}{3m - 2}$

20. $\dfrac{8m^2 + 14m - 15}{4m - 3}$

21. $\dfrac{k^3 + 3k^2 + 5k + 3}{k + 1}$

22. $\dfrac{k^3 + 2k^2 + 4k + 3}{k + 1}$

23. $\dfrac{10y^3 - 23y^2 + y - 15}{2y - 5}$

24. $\dfrac{9y^3 - 3y^2 + 4y + 4}{3y - 2}$

25. $\dfrac{4r^3 - r^2 + 17r - 15}{4r + 3}$

26. $\dfrac{5r^3 - 3r^2 - 22r + 8}{5r + 2}$

27. $\dfrac{11p + 9p^3 + 10}{1 + 3p}$

28. $\dfrac{7p + 12p^3 + 13}{1 + 2p}$

29. $\dfrac{6t^3 - 5t^2 + 18t - 15}{6t - 5}$

30. $\dfrac{8t^3 - 7t^2 + 16t - 14}{8t - 7}$

31. $\dfrac{2x^3 + 5x^2 - 3x - 21}{x^2 - x - 3}$

32. $\dfrac{3x^3 + 7x^2 - x - 8}{x^2 + x - 2}$

33. $\dfrac{6a^4 - 10a^3 - a^2 + 15a - 9}{3 + 2a^2 - 4a}$

34. $\dfrac{4a^4 + 6a^3 + 22a^2 - 15a + 75}{5 + 2a^2 - 3a}$

35. $\dfrac{20m^4 + 13m^3 - 8}{5m^2 - 3m}$

36. $\dfrac{8m^4 + 10m^3 - 7}{4m^2 - 3m}$

37. $\dfrac{28y - 26y^2 + 8y^3 - 100}{2y^2 + 7}$

38. $\dfrac{32y + 12y^3 - 39y^2 - 110}{3y^2 + 8}$

39. $\dfrac{9x^3 - 15x^2 - 24x}{3x - 5}$

40. $\dfrac{6x^3 - 27x^2 - 8x}{2x - 9}$

41. $\dfrac{20z^3 + 17z^2 + 11}{z + 4z^2}$

42. $\dfrac{24z^3 + 16z^2 + 13}{z + 6z^2}$

Discussion Exercises

43. Explain how to check the answer to a long division of polynomials problem.

44. Outline the steps in long division of polynomials.

45. When performing long division of polynomials, why do we fill in a missing power with a coefficient of 0?

46. When performing long division of polynomials, why do we write the divisor and the dividend in descending order?

Divide.

47. $(8p^3 - 125) \div (2p - 5)$ 48. $(27p^3 + 64) \div (3p + 4)$

49. $(3p^2 + \frac{5}{2}p + \frac{1}{2}) \div (2p + 1)$

50. $(2p^2 + \frac{5}{3}p + \frac{1}{3}) \div (3p + 1)$

51. $(5r^5 - 3r^3 + 6r^2) \div (4r^2 - 4)$

52. $(9r^5 - 4r^2 - 2r) \div (6r^2 + 6)$

53. $x \div (x - 1)$ 54. $x^2 \div (x^2 + 1)$

Problem Solving

55. Show that the average of any three consecutive integers is always the middle integer.

56. The total cost of producing x oven mitts is given by $0.05x^2 + 25x + 1400$. If the average cost per mitt is the total cost divided by the number of mitts x, find the average cost per mitt.

57. If a particle travels $2t^3 + 7t^2 + 8t + 28$ centimeters in $2t + 7$ seconds, find its average speed.

58. If the volume of the box in Fig. 4.2 is $m^3 + 7m^2 + 10m$, find its width.

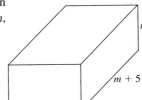

Figure 4.2

Chapter 4 Key Terms

4.1 Rational expression An expression of the form P/Q, where P and Q are polynomials and $Q \neq 0$

4.2 Reciprocal The rational expressions R/S and S/R are reciprocals of each other.

4.3 Least common denominator (LCD) The LCD of a set of rational expressions is the simplest polynomial that can be divided by the denominator of each rational expression.

4.4 Complex fraction A fraction that contains other fractions in its numerator or denominator or both

Simple fraction A fraction that is not a complex fraction

Secondary fractions (of a complex fraction) The fractions contained in the numerator or the denominator of the complex fraction

4.5 Proportion An equation of the form $\dfrac{a}{b} = \dfrac{c}{d}$, where $b \neq 0$ and $d \neq 0$

Extraneous solution (of an equation) A solution that is obtained during the solving process that does not check in the original equation

4.6 Formula An equation that relates two or more real-world quantities

Solve (a formula for a variable) Isolate that variable on one side of the equation.

4.7 Long division (of polynomials) A procedure used to divide two polynomials when the divisor contains two or more terms

Chapter 4 Key Rules/Steps

4.1 Simplifying Rational Expressions

Fundamental Principle of Rational Expressions

If P, Q, and K are polynomials and neither Q nor K is 0, then

$$\frac{P \cdot K}{Q \cdot K} = \frac{P}{Q}$$ Example: $$\frac{(x+5)(x+4)}{(x-5)(x+4)} = \frac{x+5}{x-5}$$

To Simplify a Rational Expression

1. Write the numerator and denominator in prime factored form.

$$\frac{3x^2 - 6x}{x^2 - 5x + 6} = \frac{3x(x-2)}{(x-2)(x-3)}$$

2. Divide both numerator and denominator by any factors they have in common.

$$= \frac{3x}{x-3}$$

4.2 Multiplying and Dividing Rational Expressions

If $\dfrac{P}{Q}$ and $\dfrac{R}{S}$ are rational expressions, then $\dfrac{P}{Q} \cdot \dfrac{R}{S} = \dfrac{P \cdot R}{Q \cdot S}$

and $\dfrac{P}{Q} \div \dfrac{R}{S} = \dfrac{P}{Q} \cdot \dfrac{S}{R} = \dfrac{P \cdot S}{Q \cdot R} \left(\dfrac{R}{S} \neq 0 \right).$

To Divide Two Rational Expressions

1. Invert the second fraction and multiply. Write all numerators and denominators in prime factored form.

$$\frac{m^2 - 9}{10} \div \frac{m^2 - 7m + 12}{5m^2 - 20m}$$

$$= \frac{m^2 - 9}{10} \cdot \frac{5m^2 - 20m}{m^2 - 7m + 12}$$

$$= \frac{(m-3)(m+3)}{2 \cdot 5} \cdot \frac{5m(m-4)}{(m-3)(m-4)}$$

2 & 3. Divide out common factors and multiply the remaining factors.

$$= \frac{m(m+3)}{2}$$

4.3 Adding and Subtracting Rational Expressions

If $\dfrac{P}{Q}$ and $\dfrac{R}{Q}$ are rational expressions, then

$$\frac{P}{Q} + \frac{R}{Q} = \frac{P+R}{Q} \quad \text{and} \quad \frac{P}{Q} - \frac{R}{Q} = \frac{P-R}{Q}$$

To Subtract the Expressions $\dfrac{2}{x-2} - \dfrac{x+8}{x^2 + x - 6}$

1. Write the denominators in prime factored form. Multiply together the highest power of each factor to find the LCD.

$$x - 2 = x - 2$$
$$x^2 + x - 6 = (x-2)(x+3)$$
$$\text{LCD} = (x-2)(x+3)$$

2 & 3. Write each rational expression with the LCD. Multiply out the numerators, but leave the denominators in factored form.

$$\frac{2(x+3)}{(x-2)(x+3)} - \frac{x+8}{(x-2)(x+3)}$$

$$= \frac{2x+6}{(x-2)(x+3)} - \frac{x+8}{(x-2)(x+3)}$$

4. Subtract the numerators and place the result over the LCD.

$$= \frac{2x + 6 - (x+8)}{(x-2)(x+3)}$$

$$= \frac{2x + 6 - x - 8}{(x-2)(x+3)}$$

$$= \frac{x-2}{(x-2)(x+3)}$$

5. Factor the numerator and divide out common factors.

$$= \frac{1}{x+3}$$

4.4 Complex Fractions

To Simplify a Complex Fraction

Method 1: Simplify the numerator and the denominator separately. Then invert and multiply.

$$\frac{1 - \dfrac{4}{x^2}}{1 + \dfrac{2}{x}} = \frac{\dfrac{x^2}{x^2} - \dfrac{4}{x^2}}{\dfrac{x}{x} + \dfrac{2}{x}} = \frac{\dfrac{x^2 - 4}{x^2}}{\dfrac{x+2}{x}}$$

$$= \frac{x^2 - 4}{x^2} \cdot \frac{x}{x+2}$$

$$= \frac{(x-2)(x+2)}{x^2} \cdot \frac{x}{x+2}$$

$$= \frac{x-2}{x}$$

Method 2: Multiply both numerator and denominator by the LCD of all the secondary fractions.

$$\frac{1 - \dfrac{4}{x^2}}{1 + \dfrac{2}{x}} = \frac{\left(1 - \dfrac{4}{x^2}\right)x^2}{\left(1 + \dfrac{2}{x}\right)x^2} = \frac{x^2 - 4}{x^2 + 2x}$$

$$= \frac{(x - 2)(x + 2)}{x(x + 2)}$$

$$= \frac{x - 2}{x}$$

4.5 Equations with Rational Expressions

To Solve the Equation $\dfrac{9}{r^2 + 3r} + \dfrac{3}{r + 3} = 1$

1. Factor the denominators to find the LCD.

$$\frac{9}{r(r + 3)} + \frac{3}{r + 3} = 1$$

2. Multiply every term on both sides by the LCD, $r(r + 3)$.

$$r(r + 3) \cdot \frac{9}{r(r + 3)} + r(r + 3) \cdot \frac{3}{r + 3} = r(r + 3) \cdot 1$$

$$9 + 3r = r^2 + 3r$$

3. Solve the equation resulting from Step 2.

$0 = r^2 - 9$	Subtract $3r$, subtract 9.
$0 = (r - 3)(r + 3)$	Factor.
$r - 3 = 0$ or $r + 3 = 0$	Set each factor equal to 0.
$r = 3$ $r = -3$	Solve each equation.

4. Check all solutions in the original equation. The number 3 checks, but the number -3 makes both denominators 0, so -3 is an extraneous solution and must be discarded. The solution set is $\{3\}$.

Cross-Product Rule

If $b \neq 0$ and $d \neq 0$, the proportion

$$\frac{a}{b} = \frac{c}{d} \text{ is equivalent to } ad = bc$$

Example: $\dfrac{3}{x + 4} = \dfrac{2}{x - 6}$ is equivalent to

$$3(x - 6) = 2(x + 4)$$

4.6 Problem Solving with Rational Expressions

To Solve $y = \dfrac{x + 2}{x - 3}$ **for** x

$y(x - 3) = \dfrac{x + 2}{x - 3}(x - 3)$	Multiply both sides by the LCD, $x - 3$.
$yx - 3y = x + 2$	Simplify each side.
$yx - x = 2 + 3y$	Subtract x, add $3y$.
$x(y - 1) = 2 + 3y$	Factor out x.
$x = \dfrac{2 + 3y}{y - 1}$	Divide by $y - 1$.

Work Problem

Sue can paint the garage in 6 hr, while Sam takes 8 hr. How long would it take them to paint the garage working together?

x = number of hours to paint the garage working together

$$\frac{1}{6} + \frac{1}{8} = \frac{1}{x}$$

$$24x \cdot \frac{1}{6} + 24x \cdot \frac{1}{8} = 24x \cdot \frac{1}{x}$$

$$4x + 3x = 24$$

$$7x = 24$$

$$x = 3\tfrac{3}{7} \text{ hr}$$

4.7 Dividing Polynomials

Dividing by a Monomial

To divide a polynomial by a monomial, divide each term of the polynomial by the monomial.

$$\frac{12x^2 + 6x - 3}{3x} = \frac{12x^2}{3x} + \frac{6x}{3x} - \frac{3}{3x} = 4x + 2 - \frac{1}{x}$$

Use Long Division of Polynomials to Divide $x^3 + 3x - 19$ **by** $x - 2$

1. Write the divisor and the dividend in descending order. Write any missing power with a coefficient of 0.

$$x - 2 \overline{)x^3 + 0x^2 + 3x - 19}$$

2. Divide the first term of the divisor into the first term of the dividend.

$$\begin{array}{r} x^2 \\ x - 2 \overline{)x^3 + 0x^2 + 3x - 19} \end{array}$$

3. Multiply the term of the quotient obtained in Step 2 by every term of the divisor.

$$\begin{array}{r} x^2 \\ x - 2 \overline{)x^3 + 0x^2 + 3x - 19} \\ \underline{x^3 - 2x^2} \end{array}$$

4 & 5. Subtract and bring down the next term to get the new dividend. Repeat the steps until the degree of the new dividend is less than the degree of the divisor.

$$\begin{array}{r}
x^2 + 2x\ + 7 \\
x - 2\overline{\smash{)}x^3 + 0x^2 + 3x - 19} \\
\underline{x^3 - 2x^2} \\
2x^2 + 3x \\
\underline{2x^2 - 4x} \\
7x - 19 \\
\underline{7x - 14} \\
-5 \quad \leftarrow \text{remainder}
\end{array}$$

6. Check by multiplying the divisor by the quotient and adding the remainder.

$$(x - 2)(x^2 + 2x + 7) + (-5)$$
$$= x^3 + 2x^2 + 7x - 2x^2 - 4x - 14 - 5$$
$$= x^3 + 3x - 19$$

The answer is $x^2 + 2x + 7 - \dfrac{5}{x - 2}$.

Chapter 4 Review Exercises

[4.1] Find all values of the variable that make each rational expression undefined.

1. $\dfrac{5}{x}$

2. $\dfrac{8y}{4y - 9}$

3. $\dfrac{p^2 - 6p}{p^2 - 2p - 15}$

4. $\dfrac{m + 2}{m^2 + 4}$

Simplify each rational expression.

5. $\dfrac{-15x^9 y}{12x^5 y^3}$

6. $\dfrac{6z^2 - 24z}{z - 4}$

7. $\dfrac{15t + 20}{9t^2 - 16}$

8. $\dfrac{(a - b)^2}{b^2 - a^2}$

9. $\dfrac{12p^2 + 8pq - 15q^2}{6p^2 + 7pq - 3q^2}$

10. $\dfrac{m + 2n}{m^3 + 8n^3}$

[4.2] Multiply or divide as indicated. Write your answer in simplest form.

11. $\dfrac{x^4}{12} \cdot \dfrac{4}{x}$

12. $\dfrac{a^6 b^7}{3b^3} \div \dfrac{3(a^5 b)^4}{(-a^3 b^2)^3}$

13. $\dfrac{5z - 15}{5} \cdot \dfrac{25z}{3z - 9}$

14. $\dfrac{p^2 - 4p - 21}{2p - 14} \cdot \dfrac{p^4}{p^2 - 9}$

15. $\dfrac{x^2 - 3x - 4}{x^2 - 4} \div \dfrac{x^2 + 4x + 3}{x^2 + 5x + 6}$

16. $\dfrac{c^3 - cd^2}{2c + 3d} \cdot \dfrac{2c^2 - 21cd - 36d^2}{c^2 - 11cd - 12d^2}$

17. $(y^3 + 125) \div \dfrac{y + 5}{y - 5}$

18. $\dfrac{4r^2 + 13r - 12}{20r^2 - 11r - 3} \div \dfrac{2rt + 8t - 5r - 20}{30r^2 + 6r}$

[4.3] Add or subtract as indicated. Write your answer in simplest form.

19. $\dfrac{3}{20x^2} + \dfrac{4}{20x^2} - \dfrac{2}{20x^2}$

20. $\dfrac{y + 8}{y + 3} - \dfrac{5}{y + 3}$

21. $\dfrac{5}{8a} + \dfrac{3}{2a^2}$

22. $\dfrac{3}{16x^2 y} - \dfrac{1}{20xy^3}$

23. $\dfrac{p^2}{p + 6} - p$

24. $\dfrac{8}{h - 8} - \dfrac{2}{8 - h}$

25. $\dfrac{y}{2y - 10} - \dfrac{2}{y^2 - 25}$

26. $\dfrac{a + b}{a^2 - ab} - \dfrac{a - b}{a^2 + ab}$

27. $\dfrac{k}{k^2 + k - 6} + \dfrac{1}{k^2 + 2k - 8}$

28. $\dfrac{1}{r + 1} + r - 1$

29. $\dfrac{z - 6}{z + 2} + \dfrac{16z}{z^2 - 4}$

30. $\dfrac{3m - 5n}{6m^2 - 7mn - 5n^2} - \dfrac{m - n}{12m^2 + 6mn}$

[4.4] Simplify each complex fraction.

31. $\dfrac{\dfrac{3}{x} + \dfrac{3}{y}}{\dfrac{3}{x} - \dfrac{3}{y}}$

32. $\dfrac{\dfrac{1}{a} - \dfrac{1}{ab}}{\dfrac{1}{b} + \dfrac{1}{ab}}$

33. $\dfrac{9 - \dfrac{1}{m^2}}{3 + \dfrac{1}{m}}$

34. $\dfrac{8 - \dfrac{10}{p} - \dfrac{3}{p^2}}{4 - \dfrac{4}{p} - \dfrac{3}{p^2}}$

35. $\dfrac{\dfrac{1}{x + 1} + \dfrac{1}{x - 1}}{\dfrac{x + 1}{x - 1} - \dfrac{x - 1}{x + 1}}$

36. $x - \dfrac{1}{1 - \dfrac{1}{1 + \dfrac{1}{x}}}$

Write with positive exponents and simplify.

37. $(x^{-2} - y^{-2})^{-1}$

38. $\dfrac{z + k}{z^{-1} + k^{-1}}$

[4.5] Solve each equation.

39. $\dfrac{3y + 2}{y + 2} = 0$

40. $1 - \dfrac{5}{6x} = \dfrac{4}{3} - \dfrac{1}{x}$

41. $3 + \dfrac{2}{p} - \dfrac{1}{p^2} = 0$

42. $\dfrac{w}{w - 3} = \dfrac{w}{w + 3}$

43. $\dfrac{2z}{z + 2} - \dfrac{3}{z} = 1$

44. $\dfrac{m + 1}{2m - 2} - \dfrac{m - 2}{3m - 3} = \dfrac{1}{6}$

45. $\dfrac{3}{t + 3} - \dfrac{2}{t + 5} = \dfrac{1}{t + 2}$

46. $\dfrac{3x - 2}{x + 3} - \dfrac{3x + 2}{x - 3} = \dfrac{40 - 2x}{x^2 - 9}$

[4.6] Solve each equation for the specified variable.

47. $\dfrac{P_1 V_1}{T_1} = \dfrac{P_2 V_2}{T_2}$ for P_2

48. $\dfrac{1}{f} = \dfrac{1}{f_1} + \dfrac{1}{f_2}$ for f_1

49. $S = \dfrac{a - rl}{1 - r}$ for r

50. The formula $\dfrac{1}{R} = \dfrac{1}{R_1} + \dfrac{1}{R_2}$ is used to determine the resistance in an electrical circuit. Find R_2 if $R = 8$ and $R_1 = 20$.

Solve each word problem.

51. What number must be added to the numerator and the denominator of the fraction $\frac{3}{5}$ to make the result equal $\frac{5}{6}$?

52. The sum of the reciprocals of two consecutive even integers is equal to 22 divided by the product of the two integers. Find the integers.

53. An old stamping machine takes 10 hr longer to do a job than does a new machine. If the two machines work together, the job takes 12 hr. How long does it take each machine to do the job on its own?

54. A cold-water faucet can fill an empty tub in 8 min; a hot-water faucet takes 12 min. A pump can empty a full tub in 6 min. If both faucets and the pump are on, how long will it take to fill an empty tub?

55. A plane travels 1330 mi in the same time that a motorcycle travels 175 mi. If the plane travels 330 mi/hr faster than the motorcycle, find the speed of each.

[4.7] Divide.

56. $\dfrac{9x^3 - 3x^2 + 8x}{6x}$

57. $\dfrac{8a^8 bc^2 + 12ab^4 c^5 - 36a^2 b^3 c^4}{4abc^2}$

58. $\dfrac{x^2 - 7x + 10}{x - 2}$

59. $\dfrac{6y^3 + y^2 - 8y + 6}{2y + 3}$

60. $(6p^2 - 5p + 3) \div (2p - 3)$

61. $(4z^3 - 3z + 1) \div (1 + 2z)$

62. $\dfrac{24m^4 - 14m^3 - 11}{4m^2 - 5m}$

63. $\dfrac{6r^3 - 5r^2 + 5}{2r^2 - 3r + 2}$

Chapter 4 Test

1. Find all values of x that make $\dfrac{x + 3}{4x^2 - 25}$ undefined.

2. Simplify: $\dfrac{10y^2 + 12y}{5y^2 + y - 6}$.

Multiply or divide as indicated. Write your answer in simplest form.

3. $\dfrac{3a^3}{14b^3} \div \dfrac{27a^6}{18b^5}$

4. $\dfrac{p^2 - 4}{p - 3} \cdot \dfrac{p^2 - 3p}{p + 2}$

5. $\dfrac{t^2 - t - 6}{4t^3 - 12t^2} \cdot \dfrac{2t}{t^2 - 4}$

6. $\dfrac{6x^2 - 11x - 10}{3x^2 - 4x + 1} \div \dfrac{2x^2 + x - 15}{3x^2 + 8x - 3}$

Add or subtract as indicated. Write your answer in simplest form.

7. $2p + \dfrac{5}{6p}$

8. $\dfrac{x + 1}{x - 1} - \dfrac{3}{x + 2}$

9. $\dfrac{m - 1}{m^2 - 4m} + \dfrac{4}{5m}$

10. $\dfrac{y}{y + 3} - \dfrac{18}{y^2 - 9}$

Simplify each complex fraction.

11. $\dfrac{\dfrac{m + 5}{5m}}{\dfrac{2m + 10}{m}}$

12. $\dfrac{\dfrac{x}{y} + \dfrac{x}{y^2}}{1 - \dfrac{1}{y^2}}$

Solve each equation.

13. $\dfrac{4}{x} + 1 = \dfrac{2}{x}$

14. $\dfrac{m}{m + 4} = \dfrac{2}{m}$

15. $\dfrac{p + 3}{p} - \dfrac{p - 1}{p + 2} = 0$

16. $\dfrac{x}{5x + 10} = \dfrac{x + 6}{x^2 - 4} - \dfrac{2}{x - 2}$

Solve each equation for the specified variable.

17. $F = \dfrac{gm_1 m_2}{d^2}$ for m_1

18. $d = \dfrac{r}{1 - rt}$ for t

Divide.

19. $\dfrac{24y^3 - 8y^2 - 16y}{8y^2}$ 20. $\dfrac{6x^3 - x^2 + 16}{3x + 4}$

21. $\dfrac{p^4 - 3p^3 + p^2 + 8p - 1}{p^2 - 2}$

22. Write $\dfrac{t^{-1}}{t^{-2} + 1}$ with positive exponents and simplify.

Solve each word problem.

23. One number is 4 times another. If the sum of their reciprocals is $\frac{1}{4}$, find the numbers.

24. Karen can make a quilt in 10 hr. Cathy can make the same quilt in 6 hr. How long will it take them to make the quilt if they work together?

25. A plane can travel 140 mi/hr in still air. If the plane takes the same amount of time to travel 500 mi against the wind as it does to travel 620 mi with the wind, find the speed of the wind.

Radicals, Rational Exponents, and Complex Numbers

Because of the curvature of the earth's surface, the distance that can be seen from a lighthouse depends upon the height of the lighthouse. In Sec. 5.1 we state a formula for determining the distance from an observer in a lighthouse to the horizon.

5.1 Finding Roots

TAPE AU12

The opposite of raising a number to a power is finding a root of the number. In this section we define the *n*th root of a number and learn how to find it.

Recall that to *square* 3, we write $3^2 = 9$. To find the *square root* of a number, we reverse the procedure. That is, given 9, we find the number whose square is 9. Actually there are *two* numbers whose square is 9, namely 3 and -3.

$$3^2 = 9 \quad \text{and} \quad (-3)^2 = 9$$

In general, if $b^2 = a$, then b is called a **square root** of a.

CAUTION

Do not confuse the *square* of a number with the *square root* of a number.

The **square** of 4 is 16.
The **square roots** of 4 are 2 and -2. ∎

The opposite of *cubing* a number is finding the *cube root* of a number. For example, since $2^3 = 8$, the number 2 is a cube root of 8. In general, if $b^3 = a$, then b is called a **cube root** of a.

We define the *n*th root of a number as follows:

> ### DEFINITION OF *n*th ROOT
>
> Let n be an integer greater than 1. The number b is called an **nth root** of the number a if $b^n = a$.

As we have seen, the number 9 has two square roots, 3 and -3. In many applied problems, however, we are interested only in the *positive* square root of 9, also called the **principal square root** of 9. To denote the principal square root of 9, we write $\sqrt{9}$. The symbol $\sqrt{}$ is called a **radical sign,** and the expression $\sqrt{9}$ is often read "radical 9."*

To denote the *negative* square root of 9, we write $-\sqrt{9} = -3$. To denote *both* roots of 9, we write $\pm\sqrt{9} = \pm 3$. Note that the symbol ± 3 represents the two numbers 3 and -3.

To denote the **principal nth root** of a we write $\sqrt[n]{a}$. The number n is called the **index,** and a is called the **radicand.**

$$\overset{\text{index}}{\searrow}$$
$$\sqrt[n]{a} \leftarrow \text{radicand}$$

The index 2 is usually omitted, and we simply write $\sqrt{a}$ to denote the principal square root of a.

Example 1   Find each root: (a) $\sqrt{25}$, (b) $-\sqrt{25}$, (c) $\pm\sqrt{25}$, (d) $\sqrt{-25}$.

---

*NOTE: The symbol $\sqrt{}$ is a distortion of the letter *r*, from the Latin word *radix* (meaning root).

Solution: (a) $\sqrt{25} = 5$ The positive (principal) square root of 25

(b) $-\sqrt{25} = -5$ The negative square root of 25

(c) $\pm\sqrt{25} = \pm5$ Both square roots of 25

(d) $\sqrt{-25}$ is not a real number, since there is no real number we can square to get -25.   ❑

Try Exercise 21

Find: $\sqrt{36}$.

The number $\sqrt{-25}$ is called an *imaginary number.* We shall study imaginary numbers in Sec. 5.9.

Example 2   Find each root: (a) $\sqrt[3]{216}$, (b) $-\sqrt[3]{216}$, (c) $\sqrt[3]{-216}$, (d) $-\sqrt[3]{-216}$.

Solution: (a) $\sqrt[3]{216} = 6$ Since $6^3 = 216$

(b) $-\sqrt[3]{216} = -6$

(c) $\sqrt[3]{-216} = -6$ Since $(-6)^3 = -216$

(d) $-\sqrt[3]{-216} = -(-6) = 6$ ❑

Try Exercise 35

Find: $\sqrt[3]{-8}$.

Note that the cube root of a positive number is a positive number, and the cube root of a negative number is a negative number.

Example 3   Find each root: (a) $\sqrt[4]{81}$, (b) $-\sqrt[4]{81}$, (c) $\sqrt[4]{-81}$.

Solution: (a) $\sqrt[4]{81} = 3$ Since $3^4 = 81$

(b) $-\sqrt[4]{81} = -3$

(c) $\sqrt[4]{-81}$ is not a real number.   ❑

Try Exercise 41

Find: $-\sqrt[4]{1296}$.

We summarize some of our findings about the principal nth root of a number in the following chart.

The Principal nth Root of a

Suppose a is a real number and n is an integer greater than 1.

	n is even	n is odd
a is positive	$\sqrt[n]{a}$ is positive	$\sqrt[n]{a}$ is positive
a is negative	$\sqrt[n]{a}$ is not a real number	$\sqrt[n]{a}$ is negative
a is 0	$\sqrt[n]{a}$ is 0	$\sqrt[n]{a}$ is 0

Oftentimes the principal nth root of a number is not a rational number. For example, $\sqrt{102}$, $\sqrt[3]{49}$, and $\sqrt[4]{17}$ are irrational numbers, so their decimal form neither terminates nor repeats. We can approximate their values using tables or a calculator.*

To the nearest thousandth,

$$\sqrt{102} \approx 10.100 \qquad \sqrt[3]{49} \approx 3.659 \qquad \text{and} \qquad \sqrt[4]{17} \approx 2.031$$

Since $\sqrt{a^2}$ denotes the nonnegative square root of a^2, we cannot always say that $\sqrt{a^2}$ is a, because a might be a negative number. For example $\sqrt{(-3)^2}$ does not equal

*HISTORICAL NOTE: Square root tables written on clay tablets by the early Babylonians date back as far as 2000 B.C.

-3. To ensure that $\sqrt{a^2}$ is nonnegative, we must write $\sqrt{a^2} = |a|$. In general, we have the rule below.

The Principal nth root of a^n

Suppose a is a real number and n is an integer greater than 1.

n is even	**n is odd**
$\sqrt[n]{a^n} = |a|$	$\sqrt[n]{a^n} = a$

Example 4   Find each root: (a) $\sqrt{8^2}$, (b) $\sqrt{(-8)^2}$, (c) $\sqrt[3]{(-4)^3}$, (d) $-\sqrt[4]{x^4}$, (e) $\sqrt{m^4}$.

Solution: (a) $\sqrt{8^2} = |8| = 8$ Index is even.
            (b) $\sqrt{(-8)^2} = |-8| = 8$ Index is even.
            (c) $\sqrt[3]{(-4)^3} = -4$ Index is odd.
            (d) $-\sqrt[4]{x^4} = -|x|$ Index is even.

Try Exercise 57

Find: $\sqrt{(-7)^2}$.

            (e) $\sqrt{m^4} = \sqrt{(m^2)^2} = |m^2| = m^2$ Since $m^2 \geq 0$ ❏

To avoid the necessity of writing the absolute-value symbol, we shall usually assume when we work with radicals that all variables represent positive real numbers.

Example 5   Find each root. Assume all variables represent positive real numbers.
            (a) $\sqrt{x^6}$, (b) $\sqrt[3]{y^{21}}$, (c) $\sqrt[4]{a^8 b^{20}}$.

Solution: (a) $\sqrt{x^6} = x^3$ Since $(x^3)^2 = x^6$
            (b) $\sqrt[3]{y^{21}} = y^7$ Since $(y^7)^3 = y^{21}$

Try Exercise 73

Find: $\sqrt{x^{10}}$.

            (c) $\sqrt[4]{a^8 b^{20}} = a^2 b^5$ Since $(a^2 b^5)^4 = a^8 b^{20}$ ❏

Example 6   Find each root. Assume all variables represent positive real numbers.
            (a) $\sqrt[5]{32k^{15}}$,   (b) $\sqrt[3]{-27a^6 b^{12}}$,   (c) $\sqrt[11]{(s + t)^{77}}$,   (d) $\sqrt{r^2 + 12r + 36}$.

Solution: (a) $\sqrt[5]{32k^{15}} = 2k^3$
            (b) $\sqrt[3]{-27a^6 b^{12}} = -3a^2 b^4$
            (c) $\sqrt[11]{(s + t)^{77}} = (s + t)^7$

Try Exercise 87

Find: $\sqrt[3]{-216a^9 b^{18}}$.

            (d) $\sqrt{r^2 + 12r + 36} = \sqrt{(r + 6)^2} = r + 6$ ❏

Problem Solving

Example 7   Suppose two tow trucks pull at right angles on a car stuck in the mud. If one exerts a force of 3000 lb and the other exerts a force of 4000 lb, find the magnitude of the resultant force on the car.

Solution: We use the fact that if two forces of magnitude F_1 and F_2 are exerted at right angles to one another (see Fig. 5.1), the magnitude F of the resultant force is given by the formula

$$F = \sqrt{F_1^2 + F_2^2}$$

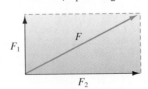

Figure 5.1

Try Exercise 97

For a motorcycle to negotiate a circular loop, its minimum velocity v in feet per second at the top of the loop must be $v = \sqrt{32r}$, where r is the radius of the loop in feet. Determine the minimum velocity required to negotiate a loop whose radius is 18 ft.

Substitute $F_1 = 3000$ and $F_2 = 4000$ into this formula.

$$F = \sqrt{(3000)^2 + (4000)^2}$$ Let $F_1 = 3000$ and $F_2 = 4000$.
$$F = \sqrt{9{,}000{,}000 + 16{,}000{,}000}$$ Square each number.
$$F = \sqrt{25{,}000{,}000}$$ Add before taking the square root.
$$F = 5000$$ Take the square root.

The magnitude of the resultant force is 5000 lb. ❏

Exercises 5.1

Completion Exercises

1. The square of 9 is _____.

2. The square roots of 9 are _____.

3. The cube root of 8 is _____.

4. The cube of 8 is _____.

Find all square roots of each number.

5. 16

6. $\dfrac{9}{100}$

7. 1.21

8. 1089

Completion Exercises

9. The symbol $\sqrt{\ }$ is called a(n) _____.

10. The expression $\sqrt{a}$ is read _____.

11. The number b is an nth root of the number a if _____.

12. The expression $\sqrt[n]{a}$ is read _____.

13. The number n in the expression $\sqrt[n]{a}$ is called the _____.

14. The number a in the equation $\sqrt[n]{a}$ is called the _____.

Discussion Exercises

15. Explain the difference between $\sqrt{25}$ and the square roots of 25.

16. Compare the solution sets of the two equations below. Are they the same? Why or why not?

    (a) $x^2 = 64$ (b) $x = \sqrt{64}$

Getting Acquainted with Your Scientific Calculator

To find $\sqrt[3]{6859}$ on your calculator, press

 (Clear) 6859 (INV) (y^x) 3 (=)   19

Find the following roots. Round approximate answers to three decimal places.

17. $\sqrt[3]{5832}$

18. $\sqrt[4]{200}$

19. $\sqrt[5]{-2744}$

20. $\sqrt[4]{-1296}$

Find each root.

21. $\sqrt{36}$

22. $\sqrt{49}$

23. $-\sqrt{4}$

24. $-\sqrt{9}$

25. $\pm\sqrt{81}$

26. $\pm\sqrt{25}$

27. $\sqrt{-100}$

28. $\sqrt{-16}$

29. $\pm\sqrt{\dfrac{4}{25}}$

30. $\pm\sqrt{\dfrac{9}{64}}$

31. $\sqrt[3]{64}$

32. $\sqrt[3]{1000}$

33. $-\sqrt[3]{125}$

34. $-\sqrt[3]{27}$

35. $\sqrt[3]{-8}$

36. $\sqrt[3]{-216}$

37. $-\sqrt[3]{-343}$

38. $-\sqrt[3]{-512}$

39. $\sqrt[4]{16}$

40. $\sqrt[4]{625}$

41. $-\sqrt[4]{1296}$

42. $-\sqrt[4]{256}$

43. $\sqrt[4]{-10{,}000}$

44. $\sqrt[4]{-81}$

45. $\sqrt[5]{243}$

46. $\sqrt[5]{32}$

47. $\sqrt[6]{1}$

48. $\sqrt[7]{0}$

Matching Exercises

Match the conditions on n and a in Exercises 49 through 52 with the values of $\sqrt[n]{a}$ in letters A through D.

49. If n is even and a is positive, then

A. $\sqrt[n]{a}$ is a positive number.

50. If n is even and a is negative, then

B. $\sqrt[n]{a}$ is a negative number.

51. If n is odd and a is positive, then

C. $\sqrt[n]{a} = 0$

52. If n is odd and a is negative, then

D. $\sqrt[n]{a}$ is not a real number.

True-False Exercises

53. If n is even, then $\sqrt[n]{a^n} = |a|$.

54. If n is odd, then $\sqrt[n]{a^n} = a$.

Find each root. Write your answer using the absolute-value symbol when necessary.

55. $\sqrt{13^2}$

56. $\sqrt{15^2}$

57. $\sqrt{(-7)^2}$

58. $\sqrt{(-9)^2}$

59. $\sqrt[3]{(-3)^3}$

60. $\sqrt[3]{(-5)^3}$

61. $\sqrt[4]{x^4}$

62. $\sqrt[6]{x^6}$

63. $\sqrt[7]{y^7}$

64. $\sqrt[5]{y^5}$

65. $\sqrt[6]{m^{12}}$

66. $\sqrt[4]{m^{16}}$

67. $\sqrt{a^8 b^2}$

68. $\sqrt{a^6 b^4}$

69. $\sqrt{p^2 - 10p + 25}$

70. $\sqrt{p^2 - 20p + 100}$

Discussion Exercises

71. Explain why $\sqrt{-9}$ could not be a real number.

72. Explain why $\sqrt{x^2}$ does not always equal x. When does $\sqrt{x^2} = x$?

Find each root. Assume all variables represent positive real numbers.

73. $\sqrt{x^{10}}$

74. $\sqrt{x^{14}}$

75. $\sqrt[3]{y^{27}}$

76. $\sqrt[3]{y^{33}}$

77. $\sqrt[4]{a^{12} b^{28}}$

78. $\sqrt[4]{a^8 b^{36}}$

79. $\sqrt{36 z^2}$

80. $\sqrt{100 z^2}$

81. $\sqrt[3]{-27 p^3}$

82. $\sqrt[3]{-64 p^3}$

83. $\sqrt[4]{81 t^8}$

84. $\sqrt[4]{16 t^{12}}$

85. $\sqrt{169 x^4 y^{16}}$

86. $\sqrt{225 x^6 y^{18}}$

87. $\sqrt[3]{-216 a^9 b^{18}}$

88. $\sqrt[3]{-125 a^{12} b^6}$

89. $-\sqrt[3]{-1000 p^{66} q^{93}}$

90. $-\sqrt[3]{-8 p^{36} q^{99}}$

91. $\sqrt[3]{\frac{8}{125} r^6 h^6}$

92. $\sqrt[3]{\frac{27}{64} r^9 h^9}$

93. $\sqrt[6]{(5x + 9)^{24}}$

94. $\sqrt[7]{(3x + 4)^{42}}$

95. $\sqrt{x^2 + 4xy + 4y^2}$

96. $\sqrt{x^2 + 8xy + 16y^2}$

Problem Solving

97. For the motorcycle in Fig. 5.2 to negotiate the circular loop successfully, its minimum velocity v in feet per second at the top of the loop must be $v = \sqrt{32r}$, where r is the radius of the loop in feet. Determine the minimum velocity required to negotiate the loop if its radius is 18 ft.

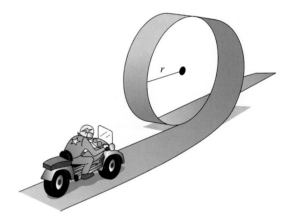

Figure 5.2

98. The distance d, in miles, to the horizon from a height of h feet (see Fig. 5.3) is given by the formula $d = \sqrt{1.5h}$. Determine the distance to the horizon from an observer in a lighthouse at a height of 54 ft.

Figure 5.3

99. The area A of a square is related to the length of its side s (Fig. 5.4) by the formula $A = s^2$. Find the length of one side of a square whose area is 324 cm^2.

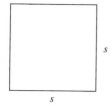

Figure 5.4

100. The volume V of a cube is related to the length of its side s (Fig. 5.5) by the formula $V = s^3$. Find the length of one side of a cube whose volume is 1331 cm^3.

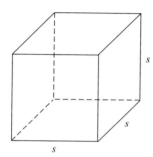

Figure 5.5

The area A of a triangle can be determined from the lengths of its sides a, b, and c using *Heron's formula*,

$$A = \sqrt{s(s - a)(s - b)(s - c)}$$

where s is one-half the perimeter of the triangle (see Fig. 5.6).*

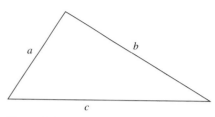

Figure 5.6

101. Find the area of a triangle with sides of length 5 m, 5 m, and 6 m.

102. Find the area of a triangle with sides of length 13 m, 13 m, and 24 m.

*HISTORICAL NOTE: Heron's formula is named after Heron of Alexandria (ca. 75 A.D.). Heron, who was probably born in Egypt but trained in Greece, helped lay the foundation for modern engineering and surveying. In his writings we find descriptions of such things as a surveyor's transit, a steam engine, mirrors for observing the back of one's head, and a device for opening temple doors using a fire on the altar.

5.2 Rational Exponents

TAPE AU13

In Sec. 3.1 we defined integer exponents. For example,

$$9^2 = 81 \qquad 9^0 = 1 \qquad \text{and} \qquad 9^{-2} = \frac{1}{9^2}$$

But what does $9^{1/2}$ mean? In this section we shall define fractional exponents.*

Consider the expression $a^{1/n}$. We want to define $a^{1/n}$ so that the rules for exponents, summarized in Sec. 3.2, apply to fractional exponents as well as to integer exponents. With that in mind, raise $a^{1/n}$ to the nth power, and apply the power-to-a-power rule.

$$(a^{1/n})^n = a^{(1/n)n} = a^1 = a$$

But if a$^{1/n}$ raised to the nth power is a, then $a^{1/n}$ is an nth root of a. This suggests the following definition.

*HISTORICAL NOTE: The first known use of fractional exponents appeared in a tract written by the French mathematician Nicole Oresme (1323–1382), the best mathematician of his day. However, the English mathematician John Wallis (1616–1703) was the first to fully explain the meaning of fractional exponents.

> **DEFINITION OF $a^{1/n}$**
>
> Suppose n is an integer greater than 1, and a is a real number such that $\sqrt[n]{a}$ is a real number. Then
>
> $$a^{1/n} = \sqrt[n]{a}$$

Example 1   Write in radical form and simplify: (a) $9^{1/2}$, (b) $-9^{1/2}$, (c) $(-9)^{1/2}$.

Solution: (a) $9^{1/2} = \sqrt{9} = 3$
(b) $-9^{1/2} = -\sqrt{9} = -3$
(c) $(-9)^{1/2} = \sqrt{-9}$, which is not a real number. ❒

Try Exercise 7

Simplify: $16^{1/2}$.

Example 2   Write in radical form and simplify:
(a) $1000^{1/3}$, (b) $(-27)^{1/3}$, (c) $625^{1/4}$, (d) $(-\frac{1}{32})^{1/5}$.

Solution: (a) $1000^{1/3} = \sqrt[3]{1000} = 10$
(b) $(-27)^{1/3} = \sqrt[3]{-27} = -3$
(c) $625^{1/4} = \sqrt[4]{625} = 5$
(d) $(-\frac{1}{32})^{1/5} = \sqrt[5]{-\frac{1}{32}} = -\frac{1}{2}$ ❒

Try Exercise 15

Simplify: $81^{1/4}$.

Now we consider the more general expression $a^{m/n}$. If we apply the power-to-a-power rule, we have

$$a^{m/n} = a^{(1/n)m} = (a^{1/n})^m = (\sqrt[n]{a})^m$$

Applying the power-to-a-power rule in a different way, we have

$$a^{m/n} = a^{m(1/n)} = (a^m)^{1/n} = \sqrt[n]{a^m}$$

These examples suggest the following definition.

> **DEFINITION OF $a^{m/n}$**
>
> Suppose m and n are positive integers ($n > 1$) with no common factor. If a is a real number such that $\sqrt[n]{a}$ and $\sqrt[n]{a^m}$ are real numbers, then
>
> $$a^{m/n} = (\sqrt[n]{a})^m = \sqrt[n]{a^m}$$

Example 3   Write in radical form and simplify: (a) $64^{2/3}$, (b) $(\frac{9}{25})^{3/2}$, (c) $(-8)^{5/3}$.

Solution: (a) $64^{2/3} = (\sqrt[3]{64})^2 = (4)^2 = 16$
(b) $(\frac{9}{25})^{3/2} = (\sqrt{\frac{9}{25}})^3 = (\frac{3}{5})^3 = \frac{27}{125}$
(c) $(-8)^{5/3} = (\sqrt[3]{-8})^5 = (-2)^5 = -32$ ❒

Try Exercise 21

Simplify: $27^{2/3}$.

Note that there are two ways to evaluate $64^{2/3}$:

$$64^{2/3} = (\sqrt[3]{64})^2 = (4)^2 = 16$$
$$64^{2/3} = \sqrt[3]{64^2} = \sqrt[3]{4096} = 16$$

In this case it was easier to use the form $(\sqrt[n]{a})^m$, rather than $\sqrt[n]{a^m}$.

We define negative fractional exponents in the same way we defined negative integer exponents.

DEFINITION OF $a^{-m/n}$

If $a^{m/n}$ is any real number except 0, then

$$a^{-m/n} = \frac{1}{a^{m/n}}$$

Example 4   Write in radical form and simplify:
(a) $100^{-1/2}$, (b) $8^{-2/3}$, (c) $(-27)^{-4/3}$.

Solution: (a) $100^{-1/2} = \dfrac{1}{100^{1/2}} = \dfrac{1}{\sqrt{100}} = \dfrac{1}{10}$

(b) $8^{-2/3} = \dfrac{1}{8^{2/3}} = \dfrac{1}{(\sqrt[3]{8})^2} = \dfrac{1}{(2)^2} = \dfrac{1}{4}$

(c) $(-27)^{-4/3} = \dfrac{1}{(-27)^{4/3}} = \dfrac{1}{(\sqrt[3]{-27})^4} = \dfrac{1}{(-3)^4} = \dfrac{1}{81}$

Try Exercise 27

Simplify: $144^{-1/2}$.

Example 5   Write in radical form. Assume all variables represent positive real numbers.
(a) $(5x)^{1/2}$, (b) $(p^2 + 1)^{1/2}$, (c) $6y^{3/4}$, (d) $(r + 8)^{-2/3}$.

Solution: (a) $(5x)^{1/2} = \sqrt{5x}$
(b) $(p^2 + 1)^{1/2} = \sqrt{p^2 + 1}$
(c) $6y^{3/4} = 6\sqrt[4]{y^3}$

(d) $(r + 8)^{-2/3} = \dfrac{1}{(r + 8)^{2/3}} = \dfrac{1}{\sqrt[3]{(r + 8)^2}}$

Try Exercise 45

Write $10y^{3/4}$ in radical form.

Since we defined fractional exponents with the rules for integer exponents in mind, these rules are true for fractional exponents as well as for integer exponents.

Rules for Exponents

Suppose r and s are rational numbers, and a and b are real numbers such that every expression below is a real number. Then

Rule 1: $a^r \cdot a^s = a^{r+s}$ **Rule 2:** $\dfrac{a^r}{a^s} = a^{r-s}$

Rule 3: $(a^r)^s = a^{rs}$ **Rule 4:** $(ab)^r = a^r b^r$ **Rule 5:** $\left(\dfrac{a}{b}\right)^r = \dfrac{a^r}{b^r}$

Example 6   Simplify. Assume all variables represent positive real numbers.

(a) $x^{7/4} \cdot x^{3/4}$,   (b) $\dfrac{y^{1/2}}{y^{1/3}}$,   (c) $(z^{-5/9})^3$.

Solution: (a) $x^{7/4} \cdot x^{3/4} = x^{7/4+3/4} = x^{10/4} = x^{5/2}$

(b) $\dfrac{y^{1/2}}{y^{1/3}} = y^{1/2-1/3} = y^{3/6-2/6} = y^{1/6}$

Try Exercise 57

Simplify: $x^{1/4} \cdot x^{5/4}$.

(c) $(z^{-5/9})^3 = z^{(-5/9)3} = z^{-5/3} = \dfrac{1}{z^{5/3}}$

Example 7 Simplify. Assume all variables represent positive real numbers.

$$\text{(a) } (2x^{1/2}y^{4/3})^6, \quad \text{(b) } \left(\frac{-8a^3b^{-12}}{c^9}\right)^{1/3}.$$

Solution:

$$\text{(a) } (2x^{1/2}y^{4/3})^6 = 2^6(x^{1/2})^6(y^{4/3})^6$$
$$= 64x^3y^8$$

$$\text{(b) } \left(\frac{-8a^3b^{-12}}{c^9}\right)^{1/3} = \frac{(-8)^{1/3}(a^3)^{1/3}(b^{-12})^{1/3}}{(c^9)^{1/3}}$$
$$= \frac{-2ab^{-4}}{c^3}$$
$$= \frac{-2a}{b^4c^3} \qquad \Box$$

Try Exercise 73

Simplify: $(2x^{3/2}y^{5/3})^6$.

Example 8 Multiply: $p^{1/2}(p^{3/2} + p)$. Assume $p > 0$.

Solution:
$$p^{1/2}(p^{3/2} + p) = p^{1/2} \cdot p^{3/2} + p^{1/2} \cdot p \qquad \text{Distributive property}$$
$$= p^{1/2+3/2} + p^{1/2+1} \qquad \text{Add the exponents.}$$
$$= p^2 + p^{3/2} \qquad \text{Simplify.} \qquad \Box$$

Try Exercise 89

Multiply: $p^{1/2}(p^{1/2} + p)$.

Example 9 Simplify: $\dfrac{x^r x}{(x^r)^{1/2}}$. Assume $x > 0$ and r is a rational number.

Solution:
$$\frac{x^r x}{(x^r)^{1/2}} = \frac{x^{r+1}}{x^{r/2}} = x^{r+1-r/2} = x^{r/2+1} \qquad \Box$$

Try Exercise 97

Simplify: $\dfrac{x^{2r}x}{(x^r)^{3/2}}$.

Problem Solving

Example 10 The year of a planet in our solar system is the time it takes that planet to make one revolution around the sun. Kepler's third law of planetary motion states that the number of Earth days, y, in a planet's year is approximately given by the formula

$$y = 0.41d^{3/2}$$

where d is the mean distance between the planet and the sun in millions of miles.* Determine the number of Earth days in the year of Venus, whose mean distance from the sun is 67,000,000 mi.

Solution: Since $d = 67,000,000 = 67$ million, substitute $d = 67$ into Kepler's formula.

$$y = 0.41d^{3/2} \qquad \text{Kepler's third law}$$
$$y = 0.41(67)^{3/2} \qquad \text{Let } d = 67.$$

Try Exercise 101

Determine the number of Earth days in the year of Mars, whose mean distance from the Sun is 141,000,000 mi.

Using a calculator, we can approximate y as follows:

$$y = 0.41(67)^{1.5} \approx 0.41(548.42) \approx 225$$

There are approximately 225 Earth days in the year of Venus. $\qquad \Box$

---

*HISTORICAL NOTE: Kepler's laws of planetary motion were named after the German mathematician and astronomer Johann Kepler (1571–1630). Kepler was the first to state the precise laws regarding the orbits of the planets. He based his lengthy calculations on the detailed observations of the Danish astronomer Tycho Brahe (1546–1601).

Exercises 5.2

For Exercises 1 through 4, assume m and n are positive integers ($n > 1$) with no common factor, and a is a real number such that $\sqrt[n]{a}$ and $\sqrt[n]{a^m}$ are real numbers.

Completion Exercises

1. The expression $a^{1/n}$ in radical form is _____.

2. The expression $a^{m/n}$ in radical form is _____ or _____.

True-False Exercises

3. The expressions $(\sqrt[n]{a})^m$ and $\sqrt[n]{a^m}$ are equal.

4. If $a^{m/n} \neq 0$, $a^{-m/n} = \dfrac{1}{a^{m/n}}$.

5. $(-64)^{1/2}$ is not a real number.

6. $(-64)^{2/3}$ is not a real number.

Write each expression in radical form and simplify.

7. $16^{1/2}$
8. $36^{1/2}$
9. $-36^{1/2}$
10. $(-16)^{1/2}$
11. $(-9)^{1/2}$
12. $8^{1/3}$
13. $(-8)^{1/3}$
14. $(-64)^{1/3}$
15. $81^{1/4}$
16. $16^{1/4}$
17. $0^{1/6}$
18. $(-1)^{1/5}$
19. $(-\frac{1}{125})^{1/3}$
20. $(\frac{25}{144})^{1/2}$
21. $27^{2/3}$
22. $125^{2/3}$
23. $(\frac{4}{25})^{3/2}$
24. $(\frac{9}{16})^{3/2}$
25. $(-8)^{2/3}$
26. $(-27)^{4/3}$
27. $144^{-1/2}$
28. $121^{-1/2}$
29. $625^{-3/4}$
30. $256^{-3/4}$
31. $(-125)^{-5/3}$
32. $(-1000)^{-5/3}$
33. $(\frac{4}{9})^{-1/2}$
34. $(\frac{16}{25})^{-1/2}$

Getting Acquainted with Your Scientific Calculator

Here are three ways to find $64^{3/2}$ on your calculator.

Method 1: (Clear) 64 (y^x) (3 ÷ 2) (=) [512]

Method 2: (Clear) 64 (INV) (y^x) 2 (y^x) 3 (=) [512]

Method 3: Use the (STO) key to store the exponent and the (RCL) key to recall the exponent from storage when needed.

(Clear) 3 ÷ 2 (=) (STO) 64 (y^x) (RCL) (=) [512]

Find each power on your calculator. Round approximate answers to three decimal places.

35. $243^{4/5}$
36. $95^{3/4}$
37. $(-169)^{2/3}$
38. $3^{-1/4}$

Write in radical form. Assume all variables represent positive real numbers.

39. $(3x)^{1/2}$
40. $(7x)^{1/2}$
41. $3x^{1/2}$
42. $7x^{1/2}$
43. $(p^2 + 4)^{1/2}$
44. $(p^2 + 9)^{1/2}$
45. $10y^{3/4}$
46. $15y^{3/5}$
47. $(2ab)^{-5/9}$
48. $(5ab)^{-4/7}$
49. $(r + 3)^{-3/8}$
50. $(r + 4)^{-5/6}$

Matching Exercises

Use the rules for exponents to match each expression in Exercises 51 through 54 with its equivalent form in letters A through D.

51. $a^m \cdot a^n$
52. $\dfrac{a^m}{a^n}$
53. $\sqrt[n]{a^m}$
54. $(a^m)^n$

A. a^{m-n}
B. a^{mn}
C. a^{m+n}
D. $a^{m/n}$

Use the rules for exponents to simplify each expression. Write your answer in exponential form with positive exponents. Assume all variables represent positive real numbers.

55. $5^{1/2} \cdot 5^{1/2}$
56. $7^{1/3} \cdot 7^{2/3}$
57. $x^{1/4} \cdot x^{5/4}$
58. $x^{1/6} \cdot x^{7/6}$
59. $\dfrac{y^{9/4}}{y^{1/4}}$
60. $\dfrac{y^{7/2}}{y^{1/2}}$
61. $\dfrac{y^{1/2}}{y^{1/5}}$
62. $\dfrac{y^{1/3}}{y^{1/4}}$
63. $\dfrac{p^{-1/2} \cdot p^{3/2}}{p^{1/2}}$
64. $\dfrac{p^{-3/2} \cdot p^{5/2}}{p^{1/2}}$

65. $\dfrac{k^{-4/3}}{k^{-2/3} \cdot k^{1/3}}$

66. $\dfrac{k^{-3/2}}{k^{-3/4} \cdot k^{1/4}}$

67. $(t^{3/2})^{-6}$

68. $(t^{2/3})^{-9}$

69. $(z^{-7/12})^{4}$

70. $(z^{-9/10})^{5}$

71. $(r^{2}s^{-5})^{1/10}$

72. $(r^{2}s^{-4})^{1/8}$

73. $(2x^{3/2}y^{5/3})^{6}$

74. $(3x^{1/2}y^{3/4})^{4}$

75. $(8p^{6}q^{9})^{2/3}$

76. $(27p^{9}q^{6})^{2/3}$

77. $\left(\dfrac{m^{10}}{n^{15}}\right)^{1/5}$

78. $\left(\dfrac{m^{20}}{n^{25}}\right)^{1/5}$

79. $\left(\dfrac{x^{1/2}y^{3/4}}{z^{-3/2}}\right)^{4}$

80. $\left(\dfrac{x^{3/2}y^{5/4}}{z^{-1/2}}\right)^{4}$

81. $\left(\dfrac{-27a^{6}b^{-15}}{c^{3}}\right)^{1/3}$

82. $\left(\dfrac{-8a^{-9}b^{3}}{c^{18}}\right)^{1/3}$

83. $\left(\dfrac{16p^{-12}}{q^{16}}\right)^{-1/4}$

84. $\left(\dfrac{81p^{-20}}{q^{8}}\right)^{-1/4}$

85. $\dfrac{h^{1/5}(h^{-2/5})^{2}}{h^{-1/2} \cdot h^{1/2} \cdot h}$

86. $\dfrac{(h^{-5/7})^{2}h^{2/7}}{h^{1/3} \cdot h \cdot h^{-1/3}}$

87. $(x^{7/2}y^{-3/2})^{4}(x^{-1/2}y^{3})^{-1}$

88. $(x^{5/2}y^{-5/3})^{6}(x^{2}y^{-1/2})^{-1}$

Multiply. Assume all variables represent positive real numbers.

89. $p^{1/2}(p^{1/2} + p)$

90. $y^{5/2}(y^{1/2} - y)$

91. $z^{-2/3}(z^{5/3} - z^{2/3})$

92. $(x^{1/4} - 3)(x^{1/4} + 5)$

93. Factor $m^{1/2}$ from $m + m^{1/2}$.

94. Factor $m^{-1/3}$ from $m^{2/3} + m^{-1/3}$.

Simplify. Assume x and y represent positive real numbers and r and s represent rational numbers.

95. $x^{3r/2}x^{r/2}$

96. $(y^{s})^{5/4}(y^{-1/3})^{3s}$

97. $\dfrac{x^{2r}x}{(x^{r})^{3/2}}$

98. $\dfrac{x^{1/r}y^{-2/s}}{x^{4/r}y^{-3/s}}$

Problem Solving

Determine the number of Earth days y in the year of each planet with the given mean distance d from the sun. Use the formula $y = 0.41d^{3/2}$, where d is in millions of miles.

99. Mercury; $d = 36{,}000{,}000$ mi

100. Earth; $d = 92{,}600{,}000$ mi

101. Mars; $d = 141{,}000{,}000$ mi

102. Jupiter; $d = 481{,}000{,}000$ mi

The sail area/displacement ratio R of a sailboat is given by the formula

$$R = \dfrac{A}{D^{2/3}}$$

where A is the area of the sail in square feet and D is the displacement of the boat in cubic feet. Calculate R for the given values of A and D.

103. $A = 375, D = 125$ 104. $A = 320, D = 64$

Discussion Exercises

105. The five laws of exponents were given in Sec. 3.2. Why were they stated again in this section?

106. Assume m and n are positive integers ($n > 1$) with no common factor, and a is a real number such that $\sqrt[n]{a}$ and $\sqrt[n]{a^{m}}$ are real numbers. Convince yourself by substituting specific values for m, n, and a that $(\sqrt[n]{a})^{m}$ and $\sqrt[n]{a^{m}}$ are equal.

107. The expression $(-1)^{2/6}$ is evaluated in two different ways below. Which way is correct and why?

(a) $(-1)^{2/6} = (-1)^{1/3} = \sqrt[3]{-1} = -1$

(b) $(-1)^{2/6} = (\sqrt[6]{-1})^{2}$, which is not a real number since $\sqrt[6]{-1}$ is not a real number.

5.3 Simplifying Radical Expressions

TAPE AU12

Since a radical can be written in exponential form using a fractional exponent, we can use the rules for exponents to prove two important rules for radicals. In this section we see how these rules can be used to simplify *radical expressions*. A **radical expression** is an algebraic expression that contains a radical. Therefore

$$2\sqrt{3} \qquad \sqrt{x} + \sqrt{y} \qquad \dfrac{1}{\sqrt[3]{m}}$$

are radical expressions.

Consider the radical $\sqrt[n]{ab}$. If we write this radical in exponential form and apply the product-to-a-power rule for exponents, we have

$$\sqrt[n]{ab} = (ab)^{1/n} = a^{1/n}b^{1/n} = \sqrt[n]{a} \cdot \sqrt[n]{b}$$

This result is called the **product rule for radicals.**

Product Rule for Radicals

If $\sqrt[n]{a}$ and $\sqrt[n]{b}$ are real numbers, then

$$\sqrt[n]{ab} = \sqrt[n]{a} \cdot \sqrt[n]{b}$$

Example 1   Simplify: $\sqrt{45}$.

Solution: The radicand 45 is not a perfect square, but 45 contains the perfect-square factor 9. Therefore factor 9 from 45 and apply the product rule.

$$\sqrt{45} = \underbrace{\sqrt{9 \cdot 5} = \sqrt{9} \cdot \sqrt{5}}_{\text{product rule}} = 3 \cdot \sqrt{5} = 3\sqrt{5}$$

❑

Try Exercise 7

Simplify: $\sqrt{12}$.

It is common to omit the multiplication symbol when multiplying a number and a radical, or when multiplying two radicals. Therefore, we often write

$$3 \cdot \sqrt{5} \text{ as } 3\sqrt{5} \qquad \text{and} \qquad \sqrt{2} \cdot \sqrt{7} \text{ as } \sqrt{2}\sqrt{7}$$

Example 2   Simplify: (a) $\sqrt{48}$, (b) $3\sqrt{600}$, (c) $\sqrt{42}$.

Solution: (a) Although 4 is a perfect-square factor of 48, the *largest* perfect-square factor of 48 is 16. Therefore, write

$$\sqrt{48} = \sqrt{16 \cdot 3} = \sqrt{16} \cdot \sqrt{3} = 4\sqrt{3}$$

If you cannot find the largest perfect-square factor of 48 by inspection, factor 48 into primes. Then

$$\sqrt{48} = \sqrt{2^4 \cdot 3} = \sqrt{2^4} \cdot \sqrt{3} = 2^2 \cdot \sqrt{3} = 4\sqrt{3}$$

(b) The largest perfect-square factor of 600 is 100. Therefore, write

$$3\sqrt{600} = 3 \cdot \sqrt{100 \cdot 6} = 3 \cdot \sqrt{100} \cdot \sqrt{6} = 3 \cdot 10 \cdot \sqrt{6} = 30\sqrt{6}$$

(c) Since $42 = 2 \cdot 3 \cdot 7$, there is no perfect-square factor of 42 (except 1). Therefore $\sqrt{42}$ cannot be simplified further.

❑

Try Exercise 11

Simplify: $4\sqrt{500}$.

Example 3   Simplify: (a) $\sqrt[3]{32}$, (b) $\sqrt[4]{162}$.

Solution: (a) The largest perfect-cube factor of 32 is 8.

$$\sqrt[3]{32} = \sqrt[3]{8 \cdot 4} \qquad \text{Factor out the perfect-cube 8.}$$
$$= \sqrt[3]{8} \cdot \sqrt[3]{4} \qquad \text{Apply the product rule.}$$
$$= 2\sqrt[3]{4} \qquad \text{Simplify.}$$

You could also factor 32 into primes. Then

$$\sqrt[3]{32} = \sqrt[3]{2^5} = \sqrt[3]{2^3} \cdot \sqrt[3]{2^2} = 2\sqrt[3]{4}$$

(b) The largest perfect-fourth-power factor of 162 is 81.

$$\sqrt[4]{162} = \sqrt[4]{81} \cdot \sqrt[4]{2} \qquad \text{Apply the product rule.}$$
$$= 3\sqrt[4]{2} \qquad \text{Simplify.} \qquad \square$$

Try Exercise 15

Simplify: $\sqrt[3]{72}$.

CAUTION

Do not forget to write the index when working with radicals that are not square roots.

Correct

$$\sqrt[3]{16} = \sqrt[3]{8} \cdot \sqrt[3]{2} = 2\sqrt[3]{2}$$

Wrong

$$\sqrt[3]{16} = \sqrt[3]{8} \cdot \sqrt[3]{2} = 2\sqrt{2} \qquad \blacksquare$$

Example 4   Simplify each radical expression. Assume all variables represent positive real numbers. (a) $\sqrt{50x^2}$, (b) $4y\sqrt{24y^3}$, (c) $\sqrt[3]{27p^6q^5}$, (d) $-\sqrt[4]{112m^{15}}$.

Solution: (a) $\sqrt{50x^2} = \sqrt{25x^2} \cdot \sqrt{2}$ $25x^2$ is the largest perfect-
$$= 5x\sqrt{2} \qquad\qquad\qquad \text{square factor of } 50x^2.$$

(b) $4y\sqrt{24y^3} = 4y\sqrt{4y^2} \cdot \sqrt{6y}$ $4y^2$ is the largest perfect-
$$= 4y \cdot 2y\sqrt{6y} \qquad\qquad \text{square factor of } 24y^3.$$
$$= 8y^2\sqrt{6y}$$

(c) $\sqrt[3]{27p^6q^5} = \sqrt[3]{27p^6q^3} \cdot \sqrt[3]{q^2}$ $27p^6q^3$ is the largest perfect-
$$= 3p^2q\sqrt[3]{q^2} \qquad\qquad \text{cube factor of } 27p^6q^5.$$

(d) $-\sqrt[4]{112m^{15}} = -\sqrt[4]{16m^{12}} \cdot \sqrt[4]{7m^3}$ $16m^{12}$ is the largest perfect-
$$= -2m^3\sqrt[4]{7m^3} \qquad\qquad \text{fourth-power factor of}$$
$$112m^{15}. \qquad \square$$

Try Exercise 19

Simplify: $\sqrt{45x^2}$.

Just as we can split a radical sign over a product, so can we split a radical sign over a quotient. If we write the radical $\sqrt[n]{a/b}$ in exponential form and apply the quotient-to-a-power rule for exponents, we have

$$\sqrt[n]{\frac{a}{b}} = \left(\frac{a}{b}\right)^{1/n} = \frac{a^{1/n}}{b^{1/n}} = \frac{\sqrt[n]{a}}{\sqrt[n]{b}}$$

This result is called the **quotient rule for radicals.**

Quotient Rule for Radicals

If $\sqrt[n]{a}$ and $\sqrt[n]{b}$ are real numbers and $b \neq 0$, then

$$\sqrt[n]{\frac{a}{b}} = \frac{\sqrt[n]{a}}{\sqrt[n]{b}}$$

Example 5   Simplify each radical expression: (a) $\sqrt{\dfrac{81}{121}}$, (b) $\sqrt{\dfrac{17}{64}}$, (c) $\sqrt[3]{\dfrac{z}{1000}}$.

Solution: (a) $\underbrace{\sqrt{\dfrac{81}{121}} = \dfrac{\sqrt{81}}{\sqrt{121}}}_{\text{quotient rule}} = \dfrac{9}{11}$

(b) $\sqrt{\dfrac{17}{64}} = \dfrac{\sqrt{17}}{\sqrt{64}} = \dfrac{\sqrt{17}}{8}$

Try Exercise 29

Simplify: $\sqrt{\dfrac{19}{100}}$.

(c) $\sqrt[3]{\dfrac{z}{1000}} = \dfrac{\sqrt[3]{z}}{\sqrt[3]{1000}} = \dfrac{\sqrt[3]{z}}{10}$ ❑

CAUTION

Although we can split a radical sign over a product or a quotient, we cannot split a radical sign over a sum or a difference.

Correct

$\sqrt{ab} = \sqrt{a}\,\sqrt{b}$

$\sqrt{\dfrac{a}{b}} = \dfrac{\sqrt{a}}{\sqrt{b}}$

$\sqrt{9+16} = \sqrt{25} = 5$

Wrong

$\sqrt{a+b} = \sqrt{a} + \sqrt{b}$

$\sqrt{a-b} = \sqrt{a} - \sqrt{b}$

$\sqrt{9+16} = \sqrt{9} + \sqrt{16} = 3 + 4 = 7$ ■

Sometimes we must use both the quotient rule and the product rule to simplify a radical expression.

Example 6 Simplify each radical expression. Assume $r > 0$.

(a) $\sqrt{\dfrac{18}{49}}$, (b) $\sqrt{\dfrac{8r^3}{25}}$, (c) $\sqrt[3]{\dfrac{128ab^8}{250a^7b^3}}$.

Solution: (a) $\sqrt{\dfrac{18}{49}} = \dfrac{\sqrt{18}}{\sqrt{49}}$ Quotient rule

$= \dfrac{\sqrt{9}\cdot\sqrt{2}}{\sqrt{49}}$ Product rule

$= \dfrac{3\sqrt{2}}{7}$ Simplify.

(b) $\sqrt{\dfrac{8r^3}{25}} = \dfrac{\sqrt{8r^3}}{\sqrt{25}}$ Quotient rule

$= \dfrac{\sqrt{4r^2}\cdot\sqrt{2r}}{\sqrt{25}}$ Product rule

$= \dfrac{2r\sqrt{2r}}{5}$ Simplify.

(c) $\sqrt[3]{\dfrac{128ab^8}{250a^7b^3}} = \sqrt[3]{\dfrac{64b^5}{125a^6}}$ Simplify the radicand.

$= \dfrac{\sqrt[3]{64b^5}}{\sqrt[3]{125a^6}}$ Quotient rule

$= \dfrac{\sqrt[3]{64b^3}\cdot\sqrt[3]{b^2}}{\sqrt[3]{125a^6}}$ Product rule

$= \dfrac{4b\sqrt[3]{b^2}}{5a^2}$ Simplify. ❑

Try Exercise 37

Simplify: $\sqrt{\dfrac{27r^3}{64}}$.

If a radical expression involves a root of a root, convert to fractional exponents.

Example 7 Simplify each radical expression. Assume $x > 0$.

(a) $\sqrt{\sqrt{5}}$, (b) $\sqrt{\sqrt[3]{x}}$, (c) $\sqrt{\sqrt{\sqrt{15}}}$.

Solution: (a) $\sqrt{\sqrt{5}} = \sqrt{5^{1/2}} = (5^{1/2})^{1/2} = 5^{(1/2)(1/2)} = 5^{1/4} = \sqrt[4]{5}$

(b) $\sqrt{\sqrt[3]{x}} = \sqrt{x^{1/3}} = (x^{1/3})^{1/2} = x^{1/6} = \sqrt[6]{x}$

Try Exercise 51

(c) $\sqrt{\sqrt{\sqrt{15}}} = ((15^{1/2})^{1/2})^{1/2} = 15^{1/8} = \sqrt[8]{15}$ ❏

Simplify: $\sqrt[3]{\sqrt{x}}$.

If the index of a radical and the exponent of the radicand contain a common factor, convert to fractional exponents.

Example 8 Simplify each radical expression. Assume all variables represent positive real numbers. (a) $\sqrt[10]{z^5}$, (b) $\sqrt[4]{9}$, (c) $\sqrt[6]{25x^4y^2}$.

Solution: (a) $\sqrt[10]{z^5} = z^{5/10} = z^{1/2} = \sqrt{z}$

(b) $\sqrt[4]{9} = \sqrt[4]{3^2} = 3^{2/4} = 3^{1/2} = \sqrt{3}$

Try Exercise 57

(c) $\sqrt[6]{25x^4y^2} = \sqrt[6]{(5x^2y)^2} = (5x^2y)^{2/6} = (5x^2y)^{1/3} = \sqrt[3]{5x^2y}$ ❏

Simplify: $\sqrt[4]{25}$.

We can summarize what it means for a radical to be in simplest form as follows.

Simplified Form of a Radical Expression

A radical expression is in simplest form when all four of the following conditions are met:

1. The radicand contains no factor raised to a power greater than or equal to the index. ($\sqrt{x^3}$ violates this rule.)

2. The radicand contains no fractions. $\left(\sqrt{\dfrac{x}{9}}\ \text{violates this rule.}\right)$

3. The exponent of the radicand and the index have no common factor. ($\sqrt[4]{x^2}$ violates this rule.)

4. No denominator contains a radical. $\left(\dfrac{1}{\sqrt{x}}\ \text{violates this rule.}\right)$

In this section we have learned how to simplify a radical expression that violates Rules 1, 2, or 3 of the simplified form of a radical expression. We shall learn how to simplify a radical expression that violates Rule 4 in Sec. 5.7.

Exercises 5.3

❖◆❖

Completion Exercises

Assume $\sqrt[n]{a}$ and $\sqrt[n]{b}$ are real numbers and $b \neq 0$.

1. The product rule for radicals states that $\sqrt[n]{ab} =$ _____.

2. The quotient rule for radicals states that $\sqrt[n]{\dfrac{a}{b}} =$

_____.

3. The first three perfect-square integers are listed below. Write the next 10 perfect-square integers.

1, 4, 9, ——, ——, ——, ——, ——, ——, ——, ——, ——, ——.

4. The first three perfect-cube integers are listed below. Write the next five perfect-cube integers.

1, 8, 27, ——, ——, ——, ——, ——.

True-False Exercises

5. The expression $\sqrt[3]{8 \cdot 3}$ simplifies to $2\sqrt{3}$.

6. The expression $\sqrt{x^2 + y^2}$ simplifies to $x + y$.

Use the product rule to simplify each radical expression. Assume all variables represent positive real numbers.

7. $\sqrt{12}$ 8. $\sqrt{18}$

9. $\sqrt{80}$ 10. $\sqrt{72}$

11. $4\sqrt{500}$ 12. $5\sqrt{700}$

13. $\sqrt{30}$ 14. $\sqrt{105}$

15. $\sqrt[3]{72}$ 16. $\sqrt[3]{200}$

17. $\sqrt[4]{48}$ 18. $\sqrt[3]{80}$

19. $\sqrt{45x^2}$ 20. $\sqrt{63x^2}$

21. $2y\sqrt{288y^3}$ 22. $3y\sqrt{363y^3}$

23. $\sqrt[3]{64p^9q^4}$ 24. $\sqrt[3]{1000p^7q^6}$

25. $-\sqrt[4]{567m^{11}}$ 26. $-\sqrt[4]{160m^{19}}$

Use the quotient rule, and the product rule when necessary, to simplify each radical expression. Assume all variables represent positive real numbers.

27. $\sqrt{\dfrac{64}{121}}$ 28. $\sqrt{\dfrac{49}{225}}$

29. $\sqrt{\dfrac{19}{100}}$ 30. $\sqrt{\dfrac{23}{36}}$

31. $\sqrt[3]{\dfrac{z}{216}}$ 32. $\sqrt[3]{\dfrac{z^2}{343}}$

33. $\sqrt{\dfrac{20}{81}}$ 34. $\sqrt{\dfrac{12}{25}}$

35. $\sqrt{\dfrac{49m}{144}}$ 36. $\sqrt{\dfrac{81m}{100}}$

37. $\sqrt{\dfrac{27r^3}{64}}$ 38. $\sqrt{\dfrac{64r^3}{81}}$

39. $\sqrt[3]{\dfrac{27t^5}{1000}}$ 40. $\sqrt[3]{\dfrac{8t^8}{125}}$

41. $\sqrt[3]{\dfrac{24ab^9}{375a^{10}b^5}}$ 42. $\sqrt[3]{\dfrac{320a^{11}b}{135a^6b^{13}}}$

True-False Exercises

43. To simplify a radical expression that involves a root of a root, convert to fractional exponents.

44. To simplify a radical expression when the index of the radical and the exponent of the radicand contain a common factor, convert to fractional exponents.

Matching Exercises

Match each radical expression in Exercises 45 through 48 with its simplified form in letters A through D. Assume $x > 0$.

45. $\sqrt{\sqrt{\sqrt{x}}}$ A. $\sqrt{x}$

46. $\sqrt{\sqrt{\sqrt{\sqrt{x}}}}$ B. $\sqrt[4]{x}$

47. $\sqrt[3]{\sqrt{\sqrt{x}}}$ C. $\sqrt[6]{x}$

48. $\sqrt[4]{x^2}$ D. $\sqrt[8]{x}$

Simplify each radical expression by first converting to fractional exponents. Assume all variables represent positive real numbers.

49. $\sqrt{\sqrt{\sqrt{3}}}$ 50. $\sqrt{\sqrt{\sqrt{7}}}$

51. $\sqrt[3]{\sqrt{\sqrt{x}}}$ 52. $\sqrt{\sqrt[4]{\sqrt{x}}}$

53. $\sqrt{\sqrt{\sqrt{26}}}$ 54. $\sqrt{\sqrt{\sqrt{35}}}$

55. $\sqrt[9]{z^3}$ 56. $\sqrt[12]{z^4}$

57. $\sqrt[4]{25}$ 58. $\sqrt[4]{49}$

59. $\sqrt[6]{81}$ 60. $\sqrt[6]{16}$

61. $\sqrt[6]{36x^2y^4}$ 62. $\sqrt[8]{100x^6y^4}$

63. $\sqrt[8]{(2t + 5)^4}$ 64. $\sqrt[6]{(3t + 4)^3}$

✎ Discussion Exercises

State why each radical expression is not in simplest form.

65. $\sqrt{9x}$ 66. $\sqrt{\dfrac{x}{4}}$

67. $\sqrt[6]{x^3}$ 68. $\dfrac{1}{\sqrt{2}}$

Simplify each radical expression. Assume all variables represent positive real numbers.

69. $\sqrt{99x^4}$ 70. $a\sqrt{a^4b^5}$

71. $\sqrt{\dfrac{24}{25}}$ 72. $14\sqrt{\dfrac{30y^2}{49}}$

73. $\sqrt[3]{\sqrt[4]{z}}$ 74. $\sqrt[3]{378m^{26}}$

75. $-\sqrt[4]{1250k^{48}}$ 76. $\sqrt[4]{9r^2t^4}$

77. $\sqrt{\dfrac{140p^3q^5}{5q^{11}}}$ 78. $\sqrt{\dfrac{2x^2y^6}{4z^4y}}$

79. $\sqrt[3]{-27k^{14}}$ 80. $\sqrt[6]{64p^3}$

81. $\sqrt[3]{\dfrac{81r^2t^{10}}{64r^5t}}$ 82. $\sqrt[4]{\dfrac{m^{25}}{625}}$

83. $\sqrt{25\sqrt{k}}$ 84. $-\sqrt[3]{-56a^{19}b^{18}c^{32}}$

Discussion Exercises

85. Does $\sqrt{a + b}$ equal $\sqrt{a} + \sqrt{b}$ for all positive numbers a and b? Substitute numbers for a and b to illustrate your answer.

86. Does $\sqrt{a - b}$ equal $\sqrt{a} - \sqrt{b}$ for all positive numbers a and b? Substitute numbers for a and b to illustrate your answer.

87. State the four conditions that must be met for a radical to be in simplest form. Illustrate with examples.

88. What is wrong with the problem below?

$$1 = \sqrt{1} = \sqrt{(-1)(-1)} = \sqrt{-1} \cdot \sqrt{-1} = (\sqrt{-1})^2 = -1$$

Problem Solving

89. The time t in seconds that it takes a pendulum l feet long to complete a cycle (see Fig. 5.7) is given by the formula

$$t = 2\pi\sqrt{\frac{l}{32}}$$

Figure 5.7

How long does it take a pendulum that is 6 ft long to complete a cycle?*

90. The speed s_a of a car that was involved in an accident can be estimated as follows: A policeman drives a car (the same car if possible) under similar conditions at some test speed s_t, and then skids to a stop. Then s_a is given by the formula

$$s_a = s_t\sqrt{\frac{l_a}{l_t}}$$

where l_a and l_t are the lengths of the skid marks from the accident and the test, respectively. Determine s_a given that $s_t = 36$ mi/hr, $l_a = 90$ ft, and $l_t = 45$ ft.

91. The length L of the longest board that can be carried horizontally around a corner joining two hallways of width w_1 and w_2 (see Fig. 5.8) is given by the formula

$$L = \sqrt{(w_1^{2/3} + w_2^{2/3})^3}$$

Find L if w_1 and w_2 are both 8 ft.

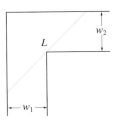

Figure 5.8

92. The amount of water a pipe can carry is determined by its cross-sectional area A. If r is the radius of the pipe (see Fig. 5.9), then $r = \sqrt{A/\pi}$. What would you need to do to the radius to enable the pipe to carry 4 times as much water? (*Hint:* Compare $r_1 = \sqrt{A/\pi}$ with $r_2 = \sqrt{4A/\pi}$.)

Figure 5.9

TAPE AU12

5.4 Multiplying and Dividing Radical Expressions

In Sec. 5.3 we used the product and quotient rules for radicals to simplify radical expressions. If we turn these rules around, we can use them to multiply and divide two radical expressions.

---

*HISTORICAL NOTE: The law for pendular motion, along with the law of centripetal force and the principle of conservation of energy, was published in 1673 by the Dutch mathematician and scientist Christiaan Huygens (1629–1695).

Assume $\sqrt[n]{a}$ and $\sqrt[n]{b}$ are real numbers.

Product Rule

$$\sqrt[n]{a}\,\sqrt[n]{b} = \sqrt[n]{ab}$$

Quotient Rule

$$\frac{\sqrt[n]{a}}{\sqrt[n]{b}} = \sqrt[n]{\frac{a}{b}} \qquad (b \neq 0)$$

We begin by illustrating how to use the product rule to multiply two radicals.

Example 1 Multiply. Assume $x > 0$. (a) $\sqrt{3} \cdot \sqrt{7}$, (b) $\sqrt{6} \cdot \sqrt{x}$, (c) $(\sqrt{5})^2$.

Solution: (a) $\underbrace{\sqrt{3} \cdot \sqrt{7} = \sqrt{3 \cdot 7}}_{\text{product rule}} = \sqrt{21}$

Try Exercise 5

Multiply: $\sqrt{3} \cdot \sqrt{5}$.

(b) $\sqrt{6} \cdot \sqrt{x} = \sqrt{6 \cdot x} = \sqrt{6x}$

(c) $(\sqrt{5})^2 = \sqrt{5} \cdot \sqrt{5} = \sqrt{5 \cdot 5} = \sqrt{25} = 5$ ❑

From Example 1(c) we see that $(\sqrt{5})^2 = 5$. This suggests the following rule.

Squaring a Radical

If a is a nonnegative real number, then

$$(\sqrt{a})^2 = \sqrt{a} \cdot \sqrt{a} = a$$

Example 2 Simplify. Assume all variables represent positive real numbers.
(a) $(\sqrt{10x})^2$, (b) $(\sqrt{3y + 4})^2$, (c) $(\sqrt{z})^3$.

Solution: (a) $(\sqrt{10x})^2 = 10x$

(b) $(\sqrt{3y + 4})^2 = 3y + 4$

Try Exercise 11

Simplify: $(\sqrt{19x})^2$.

(c) $(\sqrt{z})^3 = \sqrt{z} \cdot \sqrt{z} \cdot \sqrt{z} = z\sqrt{z}$ ❑

Example 3 Multiply. Assume $m > 0$. (a) $\sqrt[3]{4p^2} \cdot \sqrt[3]{16p}$, (b) $(3\sqrt{2m})(5\sqrt{6m})$.

Solution: (a) $\sqrt[3]{4p^2}\,\sqrt[3]{16p} = \sqrt[3]{64p^3}$ Product rule

$\qquad\qquad = 4p$ Simplify.

(b) $(3\sqrt{2m})(5\sqrt{6m}) = (3 \cdot 5)(\sqrt{2m}\sqrt{6m})$ Group the whole numbers, group the radicals.

$\qquad\qquad = 15\sqrt{12m^2}$ Multiply each group.

$\qquad\qquad = 15\sqrt{4m^2}\sqrt{3}$ Product rule

$\qquad\qquad = 15 \cdot 2m\sqrt{3}$ Simplify.

Try Exercise 31

Multiply: $\sqrt[3]{4p^2}\,\sqrt[3]{2p}$.

$\qquad\qquad = 30m\sqrt{3}$ Multiply. ❑

Example 4 Multiply: (a) $\sqrt[3]{3r^2t^2}\,\sqrt[3]{4r^5t^2}$, (b) $\sqrt[5]{9k^3}\,\sqrt[5]{27k^4}$.

Solution: (a) $\sqrt[3]{3r^2t^2}\,\sqrt[3]{4r^5t^2} = \sqrt[3]{12r^7t^4}$ Product rule

$\qquad\qquad = \sqrt[3]{r^6t^3}\,\sqrt[3]{12rt}$ Product rule

$\qquad\qquad = r^2t\sqrt[3]{12rt}$ Simplify.

(b) $\sqrt[5]{9k^3} \sqrt[5]{27k^4} = \sqrt[5]{243k^7}$ Product rule

$= \sqrt[5]{243k^5} \sqrt[5]{k^2}$ Product rule

$= 3k\sqrt[5]{k^2}$ Simplify. ◻

Try Exercise 47

Multiply: $\sqrt[5]{25k^2} \sqrt[5]{125k^4}$.

We now illustrate how to use the quotient rule to divide two radicals.

Example 5   Divide. Assume $y > 0$. (a) $\dfrac{\sqrt{45}}{\sqrt{3}}$, (b) $\dfrac{\sqrt{2y}}{\sqrt{8y^3}}$, (c) $\dfrac{\sqrt[3]{500a^2b^3}}{\sqrt[3]{4a^2}}$

Solution: (a) $\underbrace{\dfrac{\sqrt{45}}{\sqrt{3}} = \sqrt{\dfrac{45}{3}}}_{\text{quotient rule}} = \sqrt{15}$

(b) $\dfrac{\sqrt{2y}}{\sqrt{8y^3}} = \sqrt{\dfrac{2y}{8y^3}} = \sqrt{\dfrac{1}{4y^2}} = \dfrac{\sqrt{1}}{\sqrt{4y^2}} = \dfrac{1}{2y}$

Try Exercise 53

Divide: $\dfrac{\sqrt{42}}{\sqrt{7}}$.

(c) $\dfrac{\sqrt[3]{500a^2b^3}}{\sqrt[3]{4a^2}} = \sqrt[3]{\dfrac{500a^2b^3}{4a^2}} = \sqrt[3]{125b^3} = 5b$ ◻

Example 6   Divide. Assume all variables represent positive real numbers.

(a) $\dfrac{18\sqrt{90m^5}}{3\sqrt{5}}$,   (b) $\dfrac{\sqrt[4]{72p^3q^3}}{\sqrt[4]{2pq}}$.

Solution: (a) $\dfrac{18\sqrt{90m^5}}{3\sqrt{5}} = \dfrac{18}{3} \cdot \dfrac{\sqrt{90m^5}}{\sqrt{5}}$ Write as a product of two fractions.

$= 6 \cdot \sqrt{18m^5}$ Simplify $\frac{18}{3}$ and apply the quotient rule.

$= 6\sqrt{9m^4}\sqrt{2m}$ Product rule

$= 6 \cdot 3m^2\sqrt{2m}$ Simplify.

$= 18m^2\sqrt{2m}$ Multiply.

(b) $\dfrac{\sqrt[4]{72p^3q^3}}{\sqrt[4]{2pq}} = \sqrt[4]{\dfrac{72p^3q^3}{2pq}} = \sqrt[4]{36p^2q^2} = \sqrt[4]{(6pq)^2}$

$= (6pq)^{2/4}$

$= (6pq)^{1/2}$

$= \sqrt{6pq}$ ◻

Try Exercise 71

Divide: $\dfrac{28\sqrt{240m^5}}{7\sqrt{10}}$.

<u>CAUTION</u>

Neither the product rule nor the quotient rule for radicals applies when the indexes are different.

Wrong   *Wrong*   *Wrong*

$\sqrt[4]{5} \cdot \sqrt[3]{7} \ne \sqrt[4]{35}$ $\sqrt[4]{5} \cdot \sqrt[3]{7} \ne \sqrt[3]{35}$ $\sqrt[4]{5} \cdot \sqrt[3]{7} \ne \sqrt[12]{35}$ ∎

To multiply or divide two radicals having different indexes, convert to fractional exponents.

Example 7   Multiply or divide as indicated. Assume $x > 0$. (a) $\sqrt{2} \cdot \sqrt[3]{5}$, (b) $\dfrac{\sqrt{x}}{\sqrt[4]{x}}$.

Solution: (a) $\sqrt{2} \cdot \sqrt[3]{5} = 2^{1/2} \cdot 5^{1/3}$ Convert to fractional exponents.

$$= 2^{3/6} \cdot 5^{2/6}$$ Write each exponent with the LCD, 6.

$$= \sqrt[6]{2^3} \cdot \sqrt[6]{5^2}$$ Write in radical form.

$$= \sqrt[6]{2^3 \cdot 5^2}$$ Product rule

$$= \sqrt[6]{200}$$ Multiply.

Try Exercise 87

(b) $\dfrac{\sqrt{x}}{\sqrt[4]{x}} = \dfrac{x^{1/2}}{x^{1/4}} = \dfrac{x^{2/4}}{x^{1/4}} = x^{2/4 - 1/4} = x^{1/4} = \sqrt[4]{x}$ ∎

Divide: $\dfrac{\sqrt{x}}{\sqrt[6]{x}}$.

Exercises 5.4

Completion Exercises

For Exercises 1 and 2, assume $\sqrt[n]{a}$ and $\sqrt[n]{b}$ are real numbers and $b \neq 0$.

1. The quotient rule for radicals states that $\dfrac{\sqrt[n]{a}}{\sqrt[n]{b}} =$ _____.

2. The product rule for radicals states that $\sqrt[n]{a}\,\sqrt[n]{b} =$ _____.

3. If a is a nonnegative real number, then $(\sqrt{a})^2 =$ _____.

4. If $x \geq 0$, then $(\sqrt{5x})^2 =$ _____.

Multiply and simplify. Assume all variables represent positive real numbers.

5. $\sqrt{3} \cdot \sqrt{5}$
6. $\sqrt{5} \cdot \sqrt{7}$
7. $\sqrt{10} \cdot \sqrt{x}$
8. $\sqrt{15} \cdot \sqrt{x}$
9. $(\sqrt{2})^2$
10. $(\sqrt{3})^2$
11. $(\sqrt{19x})^2$
12. $(\sqrt{14x})^2$
13. $\sqrt{3p}\sqrt{6p}$
14. $\sqrt{2p}\sqrt{10p}$
15. $(\sqrt{5y+9})^2$
16. $(\sqrt{3y+7})^2$
17. $(4\sqrt{3m})^2$
18. $(9\sqrt{2m})^2$
19. $(-2\sqrt{k})^3$
20. $(-3\sqrt{k})^3$
21. $(\sqrt{z})^5$
22. $(\sqrt{z})^7$
23. $\sqrt{8}\sqrt{2a+3}$
24. $\sqrt{6}\sqrt{4a+5}$
25. $\sqrt{b+4}\sqrt{b+5}$
26. $\sqrt{b+6}\sqrt{b+3}$

27. $(\sqrt{p^2+2p+12})^2$
28. $(\sqrt{p^2+8p+11})^2$
29. $(5\sqrt{2t+1})^2$
30. $(4\sqrt{3t+2})^2$
31. $\sqrt[3]{4p^2}\,\sqrt[3]{2p}$
32. $\sqrt[3]{9p}\,\sqrt[3]{3p^2}$
33. $\sqrt{12}\sqrt{6z}\sqrt{11}$
34. $\sqrt{3}\sqrt{10z}\sqrt{20}$
35. $(4\sqrt{5m})(3\sqrt{10m})$
36. $(6\sqrt{3y})(2\sqrt{15y})$
37. $\sqrt[3]{10a}\,\sqrt[3]{5a^3}\,\sqrt[3]{20a}$
38. $\sqrt[3]{12a^2}\,\sqrt[3]{3a^3}\,\sqrt[3]{6a^2}$
39. $(-8\sqrt{40t^5})(5\sqrt{15t})$
40. $(8\sqrt{45t})(-3\sqrt{27t^3})$
41. $(-3\sqrt[3]{16})^2$
42. $(-2\sqrt[3]{9})^2$
43. $\sqrt[3]{3r^7t^2}\,\sqrt[3]{9rt^2}$
44. $\sqrt[3]{5r^5t}\,\sqrt[3]{4r^2t^4}$
45. $\sqrt[4]{28p^6}\,\sqrt[4]{4p^{10}}$
46. $\sqrt[4]{9p^5}\,\sqrt[4]{63p^7}$
47. $\sqrt[3]{25k^2}\,\sqrt[3]{125k^4}$
48. $\sqrt[5]{64k^4}\,\sqrt[5]{16k^4}$
49. $2x\sqrt{15x^{13}y^8}\,\sqrt{20x^5y^{11}}$
50. $3y\sqrt{12x^9y^6}\,\sqrt{24x^{14}y^{18}}$
51. $\sqrt{1.5 \times 10^{-2}}\sqrt{6 \times 10^{12}}$
52. $\sqrt{4.5 \times 10^{-3}}\sqrt{8 \times 10^{11}}$

Divide and simplify. Assume all variables represent positive real numbers.

53. $\dfrac{\sqrt{42}}{\sqrt{7}}$
54. $\dfrac{\sqrt{70}}{\sqrt{5}}$
55. $\dfrac{\sqrt{50}}{\sqrt{2}}$
56. $\dfrac{\sqrt{75}}{\sqrt{3}}$
57. $\dfrac{18\sqrt{17}}{\sqrt{81}}$
58. $\dfrac{14\sqrt{23}}{\sqrt{49}}$
59. $\dfrac{\sqrt{363x^2}}{\sqrt{3x}}$
60. $\dfrac{\sqrt{288x^2}}{\sqrt{2x}}$

61. $\dfrac{\sqrt{3y}}{\sqrt{27y^3}}$

62. $\dfrac{\sqrt{4y}}{\sqrt{64y^3}}$

63. $\dfrac{8\sqrt{80k^7z^5}}{24\sqrt{5kz}}$

64. $\dfrac{6\sqrt{63k^9z^3}}{30\sqrt{7kz}}$

65. $\dfrac{\sqrt{75r^3s^5t^7}}{\sqrt{15rst}}$

66. $\dfrac{\sqrt{91r^5s^9t^3}}{\sqrt{13rst}}$

67. $\dfrac{\sqrt{22xy^3}}{\sqrt{98x^9y}}$

68. $\dfrac{\sqrt{39x^5y}}{\sqrt{192xy^7}}$

69. $\dfrac{\sqrt[3]{108a^2b^6}}{\sqrt[3]{4a^2}}$

70. $\dfrac{\sqrt[3]{144a^3b^2}}{\sqrt[3]{18b^2}}$

71. $\dfrac{28\sqrt{240m^5}}{7\sqrt{10}}$

72. $\dfrac{48\sqrt{108m^3}}{12\sqrt{6}}$

73. $\dfrac{\sqrt[3]{648r^9t^2}}{\sqrt[3]{9r^2t^5}}$

74. $\dfrac{\sqrt[3]{768rt^9}}{\sqrt[3]{8r^{10}t^4}}$

75. $\dfrac{\sqrt{2x^3}\,\sqrt{7x^2y^5}}{\sqrt{504x^3y}}$

76. $\dfrac{\sqrt{3x^2y^4}\,\sqrt{5x^5y}}{\sqrt{375xy^3}}$

77. $\dfrac{\sqrt[4]{72z^{33}k^{51}}}{\sqrt[4]{6k^7}\,\sqrt[4]{4z^3}}$

78. $\dfrac{\sqrt[4]{180z^{33}k^{41}}}{\sqrt[4]{4k^5}\,\sqrt[4]{9z^9}}$

79. $\dfrac{\sqrt[4]{98p^3q^3}}{\sqrt[4]{2pq}}$

80. $\dfrac{\sqrt[3]{108p^5q^5}}{\sqrt[3]{3p^3q^3}}$

81. $\dfrac{\sqrt{6.4\times10^9}}{\sqrt{1.6\times10}}$

82. $\dfrac{\sqrt{9.9\times10^7}}{\sqrt{1.1\times10}}$

   Discussion Exercises

83. Can you use the product rule for radicals when the radicands are different? When the indexes are different? Multiply each expression.

    (a) $\sqrt{2}\cdot\sqrt{2}$ (b) $\sqrt{2}\cdot\sqrt{3}$

    (c) $\sqrt{2}\cdot\sqrt[3]{2}$ (d) $\sqrt{2}\cdot\sqrt[3]{3}$

84. Can you use the quotient rule for radicals when the radicands are different? When the indexes are different? Divide each expression.

    (a) $\dfrac{\sqrt{6}}{\sqrt{6}}$ (b) $\dfrac{\sqrt{6}}{\sqrt{2}}$

    (c) $\dfrac{\sqrt{6}}{\sqrt[3]{6}}$ (d) $\dfrac{\sqrt{6}}{\sqrt[3]{2}}$

Multiply or divide as indicated. Assume all variables represent positive real numbers.

85. $\sqrt[3]{x^2}\cdot\sqrt[4]{x^3}$ 86. $\sqrt[4]{x^3}\cdot\sqrt[5]{x^2}$

87. $\dfrac{\sqrt{x}}{\sqrt[6]{x}}$ 88. $\dfrac{\sqrt{x}}{\sqrt[8]{x}}$

89. $\sqrt{a}\,\sqrt[4]{b}$ 90. $\sqrt{a}\,\sqrt[6]{b}$

91. $\sqrt[5]{a^4}\,\sqrt[3]{b^2}$ 92. $\sqrt[4]{a^3}\,\sqrt[3]{b^2}$

5.5 Adding and Subtracting Radical Expressions

TAPE AU12

Recall that like terms are terms that have the same variables with the same exponents. **Like radicals** are radical expressions that have the same index and the same radicand.

Like radicals	*Unlike radicals*	
$3\sqrt{2},\ 5\sqrt{2}$	$3\sqrt{2},\ 5\sqrt[3]{2}$	(different indexes)
$7\sqrt{6x},\ \sqrt{6x}$	$7\sqrt{6x},\ \sqrt{6y}$	(different radicands)
$\sqrt[3]{a^2b},\ \sqrt[3]{a^2b}$	$\sqrt[3]{a^2b},\ \sqrt[4]{a^2b}$	(different indexes)

We can use the distributive property to combine (add or subtract) like radicals in essentially the same way that we used the distributive property to combine like terms.

Combining like terms:

$$3x + 5x = (3 + 5)x \qquad \text{Distributive property}$$
$$= 8x \qquad \text{Add the coefficients.}$$

Combining like radicals:

$$3\sqrt{2} + 5\sqrt{2} = (3 + 5)\sqrt{2} \qquad \text{Distributive property}$$
$$= 8\sqrt{2} \qquad \text{Add the coefficients of the radicals.}$$

Note that **we add or subtract like radicals by adding or subtracting the coefficients of the radicals.**

Example 1   Combine like radicals. Assume $y > 0$.
(a) $8\sqrt{6} + \sqrt{6}$, (b) $4\sqrt{3y} - 9\sqrt{3y}$, (c) $2\sqrt{5} + 4\sqrt{7}$, (d) $\sqrt{10} + \sqrt[3]{10}$.

Solution:

(a) $8\sqrt{6} + \sqrt{6} = 8\sqrt{6} + 1\sqrt{6}$ Since $\sqrt{6} = 1 \cdot \sqrt{6}$
                $= (8 + 1)\sqrt{6}$ Distributive property
                $= 9\sqrt{6}$ Add the coefficients.

(b) $4\sqrt{3y} - 9\sqrt{3y} = (4 - 9)\sqrt{3y}$ Distributive property
                       $= -5\sqrt{3y}$ Subtract the coefficients.

(c) $2\sqrt{5} + 4\sqrt{7}$ Cannot be simplified

(d) $\sqrt{10} + \sqrt[3]{10}$ Cannot be simplified ❏

Try Exercise 5

Combine: $3\sqrt{7} + 5\sqrt{7}$.

Sometimes we must simplify the radicals before we can combine them.

Example 2   Simplify: (a) $\sqrt{18} + \sqrt{2}$, (b) $6\sqrt{28} - \sqrt{63} - \sqrt{112}$.

Solution:

(a) $\sqrt{18} + \sqrt{2} = \sqrt{9}\sqrt{2} + \sqrt{2}$ Product rule
               $= 3\sqrt{2} + \sqrt{2}$ Simplify.
               $= 4\sqrt{2}$ Add the coefficients.

(b) $6\sqrt{28} - \sqrt{63} - \sqrt{112} = 6\sqrt{4}\sqrt{7} - \sqrt{9}\sqrt{7} - \sqrt{16}\sqrt{7}$
                       $= 12\sqrt{7} - 3\sqrt{7} - 4\sqrt{7}$
                       $= 5\sqrt{7}$ ❏

Try Exercise 23

Simplify: $\sqrt{12} + \sqrt{3}$.

Example 3   Simplify each expression. Assume all variables represent positive real numbers.
(a) $\sqrt{14m} + \sqrt{56k} - \sqrt{126m} + \sqrt{224k}$, (b) $8a\sqrt{3a^2b} - a^2\sqrt{75b}$.

Solution:

(a) $\sqrt{14m} + \sqrt{56k} - \sqrt{126m} + \sqrt{224k}$
   $= \sqrt{14m} + \sqrt{4}\sqrt{14k} - \sqrt{9}\sqrt{14m} + \sqrt{16}\sqrt{14k}$
   $= \sqrt{14m} + 2\sqrt{14k} - 3\sqrt{14m} + 4\sqrt{14k}$
   $= -2\sqrt{14m} + 6\sqrt{14k}$

(b) $8a\sqrt{3a^2b} - a^2\sqrt{75b} = 8a\sqrt{a^2}\sqrt{3b} - a^2\sqrt{25}\sqrt{3b}$
                     $= 8a^2\sqrt{3b} - 5a^2\sqrt{3b}$
                     $= (8a^2 - 5a^2)\sqrt{3b}$
                     $= 3a^2\sqrt{3b}$ ❏

Try Exercise 41

Simplify:
$\sqrt{5m} + \sqrt{80k} - \sqrt{180m} + \sqrt{245k}$.

Example 4   Simplify each expression:
(a) $\sqrt[3]{16p} - \sqrt[3]{2p} + \sqrt[3]{54p}$, (b) $\sqrt[3]{5z^3} - z\sqrt[3]{320} - \sqrt[3]{625z^3}$.

Solution:

(a) $\sqrt[3]{16p} - \sqrt[3]{2p} + \sqrt[3]{54p} = \sqrt[3]{8}\sqrt[3]{2p} - \sqrt[3]{2p} + \sqrt[3]{27}\sqrt[3]{2p}$
                         $= 2\sqrt[3]{2p} - \sqrt[3]{2p} + 3\sqrt[3]{2p}$
                         $= 4\sqrt[3]{2p}$

(b) $\sqrt[3]{5z^3} - z\sqrt[3]{320} - \sqrt[3]{625z^3}$
   $= \sqrt[3]{z^3}\sqrt[3]{5} - z\sqrt[3]{64}\sqrt[3]{5} - \sqrt[3]{125z^3}\sqrt[3]{5}$
   $= z\sqrt[3]{5} - 4z\sqrt[3]{5} - 5z\sqrt[3]{5}$
   $= -8z\sqrt[3]{5}$ ❏

Try Exercise 57

Simplify:
$\sqrt[3]{24p} - \sqrt[3]{3p} + \sqrt[3]{192p}$.

Exercises 5.5

Completion Exercises

1. Like radicals have the same _____ and the same _____ .

2. We add like radicals by adding their _____ .

True-False Exercises

3. $\sqrt{7}$ and $\sqrt[4]{7}$ are like radicals.

4. $\sqrt{5x}$ and $\sqrt{5x^2}$ are like radicals.

Combine like radicals. Assume all variables represent positive real numbers.

5. $3\sqrt{7} + 5\sqrt{7}$

6. $17\sqrt{10} - 4\sqrt{10}$

7. $6\sqrt{15} + \sqrt{15}$

8. $\sqrt{14} + 9\sqrt{14}$

9. $5\sqrt{2y} - 9\sqrt{2y}$

10. $3\sqrt{5y} - 8\sqrt{5y}$

11. $3\sqrt{7} + 5\sqrt{2}$

12. $6\sqrt{3} + 2\sqrt{5}$

13. $\sqrt{6} + \sqrt[3]{6}$

14. $\sqrt[4]{10} + \sqrt{10}$

15. $5\sqrt{19} + 2\sqrt{19} - 10\sqrt{19}$

16. $3\sqrt{23} - 8\sqrt{23} + 2\sqrt{23}$

17. $7\sqrt{m} - \sqrt{m} + 5m$

18. $9\sqrt{m} - \sqrt{m} + 3m$

19. $6a\sqrt{5} + 3a\sqrt{5}$

20. $5a\sqrt{3} + 2a\sqrt{3}$

21. $8p\sqrt{6p} - 15p\sqrt{6p}$

22. $4p\sqrt{10p} - 13p\sqrt{10p}$

Simplify each expression. Assume all variables represent positive real numbers.

23. $\sqrt{12} + \sqrt{3}$

24. $\sqrt{20} + \sqrt{5}$

25. $\sqrt{20} + \sqrt{45}$

26. $\sqrt{12} + \sqrt{48}$

27. $5\sqrt{8} - \sqrt{2}$

28. $2\sqrt{7} - 6\sqrt{28}$

29. $\sqrt{6x} + \sqrt{24x}$

30. $\sqrt{3x} + \sqrt{75x}$

31. $\frac{3}{4}\sqrt{240} + \frac{7}{2}\sqrt{540}$

32. $\frac{4}{5}\sqrt{150} + \frac{2}{3}\sqrt{486}$

33. $8\sqrt{18} - \sqrt{72} - \sqrt{50}$

34. $7\sqrt{175} - \sqrt{28} - \sqrt{63}$

35. $2\sqrt{40y} - 5\sqrt{90y} + 3\sqrt{160y}$

36. $4\sqrt{250y} - 9\sqrt{360y} + 6\sqrt{490y}$

37. $12\sqrt{p} - p + 3\sqrt{p} + 9p$

38. $18p - \sqrt{p} + 2p + 6\sqrt{p}$

39. $z\sqrt{27} + 6z\sqrt{3} - 2z\sqrt{75}$

40. $z\sqrt{8} + 3z\sqrt{2} - 5z\sqrt{18}$

41. $\sqrt{5m} + \sqrt{80k} - \sqrt{180m} + \sqrt{245k}$

42. $\sqrt{3m} - \sqrt{48k} - \sqrt{192m} + \sqrt{243k}$

43. $\sqrt{2}\sqrt{14} - 3\sqrt{98} - \sqrt{3}\sqrt{21}$

44. $\sqrt{3}\sqrt{15} + 4\sqrt{50} - \sqrt{2}\sqrt{10}$

45. $5r\sqrt{11r} + \sqrt{44r^3}$

46. $\sqrt{13r^3} + 2r\sqrt{117r}$

47. $\sqrt{216} + 7\sqrt{54} - \sqrt{\frac{6}{25}}$

48. $\sqrt{1000} - 9\sqrt{40} + \sqrt{\frac{10}{9}}$

Simplify each expression. Assume all variables represent positive real numbers.

49. $16\sqrt[3]{6} - 9\sqrt[3]{6} + 4\sqrt{6}$

50. $8\sqrt[3]{3} + 4\sqrt[3]{3} - 3\sqrt{3}$

51. $\sqrt[4]{7} + 7\sqrt[5]{7} - 6\sqrt[4]{7}$

52. $\sqrt[4]{10} - 9\sqrt[5]{10} - 11\sqrt[4]{10}$

53. $\sqrt[3]{81} + \sqrt[3]{3}$

54. $\sqrt[3]{16} + \sqrt[3]{2}$

55. $\sqrt[3]{64y} - \sqrt[3]{8y}$

56. $\sqrt[3]{125y} - \sqrt[3]{27y}$

57. $\sqrt[3]{24p} - \sqrt[3]{3p} + \sqrt[3]{192p}$

58. $\sqrt[3]{40p} - \sqrt[3]{5p} + \sqrt[3]{625p}$

59. $5\sqrt[3]{54} - 3\sqrt[3]{128} - 5\sqrt[3]{16}$

60. $2\sqrt[3]{48} - 5\sqrt[3]{162} - 2\sqrt[3]{384}$

61. $\sqrt[3]{5z^3} - z\sqrt[3]{135} - \sqrt[3]{1080z^3}$

62. $\sqrt[3]{7z^3} + \sqrt[3]{875z^3} - z\sqrt[3]{448}$

63. $\sqrt[4]{16t} + 3\sqrt[4]{t}$

64. $2\sqrt[4]{t} + \sqrt[4]{81t}$

65. $2\sqrt{3k^3} - \sqrt[4]{9k^6}$

66. $3\sqrt{2k^3} - \sqrt[4]{4k^6}$

Problem Solving

67. Find the perimeter of a rectangle with width $\sqrt{605}$ and length $3\sqrt{320}$.

68. Find the perimeter of a triangle whose sides are $\sqrt{75}$, $\sqrt{147}$, and $\sqrt{363}$.

69. The voltage V required to operate an electrical appliance is given by the formula

$$V = \sqrt{WR}$$

where W is the wattage and R is the resistance, in ohms. Determine the total voltage needed to operate two appliances connected in series, one with a wattage of 100 W and a resistance of 12 Ω, and the other with a wattage of 450 W and a resistance of 6 Ω. (*Hint:* Since the appliances are connected in series, the total voltage is the sum of the voltages of the two appliances.)

70. The speed s of a car in mph can be approximated by measuring the length L of its skid marks in feet and then using one of the formulas below.

$$s = 2\sqrt{3L} \quad \text{(wet pavement)} \quad s = 2\sqrt{5L} \quad \text{(dry pavement)}$$

Suppose a red car skidded 125 ft on wet pavement, and a blue car skidded 48 ft on dry pavement. Which car was traveling faster, and by how much?

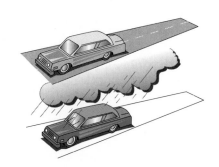

5.6 Combinations of Operations

TAPE AU12

Now that we know how to add, subtract, multiply, and divide radical expressions, we shall learn how to perform combinations of these operations.

Example 1 Multiply and simplify: (a) $\sqrt{3}(4\sqrt{7} - \sqrt{3})$, (b) $8\sqrt{2}(5\sqrt{6} + \sqrt{2})$.

Solution:

(a) $\sqrt{3}(4\sqrt{7} - \sqrt{3}) = \sqrt{3} \cdot 4\sqrt{7} - \sqrt{3} \cdot \sqrt{3}$	Distributive property
$= 4\sqrt{21} - 3$	Product rule
(b) $8\sqrt{2}(5\sqrt{6} + \sqrt{2}) = 8\sqrt{2} \cdot 5\sqrt{6} + 8\sqrt{2} \cdot \sqrt{2}$	Distributive property
$= 40\sqrt{12} + 8 \cdot 2$	Product rule
$= 40\sqrt{4}\sqrt{3} + 16$	Product rule
$= 80\sqrt{3} + 16$	Simplify. ❏

Try Exercise 11

Multiply and simplify:
$\sqrt{2}(4\sqrt{3} - \sqrt{2})$.

To multiply two radical expressions that both contain two terms, we use the FOIL method.

Example 2 Multiply and simplify:

(a) $(\sqrt{5} + 4)(\sqrt{5} + 3)$, (b) $(\sqrt{10} - 7\sqrt{2})(\sqrt{5} + 3\sqrt{2})$.

Solution:

$$
\begin{array}{cccc}
\text{F} & \text{O} & \text{I} & \text{L} \\
\downarrow & \downarrow & \downarrow & \downarrow
\end{array}
$$

$$\text{(a)}\ (\sqrt{5} + 4)(\sqrt{5} + 3) = \sqrt{5}\sqrt{5} + 3\sqrt{5} + 4\sqrt{5} + 4 \cdot 3$$
$$= 5 + 3\sqrt{5} + 4\sqrt{5} + 12$$
$$= 17 + 7\sqrt{5}$$

$$\text{(b)} \quad (\sqrt{10} - 7\sqrt{2})(\sqrt{5} + 3\sqrt{2})$$
$$= \sqrt{10}\sqrt{5} + \sqrt{10} \cdot 3\sqrt{2} - 7\sqrt{2}\sqrt{5} - 7\sqrt{2} \cdot 3\sqrt{2}$$
$$= \sqrt{50} + 3\sqrt{20} - 7\sqrt{10} - 21 \cdot 2$$
$$= \sqrt{25}\sqrt{2} + 3\sqrt{4}\sqrt{5} - 7\sqrt{10} - 42$$
$$= 5\sqrt{2} + 6\sqrt{5} - 7\sqrt{10} - 42 \qquad ❏$$

Try Exercise 25

Multiply and simplify:
$(\sqrt{5} + 6)(\sqrt{5} + 2)$.

Recall the special product rules from Sec. 3.5.
Squaring a binomial:

$$(a + b)^2 = a^2 + 2ab + b^2$$
$$(a - b)^2 = a^2 - 2ab + b^2$$

Product of two conjugates:

$$(a + b)(a - b) = a^2 - b^2$$

We use these rules in Example 3.

Example 3   Multiply and simplify: (a) $(\sqrt{11} + 9)^2$, (b) $(\sqrt{7} + \sqrt{2})(\sqrt{7} - \sqrt{2})$.

Solution: (a) Use the rule for squaring a binomial.

$$(\sqrt{11} + 9)^2 = (\sqrt{11})^2 + 2\sqrt{11} \cdot 9 + 9^2 = 11 + 18\sqrt{11} + 81$$
$$= 92 + 18\sqrt{11}$$

$$\begin{array}{ccccccc} \uparrow & \uparrow & & \uparrow & \uparrow\uparrow\uparrow & & \uparrow \\ (a & + b)^2 = & & a^2 & + 2 \cdot a \cdot b & + & b^2 \end{array}$$

(b) Use the rule for multiplying two conjugates.

Try Exercise 35

Multiply and simplify:
$(\sqrt{7} + \sqrt{3})(\sqrt{7} - \sqrt{3})$.

$$(\sqrt{7} + \sqrt{2})(\sqrt{7} - \sqrt{2}) = (\sqrt{7})^2 - (\sqrt{2})^2 = 7 - 2 = 5$$

$$\begin{array}{cccccc} \uparrow & \uparrow\uparrow & \uparrow & \uparrow & & \uparrow \\ (a & + b) & (a & - b) = & a^2 & - b^2 \end{array}$$

Example 4   Multiply and simplify. Assume $x > 0$ and $y > 0$.
(a) $(3\sqrt{x} - \sqrt{y})^2$, (b) $(\sqrt{5x} - 4)(\sqrt{5x} + 4)$.

Solution: (a) $(3\sqrt{x} - \sqrt{y})^2 = (3\sqrt{x})^2 - 2 \cdot 3\sqrt{x}\sqrt{y} + (\sqrt{y})^2$ Square of a binomial
$$= 9x - 6\sqrt{xy} + y$$ Simplify.
(b) $(\sqrt{5x} - 4)(\sqrt{5x} + 4) = (\sqrt{5x})^2 - 4^2$ Product of two conjugates
$$= 5x - 16$$ Simplify.

Try Exercise 41

Multiply and simplify:
$(4\sqrt{x} - \sqrt{y})^2$.

Example 5 illustrates how to simplify a quotient whose numerator is a radical expression with two terms.

Example 5   Simplify each expression: (a) $\dfrac{5 + 20\sqrt{7}}{15}$, (b) $\dfrac{12 \pm \sqrt{72}}{12}$.

Solution: (a) $\dfrac{5 + 20\sqrt{7}}{15} = \dfrac{5(1 + 4\sqrt{7})}{3 \cdot 5}$ Factor.

$$= \dfrac{1 + 4\sqrt{7}}{3}$$ Divide out the common factor 5.

(b) $\dfrac{12 \pm \sqrt{72}}{12} = \dfrac{12 \pm \sqrt{36}\sqrt{2}}{12}$ Product rule

$$= \dfrac{12 \pm 6\sqrt{2}}{12}$$ Simplify.

$$= \dfrac{6(2 \pm \sqrt{2})}{6 \cdot 2}$$ Factor.

Try Exercise 63

Simplify: $\dfrac{3 + 12\sqrt{22}}{9}$.

$$= \dfrac{2 \pm \sqrt{2}}{2}$$ Divide out the common factor 6.

This answer cannot be simplified further.

Exercises 5.6

Matching Exercises

Match each expression in Exercises 1 through 6 with its expanded form in letters A through F.

1. $(a + b)^2$

2. $(a + b)(a - b)$
3. $(a - b)(a + b)$
4. $(a - b)^2$
5. $(a + b)(a + 2b)$
6. $a(a + b)$

A. $a^2 + b^2$

B. $a^2 - 2ab + b^2$
C. $a^2 - b^2$
D. $a^2 + 3ab + 2b^2$
E. $a^2 + ab$
F. $a^2 + 2ab + b^2$

Multiply and simplify. Assume all variables represent positive real numbers.

7. $5(\sqrt{3} + 4)$

8. $7(\sqrt{2} + 3)$

9. $2(6\sqrt{10} - \sqrt{5})$

10. $2(4\sqrt{14} - \sqrt{7})$

11. $\sqrt{2}(4\sqrt{3} - \sqrt{2})$

12. $\sqrt{5}(6\sqrt{3} - \sqrt{5})$

13. $7\sqrt{3}(\sqrt{12} + 2\sqrt{27})$

14. $6\sqrt{2}(\sqrt{8} - 3\sqrt{18})$

15. $-\sqrt{x}(\sqrt{x} - \sqrt{xy})$

16. $-\sqrt{y}(\sqrt{y} + \sqrt{xy})$

17. $2\sqrt{3p}(5\sqrt{6p} - 4\sqrt{3})$

18. $3\sqrt{2p}(4\sqrt{6p} + 5\sqrt{2})$

19. $\sqrt[3]{2}(\sqrt[3]{4} - \sqrt[3]{7})$

20. $\sqrt[3]{3}(\sqrt[3]{9} - \sqrt[3]{11})$

21. $\sqrt{m + 5}(\sqrt{m + 5} + \sqrt{m})$

22. $\sqrt{m + 2}(\sqrt{m} - \sqrt{m + 2})$

23. $6\sqrt[3]{16}(3\sqrt[3]{4} + \sqrt[3]{6})$

24. $4\sqrt[3]{25}(3\sqrt[3]{10} + \sqrt[3]{16})$

Multiply and simplify. Assume all variables represent positive real numbers.

25. $(\sqrt{5} + 6)(\sqrt{5} + 2)$

26. $(\sqrt{3} + 4)(\sqrt{3} + 8)$

27. $(\sqrt{11} - \sqrt{5})(\sqrt{2} - \sqrt{3})$

28. $(\sqrt{13} - \sqrt{5})(\sqrt{2} - \sqrt{3})$

29. $(\sqrt{14} - 5\sqrt{2})(\sqrt{7} + 6\sqrt{2})$

30. $(\sqrt{15} - 6\sqrt{5})(\sqrt{3} + 2\sqrt{5})$

31. $(\sqrt{15} + 8)^2$

32. $(\sqrt{17} - 9)^2$

33. $(\sqrt{23} - \sqrt{2})^2$

34. $(\sqrt{29} + \sqrt{3})^2$

35. $(\sqrt{7} + \sqrt{3})(\sqrt{7} - \sqrt{3})$

36. $(\sqrt{5} + \sqrt{2})(\sqrt{5} - \sqrt{2})$

37. $(3 + \sqrt{10})(3 - \sqrt{10})$

38. $(2 - \sqrt{11})(2 + \sqrt{11})$

39. $(3\sqrt{m} + 1)(2\sqrt{m} - 5)$

40. $(5\sqrt{m} + 1)(2\sqrt{m} - 7)$

41. $(4\sqrt{x} - \sqrt{y})^2$

42. $(6\sqrt{x} + \sqrt{y})^2$

43. $(\sqrt{2p} - 10\sqrt{7})(\sqrt{2p} + 3\sqrt{5})$

44. $(\sqrt{3p} - 4\sqrt{5})(\sqrt{3p} + 2\sqrt{7})$

45. $(\sqrt{3x} + 6)(\sqrt{3x} - 6)$

46. $(\sqrt{2x} - 5)(\sqrt{2x} + 5)$

47. $(z - 2\sqrt{14})(z + 2\sqrt{14})$

48. $(z + 3\sqrt{6})(z - 3\sqrt{6})$

49. $(8 + 5\sqrt{2ab})^2$

50. $(7 - 4\sqrt{3ab})^2$

51. $(5\sqrt{r} - \sqrt{2t})(3\sqrt{r} + \sqrt{8t})$

52. $(4\sqrt{r} - \sqrt{3t})(2\sqrt{r} + \sqrt{27t})$

53. $(\sqrt{x + 8} + 4)^2$

54. $(\sqrt{x + 3} + 6)^2$

55. $(\sqrt{y} + \sqrt{y + 7})(\sqrt{y} - \sqrt{y + 7})$

56. $(\sqrt{y} + \sqrt{y + 5})(\sqrt{y} - \sqrt{y + 5})$

57. $(\sqrt[3]{5} + 9)(\sqrt[3]{25} - 4)$

58. $(\sqrt[3]{7} - 3)(\sqrt[3]{49} + 6)$

59. $(7 + 2\sqrt[3]{9})^2$

60. $(5 - 3\sqrt[3]{4})^2$

Simplify each expression. Assume $x > 0$.

61. $\dfrac{7 - 7\sqrt{13}}{7}$

62. $\dfrac{11 + 11\sqrt{19}}{11}$

63. $\dfrac{3 + 12\sqrt{22}}{9}$

64. $\dfrac{5 - 15\sqrt{6}}{20}$

65. $\dfrac{-6 - \sqrt{18}}{3}$

66. $\dfrac{-4 + \sqrt{28}}{2}$

67. $\dfrac{10 \pm \sqrt{200}}{40}$

68. $\dfrac{8 \pm \sqrt{192}}{48}$

69. $\dfrac{32 \pm 4\sqrt{44}}{24}$

70. $\dfrac{27 \pm 9\sqrt{45}}{36}$

71. $\dfrac{-25x + 6\sqrt{175x^3}}{5x}$

72. $\dfrac{-21x - 10\sqrt{98x^3}}{7x}$

Problem Solving

73. Find the area of a rectangle with width $\sqrt{10} - 3$ and length $2\sqrt{10} + 5$.

74. Find the area of a square whose side has length $1 + 4\sqrt{8}$.

5.7 Rationalizing the Denominator

In Sec. 5.3 we stated four conditions that had to be met for a radical expression to be in simplest form. The fourth condition was that no denominator contain a radical. In this section we will learn how to eliminate a radical from the denominator of a radical expression.

Example 1   Simplify: $\dfrac{3}{\sqrt{5}}$.

Solution: Eliminate the radical from the denominator by multiplying numerator and denominator by $\sqrt{5}$.

$$\frac{3}{\sqrt{5}} = \frac{3}{\sqrt{5}} \cdot \frac{\sqrt{5}}{\sqrt{5}} = \frac{3\sqrt{5}}{5}$$

Try Exercise 7

Simplify: $\dfrac{9}{\sqrt{2}}$.

Note that the denominator in Example 1 was changed from the irrational number $\sqrt{5}$ to the rational number 5. For this reason, the process of eliminating a radical from the denominator of a fraction is called **rationalizing the denominator.**

Example 2   Simplify: (a) $\sqrt{\dfrac{24}{7}}$, (b) $\dfrac{15}{2\sqrt{10}}$.

Solution: (a) $\sqrt{\dfrac{24}{7}} = \dfrac{\sqrt{24}}{\sqrt{7}}$ Quotient rule

$\qquad\qquad = \dfrac{\sqrt{4}\sqrt{6}}{\sqrt{7}}$ Product rule

$\qquad\qquad = \dfrac{2\sqrt{6}}{\sqrt{7}}$ Simplify.

$\qquad\qquad = \dfrac{2\sqrt{6}}{\sqrt{7}} \cdot \dfrac{\sqrt{7}}{\sqrt{7}}$ Rationalize the denominator.

$\qquad\qquad = \dfrac{2\sqrt{42}}{7}$ Product rule

$\qquad$ (b) $\dfrac{15}{2\sqrt{10}} = \dfrac{15}{2\sqrt{10}} \cdot \dfrac{\sqrt{10}}{\sqrt{10}}$ Rationalize the denominator.

$\qquad\qquad = \dfrac{15\sqrt{10}}{2 \cdot 10}$ Product rule

$\qquad\qquad = \dfrac{3\sqrt{10}}{4}$ Divide out the common factor 5.   ❏

Try Exercise 13

Simplify: $\sqrt{\dfrac{24}{11}}$.

Example 3   Simplify: $\dfrac{\sqrt{20x^5}}{\sqrt{15xy}}$. Assume $x > 0$ and $y > 0$.

Solution: $\dfrac{\sqrt{20x^5}}{\sqrt{15xy}} = \sqrt{\dfrac{20x^5}{15xy}}$ Quotient rule

$\qquad\qquad = \sqrt{\dfrac{4x^4}{3y}}$ Simplify the radicand.

$\qquad\qquad = \dfrac{\sqrt{4x^4}}{\sqrt{3y}}$ Quotient rule

$\qquad\qquad = \dfrac{2x^2}{\sqrt{3y}}$ Simplify the numerator.

$\qquad\qquad = \dfrac{2x^2}{\sqrt{3y}} \cdot \dfrac{\sqrt{3y}}{\sqrt{3y}}$ Rationalize the denominator.

$\qquad\qquad = \dfrac{2x^2\sqrt{3y}}{3y}$ Product rule   ❏

Try Exercise 21

Simplify: $\dfrac{\sqrt{72x^7}}{\sqrt{14xy}}$.

Example 4 illustrates how to rationalize a denominator that is a cube root.

Example 4   Simplify: $\dfrac{6}{\sqrt[3]{2}}$.

Solution: We need to make the radicand in the denominator a perfect cube. Since $2 \cdot 4 = 8$, and 8 is a perfect cube, multiply numerator and denominator by $\sqrt[3]{4}$.

Try Exercise 25

Simplify: $\dfrac{8}{\sqrt[3]{4}}$.

$$\frac{6}{\sqrt[3]{2}} = \frac{6}{\sqrt[3]{2}} \cdot \frac{\sqrt[3]{4}}{\sqrt[3]{4}} = \frac{6\sqrt[3]{4}}{\sqrt[3]{8}} = \frac{6\sqrt[3]{4}}{2} = 3\sqrt[3]{4} \qquad \square$$

To make an expression a perfect cube, you must multiply it by an expression that makes its exponent divisible by 3.

Expression	Multiply by	To obtain the perfect cube
2	2^2	2^3, or 8
$9p$	$3p^2$	$27p^3$, or 3^3p^3
$5x^5y^7$	5^2xy^2	$5^3x^6y^9$, or $125x^6y^9$

Example 5   Simplify: $\dfrac{\sqrt[3]{10}}{\sqrt[3]{18p}}$.

Solution:

$$\frac{\sqrt[3]{10}}{\sqrt[3]{18p}} = \sqrt[3]{\frac{10}{18p}} \qquad \text{Quotient rule}$$

$$= \sqrt[3]{\frac{5}{9p}} \qquad \text{Simplify the radicand.}$$

$$= \frac{\sqrt[3]{5}}{\sqrt[3]{9p}} \qquad \text{Quotient rule}$$

$$= \frac{\sqrt[3]{5}}{\sqrt[3]{9p}} \cdot \frac{\sqrt[3]{3p^2}}{\sqrt[3]{3p^2}} \qquad \text{Rationalize the denominator.}$$

$$= \frac{\sqrt[3]{15p^2}}{\sqrt[3]{27p^3}} \qquad \text{Product rule}$$

$$= \frac{\sqrt[3]{15p^2}}{3p} \qquad \text{Simplify.} \qquad \square$$

Try Exercise 29

Simplify: $\dfrac{\sqrt[3]{14}}{\sqrt[3]{63p}}$.

To rationalize a denominator like $3 - \sqrt{5}$, multiply numerator and denominator by the conjugate of the denominator. The expressions $3 + \sqrt{5}$ and $3 - \sqrt{5}$ are called **conjugates** of each other. Note that their product does not contain a radical.

$$(3 + \sqrt{5})(3 - \sqrt{5}) = 3^2 - (\sqrt{5})^2 = 9 - 5 = 4$$

$$\uparrow \qquad \uparrow \ \uparrow \qquad \uparrow \qquad \uparrow \qquad \uparrow$$

$$(a + b)\ (a - b) = a^2 - b^2$$

Example 6   Simplify: $\dfrac{2}{3 - \sqrt{5}}$.

Solution: Multiply numerator and denominator by the conjugate of the denominator.

$$\frac{2}{3 - \sqrt{5}} = \frac{2}{3 - \sqrt{5}} \cdot \frac{3 + \sqrt{5}}{3 + \sqrt{5}} \qquad \begin{array}{l}\text{The conjugate of } 3 - \sqrt{5} \text{ is} \\ 3 + \sqrt{5}.\end{array}$$

$$= \frac{2(3 + \sqrt{5})}{3^2 - (\sqrt{5})^2} \qquad \begin{array}{l}\text{Multiply numerators, multiply} \\ \text{denominators.}\end{array}$$

$$= \frac{2(3 + \sqrt{5})}{9 - 5} \ \left.\begin{array}{l}\\ \\ \\ \\ \end{array}\right\} \quad \begin{array}{l}\text{Simplify the denominator,} \\ \text{leave the numerator in fac-} \\ \text{tored form.}\end{array}$$

$$= \frac{2(3 + \sqrt{5})}{4}$$

Try Exercise 37

Simplify: $\dfrac{5}{4 - \sqrt{6}}$.

$$= \frac{3 + \sqrt{5}}{2} \qquad \begin{array}{l}\text{Divide out the common factor} \\ 2.\end{array} \qquad \square$$

Example 7   Simplify: $\dfrac{4 + \sqrt{6}}{\sqrt{2} + \sqrt{3}}$.

Solution: $\dfrac{4 + \sqrt{6}}{\sqrt{2} + \sqrt{3}} = \dfrac{4 + \sqrt{6}}{\sqrt{2} + \sqrt{3}} \cdot \dfrac{\sqrt{2} - \sqrt{3}}{\sqrt{2} - \sqrt{3}}$ The conjugate of $\sqrt{2} + \sqrt{3}$ is $\sqrt{2} - \sqrt{3}$.

$= \dfrac{4\sqrt{2} - 4\sqrt{3} + \sqrt{12} - \sqrt{18}}{(\sqrt{2})^2 - (\sqrt{3})^2}$ Multiply numerators, multiply denominators.

$\left. \begin{array}{l} = \dfrac{4\sqrt{2} - 4\sqrt{3} + 2\sqrt{3} - 3\sqrt{2}}{2 - 3} \\[2em] = \dfrac{\sqrt{2} - 2\sqrt{3}}{-1} \end{array} \right\}$ Simplify.

$= -\sqrt{2} + 2\sqrt{3}$ Divide each term by -1.   ◻

Try Exercise 47

Simplify: $\dfrac{4 + \sqrt{6}}{\sqrt{2} - \sqrt{3}}$.

In Example 8 we use the formula for squaring a binomial:

$$(a + b)^2 = a^2 + 2ab + b^2$$

Example 8   Simplify: $\dfrac{\sqrt{7x} + \sqrt{y}}{\sqrt{7x} - \sqrt{y}}$.

Solution: $\dfrac{\sqrt{7x} + \sqrt{y}}{\sqrt{7x} - \sqrt{y}} = \dfrac{\sqrt{7x} + \sqrt{y}}{\sqrt{7x} - \sqrt{y}} \cdot \dfrac{\sqrt{7x} + \sqrt{y}}{\sqrt{7x} + \sqrt{y}}$ The conjugate of $\sqrt{7x} - \sqrt{y}$ is $\sqrt{7x} + \sqrt{y}$.

Note that the product of the numerators is the square of the binomial $\sqrt{7x} + \sqrt{y}$.

$= \dfrac{(\sqrt{7x})^2 + 2\sqrt{7x}\sqrt{y} + (\sqrt{y})^2}{(\sqrt{7x})^2 - (\sqrt{y})^2}$ Multiply numerators, multiply denominators.

$= \dfrac{7x + 2\sqrt{7xy} + y}{7x - y}$ Simplify.   ◻

Try Exercise 51

Simplify: $\dfrac{\sqrt{6x} + \sqrt{y}}{\sqrt{6x} - \sqrt{y}}$.

Exercises 5.7

◆◆

True-False Exercises

1. The radical expression $\dfrac{2}{\sqrt{3}}$ is in simplest form.

2. The radical expression $\dfrac{1}{\sqrt{6} + 1}$ is in simplest form.

Completion Exercises

3. The process of eliminating a radical from the denominator of a fraction is called _____.

4. To rationalize the denominator of $\dfrac{5}{\sqrt{7}}$, multiply numerator and denominator by _____.

5. To rationalize the denominator of $\dfrac{1}{\sqrt[3]{3}}$, multiply numerator and denominator by _____.

6. To rationalize the denominator of $\dfrac{2}{\sqrt[3]{4x}}$, multiply numerator and denominator by _____.

Simplify each radical expression. Rationalize all denominators. Assume all variables represent positive real numbers.

7. $\dfrac{9}{\sqrt{2}}$

8. $\dfrac{4}{\sqrt{3}}$

9. $\dfrac{\sqrt{3}}{\sqrt{7}}$

10. $\dfrac{\sqrt{2}}{\sqrt{5}}$

11. $\dfrac{20}{\sqrt{5}}$

12. $\dfrac{28}{\sqrt{7}}$

13. $\sqrt{\dfrac{24}{11}}$

14. $\sqrt{\dfrac{32}{15}}$

15. $\dfrac{14\sqrt{3}}{\sqrt{6}}$

16. $\dfrac{25\sqrt{2}}{\sqrt{10}}$

17. $\dfrac{2\sqrt{k}}{\sqrt{10z}}$

18. $\dfrac{3\sqrt{k}}{\sqrt{6z}}$

19. $-\sqrt{\dfrac{8m^3}{75}}$

20. $-\sqrt{\dfrac{125m^5}{108}}$

21. $\dfrac{\sqrt{72x^7}}{\sqrt{14xy}}$

22. $\dfrac{\sqrt{63x^{11}}}{\sqrt{35x^3y}}$

23. $\dfrac{1}{\sqrt[3]{5}}$

24. $\dfrac{1}{\sqrt[3]{7}}$

25. $\dfrac{8}{\sqrt[3]{4}}$

26. $\dfrac{6}{\sqrt[3]{9}}$

27. $\sqrt[3]{\dfrac{m^4}{2n^2}}$

28. $\sqrt[3]{\dfrac{m^5}{5n^2}}$

29. $\dfrac{\sqrt[3]{14}}{\sqrt[3]{63p}}$

30. $\dfrac{\sqrt[3]{15}}{\sqrt[3]{20p}}$

Completion Exercises

31. The expressions $\sqrt{10} - 4$ and $\sqrt{10} + 4$ are called _____ of each other.

32. The conjugate of $\sqrt{13} + \sqrt{11}$ is _____.

33. To rationalize the denominator of $\dfrac{1}{2 + \sqrt{x}}$, multiply numerator and denominator by _____.

34. The product of $\sqrt{5} - \sqrt{2}$ and $\sqrt{5} + \sqrt{2}$ is _____.

Simplify each radical expression. Rationalize all denominators. Assume all variables represent positive real numbers.

35. $\dfrac{1}{\sqrt{2} + 1}$

36. $\dfrac{1}{2 + \sqrt{3}}$

37. $\dfrac{5}{4 - \sqrt{6}}$

38. $\dfrac{3}{5 - \sqrt{10}}$

39. $\dfrac{10}{3 + \sqrt{3}}$

40. $\dfrac{14}{\sqrt{5} + 1}$

41. $\dfrac{1}{3\sqrt{5} - 1}$

42. $\dfrac{1}{2\sqrt{7} - 1}$

43. $\dfrac{12}{\sqrt{7} + \sqrt{5}}$

44. $\dfrac{8}{\sqrt{5} + \sqrt{3}}$

45. $\dfrac{-9\sqrt{2}}{\sqrt{13} - \sqrt{2}}$

46. $\dfrac{-11\sqrt{3}}{\sqrt{17} - \sqrt{3}}$

47. $\dfrac{4 + \sqrt{6}}{\sqrt{2} - \sqrt{3}}$

48. $\dfrac{2 + \sqrt{30}}{\sqrt{5} - \sqrt{6}}$

49. $\dfrac{k - 4}{\sqrt{k} + 2}$

50. $\dfrac{k - 9}{\sqrt{k} + 3}$

51. $\dfrac{\sqrt{6x} + \sqrt{y}}{\sqrt{6x} - \sqrt{y}}$

52. $\dfrac{\sqrt{x} + \sqrt{10y}}{\sqrt{x} - \sqrt{10y}}$

53. $\dfrac{4\sqrt{a}}{4\sqrt{a} + 3\sqrt{b}}$

54. $\dfrac{3\sqrt{b}}{3\sqrt{a} + 5\sqrt{b}}$

Discussion Exercises

55. Why can you not rationalize the denominator of $\dfrac{1}{\sqrt[3]{2}}$ by multiplying numerator and denominator by $\sqrt{2}$? by $\sqrt[3]{2}$?

56. Why can you not rationalize the denominator of $\dfrac{1}{\sqrt{5} + 1}$ by multiplying numerator and denominator by $\sqrt{5}$? by $\sqrt{5} + 1$?

Rationalize the denominator of each fraction. Then add or subtract as indicated and simplify.

57. $\dfrac{\sqrt{5}}{3} - \dfrac{1}{\sqrt{5}}$

58. $\dfrac{3}{\sqrt{2}} + \sqrt{2}$

59. $\dfrac{\sqrt{8}}{\sqrt{6}} + \dfrac{\sqrt{6}}{\sqrt{8}}$

60. $\dfrac{\sqrt{8}}{\sqrt{10}} + \dfrac{\sqrt{10}}{\sqrt{8}}$

61. $\sqrt{50} + \sqrt{\dfrac{1}{2}}$

62. $\sqrt{48} + \sqrt{\dfrac{1}{3}}$

Problem Solving

The current A in amperes of an electrical appliance is given by the formula

$$A = \sqrt{\frac{W}{R}}$$

where W is the wattage of the appliance and R is the resistance of the appliance in ohms.

63. Determine the current in a hair dryer that uses 1100 W of power and has a resistance of 24 Ω.

64. Determine the current in a contact lens disinfecting unit that uses 27 W of power and has a resistance of 540 Ω.

5.8 Radical Equations

TAPE AU14

In previous chapters we learned how to solve linear equations, quadratic equations, and equations with rational expressions. In this section we learn how to solve equations with radicals.

Equations such as

$$\sqrt{x - 3} = 2 \qquad \sqrt[3]{6t - 3} = \sqrt[3]{8t + 5} \qquad \text{and} \qquad \sqrt{4p - 5} - \sqrt{2p - 5} = 2$$

are called **radical equations,** because they contain a variable in a radicand. To solve a radical equation, we use the **power rule for equality.**

Power Rule for Equality

Suppose a and b are real numbers and n is a positive integer.

$$\text{If } a = b, \text{ then } a^n = b^n$$

You will never *lose* solutions when you use the power rule to solve an equation, but you may *gain* solutions. For example, the equation $x = 5$ has only one solution, 5. But if you square both sides, you get the equation $x^2 = 25$. This equation has two solutions, namely, 5 and -5.

CAUTION

Whenever you raise both sides of an equation to an *even* power, you must check all solutions in the original equation. Those proposed solutions that do not check are **extraneous solutions** and must be discarded. Raising both sides of an equation to an *odd* power does not produce extraneous solutions, but it is still a good idea to check your answer. ■

Example 1 Solve: $\sqrt{x - 3} = 2$.

Solution: Use the power rule to square both sides. This will remove the radical.

$$\begin{aligned} \sqrt{x - 3} &= 2 && \text{Original equation} \\ (\sqrt{x - 3})^2 &= 2^2 && \text{Square both sides.} \\ x - 3 &= 4 && \text{Simplify.} \\ x &= 7 && \text{Add 3.} \end{aligned}$$

Since we raised both sides to an even power, namely, 2, we must check our solution in the original equation.

Check: $\sqrt{x-3} = 2$ Original equation

$\sqrt{7-3} = 2$ Substitute $x = 7$.

$\sqrt{4} = 2$ Simplify.

$2 = 2$ True

The number 7 checks, so the solution set is $\{7\}$.   ❐

Try Exercise 5

Solve: $\sqrt{x-1} = 2$.

Example 2 Solve: $\sqrt{4z+1} + 3 = 0$.

Solution: Isolate the radical on one side of the equation. Then square both sides.

$\sqrt{4z+1} + 3 = 0$ Original equation

$\sqrt{4z+1} = -3$ Subtract 3.

$(\sqrt{4z+1})^2 = (-3)^2$ Square both sides.

$4z + 1 = 9$ Simplify.

$4z = 8$ Subtract 1.

$z = 2$ Divide by 4.

Check this solution in the original equation.

Check: $\sqrt{4z+1} + 3 = 0$ Original equation

$\sqrt{4(2)+1} + 3 = 0$ Substitute $z = 2$.

$\sqrt{9} + 3 = 0$ Simplify.

$6 = 0$ False

Since 2 does not check, it is an extraneous solution and must be discarded. Therefore the solution set is $\varnothing$. Actually, since $\sqrt{4z+1}$ is nonnegative for *any* value of z, we could have concluded from the second step $\sqrt{4z+1} = -3$ that there was no solution.   ❐

Try Exercise 11

Solve: $\sqrt{4z-3} + 5 = 0$.

To solve the equation in Example 3 we use the rule for squaring a binomial:

$$(a-b)^2 = a^2 - 2ab + b^2$$

Example 3 Solve: $\sqrt{m+16} = m - 4$.

Solution: Square both sides. Note that you are squaring a binomial on the right side.

$$(\sqrt{m+16})^2 = (m-4)^2$$

$$m + 16 = m^2 - 8m + 16$$

This is a quadratic equation. Collect terms on the right side and solve by factoring.

$0 = m^2 - 9m$ Subtract m, subtract 16.

$0 = m(m - 9)$ Factor.

Set each factor equal to 0 and solve.

$m = 0$ or $m - 9 = 0$

$m = 9$

The number 9 checks, but 0 does not. Therefore the solution set is $\{9\}$.   ❐

Try Exercise 33

Solve: $\sqrt{m+4} = m - 2$.

CAUTION

Make sure you square each *side* of a radical equation, not each *term*.

Correct

$$\sqrt{x} = x - 2$$
$$(\sqrt{x})^2 = (x - 2)^2$$
$$x = x^2 - 4x + 4$$

Wrong

$$\sqrt{x} = x - 2$$
$$(\sqrt{x})^2 = x^2 - 2^2$$
$$x = x^2 - 4$$

∎

If the index on the radical is 3, we must *cube* both sides to remove the radical.

Example 4   Solve: $\sqrt[3]{6t - 3} = \sqrt[3]{8t + 5}$.

Solution: Use the power rule to raise both sides to a power equal to the index, 3.

$$(\sqrt[3]{6t - 3})^3 = (\sqrt[3]{8t + 5})^3 \qquad \text{Cube both sides.}$$
$$6t - 3 = 8t + 5 \qquad \text{Simplify.}$$
$$-2t = 8 \qquad \text{Subtract } 8t, \text{ add } 3.$$
$$t = -4 \qquad \text{Divide by } -2.$$

Try Exercise 39

Solve: $\sqrt[3]{2t + 5} = \sqrt[3]{4t + 11}$.

The number -4 checks, so the solution set is $\{-4\}$. ❐

When an equation contains two or more radicals, sometimes we must use the power rule more than once to remove all radicals from the equation. Here is a summary of the steps in solving a radical equation.

To Solve a Radical Equation

1. Isolate a radical on one side of the equation.
2. Raise both sides of the equation to a power equal to the index of the isolated radical.
3. If the equation still contains a radical, simplify and repeat Steps 1 and 2.
4. Solve the resulting equation.
5. Check all solutions in the original equation and discard any extraneous solutions.

We illustrate these steps in Example 5.

Example 5   Solve: $\sqrt{4p - 3} - \sqrt{2p - 5} = 2$.

Solution: Step 1:   Isolate a radical on one side of the equation.

$$\sqrt{4p - 3} - \sqrt{2p - 5} = 2 \qquad \text{Original equation}$$
$$\sqrt{4p - 3} = 2 + \sqrt{2p - 5} \qquad \text{Add } \sqrt{2p - 5}.$$

Step 2:   Since the index is 2, square both sides.

$$(\sqrt{4p - 3})^2 = (2 + \sqrt{2p - 5})^2$$

Use $(a + b)^2 = a^2 + 2ab + b^2$ to square the right side.

$$4p - 3 = 4 + 4\sqrt{2p - 5} + (2p - 5)$$

Step 3:   Since the equation still contains a radical, simplify the equation, isolate the radical, and square again.

$$4p - 3 = 4\sqrt{2p - 5} + 2p - 1 \qquad \text{Combine like terms.}$$
$$2p - 2 = 4\sqrt{2p - 5} \qquad \text{Subtract } 2p, \text{ add } 1.$$
$$p - 1 = 2\sqrt{2p - 5} \qquad \text{Divide by 2.}$$
$$(p - 1)^2 = (2\sqrt{2p - 5})^2 \qquad \text{Square both sides.}$$
$$p^2 - 2p + 1 = 4(2p - 5)$$

Step 4:   Solve the resulting equation.

$$p^2 - 2p + 1 = 8p - 20 \qquad \text{Distribute 4.}$$
$$p^2 - 10p + 21 = 0 \qquad \text{Subtract } 8p, \text{ add } 20.$$
$$(p - 3)(p - 7) = 0 \qquad \text{Factor.}$$
$$p - 3 = 0 \quad \text{or} \quad p - 7 = 0 \qquad \text{Set each factor equal to 0.}$$
$$p = 3 \qquad\qquad p = 7 \qquad \text{Solve each linear equation.}$$

Step 5:   Check all solutions in the original equation.

Check: p = 3

$$\sqrt{4(3) - 3} - \sqrt{2(3) - 5} = 2$$
$$\sqrt{9} - \sqrt{1} = 2$$
$$3 - 1 = 2$$
true

Check: p = 7

$$\sqrt{4(7) - 3} - \sqrt{2(7) - 5} = 2$$
$$\sqrt{25} - \sqrt{9} = 2$$
$$5 - 3 = 2$$
true

Try Exercise 49

Solve:
$\sqrt{3p + 4} - \sqrt{2p - 4} = 2.$

Both solutions check, so the solution set is $\{3, 7\}$.   ❒

Here is an example of an applied problem that requires you to solve a radical equation.

Problem Solving

Example 6   The time t in seconds that it takes a pendulum l feet long to complete a cycle (see Fig. 5.10) is called the period of the pendulum and is given by the formula

$$t = 2\pi\sqrt{\frac{l}{32}}$$

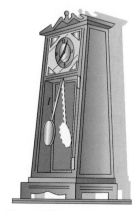

Figure 5.10

How long is a pendulum whose period is 2 sec?

Solution:

$$t = 2\pi\sqrt{\dfrac{l}{32}}$$ Write down the appropriate formula.

$$2 = 2\pi\sqrt{\dfrac{l}{32}}$$ Substitute $t = 2$.

$$\dfrac{2}{2\pi} = \sqrt{\dfrac{l}{32}}$$ Divide by 2π.

$$\dfrac{1}{\pi} = \sqrt{\dfrac{l}{32}}$$ Simplify the left side.

$$\left(\dfrac{1}{\pi}\right)^2 = \left(\sqrt{\dfrac{l}{32}}\right)^2$$ Square both sides.

$$\dfrac{1}{\pi^2} = \dfrac{l}{32}$$ Simplify.

$$\dfrac{32}{\pi^2} = l$$ Multiply by 32.

Try Exercise 57

How long is a pendulum whose period is 4 sec?

This answer checks, so the pendulum is $\dfrac{32}{\pi^2}$ (approximately 3.24) feet long. ❏

Exercises 5.8

Completion Exercises

1. A radical equation is an equation that contains a variable in a(n) _____.

2. The power rule for equality states that if $a = b$, then _____.

True-False Exercises

3. Raising both sides of an equation to an even power may introduce extraneous solutions.

4. If $\sqrt{x + 7} = x + 5$, then $x + 7 = x^2 + 25$.

Solve each equation.

5. $\sqrt{x - 1} = 2$

6. $\sqrt{x - 2} = 3$

7. $\sqrt{2y - 3} = 9$

8. $\sqrt{2y - 5} = 11$

9. $\sqrt{m - 2} = 5$

10. $\sqrt{m - 3} = 5$

11. $\sqrt{4z - 3} + 5 = 0$

12. $\sqrt{4z - 7} + 1 = 0$

13. $3 = 1 + \sqrt{5p + 4}$

14. $5 = 2 + \sqrt{5p + 9}$

15. $\sqrt{y + 2} = y$

16. $\sqrt{y + 6} = y$

17. $2z - \sqrt{5z - 1} = 0$

18. $3z - \sqrt{10z - 1} = 0$

19. $\sqrt[3]{p} - 4 = 0$

20. $\sqrt[3]{p} - 5 = 0$

21. $\sqrt[4]{m - 6} = 1$

22. $\sqrt[4]{m - 1} = 2$

23. $\sqrt{r + 1} = \sqrt{r + 2}$

24. $\sqrt{r + 4} = \sqrt{r + 3}$

25. $\sqrt[3]{y^2 - 17} = 2$

26. $\sqrt[3]{y^2 + 23} = 3$

27. $\sqrt{r^2 - 6r + 12} = r$

28. $\sqrt{r^2 - 7r + 21} = r$

29. $\sqrt{p^2 + 12p} = 3\sqrt{5}$

30. $\sqrt{p^2 + 12p} = 2\sqrt{7}$

31. $\sqrt{10z} = 2\sqrt{z + 3}$

32. $\sqrt{6z} = 2\sqrt{z + 5}$

33. $\sqrt{m + 4} = m - 2$

34. $\sqrt{m + 9} = m - 3$

35. $\sqrt{y + 6} = y$

36. $\sqrt{y + 2} = y$

37. $\sqrt{a^2 + 45} = a - 3$

38. $\sqrt{a^2 + 24} = a - 2$

39. $\sqrt[3]{2t + 5} = \sqrt[3]{4t + 11}$

40. $\sqrt[3]{3t - 2} = \sqrt[3]{5t + 2}$

41. $\sqrt{9 + 4\sqrt{x}} = 7$

42. $\sqrt{4 + 9\sqrt{x}} = 7$

43. $\sqrt{q - 3} = \sqrt{q} - 1$

44. $\sqrt{q - 5} = \sqrt{q} - 1$

45. $(3z - 2)^{1/2} - z^{1/2} = 2$

46. $(3z + 4)^{1/2} - z^{1/2} = 2$

47. $(r^{1/2} + 1)^{1/4} = 3$

48. $(r^{1/2} - 44)^{1/4} = 2$

49. $\sqrt{3p + 4} - \sqrt{2p - 4} = 2$

50. $\sqrt{5p + 6} - \sqrt{3p + 4} = 2$

51. $\sqrt{5m - 1} + \sqrt{m + 3} = 4$

52. $\sqrt{5m + 6} + \sqrt{m + 3} = 3$

Discussion Exercises

53. Without even solving the equation $\sqrt{3x - 2} = -4$, you should know that it has no real solution. Why?

54. Outline the steps in solving a radical equation.

Problem Solving

55. The principal square root of the sum of a number and 2 is 4 less than the number. Find the number.

56. The principal square root of the sum of a number and 7 is 5 less than the number. Find the number.

57. How long is a pendulum whose period is 4 sec? Use $t = 2\pi\sqrt{\dfrac{l}{32}}$.

58. How long is a pendulum whose period is 6 sec? Use $t = 2\pi\sqrt{\dfrac{l}{32}}$.

59. The distance d, in miles, to the horizon from a height of h feet is given by the formula $d = \sqrt{1.5h}$ (see Fig. 5.11). How high above the water must a blimp be to see a lifeboat 30 mi away?

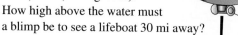

Figure 5.11

60. Ignoring air resistance, the time t, in seconds, that it takes an object to fall a distance of d feet is given by the formula

$$t = \frac{\sqrt{d}}{4}$$

A stone dropped into a well splashes into the water at the bottom of the well 3 sec later. How deep is the well?

Getting Acquainted with Your Graphing Calculator

61. Check that 11 is a solution of $\sqrt{x + 25} = x - 5$ by storing 11 in x and then calculating the value of each side of the equation.

62. Check that 81 is a solution of $\sqrt{16 + \sqrt{y}} = 5$ by storing 81 in y and calculating the value of the left side of the equation.

5.9 Complex Numbers

In Sec. 5.1 we noted that square roots of negative numbers are not real numbers. In this section we give names to these numbers, and we learn how to add, subtract, multiply, and divide them.

Imaginary Numbers

Numbers like $\sqrt{-9}$ are called **imaginary numbers.*** In order to write imaginary numbers in a simpler form, we introduce the following notation.

> **DEFINITION OF i**
>
> The **imaginary unit** is denoted i, where
> $$i = \sqrt{-1} \qquad \text{and} \qquad i^2 = -1$$

---

*HISTORICAL NOTE: René Descartes (1596–1650) introduced the term "imaginary number," thus giving them a permanent stigma. Of course, these numbers really do exist and, in fact, have a wide variety of useful applications in physics and engineering. The first application of imaginary numbers was introduced by Charles Steinmetz (1865–1923), who used them to explain the behavior of electric circuits. The Swiss mathematician Leonhard Euler (1707–1783) was the first to use the symbol i for $\sqrt{-1}$.

We now extend the product rule for radicals so that it applies when one of the radicands is negative. This allows us to write imaginary numbers more simply using the imaginary unit i.

Product Rule For Radicals

If a and b are real numbers and at least one of them is nonnegative, then

$$\sqrt{ab} = \sqrt{a}\,\sqrt{b}$$

Example 1   Write each imaginary number using i:

(a) $\sqrt{-4}$, (b) $\sqrt{-7}$, (c) $\sqrt{-18}$, (d) $-\sqrt{-\frac{1}{25}}$.

Solution: (a) $\sqrt{-4} = \sqrt{4}\,\sqrt{-1}$ Product rule

$\qquad\qquad\quad = 2i$ Since $i = \sqrt{-1}$

(b) $\sqrt{-7} = \sqrt{7}\,\sqrt{-1} = \sqrt{7}i = i\sqrt{7}$

Note that we write $\sqrt{7}i$ as $i\sqrt{7}$ to avoid confusing $\sqrt{7}i$ with $\sqrt{7i}$.

(c) $\sqrt{-18} = \sqrt{18}\,\sqrt{-1} = 3\sqrt{2}i = 3i\sqrt{2}$

(d) $-\sqrt{-\frac{1}{25}} = -\sqrt{\frac{1}{25}}\,\sqrt{-1} = -\frac{1}{5}i$ ❏

Try Exercise 13

Write $\sqrt{-9}$ using i.

The numbers $2i$, $i\sqrt{7}$, $3i\sqrt{2}$, and $-\frac{1}{5}i$ are all imaginary numbers. Other examples of imaginary numbers are

$$6 + 2i \qquad -1 - i \qquad \text{and} \qquad \frac{3}{4} + \frac{\sqrt{5}}{4}i$$

Complex Numbers

The union of the set of imaginary numbers with the set of real numbers is called the set of **complex numbers.**

DEFINITION OF A COMPLEX NUMBER

A **complex number** is a number of the form

$$a + bi$$

where a and b are real numbers and $i = \sqrt{-1}$. The form $a + bi$ is called the **standard form** of a complex number.

The number a is called the **real part** of the complex number $a + bi$. The number b is called the **imaginary part** of $a + bi$.

$$a + bi$$
$$\uparrow \qquad \uparrow$$
real part   imaginary part

If $b = 0$, the complex number $a + bi$ is simply the real number a. For example, $7 + 0i$ is simply the real number 7. If $a = 0$, the complex number $a + bi$ becomes bi, which is called a **pure imaginary number** (so long as $b \neq 0$). For example, $0 + 8i$ is the pure imaginary number $8i$.

Figure 5.12 illustrates the relationship between the set of complex numbers and its subsets. Note that **both the set of real numbers and the set of imaginary numbers are subsets of the set of complex numbers.**

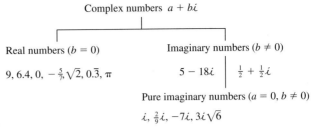

Figure 5.12

Operations on Complex Numbers

We add two complex numbers by adding their real parts and adding their imaginary parts.

> **DEFINITION OF ADDITION OF COMPLEX NUMBERS**
>
> The **sum** of the complex numbers $a + bi$ and $c + di$ is
> $$(a + bi) + (c + di) = (a + c) + (b + d)i$$

Example 2   Find the sum of $3 + 7i$ and $5 + 2i$.

Solution: $(3 + 7i) + (5 + 2i) = (3 + 5) + (7 + 2)i$ Add the real parts, add the imaginary parts.

$\qquad\qquad\qquad\qquad = 8 + 9i$ Simplify.   ❏

Try Exercise 31

Add: $(6 + 2i) + (4 + 7i)$.

An easy way to add the two numbers $3 + 7i$ and $5 + 2i$ is to think of the numbers as binomials in i. Then add like terms.

$$(3 + 7i) + (5 + 2i) = 3 + 7i + 5 + 2i \qquad \text{Remove parentheses.}$$
$$= 8 + 9i \qquad \text{Add like terms.}$$

Example 3   Add: (a) $(-6 + i) + (4 - 12i)$, (b) $(-5 + \sqrt{-16}) + 2\sqrt{-9}$.

Solution: (a) $(-6 + i) + (4 - 12i) = -6 + i + 4 - 12i$ Remove parentheses.
$\qquad\qquad\qquad\qquad\qquad\quad = -2 - 11i$ Combine like terms.

(b) Note that $\sqrt{-16} = \sqrt{16}\sqrt{-1} = 4i$ and $2\sqrt{-9} = 2\sqrt{9}\sqrt{-1} = 6i$. Therefore

Try Exercise 35

Add: $(-8 - \sqrt{-1}) + 2\sqrt{-4}$.

$$(-5 + \sqrt{-16}) + 2\sqrt{-9} = (-5 + 4i) + 6i = -5 + 10i. \qquad ❏$$

As with addition, the easiest way to subtract two complex numbers is to think of them as binomials in i.

Example 4   Subtract:
(a) $(10 + 6i) - (1 + 7i)$, (b) $(-4 + 9i) - (-4 - i)$,
(c) $\sqrt{-50} - (8 + \sqrt{-8})$.

Solution: (a) $(10 + 6i) - (1 + 7i) = 10 + 6i - 1 - 7i$ Distribute -1.
$\qquad\qquad\qquad\qquad\qquad\quad = 9 - i$ Combine like terms.

(b) $(-4 + 9i) - (-4 - i) = -4 + 9i + 4 + i$ Distribute -1.

$= 10i$ Combine like terms.

(c) $\sqrt{-50} - (8 + \sqrt{-8}) = 5i\sqrt{2} - (8 + 2i\sqrt{2})$ Write in standard form.

$= 5i\sqrt{2} - 8 - 2i\sqrt{2}$ Distribute -1.

$= -8 + 3i\sqrt{2}$ Combine like terms.

Try Exercise 41

Subtract: $(10 + 6i) - (3 + 5i)$.

DEFINITION OF MULTIPLICATION OF COMPLEX NUMBERS

The **product** of the complex numbers $a + bi$ and $c + di$ is

$$(a + bi)(c + di) = (ac - bd) + (ad + bc)i$$

Note what happens if you treat the numbers $a + bi$ and $c + di$ as binomials in i and use the FOIL method to multiply them.

$$
\begin{array}{cccc}
\text{F} & \text{O} & \text{I} & \text{L} \\
\downarrow & \downarrow & \downarrow & \downarrow
\end{array}
$$

$(a + bi)(c + di) = ac + adi + bci + bdi^2$

$= ac + adi + bci + bd(-1)$ Replace i^2 with -1.

$= (ac - bd) + (ad + bc)i$ Combine like terms.

This is the same result given in the definition of multiplication of complex numbers. Since the FOIL method is easier to remember, that is the method we use to multiply complex numbers.

Example 5   Multiply: (a) $(3 + 5i)(2 + 4i)$, (b) $(7 + 6i)(1 - 2i)$,

(c) $\sqrt{-4}(9 - \sqrt{-25})$.

Solution: (a) $(3 + 5i)(2 + 4i) = 6 + 12i + 10i + 20i^2$ FOIL method

$= 6 + 12i + 10i + 20(-1)$ Replace i^2 with -1.

$= -14 + 22i$ Combine like terms.

(b) $(7 + 6i)(1 - 2i) = 7 - 14i + 6i - 12i^2$ FOIL method

$= 7 - 14i + 6i - 12(-1)$ Replace i^2 with -1.

$= 19 - 8i$ Combine like terms.

(c) $\sqrt{-4}(-9 - \sqrt{-25}) = 2i(-9 - 5i)$ Write in standard form.

$= -18i - 10i^2$ Distribute $2i$.

$= -18i - 10(-1)$ Replace i^2 with -1.

$= 10 - 18i$ Simplify.   ❐

Try Exercise 53

Multiply: $(5 + 4i)(2 + 3i)$.

CAUTION

You cannot use the product rule for radicals to find the product $\sqrt{-4} \cdot \sqrt{-9}$, since the product rule applies only when at least one radicand is nonnegative. Instead, write each number in terms of i, and then multiply.

Correct

$\sqrt{-4} \cdot \sqrt{-9} = 2i \cdot 3i$

$= 6i^2$

$= -6$

Wrong

$\sqrt{-4} \cdot \sqrt{-9} = \sqrt{(-4)(-9)}$

$= \sqrt{36}$

$= 6$

Always write a complex number in standard form before you operate on it.   ■

Example 6   Multiply: (a) $\sqrt{-5} \cdot \sqrt{-7}$, (b) $\sqrt{3} \cdot \sqrt{-27}$, (c) $\sqrt{-12} \cdot \sqrt{-2}$.

Solution: (a) $\sqrt{-5} \cdot \sqrt{-7} = i\sqrt{5} \cdot i\sqrt{7} = i^2\sqrt{5 \cdot 7} = -\sqrt{35}$
(b) $\sqrt{3} \cdot \sqrt{-27} = \sqrt{3} \cdot i\sqrt{27} = i\sqrt{81} = 9i$
(c) $\sqrt{-12} \cdot \sqrt{-2} = i\sqrt{12} \cdot i\sqrt{2} = i^2\sqrt{24} = -\sqrt{4}\sqrt{6} = -2\sqrt{6}$ ❏

Try Exercise 63

Multiply: $\sqrt{-3} \cdot \sqrt{-7}$.

Before we can illustrate how complex numbers are divided, we need to define what is meant by the *conjugate* of a complex number. The numbers $a + bi$ and $a - bi$ are called **complex conjugates** of each other. The product of two complex conjugates is always a real number.

$$(a + bi)(a - bi) = a^2 - b^2i^2$$
$$= a^2 - b^2(-1)$$
$$= a^2 + b^2$$

Note that $a^2 + b^2$ is a real number because a and b are real numbers.
The easiest way to find the quotient of two complex numbers, such as

$$\frac{5 + 10i}{2 + i}$$

is to multiply the numerator and the denominator by the conjugate of the denominator. This produces a real number in the denominator. From there it is easy to write the fraction in the form $a + bi$ or $a - bi$.

Example 7   Divide $5 + 10i$ by $2 + i$.

Solution: Multiply numerator and denominator by the conjugate of the denominator.

$$\frac{5 + 10i}{2 + i} = \frac{5 + 10i}{2 + i} \cdot \frac{2 - i}{2 - i} \qquad \text{The conjugate of } 2 + i \text{ is } 2 - i.$$

$$= \frac{10 - 5i + 20i - 10i^2}{2^2 - i^2} \qquad \text{Multiply numerators, multiply denominators.}$$

$$= \frac{10 - 5i + 20i - 10(-1)}{4 - (-1)} \qquad \text{Replace } i^2 \text{ with } -1.$$

$$= \frac{20 + 15i}{5} \qquad \text{Combine like terms.}$$

The denominator is now a real number. Write the fraction in the standard form $a + bi$ as follows:

$$= \frac{20}{5} + \frac{15i}{5} \qquad \text{Divide each term by 5.}$$

Try Exercise 73

Divide: $\dfrac{2 + 11i}{2 + i}$.

$$= 4 + 3i \qquad \text{Simplify.} \qquad ❏$$

You can check the answer to Example 7 by multiplication as follows:

$$(2 + i)(4 + 3i) = 8 + 6i + 4i + 3i^2$$
$$= 8 + 6i + 4i + 3(-1)$$
$$= 5 + 10i$$

Example 8   Divide: (a) $\dfrac{1}{4 - 2i}$, (b) $\dfrac{7 + 6i}{3i}$.

Solution: (a) $\dfrac{1}{4-2i} = \dfrac{1}{4-2i} \cdot \dfrac{4+2i}{4+2i}$ The conjugate of $4-2i$ is $4+2i$.

$= \dfrac{4+2i}{16-4i^2}$ Multiply numerators, multiply denominators.

$= \dfrac{4+2i}{16-4(-1)}$ Replace i^2 with -1.

$= \dfrac{4+2i}{20}$ Simplify.

$= \dfrac{4}{20} + \dfrac{2i}{20}$ Divide each term by 20.

$= \dfrac{1}{5} + \dfrac{1}{10}i$ Simplify.

(b) $\dfrac{7+6i}{3i} = \dfrac{7+6i}{3i} \cdot \dfrac{-3i}{-3i}$ The conjugate of $3i$ is $-3i$.

$= \dfrac{-21i-18i^2}{-9i^2}$ Multiply numerators, multiply denominators.

$= \dfrac{-21i-18(-1)}{-9(-1)}$ Replace i^2 with -1.

$= \dfrac{18-21i}{9}$ Simplify.

$= 2 - \dfrac{7}{3}i$ Divide each term by 9.   ❏

Try Exercise 79

Divide: $\dfrac{4+10i}{5i}$.

Note the similarity between rationalizing denominators and dividing complex numbers. In one case we remove a radical from the denominator. In the other case we remove an i (which is actually the radical $\sqrt{-1}$) from the denominator.

Powers of i

We can use the fact that $i^2 = -1$ to find other powers of i.

i

$i^2 = -1$

$i^3 = i^2 \cdot i = -1 \cdot i = -i$

$i^4 = i^2 \cdot i^2 = (-1)(-1) = 1$

$i^5 = i^4 \cdot i = 1 \cdot i = i$

$i^6 = i^4 \cdot i^2 = 1 \cdot (-1) = -1$

$i^7 = i^4 \cdot i^3 = 1 \cdot (-i) = -i$

$i^8 = i^4 \cdot i^4 = 1 \cdot 1 = 1$

Note that the powers of i rotate through the four complex numbers i, -1, $-i$, and 1.

Example 9   Simplify: (a) i^{28}, (b) i^{14}.

Solution: (a) i^{28}

Since $i^4 = 1$, write i^{28} as a power of i^4.

$$i^{28} = (i^4)^7 = (1)^7 = 1$$

(b) i^{14}

Divide 4 into the exponent.

$$\begin{array}{r} 3 \\ 4\overline{)14} \\ \underline{12} \\ 2 \end{array}$$

Therefore, $14 = 4 \cdot 3 + 2$ and

$$i^{14} = i^{4 \cdot 3} \cdot i^2 = (i^4)^3 \cdot i^2 = (1)^3 \cdot (-1) = -1$$

Or simply use the remainder 2 and write

$$i^{14} = i^2 = -1$$

Try Exercise 91

Simplify: i^{30}.

Exercises 5.9

Completion Exercises

1. A number of the form $a + bi$, where a and b are real numbers, is called a(n) _____.

2. The number i is called the _____.

3. The conjugate of the complex number $a - bi$ is _____.

4. The number 8 is called the _____ of the complex number $8 + 12i$, and the number 12 is called the _____.

5. The real part of the complex number 15 is _____, and the imaginary part is _____.

6. The real part of the complex number $\dfrac{i}{7}$ is _____, and the imaginary part is _____.

Matching Exercises

Match each number in Exercises 7 through 10 with its description in letters A through D. There may be more than one answer for each exercise.

7. $3 + 4i$ A. Real number

8. -10 B. Complex number

9. $5i$ C. Imaginary number

10. $1 - \sqrt{6}$ D. Pure imaginary number

True-False Exercises

11. The number i is equal to -1.

12. The product of two complex conjugates is always a real number.

Write each complex number in standard form using i.

13. $\sqrt{-9}$ 14. $\sqrt{-36}$

15. $\sqrt{-2}$ 16. $\sqrt{-5}$

17. $\sqrt{-12}$ 18. $\sqrt{-50}$

19. $\sqrt{-\frac{7}{25}}$ 20. $\sqrt{-\frac{6}{49}}$

21. $5 + 3\sqrt{-64}$ 22. $7 - 5\sqrt{-4}$

23. $14 - \sqrt{-18}$ 24. $10 - \sqrt{-20}$

True or false.

25. Every real number is also a complex number.

26. Every imaginary number is also a complex number.

27. No real number is also an imaginary number.

28. No imaginary number is also a real number.

29. Every pure imaginary number is also an imaginary number.

30. Some imaginary numbers are not pure imaginary numbers.

Add or subtract as indicated. Write your answer in standard form.

31. $(6 + 2i) + (4 + 7i)$ 32. $(8 + 3i) + (5 + 9i)$

33. $(5 + 4i) + (-3 - 4i)$ 34. $(-8 - 14i) + (8 - 7i)$

35. $(-8 - \sqrt{-1}) + 2\sqrt{-4}$ 36. $2\sqrt{-9} + (-5 - \sqrt{-1})$

37. $9 + (16 + i)$ 38. $(17 + i) + 11$

39. $\sqrt{-36} + 7\sqrt{-4}$ 40. $\sqrt{-64} + 5\sqrt{-4}$

41. $(10 + 6i) - (3 + 5i)$ 42. $(7 + 5i) - (2 + 3i)$

43. $(-3 - 7i) - (-3 - 7i)$ 44. $(-8 - 4i) - (-8 - 4i)$

45. $11 - (4 + 5i)$ 46. $13 - (2 + 7i)$

47. $(-5 - i) - [(7 + 6i) - (8 - 2i)]$

48. $(-4 - 3i) - [(6 + 9i) - (7 - i)]$

Discussion Exercises

49. Discuss the difference between $\sqrt{2}i$ and $\sqrt{2i}$.

50. Draw a diagram of the relationships between the set of complex numbers and its subsets. Give examples of each type of number in your diagram.

51. Which method below is the correct method for finding $\sqrt{-1}\sqrt{-9}$, and why?
    Method 1: $\sqrt{-1}\sqrt{-9} = \sqrt{(-1)(-9)} = \sqrt{9} = 3$
    Method 2: $\sqrt{-1}\sqrt{-9} = i \cdot 3i = 3i^2 = -3$

52. Why is dividing two complex numbers similar to rationalizing the denominator of a fraction?

Multiply. Write your answer in standard form.

53. $(5 + 4i)(2 + 3i)$

54. $(2 + 6i)(4 + 5i)$

55. $(-8 + i)(2 - 4i)$

56. $(2 - 5i)(-7 + i)$

57. $(-1 - 9i)(-6 - 3i)$

58. $(-8 - 6i)(-1 - 4i)$

59. $(4 - 5i)(4 + 5i)$

60. $(6 + 3i)(6 - 3i)$

61. $\sqrt{-4}(5 - \sqrt{-9})$

62. $\sqrt{-16}(3 - \sqrt{-4})$

63. $\sqrt{-3} \cdot \sqrt{-7}$

64. $\sqrt{-2} \cdot \sqrt{-5}$

65. $\sqrt{2} \cdot \sqrt{-8}$

66. $\sqrt{-5} \cdot \sqrt{125}$

67. $\sqrt{-27} \cdot \sqrt{-2}$

68. $\sqrt{-3} \cdot \sqrt{-28}$

69. $(3 - 2i)^2$

70. $(5 - 3i)^2$

Divide. Write your answer in standard form.

71. $\dfrac{8}{1 - i}$

72. $\dfrac{6}{1 + i}$

73. $\dfrac{2 + 11i}{2 + i}$

74. $\dfrac{10 + 10i}{3 + i}$

75. $\dfrac{6 + 4i}{3 - 2i}$

76. $\dfrac{2 + 5i}{4 - 3i}$

77. $\dfrac{2i}{-5 + 3i}$

78. $\dfrac{5i}{-3 + 4i}$

79. $\dfrac{4 + 10i}{5i}$

80. $\dfrac{1 + 4i}{2i}$

81. $\dfrac{15i}{-3i}$

82. $\dfrac{18i}{-3i}$

83. $\dfrac{4 + \sqrt{-16}}{2 + \sqrt{-4}}$

84. $\dfrac{9 - \sqrt{-81}}{3 - \sqrt{-9}}$

Getting Acquainted with Your Scientific Calculator

Some scientific calculators can perform operations on complex numbers. Check your owner's manual. If yours does, try a few of Exercises 31 through 48 and 53 through 84 on your calculator.

Matching Exercises

Match each power of i in Exercises 85 through 88 with its simplified form in letters A through D.

85. i^2 A. 1

86. i^3 B. -1

87. i^4 C. i

88. i^5 D. $-i$

Simplify each power of i.

89. i^{16}

90. i^{21}

91. i^{30}

92. i^{25}

93. i^{11}

94. i^{10}

95. i^{-1}

96. i^{-3}

97. Show that both $2i$ and $-2i$ are square roots of -4.

98. Show that the sum of any two complex conjugates is a real number.

Chapter 5 Key Terms

5.1 **Square root** If $b^2 = a$, b is a square root of a.
Cube root If $b^3 = a$, then b is a cube root of a.
nth root If $b^n = a$, then b is an nth root of a.
Square (of a number) The square of a is a^2.
Principal square root The nonnegative square root

Radical sign The symbol $\sqrt{}$
Principal nth root ($\sqrt[n]{}$) If a number has both a positive and a negative nth root, the principal nth root is the positive nth root. If a number has only a negative nth root, then that root is the principal nth root.
Index The index of $\sqrt[n]{a}$ is n.
Radicand The radicand of $\sqrt[n]{a}$ is a.

5.3 Radical expression An algebraic expression that contains a radical

5.5 Like radicals Radical expressions that have the same index and the same radicand

5.7 Rationalizing the denominator The process of rewriting a fraction so that its denominator does not contain a radical

Conjugates The expressions $a + b$ and $a - b$ are conjugates of each other.

5.8 Radical equation An equation that contains a variable in a radicand

Extraneous solution A solution obtained when solving an equation that does not check in the original equation

5.9 Imaginary number A number of the form $a + bi$ where a and b are real numbers ($b \neq 0$) and $i = \sqrt{-1}$

Imaginary unit The number $i = \sqrt{-1}$

Complex number A number of the form $a + bi$ where a and b are real numbers and $i = \sqrt{-1}$

Standard form (of a complex number) The form $a + bi$

Real part (of a complex number) The number a in $a + bi$

Imaginary part (of a complex number) The number b in $a + bi$

Pure imaginary number A number of the form bi ($b \neq 0$)

Complex conjugates The numbers $a + bi$ and $a - bi$ are complex conjugates of each other.

Chapter 5 Key Rules/Steps

5.1 Finding Roots

Suppose a is a real number and n is an integer greater than 1.

Principal nth Root of a

n *even*	n *odd*	*Examples*
$a > 0$ $\sqrt[n]{a} > 0$	$\sqrt[n]{a} > 0$	$\sqrt[4]{81} = 3, \sqrt[3]{8} = 2$
$a < 0$ $\sqrt[n]{a}$ not real	$\sqrt[n]{a} < 0$	$\sqrt{-9}$ is not real, $\sqrt[3]{-8} = -4$
$a = 0$ $\sqrt[n]{a} = 0$	$\sqrt[n]{a} = 0$	$\sqrt{0} = 0, \sqrt[5]{0} = 0$

Principal nth Root of a^n

n *even*	n *odd*	*Examples*
$\sqrt[n]{a^n} = |a|$	$\sqrt[n]{a^n} = a$	$\sqrt{(-5)^2} = |-5| = 5$ and $\sqrt[3]{(-5)^3} = -5$

5.2 Rational Exponents

Suppose m and n are positive integers ($n > 1$) with no common factor, and a is a real number such that $\sqrt[n]{a}$ and $\sqrt[n]{a^m}$ are real numbers.

Definition of $a^{1/n}$

$$a^{1/n} = \sqrt[n]{a} \qquad\qquad 27^{1/3} = \sqrt[3]{27} = 3$$

Definition of $a^{m/n}$

$$a^{m/n} = (\sqrt[n]{a})^m = \sqrt[n]{a^m} \qquad (-8)^{4/3} = (\sqrt[3]{-8})^4 = \sqrt[3]{(-8)^4}$$
$$= \sqrt[3]{4096} = 16$$

Definition of $a^{-m/n}$

$$a^{-m/n} = \frac{1}{a^{m/n}} \ (a \neq 0) \qquad 32^{-3/5} = \frac{1}{32^{3/5}} = \frac{1}{(\sqrt[5]{32})^3} = \frac{1}{8}$$

Rules for Exponents

Suppose r and s are rational numbers, and a and b are real numbers such that every expression below is a real number.

1. $a^r \cdot a^s = a^{r+s}$ $\qquad\qquad x^{1/2} \cdot x^{3/2} = x^{1/2+3/2} = x^2$

2. $\dfrac{a^r}{a^s} = a^{r-s}$ $\qquad\qquad \dfrac{m}{m^{1/3}} = m^{1-1/3} = m^{2/3}$

3. $(a^r)^s = a^{rs}$ $\qquad\qquad (y^2)^{1/2} = y^{2 \cdot 1/2} = y$

4. $(ab)^r = a^r b^r$ $\qquad\qquad (ab)^{3/5} = a^{3/5} b^{3/5}$

5. $\left(\dfrac{a}{b}\right)^r = \dfrac{a^r}{b^r}$ $\qquad \left(\dfrac{x}{y}\right)^{-1/4} = \dfrac{x^{-1/4}}{y^{-1/4}} = \dfrac{y^{1/4}}{x^{1/4}}$

5.3 Simplifying Radical Expressions

Suppose $\sqrt[n]{a}$ and $\sqrt[n]{b}$ are real numbers.

Product Rule for Radicals

$$\sqrt[n]{ab} = \sqrt[n]{a} \cdot \sqrt[n]{b} \qquad \sqrt[3]{40} = \sqrt[3]{8 \cdot 5} = \sqrt[3]{8} \cdot \sqrt[3]{5} = 2\sqrt[3]{5}$$

Quotient Rule for Radicals

$$\sqrt[n]{\frac{a}{b}} = \frac{\sqrt[n]{a}}{\sqrt[n]{b}} \ (b \neq 0) \qquad \sqrt{\frac{12}{25}} = \frac{\sqrt{12}}{\sqrt{25}} = \frac{\sqrt{4}\sqrt{3}}{5} = \frac{2\sqrt{3}}{5}$$

Simplified Form of a Radical Expression

A radical expression is in simplest form when all four of the following conditions are met.

1. The radicand contains no factor to a power greater than or equal to the index.
$\qquad \sqrt[3]{x^4} = \sqrt[3]{x^3}\sqrt[3]{x} = x\sqrt[3]{x}$

2. The radicand contains no fractions.
$\qquad \sqrt{\dfrac{3}{4}} = \dfrac{\sqrt{3}}{\sqrt{4}} = \dfrac{\sqrt{3}}{2}$

3. The exponent of the radicand and the index have no common factor.
$\qquad \sqrt[6]{x^3} = x^{3/6} = x^{1/2} = \sqrt{x}$

4. No denominator contains a radical.
$\qquad \dfrac{1}{\sqrt{3}} = \dfrac{1}{\sqrt{3}} \cdot \dfrac{\sqrt{3}}{\sqrt{3}} = \dfrac{\sqrt{3}}{3}$

Note:

$$\sqrt{\sqrt{x}} = (x^{1/2})^{1/2} = x^{1/4} = \sqrt[4]{x}$$

5.4 Multiplying and Dividing Radical Expressions

$(3\sqrt{6})(5\sqrt{12}) = (3 \cdot 5)\sqrt{6 \cdot 12} = 15\sqrt{6^2 \cdot 2}$
$$= 15 \cdot \sqrt{6^2} \cdot \sqrt{2}$$
$$= 15 \cdot 6 \cdot \sqrt{2} = 90\sqrt{2}$$

$(\sqrt{x+5})^2 = \sqrt{x+5} \cdot \sqrt{x+5} = x+5$

$\dfrac{30\sqrt[3]{8m^5}}{3\sqrt[3]{2m}} = \dfrac{30}{3}\sqrt[3]{\dfrac{8m^5}{2m}} = 10\sqrt[3]{4m^4} = 10\sqrt[3]{m^3}\sqrt[3]{4m} = 10m\sqrt[3]{4m}$

5.5 Adding and Subtracting Radical Expressions

$3\sqrt{11} + 4\sqrt{11} = 7\sqrt{11}$

$\sqrt{45} - 2\sqrt{5} = \sqrt{9}\sqrt{5} - 2\sqrt{5} = 3\sqrt{5} - 2\sqrt{5} = \sqrt{5}$

$x\sqrt[3]{8x} - 6\sqrt[3]{x^4} = x\sqrt[3]{8}\sqrt[3]{x} - 6\sqrt[3]{x^3}\sqrt[3]{x} = 2x\sqrt[3]{x} - 6x\sqrt[3]{x}$
$$= -4x\sqrt[3]{x}$$

5.6 Combinations of Operations

$(\sqrt{3}+5)(\sqrt{2}-1) = \sqrt{6} - \sqrt{3} + 5\sqrt{2} - 5$

$(\sqrt{x}-4)^2 = (\sqrt{x})^2 - 2(\sqrt{x})4 + 4^2 = x - 8\sqrt{x} + 16$

$(a+\sqrt{b})(a-\sqrt{b}) = a^2 - (\sqrt{b})^2 = a^2 - b$

$\dfrac{2+\sqrt{12}}{2} = \dfrac{2+2\sqrt{3}}{2} = \dfrac{2(1+\sqrt{3})}{2} = 1+\sqrt{3}$

5.7 Rationalizing the Denominator

$\dfrac{10}{\sqrt{5}} = \dfrac{10}{\sqrt{5}} \cdot \dfrac{\sqrt{5}}{\sqrt{5}} = \dfrac{10\sqrt{5}}{5} = 2\sqrt{5}$

$\dfrac{1}{\sqrt[3]{4p}} = \dfrac{1}{\sqrt[3]{4p}} \cdot \dfrac{\sqrt[3]{2p^2}}{\sqrt[3]{2p^2}} = \dfrac{\sqrt[3]{2p^2}}{\sqrt[3]{8p^3}} = \dfrac{\sqrt[3]{2p^2}}{2p}$

$\dfrac{6}{\sqrt{7}+2} = \dfrac{6}{\sqrt{7}+2} \cdot \dfrac{\sqrt{7}-2}{\sqrt{7}-2} = \dfrac{6(\sqrt{7}-2)}{(\sqrt{7})^2 - 2^2}$
$$= \dfrac{6(\sqrt{7}-2)}{7-4} = \dfrac{6(\sqrt{7}-2)}{3} = 2(\sqrt{7}-2)$$

5.8 Radical Equations

Power Rule for Equality

Suppose a and b are real numbers and n is a positive integer.

If $a = b$, then $a^n = b^n$.　　　If $\sqrt{x} = 3$, then $(\sqrt{x})^2 = 3^2$.

To Solve the Radical Equation $\sqrt{x+7} - 5 = x$

1. Isolate a radical on one side of the equation.

$\sqrt{x+7} = x+5$　　Add 5.

2. Raise both sides to a power equal to the index.

$(\sqrt{x+7})^2 = (x+5)^2$　　Square both sides.

$x+7 = x^2 + 10x + 25$

3. If the equation still contains a radical, repeat Steps 1 and 2.

There is no radical, so this step is done.

4. Solve the resulting equation.

$0 = x^2 + 9x + 18$　　Subtract x, subtract 7.
$0 = (x+3)(x+6)$　　Factor.
$x+3 = 0$　or　$x+6 = 0$　　Set each factor equal to 0.
$x = -3$　　　　$x = -6$　　Solve each equation.

5. Check in the original equation.

The number -3 checks, but -6 does not. Therefore, the solution set is $\{-3\}$.

5.9 Complex Numbers

Complex numbers $a + bi$

Real numbers ($b = 0$)　　　Imaginary numbers ($b \neq 0$)

$17, 4.3, 0, -1\frac{2}{3}, \sqrt{3}, \pi$　　　$6 + 7i$　$\Big|$　$\frac{2}{3} - \frac{i}{5}$

Pure imaginary numbers ($a = 0, b \neq 0$)

$-8i, \frac{i}{2}, i\sqrt{5}$

Operations on Complex Numbers

Sum: $(4 + 5i) + (8 - 3i) = 4 + 5i + 8 - 3i = 12 + 2i$

Difference: $(-1 + 9i) - (3 - 6i) = -1 + 9i - 3 + 6i$
$$= -4 + 15i$$

Product:　$(2 - 5i)(4 + 3i) = 8 + 6i - 20i - 15i^2$
$$= 8 - 14i - 15(-1) = 23 - 14i$$
$$\sqrt{-4}\sqrt{-9} = 2i \cdot 3i = 6i^2 = 6(-1) = -6$$

Quotient:　$\dfrac{5i}{3-2i} = \dfrac{5i}{3-2i} \cdot \dfrac{3+2i}{3+2i} = \dfrac{15i + 10i^2}{3^2 - (2i)^2}$
$$= \dfrac{15i + 10(-1)}{9 - 4i^2} = \dfrac{-10 + 15i}{9 - 4(-1)} = \dfrac{-10}{13} + \dfrac{15}{13}i$$

Powers of i

i
$i^2 = -1$
$i^3 = -i$
$i^4 = 1$
Note:
$i^{27} = i^{24} \cdot i^3 = (i^4)^6 \cdot i^3 = (1)^6 \cdot (-i) = -i$

Chapter 5 Review Exercises

[5.1] Find all square roots of each number.

1. 36　　　　　　　　　　2. $\frac{49}{144}$

Find each root. Assume x and y are positive real numbers and m and n are integers.

3. $\sqrt{25}$

4. $\sqrt{-9}$

5. $-\sqrt{100}$

6. $\pm\sqrt{121}$

7. $\sqrt[3]{-27}$

8. $\sqrt[6]{64}$

9. $\sqrt[4]{81y^{12}}$

10. $\sqrt[3]{-8x^6y^9}$

11. $\sqrt[8]{(x+7)^{16}}$

12. $\sqrt{25x^{10n}y^{8m}}$

[5.2] Write in radical form and simplify.

13. $\left(\frac{4}{81}\right)^{1/2}$

14. $(-243)^{3/5}$

15. $125^{-1/3}$

16. $36^{-3/2}$

Write in radical form.

17. $4x^{2/5}$

18. $(y-7)^{1/3}$

Simplify each expression. Write your answer with positive exponents. Assume all variables represent positive real numbers.

19. $x^{5/4} \cdot x^{-1/4}$

20. $(16p^{-8}q^{16})^{-3/4}$

21. $\dfrac{p^{1/2}}{p^{1/4} \cdot p^{3/2}}$

22. $\dfrac{(x^{-3/4}y^{3/2})^{-4}}{x^{3/2}y^{-4}}$

23. Multiply: $z^{8/3}(z^{1/3} - z^{-8/3})$.

[5.3] Simplify each radical expression. Assume all variables represent positive real numbers.

24. $\sqrt{20}$

25. $5\sqrt[3]{54}$

26. $\sqrt[3]{-125a^6b^4}$

27. $\sqrt{\sqrt{64x^4}}$

28. $\sqrt{\dfrac{18m^5}{121}}$

29. $\sqrt[3]{\dfrac{81y^8}{192y}}$

30. $\sqrt[4]{\sqrt[3]{x}}$

31. $\sqrt[6]{49x^2y^4}$

32. $\sqrt{\dfrac{12zk^{15}}{36z^{11}k^3}}$

33. $\sqrt{16\sqrt{p}}$

[5.4] Multiply or divide as indicated and simplify. Assume all variables represent positive real numbers.

34. $\sqrt{5} \cdot \sqrt{10}$

35. $(\sqrt{13r})^2$

36. $\sqrt{7x}\sqrt{14x}$

37. $\sqrt[3]{5h^2t^4}\sqrt[3]{25ht^2}$

38. $\sqrt{3p^2}\sqrt{5p}\sqrt{15p^4}$

39. $(-4\sqrt{3z+1})^2$

40. $a\sqrt[5]{4a^4}\sqrt[5]{8a^3}$

41. $\dfrac{\sqrt{98}}{\sqrt{2}}$

42. $\dfrac{\sqrt{24y^5}}{\sqrt{2y}}$

43. $\dfrac{60\sqrt{6k}}{24\sqrt{294k^3}}$

44. $\dfrac{\sqrt[3]{35a^8b^{13}c^4}}{\sqrt[3]{5a^2bc}}$

45. $\dfrac{\sqrt{14p^4q^4}\sqrt{3pq^2}}{\sqrt{378p^2q^3}}$

46. $\sqrt[3]{2} \cdot \sqrt[6]{2}$

47. $\dfrac{\sqrt{m}}{\sqrt[3]{m}}$

[5.5] Simplify each expression. Assume all variables represent positive real numbers.

48. $5\sqrt{3x} - 8\sqrt{3x}$

49. $\sqrt{6} + \sqrt{24}$

50. $3\sqrt{20} + 4\sqrt{45} - 5\sqrt{80}$

51. $-y\sqrt{4xy} + 11\sqrt{xy^3}$

52. $\sqrt[3]{54p^2} - \sqrt[3]{16p^2}$

53. $-2\sqrt[4]{48} - 7\sqrt[4]{243}$

[5.6] Multiply and simplify. Assume all variables represent positive real numbers.

54. $5\sqrt{3}(\sqrt{21} - \sqrt{3})$

55. $\sqrt[3]{4}(\sqrt[3]{2} - \sqrt[3]{3})$

56. $(\sqrt{x} + \sqrt{y})(\sqrt{x} - \sqrt{y})$

57. $(t + \sqrt{10})^2$

58. $(2\sqrt{15} - 5)(3\sqrt{3} + 2)$

59. $(\sqrt{6p+5} - 4)^2$

Simplify each expression.

60. $\dfrac{6 + \sqrt{28}}{2}$

61. $\dfrac{-18 + 4\sqrt{45}}{12}$

[5.7] Simplify each radical expression. Rationalize all denominators. Assume all variables represent positive real numbers.

62. $\dfrac{-2}{\sqrt{3}}$

63. $\sqrt{\dfrac{48}{5}}$

64. $\dfrac{\sqrt{242x^9}}{\sqrt{12x^3y}}$

65. $\dfrac{14}{\sqrt[3]{2}}$

66. $\dfrac{\sqrt[3]{10}}{\sqrt[3]{9k}}$

67. $\dfrac{1}{3 - \sqrt{5}}$

68. $\dfrac{\sqrt{7} - \sqrt{5}}{\sqrt{7} + \sqrt{5}}$

69. $\dfrac{9\sqrt{2pm}}{\sqrt{2p} - \sqrt{4m}}$

70. $\dfrac{1}{\sqrt{2}} - \dfrac{1}{\sqrt{3}}$

71. $\sqrt{80z} + \dfrac{\sqrt{2}}{\sqrt{10z}}$

[5.8] Solve each equation.

72. $\sqrt{3x+1} = 5$

73. $11 = 12 + \sqrt{2y-5}$

74. $\sqrt[3]{4p^2 - 36} = 4$

75. $\sqrt{6z} = 2\sqrt{z^2 - 1}$

76. $\sqrt{m+8} = \sqrt{m} + 2$

77. $t - \sqrt{t+3} = 3$

78. $\sqrt{7r^2 - 16r + 10} = 2r - 5$

79. $\sqrt{w+2} + \sqrt{2w+5} = 1$

[5.9] Write each complex number in standard form using i. Then state the real part and the imaginary part.

80. $9 - 2\sqrt{-36}$

81. $\sqrt{-\frac{75}{16}}$

Perform the indicated operations. Write your answer in standard form.

82. $(6 + 9i) + (7 + 9i)$

83. $(8 - 4i) - (8 - 2i)$

84. $10i - (5 + 11i)$

85. $(3 + 2i)(4 - i)$

86. $3i(4 + i)$

87. $\sqrt{-4} \cdot \sqrt{-81}$

88. $\dfrac{3}{\sqrt{-64}}$

89. $\dfrac{5}{2 + 3i}$

90. $\dfrac{3 - 2i}{5 + i}$

Simplify each power of i.

91. i^7

92. i^{52}

93. i^{-2}

Chapter 5 Test

Find each root. Assume all variables represent positive real numbers.

1. $\sqrt[3]{-64}$

2. $-\sqrt[4]{81}$

3. $\sqrt{144x^{10}y^{14}}$

4. $\sqrt[5]{32p^{15}}$

Write in radical form and simplify.

5. $(-1000)^{2/3}$

6. $625^{-1/4}$

Simplify each expression. Write your answer with positive exponents. Assume m and p represent positive real numbers.

7. $(49m^4p^{-12})^{-1/2}$

8. $\left(\dfrac{3^{-1/2} \cdot 3^{3/4}}{3^{-2}}\right)^{-1}$

Simplify each radical expression. Assume all variables represent positive real numbers.

9. $6\sqrt{12x^2}$

10. $\sqrt[3]{27a^{13}b^8}$

11. $\sqrt{\dfrac{35z^8}{320z}}$

12. $\sqrt[4]{36t^6}$

Perform the indicated operations and simplify. Assume all variables represent positive real numbers.

13. $\sqrt{6} \cdot \sqrt{15} \cdot \sqrt{10}$

14. $\sqrt[4]{45r^2}\,\sqrt[4]{9r^{10}}$

15. $\dfrac{60\sqrt{3zh^{11}}}{12\sqrt{75z^3h^2}}$

16. $\dfrac{\sqrt[3]{x^2}}{\sqrt[6]{x}}$

17. $7\sqrt{3} - 2\sqrt{27}$

18. $\sqrt[3]{y^4} + y\sqrt[3]{8y} - \sqrt[3]{125y^4}$

19. $\sqrt{2}(\sqrt{18} + \sqrt{3})$

20. $(10\sqrt{p} - 3)(10\sqrt{p} + 3)$

21. $(\sqrt{6} - 5\sqrt{3})(\sqrt{2} + 7\sqrt{7})$

Simplify each expression. Rationalize all denominators.

22. $\dfrac{-14}{\sqrt{8}}$

23. $\dfrac{1}{\sqrt[3]{5}}$

24. $\dfrac{\sqrt{3} + 1}{2 - \sqrt{3}}$

Solve each equation.

25. $\sqrt{x^2 - 6x} - 4 = 0$

26. $5\sqrt{y} = \sqrt{13y + 3}$

27. $\sqrt{2m - 1} - \sqrt{m + 3} = 1$

Perform the indicated operations. Write your answer in standard form.

28. $6 + (9 - 8i) - (7 - 3i)$

29. $(5 - 2i)(5 - 6i)$

30. $\dfrac{4 + i}{3 + 2i}$

Quadratic Equations and Inequalities

Sections of railroad track are laid with a small space between them to allow for expansion when the metal becomes hot. In Sec. 6.5 we use the Pythagorean theorem to show that even a small amount of overexpansion can cause serious buckling problems.

TAPE AU15

6.1 Solving Quadratic Equations by Completing the Square

Recall that the standard form of a quadratic equation in the variable x is

$$ax^2 + bx + c = 0 \qquad (a \neq 0)$$

In Sec. 3.10 we learned to solve a quadratic equation by factoring. For example, to solve $2x^2 + 11x = 6$ by factoring we proceed as follows:

$2x^2 + 11x - 6 = 0$		Write in standard form.
$(2x - 1)(x + 6) = 0$		Factor the left side.
$2x - 1 = 0$ or $x + 6 = 0$		Set each factor equal to 0.
$x = \frac{1}{2}$ $x = -6$		Solve each linear equation.

Although the factoring method is the easiest way to solve a quadratic equation, not all quadratic equations can be solved by factoring. For example, we cannot solve the equation $x^2 - 6x + 2 = 0$ by factoring because we cannot factor the left side.

In this section we learn a method that can be used to solve any quadratic equation. This method is based on the **square root property.**

The Square Root Property

Suppose k is a real number.

$$\text{If } x^2 = k, \text{ then } x = \sqrt{k} \text{ or } x = -\sqrt{k}$$

We usually write the two equations $x = \sqrt{k}$ or $x = -\sqrt{k}$ more simply as $x = \pm\sqrt{k}$.

Example 1   Solve: $x^2 - 11 = 0$.

Solution:
$x^2 - 11 = 0$	Original equation
$x^2 = 11$	Solve for x^2.
$x = \pm\sqrt{11}$	Square root property

We could approximate these solutions as decimals, but normally we leave them in radical form. Therefore the solution set is $\{\sqrt{11}, -\sqrt{11}\}$. Check each solution in the original equation.   ❐

Try Exercise 7

Solve: $x^2 - 13 = 0$.

CAUTION

Make sure you take *both* the positive and the negative square roots when you use the square root property.

Correct

$x^2 = 9$

$x = \pm 3$

Wrong

~~$x^2 = 9$~~

~~$x = 3$~~   ■

The solutions to a quadratic equation may be imaginary numbers.

Example 2   Solve: $y^2 + 25 = 0$.

Solution:

$$y^2 + 25 = 0 \qquad \text{Original equation}$$
$$y^2 = -25 \qquad \text{Solve for } y^2.$$
$$y = \pm\sqrt{-25} \qquad \text{Square root property}$$
$$y = \pm 5i \qquad \text{Simplify the radical.}$$

Try Exercise 13

Solve: $y^2 + 100 = 0$.

The solution set is $\{5i, -5i\}$. Check in the original equation.   ◻

We can also use the square root property to solve equations like

$$(x - 1)^2 = 12$$

Example 3 Solve: $(x - 1)^2 = 12$.

Solution:

$$(x - 1)^2 = 12 \qquad \text{Original equation}$$
$$x - 1 = \pm\sqrt{12} \qquad \text{Square root property}$$
$$x - 1 = \pm 2\sqrt{3} \qquad \text{Simplify the radical.}$$
$$x = 1 \pm 2\sqrt{3} \qquad \text{Add 1 to both sides.}$$

Try Exercise 19

Solve: $(x - 9)^2 = 28$.

The solution set is $\{1 + 2\sqrt{3},\, 1 - 2\sqrt{3}\}$.   ◻

Example 4 Solve: $(2p + 7)^2 = -9$.

Solution:

$$(2p + 7)^2 = -9 \qquad \text{Original equation}$$
$$2p + 7 = \pm\sqrt{-9} \qquad \text{Square root property}$$
$$2p + 7 = \pm 3i \qquad \text{Simplify the radical.}$$
$$2p = -7 \pm 3i \qquad \text{Subtract 7 from both sides.}$$
$$p = \frac{-7 \pm 3i}{2} \qquad \text{Divide both sides by 2.}$$

Try Exercise 27

Solve: $(3p + 7)^2 = -4$.

The solution set is $\{-\frac{7}{2} + \frac{3}{2}i,\, -\frac{7}{2} - \frac{3}{2}i\}$.   ◻

We can use the square root property to solve *any* quadratic equation, because we can write any quadratic equation in the form

$$(x + d)^2 = k$$

To write the left side of the equation in the form $(x + d)^2$, first write the left side as the perfect-square trinomial

$$x^2 + 2dx + d^2$$

Note that the last term d^2 is the square of half the coefficient of x. That is,

$$\left(\frac{\text{coefficient of } x}{2}\right)^2 = \left(\frac{2d}{2}\right)^2 = d^2$$

For example, to write $x^2 + 6x$ as a perfect-square trinomial, add the square of half the coefficient of x. Note that the coefficient of x is 6, half of 6 is 3, and the square of 3 is 9. Therefore add 9 to $x^2 + 6x$. This gives the perfect-square trinomial

$$x^2 + 6x + 9$$

Since adding 9 to $x^2 + 6x$ produces a perfect-square trinomial, adding 9 is called **completing the square** on $x^2 + 6x$. Note that

$$x^2 + 6x + 9 = (x + 3)^2$$

Example 5   Solve $x^2 - 6x + 2 = 0$ by completing the square.

Solution:
$$x^2 - 6x + 2 = 0 \qquad \text{Original equation}$$
$$x^2 - 6x = -2 \qquad \text{Subtract 2.}$$

To make the left side a perfect-square trinomial, add the square of half the coefficient of x to both sides. The coefficient of x is -6, half of -6 is -3, and $(-3)^2 = 9$.

$$x^2 - 6x + 9 = -2 + 9 \qquad \text{Add 9 to both sides.}$$

Factor the left side and simplify the right side.

$$(x - 3)^2 = 7$$

The equation is now in the form $(x + d)^2 = k$. Solve using the square root property.

$$x - 3 = \pm\sqrt{7} \qquad \text{Square root property}$$
$$x = 3 \pm \sqrt{7} \qquad \text{Add 3 to both sides.}$$

The solution set is $\{3 + \sqrt{7}, 3 - \sqrt{7}\}$. ❐

Try Exercise 39

Solve: $x^2 - 6x + 3 = 0$.

If $a \neq 1$ in the equation $ax^2 + bx + c = 0$, divide both sides by a before completing the square.

Example 6   Solve $2m^2 - 2m = 1$ by completing the square.

Solution: To make the coefficient of m^2 equal 1, divide both sides by 2.

$$\frac{2m^2}{2} - \frac{2m}{2} = \frac{1}{2} \qquad \text{Divide each term by 2.}$$

$$m^2 - m = \frac{1}{2} \qquad \text{Simplify.}$$

The coefficient of m is -1, half of -1 is $-\frac{1}{2}$, and $(-\frac{1}{2})^2 = \frac{1}{4}$. Therefore complete the square by adding $\frac{1}{4}$ to both sides.

$$m^2 - m + \frac{1}{4} = \frac{1}{2} + \frac{1}{4} \qquad \text{Add } \tfrac{1}{4} \text{ to both sides.}$$

$$\left(m - \frac{1}{2}\right)^2 = \frac{3}{4} \qquad \begin{array}{l}\text{Factor the left side, simplify} \\ \text{the right side.}\end{array}$$

$$m - \frac{1}{2} = \pm\sqrt{\frac{3}{4}} \qquad \text{Square root property}$$

$$m - \frac{1}{2} = \pm\frac{\sqrt{3}}{2} \qquad \text{Since } \sqrt{\frac{3}{4}} = \frac{\sqrt{3}}{\sqrt{4}} = \frac{\sqrt{3}}{2}$$

$$m = \frac{1}{2} \pm \frac{\sqrt{3}}{2} \qquad \text{Add } \tfrac{1}{2} \text{ to both sides.}$$

The solution set is $\left\{\dfrac{1}{2} + \dfrac{\sqrt{3}}{2}, \dfrac{1}{2} - \dfrac{\sqrt{3}}{2}\right\}$. ❐

Try Exercise 41

Solve: $2m^2 - 2m = 3$.

We can summarize the steps in solving a quadratic equation by completing the square as follows.

To Solve a Quadratic Equation by Completing the Square

1. Write the equation in the form $ax^2 + bx = -c$.
2. If $a \neq 1$, divide both sides by a.
3. Add the square of half the coefficient of x to both sides.
4. Factor the left side and simplify the right side.
5. Use the square root property to solve the equation resulting from Step 4.

Note:

You can do Steps 1 and 2 in either order.

We illustrate these steps in Example 7.

Example 7   Solve $3x^2 - 4x + 18 = 0$ by completing the square.

Solution: Step 1:   Write the equation in the form $ax^2 + bx = -c$.

$$3x^2 - 4x + 18 = 0$$
$$3x^2 - 4x = -18$$

Step 2:   Since $a \neq 1$, divide both sides by a.

$$\frac{3x^2}{3} - \frac{4x}{3} = \frac{-18}{3}$$

$$x^2 - \frac{4}{3}x = -6$$

Step 3:   Add the square of half the coefficient of x to both sides.

$$x^2 - \frac{4}{3}x + \frac{4}{9} = -6 + \frac{4}{9}$$

Step 4:   Factor the left side and simplify the right side.

$$\left(x - \frac{2}{3}\right)^2 = \frac{-50}{9}$$

Step 5:   Use the square root property to solve.

$$x - \frac{2}{3} = \pm\sqrt{\frac{-50}{9}}$$

$$x - \frac{2}{3} = \pm\frac{5i\sqrt{2}}{3}$$

$$x = \frac{2}{3} \pm \frac{5i\sqrt{2}}{3}$$

The solution set is $\left\{\dfrac{2}{3} + \dfrac{5i\sqrt{2}}{3}, \dfrac{2}{3} - \dfrac{5i\sqrt{2}}{3}\right\}$.   ❐

Try Exercise 63

Solve: $3x^2 - 8x + 12 = 0$.

Problem Solving

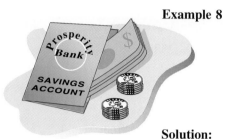

Example 8   The value V of a savings account that compounds interest annually is given by the formula

$$V = P(1 + r)^t$$

where P is the principal, r is the interest rate in decimal form, and t is the time in years. What interest rate, compounded annually, will make $100 grow to $121 in 2 yr?

Solution: Substitute $V = 121$, $P = 100$, and $t = 2$ into the formula above. Then solve for r.

$$121 = 100(1 + r)^2$$
$$1.21 = (1 + r)^2 \qquad \text{Divide both sides by 100.}$$
$$\pm 1.1 = 1 + r \qquad \text{Square root property}$$
$$\pm 1.1 - 1 = r \qquad \text{Subtract 1 from both sides.}$$
$$r = 1.1 - 1 \quad \text{or} \quad r = -1.1 - 1 \qquad \text{Split into two equations.}$$
$$r = 0.1 \qquad\qquad\quad r = -2.1$$

The interest rate could not be -2.1. Therefore, the interest rate is 0.1, or 10 percent.   ∏

Try Exercise 73

What interest rate compounded annually will make $100 grow to $144 in 2 yr?

Exercises 6.1

Completion Exercises

1. The square root property states that if $x^2 = k$, then _____ or _____.

2. To complete the square on $y^2 + 6y$, add _____.

True-False Exercises

3. If $m^2 = 25$, then $m = 5$.

4. The solutions of $p^2 = -9$ are $p = \pm 3$.

Use the square root property to solve each equation. All of the solutions are real numbers.

5. $y^2 = 64$
6. $y^2 = 81$
7. $x^2 - 13 = 0$
8. $x^2 - 17 = 0$
9. $4z^2 - 25 = 0$
10. $9z^2 - 16 = 0$

Use the square root property to solve each equation. All of the solutions are imaginary numbers.

11. $x^2 = -36$
12. $x^2 = -49$
13. $y^2 + 100 = 0$
14. $y^2 + 121 = 0$
15. $3t^2 + 24 = 0$
16. $2t^2 + 54 = 0$

Use the square root property to solve each equation. All of the solutions are real numbers.

17. $(y - 2)^2 = 81$
18. $(y - 1)^2 = 64$
19. $(x - 9)^2 = 28$
20. $(x - 3)^2 = 45$
21. $(2m + 4)^2 = 80$
22. $(3m + 6)^2 = 72$

Use the square root property to solve each equation. All of the solutions are imaginary numbers.

23. $(z - 1)^2 = -16$
24. $(z - 4)^2 = -1$
25. $(x + 11)^2 = -3$
26. $(x + 13)^2 = -2$
27. $(3p + 7)^2 = -4$
28. $(5p + 9)^2 = -36$

Complete the square on each expression. Then factor the resulting perfect-square trinomial.

29. $x^2 + 8x$

30. $x^2 + 10x$

31. $y^2 - 4y$

32. $y^2 - 6y$

33. $p^2 + 3p$

34. $p^2 - p$

35. $r^2 - \frac{4}{3}r$

36. $r^2 + \frac{6}{5}r$

Solve each equation by completing the square. All of the solutions are real numbers.

37. $p^2 - 4p - 5 = 0$

38. $p^2 - 4p - 12 = 0$

39. $x^2 - 6x + 3 = 0$

40. $x^2 - 8x + 9 = 0$

41. $2m^2 - 2m = 3$

42. $2m^2 - 2m = 5$

43. $5k^2 = 4k + 1$

44. $3k^2 = 4k + 4$

45. $3z^2 + 6z - 2 = 0$

46. $5z^2 + 10z - 1 = 0$

47. $6y = 3y^2 - 5$

48. $12y = 3y^2 - 8$

Discussion Exercises

49. Compare the advantages and disadvantages of the following two methods for solving a quadratic equation: (a) factoring, (b) completing the square.

50. Describe, in words, how to complete the square on the expression $x^2 + bx$.

51. Solve $2x^2 + x - 2 = 0$ by first (a) adding 2 to both sides, (b) dividing both sides by 2. Is there a significant difference in the two methods?

52. Outline the steps in solving a quadratic equation by completing the square.

Solve each equation by completing the square. Some of the solutions are imaginary numbers.

53. $y^2 - 6y + 34 = 0$

54. $y^2 - 4y + 53 = 0$

55. $z^2 + 4z - 7 = 0$

56. $z^2 + 2z - 9 = 0$

57. $r^2 - 10r + 13 = 0$

58. $r^2 - 12r + 8 = 0$

59. $t^2 + 14t + 59 = 0$

60. $t^2 + 18t + 87 = 0$

61. $4m^2 - 4m + 65 = 0$

62. $9m^2 - 6m + 37 = 0$

63. $3x^2 - 8x + 12 = 0$

64. $3x^2 - 2x + 17 = 0$

Solve each equation for x.

65. $x^2 - 4a^2 = 0$

66. $2x^2 - 72b^2 = 0$

67. $4x^2 = a^2 + 81$

68. $(5x - 3a)^2 = 64b^2$

69. $x^2 - 2ax + a^2 = 1$

70. $x^2 - 4ax + 4a^2 = 9$

Problem Solving

71. Twenty-one less than 5 times the square of a number is 699. Find the number.

72. If twice a positive number is added to its square, the result is 2. Find the number.

73. What interest rate compounded annually will make $100 grow to $144 in 2 yr?

74. The speed of the current in a stream can be determined using an open-ended, L-shaped tube like the one shown in Fig. 6.1. The current speed s in feet per second is given by *Torricelli's law** as

$$s^2 = 64h$$

where h is the height, in feet, of the water in the tube. Determine the current speed if $h = 9$ in.

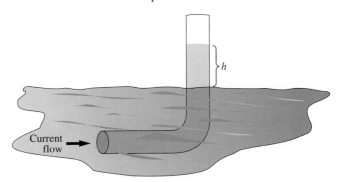

Figure 6.1

Getting Acquainted with Your Graphing Calculator

75. Check that $3 + \sqrt{6}$ is a solution of $x^2 - 6x + 3 = 0$ by storing $3 + \sqrt{6}$ in x and then calculating the value of $x^2 - 6x + 3$ (it should be 0). Then check $3 - \sqrt{6}$ in the same way. *Note:* Your calculator will store a decimal approximation of $3 + \sqrt{6}$ in x. Therefore, it may calculate $x^2 - 6x + 3$ to be a value of approximately 0, such as 1×10^{-12} (0.000000000001).

---

*HISTORICAL NOTE: Torricelli's law is named after the Italian mathematician and scientist Evangelista Torricelli (1608–1647). Torricelli's invention of the mercury barometer was, after Galileo's planetary observations, considered to be the second greatest feat of the scientific revolution.

6.2 The Quadratic Formula

Although we can solve any quadratic equation by completing the square, this method is sometimes tedious. In this section we will use the method of completing the square to develop a formula that will enable us to solve any quadratic equation.

Consider the standard form of a quadratic equation.

$$ax^2 + bx + c = 0 \qquad (a \neq 0)$$

We shall solve this equation for x by completing the square.

$$ax^2 + bx = -c \qquad \text{Subtract } c \text{ from both sides.}$$

$$x^2 + \frac{b}{a}x = -\frac{c}{a} \qquad \text{Divide both sides by } a.$$

The coefficient of x is $\frac{b}{a}$. Half of $\frac{b}{a}$ is $\frac{1}{2} \cdot \frac{b}{a} = \frac{b}{2a}$. Finally, $\left(\frac{b}{2a}\right)^2 = \frac{b^2}{4a^2}$. Therefore, complete the square by adding $\frac{b^2}{4a^2}$ to both sides.

$$x^2 + \frac{b}{a}x + \frac{b^2}{4a^2} = -\frac{c}{a} + \frac{b^2}{4a^2}$$

Factor the left side and simplify the right side.

$$\left(x + \frac{b}{2a}\right)^2 = \frac{-c}{a} \cdot \frac{4a}{4a} + \frac{b^2}{4a^2}$$

$$\left(x + \frac{b}{2a}\right)^2 = \frac{-4ac}{4a^2} + \frac{b^2}{4a^2}$$

$$\left(x + \frac{b}{2a}\right)^2 = \frac{b^2 - 4ac}{4a^2}$$

Apply the square root property.

$$x + \frac{b}{2a} = \pm\sqrt{\frac{b^2 - 4ac}{4a^2}}$$

Simplify the radical

$$x + \frac{b}{2a} = \pm\frac{\sqrt{b^2 - 4ac}}{\sqrt{4a^2}} \qquad \text{Quotient rule}$$

$$x + \frac{b}{2a} = \pm\frac{\sqrt{b^2 - 4ac}}{2a}$$

Subtract $\frac{b}{2a}$ from both sides.

$$x = -\frac{b}{2a} \pm \frac{\sqrt{b^2 - 4ac}}{2a}$$

Combine the fractions.

$$x = \frac{-b \pm \sqrt{b^2 - 4ac}}{2a}$$

This last equation is called the **quadratic formula** and you should memorize it.*

The Quadratic Formula

The solutions to the quadratic equation $ax^2 + bx + c = 0$ are

$$x = \frac{-b \pm \sqrt{b^2 - 4ac}}{2a}$$

The following examples show you how to use the quadratic formula. Note that you must write the equation in the standard form $ax^2 + bx + c = 0$ before you identify a, b, and c.

Example 1 Solve $5x^2 + 3x = 2$ using the quadratic formula.

Solution: Write the equation in standard form.

$$\begin{array}{ll} 5x^2 + 3x = 2 & \text{Original equation} \\ 5x^2 + 3x - 2 = 0 & \text{Subtract 2.} \end{array}$$

Identify a, b, and c.

$$a = 5 \qquad b = 3 \qquad c = -2$$

Substitute the values for a, b, and c into the quadratic formula.

$$x = \frac{-b \pm \sqrt{b^2 - 4ac}}{2a} \qquad \text{Quadratic formula}$$

$$x = \frac{-3 \pm \sqrt{3^2 - 4(5)(-2)}}{2(5)} \qquad \text{Let } a = 5, b = 3, \text{ and } c = -2.$$

Simplify the right side.

$$x = \frac{-3 \pm \sqrt{9 + 40}}{10} = \frac{-3 \pm \sqrt{49}}{10} = \frac{-3 \pm 7}{10}$$

Separate into two equations and simplify the right side of each equation.

$$x = \frac{-3 + 7}{10} = \frac{4}{10} = \frac{2}{5} \qquad \text{or} \qquad x = \frac{-3 - 7}{10} = \frac{-10}{10} = -1$$

Try Exercise 9

Solve: $3x^2 + 2x = 1$.

Both numbers check, so the solution set is $\{\frac{2}{5}, -1\}$. ❒

The fact that the solutions to Example 1 were *rational* numbers means that we could have solved the equation by factoring. This is not the case in Example 2.

---

*HISTORICAL NOTE: Both the quadratic formula and the method of completing the square were known by the Babylonians, perhaps as early as 2000 B.C. The Babylonians, however, did not accept quadratic equations whose solutions were negative numbers.

Example 2   Solve: $4y^2 = -2 - 7y$.

Solution: Write the equation in standard form.

$$4y^2 = -2 - 7y \qquad \text{Original equation}$$
$$4y^2 + 7y + 2 = 0 \qquad \text{Add } 7y, \text{ add } 2.$$

Substitute $a = 4$, $b = 7$, and $c = 2$ into the quadratic formula.

$$y = \frac{-b \pm \sqrt{b^2 - 4ac}}{2a}$$

$$y = \frac{-7 \pm \sqrt{7^2 - 4(4)(2)}}{2(4)}$$

Simplify the right side.

$$y = \frac{-7 \pm \sqrt{49 - 32}}{8} = \frac{-7 \pm \sqrt{17}}{8}$$

The solution set is $\left\{ \dfrac{-7 + \sqrt{17}}{8}, \dfrac{-7 - \sqrt{17}}{8} \right\}$.   ❑

Try Exercise 17

Solve: $4y^2 = -3 - 9y$.

We can summarize the steps in solving a quadratic equation using the quadratic formula as follows.

To Solve a Quadratic Equation Using the Quadratic Formula

1. Write the equation in the standard form $ax^2 + bx + c = 0$.
2. Identify a, b, and c.
3. Substitute the values for a, b, and c into the quadratic formula

$$x = \frac{-b \pm \sqrt{b^2 - 4ac}}{2a}$$

4. Simplify the right side.

We illustrate these steps in Example 3.

Example 3   Solve: $\dfrac{x^2}{4} - \dfrac{x}{3} - \dfrac{5}{12} = 0$.

Solution: Step 1:   Write the equation in standard form. The equation is already in standard form. We could use $a = \frac{1}{4}$, $b = -\frac{1}{3}$, and $c = -\frac{5}{12}$. However, our calculations will be simpler if we clear fractions by multiplying both sides by the LCD, 12.

$$12 \cdot \frac{x^2}{4} - 12 \cdot \frac{x}{3} - 12 \cdot \frac{5}{12} = 12 \cdot 0 \qquad \text{Multiply every term by 12.}$$
$$3x^2 - 4x - 5 = 0 \qquad \text{Simplify.}$$

Step 2:   Identify a, b, and c.

$$a = 3 \qquad b = -4 \qquad c = -5$$

Step 3: Substitute the values for a, b, and c into the quadratic formula.

$$x = \frac{-b \pm \sqrt{b^2 - 4ac}}{2a} \qquad \text{Quadratic formula}$$

$$x = \frac{-(-4) \pm \sqrt{(-4)^2 - 4(3)(-5)}}{2(3)} \qquad \text{Substitute } a = 3, b = -4, \text{ and } c = -5.$$

Step 4: Simplify the right side.

$$x = \frac{4 \pm \sqrt{16 + 60}}{6} = \frac{4 \pm \sqrt{76}}{6}$$

But $\sqrt{76} = \sqrt{4} \cdot \sqrt{19} = 2\sqrt{19}$. Therefore,

$$x = \frac{4 \pm 2\sqrt{19}}{6} = \frac{2(2 \pm \sqrt{19})}{6} = \frac{2 \pm \sqrt{19}}{3}$$

The solution set is $\left\{ \dfrac{2 + \sqrt{19}}{3}, \dfrac{2 - \sqrt{19}}{3} \right\}$.   ❑

Try Exercise 27

Solve: $\dfrac{x^2}{4} - \dfrac{x}{3} - \dfrac{1}{12} = 0.$

CAUTION

Be careful when you are simplifying your answer after using the quadratic formula.

Correct   *Wrong*

$$\frac{6 + 3\sqrt{5}}{6} = \frac{3(2 + \sqrt{5})}{3 \cdot 2}$$
$$= \frac{2 + \sqrt{5}}{2}$$

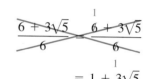

$$= 1 + 3\sqrt{5}$$

   ■

If the radicand $b^2 - 4ac$ in the quadratic formula is a *negative* number, the solutions to the quadratic equation are imaginary numbers. This is illustrated in Example 4.

Example 4 Solve: $r^2 - 6r + 25 = 0.$

Solution: Substitute $a = 1$, $b = -6$, and $c = 25$ into the quadratic formula.

$$r = \frac{-(-6) \pm \sqrt{(-6)^2 - 4(1)(25)}}{2(1)}$$

$$r = \frac{6 \pm \sqrt{36 - 100}}{2} = \frac{6 \pm \sqrt{-64}}{2}$$

But $\sqrt{-64} = \sqrt{64}\,\sqrt{-1} = 8i$. Therefore,

$$r = \frac{6 \pm 8i}{2} = \frac{6}{2} \pm \frac{8i}{2} = 3 \pm 4i$$

Try Exercise 39

Solve: $r^2 - 6r + 13 = 0.$

The solution set is $\{3 + 4i, 3 - 4i\}$.   ❑

Here is an example of an applied problem that we can solve using the quadratic formula.

Problem Solving

Example 5 A kennel owner has 40 m of fencing to enclose a rectangular exercise pen. One side of the pen will lie along the kennel and needs no additional fencing (see Fig. 6.2). What should the dimensions of the pen be if the area is to be 100 m²?

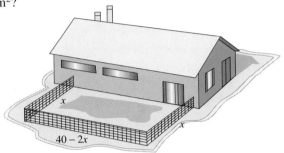

Figure 6.2

Solution: Let x = the width of the pen. Since there are 40 m of fencing, and the two widths use up $2x$ meters of the 40 m, the length of the pen is $40 - 2x$. Therefore we can write the following equation:

$$\begin{array}{ccc} \text{Area of} & & \\ \text{pen} & \text{is} & 100 \\ \downarrow & \downarrow & \downarrow \end{array}$$

$$x(40 - 2x) = 100$$

Solve for x.

$$40x - 2x^2 = 100 \qquad \text{Distribute } x.$$
$$-2x^2 + 40x - 100 = 0 \qquad \text{Subtract 100.}$$

Simplify the equation by dividing both sides by -2.

$$\frac{-2x^2}{-2} + \frac{40x}{-2} - \frac{100}{-2} = \frac{0}{-2}$$
$$x^2 - 20x + 50 = 0$$

Substitute $a = 1$, $b = -20$, and $c = 50$ into the quadratic formula and simplify.

$$x = \frac{-(-20) \pm \sqrt{(-20)^2 - 4(1)(50)}}{2(1)}$$

$$x = \frac{20 \pm \sqrt{400 - 200}}{2} = \frac{20 \pm \sqrt{200}}{2}$$

But $\sqrt{200} = \sqrt{100}\sqrt{2} = 10\sqrt{2}$. Therefore,

$$x = \frac{20 \pm 10\sqrt{2}}{2} = \frac{20}{2} \pm \frac{10\sqrt{2}}{2} = 10 \pm 5\sqrt{2}$$

Try Exercise 51

What should the dimensions of the pen in Fig. 6.3 be if the owner has 30 m of fencing and the area is to be 110 m²?

One possible width for the pen is $x = 10 + 5\sqrt{2} \approx 17.1$ m. In that case the length would be $40 - 2x \approx 5.9$ m. A second possible width for the pen is $x = 10 - 5\sqrt{2} \approx 2.9$ m. In that case the length would be $40 - 2x \approx 34.1$ m.   ☐

Exercises 6.2

Completion Exercises

1. By the quadratic formula, the solutions to the quadratic equation $ax^2 + bx + c = 0$ are $x =$ _____.

2. For the quadratic equation $x^2 - 5x + 2 = 0$, $a =$ _____, $b =$ _____, and $c =$ _____.

True-False Exercises

3. The quadratic formula can be used to solve any quadratic equation.

4. If the solutions to a quadratic equation with integer coefficients are rational numbers, the equation can be solved by factoring.

5. The expression $\dfrac{2 + \sqrt{7}}{2}$ simplifies to $1 + \sqrt{7}$.

6. The quadratic formula can be written as
$$x = -b \pm \frac{\sqrt{b^2 - 4ac}}{2a}.$$

7. The best first step in solving $\dfrac{x^2}{4} - \dfrac{x}{3} - \dfrac{1}{12} = 0$ is to multiply both sides by the LCD, 12.

8. If $b^2 - 4ac$ is a negative number, the solutions to the quadratic equation $ax^2 + bx + c = 0$ are imaginary numbers.

Solve each equation using the quadratic formula. All of the solutions are real numbers.

9. $3x^2 + 2x = 1$

10. $5x^2 + 2x = 3$

11. $x^2 + x - 3 = 0$

12. $x^2 + x - 4 = 0$

13. $p^2 + 7p + 1 = 0$

14. $p^2 + 3p - 9 = 0$

15. $r^2 = 4r - 1$

16. $r^2 = 4r + 7$

17. $4y^2 = -3 - 9y$

18. $5y^2 = -1 - 7y$

19. $k^2 - 8k + 16 = 0$

20. $k^2 + 12k + 36 = 0$

21. $4p(p - 2) = -1$

22. $(2p + 3)(2p + 1) = 2$

23. $(5r + 7)(r + 1) = 1$

24. $5r(r + 2) = -4$

25. $90y^2 = 60y + 20$

26. $140y^2 = 200y + 20$

27. $\dfrac{x^2}{4} - \dfrac{x}{3} - \dfrac{1}{12} = 0$

28. $\dfrac{x^2}{4} + \dfrac{x}{3} - \dfrac{5}{12} = 0$

29. $\frac{1}{2}k^2 - \frac{5}{4} = k$

30. $\frac{3}{2}k^2 - 2 = k$

31. $0.2z^2 + 1.2z - 0.5 = 0$

32. $0.3z^2 - 1.4z - 0.4 = 0$

Discussion Exercises

33. Describe, in words, the standard form of a quadratic equation.

34. Can you think of a situation where you would use the quadratic formula to solve a quadratic equation, even though the equation could be solved by factoring?

35. Which of the equations below could be solved using the quadratic formula? Would you use the quadratic formula to solve any of the three equations? Why or why not?

    (a) $x^2 = 0$ (b) $x^2 - 4 = 0$ (c) $x^2 - 3x = 0$

36. Outline the steps in solving a quadratic equation using the quadratic formula.

Solve each equation using the quadratic formula. All of the solutions are imaginary numbers.

37. $x^2 + 2x + 2 = 0$

38. $x^2 + 4x + 5 = 0$

39. $r^2 - 6r + 13 = 0$

40. $r^2 - 6r + 10 = 0$

41. $2y^2 + 5y + 4 = 0$

42. $3y^2 + 9y + 7 = 0$

43. $t^2 = 2t - 7$

44. $t^2 = 4t - 10$

45. $10p = 9 + 5p^2$

46. $12p = 11 + 6p^2$

47. $2ix^2 + 5x - 3i = 0$

48. $ix^2 + 7x - 6i = 0$

Problem Solving

Round approximate answers to the nearest tenth.

49. The square of a positive number is 4 more than twice the number. Find the number.

50. The square of a positive number is 2 more than 6 times the number. Find the number.

A farmer has 30 m of fencing to enclose a rectangular pen. One side of the pen will lie along the barn and needs no additional fencing (see Fig. 6.3).

Figure 6.3

51. What should the dimensions of the pen in Fig. 6.3 be if the area is to be 110 m²?

52. What should the dimensions of the pen in Fig. 6.3 be if the area is to be 90 m²?

53. A piece of sheet metal 16 in. wide is to be made into a trough with a rectangular cross section by folding up its edges (see Fig. 6.4). What should the height x of the trough in Fig. 6.4 be if the area of the cross section is to be 12 in.²

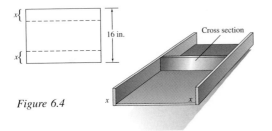

Figure 6.4

54. A slingshot propels a stone upward with an initial velocity of 80 ft/sec from a balcony that is 20 ft high (see Fig. 6.5). If air resistance is ignored, the height h, in feet, of the stone after t seconds is given by the formula

$$h = -16t^2 + 80t + 20$$

(a) When will the height h be 108 ft?

(b) When will the height h be 124 ft?

(c) When will the stone hit the ground?

Figure 6.5

 Getting Acquainted with Your Graphing Calculator

PROGRAMMING YOUR CALCULATOR   You can program the quadratic formula into your calculator. Check your owner's manual to see how to do this. Then use this program to solve each of the following quadratic equations.

55. $2x^2 + x - 15 = 0$

56. $3x^2 - 4x - 5 = 0$

57. $x^2 + x + 1 = 0$

58. $16x^2 - 72x + 81 = 0$

6.3 More about Quadratic Equations

TAPE AU15

We have discussed four methods for solving a quadratic equation—factoring, the square root property, completing the square, and the quadratic formula. Here is a summary of the advantages and disadvantages of each method.

Choosing a Method to Solve a Quadratic Equation

Method	Advantages	Disadvantages
1. Factoring	Easy and fast	Not all equations factorable; some equations difficult to factor because their coefficients are large
2. Square root property	Best method for solving equations of the form $x^2 = k$ and $(ax + b)^2 = k$	Many equations not of this form
3. Completing the square	Will solve any quadratic equation; useful in other areas of mathematics	Sometimes tedious to use
4. Quadratic formula	Will solve any quadratic equation; useful in proving properties involving the solutions of a quadratic equation	Not as easy to use as factoring or the square root property

There are certain special types of equations that are easiest to solve by completing the square. For example, completing the square is probably the easiest method to use in solving the equation $x^2 - 12x - 121 = 0$, where $a = 1$, b is even, and c is large. However, for most quadratic equations, first try factoring or the square root property. If you are having no luck, use the quadratic formula.

Example 1   Solve: $3x^2 + 7x = 20$.

Solution: If we write the equation in standard form, the left side is easily factored. Therefore, solve by factoring.

$$3x^2 + 7x = 20 \qquad \text{Original equation}$$
$$3x^2 + 7x - 20 = 0 \qquad \text{Subtract 20.}$$
$$(3x - 5)(x + 4) = 0 \qquad \text{Factor.}$$
$$3x - 5 = 0 \quad \text{or} \quad x + 4 = 0 \qquad \text{Set each factor equal to 0.}$$
$$x = \tfrac{5}{3} \qquad\qquad x = -4 \qquad \text{Solve each linear equation.}$$

Try Exercise 5

Solve: $5x^2 + 13x = 6$.

The solution set is $\{\tfrac{5}{3}, -4\}$.   ❑

Example 2   Solve: $3y^2 - 144 = 0$.

Solution: The equation is easily written in the form $y^2 = k$. Therefore, solve using the square root property.

$$3y^2 - 144 = 0 \qquad \text{Original equation}$$
$$3y^2 = 144 \qquad \text{Add 144.}$$
$$y^2 = 48 \qquad \text{Divide by 3.}$$
$$y = \pm\sqrt{48} \qquad \text{Square root property}$$
$$y = \pm 4\sqrt{3} \qquad \text{Simplify the radical}$$

Try Exercise 9

Solve: $2y^2 - 36 = 0$.

The solution set is $\{4\sqrt{3}, -4\sqrt{3}\}$.   ❑

Example 3   Solve: $5p^2 - 9p + 2 = 0$.

Solution: The left side cannot be factored using integers. Therefore, use the quadratic formula with $a = 5$, $b = -9$, and $c = 2$.

$$p = \frac{-b \pm \sqrt{b^2 - 4ac}}{2a} \qquad \text{Quadratic formula}$$

$$p = \frac{-(-9) \pm \sqrt{(-9)^2 - 4(5)(2)}}{2(5)} \qquad \text{Let } a = 5, b = -9, \text{ and } c = 2.$$

$$p = \frac{9 \pm \sqrt{81 - 40}}{10} = \frac{9 \pm \sqrt{41}}{10} \qquad \text{Simplify.}$$

The solution set is $\left\{ \dfrac{9 + \sqrt{41}}{10}, \dfrac{9 - \sqrt{41}}{10} \right\}$.

Try Exercise 13

Solve: $3p^2 - 11p + 7 = 0$.   ❑

The Discriminant

Recall that the solutions of the quadratic equation $ax^2 + bx + c = 0$ are given by the quadratic formula as

$$x = \frac{-b \pm \sqrt{b^2 - 4ac}}{2a}$$

By calculating the value of the radicand $b^2 - 4ac$, we can determine whether the solutions are rational numbers, irrational numbers, or imaginary numbers. We call $b^2 - 4ac$ the **discriminant** of the quadratic equation $ax^2 + bx + c = 0$.

Example 4 Use the discriminant to determine whether the solutions of each equation are rational numbers, irrational numbers, or imaginary numbers.

(a) $6x^2 - 5x - 4 = 0$ (b) $2m^2 + 5 = 3m$

Solution: (a) Substitute $a = 6$, $b = -5$, and $c = -4$ into the discriminant.

$$b^2 - 4ac = (-5)^2 - 4(6)(-4) = 25 + 96 = 121$$

Note that 121 is a perfect square ($121 = 11^2$), so $\sqrt{121}$ is an integer. Therefore the two solutions will be rational numbers.

(b) First write the equation in the standard form $2m^2 - 3m + 5 = 0$. Then $a = 2$, $b = -3$, $c = 5$, and the discriminant is

$$b^2 - 4ac = (-3)^2 - 4(2)(5) = 9 - 40 = -31$$

Since $\sqrt{-31}$ is an imaginary number, the equation will have two imaginary solutions. ❏

Try Exercise 23

Use the discriminant to determine the type of solutions to $4x^2 - 3x - 10 = 0$.

Here is a summary of what the discriminant tells us about the solutions of a quadratic equation.

The Discriminant

The discriminant of the quadratic equation $ax^2 + bx + c = 0$ is $b^2 - 4ac$. If a, b, and c are integers, then the type of solutions can be determined as follows:

Discriminant	Solutions
Positive, and a perfect square	Two rational solutions
Positive, but not a perfect square	Two irrational solutions
Zero	One rational solution
Negative	Two imaginary solutions

We can also use the discriminant to determine whether a given trinomial is factorable. **A trinomial with integer coefficients can be factored using integers if and only if its discriminant is a perfect square.**

Example 5 Use the discriminant to determine whether each trinomial can be factored using integers: (a) $54x^2 + 69x + 20$, (b) $6z^2 + 3z - 8$.

Solution: (a) Since $a = 54$, $b = 69$ and $c = 20$,

$$b^2 - 4ac = 69^2 - 4(54)(20) = 4761 - 4320 = 441$$

Since 441 is a perfect square ($441 = 21^2$), the trinomial can be factored using integers. Note that

$$54x^2 + 69x + 20 = (9x + 4)(6x + 5)$$

(b) Since $a = 6$, $b = 3$, and $c = -8$,

$$b^2 - 4ac = 3^2 - 4(6)(-8) = 9 + 192 = 201$$

Since 201 is *not* a perfect square, the trinomial cannot be factored using integers.   ❐

Try Exercise 37

Use the discriminant to determine whether $9z^2 + 8z - 3$ can be factored using integers.

Sum and Product of the Solutions of $ax^2 + bx + c = 0$

Suppose the two solutions of the quadratic equation $ax^2 + bx + c = 0$ are called x_1 and x_2. Using the quadratic formula, we can write

$$x_1 = \frac{-b + \sqrt{b^2 - 4ac}}{2a} \quad \text{and} \quad x_2 = \frac{-b - \sqrt{b^2 - 4ac}}{2a}$$

The *sum* of the two solutions is

$$x_1 + x_2 = \frac{-b + \sqrt{b^2 - 4ac}}{2a} + \frac{-b - \sqrt{b^2 - 4ac}}{2a}$$

Add the numerators and simplify.

$$x_1 + x_2 = \frac{-b + \sqrt{b^2 - 4ac} - b - \sqrt{b^2 - 4ac}}{2a} = \frac{-2b}{2a} = -\frac{b}{a}$$

The *product* of the two solutions is

$$x_1 \cdot x_2 = \frac{-b + \sqrt{b^2 - 4ac}}{2a} \cdot \frac{-b - \sqrt{b^2 - 4ac}}{2a}$$

Multiply the numerators and multiply the denominators. Note that when you multiply the numerators you are multiplying two conjugates. Therefore the product is the difference of two squares.

$$x_1 \cdot x_2 = \frac{(-b)^2 - (\sqrt{b^2 - 4ac})^2}{(2a)^2} = \frac{b^2 - (b^2 - 4ac)}{4a^2}$$

$$x_1 \cdot x_2 = \frac{b^2 - b^2 + 4ac}{4a^2} = \frac{4ac}{4a^2} = \frac{c}{a}$$

We can summarize these results as follows.

Sum and Product of the Solutions of $ax^2 + bx + c = 0$

If x_1 and x_2 are the solutions of the quadratic equation $ax^2 + bx + c = 0$, then

$$x_1 + x_2 = -\frac{b}{a} \quad \text{and} \quad x_1 \cdot x_2 = \frac{c}{a}$$

We can now find the sum and the product of the solutions of a quadratic equation without actually solving the equation.

Example 6   Find the sum and the product of the solutions of $3x^2 - 4x - 7 = 0$.

Solution: The sum of the solutions is

$$x_1 + x_2 = -\frac{b}{a} = -\frac{-4}{3} = \frac{4}{3}$$

The product of the solutions is

$$x_1 \cdot x_2 = \frac{c}{a} = \frac{-7}{3}$$

Try Exercise 43

Find the sum and product of the solutions of the equation $4x^2 + 3x - 10 = 0$.

It can also be shown that if x_1 and x_2 are numbers such that both $x_1 + x_2 = -\frac{b}{a}$ and $x_1 \cdot x_2 = \frac{c}{a}$, then x_1 and x_2 are solutions of $ax^2 + bx + c = 0$. Therefore, the sum and product formulas can be used to check the solutions of a quadratic equation. This is often easier than substituting the solutions back into the original equation.

Example 7 Use the sum and product formulas to determine whether the numbers $\frac{3 \pm \sqrt{29}}{2}$ are the solutions of the equation $x^2 - 3x - 5 = 0$.

Solution: The sum of the solutions of $x^2 - 3x - 5 = 0$ should be

$$x_1 + x_2 = -\frac{b}{a} = -\frac{-3}{1} = 3$$

and the product should be

$$x_1 \cdot x_2 = \frac{c}{a} = \frac{-5}{1} = -5$$

The sum of the given numbers is

$$\frac{3 + \sqrt{29}}{2} + \frac{3 - \sqrt{29}}{2} = \frac{(3 + \sqrt{29}) + (3 - \sqrt{29})}{2} = \frac{6}{2} = 3$$

and the product is

$$\frac{3 + \sqrt{29}}{2} \cdot \frac{3 - \sqrt{29}}{2} = \frac{3^2 - (\sqrt{29})^2}{4} = \frac{9 - 29}{4} = \frac{-20}{4} = -5$$

Since both the sum and product of the given numbers agree with the sum and product of the solutions of the equation $x^2 - 3x - 5 = 0$, the given numbers are solutions.

Try Exercise 53

Use the sum and product formulas to determine whether $\frac{5 \pm \sqrt{13}}{6}$ are the solutions of $3x^2 - 5x + 1 = 0$.

Determining a Quadratic Equation from Its Solutions

Given any two complex numbers, we can write a quadratic equation having the given numbers as its solutions. We do this by reversing the steps for solving a quadratic equation.

Example 8 Write a quadratic equation with integer coefficients having the solution set $\{\frac{3}{5}, -2\}$.

Solution: Set x equal to each solution.

$$x = \frac{3}{5} \quad \text{or} \quad x = -2$$

Isolate 0 in both equations.

$$x - \frac{3}{5} = 0 \quad \text{or} \quad x + 2 = 0$$

Since both $x - \frac{3}{5}$ and $x + 2$ equal 0, their product must be 0.

$$\left(x - \frac{3}{5}\right)(x + 2) = 0$$

$$x^2 + 2x - \frac{3}{5}x - \frac{6}{5} = 0 \qquad \text{FOIL method}$$

Multiply by the LCD 5 to clear fractions.

$$5x^2 + 10x - 3x - 6 = 0$$
$$5x^2 + 7x - 6 = 0 \qquad \text{Simplify.}$$

Check by solving the equation, or by substituting $x = \frac{3}{5}$ and $x = -2$ into the equation, or by using the sum and product formulas.   ☐

Try Exercise 59

Write a quadratic equation with integer coefficients having the solution set $\{\frac{2}{7}, -1\}$.

Exercises 6.3

Solve each quadratic equation. Some of the solutions are imaginary numbers.

1. $y^2 = 121$
2. $y^2 = 100$
3. $8z^2 = 20z$
4. $6z^2 = 21z$
5. $5x^2 + 13x = 6$
6. $4x^2 + 7x = 15$
7. $(m - 4)^2 = 25$
8. $(m + 6)^2 = 16$
9. $2y^2 - 36 = 0$
10. $3y^2 - 72 = 0$
11. $t^2 + 81 = 0$
12. $t^2 + 64 = 0$
13. $3p^2 - 11p + 7 = 0$
14. $5p^2 - 9p + 3 = 0$
15. $(5k + 7)(k - 3) = -22$
16. $(k - 1)(3k - 9) = 7$
17. $r^2 + 2r + \frac{9}{4} = 0$
18. $r^2 - 2r + \frac{11}{9} = 0$

22. What is the discriminant of the quadratic equation $ax^2 + bx + c = 0$? If a, b, and c are integers, how is the discriminant used to identify the type of solutions of $ax^2 + bx + c = 0$?

Use the discriminant to determine whether the solutions of each equation are (a) two rational numbers, (b) two irrational numbers, (c) one rational number, or (d) two imaginary numbers. Do not solve the equation.

23. $4x^2 - 3x - 10 = 0$
24. $x^2 + 5x + 4 = 0$
25. $x^2 + x - 7 = 0$
26. $x^2 + x - 9 = 0$
27. $2m^2 + 4 = 5m$
28. $2m^2 + 3 = 4m$
29. $9p^2 = 6p - 1$
30. $9p^2 = 12p - 4$

Completion Exercises

19. The expression $b^2 - 4ac$ is called the _____ of the quadratic equation $ax^2 + bx + c = 0$.

20. The sum of the solutions of $ax^2 + bx + c = 0$ is given by the formula $x_1 + x_2 = $ _____, and the product by $x_1 \cdot x_2 = $ _____.

Discussion Exercises

21. State the four methods of solving a quadratic equation and discuss the advantages and disadvantages of each method.

True-False Exercises

31. A trinomial with integer coefficients can be factored using integers if and only if its discriminant is a perfect square.

32. The solutions of a quadratic equation with integer coefficients could be one real number and one imaginary number.

33. The solutions of a quadratic equation with integer coefficients could be one rational number and one irrational number.

34. We can use the sum and product formulas to check the solutions of a quadratic equation.

Use the discriminant to determine whether each trinomial can be factored using integers. Factor those trinomials that can be factored using integers.

35. $12x^2 + 23x + 10$

36. $16x^2 + 46x + 15$

37. $9z^2 + 8z - 3$

38. $6z^2 - 7z - 4$

39. $8m^2 - 12m + 5$

40. $10m^2 - 14m + 5$

41. $4r^2 - 28r + 49$

42. $9r^2 - 30r + 25$

Find the sum and the product of the solutions of each equation. Do not solve the equation.

43. $4x^2 + 3x - 10 = 0$

44. $2x^2 - 5x - 18 = 0$

45. $y^2 - 5y + 2 = 0$

46. $y^2 + 7y + 4 = 0$

47. $6z^2 = 9z$

48. $8z^2 = -10z$

49. $16p^2 + 81 = 0$

50. $25p^2 - 49 = 0$

Use the sum and product formulas to determine whether the given numbers are the solutions of the given equation.

51. $x^2 + 3x - 4 = 0$; $1, -4$

52. $2x^2 + 7x - 15 = 0$; $\frac{3}{2}, -5$

53. $3x^2 - 5x + 1 = 0$; $\dfrac{5 \pm \sqrt{13}}{6}$

54. $r^2 - r - 1 = 0$; $\dfrac{1 \pm \sqrt{5}}{2}$

55. $4x^2 - 8x - 9 = 0$; $\dfrac{2 \pm \sqrt{11}}{2}$

56. $4x^2 - 16x + 7 = 0$; $\dfrac{4 \pm 3i}{2}$

Write a quadratic equation with integer coefficients having the given solution set. Write your final answer in standard form.

57. $\{9, 4\}$

58. $\{1, -10\}$

59. $\{\frac{2}{7}, -1\}$

60. $\{\frac{3}{4}, \frac{1}{5}\}$

61. $\{-\frac{5}{6}, 0\}$

62. $\{8\}$

63. $\{3\sqrt{5}, -3\sqrt{5}\}$

64. $\{2 - \sqrt{6}, 2 + \sqrt{6}\}$

65. $\{3 + 4i, 3 - 4i\}$

66. $\{10i, -10i\}$

Determine the value(s) of k that cause the given equation to have exactly one solution.

67. $x^2 + 6x + k = 0$

68. $x^2 + 8x + k = 0$

69. $2x^2 - kx + 8 = 0$

70. $2x^2 - kx + 2 = 0$

6.4 Equations That Lead to Quadratic Equations

TAPE AU14

There are many types of nonquadratic equations that *lead* to quadratic equations. In this section we learn to solve several of these types of equations.

Equations with Rational Expressions

Sometimes an equation with rational expressions will lead to a quadratic equation when we multiply both sides by the LCD.

Example 1   Solve: $4 + \dfrac{3}{m} - \dfrac{135}{2m^2} = 0$.

Solution: Multiply every term by the LCD, $2m^2$, to clear the fractions.

$$2m^2 \cdot 4 + 2m^2 \cdot \frac{3}{m} - 2m^2 \cdot \frac{135}{2m^2} = 2m^2 \cdot 0$$

$$8m^2 + 6m - 135 = 0$$

Instead of wasting time trying to factor the left side, use the quadratic formula with $a = 8$, $b = 6$, and $c = -135$.

$$m = \frac{-6 \pm \sqrt{6^2 - 4(8)(-135)}}{2(8)}$$

$$m = \frac{-6 \pm \sqrt{36 + 4320}}{16} = \frac{-6 \pm \sqrt{4356}}{16}$$

Using a calculator we find that $\sqrt{4356} = 66$. Therefore

$$m = \frac{-6 + 66}{16} \qquad \text{or} \qquad m = \frac{-6 - 66}{16}$$

$$m = \frac{60}{16} = \frac{15}{4} \qquad\qquad m = \frac{-72}{16} = -\frac{9}{2}$$

The original equation contains a variable in the denominator, so we must check to make sure that no solution causes a denominator to equal 0. Since neither $\frac{15}{4}$ nor $-\frac{9}{2}$ makes a denominator 0, the solution set is $\{\frac{15}{4}, -\frac{9}{2}\}$. ❏

Try Exercise 5

Solve: $3 - \dfrac{17}{4m} - \dfrac{10}{m^2} = 0$.

Radical Equations

In Sec. 5.8 we learned to solve equations with radicals. We can now solve radical equations that lead to quadratic equations that are not factorable.

Example 2   Solve: $\sqrt{2y - 1} - \sqrt{y} = 1$.

Solution: Isolate the first radical by adding $\sqrt{y}$ to both sides.

$$\sqrt{2y - 1} = 1 + \sqrt{y}$$

Square both sides.

$$(\sqrt{2y - 1})^2 = (1 + \sqrt{y})^2$$
$$2y - 1 = 1 + 2\sqrt{y} + y$$

Isolate the remaining radical by subtracting y and subtracting 1.

$$y - 2 = 2\sqrt{y}$$

Square again and simplify.

$$(y - 2)^2 = (2\sqrt{y})^2$$
$$y^2 - 4y + 4 = 4y$$
$$y^2 - 8y + 4 = 0$$

Use the quadratic formula with $a = 1$, $b = -8$, and $c = 4$.

$$y = \frac{-(-8) \pm \sqrt{(-8)^2 - 4(1)(4)}}{2(1)}$$

$$y = \frac{8 \pm \sqrt{64 - 16}}{2} = \frac{8 \pm \sqrt{48}}{2}$$

$$y = \frac{8 \pm 4\sqrt{3}}{2} = 4 \pm 2\sqrt{3}$$

Since we squared both sides of the equation, we must check our solutions in the original equation. We will do this with the aid of a calculator.

Check: $x = 4 + 2\sqrt{3} \approx 7.464$ *Check:* $x = 4 - 2\sqrt{3} \approx 0.536$

$\sqrt{2(7.464) - 1} - \sqrt{7.464} \approx 1$ $\sqrt{2(0.536) - 1} - \sqrt{0.536} \approx 1$

$3.732 - 2.732 \approx 1$ $0.268 - 0.732 \approx 1$

$1 \approx 1$ $-0.464 \approx 1$

         true   false

The number $4 - 2\sqrt{3}$ is an extraneous solution. Therefore the solution set is $\{4 + 2\sqrt{3}\}$.   ❏

Try Exercise 21

Solve: $\sqrt{2y - 3} - \sqrt{y} = 1$.

Equations of Quadratic Form

Consider the three following equations.

$$x^4 - 10x^2 + 9 = 0*$$
$$p^{2/3} - 3p^{1/3} - 4 = 0$$
$$3(3r - 4)^2 + 2(3r - 4) - 1 = 0$$

Note that in each case the variable factor of the first term is the square of the variable factor of the middle term. That is,

$$x^4 = (x^2)^2 \qquad p^{2/3} = (p^{1/3})^2 \qquad \text{and} \qquad (3r - 4)^2 = (3r - 4)^2$$

If we let u equal the variable factor of the middle term, each equation will take the form $au^2 + bu + c = 0$. An equation of the form $au^2 + bu + c = 0$, where u is some algebraic expression and $a \neq 0$, is called an **equation of quadratic form.**

Example 3   Solve: $x^4 - 10x^2 + 9 = 0$.

Solution: Let $u = x^2$. Then $u^2 = (x^2)^2 = x^4$.

$x^4 - 10x^2 + 9 = 0$		Original equation
$u^2 - 10u + 9 = 0$		Let $x^4 = u^2$ and $x^2 = u$.
$(u - 1)(u - 9) = 0$		Factor.
$u - 1 = 0$ or $u - 9 = 0$		Set each factor equal to 0.
$u = 1$ $\qquad$ $u = 9$		Solve each linear equation.
$x^2 = 1$ $\qquad$ $x^2 = 9$		Let $u = x^2$.
$x = \pm 1$ $\qquad$ $x = \pm 3$		Square root property

Try Exercise 27

Solve: $x^4 - 5x^2 + 4 = 0$.

The solution set is $\{1, -1, 3, -3\}$. Check in the original equation. ❏

Example 4   Solve: $p^{2/3} - 3p^{1/3} - 4 = 0$.

Solution: Let $u = p^{1/3}$. Then $u^2 = (p^{1/3})^2 = p^{2/3}$, and the original equation becomes

$$u^2 - 3u - 4 = 0$$

Solve by factoring.

$$(u + 1)(u - 4) = 0$$
$$u = -1 \qquad \text{or} \qquad u = 4$$

Since $u = p^{1/3}$, the last two equations become

$$p^{1/3} = -1 \qquad \text{or} \qquad p^{1/3} = 4$$

To find p, cube each side.

$$(p^{1/3})^3 = (-1)^3 \quad \text{or} \quad (p^{1/3})^3 = 4^3$$
$$p = -1 \qquad \qquad p = 64$$

Try Exercise 35

Solve: $p^{2/3} - 2p^{1/3} - 3 = 0$.

The solution set is $\{-1, 64\}$. Check in the original equation. ❏

---

*HISTORICAL NOTE: The Arab mathematician al-Karkhî (ca. 1029) was the first to use numeric methods to find the solutions (including nonrational solutions) of equations of the form $ax^{2n} + bx^n + c = 0$.

Many types of applied problems result in equations that lead to quadratic equations. To solve Example 5, we need to recall the rule for solving work problems from Sec. 4.6:

If it takes x hours to do a job, then the portion of the job done in 1 hr is $\dfrac{1}{x}$.

Problem Solving

Example 5 One pipe can fill an oil tank in 3 hr less time than it takes a smaller pipe. The two pipes working together can fill the tank in 6 hr. How long does it take each pipe to fill the tank on its own?

Solution:

$$x = \text{number of hours it takes the smaller pipe}$$
$$x - 3 = \text{number of hours it takes the larger pipe}$$

Portion of tank filled by smaller pipe in 1 hr	+	portion of tank filled by larger pipe in 1 hr	=	portion of tank filled by both pipes in 1 hr
↓	↓	↓	↓	↓
$\dfrac{1}{x}$	+	$\dfrac{1}{x-3}$	=	$\dfrac{1}{6}$

Multiply every term by the LCD, $6x(x - 3)$. Then simplify.

$$6x(x - 3)\frac{1}{x} + 6x(x - 3)\frac{1}{x-3} = 6x(x - 3)\frac{1}{6}$$
$$6(x - 3) + 6x = x(x - 3)$$
$$6x - 18 + 6x = x^2 - 3x$$
$$0 = x^2 - 15x + 18$$

Use the quadratic formula with $a = 1$, $b = -15$, and $c = 18$.

$$x = \frac{-(-15) \pm \sqrt{(-15)^2 - 4(1)(18)}}{2(1)}$$
$$x = \frac{15 \pm \sqrt{225 - 72}}{2} = \frac{15 \pm \sqrt{153}}{2}$$

Since $\sqrt{153} \approx 12.37$, we have

$$x \approx \frac{15 + 12.37}{2} \qquad \text{or} \qquad x \approx \frac{15 - 12.37}{2}$$
$$x \approx 13.7 \qquad\qquad\qquad x \approx 1.3$$

The smaller pipe's time cannot be 1.3 hr, because that would make the larger pipe's time $1.3 - 3 = -1.7$ hr. Therefore the smaller pipe's time is approximately 13.7 hr, and the larger pipe's time is approximately $13.7 - 3 = 10.7$ hr. ❑

Try Exercise 51

One pump can fill an underground gasoline tank in 1 hr less time than it takes a slower pump. The two pumps working together can fill the tank in 4 hr. How long does it take each pump to fill the tank on its own?

Exercises 6.4

True-False Exercises

1. Usually the best first step in solving an equation with fractions is to multiply both sides by the LCD.

2. If an equation contains a variable in a denominator, you must check to make sure that no proposed solution makes a denominator 0.

Solve each equation with rational expressions.

3. $1 + \dfrac{15}{x} + \dfrac{54}{x^2} = 0$ 4. $1 - \dfrac{12}{x} + \dfrac{32}{x^2} = 0$

5. $3 - \dfrac{17}{4m} - \dfrac{10}{m^2} = 0$ 6. $9 + \dfrac{33}{2m} - \dfrac{20}{m^2} = 0$

7. $2y - 3 = \dfrac{1}{y + 1}$ 8. $3y - 2 = \dfrac{1}{y - 1}$

9. $\dfrac{1}{r} - \dfrac{2}{r - 4} = 2$ 10. $\dfrac{2}{r} + \dfrac{1}{r + 2} = 3$

11. $\dfrac{3z}{z + 6} - \dfrac{6}{(z + 6)^2} = 4$ 12. $\dfrac{4z}{z - 4} - \dfrac{14}{(z - 4)^2} = 5$

True-False Exercises

13. If you square both sides of an equation, you must check your answers in the original equation.

14. The equation $x^4 - 2x^2 + 1 = 0$ is called an equation of quadratic form.

Solve each radical equation.

15. $\sqrt{p + 4} - p\sqrt{2} = 0$ 16. $\sqrt{p + 1} - p\sqrt{3} = 0$

17. $m - 4 = \sqrt{5m - 1}$ 18. $m - 5 = \sqrt{3m + 7}$

19. $\sqrt{3z - 8} = 2 + \sqrt{z}$ 20. $\sqrt{4z + 6} = 3 + \sqrt{z}$

21. $\sqrt{2y - 3} - \sqrt{y} = 1$ 22. $\sqrt{3y - 1} - \sqrt{y} = 1$

23. $\sqrt{7r + 2} = \sqrt{r - 1} + 3$

24. $\sqrt{7r + 23} = \sqrt{3r + 7} + 2$

Completion Exercises

25. To solve the equation $x^4 - 5x^2 + 4 = 0$, you can make the substitution $u = $ _____.

26. To solve the equation $m^{1/2} - 3m^{1/4} + 2 = 0$, you can make the substitution $u = $ _____.

Solve each equation of quadratic form.

27. $x^4 - 5x^2 + 4 = 0$ 28. $x^4 - 13x^2 + 36 = 0$

29. $t^4 - 11t^2 + 18 = 0$ 30. $t^4 - 7t^2 + 12 = 0$

31. $4y^4 - 13y^2 + 9 = 0$ 32. $4y^4 - 29y^2 + 25 = 0$

33. $m^{1/2} - 5m^{1/4} + 6 = 0$ 34. $m^{1/2} - 3m^{1/4} + 2 = 0$

35. $p^{2/3} - 2p^{1/3} - 3 = 0$ 36. $p^{2/3} - p^{1/3} - 6 = 0$

37. $6r^{-2} + 29r^{-1} - 5 = 0$ 38. $7r^{-2} + 41r^{-1} - 6 = 0$

39. $t^4 + 3 = 6t^2$ 40. $t^4 + 4 = 6t^2$

41. $x^4 + 35x^2 = 36$ 42. $x^4 + 21x^2 = 100$

43. $y^4 + 29y^2 + 100 = 0$ 44. $y^4 + 25y^2 + 144 = 0$

45. $(2m + 3)^2 - 2(2m + 3) - 63 = 0$

46. $(2m - 5)^2 - 8(2m - 5) - 33 = 0$

Discussion Exercises

47. Explain how to recognize an equation of quadratic form.

48. Discuss the procedure for solving an equation of quadratic form by substitution. Do you think you could solve such an equation without making a substitution? Try this with Examples 3 and 4.

49. What is the basic rule to remember when solving work problems?

50. Why is it important to check your solutions when you solve a radical equation or an equation with rational exponents?

Problem Solving

51. One pump can fill an underground gasoline tank in 1 hr less time than it takes a slower pump. The two pumps working together can fill the tank in 4 hr. How long does it take each pump to fill the tank on its own?

52. One worker can unload a boxcar in 2 hr less time than it takes a slower worker. Working together, the workers can unload the boxcar in 5 hr. How long does it take each worker to unload the boxcar working alone?

53. An old copier takes 1 hr longer to copy the annual report than does a new copier. If both copiers are used, the job takes $1\frac{1}{2}$ hr. How long does it take each copier to do the job on its own?

54. An old loom takes 3 hr longer to produce a bolt of cloth than does a new loom. If both looms are used, the job takes $2\frac{1}{2}$ hr. How long does it take each loom to do the job on its own?

55. A worker with a shovel takes 2 hr longer to dig a trench than does a worker with a trenching machine. If both work together, they can dig the trench in 3 hr and 15 min. Determine how long it takes each worker to dig the trench working alone.

56. A small hose takes 1 hr longer to fill a wading pool than does a larger hose. If both hoses are used, the pool can be filled in 2 hr and 20 min. Determine how long it takes each hose to fill the pool by itself.

 Getting Acquainted with Your Graphing Calculator

57. Check that $6 + 2\sqrt{5}$ is a solution of $\sqrt{2x - 3} - \sqrt{x} = 1$ by storing $6 + 2\sqrt{5}$ in x and then calculating the value of the left side of the equation (it should be 1, or approximately 1). Then check $6 - 2\sqrt{5}$ in the same way.

58. Check that 1 is a solution of $x^{2/3} - 4x^{1/3} + 3 = 0$ by storing 1 in x and then calculating the value of the left side of the equation (it should be 0, or approximately 0). Then check 27 in the same way.

6.5 Problem Solving with Quadratic Equations

TAPE AU9

In this section we learn how to solve applied problems that lead to quadratic equations.

Formulas

One of the most important formulas in plane geometry is the **Pythagorean theorem.***

The Pythagorean Theorem

In any right triangle (see Fig. 6.6), the square of the hypotenuse (the longest side) equals the sum of the squares of the two legs,

$$c^2 = a^2 + b^2$$

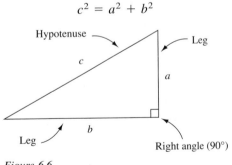

Figure 6.6

Note:

The hypotenuse is always the side opposite the right angle.

Example 1   Sections of railroad track are laid with a small space between them to allow for expansion when the metal becomes hot. Suppose two 10-ft sections are

---

*HISTORICAL NOTE: The Pythagorean theorem is named after the Greek mathematician Pythagoras (ca. 572–495 B.C.) and his secret mathematical and philosophical society of followers, the Pythagoreans. However, the theorem was known to the Babylonians more than a thousand years earlier.

laid so that the outer ends are fixed. If each section expands 1 in., and no space is allowed for this expansion, by how much, x, will the junction be raised (see Fig. 6.7)?

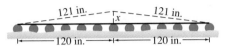

Figure 6.7

Solution: The left side of Fig. 6.7 (as well as the right side) forms a right triangle. Since we know the lengths of the hypotenuse and one of the legs, we use the Pythagorean theorem to find the unknown leg.

$$a^2 + b^2 = c^2 \qquad \text{Pythagorean theorem}$$

Convert 10 ft to $10 \cdot 12 = 120$ in., and 10 ft + 1 in. to 121 in.

$$
\begin{aligned}
x^2 + 120^2 &= 121^2 &&\text{Let } a = x, b = 120, \text{ and } c = 121.\\
x^2 + 14{,}400 &= 14{,}641 &&\text{Square each number.}\\
x^2 &= 241 &&\text{Subtract 14,400.}\\
x &= \pm\sqrt{241} &&\text{Square root property}
\end{aligned}
$$

The distance x cannot be $-\sqrt{241}$. Therefore the junction is raised $\sqrt{241} \approx 15.5$ in.   ❏

Solving for a Specified Variable

Recall that we solve an equation for a specified variable by isolating that variable on one side of the equation.

Example 2   Solve $x^2 + y^2 = r^2$ for x.

Solution:
$$
\begin{aligned}
x^2 + y^2 &= r^2 &&\text{Original equation}\\
x^2 &= r^2 - y^2 &&\text{Subtract } y^2.\\
x &= \pm\sqrt{r^2 - y^2} &&\text{Square root property}
\end{aligned}
$$
❏

Example 3   The formula $T = \pi r l + \pi r^2$ gives the total surface area T of a right circular cone (Fig. 6.8). Solve this formula for r.

Figure 6.8

Solution: Since one term contains r^2 and another contains r, this is a quadratic equation in r. Write the equation in standard form.

$$
\begin{aligned}
T &= \pi r l + \pi r^2 &&\text{Original equation}\\
0 &= \pi r^2 + \pi l r - T &&\text{Subtract } T.
\end{aligned}
$$

Use the quadratic formula with $a = \pi$, $b = \pi l$, and $c = -T$.

Try Exercise 13

A plane with an air speed of 120 mi/hr flies on a heading of due north for 1 hr. During this time, a 22-mi/hr wind blowing from west to east causes the plane to travel in a northeastern direction, as shown in Fig. 6.11. Find the total distance traveled by the plane.

Try Exercise 19

Solve $c^2 = b^2 - a^2$ for a.

$$r = \frac{-\pi l \pm \sqrt{(\pi l)^2 - 4(\pi)(-T)}}{2(\pi)}$$

$$r = \frac{-\pi l \pm \sqrt{\pi^2 l^2 + 4\pi T}}{2\pi}$$ ❑

Try Exercise 27

Solve $T = 2\pi rh + 2\pi r^2$ for r.

Word Problems

You can solve the word problems in this section using the same four steps you used in Sec. 2.3.

Example 4   A landscaper wants to surround a flower bed that is 4 ft by 8 ft with a border of mulch of uniform width (see Fig. 6.9). If there is enough mulch to cover 28 ft², how wide should the border be?

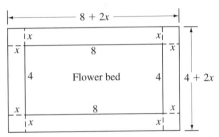

Figure 6.9

Solution: Let $x =$ the width of the border. Since the area of the border is 28 ft², we can write the following equation.

$$\text{Total area} - \text{area of bed} = \text{area of border}$$
$$\downarrow \qquad\qquad \downarrow \qquad \downarrow$$
$$(8 + 2x)(4 + 2x) - 4 \cdot 8 = 28$$
$$32 + 24x + 4x^2 - 32 = 28$$
$$4x^2 + 24x - 28 = 0$$

This is a quadratic equation. Solve by factoring.

$$x^2 + 6x - 7 = 0 \qquad \text{Divide by 4.}$$
$$(x - 1)(x + 7) = 0 \qquad \text{Factor.}$$
$$x = 1 \quad \text{or} \quad x = -7$$

The border cannot be -7 ft wide. Therefore the border should be 1 ft wide.   ❑

Try Exercise 31

Sue has a flower bed that is 4 ft by 8 ft. She wants to surround the bed with a border of mulch of uniform width. She has enough mulch to cover 64 ft². How wide should the border be?

When solving a motion problem we often use the formula $d = rt$, where d is distance, r is rate, and t is time.

Example 5   A jogger and a cyclist leave the same point simultaneously. The jogger travels north and the cyclist travels east. After 1 hr, they are 26 mi apart. If the cyclist travels 14 mi/hr faster than the jogger, find the rate of each.

Solution: Let $x =$ the rate of the jogger. Then $x + 14 =$ the rate of the cyclist. Since $d = rt$, after 1 hr the jogger's distance is $x \cdot 1 = x$, and the cyclist's distance is $(x + 14) \cdot 1 = x + 14$ (see Fig. 6.10).

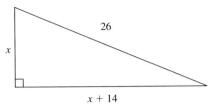

Figure 6.10

Using the Pythagorean theorem, we can write the following equation:

$$\begin{array}{ccccc} a^2 & + & b^2 & = & c^2 \\ \downarrow & & \downarrow \quad \downarrow & & \downarrow \quad \downarrow \\ x^2 & + & (x + 14)^2 & = & 26^2 \end{array}$$

Write in standard form and solve by factoring.

$$\begin{array}{ll} x^2 + x^2 + 28x + 196 = 676 & \text{Square } x + 14, \text{ square } 26. \\ 2x^2 + 28x - 480 = 0 & \text{Write in standard form.} \\ x^2 + 14x - 240 = 0 & \text{Divide by 2.} \\ (x - 10)(x + 24) = 0 & \text{Factor.} \\ x = 10 \quad \text{or} \quad x = -24 \end{array}$$

The jogger's rate cannot be -24 mi/hr. Therefore the jogger's rate is 10 mi/hr, and the cyclist's rate is $10 + 14 = 24$ mi/hr. ❏

Try Exercise 39

Two ships leave the same port simultaneously. Ship A travels north and ship B travels east. After 1 hr, the ships are 25 mi apart. If ship B travels 5 mi/hr faster than ship A, find the rate of each.

Sometimes it is helpful to make a chart when solving a motion problem.

Example 6 On the third Sunday of each month, Violet drives 70 mi to visit her grand-daughter at college. If she drove 5 mi/hr faster than her normal speed, she would arrive 15 min sooner. Find Violet's normal speed.

Solution: Let $x =$ Violet's normal speed. Then construct the chart below.

	d	r	t
Normal trip	70	x	$\dfrac{70}{x}$
Fast trip	70	$x + 5$	$\dfrac{70}{x + 5}$

Since $t = \dfrac{d}{r}$

Since the rates are in miles per hour, convert 15 min to hours.

$$15 \text{ min} = \frac{15}{60} \text{ hr} = \frac{1}{4} \text{ hr}$$

Using the fact that the difference in time between a normal trip and a fast trip is $\frac{1}{4}$ hr, we can write the following equation:

$$\begin{array}{ccccc} \text{Time for a} & & \text{time for a} & = & \dfrac{1}{4} \\ \text{normal trip} & - & \text{fast trip} & & \\ \downarrow & \downarrow & \downarrow & & \downarrow \quad \downarrow \\ \dfrac{70}{x} & - & \dfrac{70}{x + 5} & = & \dfrac{1}{4} \end{array}$$

Multiply each term by the LCD, $4x(x + 5)$, and simplify.

$$4x(x + 5)\frac{70}{x} - 4x(x + 5)\frac{70}{x + 5} = 4x(x + 5)\frac{1}{4}$$
$$280(x + 5) - 280x = x(x + 5)$$
$$280x + 1400 - 280x = x^2 + 5x$$
$$0 = x^2 + 5x - 1400$$

Instead of trying to factor the right side, use the quadratic formula with $a = 1$, $b = 5$, and $c = -1400$.

$$x = \frac{-5 \pm \sqrt{5^2 - 4(1)(-1400)}}{2(1)}$$
$$x = \frac{-5 \pm \sqrt{25 + 5600}}{2} = \frac{-5 \pm \sqrt{5625}}{2}$$

Since $\sqrt{5625} = 75$, we have

$$x = \frac{-5 + 75}{2} = 35 \qquad \text{or} \qquad x = \frac{-5 - 75}{2} = -40$$

Violet's normal speed cannot be -40 mi/hr, so it must be 35 mi/hr. ❒

Try Exercise 41

A family moves to a new home 150 mi away. The parents drive a rented truck while their children drive the family car. Find the speed of each vehicle if the kids drive 10 mi/hr faster than their parents and arrive 30 min sooner.

Exercises 6.5

Completion Exercises

1. The Pythagorean theorem states that if a and b are the legs of a right triangle and c is the hypotenuse, then _____.

2. To solve an equation for a variable means to _____.

Discussion Exercises

3. What is a right triangle?

4. What is the hypotenuse and what are the legs of a right triangle?

5. State the Pythagorean theorem in words.

6. Outline the four steps in solving a word problem.

Use the Pythagorean theorem to find the unknown side of each right triangle.

7.

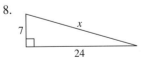

8.

9.

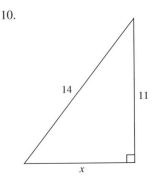

10.

11.

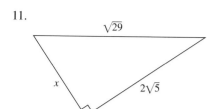

12.

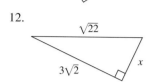

Problem Solving

13. A plane with an air speed of 120 mi/hr flies on a heading of due north for 1 hr. During this time, a 22-mi/hr wind blowing from west to east causes the plane to travel in a northeastern direction, as shown in Fig. 6.11. Find the total distance traveled by the plane

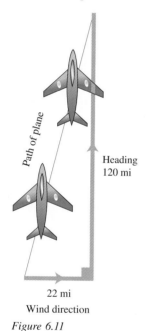

Figure 6.11

14. How long must a bracing wire be to reach the top of a 15-ft antenna from a point on the ground 8 ft from the base of the antenna?

15. Determine the distance from one corner to the opposite corner of a square boxing ring that is 20 ft by 20 ft.

16. A baseball infield is a square 90 ft by 90 ft (see Fig. 6.12). Determine the distance of a throw from home plate to second base.

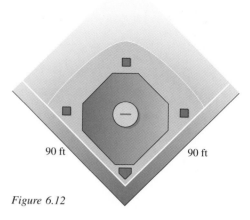

Figure 6.12

17. Find x in Fig. 6.13.

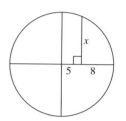

Figure 6.13

18. Determine the height of the pile of logs in Fig. 6.14 if the diameter of each log is 6 in.

Figure 6.14

Solve each equation for the specified variable.

19. $c^2 = b^2 - a^2$ for a

20. $V = \frac{1}{3}\pi r^2 h$ for r

21. $V = \sqrt{WR}$ for W

22. $A = \sqrt{\dfrac{W}{R}}$ for W

23. $F = \dfrac{gm_1 m_2}{d^2}$ for d

24. $I = \dfrac{W}{4\pi r^2}$ for r

25. $V = P(1 + r)^2$ for r

26. $T = 2\pi\sqrt{\dfrac{l}{32}}$ for l

27. $T = 2\pi rh + 2\pi r^2$ for r

28. $h = -16t^2 + vt$ for t

29. $S = \dfrac{n(n + 1)}{2}$ for n

30. $9x^2 + 6xy + y^2 - 16 = 0$ for x

Problem Solving

31. Sue has a flower bed that is 4 ft by 8 ft. She wants to surround the bed with a border of mulch of uniform width. She has enough mulch to cover 64 ft². How wide should the border of mulch be?

32. Dave has a swimming pool that is 20 ft by 30 ft. He wants to surround the pool with a concrete border of uniform width. He has enough concrete to cover 336 ft². How wide should he make the border?

33. An open pan is to be made from a square piece of sheet metal by cutting a 2-in. square from each corner and then turning up the sides (see Fig. 6.15). What size piece of sheet metal is needed if the volume of the pan is to be 242 in³?

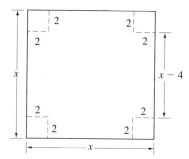

Figure 6.15

34. A box without a top is to be constructed from a square piece of cardboard by cutting a 5-in. square from each corner and then turning up the sides (see Fig. 6.16). What size piece of cardboard should be used if the volume of the box is to be 720 in³?

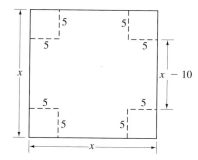

Figure 6.16

35. Find the lengths of the sides of a right triangle if one leg is 2 in. longer than the other leg, and the hypotenuse is 2 in. shorter than twice the shorter leg.

36. Find the lengths of the sides of a right triangle if one leg is 2 in. longer than twice the other leg, and the hypotenuse is 2 in. shorter than three times the shorter leg.

37. Determine the length of one side of a square whose diagonal is 16 ft long.

38. Determine the length of one side of a square whose diagonal is 24 ft long.

39. Two ships leave the same port simultaneously. Ship A travels north and ship B travels east. After 1 hr, the ships are 25 mi apart. If ship B travels 5 mi/hr faster than ship A, find the rate of each.

40. Two traffic helicopters leave the top of a building simultaneously. Helicopter A travels south and helicopter B travels west. After 1 hr, the helicopters are 50 mi apart. If helicopter B travels 10 mi/hr faster than helicopter A, find the rate of each.

41. A family moves to a new home 150 mi away. The parents drive a rented truck while their children drive the family car. Find the speed of each vehicle if the children drive 10 mi/hr faster than their parents and arrive 30 min sooner.

42. A trucker drives to a city 60 mi away, unloads, and then returns. The trucker drives 20 mi/hr faster on the return trip, which takes 30 min less time. How fast did the trucker drive each way?

43. A kayaker who paddles 5 mi/hr in still water takes 1 hr longer to paddle 12 mi upstream to a camping ground than she does to make the return trip. Find the speed of the current.

	d	r	t
Upstream	12	$5 - x$	
Downstream	12	$5 + x$	

44. A pontoon boat that travels 10 mi/hr in still water takes 1 hr longer to travel 24 mi against the current than it does to travel the same distance with the current. Find the speed of the current.

	d	r	t
Upstream	24	$10 - x$	
Downstream	24	$10 + x$	

45. A boat can travel 6 mi upstream and return in a total time of $3\frac{1}{2}$ hr. If the speed of the current is 3 mi/hr, find the speed of the boat in still water.

46. A plane can fly 100 mi against the wind and return with the wind in a total time of $1\frac{1}{2}$ hr. If the speed of the wind is 40 mi/hr, find the speed of the plane in still air.

6.6 Nonlinear Inequalities

TAPE AU9
TAPE AU11

In Sec. 2.5 we solved *linear* inequalities, such as $2x - 5 < 0$, by isolating the variable on one side of the inequality.

$$2x - 5 < 0 \qquad \text{Original inequality}$$
$$2x < 5 \qquad \text{Add 5.}$$
$$x < \tfrac{5}{2} \qquad \text{Divide by 2.}$$

We wrote the solution set in interval notation as $(-\infty, \tfrac{5}{2})$. Then we graphed the solution set on a number line as shown in Fig. 6.17.

Figure 6.17

In this section we learn to solve *nonlinear* inequalities.

Quadratic Inequalities

If the equals sign in a quadratic equation is replaced by an inequality symbol, the result is a **quadratic inequality.** Here are three examples of quadratic inequalities:

$$(x - 2)(x + 4) < 0 \qquad y^2 - 2y - 3 > 0 \qquad r^2 \leq 25$$

DEFINITION OF A QUADRATIC INEQUALITY

A **quadratic inequality** in the variable x is an inequality that can be written in the form

$$ax^2 + bx + c < 0$$

where a, b, and c are real numbers and $a \neq 0$.

We can replace the symbol $<$ in the definition of a quadratic inequality with any of the symbols $>$, $\leq$, or $\geq$.

One way to solve a quadratic inequality is to make a **truth diagram.** To illustrate why this method works, consider the quadratic inequality $x^2 + 2x - 8 < 0$. We can write this inequality in factored form as

$$(x - 2)(x + 4) < 0$$

Now for x values greater than 2, the factor $x - 2$ is positive; for x values less than 2, the factor $x - 2$ is negative (see Fig. 6.18).

Figure 6.18

Similarly, the factor $x + 4$ is positive when $x > -4$ and negative when $x < -4$ (see Fig. 6.19).

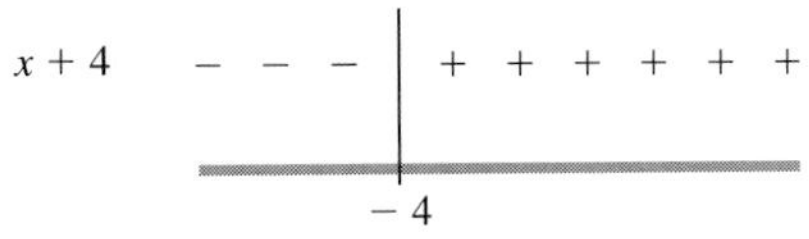

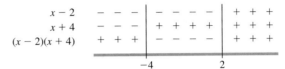

Figure 6.19

Since the factors $x - 2$ and $x + 4$ have opposite signs when x is between -4 and 2, their product is negative when x is between -4 and 2 (see Fig. 6.20). That is, $(x - 2)(x + 4) < 0$ when $-4 < x < 2$.

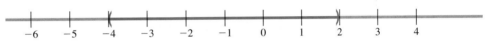

Figure 6.20

Therefore the solution set of the inequality $x^2 + 2x - 8 < 0$ is the interval $(-4, 2)$. The graph of the solution set is shown in Fig. 6.21.

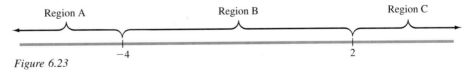

Figure 6.21

Figure 6.22 shows how the solution set of $(x - 2)(x + 4) < 0$ changes when $<$ is replaced by $\leq$, $>$, or $\geq$.

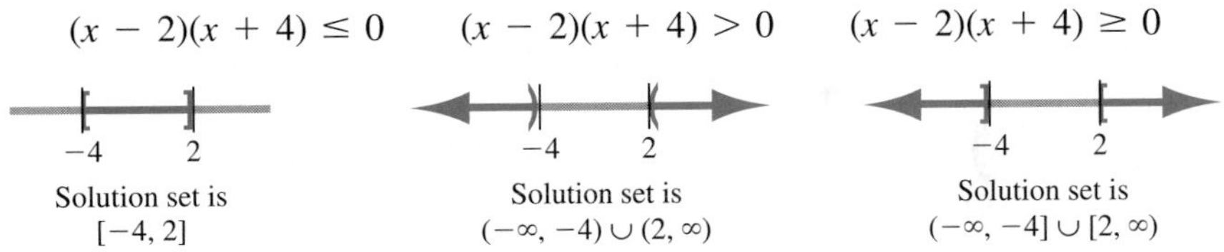

$(x - 2)(x + 4) \leq 0$ $(x - 2)(x + 4) > 0$ $(x - 2)(x + 4) \geq 0$

Solution set is
$[-4, 2]$

Solution set is
$(-\infty, -4) \cup (2, \infty)$

Solution set is
$(-\infty, -4] \cup [2, \infty)$

Figure 6.22

Note that -4 and 2 are the solutions of the equation $x^2 + 2x - 8 = 0$. Also, the numbers -4 and 2, which we shall call **boundary values,** divide the number line into three regions (see Fig. 6.23). The product $(x - 2)(x + 4)$ is either positive throughout an entire region or negative throughout an entire region (Fig. 6.23).

Region A   Region B   Region C

-4 2

Figure 6.23

This means that we can determine the sign of the product $(x - 2)(x + 4)$ simply by testing one number from each of the three regions. We illustrate how this is done in Example 1.

Example 1   Solve: $y^2 - 2y - 3 > 0$. Then graph the solution set.

Solution: First, solve the *equation* $y^2 - 2y - 3 = 0$ to find the boundary values.

$$y^2 - 2y - 3 = 0$$
$$(y - 3)(y + 1) = 0$$
$$y = 3 \quad \text{or} \quad y = -1$$

Graph the boundary values on a number line (Fig. 6.24).

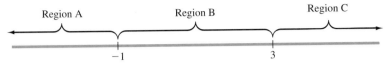

Figure 6.24

Choose a number (any number) from each of the three regions and test it in the original inequality.

Test $x = -2$ Test $x = 0$ Test $x = 5$

$$(-2)^2 - 2(-2) - 3 > 0 \qquad 0^2 - 2(0) - 3 > 0 \qquad 5^2 - 2(5) - 3 > 0$$
$$4 + 4 - 3 > 0 \qquad 0 - 0 - 3 > 0 \qquad 25 - 10 - 3 > 0$$
$$5 > 0 \qquad\qquad -3 > 0 \qquad\qquad 12 > 0$$
$$\text{True} \qquad\qquad \text{False} \qquad\qquad \text{True}$$

Write the results of the three tests on the number line (Fig. 6.25).

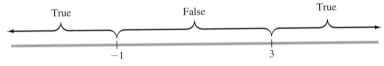

Figure 6.25

Since the inequality is true when $x < -1$ or when $x > 3$ (Fig. 6.25), the solution set is $(-\infty, -1) \cup (3, \infty)$. The graph of the solution set is shown in Fig. 6.26. ❏

Figure 6.26

Try Exercise 9

Solve: $y^2 - 4y - 5 > 0$. Then graph the solution set.

CAUTION

Do not choose a boundary value as your test point.

Wrong   *Wrong*

~~Test $x = -1$~~   ~~Test $x = 3$~~ ∎

Here is a summary of the steps in solving a quadratic inequality.

To Solve a Quadratic Inequality

1. Replace the inequality symbol with an equals sign. Then solve the resulting equation to find the boundary values.
2. Graph the boundary values on a number line. This will divide the number line into regions.
3. Choose a number from each region and test it in the original inequality. Using the results of these tests, label each region "True" or "False."
4. The numbers in those regions labeled "True" form the solution set. Include the boundary values in the solution set if the original inequality symbol is $\leq$ or $\geq$.

We illustrate these steps in Example 2.

Example 2   Solve: $r^2 \leq 25$. Then graph the solution set.

Solution: Step 1:   Replace the inequality symbol with an equals sign. Then solve the resulting equation to find the boundary values.

$$r^2 = 25 \qquad \text{Replace} \leq \text{with} =.$$
$$r = \pm 5 \qquad \text{Square root property}$$

Step 2:   Graph the boundary values on a number line. This will divide the number line into regions.

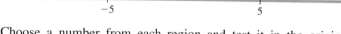

Step 3:   Choose a number from each region and test it in the original inequality.

Test x = −6	*Test x = 0*	*Test x = 6*
$(-6)^2 \leq 25$	$0^2 \leq 25$	$6^2 \leq 25$
$36 \leq 25$	$0 \leq 25$	$36 \leq 25$
False	True	False

Using the results of these tests, label each region "True" or "False."

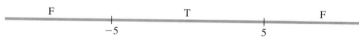

Step 4:   The numbers in those regions labeled "True" form the solution set. Include the boundary values since the original inequality symbol is $\leq$, rather than $<$.

$$\text{Solution set} = [-5, 5]$$

The graph of the solution set is shown in Fig. 6.27.   ❑

Try Exercise 15

Solve: $r^2 \leq 16$. Then graph the solution set.

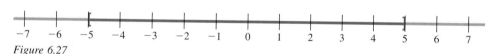

Figure 6.27

CAUTION

Do not try to solve a quadratic inequality in the same way that you solve a quadratic equation.

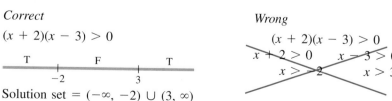

Correct

$(x + 2)(x - 3) > 0$

Solution set $= (-\infty, -2) \cup (3, \infty)$

Wrong

$(x + 2)(x - 3) > 0$
$x + 2 > 0 \qquad x - 3 > 0$
$x > 2 \qquad x > 3$

NOTE:

The inequality $a \cdot b > 0$ does not necessarily mean that both $a > 0$ and $b > 0$. The inequality $a \cdot b > 0$ is also true when $a < 0$ and $b < 0$.   ∎

Higher-Degree Inequalities

We can solve an inequality of degree 3 or higher in much the same way that we solve a quadratic inequality.

Example 3   Solve: $x^3 - 7x^2 + 12x > 0$. Then graph the solution set.

Solution: Replace $>$ with $=$ and solve the resulting equation.

$$x^3 - 7x^2 + 12x = 0$$
$$x(x^2 - 7x + 12) = 0$$
$$x(x - 3)(x - 4) = 0$$
$$x = 0 \quad \text{or} \quad x = 3 \quad \text{or} \quad x = 4$$

Graph the boundary values 0, 3, and 4. Then test a number from each region. This will produce the truth diagram shown in Fig. 6.28.

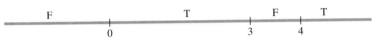

Figure 6.28

Try Exercise 27

Solve: $x^3 - 9x^2 + 20x > 0$.
Then graph the solution set.

From Fig. 6.28 we see that the solution set is $(0, 3) \cup (4, \infty)$. The graph of the solution set is shown in Fig. 6.29.   ❑

Figure 6.29

Rational Inequalities

Although we can solve an inequality with rational expressions in much the same way that we solve a quadratic inequality, there is one significant difference. To see this difference, consider the inequality

$$\frac{x - 2}{x - 5} < 0$$

In this case, the boundary values occur not only when $\dfrac{x - 2}{x - 5} = 0$, but also when $x - 5 = 0$ (see Fig. 6.30).

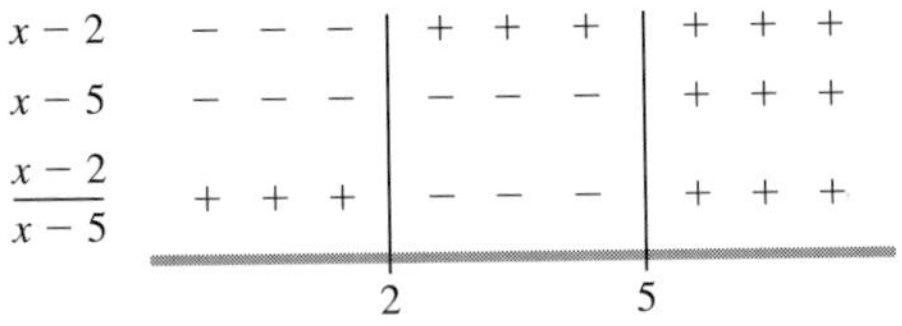

<table>
<tr><td>$x - 2$</td><td>$-\ -\ -$</td><td>$+\ +\ +$</td><td>$+\ +\ +$</td></tr>
<tr><td>$x - 5$</td><td>$-\ -\ -$</td><td>$-\ -\ -$</td><td>$+\ +\ +$</td></tr>
<tr><td>$\dfrac{x - 2}{x - 5}$</td><td>$+\ +\ +$</td><td>$-\ -\ -$</td><td>$+\ +\ +$</td></tr>
</table>

2   5

Figure 6.30

Therefore, to find the boundary values we must not only replace $<$ with $=$, but we must also set the denominator equal to 0 and solve.

Example 4   Solve: $\dfrac{x - 2}{x - 5} < 0$. Then graph the solution set.

Solution: Replace $<$ with $=$ and solve the resulting equation.

$$\frac{x - 2}{x - 5} = 0$$

$$(x - 5)\frac{x - 2}{x - 5} = (x - 5)0 \qquad \text{Multiply by the LCD, } x - 5.$$

$$x - 2 = 0 \qquad \text{Simplify.}$$

$$x = 2 \qquad \text{Add 2.}$$

Set the denominator equal to 0 and solve.

$$x - 5 = 0$$
$$x = 5$$

Graph the boundary values 2 and 5 on a number line and test a number from each region. This will produce the truth diagram shown in Fig. 6.31.

Figure 6.31

Try Exercise 37

Solve: $\dfrac{x - 1}{x - 6} < 0$. Then graph the solution set.

From Fig. 6.31 we see that the solution set is (2, 5). The graph of the solution set is shown in Fig. 6.32. ❑

Figure 6.32

Here is a summary of the steps in solving an inequality with rational expressions.

To Solve a Rational Inequality

1. Replace the inequality symbol with an equals sign and solve the resulting equation.
2. Set the denominator equal to 0 and solve the resulting equation.
3. Use the boundary values found in Steps 1 and 2 to divide the number line into regions.
4. Choose a number from each region and test it in the original inequality. Label each region "True" or "False."
5. The numbers in those regions labeled "True" form the solution set. Include the boundary values if the original inequality is $\leq$ or $\geq$, but **always exclude those boundary values that make the denominator 0.**

Example 5 Solve: $\dfrac{m^2 + 5}{m + 1} \geq 3$. Then graph the solution set.

Solution: Step 1:   Replace $\geq$ with $=$ and solve the resulting equation.

$$\frac{m^2 + 5}{m + 1} = 3$$

$$m^2 + 5 = 3(m + 1) \qquad \text{Multiply by } m + 1.$$

$$m^2 + 5 = 3m + 3$$

$$m^2 - 3m + 2 = 0$$

$$(m - 1)(m - 2) = 0$$

$$m = 1 \qquad \text{or} \qquad m = 2$$

Step 2:   Set the denominator equal to 0 and solve the resulting equation.

$$m + 1 = 0$$
$$m = -1$$

Step 3:   Use the boundary values found in Steps 1 and 2 to divide the number line into regions.

Step 4:   Choose a number from each region and test it in the original inequality. Label each region "True" or "False."

Step 5:   The numbers in those regions labeled "True" form the solution set. Include 1 and 2 since the original inequality symbol is $\geq$ rather than $>$. Exclude -1 since it makes the denominator 0.

$$\text{Solution set} = (-1, 1] \cup [2, \infty)$$

Try Exercise 43

Solve: $\dfrac{m^2 + 7}{m + 1} \geq 4$. Then graph the solution set.

The graph of the solution set is shown in Fig. 6.33.

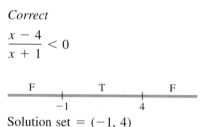

Figure 6.33

CAUTION

You must be careful when you multiply both sides of an inequality by a variable expression. You do not know whether that expression is positive or negative, so you do not know whether to reverse the direction of the inequality symbol or not.

Correct

$$\frac{x - 4}{x + 1} < 0$$

Solution set $= (-1, 4)$

Wrong

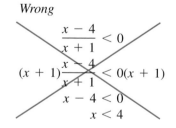

Exercises 6.6

True-False Exercises

1. The inequality $x^2 - 7x + 12 < 0$ is a quadratic inequality.

2. Do not choose a boundary value as a test point when constructing a truth diagram.

Completion Exercises

3. The boundary values for the inequality $(x - 9)(x + 6) > 0$ are _____.

4. The boundary values for the inequality $\dfrac{x + 7}{x - 5} < 0$ are _____.

Matching Exercises

Match each quadratic inequality in Exercises 5 through 8 with the graph of its solution set in letters *A* through *D*.

5. $(x - 3)(x + 2) \geq 0$ A.

6. $(x - 3)(x + 2) > 0$ B.

7. $(x - 3)(x + 2) < 0$ C.

8. $(x - 3)(x + 2) \leq 0$ D.

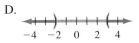

Solve each quadratic inequality. Then graph the solution set.

9. $y^2 - 4y - 5 > 0$ 10. $y^2 - 4y - 12 > 0$

11. $p^2 + 6p + 8 \leq 0$ 12. $p^2 + 7p + 12 \leq 0$

13. $2m^2 + 13m \geq 7$ 14. $2m^2 + 7m \geq 15$

15. $r^2 \leq 16$ 16. $r^2 \geq 49$

17. $t^2 \geq 8t$ 18. $t^2 \leq 10t$

19. $x^2 > 2x + 2$ 20. $x^2 < 2x + 4$

21. $2y^2 + 4y - 1 \leq 0$ 22. $4y^2 + 8y + 1 \geq 0$

Solve each higher-order inequality. Then graph the solution set.

23. $(p + 2)(p + 6)(p - 4) < 0$

24. $(p - 1)(p + 3)(p + 5) > 0$

25. $(y - 9)(y + 1)(4y - 3) \geq 0$

26. $(y - 4)(y - 8)(5y + 2) \leq 0$

27. $x^3 - 9x^2 + 20x > 0$ 28. $x^3 + 5x^2 + 6x < 0$

29. $m^3 - 4m^2 + 4m \leq 0$ 30. $m^3 + 6m^2 + 9m \geq 0$

True-False Exercises

31. The inequality $\dfrac{x + 1}{x + 6} > 0$ is a linear inequality.

32. When solving a rational inequality, always exclude from the solution set any numbers that make the denominator 0.

Matching Exercises

Match each rational inequality in Exercises 33 through 36 with the graph of its solution set in letters *A* through *D*.

33. $\dfrac{x - 4}{x + 1} > 0$ A.

34. $\dfrac{x - 4}{x + 1} \geq 0$ B.

35. $\dfrac{x - 4}{x + 1} \leq 0$ C.

36. $\dfrac{x - 4}{x + 1} < 0$ D.

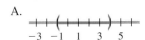
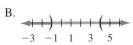

Solve each rational inequality. Then graph the solution set.

37. $\dfrac{x - 1}{x - 6} < 0$ 38. $\dfrac{x - 4}{x + 2} > 0$

39. $\dfrac{3y - 1}{y + 5} \geq 0$ 40. $\dfrac{4y + 5}{y - 1} \leq 0$

41. $\dfrac{-2}{z + 4} \leq 1$ 42. $\dfrac{-3}{z + 3} \geq 1$

43. $\dfrac{m^2 + 7}{m + 1} \geq 4$ 44. $\dfrac{m^2 + 2}{m + 4} \leq 3$

45. $\dfrac{3}{2r - 1} < r$ 46. $\dfrac{2}{3r + 1} > r$

47. $\dfrac{x^2 + x}{x - 11} < 0$ 48. $\dfrac{x^2 - 5x}{x + 13} > 0$

49. $\dfrac{16}{y^2 - 4} \le 0$

50. $\dfrac{18}{y^2 - 9} \ge 0$

51. $\dfrac{z^2 + 2z + 1}{z - 7} \ge 0$

52. $\dfrac{z^2 - 10z + 25}{z + 9} \le 0$

$$\dfrac{x - 1}{x + 4} > 0$$

$$(x + 4)\dfrac{x - 1}{x + 4} > (x + 4)0$$

$$x - 1 > 0$$

$$x > 1$$

Discussion Exercises

53. Outline the steps in solving a quadratic inequality.

54. Outline the steps in solving a rational inequality.

55. The following method for solving a quadratic inequality is **wrong.** Explain why.

$$x^2 - 5x + 6 > 0$$
$$(x - 3)(x - 2) > 0$$
$$x - 3 > 0 \quad \text{or} \quad x - 2 > 0$$
$$x > 3 \qquad\qquad x > 2$$

56. The following method for solving a rational inequality is **wrong.** Explain why.

Problem Solving

57. A company's daily profit P in thousands of dollars is given by $P = -4x^2 + 20x$, where x is the number of units produced in hundreds. Find those values of x for which P is at least 16.

58. If we ignore air resistance, the height h, in feet, of an object projected vertically upward with an initial velocity of 176 ft/sec is given by $h = -16t^2 + 176t$. Find those values of t for which h is at least 288. (t is time in seconds.)

Chapter 6 Key Terms

6.3 **Discriminant**   The discriminant of the quadratic equation $ax^2 + bx + c = 0$ is $b^2 - 4ac$.

6.4 **Equation of quadratic form**   An equation of the form $au^2 + bu + c = 0$ $(a \ne 0)$ where u is some algebraic expression

6.6 **Quadratic inequality**   An inequality that can be written in the form $ax^2 + bx + c < 0$ or $ax^2 + bx + c \le 0$, where $a \ne 0$

Chapter 6 Key Rules/Steps

6.1 Solving Quadratic Equations by Completing the Square

The Square Root Property

Suppose k is a real number.

If $x^2 = k$, then $x = \sqrt{k}$ or $x = -\sqrt{k}$.

Example: If $x^2 = 9$, then $x = \pm\sqrt{9} = \pm 3$.

To Solve $4x^2 + 24x + 5 = 0$ by Completing the Square

1. Write the equation in the form $ax^2 + bx = -c$.

$$4x^2 + 24x = -5 \qquad \text{Subtract 5.}$$

2. If $a \ne 1$, divide both sides by a.

$$x^2 + 6x = -\tfrac{5}{4} \qquad \text{Divide by 4.}$$

3. Add the square of half the coefficient of x to both sides.

$$x^2 + 6x + 9 = -\tfrac{5}{4} + 9 \qquad \text{Add } (\tfrac{6}{2})^2 = (3)^2 = 9 \text{ to both sides.}$$

4. Factor the left side and simplify the right side.

$$(x + 3)^2 = \tfrac{31}{4} \qquad \text{Since } -\tfrac{5}{4} + 9 = -\tfrac{5}{4} + \tfrac{36}{4} = \tfrac{31}{4}$$

5. Use the square root property to solve.

$$x + 3 = \pm\sqrt{\dfrac{31}{4}}$$

$$x = -3 \pm \dfrac{\sqrt{31}}{2}$$

6.2 The Quadratic Formula

The solutions of the quadratic equation $ax^2 + bx + c = 0$ are

$$x = \dfrac{-b \pm \sqrt{b^2 - 4ac}}{2a}$$

Example: The solutions of $x^2 + 3x - 2 = 0$ are

$$x = \frac{-3 \pm \sqrt{3^2 - 4(1)(-2)}}{2(1)} = \frac{-3 \pm \sqrt{17}}{2}$$

6.3 More about Quadratic Equations

The Discriminant

The discriminant of the quadratic equation $ax^2 + bx + c = 0$ is $b^2 - 4ac$. If a, b, and c are integers, then the type of solutions can be determined as follows:

Discriminant	Solutions
Positive, and a perfect square	Two rational solutions
Positive, but not a perfect square	Two irrational solutions
Zero	One rational solution
Negative	Two imaginary solutions

Example: There are two rational solutions of $3x^2 + 5x - 2 = 0$, since $5^2 - 4(3)(-2) = 25 + 24 = 49$.

Example: There are two irrational solutions of $x^2 + 3x + 1 = 0$, since $3^2 - 4(1)(1) = 9 - 4 = 5$.

Example: There is one rational solution of $x^2 - 6x + 9 = 0$, since $(-6)^2 - 4(1)(9) = 36 - 36 = 0$.

Example: There are two imaginary solutions of $x^2 + x + 1 = 0$, since $1^2 - 4(1)(1) = 1 - 4 = -3$.

Note:

A trinomial with integer coefficients can be factored using integers if and only if its discriminant is a perfect square.

Sum and Product of the Solutions of $ax^2 + bx + c = 0$

If x_1 and x_2 are the two solutions of the quadratic equation $ax^2 + bx + c = 0$, then

$$x_1 + x_2 = -\frac{b}{a} \text{ and } x_1 \cdot x_2 = \frac{c}{a}.$$

Example: The sum of the solutions of $2x^2 - 3x - 5 = 0$ is $x_1 + x_2 = -\frac{-3}{2} = \frac{3}{2}$, and the product of the solutions is

$$x_1 \cdot x_2 = \frac{-5}{2}.$$

6.4 Equations That Lead to Quadratic Equations

To Solve $x^4 - 5x^2 + 4 = 0$

$u^2 - 5u + 4 = 0$		Let $x^2 = u$ and $x^4 = u^2$.
$(u - 4)(u - 1) = 0$		Factor.
$u - 4 = 0$ or $u - 1 = 0$		Set each factor equal to 0.
$u = 4$	$u = 1$	Solve each linear equation.
$x^2 = 4$	$x^2 = 1$	Let $u = x^2$.
$x = \pm 2$	$x = \pm 1$	Square root property

6.5 Problem Solving with Quadratic Equations

The Pythagorean Theorem

In any right triangle, the square of the hypotenuse (the longest side) equals the sum of the squares of the two legs (Fig. 6.34).

$$c^2 = a^2 + b^2$$

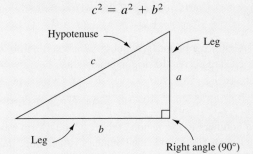

Figure 6.34

Solve $V = \pi r^2 h$ for r

$$\frac{V}{\pi h} = r^2 \qquad \text{Divide by } \pi h.$$

$$r = \pm \sqrt{\frac{V}{\pi h}} \qquad \text{Square root property}$$

Solve $h = -16t^2 + vt + s$ for t

Write in standard form and use the quadratic formula.

$$16t^2 - vt + (h - s) = 0$$

$$t = \frac{-(-v) \pm \sqrt{(-v)^2 - 4(16)(h - s)}}{2(16)}$$

$$t = \frac{v \pm \sqrt{v^2 - 64(h - s)}}{32}$$

Find the lengths of the sides of a right triangle if one leg is 1 cm longer than the other leg, and the hypotenuse is 2 cm longer than the shorter leg (Fig. 6.35).

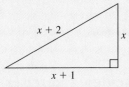

Figure 6.35

$x^2 + (x + 1)^2 = (x + 2)^2$	Pythagorean theorem
$x^2 + x^2 + 2x + 1 = x^2 + 4x + 4$	Square $x + 1$, square $x + 2$.
$x^2 - 2x - 3 = 0$	Standard form
$(x + 1)(x - 3) = 0$	Factor.
$x = -1$ or $x = 3$ cm, $x + 1 = 4$ cm, $x + 2 = 5$ cm	

6.6 Nonlinear Inequalities

To Solve the Quadratic Inequality $x^2 - 2x - 8 < 0$

1. Replace $<$ with $=$. Then solve to find the boundary values.

$$x^2 - 2x - 8 = 0$$
$$(x + 2)(x - 4) = 0$$
$$x = -2 \quad \text{or} \quad x = 4$$

2 & 3. Graph the boundary values. Then test a number from each region in the original inequality.

```
        F         T          F
    ────┼──────────┼────
        -2          4
```

4. The numbers in the region labeled "True" form the solution set. Do not include the boundary values since the original inequality symbol is $<$ (not $\leq$).

$$\text{Solution set} = (-2, 4)$$

To Solve the Rational Inequality $\dfrac{x + 3}{x - 1} \geq 0$

1 & 2. Replace $\geq$ with $=$ and solve. Also, set the denominator equal to 0 and solve.

$$\frac{x + 3}{x - 1} = 0 \qquad x - 1 = 0$$
$$x + 3 = 0 \qquad x = 1$$
$$x = -3$$

3 & 4. Graph the boundary values found in steps 1 and 2. Then test a number from each region in the original inequality.

```
        T         F          T
    ════┼──────────┼════
        -3          1
```

5. The numbers in those regions labeled "True" form the solution set. Include $x = -3$ since the original inequality symbol is $\geq$ (not $>$), but exclude $x = 1$ since 1 makes the denominator 0.

$$\text{Solution set} = (-\infty, -3] \cup (1, \infty)$$

Chapter 6 Review Exercises

[6.1] Use the square root property to solve each equation.

1. $x^2 = 100$

2. $25y^2 - 16 = 0$

3. $z^2 + 81 = 0$

4. $4p^2 + 48 = 0$

5. $(3t + 10)^2 = 45$

6. $(m - \frac{2}{3})^2 = -\frac{1}{9}$

Complete the square on each expression. Then factor the resulting perfect-square trinomial.

7. $x^2 - 12x$

8. $y^2 + \frac{2}{3}y$

Solve each equation by completing the square.

9. $x^2 - 2x - 2 = 0$

10. $3y^2 - 18y - 1 = 0$

11. $2m^2 - 3m + 2 = 0$

12. Solve $(5x + 6a)^2 = 64b^2$ for x.

13. If $V = P(1 + r)^t$, $V = 169$, $P = 100$, and $t = 2$, find r.

[6.2] Solve each equation using the quadratic formula.

14. $8x^2 + 2x = 3$

15. $2y^2 + y - 2 = 0$

16. $\dfrac{z^2}{3} + \dfrac{z}{6} - 3 = 0$

17. $4t^2 - 4t - 11 = 0$

18. $r^2 + 2r = 5r + 12$

19. $4p(1 - p) = 1$

20. $y^2 = 12y - 37$

21. $3x(x - 2) + 5 = 0$

22. What should the height x of the trough be in Fig. 6.36 if the area of the cross section is to be 18 in.^{2}?

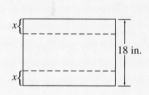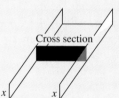

Figure 6.36

[6.3] Solve each quadratic equation.

23. $(t + 2)(t - 2) = 5$

24. $6p^2 - 17p - 14 = 0$

25. $2y^2 - 3y = 7$

26. $(4m + 5)(m - 1) = m - 9$

Use the discriminant to determine whether the solutions of each equation are (a) two rational numbers, (b) two irrational numbers, (c) one rational number, or (d) two imaginary numbers. Do not solve the equation.

27. $3t^2 - 2t + 4 = 0$

28. $9p^2 = 24p - 16$

Use the discriminant to determine whether each trinomial can be factored using integers. Factor those trinomials that can be factored using integers.

29. $8x^2 - 14x - 9$

30. $6y^2 - 8y - 15$

Find the sum and the product of the solutions of each equation. Do not solve the equation.

31. $4z^2 + z - 6 = 0$

32. $2m^2 = 12m$

Write a quadratic equation with integer coefficients having the given solution set. Write your final answer in standard form.

33. $\left\{3, -\dfrac{4}{9}\right\}$ 34. $\left\{\dfrac{\sqrt{3}}{2}, -\dfrac{\sqrt{3}}{2}\right\}$

[6.4] Solve each equation.

35. $\dfrac{1}{p - 2} = 3p + 5$ 36. $\dfrac{5}{3m - 4} = \dfrac{m}{m + 1} - 2$

37. $\sqrt{2r + 5} = r - 1$ 38. $\sqrt{7t - 4} - \sqrt{t - 1} = 3$

39. $y^4 - 29y^2 + 100 = 0$ 40. $x^4 + 15x^2 - 16 = 0$

41. $z^{2/3} + 2z^{1/3} - 3 = 0$

42. $(k + 7)^2 - 8(k + 7) + 4 = 0$

43. An old computer takes 2 hr longer to do a job than does a new computer. Working together, the computers can do the job in 4 hr. How long does it take each computer to do the job on its own?

[6.5] Use the Pythagorean theorem to find the unknown side of the right triangles in Figs. 6.37 and 6.38.

44.   45.

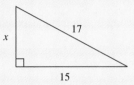   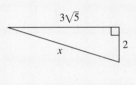

Figure 6.37   Figure 6.38

46. Two ships steam out of port, one traveling north at a rate of 20 mi/hr and the other traveling west at 21 mi/hr. How far apart are the ships after 2 hr?

Solve each equation for the specified variable.

47. $v = \sqrt{2gr}$ for g 48. $A = \pi(R^2 - r^2)h$ for r

49. $x^2 - 2xy + y^2 = 36$ for x

Solve each word problem.

50. The length of a rectangle is 7 m more than twice the width. The area of the rectangle is 319 m². Find the width and the length.

51. Kerry has a garden that is 6 ft by 4 ft. She wants to surround the garden with a brick walk of uniform width. She has enough bricks to cover 24 ft². How wide should the walk be?

52. Find the lengths of the sides of a right triangle if one leg is 1 in. less than twice the shorter leg, and the hypotenuse is 1 in. more than twice the shorter leg.

53. When Maria rides her moped to her cabin 30 mi away, it takes $\frac{1}{2}$ hr less time than when she rides her bicycle. If her moped travels 10 mi/hr faster than her bike, find the speed of each.

[6.6] Solve each inequality. Then graph the solution set.

54. $x^2 - 9 > 0$ 55. $3y^2 + y \le 10$

56. $4z^2 + 6z \ge 0$ 57. $p^2 - 8p + 8 < 0$

58. $(m + 1)(m + 4)(m - 6) \ge 0$

59. $r^3 + 4r^2 + 4r < 0$

60. $\dfrac{x - 5}{x + 2} \ge 0$ 61. $\dfrac{3y}{2y - 3} < -1$

62. $\dfrac{5}{r - 4} > r$ 63. $\dfrac{m^2 - 7m}{m - 3} \le 0$

Chapter 6 Test

1. Complete the square on $x^2 - 8x$. Then factor the resulting trinomial.

2. Use the discriminant to determine the number and type of solutions to $3p^2 = 6p - 2$.

3. Without solving, find the sum and the product of the solutions to $6r^2 - 2r + 9 = 0$.

4. Write a quadratic equation whose solution set is $\{5\}$. Write your final answer in standard form.

Use the square root property to solve.

5. $x^2 - 20 = 0$ 6. $(2y + 5)^2 = 121$

Solve by completing the square.

7. $z^2 + 6z + 3 = 0$ 8. $4m^2 + 12m = -7$

Solve using the quadratic formula.

9. $3x^2 - 4x + 1 = 0$ 10. $p^2 + p - 1 = 0$

11. $y^2 + 8y = 1$ 12. $2z^2 = 5z - 6$

Solve each equation.

13. $6 + \dfrac{37}{x} - \dfrac{60}{x^2} = 0$ 14. $\sqrt{y + 5} = y\sqrt{3}$

15. $z^4 - 13z^2 + 36 = 0$ 16. $p^4 - 2p^2 - 3 = 0$

Solve each inequality. Then graph the solution set.

17. $2x^2 - 5x > 7$

18. $(y - 3)(y + 3)(y - 5) < 0$ 19. $\dfrac{r^2 + 2}{r - 1} \ge 6$

Solve for the specified variable.

20. $x^2 - 16a^2 = 0$ for x

21. $S = \pi d\sqrt{y^2 + 1}$ for y

Solve each word problem.

22. If twice a positive number is added to its square, the result is 5. Find the number.

23. One leg of a right triangle is 1 in. longer than the other. The hypotenuse is 1 in. shorter than twice the shorter leg. Find the length of each side.

24. One combine machine can harvest a field in 2 hr less time than another. Working together, the machines can harvest the field in 4 hr. How long does it take each machine to harvest the field on its own?

25. A box without a top is to be constructed from a square piece of cardboard by cutting a 3-in. square from each corner and then turning up the sides (see Fig. 6.39). What size piece of cardboard should be used if the volume of the box is to be 300 in.^{3}?

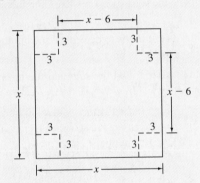

Figure 6.39

Cumulative Test for Chapters 4, 5, and 6

1. Simplify: $\dfrac{1 - \dfrac{16}{x^2}}{\dfrac{1}{x} + \dfrac{4}{x^2}}$.

2. Solve $y = \dfrac{x - 1}{x + 2}$ for x.

3. Divide and write your answer in standard form: $\dfrac{1 + 5i}{4 - 2i}$.

4. Use the discriminant to determine the number and type of solutions to $x^2 - 3x + 5 = 0$.

5. Without solving, find the sum and the product of the solutions to $3x^2 + x = 7$.

Perform the indicated operations. Write your answer in simplest form.

6. $\dfrac{y^2 + 4y - 21}{y^2 - 8y + 15} \div \dfrac{4y^2 + 28y}{y^2 - 4y - 5}$

7. $\dfrac{p}{p - 3} - \dfrac{6p}{p^2 - 9}$

8. $\dfrac{16m^3 + 20m^2 - 4m}{4m^2}$

9. $\dfrac{5x^2 + 2x^3 - 3x - 21}{x^2 - x - 3}$

Simplify. Rationalize all denominators. Assume all variables represent positive real numbers.

10. $(-27)^{2/3}$

11. $(36m^{12})^{-1/2}$

12. $5\sqrt{80x^4y^5}$

13. $\sqrt[3]{24a^8}$

14. $\sqrt{20m} + \sqrt{5m}$

15. $(2\sqrt{t} + 5)(3\sqrt{t} - 8)$

16. $\dfrac{10}{\sqrt[3]{4}}$

17. $\dfrac{1}{\sqrt{5} - 2}$

Solve each equation.

18. $\dfrac{6y}{y + 3} - \dfrac{4}{y} = 1$

19. $\sqrt{x + 7} - \sqrt{x} = 1$

Solve each quadratic equation using the indicated method.

20. $3p^2 + 10p = 8$ (factoring)

21. $2y^2 - 24 = 0$ (square root property)

22. $x^2 - 8x + 5 = 0$ (completing the square)

23. $x^2 + 12x + 25 = 0$ (quadratic formula)

Solve each inequality. Then graph the solution set.

24. $x^2 + 3x \le 10$

25. $\dfrac{x + 2}{x - 3} > 0$

Solve each word problem.

26. One leg of a right triangle is 4 in. more than twice the other leg. The hypotenuse is 6 in. more than twice the shorter leg. Find the length of each side.

27. Tony can cycle 8 mi in the same time that Anita can jog 3 mi. If Tony travels 15 mi/hr faster than Anita, find the speed of each.

28. One machine takes an hour longer to print 500 T-shirts than does another machine. If the two machines work together, they can print the T-shirts in 2 hr. How long would it take each machine to print the T-shirts on its own?

Linear Equations and Inequalities in Two Variables

The profit y of a tie shop is related to the number of ties sold, x. In Sec. 7.2 we graph the relationship between x and y. Then we use this graph to determine the number of ties that must be sold for the shop to break even.

7.1 The Rectangular Coordinate System

TAPE AU16
TAPE AU17

You have probably heard the expression "A picture is worth a thousand words." In mathematics, we often draw pictures called *graphs* to illustrate the relationship between two variable quantities. Graphs paved the way for, among other things, the development of calculus during the latter part of the 17th century. In short, graphing marked the birth of modern mathematics.

The Rectangular Coordinate System

In Chap. 2 we graphed equations and inequalities in one variable using a number line.

Graph of x = 3   *Graph of x > −2*

To graph an equation in *two* variables, such as $2x + 3y = 12$, we need *two* number lines, one for x and one for y. The two number lines are drawn perpendicular to each other as shown in Fig. 7.1. The horizontal number line is called the **x-axis,** and the vertical number line is called the **y-axis.** The point of intersection of the two axes is called the **origin.**

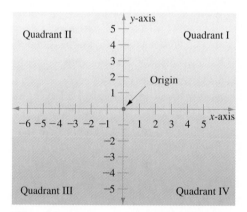

Figure 7.1

The axes divide the plane on which they are drawn into four separate regions, called **quadrants.** The quadrants are always numbered as shown in Fig. 7.1. **The axes themselves are not part of any quadrant.**

To identify a point in the plane, we must give both a value for x and a value for y. For simplicity, we write the pair of values $x = 4$ and $y = 2$ as (4, 2). The expression (4, 2) is called an **ordered pair** of numbers because it consists of two numbers in a specified order. The first number in an ordered pair is called the **x-coordinate,** or **abscissa.** The second number is called the **y-coordinate,** or **ordinate.**

(4, 2)

x-coordinate   y-coordinate
abscissa   ordinate

To **plot** the point that corresponds to the ordered pair (4, 2), start at the origin, go 4 units to the right and 2 units up (see Fig. 7.2). For simplicity, we usually say "the point (4, 2)" rather than "the point that corresponds to (4, 2)."

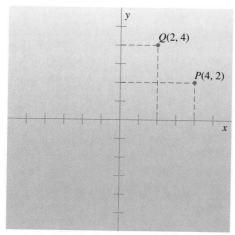

Figure 7.2

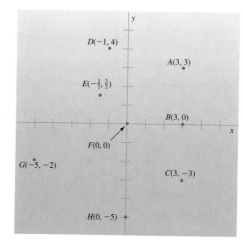

Figure 7.3

The point P is called the **graph** of the ordered pair (4, 2). Note that the point Q, whose coordinates are (2, 4), is different from point P (see Fig. 7.2).

The graphs of several other ordered pairs are shown in Fig. 7.3. Note that the point (3, 3) lies in quadrant 1, the point (3, −3) lies in quadrant IV, and the point (3, 0) lies on the positive part of the x-axis between quadrants I and IV.

This system of assigning ordered pairs of numbers to points in the plane is called a **rectangular coordinate system.***

CAUTION

The expression (1, 3) now has two entirely different meanings:

(1, 3) is the *ordered pair* whose coordinates are 1 and 3.
(1, 3) is the *open interval* of numbers from 1 to 3.

We rely on the context of the problem to tell us which meaning is intended.   ∎

The Distance Formula

We can use the *Pythagorean theorem,* discussed in Sec. 6.5, to develop a formula that will enable us to find the distance between any two points in the plane. Consider the two points $P(x_1, y_1)$ and $Q(x_2, y_2)$ shown in Fig. 7.4. To find the distance d between P and Q, construct the right triangle PQR.

Note that the coordinates of R must be (x_2, y_1). Therefore the distance between P and R is $x_2 - x_1$, and the distance between Q and R is $y_2 - y_1$. By the Pythagorean theorem,

$$d^2 = (x_2 - x_1)^2 + (y_2 - y_1)^2$$

---

*HISTORICAL NOTE: A rectangular coordinate system is also called a **Cartesian coordinate system** in honor of its inventor, René Descartes (1596–1650). Descartes was the leading mathematician and philosopher of his day, and as such was invited by Queen Christina of Sweden to tutor her. At first he declined, but when Christina sent her entire fleet of ships to get him, he was so impressed he agreed to go. However, Descartes had been in ill health for most of his life, often staying in bed until noon, and getting up early in the morning in an unheated castle to tutor her took its toll. He soon developed pneumonia and died. Descartes supposedly got the idea for his coordinate system while watching a fly crawl on the ceiling of his bedroom.

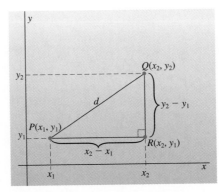

Figure 7.4

Now use the square root property to solve for d. Since d represents a positive number, take only the positive square root of the right side.

The Distance Formula

The **distance** d between the two points $P(x_1, y_1)$ and $Q(x_2, y_2)$ is given by
$$d = \sqrt{(x_2 - x_1)^2 + (y_2 - y_1)^2}$$
We shall also denote the distance d by PQ.

Although we chose points P and Q in the first quadrant in Fig. 7.4, it can be shown that the distance formula is true for *any* two points in the plane.

Since $x_2 - x_1$ and $x_1 - x_2$ are opposites, their squares are equal. Similarly, $(y_2 - y_1)^2 = (y_1 - y_2)^2$. Therefore it does not matter which point we call (x_1, y_1) and which point we call (x_2, y_2).

Example 1   Find the distance between the two points $(2, 3)$ and $(5, 7)$.

Solution: Choose $(x_1, y_1) = (2, 3)$ and $(x_2, y_2) = (5, 7)$. Then
$$\begin{aligned} d &= \sqrt{(x_2 - x_1)^2 + (y_2 - y_1)^2} \\ &= \sqrt{(5 - 2)^2 + (7 - 3)^2} \\ &= \sqrt{3^2 + 4^2} = \sqrt{9 + 16} = \sqrt{25} = 5 \end{aligned}$$
❒

Try Exercise 23

Find the distance between $(2, 1)$ and $(6, 4)$.

We can use the distance formula to determine whether three points lie on the same line. Points that lie on the same straight line are said to be **collinear.**

Example 2   Determine whether the three points $A(-6, -5)$, $B(-2, 1)$, and $C(0, 4)$ are collinear.

Solution: Find the distance between each pair of points.
$$\begin{aligned} AB &= \sqrt{(-6 - (-2))^2 + (-5 - 1)^2} = \sqrt{(-4)^2 + (-6)^2} = \sqrt{52} = 2\sqrt{13} \\ BC &= \sqrt{(-2 - 0)^2 + (1 - 4)^2} = \sqrt{(-2)^2 + (-3)^2} = \sqrt{13} \\ AC &= \sqrt{(-6 - 0)^2 + (-5 - 4)^2} = \sqrt{(-6)^2 + (-9)^2} = \sqrt{117} = 3\sqrt{13} \end{aligned}$$

Try Exercise 39

Determine whether $A(2, 1)$, $B(-4, 3)$, and $C(8, -1)$ are collinear.

Since $AB + BC = AC$, the three points are collinear (see Fig. 7.5).   ❐

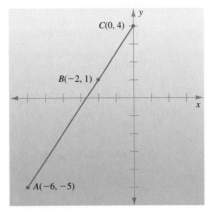

Figure 7.5

The Midpoint Formula

To find the midpoint of the line segment joining two points, average the x-coordinates and average the y-coordinates (see Fig. 7.6).

The Midpoint Formula

The **midpoint** of the line segment joining the two points $P(x_1, y_1)$ and $Q(x_2, y_2)$ is denoted $M(\bar{x}, \bar{y})$, where

$$\bar{x} = \frac{x_1 + x_2}{2} \quad \text{and} \quad \bar{y} = \frac{y_1 + y_2}{2}$$

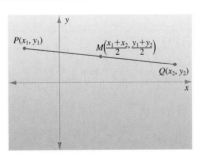

Figure 7.6

Example 3   Determine the midpoint of the line segment joining $(-5, -4)$ and $(11, -3)$.

Solution: Substituting $(x_1, y_1) = (-5, -4)$ and $(x_2, y_2) = (11, -3)$ into the midpoint formula, we have

$$\bar{x} = \frac{-5 + 11}{2} = 3 \quad \text{and} \quad \bar{y} = \frac{-4 + (-3)}{2} = -\frac{7}{2}$$

The midpoint is $M(3, -\frac{7}{2})$.   ❐

Try Exercise 45

Determine the midpoint of the line segment joining $(-7, -4)$ and $(13, -5)$.

Often the units in an applied problem are large, so we must make appropriate adjustments when we label the tick marks on the axes.

Problem Solving

Example 4 Suppose x represents a single person's taxable income, and y represents the federal tax owed on that income. Interpret each ordered pair graphed in Fig. 7.7.

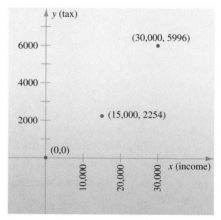

Figure 7.7

Solution:

Ordered pair	Interpretation
(0, 0)	If taxable income is \$0, tax owed is \$0.
(15,000, 2254)	If taxable income is \$15,000, tax owed is \$2254.
(30,000, 5996)	If taxable income is \$30,000, tax owed is \$5996.

Try Exercise 63

Suppose x represents taxable income and y represents the Social Security tax on that income. Interpret each ordered pair graphed in Fig. 7.9.

Exercises 7.1

Completion Exercises

1. The point of intersection of the x-axis and the y-axis is called the _____.

2. The x- and y-axes divide the plane into four separate regions, called _____.

3. The x-coordinate of the ordered pair $(3, -5)$ is also called the _____.

4. The y-coordinate of $(3, -5)$ is also called the _____.

5. The distance between the two points (x_1, y_1) and (x_2, y_2) is given by the formula $d =$ _____.

6. The midpoint $(\bar{x}, \bar{y})$ of the line segment joining the two points (x_1, y_1) and (x_2, y_2) is given by the formulas $\bar{x}$ _____ and $\bar{y} =$ _____.

True-False Exercises

7. The axes are not part of any quadrant.

8. The expressions $(x_1 - x_2)^2$ and $(x_2 - x_1)^2$ are equal.

9. Give the coordinates of each point graphed in Fig. 7.8. Then state the quadrant in which each point lies.

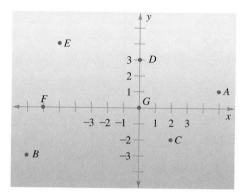

Figure 7.8

Discussion Exercise

10. Discuss the two meanings of the expression (2, 5). How do you know which meaning is intended?

Plot each point on a rectangular coordinate system.

11. (4, 3)

12. (−2, 5)

13. (−6, −1)

14. (1, −4)

15. (0, −5)

16. (−8, 0)

17. ($\frac{7}{2}$, 0)

18. (0, $\frac{9}{2}$)

19. (−5, 25)

20. (10, 35)

21. (2, −70)

22. (−3, −80)

Find the distance between each pair of points.

23. (2, 1) and (6, 4)

24. (1, 2) and (9, 8)

25. (−2, 3) and (1, 1)

26. (3, −5) and (4, 3)

27. (−3, 9) and (3, −1)

28. (−1, 3) and (2, −6)

29. (7.1, −8.3) and (6.6, −9.5)

30. (0, $\frac{1}{2}$) and (−$\frac{1}{3}$, $\frac{1}{4}$)

31. (a^2, ab) and (b^2, −ab)

32. (−ab, a^2) and (ab, b^2)

Use the distance formula to find the distance between each pair of points. Then graph the points and determine the distance using the graph. Why is it not necessary to use the distance formula for these problems?

33. (5, 2) and (8, 2)

34. (−3, 1) and (5, 1)

35. (−4, −1) and (−4, 6)

36. (−5, −9) and (−5, −2)

Use the distance formula to determine whether the given three points are collinear.

37. A(−11, −5), B(−7, −2), C(1, 4)

38. A(−13, 7), B(−1, 2), C(11, −3)

39. A(2, 1), B(−4, 3), C(8, −1)

40. A(2, 4), B(0, 0), C(3, 6)

41. A(0, 0), B(4, −9), C(−4, 6)

42. A(4, −5), B(1, −2), C(−2, 3)

Determine the midpoint of the line segment joining each pair of points.

43. (2, 3) and (6, 9)

44. (−5, 4) and (1, 10)

45. (−7, −4) and (13, −5)

46. (14, 8) and (−9, −6)

47. (0, −10) and (−3, −1)

48. (−3, −13) and (−4, 0)

49. ($\frac{2}{3}$, −$\frac{3}{4}$) and ($\frac{4}{5}$, $\frac{1}{8}$)

50. (−9.1, −2.8) and (1.7, −5.4)

Given that $a > 0$ and $b < 0$, determine the quadrant in which each point lies.

51. (a, b)

52. (b, a)

53. (b, −a)

54. (−a, −b)

55. (ab, $a − b$)

56. $\left(b - a, \dfrac{a}{b} \right)$

Problem Solving

57. An isosceles triangle is a triangle with two equal sides. Show that the triangle with vertices at A(−1, 3), B(2, 6), and C(3, 2) is an isosceles triangle.

58. An equilateral triangle is a triangle with three equal sides. Show that the triangle with vertices at A(0, $\sqrt{3}$), B(1, 0), and C(−1, 0) is an equilateral triangle.

59. Find x if the distance between (3, 2) and (x, 1) is $\sqrt{5}$.

60. Find y if the distance between (1, 4) and (5, y) is 5.

61. If M(−1, 2) is the midpoint of the line segment joining points P(6, −3) and Q, find the coordinates of Q.

62. If M($\frac{1}{2}$, $\frac{11}{2}$) is the midpoint of the line segment joining points P(4, 2) and Q, find the coordinates of Q.

63. Suppose x represents taxable income and y represents the Social Security tax on that income. Interpret each ordered pair graphed in Fig. 7.9.

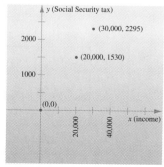

Figure 7.9

64. Suppose x represents the year and y represents the prime interest rate. Interpret each ordered pair graphed in Fig. 7.10. (The break in the x-axis means that the years 1 through 1979 have been omitted.)

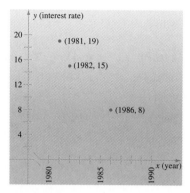

Figure 7.10

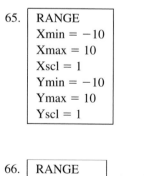 *Getting Acquainted with Your Graphing Calculator*

SETTING THE VIEWING RECTANGLE   Set the **Viewing Rectangle** for a graph so that it appears as shown in Exercises 65 through 68. The Viewing Rectangle is determined by six values: the minimum x-value, the maximum x-value, the scale on the x-axis, the minimum y-value, the maximum y-value, and the scale on the y-axis. By entering these six values into your calculator, you are setting the **RANGE** of the Viewing Rectangle. Your calculator probably has a command that automatically returns the RANGE values to a standard setting.

65. | RANGE |
     Xmin = −10
     Xmax = 10
     Xscl = 1
     Ymin = −10
     Ymax = 10
     Yscl = 1

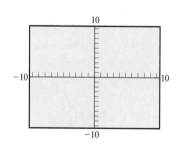

66. | RANGE |
     Xmin = −6
     Xmax = 6
     Xscl = 1
     Ymin = −5
     Ymax = 5
     Yscl = 1

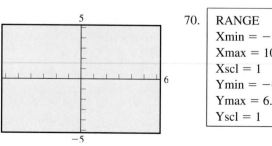

67. | RANGE |
     Xmin = 0
     Xmax = 5
     Xscl = 1
     Ymin = −3
     Ymax = 4
     Yscl = 1

68. | RANGE |
     Xmin = 0
     Xmax = 7
     Xscl = 1
     Ymin = 0
     Ymax = 2
     Yscl = 1

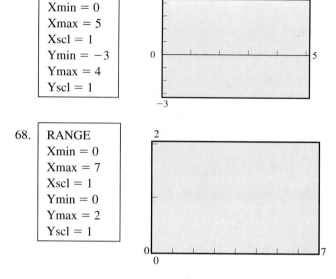

DRAWING A POINT   Set the Viewing Rectangle so that it appears as shown in Exercises 69 and 70. Then draw the points shown.

READING THE COORDINATES OF A POINT   Move the cursor to each point and check the coordinates of the point.

Note:

Your calculator screen consists of thousands of small squares, called **pixels** (short for picture elements). Therefore, your calculator will give the coordinates of each point accurate to within the width of one pixel. These coordinates may or may not be the exact coordinates of the point.

69. | RANGE |
     Xmin = −5
     Xmax = 7
     Xscl = 1
     Ymin = −3
     Ymax = 4
     Yscl = 1

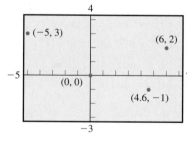

70. | RANGE |
     Xmin = −9
     Xmax = 10
     Xscl = 1
     Ymin = −6
     Ymax = 6.6
     Yscl = 1

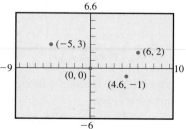

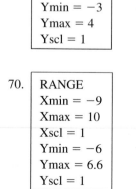

71. **CHANGING THE SCALE** Set the Viewing Rectangle as shown and draw the two points shown. Use the cursor to check the coordinates of each point.

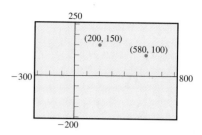

RANGE
Xmin = −300
Xmax = 800
Xscl = 100
Ymin = −200
Ymax = 250
Yscl = 50

Note:

Since the coordinates of the points shown are large, we set the scale on the *x*-axis to 100 and the scale on the *y*-axis to 50. This makes the distance between tick marks on the *x*-axis 100 units and the distance between tick marks on the *y*-axis 50 units.

TAPE AU16

7.2 Graphing Linear Equations

In this section you will learn to graph linear equations on a rectangular coordinate system.

> ### DEFINITION OF A LINEAR EQUATION
>
> A **linear equation** in two variables is an equation that can be written in the form
>
> $$ax + by = c$$
>
> where *a* and *b* are not both 0. This form is called the **standard form** of a linear equation.

Here are some examples that illustrate the definition of a linear equation:

Linear equations	*Standard form*	*Nonlinear equations*
$3x + 4y = 12$	Already in standard form	$\frac{3}{x} + \frac{4}{y} = 12$
$y = 2x - 6$	$-2x + y = -6$	$y = 2x^2 - 6$
$y = 5$	$0x + y = 5$	$\sqrt{x} + \sqrt{y} = 5$

Note that the key feature of a linear equation is that the variables are all to the first power.

Since the graph of an equation is actually a picture of its solution set, we begin by defining a solution of an equation in two variables. A **solution** of an equation in two variables is an ordered pair of numbers that satisfies the equation. For example $(1, -4)$ is a solution of $y = 2x - 6$, since the equation becomes a true statement when $x = 1$ and $y = -4$.

$$y = 2x - 6 \qquad \text{Original equation}$$
$$-4 = 2(1) - 6 \qquad \text{Let } x = 1 \text{ and } y = -4.$$
$$-4 = -4 \qquad \text{True}$$

Actually there are an *infinite* number of ordered pairs that are solutions of $y = 2x - 6$. The **graph** of $y = 2x - 6$ is the set of all points (x, y) in the plane that are solutions of the equation. Since there are an infinite number of solutions, we graph several solutions and look for a pattern.

Example 1 Graph: $y = 2x - 6$.

Solution: To find solutions, substitute any convenient value for either variable. Then solve for the other variable.

If $x = 0$, then $y = 2(0) - 6 = -6$

If $x = 5$, then $y = 2(5) - 6 = 4$

If $y = 0$, then $0 = 2x - 6$

$$6 = 2x$$
$$x = 3$$

Therefore, three solutions of $y = 2x - 6$ are $(0, -6)$, $(5, 4)$, and $(3, 0)$. You can verify that $(1, -4)$, $(2, -2)$, and $(4, 2)$ are also solutions. The easiest way to collect solutions is in a *table of values,* like the one below:

x	y
0	-6
1	-4
2	-2
3	0
4	2
5	4

If you plot the six ordered pairs from the table, you obtain the six points shown in Fig. 7.11. Note that the six points lie on the same straight line. When *all* of the solutions of $y = 2x - 6$ are graphed, they form the straight line shown in Fig. 7.12. ❑

Try Exercise 7

Complete the table of values below using the given equation. Then plot the points and draw a straight line through them.

$$y = 2x - 4$$

x	y
0	
1	
3	
	0

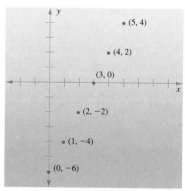

Figure 7.11

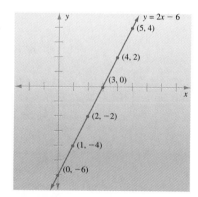

Figure 7.12

The graph of every linear equation in two variables is a straight line. Since a line is determined when any two points on the line are known, you need to find only two solutions of the equation to graph the entire line. However, it is a good idea to find a third solution as a check. If the three points do not line up, you have made a mistake.

Two points which we usually include in the graph of a line are the points where the line crosses the axes. The **x-intercept** is the x-value of the point where the line crosses the x-axis. The **y-intercept** is the y-value of the point where the line crosses the y-axis. For example, the x-intercept of the line in Fig. 7.12 is 3, and the y-intercept is -6.

To Find the Intercepts of a Line

Let $x = 0$ to find the y-intercept.

Let $y = 0$ to find the x-intercept.

Example 2   Graph: $2x + 3y = 12$. Label the x-intercept and the y-intercept.

Solution: Let $x = 0$ to find the y-intercept. Let $y = 0$ to find the x-intercept.

$$\text{If } x = 0, \text{ then} \qquad\qquad \text{If } y = 0, \text{ then}$$
$$2(0) + 3y = 12 \qquad\qquad 2x + 3(0) = 12$$
$$3y = 12 \qquad\qquad\qquad 2x = 12$$
$$y = 4 \qquad\qquad\qquad\qquad x = 6$$

Then find a check point.

$$\text{If } x = 2, \text{ then } 2(2) + 3y = 12$$
$$4 + 3y = 12$$
$$3y = 8$$
$$y = \tfrac{8}{3}$$

Collect the three ordered pairs in the table of values below:

x	y
0	4
6	0
2	$\tfrac{8}{3}$

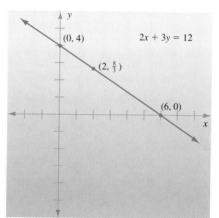

Figure 7.13

Plot the points and draw a straight line through them to produce the graph shown in Fig. 7.13.   ❑

Try Exercise 23

Graph: $5x + 7y = 35$. Label the intercepts.

Example 3   Graph: $y = \tfrac{1}{2}x$. Label the x-intercept and the y-intercept.

Solution: If $x = 0$, then $y = 0$. This gives the ordered pair $(0, 0)$. If $y = 0$, then $x = 0$. This gives the same ordered pair $(0, 0)$. This means that both intercepts occur at the same point, $(0, 0)$. Therefore we need to find *two* other points on the line. If $x = 2$, then $y = 1$. If $x = 4$, then $y = 2$. Collect the three ordered pairs in the table below:

x	y
0	0
2	1
4	2

Try Exercise 27

Graph: $y = \frac{1}{3}x$. Label the intercepts.

Then plot the points and draw a straight line through them to produce the graph shown in Fig. 7.14.   ❏

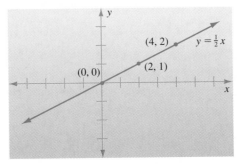

Figure 7.14

We can summarize the steps in graphing a linear equation in two variables as follows.

To Graph a Linear Equation in Two Variables

1. Let $x = 0$ to find the y-intercept. Let $y = 0$ to find the x-intercept.
2. To find a check point, substitute any convenient value for one of the variables and solve for the other variable.
3. Plot the points found in Steps 1 and 2, and draw a straight line through them.

Note:

If the intercepts coincide at $(0, 0)$, you will need to find *two* other points in Step 2.

Horizontal and Vertical Lines

If one of the two variables in a linear equation is missing, the graph is a horizontal line or a vertical line.

Example 4   Graph: $y = 4$.

Solution: We can write the equation as

$$0x + 1y = 4$$

For any x-value we choose, the corresponding y-value is 4. For example, if $x = 3$,

$$0 \cdot 3 + 1y = 4$$
$$y = 4$$

Two other solutions are $(0, 4)$ and $(-4, 4)$. Plotting these points and drawing a straight line through them produces the horizontal line shown in Fig. 7.15. Note that a horizontal line does not have an x-intercept (unless the horizontal line is the x-axis; then *every* x-value is an x-intercept).   ❏

Try Exercise 31

Graph: $y = 5$. Label the intercepts.

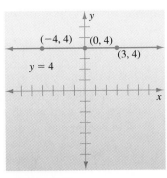

Figure 7.15

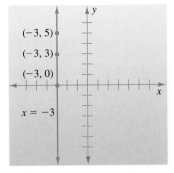

Figure 7.16

Example 5   Graph: $x = -3$.

Solution: We can write this equation as $1x + 0y = -3$. For any y-value we choose, the corresponding x-value is -3. Therefore $(-3, 0)$, $(-3, 3)$, and $(-3, 5)$ are three solutions. Plotting these points and drawing a straight line through them produces the graph shown in Fig. 7.16. Note that a vertical line does not have a y-intercept (unless the vertical line is the y-axis). ❏

Try Exercise 33

Graph: $x = -4$. Label the intercepts.

From Example 4 we see that the graph of $y = 4$ is a horizontal line with y-intercept 4. From Example 5 we see that the graph of $x = -3$ is a vertical line with x-intercept -3. These two observations suggest the following rules.

Horizontal and Vertical Lines

The graph of $y = k$ is a horizontal line with y-intercept k.
The graph of $x = k$ is a vertical line with x-intercept k.

Using these rules, we can graph a horizontal line or a vertical line without constructing a table of values.

Example 6   Graph: $2y + 5 = 0$.

Solution: Write the equation in the form $y = k$.

$$2y + 5 = 0$$
$$2y = -5$$
$$y = -\tfrac{5}{2}$$

The graph is the horizontal line with y-intercept $-\tfrac{5}{2}$ shown in Fig. 7.17. ❏

Try Exercise 39

Graph: $2y + 9 = 0$. Label the intercepts.

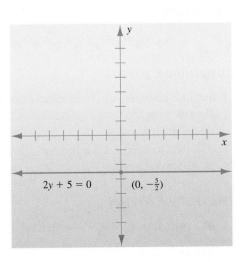

Figure 7.17

Problem Solving

Example 7   The monthly profit y of a tie shop that sells x ties per month is $5 per tie minus a monthly overhead of $600. The relationship between x and y is graphed in Fig. 7.18. Use this graph to estimate each of the following: (a) the monthly profit when no ties are sold, (b) the number of ties that must be sold each month to break even, (c) the number of ties that must be sold each month to produce a monthly profit of $1000.

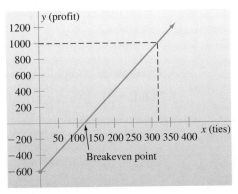

Figure 7.18

Try Exercise 49

The monthly profit y of a pipe shop that sells x pipes per month is $5 per pipe minus a monthly overhead of $400. The relationship between x and y is graphed in Fig. 7.19. Use this graph to determine (a) the monthly profit when no pipes are sold, (b) the number of pipes that must be sold each month to break even, and (c) the number of pipes that must be sold each month to produce a monthly profit of $1000.

Solution: (a) If no ties are sold, then $x = 0$. From the graph, if $x = 0$, then $y = -600$. Therefore if no ties are sold, the monthly profit is $-\$600$.

(b) The breakeven point occurs when profit y is 0. If $y = 0$, we estimate from the graph that $x = 120$. Therefore 120 ties must be sold each month to break even.

(c) From the graph, we estimate that $x = 320$ when $y = 1000$. Therefore 320 ties must be sold each month to produce a monthly profit of $1000.   ❏

Exercises 7.2

◆◇◆

Completion Exercises

1. A linear equation in two variables is an equation that can be written in the form _____, where a and b are not both 0.

2. The form $ax + by = c$ is called the _____ form of a linear equation.

True-False Exercises

3. The equation $x + y^2 = 9$ is a linear equation.

4. The equation $y = \dfrac{1}{x}$ is a linear equation.

5. The graph of every linear equation in two variables is a straight line.

6. The graph of $x = 8$ is a horizontal line.

Complete each table of values using the given equation. Then plot the points and draw a straight line through them.

7. $y = 2x - 4$

x	y
0	
1	
3	
	0

8. $y = -3x + 6$

x	y
0	
1	
3	
	0

9. $4x - 3y = 12$

x	y
0	
	0
1	

10. $3x + 5y = -15$

x	y
0	
	0
2	

11. $y = -3$

x	y
-3	
0	
4	

12. $x = 4$

x	y
	-2
	0
	4

✏ Discussion Exercises

13. What is a solution of a linear equation in two variables? How many solutions does a linear equation in two variables have?

14. What are the x- and y-intercepts of a line? Explain how to find the intercepts.

15. If the x-intercept of a line is 0, must the y-intercept be 0 as well? Explain.

16. How can you tell just by looking at a linear equation in two variables whether its graph is a horizontal line, a vertical line, or a slant line?

17. Why do we usually find three points when we graph a line, when only two points are needed to draw a line?

18. Outline the steps in graphing a linear equation in two variables.

Graph each equation. Label the x-intercept and the y-intercept.

19. $x + y = 4$

20. $x - y = 5$

21. $x - 2y = -10$

22. $3x + y = 9$

23. $5x + 7y = 35$

24. $-6x + 5y = -30$

25. $y = -3x - 6$

26. $y = -2x + 4$

27. $y = \frac{1}{3}x$

28. $y = -\frac{1}{4}x$

29. $-4x + 9y = 27$

30. $7x - 3y = 14$

31. $y = 5$

32. $x = 3$

33. $x = -4$

34. $y = -2$

35. $y = 2x$

36. $y = -3x$

37. $x - 6 = 0$

38. $y - 1 = 0$

39. $2y + 9 = 0$

40. $2x + 7 = 0$

41. $x = 0$

42. $y = 0$

Write an equation that describes each statement. Then graph the equation.

43. The y-value is 8 more than 4 times the x-value.

44. The y-value is 10 less than 5 times the x-value.

45. The sum of the x-value and twice the y-value is 6.

46. The difference of the x-value and 3 times the y-value is 9.

47. Nineteen less than 3 times the x-value is 2.

48. Thirteen more than twice the y-value is 1.

Problem Solving

49. The monthly profit y of a pipe shop that sells x pipes per month is $5 per pipe minus a monthly overhead of $400. The relationship between x and y is graphed in Fig. 7.19. Use this graph to determine (a) the monthly profit when no pipes are sold, (b) the number of pipes that must be sold each month to break even, and (c) the number of pipes that must be sold each month to produce a monthly profit of $1000.

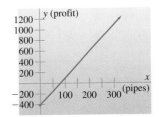

Figure 7.19

50. The daily cost y of a company that manufactures x sweatshirts each day is $4 per sweatshirt plus a daily overhead of $50. The relationship between x and y is graphed in Fig. 7.20. Use this graph to determine (a) the daily cost of manufacturing no sweatshirts, (b) the daily cost of manufacturing 25 sweatshirts per day, and (c) the number of sweatshirts that must be manufactured each day to produce a daily cost of $250.

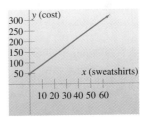

Figure 7.20

51. Suppose the demand for a certain exercise machine is 4500 when the price is $30, but only 2000 when the price is $40. Let y represent the number of exercisers demanded and let x represent the price. (a) Graph the two ordered pairs described, and draw a line through the two points. (b) Use this line to estimate the demand if the price increases to $45.

52. Suppose the demand for a certain calculator is 5000 when the price is $25, but only 1500 when the price is $35. Let y represent the number of calculators demanded and let x represent the price. (a) Graph the two ordered pairs described, and draw a line through the two points. (b) Use this line to estimate the demand if the price drops to $20.

Getting Acquainted with Your Graphing Calculator

The Viewing Rectangle should be chosen to display the important features of a graph. For example, if the graph is a straight line, both the x-intercept and the y-intercept of the line should be displayed. Here is the graph of the line $y = 2x - 6$ as seen in three different Viewing Rectangles.

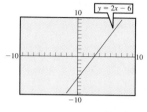

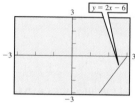

GRAPHING AN EQUATION   Set the Viewing Rectangle as described by the RANGE values given in Exercises 53 through 58. Then graph the given equation.

53. $y = x + 3$ 54. $y = -2x + 4$ 55. $y = 4$

RANGE
Xmin = −6
Xmax = 4
Xscl = 1
Ymin = −3
Ymax = 7
Yscl = 1

RANGE
Xmin = −5
Xmax = 8
Xscl = 1
Ymin = −4
Ymax = 6
Yscl = 1

RANGE
Xmin = −10
Xmax = 10
Xscl = 1
Ymin = −5
Ymax = 10
Yscl = 1

56.

RANGE
Xmin = −4
Xmax = 4
Xscl = 1
Ymin = −4
Ymax = 4
Yscl = 1

$y = \frac{1}{2}x$ (*Hint:* Write the equation as $y = (1/2)x$ or as $y = 0.5x$. The equation $y = 1/2x$ is actually $y = \dfrac{1}{2x}$.)

57.

RANGE
Xmin = −150
Xmax = 400
Xscl = 50
Ymin = −80
Ymax = 100
Yscl = 10

$y = \frac{1}{4}x - 50$ (See hint for Exercise 56.)

58.

RANGE
Xmin = −4
Xmax = 8
Xscl = 1
Ymin = −5
Ymax = 10
Yscl = 1

$3x + 2y = 12$ (*Hint:* Solve for y.)

DRAWING A LINE SEGMENT BETWEEN TWO POINTS   Set the Viewing Rectangle described by the given RANGE values. Then draw a line segment connecting the given two points.

RANGE
Xmin = −6
Xmax = 6
Xscl = 1
Ymin = −4
Ymax = 4
Yscl = 1

59. (2, 1), (4, 3)

60. (−3, 3), (5, −2)

7.3 The Slope of a Line

TAPE AU18

One characteristic of a line that we would like to measure is its "steepness." We refer to the steepness of a line as the *slope* of the line. In this section we define the slope of a line and learn how to use slope to graph a line.

Calculating the Slope of a Line

Consider the line shown in Fig. 7.21. The change in y (the *rise*) between points P and Q is $y_2 - y_1$. The change in x (the *run*) between P and Q is $x_2 - x_1$.

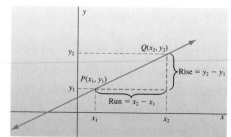

Figure 7.21

The slope of the line is the ratio of rise to run.

DEFINITION OF SLOPE

The **slope** of the nonvertical line through the two points (x_1, y_1) and (x_2, y_2) is denoted m,* and is given by

$$m = \frac{\text{change in } y \text{ (rise)}}{\text{change in } x \text{ (run)}} = \frac{y_2 - y_1}{x_2 - x_1}$$

It can be shown using similar triangles that the slope of a line is the same regardless of which two points on the line are used to calculate the slope.

Example 1   Find the slope of the line through $(-2, 1)$ and $(5, -3)$.

Solution: Let $(x_1, y_1) = (-2, 1)$ and $(x_2, y_2) = (5, -3)$. Then

$$m = \frac{y_2 - y_1}{x_2 - x_1} = \frac{-3 - 1}{5 - (-2)} = \frac{-4}{7}$$

Try Exercise 15

Find the slope of the line through $(-3, 5)$ and $(7, -2)$.

A slope of $\dfrac{-4}{7}$ means that a run of 7 units produces a rise of -4 units (see Fig. 7.22).

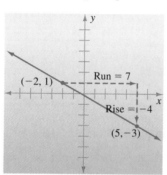

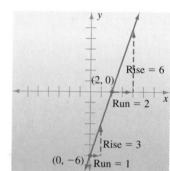

Figure 7.22   *Figure 7.23*

CAUTION

When calculating slope, you can choose either point as (x_1, y_1). However, do not start with the y-value of one point and the x-value of the other point. This gives the opposite of the correct slope.

Correct   *Correct*   *Wrong*

$$m = \frac{y_2 - y_1}{x_2 - x_1} \qquad m = \frac{y_1 - y_2}{x_1 - x_2} \qquad m \neq \frac{y_2 - y_1}{x_1 - x_2}$$

For simplicity, we often shorten the phrase "the line whose equation is $3x - y = 6$" to "the line $3x - y = 6$."

Example 2   Find the slope of the line $3x - y = 6$.

Solution: To find the slope we need two points on the line. The intercepts are easy points to calculate. If $x = 0$, then $y = -6$. If $y = 0$, then $x = 2$. Therefore two points on the line are $(0, -6)$ and $(2, 0)$.

Try Exercise 29

Find the slope of the line $4x - y = 8$.

$$m = \frac{-6 - 0}{0 - 2} = \frac{-6}{-2} = 3$$

---

*The letter m (for slope) is taken from the French word "monter," meaning "to climb."

A slope of 3, or $\frac{3}{1}$, means that a run of 1 unit produces a rise of 3 units (see Fig. 7.23). Since $\frac{3}{1} = \frac{6}{2}$, it also means a run of 2 units produces a rise of 6 units (see Fig. 7.23).

Example 3   Find the slope of the line $y = -3$.

Solution: Two points on the line (see Fig. 7.24) are $(0, -3)$ and $(4, -3)$. Therefore

$$m = \frac{-3 - (-3)}{0 - 4} = \frac{0}{-4} = 0 \qquad \square$$

Try Exercise 33

Find the slope of the line $y = -4$.

Since all points on any horizontal line have the same y-value, the difference $y_2 - y_1$ in the slope formula is 0. Therefore **the slope of any horizontal line is 0.**

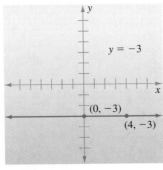

Figure 7.24   *Figure 7.25*

Example 4   Find the slope of the line $x = 4$.

Solution: Two points on the line (see Fig. 7.25) are $(4, 0)$ and $(4, 5)$. Therefore

$$m = \frac{0 - 5}{4 - 4} = \frac{-5}{0} \qquad \text{(which is undefined)} \qquad \square$$

Try Exercise 35

Find the slope of the line $x + 6 = 0$.

Since all points on any vertical line have the same x-value, the difference $x_2 - x_1$ in the slope formula will be 0. Therefore **the slope of any vertical line is undefined.** This is why we specified that the line be *nonvertical* in our definition of slope.

We summarize some of our findings about slope in Fig. 7.26.

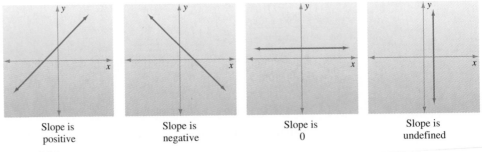

Slope is positive   Slope is negative   Slope is 0   Slope is undefined

Figure 7.26

In Sec. 7.1 we used the distance formula to determine whether three points were collinear (on the same straight line). Example 5 illustrates how to use slope to determine whether three points are collinear.

Example 5   Determine whether the three points $A(5, 6)$, $B(-1, 3)$, and $C(-3, 2)$ are collinear.

Solution: If the three points lie on the same line, the slope will be the same no matter which two points are used to calculate the slope.

$$\text{Slope of line } AB = \frac{6 - 3}{5 - (-1)} = \frac{3}{6} = \frac{1}{2}$$

$$\text{Slope of line } BC = \frac{3 - 2}{-1 - (-3)} = \frac{1}{2}$$

Since the slopes of the two lines are the same, and since the lines share the common point B, the three points A, B, and C are collinear.   ❏

Try Exercise 37

Determine whether $A(6, 5)$, $B(-2, 3)$, and $C(-6, 2)$ are collinear.

Graphing a Line Using One Point and the Slope

In Sec. 7.2 we graphed lines by finding two points on the line. We can also graph a line if we know one point on the line and the slope of the line.

Example 6   Graph the line that passes through $(0, 2)$ and has slope $\frac{3}{4}$.

Solution: First plot the point $(0, 2)$ as shown in Fig. 7.27. Then note that

$$m = \frac{3}{4} = \frac{\text{rise}}{\text{run}}$$

Therefore, from $(0, 2)$ go *up* 3 units (since the rise is 3) and *right* 4 units (since the run is 4). This brings you to the point $(4, 5)$. Draw a line through $(0, 2)$ and $(4, 5)$, as shown in Fig. 7.27.   ❏

Try Exercise 43

Graph the line through $(0, 2)$ with slope $\frac{3}{5}$.

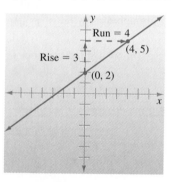

Figure 7.27   *Figure 7.28*

Example 7   Graph the line that passes through $(-1, 2)$ and has slope $-\frac{3}{5}$.

Solution: First plot the point $(-1, 2)$ as shown in Fig. 7.28. Then note that

$$m = -\frac{3}{5} = \frac{-3}{5} = \frac{\text{rise}}{\text{run}}$$

Therefore, from $(-1, 2)$ go *down* 3 units (since the rise is -3) and *right* 5 units (since the run is 5). This brings you to the point $(4, -1)$. Draw a line through $(-1, 2)$ and $(4, -1)$, as shown in Fig. 7.28.   ❏

Try Exercise 49

Graph the line through $(-2, 1)$ with slope $-\frac{4}{7}$.

To Graph a Line Using One Point and the Slope

1. Plot the known point.
2. Starting at the known point, travel *up* (if the slope is positive) or *down* (if the slope is negative) a distance given by the rise. Then travel *right* a distance given by the run. Plot a second point at this new location.
3. Draw a line through the two points plotted in Steps 1 and 2.

Exercises 7.3

Completion Exercises

1. The slope of the nonvertical line through the two points (x_1, y_1) and (x_2, y_2) is given by the formula $m = $ _____.

2. A slope of $\frac{3}{5}$ means that a run of _____ units produces a rise of _____ units.

True-False Exercises

3. The slope of a line measures the steepness of the line.

4. The slope of a line is the same regardless of which two points on the line are used to calculate the slope.

Matching Exercises

Match each description of the slope in Exercises 5 through 8 with a line it might represent in letters *A* through *D*.

5. $m > 0$

A.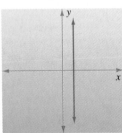

6. $m < 0$

B.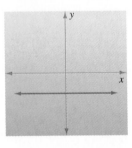

7. $m = 0$

C.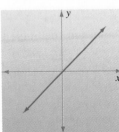

8. *m* is undefined

D.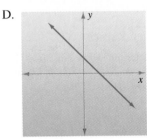

Find the slope of the line through each pair of points.

9. (2, 4) and (4, 8)

10. (2, 6) and (3, 9)

11. (−4, 2) and (−1, −16)

12. (−3, 2) and (−1, −8)

13. (−1, −3) and (7, 3)

14. (−4, −3) and (8, 7)

15. (−3, 5) and (7, −2)

16. (6, −1) and (−3, 4)

17. (5, 9) and (6, 9)

18. (4, −11) and (5, −11)

19. (−7, 1) and (−7, −6)

20. (1, −2) and (1, 3)

Find the slope of each line in Exercises 21 through 26.

21.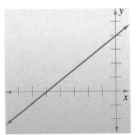

22.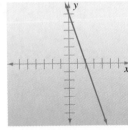

23.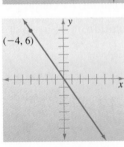

(−4, 6)

24.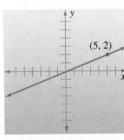

(5, 2)

25.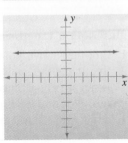

26.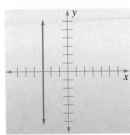

Find the slope of each line.

27. $y = -3x + 9$ 28. $y = 2x - 6$

29. $4x - y = 8$ 30. $x + 5y = 10$

31. $2x + 7y = 28$ 32. $4x - 9y = 36$

33. $y = -4$ 34. $y = 7$

35. $x + 6 = 0$ 36. $x - 1 = 0$

Use slope to determine whether the given three points are collinear.

37. $A(6, 5)$, $B(-2, 3)$, $C(-6, 2)$

38. $A(7, 1)$, $B(-5, -3)$, $C(-8, -4)$

39. $A(-4, 1)$, $B(0, 0)$, $C(7, -2)$

40. $A(-2, -4)$, $B(0, 0)$, $C(3, 8)$

41. $A(-7, -1)$, $B(-2, -4)$, $C(3, -7)$

42. $A(-6, 1)$, $B(-1, -3)$, $C(4, -7)$

Graph the line that passes through the given point and has the given slope.

43. $(0, 2)$, $m = \frac{3}{5}$ 44. $(0, 1)$, $m = -\frac{2}{3}$

45. $(0, 0)$, $m = -\frac{1}{2}$ 46. $(0, 0)$, $m = \frac{1}{4}$

47. $(0, 0)$, $m = 8$ 48. $(0, 0)$, $m = 10$

49. $(-2, 1)$, $m = -\frac{4}{7}$ 50. $(-3, -2)$, $m = -\frac{5}{8}$

51. $(5, -3)$, $m = 1$ 52. $(-4, 5)$, $m = -1$

53. $(-6, -4)$, $m = 0$ 54. $(7, 3)$, $m = 0$

55. $(3, 1)$, m undefined   56. $(2, -2)$, m undefined

✎ Discussion Exercises

57. Which of the following are correct ways to calculate the slope of the line through $(1, 3)$ and $(4, 5)$, and which are incorrect? Explain why.

    (a) $\dfrac{5 - 3}{4 - 1}$ (b) $\dfrac{5 - 3}{1 - 4}$ (c) $\dfrac{4 - 1}{5 - 3}$ (d) $\dfrac{3 - 5}{1 - 4}$

58. (a) Why is the slope of any horizontal line 0? (b) Why is the slope of any vertical line undefined?

59. (a) Why is the slope of a line that is rising (from left to right) always a positive number? (b) Why is the slope of a line that is falling (from left to right) always a negative number?

60. Outline the steps in graphing a line using one point and the slope.

Problem Solving

61. A parallelogram is a four-sided figure with opposite sides parallel. Show that the four-sided figure with vertices at $A(1, 5)$, $B(3, 8)$, $C(4, 1)$, and $D(6, 4)$ is a parallelogram.

62. A rhombus is a four-sided plane figure with opposite sides parallel and all sides equal. Show that the four-sided figure with vertices at $A(3, 2)$, $B(5, 5)$, $C(3, 8)$, and $D(1, 5)$ is a rhombus.

63. Find t if the line through $(7, t^2)$ and $(4, 2t)$ has slope 5.

64. Find t if the line through $(9, t^2)$ and $(2, 3t)$ has slope 4.

65. Determine the total area of the roof in Fig. 7.29 given that the slope of the roof is $\frac{5}{12}$.

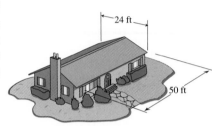

Figure 7.29

66. The grade of a road is the ratio of the change in elevation of the road to the corresponding horizontal change (see Fig. 7.30). If you drive on a road that has a constant grade of 2%, by how much will you increase your elevation if your horizontal change is 6000 ft?

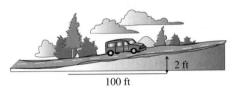

Figure 7.30   Grade $= \frac{2}{100} = 0.02 = 2\%$

📉 Getting Acquainted with Your Graphing Calculator

DRAWING A LINE USING THE CURSOR   Set the Viewing Rectangle as described and draw the given point. Then use the cursor to draw a line from the given point to another point on the line, using the given rise and run.

RANGE
Xmin $= -6$
Xmax $= 6$
Xscl $= 1$
Ymin $= -4$
Ymax $= 4$
Yscl $= 1$

67. $(1, 1)$, $m = \frac{2}{3}$
    (rise $= 2$, run $= 3$)

68. $(-3, 2)$, $m = \frac{-5}{4}$
    (rise $= -5$, run $= 4$)

7.4 Forms of a Linear Equation

TAPE AU18

In Sec. 7.2 we defined the standard form of a linear equation as $ax + by = c$. In this section we discuss two more forms of a linear equation.

Point-Slope Form

Consider the line that passes through the point (x_1, y_1) and has slope m, as shown in Fig. 7.31.

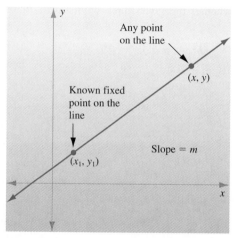

Figure 7.31

If (x, y) is any other point on the line, we can use the slope formula to write the following equation:

$$\frac{y - y_1}{x - x_1} = m$$

Multiply both sides by $x - x_1$ and simplify.

$$(x - x_1)\frac{y - y_1}{x - x_1} = m(x - x_1)$$

$$y - y_1 = m(x - x_1)$$

Since (x_1, y_1) is a known point on the line and m is the known slope of the line, this last equation is called the **point-slope form** of a linear equation.

Point-Slope Form of a Linear Equation

An equation of the line through (x_1, y_1) with slope m is

$$y - y_1 = m(x - x_1)$$

To write an equation of a line using the point-slope form, you must know one point on the line and the slope of the line.

Example 1 Find an equation of the line through $(2, 7)$ with slope 3. Then write your answer in standard form.

Solution: Since you are given one pont and the slope, use the point-slope form.

$$y - y_1 = m(x - x_1)$$ Point-slope form

$$y - 7 = 3(x - 2)$$ Substitute $y_1 = 7$, $m = 3$, and $x_1 = 2$.

$$y - 7 = 3x - 6$$ Distributive property

$$-3x + y = 1$$ Subtract $3x$, add 7.

or $$3x - y = -1$$ Multiply by -1.   ❐

Try Exercise 3

Find an equation of the line through (3, 5) with slope 2. Then write your answer in standard form.

CAUTION

Normally, writing a linear equation in the standard form $ax + by = c$ means that a, b, and c are integers with no common factor, and $a \geq 0$.

Best	*Not as good*	*Not as good*	*Not as good*
$3x - y = -1$	$-3x + y = 1$	$6x - 2y = -2$	$x - \frac{1}{3}y = -\frac{1}{3}$	■

Example 2   Find an equation of the line through $(3, -4)$ and $(8, -6)$. Then write your answer in standard form.

Solution: First, find the slope of the line.

$$m = \frac{-4 - (-6)}{3 - 8} = \frac{2}{-5} = -\frac{2}{5}$$

Then substitute the slope and either of the points into the point-slope form.

$$y - y_1 = m(x - x_1)$$ Point-slope form

$$y - (-4) = -\frac{2}{5}(x - 3)$$ Let $y_1 = -4$, $m = -\frac{2}{5}$, and $x_1 = 3$.

$$y + 4 = -\frac{2}{5}(x - 3)$$ Simplify.

$$5y + 20 = -2(x - 3)$$ Multiply by 5.

$$5y + 20 = -2x + 6$$ Distributive property

$$2x + 5y = -14$$ Add $2x$, subtract 20.

You should verify that substituting $(8, -6)$ into the point-slope form instead of $(3, -4)$ produces the same answer.   ❐

Try Exercise 17

Find an equation of the line through $(1, -6)$ and $(4, -11)$. Then write your answer in standard form.

Problem Solving

Example 3   The graph in Fig. 7.32 shows the sales of a firm in its first and fifth years of business. (a) Write a linear equation relating x and y. (b) Use this equation to predict sales in the eighth year.

Solution: (a) Two points on the line are (1, 6000) and (5, 9000). Use these points to find the slope.

$$m = \frac{9000 - 6000}{5 - 1} = \frac{3000}{4} = 750$$

Therefore the equation is

$$y - y_1 = m(x - x_1) \qquad \text{Point-slope form}$$

$$y - 6000 = 750(x - 1) \qquad \text{Let } y_1 = 6000, \, m = 750, \text{ and } x_1 = 1.$$

$$y - 6000 = 750x - 750 \qquad \text{Distribute 750.}$$

$$y = 750x + 5250 \qquad \text{Add 6000.}$$

(b) Substitute $x = 8$ into the equation from part (a).

$$y = 750(8) + 5250 = 11{,}250$$

Sales in the eighth year are projected to be $11,250.   ❒

Try Exercise 67

The graph in Fig. 7.37 shows the sales of a firm in its first and fifth years of business. (a) Write a linear equation relating x and y. (b) Use this equation to predict sales in the seventh year.

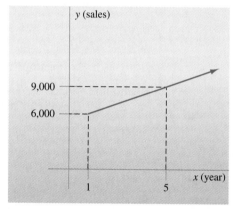

Figure 7.32

Slope-Intercept Form

Since the y-intercept of a line identifies a point on the line, we can write an equation of a line if we know the y-intercept and the slope. Consider the line with slope m and y-intercept b, as shown in Fig. 7.33.

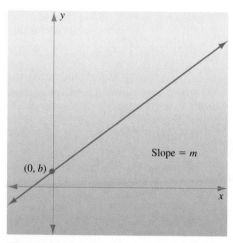

Figure 7.33

We use the point-slope form to write an equation for the line.

$$y - y_1 = m(x - x_1) \qquad \text{Point-slope form}$$

$$y - b = m(x - 0) \qquad \text{Let } y_1 = b, m = m, \text{ and}$$
$$x_1 = 0.$$

$$y - b = mx \qquad \text{Simplify.}$$

$$y = mx + b \qquad \text{Add } b.$$

This last equation is called the **slope-intercept form** of a linear equation.

Slope-Intercept Form of a Linear Equation

An equation of the line with slope m and y-intercept b is
$$y = mx + b$$

In other words, **when a linear equation in two variables is solved for y, the coefficient of x is the slope and the constant term is the y-intercept.**

Example 4 Find an equation of the line with slope 4 and y-intercept -9.

Solution:

$$y = mx + b \qquad \text{Slope-intercept form}$$
$$y = 4x + (-9) \qquad \text{Let } m = 4 \text{ and } b = -9.$$
$$y = 4x - 9 \qquad \text{Simplify.} \qquad \square$$

Try Exercise 23

Find an equation of the line with slope 6 and y-intercept -3.

Example 5 Find the slope and the y-intercept of the line $8x + 5y = 11$.

Solution: Solve for y to put the equation in slope-intercept form.

$$8x + 5y = 11 \qquad \text{Original equation}$$
$$5y = -8x + 11 \qquad \text{Subtract } 8x.$$
$$y = -\frac{8}{5}x + \frac{11}{5} \qquad \text{Divide by 5.}$$

Try Exercise 39

Find the slope and y-intercept of the line $4x - 5y = 5$.

The slope is $-\frac{8}{5}$ and the y-intercept is $\frac{11}{5}$. $\square$

Example 6 Find the slope and the y-intercept of each line: (a) $y = 6$, (b) $x = -3$.

Solution: (a) We can write the equation $y = 6$ in slope-intercept form as

$$y = 0x + 6$$

Therefore the slope is 0 (which means the line is horizontal), and the y-intercept is 6.

(b) We cannot write $x = -3$ in slope-intercept form. The graph is a vertical line, so the slope is undefined and there is no y-intercept. $\square$

Try Exercise 43

Find the slope and y-intercept of the line $y = 8$.

If we write a linear equation in slope-intercept form, we can graph the line without constructing a table of values.

Example 7 Find the slope and the y-intercept of the line $x - y = 0$. Then use this information to graph the line.

Solution: Write the equation in slope-intercept form.

$$x - y = 0 \qquad \text{Original equation}$$
$$-y = -x \qquad \text{Subtract } x.$$
$$y = x \qquad \text{Multiply by } -1.$$

Since the equation can be written as $y = 1x + 0$, the slope is 1 and the y-intercept is 0. Therefore plot the point $(0, 0)$ as shown in Fig. 7.34. Since the slope is 1, or $\frac{1}{1}$, start at $(0, 0)$ and go 1 unit up (the rise) and 1 unit right (the run). This brings you to the point $(1, 1)$. Draw a line through $(0, 0)$ and $(1, 1)$, as shown in Fig. 7.34. ❏

Try Exercise 47

Find the slope and y-intercept of the line $x - 2y = 0$. Then use this information to graph the line.

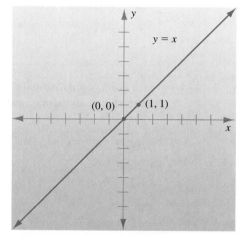

Figure 7.34

Here is a summary of the various forms of a linear equation.

Forms of a Linear Equation

Equation	Description
$ax + by = c$	Standard form (a and b not both 0): Let $x = 0$ to find the y-intercept; let $y = 0$ to find the x-intercept.
$y = k$	Horizontal line: Slope is 0, y-intercept is k.
$x = k$	Vertical line: Slope is undefined, x-intercept is k.
$y - y_1 = m(x - x_1)$	Point-slope form: Slope is m; passes through (x_1, y_1).
$y = mx + b$	Slope-intercept form: Slope is m, y-intercept is b.

Parallel and Perpendicular Lines

Given two lines in the plane, we can use their slopes to determine whether the lines are parallel, perpendicular, or neither parallel nor perpendicular.

Consider the two lines graphed in Fig. 7.35. The lines are parallel since they do not intersect no matter how far they are extended. Since parallel lines have the same "steepness," we would expect their slopes to be equal. Note that both lines have slope 2. This suggests the following rule.

Parallel Lines

Suppose line L_1 has slope m_1 and line L_2 has slope m_2.

1. If L_1 is parallel to L_2, then $m_1 = m_2$.
2. If $m_1 = m_2$, then L_1 is parallel to L_2.

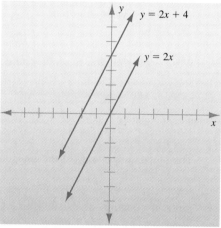

Figure 7.35

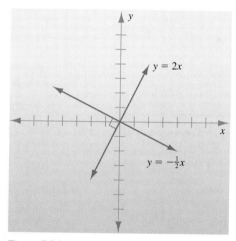

Figure 7.36

   Now consider the two lines graphed in Fig. 7.36. The lines are perpendicular since they intersect at right angles. Note that the product of their slopes is -1; that is, $2(-\frac{1}{2}) = -1$. This suggests the following rule.

Perpendicular Lines

Suppose line L_1 has slope m_1 and line L_2 has slope m_2.

1. If L_1 is perpendicular to L_2, then $m_1 \cdot m_2 = -1$.
2. If $m_1 \cdot m_2 = -1$, then L_1 is perpendicular to L_2.

Note that the equation $m_1 \cdot m_2 = -1$ can also be written as $m_1 = -\dfrac{1}{m_2}$, or as $m_2 = -\dfrac{1}{m_1}$. Therefore **two (nonvertical) lines are perpendicular if and only if their slopes are negative reciprocals of each other.**

Example 8   Determine whether the given lines are parallel, perpendicular, or neither.

$$4x - y = 7$$
$$-12x + 3y = 8$$

Solution: Write each equation in slope-intercept form to determine the slope.

$$\begin{array}{ll} 4x - y = 7 & -12x + 3y = 8 \\ -y = -4x + 7 & 3y = 12x + 8 \\ y = 4x - 7 & y = 4x + \frac{8}{3} \\ m = 4 & m = 4 \end{array}$$

The slopes are equal, so the lines are parallel.   ❏

Try Exercise 51

Determine whether the lines $4x - y = 6$ and $-8x + 2y = 7$ are parallel, perpendicular, or neither.

Example 9   Find an equation of the line through $(-2, 1)$ that is perpendicular to the line $8x - 6y = 15$. Then write your answer in standard form.

Solution: Write $8x - 6y = 15$ in slope-intercept form to find the slope.

$$\begin{array}{ll} 8x - 6y = 15 & \text{Original equation} \\ -6y = -8x + 15 & \text{Subtract } 8x. \\ y = \dfrac{-8}{-6}x + \dfrac{15}{-6} & \text{Divide by } -6. \\ y = \dfrac{4}{3}x - \dfrac{5}{2} & \text{Simplify.} \end{array}$$

The slope of the line $8x - 6y = 15$ is $\frac{4}{3}$. Therefore the slope of any line perpendicular to this line is $-\frac{3}{4}$ (the negative reciprocal of $\frac{4}{3}$). Use the point-slope form with $(x_1, y_1) = (-2, 1)$ and $m = -\frac{3}{4}$ to write the desired equation.

$$\begin{array}{ll} y - y_1 = m(x - x_1) & \text{Point-slope form} \\ y - 1 = -\dfrac{3}{4}(x - (-2)) & \text{Let } y_1 = 1, m = -\frac{3}{4}, \text{ and} \\ & \quad x_1 = -2. \\ 4y - 4 = -3(x + 2) & \text{Multiply by 4.} \\ 4y - 4 = -3x - 6 & \text{Distributive property} \\ 3x + 4y = -2 & \text{Add } 3x, \text{ add 4.} \end{array}$$ ❏

Try Exercise 59

Find an equation of the line through $(-3, 1)$ that is perpendicular to $6x - 4y = 10$. Then write your answer in standard form.

Exercises 7.4

Completion Exercises

1. An equation of the line through (x_1, y_1) with slope m is _____. This is called the _____ form of a linear equation.

2. An equation of the line with slope m and y-intercept b is _____. This is called the _____ form of a linear equation.

Find an equation of the line through the given point and with the given slope. Then write your answer in standard form.

3. $(3, 5)$, $m = 2$ 4. $(7, 4)$, $m = 3$

5. $(-1, -8)$, $m = -\frac{4}{5}$ 6. $(6, -1)$, $m = -\frac{7}{3}$

7. $(0, 0)$, $m = \frac{7}{2}$ 8. $(0, 0)$, $m = -\frac{4}{9}$

9. $(5, -10)$, $m = 0$ 10. $(-3, 9)$, $m = 0$

11. $(-4, 1)$, m undefined   12. $(8, -2)$, m undefined

Find an equation of the line through the given pair of points. Then write your answer in standard form.

13. $(8, -4)$, $(5, -7)$ 14. $(-2, 6)$, $(3, 1)$

15. $(7, 7)$, $(3, 0)$ 16. $(-7, -9)$, $(-4, -1)$

17. $(1, -6)$, $(4, -11)$ 18. $(0, 8)$, $(5, 5)$

19. $(0, 0)$, $(0, 1)$ 20. $(-5, 11)$, $(-5, 6)$

21. $(-13, 2)$, $(9, 2)$ 22. $(0, 0)$, $(12, 0)$

Find an equation of the line with the given slope and the given y-intercept.

23. $m = 6$, $b = -3$ 24. $m = -5$, $b = 7$

25. $m = -1$, $b = 15$ 26. $m = 1$, $b = -10$

27. $m = \frac{1}{4}$, $b = 0$ 28. $m = -\frac{1}{6}$, $b = 8$

29. $m = -\frac{5}{8}$, $b = -\frac{2}{5}$ 30. $m = \frac{4}{9}$, $b = -\frac{3}{4}$

31. $m = 0$, $b = 7$ 32. $m = 0$, $b = -1$

Discussion Exercises

33. When would you use the point-slope form to write an equation for a line?

34. When would you use the slope-intercept form to write an equation for a line?

35. Why can you not use the point-slope form to write an equation for a vertical line? Can you use the point-slope form to write an equation for a horizontal line?

36. Summarize the five forms of a linear equation.

Find the slope and the y-intercept of each line. Then use this information to graph the line.

37. $y = 3x + 2$ 38. $y = 2x + 1$

39. $4x - 5y = 5$ 40. $3x - 5y = 10$

41. $7x + 2y = 11$ 42. $-8x + 3y = -16$

43. $y = 8$ 44. $y = -5$

45. $x = -1$ 46. $x = 7$

47. $x - 2y = 0$ 48. $x + y = 0$

Completion Exercises

49. When a linear equation in x and y is solved for y, the coefficient of x is the _____, and the constant term is the _____.

50. Suppose m_1 and m_2 are the slopes of lines L_1 and L_2, respectively. Line L_1 is parallel to line L_2 if and only if $m_1 = $ _____. L_1 is perpendicular to L_2 if and only if $m_1 = $ _____.

Determine whether the given lines are parallel, perpendicular, or neither.

51. $4x - y = 6$
    $-8x + 2y = 7$ 52. $-2x + y = 3$
    $6x - 3y = -5$

53. $4x - 3y = 12$
    $8x + 6y = 27$ 54. $3x - 5y = 10$
    $6x + 10y = 25$

55. $y = 5x - 1$
    $2x + 10y = 15$ 56. $y = 4x + 1$
    $3x + 12y = 8$

Find an equation of the line satisfying the given conditions. Then write your answer in standard form.

57. Through $(4, -6)$ and parallel to $3x + y = 7$

58. Through $(-1, 5)$ and parallel to $4x - y = 2$

59. Through $(-3, 1)$ and perpendicular to $6x - 4y = 10$

60. Through $(-2, -8)$ and perpendicular to $9x + 6y = 14$

61. Parallel to $x = 5y$ with y-intercept -2

62. Parallel to $x = -6y$ with y-intercept 1

63. Perpendicular to $y = 9$ with x-intercept 8

64. Perpendicular to $y = 7$ with x-intercept -3

✎ Discussion Exercises

65. Discuss three different ways to find the slope of the line $3x + 4y = 24$.

66. Suppose you are given the equations of two lines, one of which is vertical. Since the slope of a vertical line is undefined, how can you tell whether the lines represented by the two equations are (a) parallel? (b) perpendicular?

Problem Solving

67. The graph in Fig. 7.37 shows the sales of a firm in its first and fifth years of business. (a) Write a linear equation relating x and y. (b) Use this equation to predict sales in the seventh year.

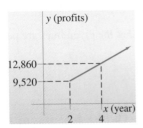

Figure 7.37

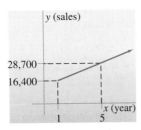

Figure 7.38

68. The graph in Fig. 7.38 shows the profits of a company in its second and fourth years of business. (a) Write a linear equation relating x and y. (b) Use this equation to predict profits in the fifth year.

69. When monthly dues at a tanning salon are $20, the salon has 450 customers. When dues are $25, the demand for memberships is only 375. (a) Write a linear equation relating x (dues) and y (demand). (b) Use this equation to predict the demand if dues are raised to $35.

70. The demand for a certain watch is 3000 when the price is $20, but only 1000 when the price is $25. (a) Write a linear equation relating x (price) and y (demand). (b) Use this equation to predict the demand if the price drops to $10.

📈 Getting Acquainted with Your Graphing Calculator

SQUARING A GRAPH   The graph of the line $y = x$ is shown in three different Viewing Rectangles below.

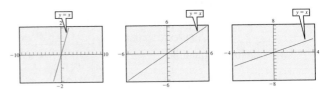

Note:

The slope of the line $y = x$ appears different in the three Viewing Rectangles. This is because the spacing between the tick marks on the x-axis and the y-axis is different in the three Viewing Rectangles. Since the ratio of screen height to screen width on most graphing calculators is $\frac{2}{3}$, you will get a true geometric picture of the graph if you use a RANGE setting like one of the three given below.

RANGE	RANGE	RANGE
Xmin = −3	Xmin = −6	Xmin = 0
Xmax = 3	Xmax = 6	Xmax = 15
Xscl = 1	Xscl = 1	Xscl = 1
Ymin = −2	Ymin = −4	Ymin = −2
Ymax = 2	Ymax = 4	Ymax = 8
Yscl = 1	Yscl = 1	Yscl = 1

Note that in each of the three RANGE settings above, Ymax − Ymin = $\frac{2}{3}$(Xmax − Xmin). For example, in the first RANGE setting, Ymax − Ymin = $2 - (-2) = 4$ and $\frac{2}{3}$(Xmax − Xmin) = $\frac{2}{3}(3 - (-3)) = \frac{2}{3}(6) = 4$. Your calculator may have a **SQUARE** setting that automatically spaces the tick marks equally on both axes.

71. Graph the perpendicular lines $y = 2x$ and $y = -\frac{1}{2}x$ in both Viewing Rectangles described below, and note that the lines *appear* perpendicular only in Viewing Rectangle (b).

(a)	RANGE	(b)	RANGE
	Xmin = −4		Xmin = −3
	Xmax = 4		Xmax = 3
	Xscl = 1		Xscl = 1
	Ymin = −4		Ymin = −2
	Ymax = 4		Ymax = 2
	Yscl = 1		Yscl = 1

72. Graph the parallel lines $y = 3x - 5$ and $y = 3x + 2$ in any Viewing Rectangle and note that the lines appear to be parallel.

73. Graph the three lines $y = 4x$, $y = x$, and $y = \frac{1}{4}x$ in the Viewing Rectangle described below and compare the slopes of the three lines.

```
RANGE
Xmin = -6
Xmax = 6
Xscl = 1
Ymin = -4
Ymax = 4
Yscl = 1
```

74. Graph the three lines $y = -4x$, $y = -x$, and $y = -\frac{1}{4}x$ in the Viewing Rectangle described above and compare the slopes of the three lines.

75. **TRACING A GRAPH** A salesperson earns $150 per week plus 35% of all sales the salesperson makes that week. If y represents weekly earnings and x represents weekly sales, then $y = 0.35x + 150$. The graph of this equation is shown in the first figure.

```
RANGE
Xmin = 0
Xmax = 1000
Xscl = 50
Ymin = -100
Ymax = 800
Yscl = 50
```

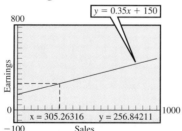

Use the trace feature to move the cursor along the graph and read the coordinates of the cursor location. The first figure shows that when weekly sales x are approximately $305, weekly earnings y are approximately $257.

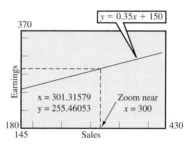

ZOOMING IN ON A POINT To determine weekly earnings when weekly sales are $300, locate the cursor as close to $x = 300$ as you can and zoom in on that point (see second figure). The RANGE values will be updated automatically to reflect the change in the Viewing Rectangle. (*Note:* The values shown in the Viewing Rectangle of your calculator may be different than the values shown in these two figures.) Continue to zoom in until you get to within 1 cent of $x = \$300$. Then read the corresponding y-value, $255.

(a) Zoom near $x = 600$ to determine weekly earnings if $600 worth of goods are sold per week.

(b) Zoom near $y = 430$ to determine weekly sales needed to produce weekly earnings of $430.

(c) **PANNING** If you continue to trace the graph to the right, the Viewing Rectangle will pan to the right so that you can continue to view the graph. The RANGE values will be updated automatically to reflect the change in the Viewing Rectangle. Pan right to determine weekly earnings if $1400 worth of goods are sold per week.

76. Use your answer to part (a) of Exercise 67 and your graphing calculator to answer part (b).

77. Use your answer to part (a) of Exercise 69 and your graphing calculator to answer part (b).

7.5 Graphing Linear Inequalities

As we have seen, the graph of a linear equation in two variables is a straight line. In this section we learn to graph linear *inequalities* in two variables.

Examples of linear inequalities are

$$y > x - 2 \qquad 3x + 5y \le 0 \qquad \text{and} \qquad y \ge 2$$

Note that a linear inequality is simply a linear equation whose equals sign has been replaced with one of the symbols $<$, $>$, $\le$, or $\ge$.

> **DEFINITION OF A LINEAR INEQUALITY IN TWO VARIABLES**
>
> A **linear inequality** in the two variables x and y is an inequality that can be written in the form
>
> $$ax + by < c$$
>
> where a and b are not both 0.

We can replace the symbol $<$ in the definition of a linear inequality with any of the symbols $>$, $\leq$, or $\geq$.

The **graph** of a linear inequality in two variables is the set of all points (x, y) in the plane that are solutions of the inequality. For example, the set of all points that are solutions of the inequality $x > 2$ is shown in Fig. 7.39.

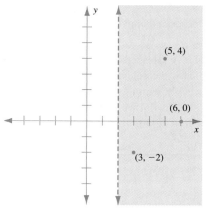

Figure 7.39

Note that every point in the shaded region of Fig. 7.39 has an x-value greater than 2.

Example 1   Graph: $y > x - 2$.

Solution: First graph the line $y = x - 2$, as shown in Fig. 7.40. Note that this line, called the *boundary line,* divides the plane into two regions, called **half-planes.** One half-plane is the graph of $y > x - 2$. The other half-plane is the graph of $y < x - 2$. The boundary line is drawn as a dashed line to indicate that the boundary line is not part of the graph of $y > x - 2$.

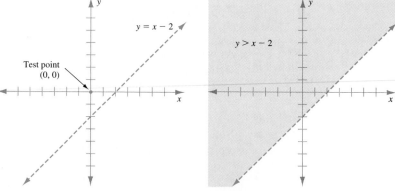

Figure 7.40   *Figure 7.41*

To determine which half-plane is the graph of $y > x - 2$, choose a *test point* not on the boundary line. We usually choose $(0, 0)$ to make the arithmetic easy, but any point not on the boundary line can be used. Substitute the test point into the original inequality.

$$y > x - 2 \qquad \text{Original inequality}$$
$$0 > 0 - 2 \qquad \text{Let } x = 0 \text{ and } y = 0.$$
$$0 > -2 \qquad \text{True}$$

Since $(0, 0)$ is a solution of $y > x - 2$, shade the half-plane that contains $(0, 0)$. The graph of $y > x - 2$ is shown in Fig. 7.41. ❏

Try Exercise 9

Graph: $y > x - 3$.

A half-plane that includes its boundary line is called a **closed half-plane**. A half-plane that does not include its boundary line is an **open half-plane**. The graph in Fig. 7.41 is an open half-plane.

Here are the steps in graphing a linear inequality in two variables.

To Graph a Linear Inequality in Two Variables

1. Replace the inequality symbol with an equals sign and graph the resulting boundary line. Draw a solid boundary line if the inequality symbol is $\leq$ or $\geq$. Draw a dashed boundary line if the symbol is $<$ or $>$.
2. Choose a test point not on the boundary line and substitute it into the original inequality. If the result is a true statement, shade the half-plane that contains the test point. If the result is a false statement, shade the half-plane that does not contain the test point.

We illustrate these steps in Example 2.

Example 2   Graph: $3x + 5y \leq 0$.

Solution: Step 1:   Replace $\leq$ with $=$ and graph the resulting boundary line. Draw a solid line since the original inequality symbol is $\leq$, not $<$ (see Fig. 7.42).

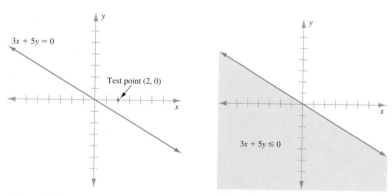

Figure 7.42   *Figure 7.43*

Step 2:   Choose a test point not on the boundary line, say $(2, 0)$, and substitute it into the original inequality.

$$3x + 5y \leq 0 \qquad \text{Original inequality}$$
$$3(2) + 5(0) \leq 0 \qquad \text{Let } x = 2 \text{ and } y = 0.$$
$$6 \leq 0 \qquad \text{False}$$

Since the result is a false statement, shade the half plane that does not contain (2, 0). The graph of $3x + 5y \leq 0$ is shown in Fig. 7.43.   ❐

Try Exercise 17

Graph: $2x + 7y \leq 0$.

The graph of the **intersection** of two linear inequalities is the set of all points in the plane that satisfy *both* inequalities.

Example 3   Graph the intersection of $3x + 2y \geq 6$ and $y \geq 2$.

Solution: First graph $3x + 2y \geq 6$, as shown in Fig. 7.44. Then graph $y \geq 2$, as shown in Fig. 7.45.

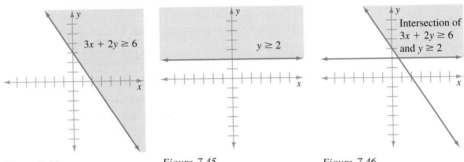

Figure 7.44   Figure 7.45   Figure 7.46

The graph of the intersection of $3x + 2y \geq 6$ and $y \geq 2$ is the region common to the half-planes shown in Figs. 7.44 and 7.45. This graph is shown in Fig. 7.46.   ❐

Try Exercise 27

Graph the intersection of
$7x + 3y \geq 21$ and $y \geq 3$.

The graph of the **union** of two linear inequalities is the set of all points in the plane that satisfy *at least one* of the inequalities.

Example 4   Graph the union of $y \leq x$ and $x < -2$.

Solution: The graph of $y \leq x$ is shown in Fig. 7.47. The graph of $x < -2$ is shown in Fig. 7.48.

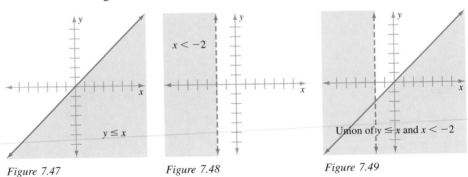

Figure 7.47   Figure 7.48   Figure 7.49

Try Exercise 33

Graph the union of $y \leq 3x$ and
$x < -1$.

The graph of the union of $y \leq x$ and $x < -2$ consists of all those points that satisfy $y \leq x$ or $x < -2$ (or both). This graph is shown in Fig. 7.49.   ❐

Absolute-Value Inequalities

Recall from Sec. 2.7 that if a is a positive number, then

$$|x| < a \text{ is equivalent to } -a < x < a$$
$$|x| > a \text{ is equivalent to } x < -a \text{ or } x > a$$

We use these rules to graph absolute-value inequalities.

Example 5 Graph: $|x| < 3$.

Solution: Since $|x| < 3$ is equivalent to $-3 < x < 3$, the graph of $|x| < 3$ consists of all those points between the vertical boundary lines $x = -3$ and $x = 3$ (see Fig. 7.50). ❏

Try Exercise 43

Graph: $|x| < 4$.

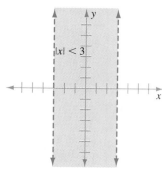

Figure 7.50

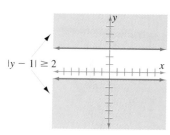

Figure 7.51

Example 6 Graph: $|y - 1| \geq 2$.

Solution: First note that $|y - 1| \geq 2$ is equivalent to

$$y - 1 \leq -2 \quad \text{or} \quad y - 1 \geq 2$$

Solve both inequalities for y.

$$y \leq -1 \quad \text{or} \quad y \geq 3$$

Therefore the graph of $|y - 1| \geq 2$ is the graph of the union of $y \leq -1$ and $y \geq 3$ (see Fig. 7.51). ❏

Try Exercise 49

Graph: $|y - 4| \geq 3$.

Exercises 7.5

True-False Exercises

1. A linear inequality is simply a linear equation whose equals sign has been replaced by an inequality symbol.

2. The graph of a linear inequality is a half-plane.

Matching Exercises

Match each linear inequality in Exercises 3 through 6 with its graph in letters A through D.

3. $x > 2$

A.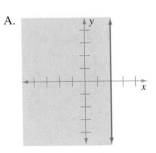

4. $x \geq 2$

B.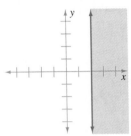

5. $x \leq 2$

C.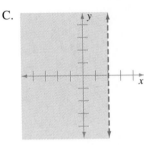

6. $x < 2$

D.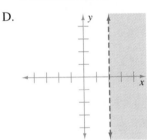

Discussion Exercises

7. What is a half-plane? Discuss the difference between a closed half-plane and an open half-plane.

8. What is meant by the graph of a linear inequality?

Graph each linear inequality.

9. $y > x - 3$

10. $y < x + 2$

11. $y < 2x + 6$

12. $y > 3x - 6$

13. $x + y \leq 4$

14. $x + y \geq -5$

15. $3x - 4y \geq 12$

16. $4x - 5y \leq 20$

17. $2x + 7y \leq 0$

18. $3x + 8y \geq 0$

19. $y < 4$

20. $y \geq 1$

Discussion Exercises

21. How do you find the boundary line when you graph a linear inequality? How do you know whether to draw a dashed boundary line or a solid one?

22. Outline the steps in graphing a linear inequality in two variables.

Completion Exercises

23. The graph of the intersection of two linear inequalities is the set of all points in the plane that satisfy _____ inequalities.

24. The graph of the union of two linear inequalities is the set of all points in the plane that satisfy _____ inequalities.

Graph the intersection of each pair of inequalities.

25. $5x - y > 10, x > 2$

26. $x + 6y < 6, x < 4$

27. $7x + 3y \geq 21, y \geq 3$

28. $3x - 5y \leq 15, y \leq 1$

29. $y > x, x + y \leq -4$

30. $y \geq -x, x - y < -3$

Graph the union of each pair of inequalities.

31. $-2x + y \leq 8, y \leq -5$

32. $-4x - y \geq 12, x \geq -2$

33. $y \leq 3x, x < -1$

34. $y \geq 2x, y > 3$

35. $x + 4y > 8, 2x - 3y < 9$

36. $x + 2y < 4, 3x - 5y > 20$

Completion Exercises

37. If $a > 0$, then $|x| < a$ is equivalent to _____.

38. If $a > 0$, then $|x| > a$ is equivalent to _____.

Graph each inequality.

39. $-2 < x \leq 5$

40. $-3 \leq x < 4$

41. $-1 \leq y < 3$

42. $-2 < y \leq 6$

43. $|x| < 4$

44. $|x| \leq 3$

45. $|x + 3| > 3$

46. $|x - 3| \geq 5$

47. $|y| \leq 2$

48. $|y| > 1$

49. $|y - 4| \geq 3$

50. $|y + 2| < 2$

For each pair of inequalities, graph (a) the intersection, (b) the union.

51. $|x| \leq 2, |y| \leq 1$

52. $|x| \geq 3, |y| \geq 2$

53. $|x - 1| < 3, |y + 2| > 5$

54. $|x + 4| > 1, |y - 3| < 2$

Problem Solving

55. A dieter wants to eat no more than 1800 calories of protein and fat combined each day. Protein contains 4 cal/g

and fat contains 9 cal/g. If the dieter eats x grams of protein and y grams of fat each day, then $4x + 9y \leq 1800$. Since neither x nor y can be negative, we can also write that $x \geq 0$ and $y \geq 0$. Graph the intersection of these three inequalities. Then choose an ordered pair from the graphed region and interpret it.

56. A shop orders x swimsuits for men at \$8 per suit and y swimsuits for women at \$15 per suit. If the shop must order at least \$600 worth of swimsuits from this supplier to keep the account open, then $8x + 15y \geq 600$. Since neither x nor y can be negative, we can also write that $x \geq 0$ and $y \geq 0$. Graph the intersection of these three inequalities. Then choose an ordered pair from the graphed region and interpret it.

 Getting Acquainted with Your Graphing Calculator

SHADING A GRAPH   Use the Viewing Rectangle described at the top of the next column to graph each inequality in Exercises 57 through 64.

RANGE
Xmin = -9
Xmax = 9
Xscl = 1
Ymin = -6
Ymax = 6
Yscl = 1

57. $y > 3$
58. $x < -1$
59. $y < x - 4$
60. $6x + 2y < 5$

Graph the intersection of each pair of inequalities.

61. $y > x - 2, y < x + 2$
62. $|x| < 3, |y| < 1$

Graph the union of each pair of inequalities.

63. $|x| < 4, |y| > 2$
64. $x + y > 3, x - 2y < -6$

 ## 7.6 Variation

TAPE AU19

Many types of applied problems can be solved using *variation*. In this section we study three basic types of variation—*direct variation, inverse variation,* and *joint variation.*

Direct Variation

The relationship between the circumference C of a circle and the diameter d is given by the equation

$$C = \pi d$$

Note that, if we double the diameter, the circumference is also doubled. If we triple the diameter, the circumference is tripled. We say that C *varies directly as d*. The constant π is called the *constant of variation.*

> ### DEFINITION OF DIRECT VARIATION
>
> The statement **y varies directly as x** means that
>
> $$y = kx$$
>
> for some real constant k, called the **constant of variation.**

If $y = kx$, we also say that **y is directly proportional to x** and we call k the **constant of proportionality.**

The equation $y = kx$ is called a *general variation equation*. If we substitute a specific value of k into the general variation equation, we obtain a *specific variation equation*.

Example 1   Suppose y varies directly as x, and $y = 30$ when $x = 5$. (a) Find the constant of variation, k. (b) Find the specific variation equation. (c) Find y when $x = \frac{3}{2}$.

Solution: (a) The statement y varies directly as x means that

$$y = kx$$

Since $y = 30$ when $x = 5$, substitute and get

$$30 = k \cdot 5$$

Solving for k gives

$$k = 6$$

(b) Substitute $k = 6$ into the general variation equation $y = kx$ to obtain the specific variation equation $y = 6x$.

(c) Substitute $x = \frac{3}{2}$ into the specific variation equation $y = 6x$.

$$y = 6\left(\tfrac{3}{2}\right) = 9 \qquad \square$$

Try Exercise 15

Suppose y varies directly as x, and $y = 20$ when $x = 5$. (a) Find the constant of variation, k. (b) Find the specific variation equation. (c) Find y when $x = \frac{5}{2}$.

Example 2   Under normal conditions the price of gold, y, is directly proportional to the price of silver, x. If gold is \$320 an ounce when silver is \$8 an ounce, find the price of gold when silver is \$11 an ounce.

Solution: Write the general variation equation.

$$y = kx$$

Substitute $y = 320$ and $x = 8$.

$$320 = k \cdot 8$$

Solve for k.

$$k = 40$$

Substitute $k = 40$ into $y = kx$ to get the specific variation equation.

$$y = 40x$$

Substitute $x = 11$ into $y = 40x$.

$$y = 40(11) = 440$$

The price of gold will be \$440 an ounce when silver is \$11 an ounce. $\square$

Try Exercise 25

Suppose the price of gold, y, is directly proportional to the price of silver, x. If gold is \$350 an ounce when silver is \$10 an ounce, find the price of gold when silver is \$9 an ounce.

Sometimes one quantity varies directly as the square of another quantity. For example, the area A of a circle is related to the radius r by the equation

$$A = \pi r^2$$

We say that A *varies directly as the square of r*. This means that if the radius is doubled, the area is multiplied by 4 (see Fig. 7.52). In other words, a pipe whose radius is twice the radius of another pipe can carry *four* times as much water as the smaller pipe.

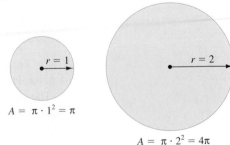

$A = \pi \cdot 1^2 = \pi$

$A = \pi \cdot 2^2 = 4\pi$

Figure 7.52

The statement *y* **varies directly as the *n*th power of *x*** means that $y = kx^n$, for some real constant *k*. Note that $y = kx$ is the special case of $y = kx^n$ that occurs when $n = 1$.

Inverse Variation

Sometimes one quantity gets *smaller* as another quantity gets *larger*. For example, the resistance *R* of a fixed length of wire is related to the diameter *d* of the wire by the equation

$$R = \frac{k}{d^2}$$

In this case *k* is determined by such factors as the material used to make the wire and the temperature of the wire. Note that *R* gets smaller as *d* gets larger. We say that *R varies inversely as the square of d,* or *R is inversely proportional to d².*

DEFINITION OF INVERSE VARIATION

The statement *y* **varies inversely as *x*** means that

$$y = \frac{k}{x}$$

for some real constant *k*.

The statement *y* **varies inversely as the *n*th power of *x*** means that

$$y = \frac{k}{x^n}$$

for some real constant *k*.

Example 3   A wire 0.06 cm in diameter has a resistance of 1 Ω. Find the resistance of a wire of the same length and material but with a diameter of 0.03 cm.

Solution: Write the general variation equation.

$$R = \frac{k}{d^2}$$

Since $R = 1$ when $d = 0.06$, we have

$$1 = \frac{k}{(0.06)^2}$$

$$1 = \frac{k}{0.0036}$$

$$k = 0.0036$$

Therefore the specific variation equation for the wire is

$$R = \frac{0.0036}{d^2}$$

When $d = 0.03$,

$$R = \frac{0.0036}{(0.03)^2} = \frac{0.0036}{0.0009} = 4$$

The resistance is 4 Ω.

Try Exercise 31

The resistance *R* of a wire varies inversely as the square of the diameter *d*. A wire 0.06 cm in diameter has a resistance of 1 Ω. Find the resistance of a wire of the same length and material but with a diameter of 0.02 cm.

Joint Variation

Sometimes one quantity varies directly as the product of two or more other quantities. For example, the maximum uniformly distributed load L that a rectangular beam (see Fig. 7.53) will support is related to the width w and the depth d of the beam by the equation

$$L = kwd^2$$

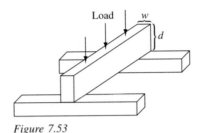

Figure 7.53

We say that L *varies jointly as w and the square of d.*

DEFINITION OF JOINT VARIATION

The statement **y varies jointly as x and z** means that

$$y = kxz$$

for some real constant k.

All of the variation problems in this section can be solved using the same five steps. These steps are summarized below.

To Solve a Variation Problem

1. Write the general variation equation.
2. Substitute the original values of the variables into the general variation equation.
3. Solve the equation resulting from Step 2 for k.
4. Substitute the value of k found in Step 3 into the general variation equation. This gives the specific variation equation.
5. Substitute the new value(s) into the specific variation equation to find the unknown value.

We illustrate these steps in Example 4.

Example 4 A beam 4 in. wide and 10 in. deep will support a load of 1200 lb. Determine the load that a beam 2 in. wide and 8 in. deep, of the same length and material, will support.

Solution: Step 1: Write the general variation equation.

$$L = kwd^2 \qquad \text{See discussion for Fig. 7.53.}$$

Step 2: Substitute the original values of the variables into the general variation equation.

$$1200 = k(4)(10)^2 \qquad \text{Let } L = 1200, w = 4, \text{ and } d = 10.$$

Step 3: Solve for k.

$$1200 = k \cdot 400 \qquad \text{Simplify the right side.}$$
$$k = 3 \qquad \text{Divide by 400.}$$

Try Exercise 33

The maximum load L that a rectangular beam will support varies jointly as the width and the square of the depth. A beam 2 in. wide and 10 in. deep will support a load of 800 lb. Determine the load that a beam 6 in. wide and 8 in. deep, of the same length and material, will support.

Step 4: Substitute $k = 3$ into the general variation equation. This gives the specific variation equation.

$$L = 3wd^2 \qquad \text{Let } k = 3 \text{ in } L = kwd^2.$$

Step 5: Substitute the new values into the specific variation equation to find the unknown value.

$$L = 3(2)(8)^2 = 384 \qquad \text{Let } w = 2 \text{ and } d = 8.$$

The beam will support 384 lb.   ❑

Exercises 7.6

Completion Exercises

1. The equation $y = kx$ is called a(n) _____ variation equation, and the equation $y = 5x$ is called a(n) _____ variation equation.

2. In a general variation equation, such as $y = kx$, the constant k is called the _____ or the _____.

Write each statement as a general variation equation.

3. p varies directly as q.

4. r is directly proportional to s^3.

5. r varies directly as s^2.

6. u varies inversely as v.

7. w is inversely proportional to the square of z.

8. w varies inversely as the cube of z.

9. y varies jointly as x and z^4.

10. y varies jointly as x^5 and z.

11. T is directly proportional to r^2 and inversely proportional to s^7.

12. T is directly proportional to r^4 and inversely proportional to s^6.

13. Newton's universal law of gravitation states that any two bodies in the universe, whether two planets, two mustard seeds, or a planet and a mustard seed, attract one another with a force F that is directly proportional to the product of their masses m_1 and m_2, and inversely proportional to the square of the distance d between them.

14. A sociologist, Joseph Cavanaugh, found that the number N of long-distance phone calls between two cities varies jointly with their populations p_1 and p_2, and inversely as the square of the distance d between them.

For each problem, (a) find the constant of variation, k; (b) find the specific variation equation; and (c) find y when $x = \frac{5}{2}$.

15. y varies directly as x, and $y = 20$ when $x = 5$.

16. y varies directly as x, and $y = 48$ when $x = 6$.

17. y varies directly as x^2, and $y = 12$ when $x = 2$.

18. y varies directly as x^2, and $y = 18$ when $x = 3$.

19. y varies inversely as x, and $y = 4$ when $x = 10$.

20. y varies inversely as x, and $y = 2$ when $x = 15$.

Problem Solving

21. Suppose y varies jointly as x and z, and $y = 18$ when $x = 6$ and $z = 2$. Find y when $x = 4$ and $z = 5$.

22. Suppose y varies jointly as x and z, and $y = 16$ when $x = 4$ and $z = 3$. Find y when $x = 6$ and $z = 5$.

23. Suppose y varies jointly as x and the square of z, and $y = 48$ when $x = \frac{2}{3}$ and $z = 6$. Find y when $x = 8$ and $z = \frac{1}{2}$.

24. Suppose y varies jointly as x and the square of z, and $y = 96$ when $x = \frac{3}{4}$ and $z = 8$. Find y when $x = 6$ and $z = \frac{1}{2}$.

25. Suppose the price of gold, y, is directly proportional to the price of silver, x. If gold is \$350 an ounce when silver is \$10 an ounce, find the price of gold when silver is \$9 an ounce.

26. According to Hooke's law,* the distance y that a spring is stretched varies directly as the weight x on the spring. If a weight of 100 kg stretches a spring 4 cm, how far will a weight of 175 kg stretch the spring?

27. Ignoring air resistance, the distance d that an object falls varies directly as the square of the time t that it falls. If an object falls 64 ft in 2 sec, how far will it fall in 4 sec?

28. The pressure p in pounds per square foot of a wind is directly proportional to the square of the velocity v of the wind. If a 10-mi/hr wind produces a pressure of 0.3 lb/ft^2, what pressure will a 100-mi/hr wind produce?

29. Suppose the time T required to complete a task is inversely proportional to the number of people, N, working on the task. If it takes 12 people 15 hr to construct the stage for a rock concert, how long will it take 10 people?

30. Boyle's law** states that the volume V of a gas varies inversely as the pressure P. If a gas has a volume of 100 in^3. when the pressure is 15 lb/in^2., what is the volume when the pressure is 10 lb/in^2?

31. The resistance R of a wire varies inversely as the square of the diameter d. A wire 0.06 cm in diameter has a resistance of 1 Ω. Find the resistance of a wire of the same length and material but with a diameter of 0.02 cm.

32. The resistance R of a wire is inversely proportional to the square of the diameter d. A wire 0.03 cm in diameter has a resistance of 2 Ω. Find the resistance of a wire of the same length and material but with a diameter of 0.01 cm.

33. The maximum load L that a rectangular beam will support varies jointly as the width and the square of the depth. A beam 2 in. wide and 10 in. deep will support a load of 800 lb. Determine the load that a beam 6 in. wide and 8 in. deep, of the same length and material, will support.

34. The crushing load L that a cylindrical pillar will support is directly proportional to the fourth power of its diameter d, and inversely proportional to the square of its height h. A column 10 ft high and 2 ft in diameter will support a load of 8 tons. Determine the load that a column 15 ft high and 3 ft in diameter will support.

35. The effectiveness of insulation is measured in R-values. If 6 in. of fiberglass insulation give an R-value of 19, how many inches of fiberglass give an R-value of 40?

36. We measure the effectiveness of insulation using R-values. If 0.5 in. of Styrofoam insulation gives an R-value of 3.5, how many inches of Styrofoam give an R-value of 20?

37. The distance d required to stop an automobile is directly proportional to the square of its speed s. How many times greater is the distance required to stop an auto traveling at 60 mi/hr than an auto traveling at 30 mi/hr?

38. The illumination I from a light source is inversely proportional to the square of the distance d from the source. How many times greater is the illumination at 10 ft from the source than at 20 ft from the source?

Discussion Exercises

39. What is the difference between a general variation equation and a specific variation equation? Give examples.

40. Discuss the difference between direct variation, inverse variation, and joint variation. Give examples from daily life.

41. Outline the steps in solving a variation problem.

Getting Acquainted with Your Graphing Calculator

42. If y varies directly as x, and $y = 20$ when $x = 15$, write the specific variation equation. Then graph the specific variation equation and use the zoom feature to find y when $x = 30$.

---

*HISTORICAL NOTE: Hooke's law is named after the English mathematician and physicist Robert Hooke (1635–1703). Hooke, a professor of geometry at Gresham College, also designed pendulums and watches, and anticipated the inverse square law for planetary motion (later proved by Newton.)
**HISTORICAL NOTE: Boyle's law is named after the English chemist Robert Boyle (1627–1691).

43. The distance y that a spring is stretched varies directly as the weight x on the spring. If a weight of 40 kg stretches a spring 16 cm, write the specific variation equation. Then graph the specific variation equation and use the zoom feature to find (a) the distance a weight of 30 kg would stretch the spring and (b) the weight needed to stretch the spring a distance of 25 cm.

Chapter 7 Key Terms

7.1 x-axis The horizontal number line in a rectangular coordinate system

y-axis The vertical number line in a rectangular coordinate system

Origin The point of intersection of the two axes in a rectangular coordinate system

Quadrants The four regions separated by the axes in a rectangular coordinate system

Ordered pair (of numbers) Two numbers written in a specified order

x-coordinate (abscissa) The first number in an ordered pair

y-coordinate (ordinate) The second number in an ordered pair

Plot (an ordered pair) To locate the point in a rectangular coordinate system that corresponds to the ordered pair

Graph (of an ordered pair) The point in a rectangular coordinate system that corresponds to the ordered pair

Rectangular (Cartesian) coordinate system A system of assigning ordered pairs of numbers to points in the plane

Collinear points Points that lie on the same straight line

7.2 Linear equation (in two variables) An equation that can be written in the form $ax + by = c$, where a and b are not both 0

Solution (of an equation in two variables) An ordered pair of numbers that satisfies the equation

Graph (of an equation in two variables) The set of all points (x, y) that are solutions of the equation

x-intercept (of a line) The x-value of the point where the line crosses the x-axis

y-intercept (of a line) The y-value of the point where the line crosses the y-axis

7.3 Slope (of a line) The ratio of rise (change in y) to run (change in x)

7.5 Linear inequality (in two variables) An inequality that can be written in the form $ax + by < c$ (or $ax + by \le c$), where a and b are not both 0

Graph (of an inequality in two variables) The set of all points (x, y) that are solutions of the inequality

Half-plane The plane region on one side of a boundary line. If the region includes the boundary line, it is a closed half-plane; if it does not include the boundary line, it is an open half-plane.

Intersection (of two inequalities) The set of all points that satisfy both inequalities

Union (of two inequalities) The set of all points that satisfy at least one of the inequalities

Chapter 7 Key Rules/Steps

7.1 The Rectangular Coordinate System

Distance Formula

The distance between (x_1, y_1) and (x_2, y_2) is

$$d = \sqrt{(x_2 - x_1)^2 + (y_2 - y_2)^2}$$

Example: The distance between $(-2, 3)$ and $(1, -1)$ is

$$d = \sqrt{(1 - (-2))^2 + (-1 - 3)^2} = \sqrt{3^2 + (-4)^2}$$
$$= \sqrt{9 + 16} = \sqrt{25} = 5$$

Midpoint Formula

The midpoint of the line segment joining (x_1, y_1) and (x_2, y_2) is $M(\bar{x}, \bar{y})$, where

$$\bar{x} = \frac{x_1 + x_2}{2} \text{ and } \bar{y} = \frac{y_1 + y_2}{2}$$

Example: For the line segment joining $(-4, 9)$ and $(-2, 3)$,

$$\bar{x} = \frac{-4 + (-2)}{2} = -3 \text{ and } \bar{y} = \frac{9 + 3}{2} = 6$$

The midpoint is $M(-3, 6)$.

7.2 Graphing Linear Equations

To Find the Intercepts of the Line $2x - 3y = 12$

Let $x = 0$ to find the y-intercept.

Let $y = 0$ to find the x-intercept.

y-intercept

$2(0) - 3y = 12$
$y = -4$

x-intercept

$2x - 3(0) = 12$
$x = 6$

To Graph the Linear Equation $2x - 3y = 12$

1. Find the y-intercept -4 and the x-intercept 6.

2. Let $x = 3$ to find a third (check) point.

$$2(3) - 3y = 12$$
$$-3y = 6$$
$$y = -2$$

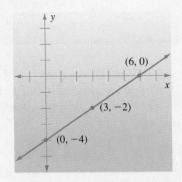

3. Plot the points $(0, -4)$, $(6, 0)$, and $(3, -2)$, and draw a straight line through them.

Horizontal and Vertical Lines

The graph of $x = k$ is a vertical line with x-intercept k.

The graph of $y = k$ is a horizontal line with y-intercept k.

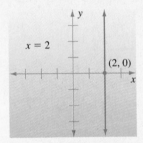

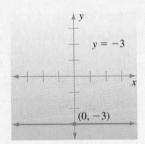

7.3 The Slope of a Line

Slope

The slope of the nonvertical line through (x_1, y_1) and (x_2, y_2) is

$$m = \frac{y_2 - y_1}{x_2 - x_1}$$

Example: The slope of the line through $(-1, 3)$ and $(4, 7)$ is

$$m = \frac{7 - 3}{4 - (-1)} = \frac{4}{5}$$

To Graph the Line through $(0, 2)$ with Slope $\dfrac{-3}{4}$

1. Plot the point $(0, 2)$.

2. Travel down 3 units and right 4 units. Plot the point $(4, -1)$.

3. Draw a line through $(0, 2)$ and $(4, -1)$.

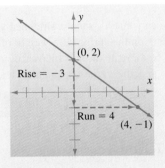

7.4 Forms of a Linear Equation

Standard Form

Assume a and b are not both 0.

$$ax + by = c$$

Example: $2x + 3y = 6$

Point-Slope Form

An equation of the line through (x_1, y_1) with slope m is

$$y - y_1 = m(x - x_1)$$

Example: An equation of the line through $(-2, 5)$ with slope 3 is

$$y - 5 = 3(x + 2)$$

Slope-Intercept Form

An equation of the line with slope m and y-intercept b is

$$y = mx + b$$

Example: An equation of the line with slope -1 and y-intercept 6 is

$$y = -x + 6$$

Parallel/Perpendicular Lines

The slopes of parallel lines are equal.

Example: $y = 4x + 1$ and $y = 4x - 5$ are parallel lines, since both have a slope of 4.

The slopes of perpendicular lines are negative reciprocals.

Example: $y = 2x - 7$ and $y = -\frac{1}{2}x + 3$ are perpendicular lines, since 2 and $-\frac{1}{2}$ are negative reciprocals.

7.5 Graphing Linear Inequalities

To Graph the Linear Inequality $x + 2y < 6$

1. Draw the (dashed) boundary line $x + 2y = 6$.

2. Test $(0, 0)$ in $x + 2y < 6$.

$$0 + 2(0) < 6 \quad \text{True}$$

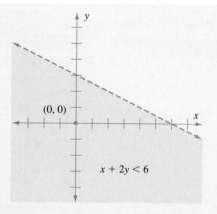

Since the result is a true statement, shade the half-plane that contains $(0, 0)$.

Absolute-Value Inequalities

Let a be a positive number.

$|x| < a$ is equivalent to $-a < x < a$.

$|x| > a$ is equivalent to $x < -a$ or $x > a$.

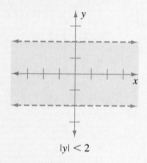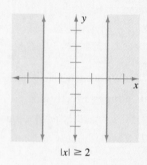

$|y| < 2$ $|x| \geq 2$

7.6 Variation

Direct Variation

$$y = kx^n$$

k is the constant of variation, or the constant of proportionality.

Example: The area A of a circle varies directly as the square of the radius r.

$$A = \pi r^2$$

Inverse Variation

$$y = \frac{k}{x^n}$$

k is the constant of variation, or the constant of proportionality.

Example: The resistance R of a wire varies inversely as the square of the diameter d.

$$R = \frac{k}{d^2}$$

Joint Variation

$$y = kxz$$

k is the constant of variation, or the constant of proportionality.

Example: The load L a beam will support varies jointly as its width w and the square of its depth d.

$$L = kwd^2$$

To Solve a Variation Problem

Suppose y varies directly as x, and $y = 24$ when $x = 6$. Find y when $x = 10$.

1. Write the general variation equation. $y = kx$
2. Substitute $y = 24$ and $x = 6$. $24 = k(6)$
3. Solve for k. $k = 4$
4. Substitute $k = 4$ into $y = kx$ to get the specific variation equation. $y = 4x$
5. Substitute $x = 10$. $y = 4(10) = 40$

Chapter 7 Review Exercises

[7.1] Graph each pair of points. Then find the distance between them.

1. $(2, 8)$ and $(7, -4)$ 2. $(-6, -1)$ and $(-4, 5)$

Determine the midpoint of the line segment joining each pair of points.

3. $(9, -6)$ and $(1, 3)$ 4. $(\frac{3}{4}, -\frac{5}{4})$ and $(-\frac{1}{2}, -\frac{7}{8})$

5. Use the distance formula to determine whether $A(2, 5)$, $B(3, 0)$, and $C(1, 10)$ are collinear.

6. If $a < 0$ and $b < 0$, in what quadrant does (ab, b) lie?

7. Find y if the distance between $(2, 5)$ and $(0, y)$ is 2.

[7.2] Determine which of the ordered pairs listed are solutions of the given equation.

8. $3x - 4y = 24$: (a) $(4, 3)$, (b) $(-4, -9)$

9. $y = -6x + 1$: (a) $(0, 1)$, (b) $(-1, 7)$

Complete each table of values using the given equation. Then plot the points and draw a straight line through them.

10. $y = \frac{2}{5}x$

x	y
0	
-5	
	2

11. $x = 8$

x	y
	-3
	0
	4

Graph each equation. Label the *x*-intercept and the *y*-intercept.

12. $3x - 5y = 15$

13. $y = -4x + 7$

14. $2y + 11 = 0$

15. $x + y = 0$

16. Write an equation that describes the statement "The sum of twice the *x*-value and 6 times the *y*-value is -10." Then graph the equation.

[7.3] Find the slope of the line through each pair of points.

17. $(3, 14)$ and $(1, 2)$

18. $(3, -2)$ and $(-6, 4)$

19. $(0, \frac{1}{2})$ and $(-\frac{1}{3}, 0)$

Find the slope of each line graphed in Exercises 20 and 21.

20.

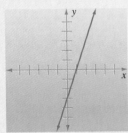

21.

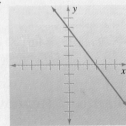

Find the slope of each line.

22. $3x + y = 6$

23. $4x - 7y = 19$

24. $y = 9$

Use the slope to determine whether the given three points are collinear.

25. $A(1, 8)$, $B(9, 11)$, $C(2, 7)$

26. $A(-1, 8)$, $B(3, 2)$, $C(7, -4)$

Graph the line that passes through the given point and has the given slope.

27. $(0, 3)$, $m = -2$

28. $(-4, -2)$, $m = \frac{5}{9}$

29. $(0, 0)$, m undefined

30. Find t if the line through $(8, t^2)$ and $(9, 6t)$ has slope 9.

[7.4] Find an equation of the line satisfying the given conditions. Then write your answer in standard form.

31. Slope 2, *y*-intercept -5

32. Through $(-4, 6)$, slope -1

33. Through $(3, -7)$ and $(-1, -4)$

34. Parallel to $8x - 6y = 9$, *y*-intercept 10

35. Perpendicular to $3x - y = -14$, *y*-intercept 0

36. Through $(5, -3)$ and $(5, 7)$

Find the slope and the *y*-intercept of each line. Then use this information to graph the line.

37. $y = 2x - 6$

38. $4x + 7y = 21$

Determine whether the given lines are parallel, perpendicular, or neither.

39. $2x + y = 7$
    $4x = 8y$

40. $6x + 4y = -11$
    $-9x - 6y = 20$

41. $y = 4 - \dfrac{x}{5}$
    $3x - 15y = 30$

[7.5] Graph each inequality.

42. $y > 2x + 1$

43. $5x + 7y \le 35$

44. $y < 3$

45. $x \ge -4$

46. $-2 \le x < 2$

47. $|y - 5| \ge 1$

For each pair of inequalities, graph (a) the intersection, (b) the union.

48. $x - 4y > 8$
    $y < -\frac{3}{4}x$

49. $|x + 1| > 1$
    $|y| \le 2$

[7.6] Write each statement as a general variation equation.

50. *s* varies directly as *t*.

51. *A* is inversely proportional to the cube of *r*.

52. *T* varies jointly as *x* and *y* and inversely as z^2.

Solve each word problem.

53. Suppose *y* varies directly as x^2, and $y = 24$ when $x = 4$. Find *y* when $x = 3$.

54. Suppose *z* is directly proportional to *x* and inversely proportional to *y*, and $z = 10$ when $x = 4$ and $y = 2$. Find *z* when $x = 8$ and $y = 10$.

55. The volume *V* of a gas is inversely proportional to the pressure *P*. If a gas has a volume of 8 m³ when the pressure is 30 g/m², what is the volume when the pressure is 80 g/m²?

56. The maximum load *L* that a rectangular beam will support varies jointly as the width *w* and the square of the depth *d*. A beam 4 in. wide and 5 in. deep will support a load of 600 lb. Determine the load that a beam 3 in. wide and 12 in. deep, of the same length and material, will support.

Chapter 7 Test

1. Find the slope of the line through $(3, -4)$ and $(9, -1)$.

2. Determine the midpoint of the line segment joining

$(-11, 6)$ and $(3, 7)$.

3. Find the distance between $(-7, -2)$ and $(-4, 3)$.

4. Determine whether $A(-4, 6)$, $B(-1, -1)$, and $C(3, -8)$ are collinear.

5. Determine whether the lines represented by $y = 3x - 4$ and $3x + 9y = -24$ are parallel, perpendicular, or neither.

6. Graph the line through $(1, 5)$ with slope $-\frac{1}{4}$.

7. Find the slope and y-intercept of the line $3x - 2y = 12$. Then use this information to graph the line.

Find an equation of the line satisfying the given conditions. Then write your answer in standard form.

8. Through $(3, 4)$ and $(-1, -2)$

9. Through $(-6, 8)$ and $(-5, 8)$

10. Through $(-7, -11)$, slope undefined

11. Parallel to $4x + 9y = 18$, y-intercept -3

Graph each equation. Label the x-intercept and the y-intercept.

12. $4x + 3y = 20$

13. $5y - 8x = 0$

14. $y = -7$

15. $3x - 9 = 0$

Graph each inequality.

16. $5x - 2y > 10$

17. $y \leq -x - 4$

Consider the inequalities $x \leq -3$ and $|y| < 1$.

18. Graph the intersection.

19. Graph the union.

Solve each word problem.

20. Suppose y varies directly as x, and $y = 12$ when $x = 8$. Find y when $x = 6$.

21. Suppose z varies jointly as x and y^3, and $z = 32$ when $x = 5$ and $y = 2$. Find z when $x = 15$ and $y = 3$.

22. The resistance R of a wire is inversely proportional to the square of the diameter d. A wire 0.05 cm in diameter has a resistance of 8 Ω. Find the resistance of a wire of the same length and material but with a diameter of 0.02 cm.

Relations and Functions

The maximum pulse rate that a healthy person should attain while exercising is a function of the person's age. In this chapter we study various types of relations and functions and their applications.

8.1 Defining Relations and Functions

TAPE AU19

We begin this section by defining a relation. Then we define a function as a special kind of relation. Functions occupy a central position in mathematics, so it is important that you know how to recognize them, how to write them, and how to use them.

Relations

Suppose you earn $6 an hour. Then your earnings are related to the number of hours that you work. One way to express this relationship is to write a set of ordered pairs. For example, the ordered pair (3, 18) represents the fact that if you work 3 hr you will earn $18. The ordered pair (8, 48) represents the fact that if you work 8 hr you will earn $48. In this way, a set of ordered pairs defines a *relation* between two variable quantities—the number of hours you work and your earnings.

> **DEFINITION OF A RELATION**
>
> A **relation** is a set of ordered pairs. The set of all first coordinates is called the **domain** of the relation. The set of all second coordinates is the **range.**

The set of ordered pairs {(2, 6), (−4, 9), (6, 0)} is an example of a relation. The domain of this relation is the set {2, −4, 6}. The range is {6, 9, 0}. One way to picture this relation is shown in Fig. 8.1.

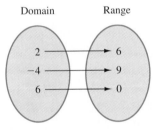

Figure 8.1

Note that the relation assigns to each value in the domain a value in the range.

Functions

A special kind of relation called a *function* is useful in solving applied problems.

> **DEFINITION OF A FUNCTION**
>
> A **function** is a relation in which each first coordinate corresponds to *exactly one* second coordinate.

Example 1 State the domain and the range of each relation. Determine whether the relation is a function.
(a) {(1, 4), (−3, 2), (0, 7), (8, 4)}   (b) {(5, 6), (1, 3), (5, −9)}

Solution: (a) The domain is {1, −3, 0, 8}, and the range is {4, 2, 7}. Since each first coordinate corresponds to exactly one second coordinate, the relation is a function (see Fig. 8.2). The fact that two first coordinates, namely, 1 and 8, correspond to the same second coordinate 4 does not violate the definition of a function.

(b) The domain is {5, 1}, and the range is {6, 3, −9}. Since the first coordinate 5 corresponds to two different second coordinates, namely, 6 and −9, the relation is *not* a function (see Fig. 8.3). ❑

Try Exercise 5

State the domain and range of the relation {(2, 5), (−6, 4), (0, 3), (9, 5)}. Is this relation a function?

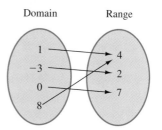

Figure 8.2

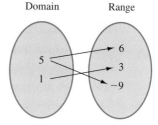

Figure 8.3

Since a function is a special kind of relation, **a function is always a relation, but a relation is not necessarily a function.**

Many functions have too many ordered pairs to list. In those cases we define the function by writing a rule, usually in the form of an equation. The rule tells us how to get the second coordinate of the ordered pair when we are given the first coordinate. For example, if you earn $6 an hour, then your daily earnings y are related to the number of hours x that you work each day by the rule

$$y = 6x$$

Therefore the function is the set of ordered pairs $\{(x, y)|y = 6x\}$, but it is common to refer to the function as simply the equation $y = 6x$.

Since your earnings y depend upon the number of hours you work, x, we call y the **dependent variable.** On the other hand, the values of x may be chosen at will, so we call x the **independent variable.**

$$y = 6x$$
↑   ↑
Dependent   Independent
variable   variable

Since the equation $y = 6x$ is a function, we say that *y is a function of x.* In other words, your earnings are a function of the number of hours you work.

When a function is written as an equation, the domain is the set of all possible values of the independent variable, and the range is the set of all possible values of the dependent variable. Since you cannot work a negative number of hours, nor can you work more than 24 hr in a day, the domain of the function $y = 6x$ is $\{x|0 \le x \le 24\}$. If you work 0 hr, your daily earnings are $0, and if you work 24 hr, your daily earnings are $144. Therefore the range of the function $y = 6x$ is $\{y|0 \le y \le 144\}$.

Example 2   For each rule, determine whether y is a function of x.

(a) $y = x + 1$ (b) $x^2 + y^2 = 25$

Solution: (a) If you substitute any value for x and add 1, you get exactly one value for y. Therefore, $y = x + 1$ is a function.
(b) Substitute a value for x, say $x = 4$, into $x^2 + y^2 = 25$.

$$4^2 + y^2 = 25 \qquad \text{Let } x = 4.$$
$$16 + y^2 = 25 \qquad \text{Simplify.}$$
$$y^2 = 9 \qquad \text{Subtract 16.}$$
$$y = \pm 3 \qquad \text{Square root property}$$

Since there are two y-values, namely, 3 and -3, that correspond to the x-value 4, the equation $x^2 + y^2 = 25$ is not a function.   ❒

Try Exercise 19

If $y = x + 3$, determine whether y is a function of x.

We can determine whether a graph is the graph of a function by applying the **vertical line test.**

Vertical Line Test

A graph in the plane represents a function provided that no vertical line intersects the graph at more than one point.

The vertical line test is illustrated in Figs. 8.4 and 8.5. Since no vertical line intersects the graph in Fig. 8.4 at more than one point, this graph represents a function. In Fig. 8.5, the vertical line $x = 4$ intersects the graph at two points, namely, $(4, 3)$ and $(4, -3)$. This means that the x-value 4 corresponds to two y-values, namely, 3 and -3. Therefore this graph does not represent a function.

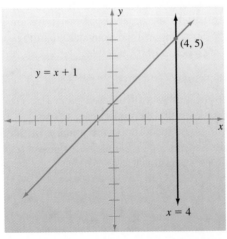

Figure 8.4

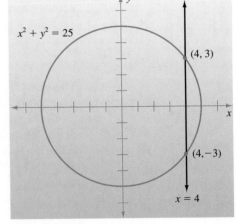

Figure 8.5

If y is a function of x, and there is no information to the contrary, we shall make the following agreement: **The domain of the function is the set of all those real values of x that produce a real value for y.**

Example 3   Find the domain of each function.

(a) $y = 3x - 8$ (b) $y = \dfrac{x}{x^2 - 6x + 5}$ (c) $y = \sqrt{2x + 7}$

Solution: (a) If you substitute any real value for x, multiply by 3, and subtract 8, the result is a real value for y. Therefore the domain is the set of all real numbers. Using interval notation, we can write the domain as $(-\infty, \infty)$.

(b) Find those values that make the denominator 0.

$$x^2 - 6x + 5 = 0 \qquad \text{Set the denominator equal to 0.}$$

$$(x - 5)(x - 1) = 0 \qquad \text{Factor.}$$

$$x = 5 \quad \text{or} \quad x = 1 \qquad \text{Set each factor equal to 0 and solve.}$$

The domain consists of all real numbers except 5 and 1, written $\{x \mid x \neq 5, 1\}$.

(c) The radicand must be nonnegative to produce a real value for y.

$$2x + 7 \geq 0 \qquad \text{Radicand is positive or 0.}$$

$$2x \geq -7 \qquad \text{Subtract 7.}$$

$$x \geq -\frac{7}{2} \qquad \text{Divide by 2.}$$

The domain is $\left[-\frac{7}{2}, \infty\right)$. ❏

Try Exercise 45

Find the domain of the function $y = \sqrt{2x + 9}$.

Example 4 Find the range of each function.
(a) $y = 4x$ (b) $y = x^2$

Solution: (a) Since y can be any real number, the range is $(-\infty, \infty)$.
(b) Since y is the square of a real number, y must be nonnegative. Therefore the range is $[0, \infty)$. ❏

Try Exercise 51

Find the range of the function $y = 6x$.

Finding the range of a relation from its equation is usually more difficult than finding the domain. However, we can determine both the domain and the range of a relation from its graph.

Example 5 Find the domain and the range of the relation graphed in Fig. 8.6. Determine whether the relation is a function.

Solution: The graph in Fig. 8.6 consists of points whose x-values go from -4 to 4 and whose y-values go from -3 to 3. Therefore the domain is $[-4, 4]$, and the range is $[-3, 3]$. The graph fails the vertical line test, so the relation is not a function. ❏

Try Exercise 57

Find the domain and range of the relation graphed below. Is the relation a function?

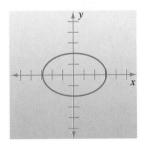

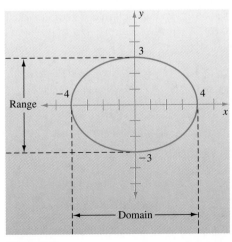

Figure 8.6

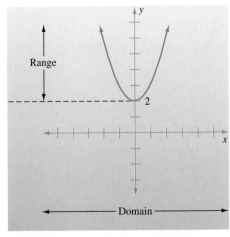

Figure 8.7

Try Exercise 59

Find the domain and range of the relation graphed below. Is the relation a function?

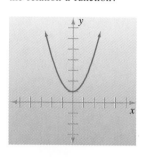

Example 6   Find the domain and the range of the relation graphed in Fig. 8.7. Determine whether the relation is a function.

Solution: The graph in Fig. 8.7 consists of points whose x-values go from $-\infty$ to ∞ and whose y-values go from 2 to ∞. Therefore the domain is $(-\infty, \infty)$, and the range is $[2, \infty)$. The graph passes the vertical line test, so the relation is a function.   ◻

Exercises 8.1

Completion Exercises

1. A set of ordered pairs is called a(n) _____.

2. The set of all first coordinates of a relation is called the _____, and the set of all second coordinates is called the _____.

3. A relation in which each first coordinate corresponds to exactly one second coordinate is called a(n) _____.

4. In the equation $y = x^2$, y is called the _____ variable, and x is called the _____ variable.

State the domain and the range of each relation. Determine whether the relation is a function.

5. $\{(2, 5), (-6, 4), (0, 3), (9, 5)\}$

6. $\{(8, 1), (-5, 7), (10, 1), (0, 2)\}$

7. $\{(3, 9), (1, 7), (3, -8)\}$ 8. $\{(6, 9), (4, 8), (4, -3)\}$

9. $\{(1, 1), (2, 2), (3, 3)\}$ 10. $\{(1, 2), (2, 3), (3, 4)\}$

11. $\{(-5, 0), (-6, 0), (-7, 0)\}$

12. $\{(8, 11), (9, 11), (10, 11)\}$

13.

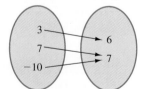

Domain   Range

14.

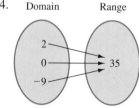

Domain   Range

15.

Domain   Range

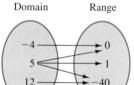

16.

Domain   Range

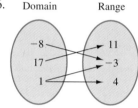

Discussion Exercises

17. Is every function a relation? Is every relation a function? Explain and give examples.

18. State the vertical line test and explain why it works. Use examples to illustrate your point.

For each rule, determine whether y is a function of x.

19. $y = x + 3$ 20. $y = 2x - 10$

21. $y = \sqrt{5x}$ 22. $y = \sqrt{x + 1}$

23. $y = x^2$ 24. $y = |x|$

25. $y = \dfrac{1}{x - 7}$ 26. $y = \dfrac{1}{x + 4}$

27. $x^2 + y^2 = 9$ 28. $x^2 + y^2 = 16$

29. $y < 3x - 1$ 30. $y > x + 8$

31. $y^2 = x + 2$ 32. $y^2 = 4x$

Matching Exercises

Match each function in Exercises 33 through 36 with its domain in letters A through D.

33. $y = x - 2$ A. $\{x \mid x \neq 2\}$

34. $y = \sqrt{x - 2}$ B. $\{x \mid x \neq 2, -2\}$

35. $y = \dfrac{1}{x - 2}$ C. $[2, \infty)$

36. $y = \dfrac{1}{x^2 - 4}$ D. $(-\infty, \infty)$

Find the domain of each function.

37. $y = 4x^2 + 1$ 38. $y = 6x + 8$

39. $y = \dfrac{x + 6}{x^2 - 9}$ 40. $y = \dfrac{9}{x + 2}$

41. $y = \dfrac{-5x}{x^2 + 49}$ 42. $y = \dfrac{3x}{x^2 + 1}$

43. $y = \dfrac{x}{x^2 - 7x + 12}$ 44. $y = \dfrac{-x}{x^2 + 9x}$

45. $y = \sqrt{2x + 9}$ 46. $y = \sqrt{3x - 10}$

47. $y = \sqrt[3]{x - 3}$ 48. $y = \sqrt[3]{x + 5}$

49. $y = \dfrac{x - 4}{\sqrt{15 - 3x}}$ 50. $y = \dfrac{x + 6}{\sqrt{24 + 4x}}$

Find the range of each function.

51. $y = 6x$ 52. $y = 3x + 7$

53. $y = (x + 2)^2$ 54. $y = x^2 + 2$

55. $y = |x| - 5$ 56. $y = |x|$

Find the domain and the range of each relation graphed in Exercises 57 through 68. Determine whether the relation is a function.

57.
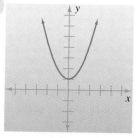

58.
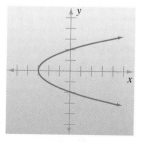

59.
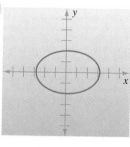

60.
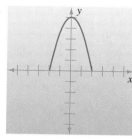

61.
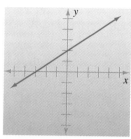

62.
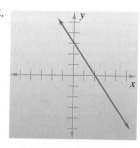

63.
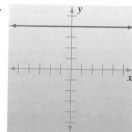

64.
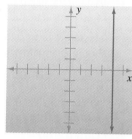

65.
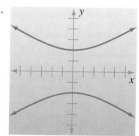

66.
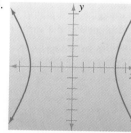

67.
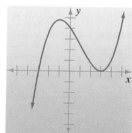

68.
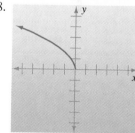

Discussion Exercises

69. Can the graph of a function have more than one y-intercept? Why or why not?

70. Can the graph of a function have more than one x-intercept? Why or why not?

Problem Solving

71. Write an equation that gives the area A of a square as a function of its side x. Then state the domain and range.

72. Write an equation that gives the perimeter P of a square as a function of its side x. Then state the domain and range.

73. You earn $4.75 an hour. Write your daily earnings y as a function of the number of hours x that you work each day. Then state the domain and range.

74. The sales tax rate in Ohio is 5 percent. Write the sales tax y that you would pay on an item as a function of the purchase price x of the item. Then state the domain and range.

75. The maximum pulse rate R that a healthy person aged 18 to 55 should attain while exercising is determined by subtracting the person's age a from 220. Write R as a function of a. Then state the domain and range.

76. The target heart rate T that a healthy person aged 18 to 55 should maintain while exercising is determined by subtracting 80 percent of the person's age a from 176. Write T as a function of a. Then state the domain and range.

 Getting Acquainted with Your Graphing Calculator

Graph each function on your calculator and note that each graph passes the vertical line test. Then determine the domain and the range from the graph. Use the Viewing Rectangle described in the next column.

RANGE
Xmin = −9
Xmax = 9
Xscl = 1
Ymin = −6
Ymax = 6
Yscl = 1

77. $y = 3$

78. $y = 2x + 1$

79. $y = x^3 - 4x$

80. $y = x^2 - 3$

81. $y = |x| - 2$

82. $y = \dfrac{6}{x}$

83. $y = \sqrt{x + 4}$ [*Hint:* You may need to write $y = \sqrt{(x + 4)}$. Otherwise, your calculator will graph $y = \sqrt{x} + 4$.]

84. $y = \sqrt{10 - 2x}$ [*Hint:* You may need to write $y = \sqrt{(10 - 2x)}$. Otherwise, your calculator will graph $y = \sqrt{10} - 2x$.]

85. $y = \sqrt{9 - x^2}$ [*Hint:* Write $y = \sqrt{(9 - x^2)}$.]

86. $y = \sqrt[3]{x}$ (*Hint:* $\sqrt[3]{x} = x^{1/3}$.)

 ## 8.2 Function Notation and Operations on Functions

In this section we introduce a notation used to write functions. We also learn how to add, subtract, multiply, and divide functions, and we learn an operation called *composition of functions*.

Function Notation

To denote that y is a function of x, we write

$$y = f(x)$$

The symbol $f(x)$ is read "f of x." It does *not* mean f times x.

Since y and $f(x)$ are equal, they can be used interchangeably. This means that we can use either of the notations below to describe the function whose second coordinate is the square of the first coordinate:

$$y = x^2 \qquad \text{or} \qquad f(x) = x^2$$

The **function notation** $f(x) = x^2$ has certain advantages over the notation $y = x^2$. For example, using $y = x^2$ we must write

$$\text{if } x = 3, \text{ then } y = 3^2 = 9$$

Using $f(x) = x^2$ we simply write

$$f(3) = 3^2 = 9$$

The equation $f(3) = 9$ is read "f of 3 equals 9." Note that $f(3)$ is the range value that corresponds to the domain value 3 (see Fig. 8.8). As such, $f(3)$ is called the **value of the function f at 3.**

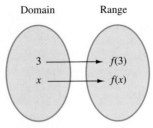

Figure 8.8

We can think of the function $f(x) = x^2$ as a "machine" that squares each number it is given. The inputs to the machine constitute the domain, and the outputs the range (see Fig. 8.9).

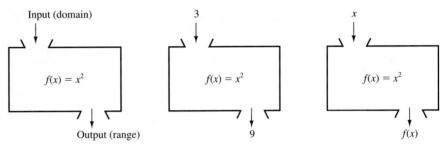

Figure 8.9

Function Notation*

The equation $y = f(x)$ is read "*y* equals *f* of *x*."

1. f is the name of the function.
2. x is the domain value.
3. $f(x)$ is the range value (or the value of the function) at the domain value x.

Example 1   Suppose $f(x) = 3x^2 - 5x + 8$. Determine each of the following:
(a) $f(4)$, (b) $f(\frac{1}{3})$, (c) $f(-a)$, (d) $f(x + h)$.

Solution: (a) $f(x) = 3x^2 - 5x + 8$ Original function

$\quad\quad\quad f(4) = 3(4)^2 - 5(4) + 8$ Replace x with 4.

$\quad\quad\quad\quad\quad = 48 - 20 + 8$

$\quad\quad\quad\quad\quad = 36$ Simplify.

$\quad$ (b) $f(\tfrac{1}{3}) = 3(\tfrac{1}{3})^2 - 5(\tfrac{1}{3}) + 8$ Replace x with $\tfrac{1}{3}$.

$\quad\quad\quad\quad = \tfrac{1}{3} - \tfrac{5}{3} + \tfrac{24}{3}$

$\quad\quad\quad\quad = \tfrac{20}{3}$ Simplify.

$\quad\quad f(-a) = 3(-a)^2 - 5(-a) + 8$ Replace x with $-a$.

$\quad\quad\quad\quad = 3a^2 + 5a + 8$ Simplify.

---

*HISTORICAL NOTE: The Swiss mathematician Leonhard Euler (1707–1783) was the first to formulate a definition of the term "function" that is somewhat similar to the definition used today. He was also the first to use the function notation $f(x)$. Euler wrote more on the subject of mathematics than any other writer in history. He continued writing even after he became totally blind in 1768. Euler's definition of a function was later modified by the German mathematician Lejeune Dirichlet (1805–1859), and it is Dirichlet's definition that is essentially in use today.

(d) $f(x + h) = 3(x + h)^2 - 5(x + h) + 8$ Replace x with $x + h$.
$\left. \begin{aligned} &= 3(x^2 + 2xh + h^2) - 5x - 5h + 8 \\ &= 3x^2 + 6xh + 3h^2 - 5x - 5h + 8 \end{aligned} \right\}$ Simplify.

∎

Try Exercise 13

If $f(x) = 6x^2 - 4x + 3$, find (a) $f(3)$, (b) $f(\frac{1}{2})$, (c) $f(-a)$, (d) $f(x + h)$.

Although we usually use the letter f to name a function, any letter may be used. Therefore each of the functions

$$g(x) = x^2 \qquad h(x) = x^2 \qquad \text{and} \qquad P(t) = t^2$$

is equivalent to the function $f(x) = x^2$, since each has the range value a^2 when the domain value is a.

Example 2   Suppose $g(x) = 5$. Determine each of the following: (a) $g(1)$, (b) $g(5)$, (c) $g(0)$.

Solution: We can write the function as $g(x) = 0 \cdot x + 5$.

(a) $g(1) = 0 \cdot 1 + 5 = 5$ Replace x with 1.
(b) $g(5) = 0 \cdot 5 + 5 = 5$ Replace x with 5.
(c) $g(0) = 0 \cdot 0 + 5 = 5$ Replace x with 0.

The function $g(x) = 5$ is called a *constant function*, because its value is constant for any x-value. ∎

Try Exercise 15

If $g(x) = 4$, find (a) $g(4)$, (b) $g(-2)$, (c) $g(0)$, (d) $g(3b)$.

Figure 8.10 illustrates the results of Example 2.

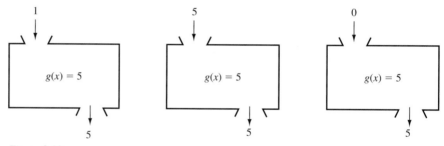

Figure 8.10

Operations on Functions

We can add, subtract, multiply, and divide two functions to produce a new function.

DEFINITION OF THE SUM, DIFFERENCE, PRODUCT, AND QUOTIENT OF TWO FUNCTIONS

Suppose f and g are functions, and x is a number in the domain of both f and g.

Sum: $(f + g)(x) = f(x) + g(x)$

Difference: $(f - g)(x) = f(x) - g(x)$

Product: $(fg)(x) = f(x) \cdot g(x)$

Quotient: $\left(\dfrac{f}{g}\right)(x) = \dfrac{f(x)}{g(x)} \qquad (g(x) \neq 0)$

Example 3   Suppose $f(x) = 3x - 5$ and $g(x) = x^2 + 4$. Determine each of the following: (a) $(f + g)(1)$, (b) $(fg)(-2)$, (c) $\left(\dfrac{f}{g}\right)(0)$.

Solution: (a) Since

$$f(x) = 3x - 5 \quad \text{and} \quad g(x) = x^2 + 4,$$
$$f(1) = 3(1) - 5 \qquad\qquad g(1) = 1^2 + 4$$
$$= -2 \qquad\qquad\qquad = 5$$

Therefore,

$$(f + g)(1) = f(1) + g(1) = -2 + 5 = 3$$

(b) Using the definition of the product of two functions, we have

$$(fg)(-2) = f(-2) \cdot g(-2) = [3(-2) - 5] \cdot [(-2)^2 + 4]$$
$$= (-11) \cdot (8) = -88$$

(c) Using the definition of the quotient of two functions, we have

$$\left(\frac{f}{g}\right)(0) = \frac{f(0)}{g(0)} = \frac{3(0) - 5}{0^2 + 4} = -\frac{5}{4} \qquad \square$$

Try Exercise 19

If $f(x) = 3x + 2$ and $g(x) = x^2 - 9$, find $(f + g)(2)$.

Example 4   Suppose $f(x) = x^2$ and $g(x) = x - 2$. Determine (a) $(f - g)(x)$, (b) $(fg)(x)$, and (c) $\left(\dfrac{f}{g}\right)(x)$. State the domain of each.

Solution: (a) $(f - g)(x) = f(x) - g(x) = x^2 - (x - 2) = x^2 - x + 2$

(b) $(fg)(x) = f(x) \cdot g(x) = x^2 \cdot (x - 2) = x^3 - 2x^2$

(c) $\left(\dfrac{f}{g}\right)(x) = \dfrac{f(x)}{g(x)} = \dfrac{x^2}{x - 2}$

Since the domain of both f and g is $(-\infty, \infty)$, the domain of both $f - g$ and fg is $(-\infty, \infty)$. The domain of $\dfrac{f}{g}$ is $\{x \mid x \neq 2\}$, since $x = 2$ makes the

Try Exercise 25

If $f(x) = 3x + 2$ and $g(x) = x^2 - 9$, find $(fg)(x)$ and state the domain.

denominator $g(x) = x - 2$ equal to 0. $\qquad \square$

CAUTION

If $f(x) = x$ and $g(x) = \dfrac{1}{x}$, then

$$(fg)(x) = f(x) \cdot g(x) = x \cdot \frac{1}{x} = 1$$

The domain of the function $h(x) = 1$ is $(-\infty, \infty)$. However, the domain of the product function $(fg)(x) = 1$ shown above is $\{x \mid x \neq 0\}$, because $x = 0$ is not in the domain of g. Therefore, while you can often determine the domain of fg, $f + g$, $f - g$, or $\dfrac{f}{g}$ by looking at the final function, sometimes you must investigate the domains of the original functions f and g. $\qquad \blacksquare$

Another operation between two functions involves substituting a domain value into one of the functions, and then substituting the resulting range value into the other function. We call this operation *composition of functions.*

DEFINITION OF COMPOSITION OF FUNCTIONS

Suppose f and g are functions such that x is in the domain of g and $g(x)$ is in the domain of f. Then the **composite of f and g** is denoted $f \circ g$, and it is given by

$$(f \circ g)(x) = f(g(x))$$

Note:

$(f \circ g)(x)$ is read "f composite g of x," and $f(g(x))$ is read "f of g of x."

The composite of f and g is illustrated in Fig. 8.11.

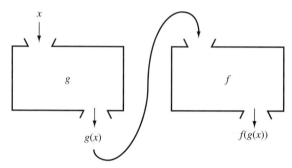

Figure 8.11

Example 5   Suppose $f(x) = 2x - x^2$ and $g(x) = x + 1$. Determine each of the following: (a) $(f \circ g)(3)$, (b) $(g \circ f)(3)$.

Solution: (a) See Fig. 8.12.

$$
\begin{aligned}
(f \circ g)(3) &= f(g(3)) && \text{Definition of composite of } f \\
&&& \text{and } g \\
&= f(4) && \text{Since } g(3) = 3 + 1 = 4 \\
&= -8 && \text{Since } f(4) = 2(4) - 4^2 = -8
\end{aligned}
$$

(b) See Fig. 8.13.

$$
\begin{aligned}
(g \circ f)(3) &= g(f(3)) && \text{Composite of } g \text{ and } f \\
&= g(-3) && \text{Since } f(3) = 2(3) - 3^2 = -3 \\
&= -2 && \text{Since } g(-3) = -3 + 1 = -2
\end{aligned}
$$

Try Exercise 33(a) and (b)

If $f(x) = 4x - x^2$ and $g(x) = x + 2$, find $(f \circ g)(2)$, and $(g \circ f)(2)$.

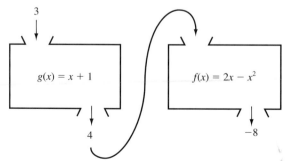

Figure 8.12

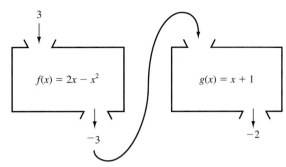

Figure 8.13

Example 5 illustrates that $(f \circ g)(3) \neq (g \circ f)(3)$. That is, **composition of functions is not a commutative operation.**

Example 6   Suppose $f(x) = 2x - x^2$ and $g(x) = x + 1$. Determine (a) $(f \circ g)(x)$, (b) $(g \circ f)(x)$. State the domain of each.

Solution: (a) $(f \circ g)(x) = f(g(x))$ Composite of f and g

$\qquad\qquad\qquad = f(x + 1)$ Since $g(x) = x + 1$

$\qquad\qquad\qquad = 2(x + 1) - (x + 1)^2$ Since $f(x) = 2x - x^2$

$\left.\begin{array}{l}\qquad\qquad\qquad = 2x + 2 - (x^2 + 2x + 1) \\ \qquad\qquad\qquad = -x^2 + 1\end{array}\right\}$ Simplify.

(b) $(g \circ f)(x) = g(f(x))$ Composite of g and f

$\qquad\qquad\qquad = g(2x - x^2)$ Since $f(x) = 2x - x^2$

$\qquad\qquad\qquad = (2x - x^2) + 1$ Since $g(x) = x + 1$

$\qquad\qquad\qquad = -x^2 + 2x + 1$ Descending order

Since the domain of both f and g is $(-\infty, \infty)$, the domain of both $f \circ g$ and $g \circ f$ is $(-\infty, \infty)$.   ❐

Try Exercise 33(c) and (d)

If $f(x) = 4x - x^2$ and $g(x) = x + 2$, find $(f \circ g)(x)$ and $(g \circ f)(x)$. State the domain of each.

Example 7   Suppose $f(x) = \dfrac{1}{x}$ and $g(x) = \sqrt{x + 1}$. Find (a) $(f + g)(x)$, (b) $(f \circ g)(x)$, and (c) $(g \circ f)(x)$. State the domain of each.

Solution: (a) $(f + g)(x) = f(x) + g(x) = \dfrac{1}{x} + \sqrt{x + 1}$

Since the domain of f is $\{x \mid x \neq 0\}$ and the domain of g is $[-1, \infty)$, the domain of $f + g$ is $[-1, 0) \cup (0, \infty)$.

(b) $(f \circ g)(x) = f(g(x)) = f(\sqrt{x + 1}) = \dfrac{1}{\sqrt{x + 1}}$

The domain of $f \circ g$ consists of all those numbers x in the domain of g (namely, $x \geq -1$) such that $g(x) = \sqrt{x + 1}$ is in the domain of f (that is, such that $\sqrt{x + 1} \neq 0$, or such that $x \neq -1$). Therefore, the domain of $f \circ g$ is $(-1, \infty)$.

(c) $(g \circ f)(x) = g(f(x)) = g\left(\dfrac{1}{x}\right) = \sqrt{\dfrac{1}{x} + 1} = \sqrt{\dfrac{1 + x}{x}}$

To find the domain, use the methods of Sec. 6.6 to solve the inequality $\dfrac{1 + x}{x} \geq 0$.

Try Exercise 37(c)

If $f(x) = \dfrac{6}{x}$ and $g(x) = \sqrt{x + 2}$, find $(f \circ g)(x)$ and state the domain.

True   False   True

——+————+——
$\qquad$ -1 $\qquad$ 0

Therefore, the domain of $g \circ f$ is $(-\infty, -1] \cup (0, \infty)$.   ❐

CAUTION

If $f(x) = \dfrac{1}{x}$ and $g(x) = \dfrac{1}{x}$, then

$$(f \circ g)(x) = f(g(x)) = f\left(\dfrac{1}{x}\right) = \dfrac{1}{1/x} = x$$

Even though the domain of $h(x) = x$ is $(-\infty, \infty)$, the domain of the composite function $(f \circ g)(x) = x$ shown above is $\{x \mid x \neq 0\}$, because $x = 0$ is not in the domain of g. ■

Sometimes it is important to be able to visualize a single function as a composite of two functions.

Example 8 If $h(x) = (x^2 + 5)^{10}$, find two functions f and g such that $(f \circ g)(x) = h(x)$.

Solution: One choice is $g(x) = x^2 + 5$ and $f(x) = x^{10}$. Then

$$(f \circ g)(x) = f(g(x)) = f(x^2 + 5) = (x^2 + 5)^{10} = h(x)$$

You can verify that $g(x) = x^2$ and $f(x) = (x + 5)^{10}$ will also work. ❑

Try Exercise 39

If $h(x) = (x^2 - 7)^{15}$, find two functions f and g such that $(f \circ g)(x) = h(x)$.

Example 9 Suppose $f = \{(8, 9), (1, 2), (0, 3)\}$ and $g = \{(6, -4), (-7, 1), (4, -5)\}$. Determine each of the following:

(a) $f(8) - g(4)$, (b) $f(0) \cdot g(6)$, (c) $f(g(-7))$.

Solution: (a) Since $f(8) = 9$ and $g(4) = -5$, we have

$$f(8) - g(4) = 9 - (-5) = 14$$

(b) Since $f(0) = 3$ and $g(6) = -4$, we have

$$f(0) \cdot g(6) = 3 \cdot (-4) = -12$$

(c) Since $g(-7) = 1$, we have

$$f(g(-7)) = f(1) = 2$$ ❑

Try Exercise 45

Suppose $f = \{(7, 9), (0, -4), (12, 7), (2, 6), (-7, 4)\}$ and $g = \{(6, 7), (8, -8), (-8, 2), (4, 2), (0, -3)\}$. Determine $f(7) \cdot g(0)$.

The expression

$$\dfrac{f(t) - f(a)}{t - a}$$

is called a *difference quotient*. If you go on to study calculus, you will need to know how to work with difference quotients.

Example 10 Find $\dfrac{f(t) - f(a)}{t - a}$ for the function $f(x) = x^2 + 3$. Then simplify.

Solution: Since $f(t) = t^2 + 3$ and $f(a) = a^2 + 3$, we have

Try Exercise 53

If $f(x) = x^2 + 4$, find $\dfrac{f(t) - f(a)}{t - a}$ and simplify.

$$\dfrac{f(t) - f(a)}{t - a} = \dfrac{(t^2 + 3) - (a^2 + 3)}{t - a}$$

$$= \dfrac{t^2 - a^2}{t - a} = \dfrac{(t + a)(t - a)}{t - a} = t + a$$ ❑

Many types of applied problems involve putting two functions together in some way to form a third function. Here is an example that involves putting two functions together using the operation of function composition.

Problem Solving

Example 11   The radius r of a circular oil slick is increasing at the rate of 10 ft/hr. Write the area A of the oil slick as a function of time t.

Solution: Since the radius is increasing 10 ft/hr, we can write r as a function of t as follows.

$$r = f(t) = 10t$$

Since the oil slick is circular, we can write A as a function of r as follows:

$$A = g(r) = \pi r^2$$

Therefore, the composite of g and f gives A as a function of t.

$$A = (g \circ f)(t) = g(f(t)) = g(10t) = \pi(10t)^2 = 100\pi t^2 \qquad \square$$

Try Exercise 79

The radius r of a circular oil slick is increasing at the rate of 5 ft/hr. Write the area A of the oil slick as a function of time t.

The relationship between time t, radius r, and area A in Example 11, is shown in the following chart:

Time t	Radius $r = 10t$	Area $A = 100\pi t^2$
0 hr	$10(0) = 0$ ft	$100\pi(0)^2 = 0$ ft^2
1 hr	$10(1) = 10$ ft	$100\pi(1)^2 = 100\pi$ ft^2
2 hr	$10(2) = 20$ ft	$100\pi(2)^2 = 400\pi$ ft^2
3 hr	$10(3) = 30$ ft	$100\pi(3)^2 = 900\pi$ ft^2

Exercises 8.2

Completion Exercises

1. To denote that y is a function of x, we often write $y = $ _____.

2. Using function notation, we can write the function $y = 2x$ as _____ $= 2x$.

True-False Exercises

3. If f is a function, the expression $f(2)$ means f times 2.

4. The expression $f(2)$ is the y-value that corresponds to an x-value of 2.

5. In general, $(fg)(x) = (gf)(x)$.

6. In general, $(f \circ g)(x) = (g \circ f)(x)$.

Matching Exercises

If $y = g(x)$ is a function, match each phrase in Exercises 7 through 10 with its symbol in letters A through D. There may be more than one answer.

7. Name of the function   A. y

8. Domain value   B. x

9. Range value   C. g

10. Value of the function   D. $g(x)$

For each function, determine (a) $f(3)$, (b) $f(\frac{1}{2})$, (c) $f(-a)$, and (d) $f(x + h)$.

11. $f(x) = 4x$ 12. $f(x) = 3 - 5x$

13. $f(x) = 6x^2 - 4x + 3$ 14. $f(x) = x^2 + x$

For each function, determine (a) $g(4)$, (b) $g(-2)$, (c) $g(0)$, and (d) $g(3b)$.

15. $g(x) = 4$

16. $g(x) = -x^2 - x + 7$

17. $g(x) = x^3 - 2x^2 + 5x - 9$

18. $g(x) = \dfrac{x + 2}{x - 4}$

Suppose $f(x) = 3x + 2$ and $g(x) = x^2 - 9$. Determine each of the following. State the domain for Exercises 20, 21, 25, 26, 29, and 30.

19. $(f + g)(2)$

20. $(f + g)(x)$

21. $(f - g)(x)$

22. $(f - g)(7)$

23. $(fg)(-1)$

24. $(gf)(5)$

25. $(fg)(x)$

26. $(gf)(x)$

27. $\left(\dfrac{f}{g}\right)(0)$

28. $\left(\dfrac{g}{f}\right)(1)$

29. $\left(\dfrac{f}{g}\right)(x)$

30. $\left(\dfrac{g}{f}\right)(x)$

For each pair of functions, determine (a) $(f \circ g)(2)$, (b) $(g \circ f)(2)$, (c) $(f \circ g)(x)$, (d) $(g \circ f)(x)$. State the domain for parts (c) and (d).

31. $f(x) = 2x + 7, g(x) = 5x - 1$

32. $f(x) = 5x, g(x) = 10$

33. $f(x) = 4x - x^2, g(x) = x + 2$

34. $f(x) = x - 2, g(x) = x^2 + 2x$

Find (a) $(f + g)(x)$, (b) $(fg)(x)$, (c) $(f \circ g)(x)$, and (d) $(g \circ f)(x)$ for each pair of functions. State the domain of each.

35. $f(x) = \dfrac{x}{x - 1}, g(x) = x + 5$

36. $f(x) = \dfrac{1}{x}, g(x) = \dfrac{x + 1}{x + 3}$

37. $f(x) = \dfrac{6}{x}, g(x) = \sqrt{x + 2}$

38. $f(x) = \sqrt{x - 1}, g(x) = \sqrt{4 - x}$

Find two functions f and g such that $(f \circ g)(x) = h(x)$. There is more than one answer.

39. $h(x) = (x^2 - 7)^{15}$

40. $h(x) = \sqrt{x^3 + 1}$

41. $h(x) = \dfrac{1}{4x + 5}$

42. $h(x) = (x - 3)^2 + 6(x - 3)$

Suppose $f = \{(7, 9), (0, -4), (12, 7), (2, 6), (-7, 4)\}$ and $g = \{(6, 7), (8, -8), (-8, 2), (4, 2), (0, -3)\}$. Determine each of the following.

43. $f(12) - g(4)$

44. $f(2) + g(8)$

45. $f(7) \cdot g(0)$

46. $f(0) \cdot g(6)$

47. $f(g(-8))$

48. $g(f(-7))$

49. $f(6)$

50. $g(2)$

Find $\dfrac{f(t) - f(a)}{t - a}$ for each function. Then simplify.

51. $f(x) = 8$

52. $f(x) = 5x - 1$

53. $f(x) = x^2 + 4$

54. $f(x) = 3x^2$

55. $f(x) = x^2 - 7x$

56. $f(x) = 2x^2 + x$

Use the graphs in Figs. 8.14 and 8.15 to determine each of the following.

57. $(f + g)(0)$

58. $(f - g)(-4)$

59. $(fg)(2)$

60. $\left(\dfrac{f}{g}\right)(0)$

61. $(f \circ g)(-1)$

62. $(g \circ f)(-1)$

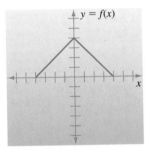

Figure 8.14   *Figure 8.15*

Discussion Exercises

63. Discuss some advantages the notation $f(x) = 5x + 3$ has over the notation $y = 5x + 3$.

64. Explain the difference between $(fg)(x)$ and $(f \circ g)(x)$. Use examples to illustrate your point.

The starting roster of a coed volleyball team is given below.

Player	Age in years	Weight in pounds	Height in inches
Sean	18	160	70
Cindy	19	110	64
Dan	20	165	68
Kim	18	100	62
Mike	21	195	73
Beth	19	145	68

If $f(x)$ = age of x, $g(x)$ = weight of x, and $h(x)$ = height of x, determine each of the following.

65. f(Kim)   66. h(Dan)
67. g(Kim)   68. f(Dan)
69. h(Cindy)   70. g(Mike)

Determine whether each relation is a function. In each case the domain is the set of all United States citizens who have Social Security numbers.

71. $f(x)$ = Social Security number of x
72. $g(x)$ = phone number of x
73. $h(x)$ = child of x
74. $p(x)$ = mother of x

For Exercises 75 and 76, let $f(x) = 3x + 5$.

75. Show that $f(a + b) \neq f(a) + f(b)$.
76. Show that $f(2a) \neq 2f(a)$.

Problem Solving

77. A company can produce shower curtains at a cost of $4 per curtain plus a weekly overhead of $275. If $C(x)$ denotes the total weekly cost of producing x shower curtains, write a function that expresses $C(x)$ in terms of x. Then find and interpret $C(0)$ and $C(235)$.

78. A sales representative earns $200 per week plus 15 percent of the value of her sales that week. If $E(x)$ denotes her total weekly earnings based on sales totaling x dollars,

write a function that expresses $E(x)$ in terms of x. Then find and interpret $E(0)$ and $E(2800)$.

79. The radius r of a circular oil slick is increasing at the rate of 5 ft/hr. Write the area A of the oil slick as a function of time t.

80. A pebble is dropped into a pond, and the radius r of the circular ripple it produces increases at the rate of 3 in./sec. Write the area A inside the ripple as function of time t.

81. Jeremy can run at a speed of 4 yd/sec. His father can run twice as fast. Use function composition to write his father's distance D as a function of time t.

82. Amanda can type 30 words per minute. Her teacher can type 3 times as fast. Use function composition to write the number of words, W, per minute her teacher can type as a function of time t.

Getting Acquainted with Your Graphing Calculator

COMBINING FUNCTIONS   For Exercises 83 and 84, enter $f(x) = x + 2$ and $g(x) = 2 - x$ into your calculator. You can do this on some calculators by typing $y_1 = x + 2$ and $y_2 = 2 - x$.

83. Enter $y_3 = y_1 + y_2$ into your calculator. Then graph the three functions $y_1 = f(x)$, $y_2 = g(x)$, and $y_3 = (f + g)(x)$ in the same Viewing Rectangle.

84. Enter $y_4 = y_1 \cdot y_2$ into your calculator. Then graph the three functions $y_1 = f(x)$, $y_2 = g(x)$, and $y_4 = (fg)(x)$ in the same Viewing Rectangle.

8.3 Constant, Linear, and Quadratic Functions

TAPE AU20

In this section we examine three specific types of functions and their graphs. They are the *constant function,* the *linear function,* and the *quadratic function.*

Constant Functions

The constant function is so named because it has a constant range value.

DEFINITION OF A CONSTANT FUNCTION

A **constant function** is a function that can be written in the form

$$f(x) = k$$

where k is a real number.

The constant function $f(x) = k$ is equivalent to $y = k$. From Sec. 7.2, we know that the graph of this equation is a horizontal line through the point $(0, k)$.

Example 1   Graph the constant function $f(x) = 5$.

Solution: This equation is equivalent to $y = 5$. The graph is the horizontal line through $(0, 5)$ shown in Fig. 8.16. We can label the vertical axis as the y-axis or the $f(x)$-axis.   ❏

Try Exercise 3

Graph: $f(x) = 3$.

Linear Functions

The constant function $f(x) = k$ is a special case of a linear function.

> ### DEFINITION OF A LINEAR FUNCTION
>
> A **linear function** is a function that can be written in the form
>
> $$f(x) = mx + b$$
>
> where m and b are real numbers.

The linear function $f(x) = mx + b$ is equivalent to $y = mx + b$. From Sec. 7.4, we know that the graph of this equation is a line with slope m and y-intercept b.

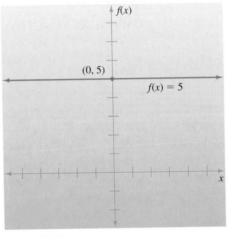

Figure 8.16

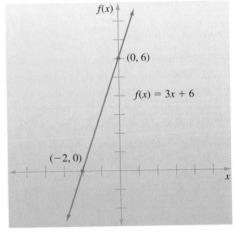

Figure 8.17

Example 2   Graph the linear function $f(x) = 3x + 6$.

Solution: The graph is the line with slope 3 and y-intercept 6 shown in Fig. 8.17.   ❏

Try Exercise 9

Graph: $f(x) = 2x + 2$.

Quadratic Functions

When $f(x)$ equals a polynomial of the second degree, the result is a quadratic function.

> ### DEFINITION OF A QUADRATIC FUNCTION
>
> A **quadratic function** is a function that can be written in the form
>
> $$f(x) = ax^2 + bx + c$$
>
> where a, b, and c are real numbers and $a \neq 0$.

The graph of any quadratic function is a *parabola*.

DEFINITION OF A PARABOLA

A **parabola** is the set of all points in the plane that are equidistant from a fixed line and a fixed point not on the line. The fixed line is called the **directrix** and the fixed point is the **focus** (see Fig. 8.18).

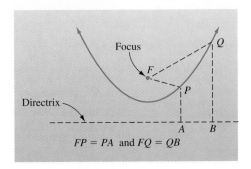

Figure 8.18

When an object is projected upward (but not vertically) and air resistance is ignored, the trajectory of the object is a parabola. Solar furnaces use parabolas as an integral part of their design. The reflectors in spotlights and satellite dishes have parabolic cross sections, and the 200-in. mirror in the telescope at Mount Palomar Observatory uses its parabolic shape to reflect light to its focus 55 ft away. See Fig. 8.19.

Ignoring air resistance, the path of this cannonball is a portion of a parabola

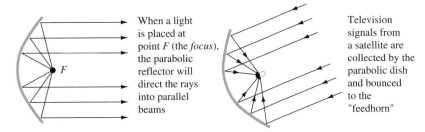

Figure 8.19

If we let $a = 1$, $b = 0$, and $c = 0$ in the quadratic function $f(x) = ax^2 + bx + c$, we obtain the simplest equation of a parabola, namely, $f(x) = x^2$.

Example 3   Graph the quadratic function $f(x) = x^2$.

Solution: Substitute any convenient values for x into $f(x) = x^2$, and compute the corresponding values of $f(x)$. For example, if $x = 3$, then $f(3) = 3^2 = 9$. Continuing to substitute values for x, we obtain the table of values below:

x	3	2	1	0	−1	−2	−3
$f(x)$	9	4	1	0	1	4	9

Plot the points from this table and draw a smooth curve through them. This produces the parabola shown in Fig. 8.20.   ❏

Try Exercise 27

Use $f(x) = x^2 + 1$ to complete the table of values below. Then plot the points and draw a smooth curve through them.

x	3	2	1	0	−1	−2	−3
$f(x)$							

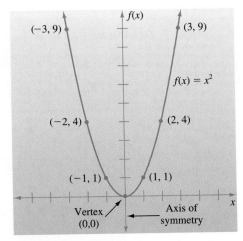

Figure 8.20

If we "fold" Fig. 8.20 along the $f(x)$-axis, the two halves of the parabola will coincide. Therefore the $f(x)$-axis is called the **axis of symmetry.** The point at which the axis of symmetry intersects the parabola is called the **vertex.** The vertex of the parabola in Fig. 8.20 is (0, 0), and it is the lowest point of the parabola.

CAUTION

Students often think that every U-shaped curve is a parabola. This is not the case. For example, the graph of $y = x^4$ shown in Fig. 8.21 is a U-shaped curve, but it is not a parabola.

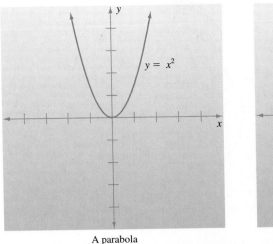

Figure 8.21

If a is a *negative* number, the parabola $y = ax^2 + bx + c$ will open *downward*. For example, if $a = -1$, $b = 0$, and $c = 1$, we have the parabola $y = -x^2 + 1$. Example 4 shows that this parabola opens downward.

Example 4   Graph the quadratic function $y = -x^2 + 1$.

Solution: Substitute values for x and compute the corresponding values of y. For example, if $x = 3$, then $y = -3^2 + 1 = -8$. Continuing to substitute values for x, we obtain the table of values below:

x	3	2	1	0	−1	−2	−3
y	−8	−3	0	1	0	−3	−8

Plot these points and draw a smooth curve through them to produce the parabola shown in Fig. 8.22. Note that when the parabola opens downward, the vertex, in this case (0, 1), is the highest point of the parabola. Note also that the vertex is halfway between the two x-intercepts.   ❐

Try Exercise 33

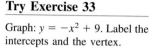

Graph: $y = -x^2 + 9$. Label the intercepts and the vertex.

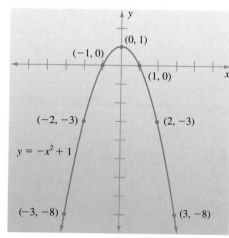

Figure 8.22

In order to graph more complicated quadratic functions, we now discuss how to find the y-intercept, the x-intercepts, and the vertex. To find the y-intercept, let $x = 0$.

$$y = ax^2 + bx + c \qquad \text{General quadratic function}$$
$$y = a(0)^2 + b(0) + c \qquad \text{Let } x = 0.$$
$$y = c \qquad \text{Simplify.}$$

Therefore the parabola $y = ax^2 + bx + c$ has one y-intercept and it is c.
To find the x-intercepts, let $y = 0$.

$$y = ax^2 + bx + c \qquad \text{General quadratic function}$$
$$0 = ax^2 + bx + c \qquad \text{Let } y = 0.$$

Since this is a quadratic equation, it will have two real solutions, one real solution, or no real solutions. Therefore the parabola $y = ax^2 + bx + c$ will have 2, 1, or 0 x-intercepts (see Fig. 8.23).

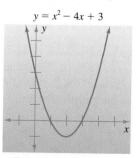

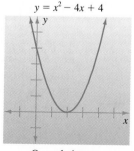

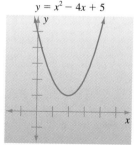

Two solutions to
$0 = x^2 - 4x + 3$;
namely, $x = 1$ and $x = 3$

One solution to
$0 = x^2 - 4x + 4$;
namely, $x = 2$

No real solution to
$0 = x^2 - 4x + 5$

Two x-intercepts

One x-intercept

No x-intercept

Figure 8.23

To develop a formula for the vertex, suppose the parabola $y = ax^2 + bx + c$ has two x-intercepts. These intercepts are the solutions of the equation $0 = ax^2 + bx + c$, namely,

$$x = \frac{-b + \sqrt{b^2 - 4ac}}{2a} \quad \text{and} \quad x = \frac{-b - \sqrt{b^2 - 4ac}}{2a}$$

Since the x-value of the vertex is halfway between the two x-intercepts, the x-value of the vertex is

$$x = \frac{1}{2}\left(\frac{-b + \sqrt{b^2 - 4ac}}{2a} + \frac{-b - \sqrt{b^2 - 4ac}}{2a} \right) \qquad \text{Average the } x\text{-intercepts.}$$

$$= \frac{1}{2}\left(\frac{-2b}{2a} \right) \qquad \text{Add the numerators.}$$

$$= \frac{-b}{2a} \qquad \text{Simplify.}$$

This formula gives the x-value of the vertex even when the parabola does *not* have two x-intercepts. To find the y-value of the vertex, simply substitute the x-value of the vertex into the equation $y = ax^2 + bx + c$.

Example 5   Graph $y = x^2 - 6x + 8$. Label the y-intercept, the x-intercepts, and the vertex.

Solution: Since $c = 8$, the y-intercept is 8. To find the x-intercepts, let $y = 0$.

$y = x^2 - 6x + 8$	Original equation
$0 = x^2 - 6x + 8$	Let $y = 0$.
$0 = (x - 2)(x - 4)$	Factor.
$x - 2 = 0 \quad \text{or} \quad x - 4 = 0$	Set each factor equal to 0.
$x = 2 \qquad\qquad\quad x = 4$	Two x-intercepts

To find the x-value of the vertex, substitute $b = -6$ and $a = 1$ into the vertex formula.

$$x = \frac{-b}{2a} \qquad \text{Vertex formula}$$

$$x = \frac{-(-6)}{2(1)} \qquad \text{Let } b = -6 \text{ and } a = 1.$$

$$x = 3 \qquad \text{Simplify.}$$

To find the y-value of the vertex, substitute $x = 3$ into $y = x^2 - 6x + 8$.

$$y = 3^2 - 6(3) + 8 = 9 - 18 + 8 = -1$$

Since $a = 1$ (a positive number), the parabola opens upward as shown in Fig. 8.24. The axis of symmetry of this parabola is the vertical line $x = 3$. Using symmetry, we conclude that the point $(6, 8)$ must be on the parabola since the point $(0, 8)$ is on the parabola. ❑

Try Exercise 39

Graph: $y = x^2 - 6x + 5$. Label the intercepts and the vertex.

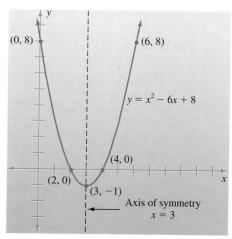

Figure 8.24

We can summarize the procedure for graphing $y = ax^2 + bx + c$ as follows.

To Graph the Parabola $y = ax^2 + bx + c$

1. Note that the y-intercept is c.
2. Let $y = 0$ to find the x-intercepts. There will be two, one, or no x-intercepts.
3. Use the formula $x = \dfrac{-b}{2a}$ to find the x-value of the vertex.
4. Substitute the x-value found in step 3 into the original equation to find the y-value of the vertex.
5. If $a > 0$, the parabola opens upward. If $a < 0$, the parabola opens downward.
6. Draw a smooth curve through the y-intercept, the x-intercepts, and the vertex. Plot additional points if necessary.

We illustrate these steps in Example 6.

Example 6 Graph $y = -x^2 + 4x - 4$. Label the intercepts and the vertex.

Solution: Step 1: Note that the y-intercept is c.

$$\text{The } y\text{-intercept is } -4.$$

Step 2: Let $y = 0$ to find the x-intercepts.

$$0 = -x^2 + 4x - 4 \qquad \text{Let } y = 0.$$
$$0 = x^2 - 4x + 4 \qquad \text{Multiply by } -1.$$
$$0 = (x - 2)^2 \qquad \text{Factor.}$$
$$x = 2 \qquad \text{One } x\text{-intercept}$$

Step 3: Use the formula $x = \dfrac{-b}{2a}$ to find the x-value of the vertex.

$$x = \frac{-4}{2(-1)} \qquad \text{Let } b = 4 \text{ and } a = -1.$$
$$x = 2 \qquad \text{Simplify.}$$

Step 4: Substitute the x-value found in step 3 into the original equation to find the y-value of the vertex.

$$y = -x^2 + 4x - 4 \qquad \text{Original equation}$$
$$y = -2^2 + 4(2) - 4 \qquad \text{Let } x = 2.$$
$$y = 0 \qquad \text{Simplify.}$$

Step 5: If $a > 0$ the parabola opens upward. If $a < 0$ the parabola opens downward.

Since $a = -1$ (that is, $a < 0$), the parabola opens downward.

Step 6: Draw a smooth curve through the y-intercept, the x-intercept, and the vertex. Plot additional points if necessary.

Since $(0, -4)$ is on the parabola, by symmetry $(4, -4)$ must be on the parabola. Therefore we have the graph shown in Fig. 8.25. ❐

Try Exercise 41

Graph: $y = -x^2 + 2x - 1$. Label the intercepts and the vertex.

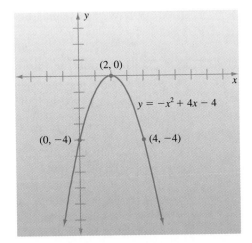

Figure 8.25

CAUTION

Here are some common errors that students make when graphing parabolas.

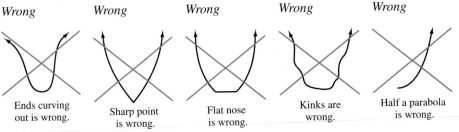

Here is a summary of constant, linear, and quadratic functions.

Constant Function

$f(x) = k$

y-intercept is k.

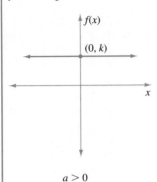

Linear Function

$f(x) = mx + b$

Slope is m, y-intercept is b.

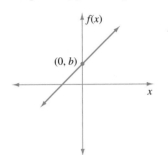

Quadratic Function

$f(x) = ax^2 + bx + c$

y-intercept is c.

Let $y = 0$ to find x-intercepts.

Vertex at $x = \dfrac{-b}{2a}$

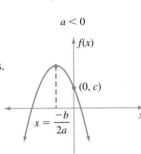

Exercises 8.3

Completion Exercises

1. $f(x) = k$ is called a(n) _____ function. Its graph is a horizontal line through the point _____.

2. $f(x) = mx + b$ is called a(n) _____ function. Its graph is a line with slope _____ and y-intercept _____.

Graph each constant function.

3. $f(x) = 3$

4. $f(x) = -4$

5. $g(x) = -2$

6. $g(x) = 1$

7. $h(x) = \frac{7}{2}$

8. $h(x) = 0$

Graph each linear function.

9. $f(x) = 2x + 2$

10. $f(x) = 3x - 3$

11. $g(x) = -4 - x$

12. $g(x) = 5 + x$

13. $h(x) = \frac{2}{3}x$

14. $h(x) = -\frac{3}{4}x$

Completion Exercises

15. $f(x) = ax^2 + bx + c$ is called a(n) _____ function. Its graph is a _____, with y-intercept _____, and vertex at $x =$ _____. To find the y-value of the vertex, _____.

16. The vertical line through the vertex of the parabola $y = ax^2 + bx + c$ is called the _____.

17. If $a > 0$, the parabola $y = ax^2 + bx + c$ opens _____ (up, down); if $a < 0$, the parabola $y = ax^2 + bx + c$ opens _____ (up, down).

18. To find the x-intercepts of the parabola $y = ax^2 + bx + c$, let $y =$ _____.

True-False Exercises

19. $f(x) = 3x$ is a constant function.

20. $f(x) = \sqrt{x} + 4$ is a linear function.

21. $f(x) = \frac{\cdot 1}{x^2}$ is a quadratic function.

22. Every U-shaped curve is a parabola.

Matching Exercises

Match each function in Exercises 23 through 26 with its graph in letters A through D.

23. $f(x) = 2$ A.
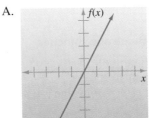

24. $f(x) = 2x$ B.
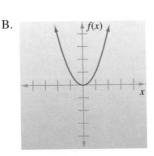

25. $f(x) = x^2$ C.
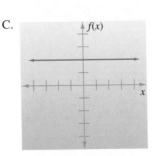

26. $f(x) = -x^2$ D.
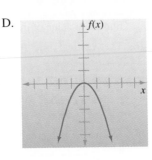

Complete each table of values using the given quadratic function. Then plot the points and draw a smooth curve through them.

27. $f(x) = x^2 + 1$

x	$f(x)$
3	
2	
1	
0	
−1	
−2	
−3	

28. $f(x) = -x^2$

x	$f(x)$
3	
2	
1	
0	
−1	
−2	
−3	

29. $g(x) = -2x^2$

x	$g(x)$
2	
1	
$\frac{1}{2}$	
0	
$-\frac{1}{2}$	
−1	
−2	

30. $g(x) = \frac{1}{4}x^2$

x	$g(x)$
6	
4	
2	
0	
−2	
−4	
−6	

Graph each quadratic function. Label the intercepts and the vertex.

31. $y = x^2 - 4$ 32. $y = x^2 - 9$

33. $y = -x^2 + 9$ 34. $y = -x^2 + 4$

35. $y = x^2 - 4x$ 36. $y = x^2 + 2x$

37. $y = 5 - x^2$ 38. $y = 7 - x^2$

39. $y = x^2 - 6x + 5$ 40. $y = x^2 - 4x + 3$

41. $y = -x^2 + 2x - 1$ 42. $y = -x^2 + 6x - 9$

43. $y = x^2 + 3x + 1$ 44. $y = x^2 - 3x + 1$

45. $y = \frac{1}{2}x^2 + x + 1$ 46. $y = \frac{1}{3}x^2 + x + 1$

Discussion Exercises

47. How many x-intercepts does the parabola $y = ax^2 + bx + c$ have? How many y-intercepts? Use graphs to illustrate your point.

48. Outline the steps to graph the parabola $y = ax^2 + bx + c$.

Identify as a constant function, a linear function, or a quadratic function. Then graph the function. Label all intercepts and vertices.

49. $f(x) = x$ 50. $f(x) = 9$

51. $g(x) = \frac{1}{3}x^2$ 52. $g(x) = 2x^2 - 8$

53. $2y + 3 = 0$

54. $6x - 2y = 12$

55. $x^2 + y = 6x$

56. $x^2 + 4x = y + 5$

Use the definition of a parabola to sketch, as accurately as you can, the parabola that satisfies the conditions of each exercise.

57. Directrix is the x-axis, focus is the point $(0, 4)$

58. Directrix is the line $y = 3$, focus is the point $(2, 1)$

  Getting Acquainted with Your Graphing Calculator

Graph each function.

59. $f(x) = 5$

60. $f(x) = -6$

61. $g(x) = 4x + 1$

62. $g(x) = \frac{1}{2}x - 3$

63. $h(x) = x^2 + x - 6$

64. $h(x) = -\frac{1}{2}x^2 + x + 4$

Hint: For Exercises 65 and 66, the parabola may appear to be almost flat when you zoom in on the vertex. To compensate for this problem, make Ymin and Ymax very close together while still containing the vertex.

65. A frog leap follows a path given by $y = 2x - \frac{1}{4}x^2$, where x and y are in feet and $0 \le x \le 8$. Graph this path. Then zoom in on the vertex to determine the maximum height attained by the frog.

66. A football punt follows a path given by $y = 2x - \frac{1}{30}x^2$, where x and y are in yards and $0 \le x \le 60$. Graph this path. Then zoom in on the vertex to determine the maximum height attained by the football.

8.4 More about Parabolas

TAPE AU20

In this section we illustrate how to use the graph of a parabola to solve an applied problem. We also discuss shifting, flipping, and stretching graphs, and we see how these ideas lead to another method for graphing a parabola. The concepts of shifting, flipping, and stretching are especially important because they apply to graphs in general, and not just the graphs of parabolas. Finally, we graph parabolas that open left or right.

Maximum and Minimum Problems

From Sec. 8.3, we know that the vertex of the parabola $y = ax^2 + bx + c$ is either the highest point or the lowest point of the graph. Therefore, we can use the formula for the vertex, $x = \dfrac{-b}{2a}$, to solve certain types of maximum and minimum problems.

Problem Solving

Example 1   A farmer plans to use 12 mi of fencing to construct a rectangular pasture. One side of the farmer's property lies along a straight river and needs no fencing (see Fig. 8.26). What should the width and length of the pasture be to maximize its area? What is the maximum area?

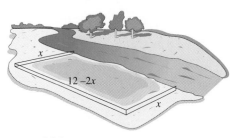

Figure 8.26

Solution: Let x be the width of the pasture. Since there are 12 mi of fencing, and the two widths use up $2x$ miles of the 12 mi, the length of the pasture must be $12 - 2x$ (see Fig. 8.26). Therefore the area of the pasture is

$$A = wl \qquad \text{Area of a rectangle}$$
$$A = x(12 - 2x) \qquad \text{Let } w = x \text{ and } l = 12 - 2x.$$
$$A = 12x - 2x^2 \qquad \text{Distributive property}$$
$$A = -2x^2 + 12x \qquad \text{Descending order}$$

From Sec. 8.3, we know that the graph of this equation is a parabola with A-intercept 0. Let $A = 0$ to find the x-intercepts.

$$0 = -2x^2 + 12x \qquad \text{Let } A = 0.$$
$$0 = -2x(x - 6) \qquad \text{Factor.}$$
$$-2x = 0 \quad \text{or} \quad x - 6 = 0 \qquad \text{Set each factor equal to 0.}$$
$$x = 0 \qquad\qquad x = 6 \qquad \text{Two } x\text{-intercepts}$$

The x-value of the vertex is

$$x = \frac{-b}{2a} = \frac{-12}{2(-2)} = \frac{-12}{-4} = 3$$

The A-value of the vertex is

$$A = -2(3)^2 + 12(3) = -2(9) + 36 = 18$$

Since $a = -2$ (a negative number), the parabola opens downward as shown in Fig. 8.27. Therefore the vertex is the highest point of the parabola, and it represents the maximum value of the area A. The maximum area of 18 mi^2 occurs when the width is $x = 3$ mi and the length is $12 - 2x = 12 - 2(3) = 6$ mi. ∎

Try Exercise 3

A rancher plans to use 8 mi of fencing to construct a rectangular pasture. One side of the rancher's property lies along an existing fence and needs no additional fencing (see Fig. 8.41). What should the width and length of the pasture be to maximize its area? What is the maximum area?

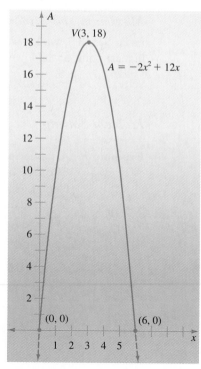

Figure 8.27

Shifting a Graph

Consider the three graphs shown in Fig. 8.28. Note that the graph of $y = x^2 + 2$ is simply the graph of $y = x^2$ shifted up 2 units. Similarly, the graph of $y = x^2 - 3$ is the graph of $y = x^2$ shifted down 3 units. This suggests the following rule.

Vertical Shifts

Assume k is a positive number and f is a function.

1. The graph of $y = f(x) + k$ is the graph of $y = f(x)$ shifted up k units.
2. The graph of $y = f(x) - k$ is the graph of $y = f(x)$ shifted down k units.

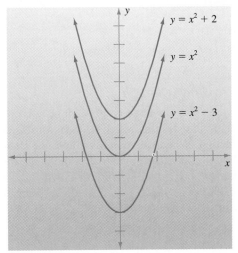

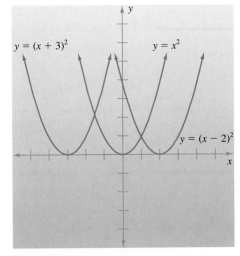

Figure 8.28   *Figure 8.29*

Now consider the three graphs shown in Fig. 8.29. The graph of $y = (x - 2)^2$ is the graph of $y = x^2$ shifted to the right 2 units. The graph of $y = (x + 3)^2$ is the graph of $y = x^2$ shifted to the left 3 units. This suggests the following rule.

Horizontal Shifts

Assume h is a positive number and f is a function.

1. The graph of $y = f(x - h)$ is the graph of $y = f(x)$ shifted right h units.
2. The graph of $y = f(x + h)$ is the graph of $y = f(x)$ shifted left h units.

Example 2 illustrates graphs that have both a vertical and a horizontal shift.

Example 2   Graph each quadratic function.
   (a) $y = (x - 4)^2 + 3$ (b) $y = (x + 2)^2 - 4$

Solution: (a) Shift the graph of $y = x^2$ right 4 units and up 3 units to produce the graph shown in Fig. 8.30.
   (b) Shift the graph of $y = x^2$ left 2 units and down 4 units to produce the graph shown in Fig. 8.31.   ❐

Try Exercise 23

Graph: $y = (x - 1)^2 + 2$.

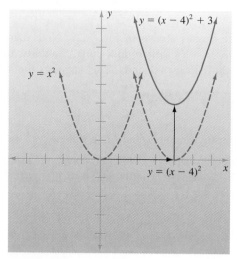

Figure 8.30

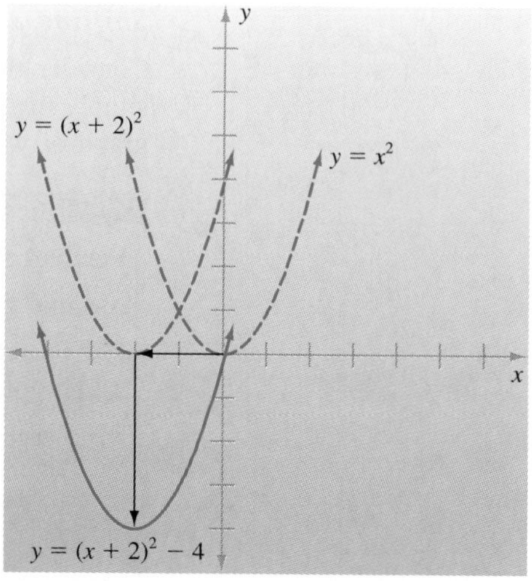

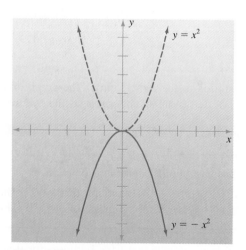

Figure 8.31

Flipping and Stretching a Graph

When we shift a graph, we move the graph right or left, or up or down, without changing its size, shape, or the direction in which it opens (up or down). When we flip a graph, we reverse its direction. When we stretch a graph, we change its size or shape (or both).

Recall from Sec. 8.3 that the number a in the quadratic function $y = ax^2 + bx + c$ determines whether the parabola opens up or down. If $a > 0$, the parabola opens up; if $a < 0$, the parabola opens down. Therefore, the graph of $y = -x^2$ is the graph of $y = x^2$ flipped over the x-axis (see Fig. 8.32).

Figure 8.32

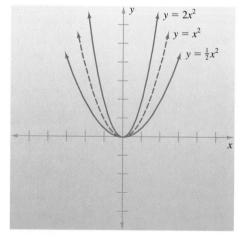

Figure 8.33

The number a in $y = ax^2 + bx + c$ also determines how "fat" or how "thin" the parabola is. As the absolute value of a gets larger, the parabola gets thinner. Therefore, $y = 2x^2$ is thinner than $y = x^2$, which is thinner than $y = \frac{1}{2}x^2$ (see Fig. 8.33).

Example 3   Graph each quadratic function.
(a) $y = 2x^2 - 4$ (b) $y = -\frac{1}{2}(x - 5)^2$

Solution: (a) Shift the graph of $y = 2x^2$ (which is thinner than the graph of $y = x^2$) down 4 units to produce the graph shown in Fig. 8.34.

                (b) Since a is negative, flip the graph of $y = \frac{1}{2}x^2$ (which is fatter than the graph of $y = x^2$) over the x-axis. Then shift this graph to the right 5 units to produce the graph shown in Fig. 8.35.   ❐

Try Exercise 29

Graph: $y = \frac{1}{2}x^2 - 4$.

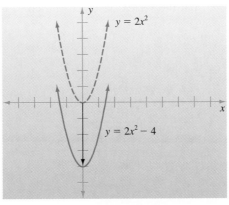

Figure 8.34   *Figure 8.35*

We summarize the concepts of shifting, flipping, and stretching a parabola below:

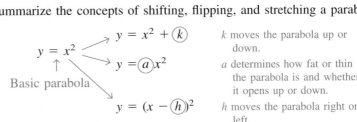

$y = x^2 + \boxed{k}$ k moves the parabola up or down.

$y = \boxed{a}x^2$ a determines how fat or thin the parabola is and whether it opens up or down.

$y = (x - \boxed{h})^2$ h moves the parabola right or left.

The parabola $y = ax^2 + bx + c$ is called a **vertical parabola** because the axis of symmetry is a vertical line. We can collect the three ideas discussed above into one equation, called the **standard form** for the equation of a vertical parabola.

Standard Form of a Vertical Parabola

The graph of the quadratic function
$$y = a(x - h)^2 + k \qquad (a \neq 0)$$
is a vertical parabola with vertex at (h, k). The parabola opens up if $a > 0$ and down if $a < 0$. The axis of symmetry is the vertical line $x = h$.

Examples 4 and 5 show how to write a quadratic function in standard form by completing the square. If you need to review how to complete the square, see Sec. 6.1.

Example 4   Graph $y = x^2 - 2x - 3$ by writing the equation in standard form.

Solution: First, move the constant to the other side.

$$y = x^2 - 2x - 3 \qquad \text{Original equation}$$
$$y + 3 = x^2 - 2x \qquad \text{Add 3 to both sides.}$$

Complete the square on the right side by adding the square of one-half the coefficient of x to both sides. The coefficient of x is -2, one-half of -2 is -1, and the square of -1 is 1. Therefore add 1 to both sides.

$$y + 3 + 1 = x^2 - 2x + 1 \quad \text{Add 1 to both sides.}$$

$$y + 4 = (x - 1)^2 \quad \text{Simplify the left side, factor the right side.}$$

$$y = (x - 1)^2 - 4 \quad \text{Subtract 4 from both sides.}$$

This parabola has the same size, shape, and direction as the parabola $y = x^2$, but the vertex has been shifted to the point $(1, -4)$, as shown in Fig. 8.36.

Try Exercise 37

Graph $y = x^2 - 4x - 5$ by writing the equation in standard form.

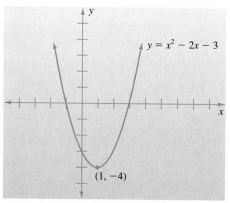

Figure 8.36

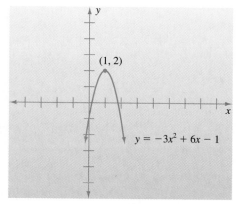

Figure 8.37

Example 5   Graph $y = -3x^2 + 6x - 1$ by writing the equation in standard form.

Solution:

$$y = -3x^2 + 6x - 1 \quad \text{Original equation}$$

$$\frac{y}{-3} = x^2 - 2x + \frac{1}{3} \quad \text{Since } a \neq 1, \text{ divide both sides by } a.$$

$$\frac{y}{-3} - \frac{1}{3} = x^2 - 2x \quad \text{Subtract } \tfrac{1}{3} \text{ from both sides.}$$

$$\frac{y}{-3} - \frac{1}{3} + 1 = x^2 - 2x + 1 \quad \text{Add } (\tfrac{-2}{2})^2 = 1 \text{ to both sides.}$$

$$\frac{y}{-3} + \frac{2}{3} = (x - 1)^2 \quad \text{Simplify the left side, factor the right side.}$$

$$\frac{y}{-3} = (x - 1)^2 - \frac{2}{3} \quad \text{Subtract } \tfrac{2}{3} \text{ from both sides.}$$

$$y = -3(x - 1)^2 + 2 \quad \text{Multiply both sides by } -3.$$

Flip the graph of $y = 3x^2$ (which is thinner than the graph of $y = x^2$) over the x-axis, and move the vertex to the point $(1, 2)$, to produce the parabola shown in Fig. 8.37.

Try Exercise 41

Graph $y = -2x^2 + 4x + 1$ by writing the equation in standard form.

Horizontal Parabolas

If we interchange the variables x and y in the equation $y = ax^2 + bx + c$, we obtain the equation

$$x = ay^2 + by + c$$

This last equation represents a parabola that opens right (if $a > 0$) or left (if $a < 0$). A parabola that opens right or left is called a **horizontal parabola** because its axis of symmetry is a horizontal line.

Example 6   Graph: $x = y^2$.

Solution: This parabola has the same shape and the same vertex, namely, $(0, 0)$, as the parabola $y = x^2$. The only difference is that the parabola $x = y^2$ opens right, rather than up (see Fig. 8.38.).   ◗

Try Exercise 45

Graph: $x = 2y^2$.

Since the graph in Fig. 8.38 fails the vertical line test, the relation $x = y^2$ is *not* a function. In general, **horizontal parabolas do not represent functions.**

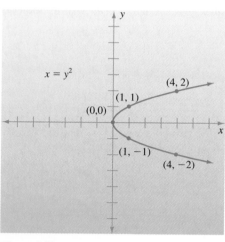

Figure 8.38   *Figure 8.39*

One way to graph a horizontal parabola is to follow the steps for graphing a vertical parabola given in Sec. 8.3, but switch the roles of x and y.

Example 7   Graph: $x = -y^2 + 4$. Label the intercepts and the vertex.

Solution: The x-intercept is 4. Let $x = 0$ to find the y-intercepts.

$$0 = -y^2 + 4 \qquad \text{Let } x = 0.$$
$$y^2 = 4 \qquad \text{Add } y^2.$$
$$y = \pm 2 \qquad \text{Square root property}$$

The two y-intercepts are 2 and -2. The y-value of the vertex is

$$y = \frac{-b}{2a} = \frac{-0}{2(-1)} = 0$$

The x-value of the vertex is

$$x = -0^2 + 4 = 4$$

Try Exercise 47

Graph: $x = -y^2 + 9$.

Since $a = -1$ (a negative number), the parabola opens left, as shown in Fig. 8.39.   ◗

Note that the graph of $x = -y^2 + 4$ is simply the graph of $x = y^2$ flipped over the y-axis and shifted 4 units to the right. That is, the concepts of shifting, flipping, and stretching apply to horizontal parabolas as well as to vertical parabolas.

Another way to graph a horizontal parabola is to write the equation in **standard form.**

Standard Form of a Horizontal Parabola

The graph of

$$x = a(y - k)^2 + h \qquad (a \neq 0)$$

is a horizontal parabola with vertex at (h, k). The parabola opens right if $a > 0$ and left if $a < 0$. The axis of symmetry is the horizontal line $y = k$.

Example 8   Graph $x = y^2 + 4y - 5$ by writing the equation in standard form.

Solution:

$$x + 5 = y^2 + 4y \qquad \text{Add 5 to both sides.}$$
$$x + 5 + 4 = y^2 + 4y + 4 \qquad \text{Add } (\tfrac{4}{2})^2 = 4 \text{ to both sides to complete the square.}$$
$$x + 9 = (y + 2)^2 \qquad \text{Simplify the left side, factor the right side.}$$
$$x = (y + 2)^2 - 9 \qquad \text{Subtract 9 from both sides.}$$

Shift the parabola $x = y^2$ so that the vertex is at $(-9, -2)$ to produce the parabola shown in Fig. 8.40.   ❏

Try Exercise 53

Graph: $x = y^2 - 6y + 5$.

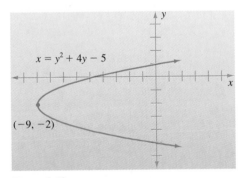

$x = y^2 + 4y - 5$

$(-9, -2)$

Figure 8.40

Exercises 8.4

Problem Solving

1. A ball is thrown vertically upward with an initial velocity of 48 ft/sec. If we ignore air resistance, the ball's height h after t seconds is given by the quadratic function $h = -16t^2 + 48t$. At what time t does the ball reach its maximum height? What is the maximum height?

2. A pharmaceutical firm can manufacture x liters of a drug for an hourly profit of P dollars, given by the quadratic function $P = -2x^2 + 40x - 150$. How many liters x produce maximum hourly profit? What is the maximum hourly profit?

3. A rancher plans to use 8 mi of fencing to construct a rectangular pasture. One side of the rancher's property lies along an existing fence and needs no additional fencing (see Fig. 8.41). What should the width and length of the pasture be to maximize its area? What is the maximum area?

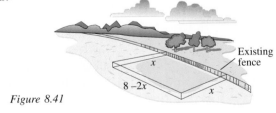

Existing fence

x

$8 - 2x$

x

Figure 8.41

4. The outside edges of a piece of sheet metal that is 16 in. wide are folded up to make a trough (see Fig. 8.42). What should x be to maximize the cross-sectional area of the trough? What is the maximum area?

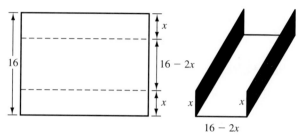

Figure 8.42

5. A car rental company has 50 cars. All of the cars are rented when the rate is $17 per day, but for each $1 increase in the rental rate, two cars are not rented. What daily rental rate produces maximum income for the company?

6. Of all pairs of numbers whose sum is 30, find the pair that has the largest product.

Matching Exercises

Match each quadratic function in Exercises 7 through 12 with its graph in letters A through F.

7. $y = x^2 + 2$

8. $y = x^2 - 2$

9. $y = (x - 2)^2$

10. $y = (x + 2)^2$

11. $y = (x + 2)^2 - 2$

12. $y = (x - 2)^2 + 2$

A.
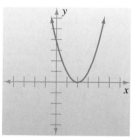

B.
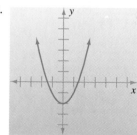

C.
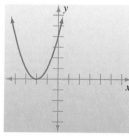

D.
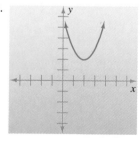

E.
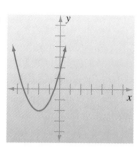

F.
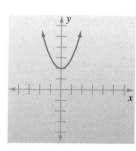

Match each quadratic function in Exercises 13 through 16 with its graph in letters A through D.

13. $y = x^2$

14. $y = -x^2$

15. $y = -2x^2$

16. $y = \frac{1}{2}x^2$

A.
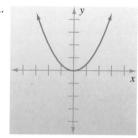

B.
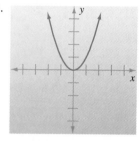

C.
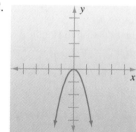

D.
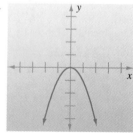

Completion Exercises

17. The graph of $y = a(x - h)^2 + k$ $(a \neq 0)$ is a _____ (horizontal, vertical) parabola with vertex at _____. The parabola opens up if _____ and down if _____. The axis of symmetry is the vertical line _____.

18. The graph of $x = a(y - k)^2 + h$ $(a \neq 0)$ is a _____ (horizontal, vertical) parabola with vertex at _____. The parabola opens right if _____ and left if _____. The axis of symmetry is the horizontal line _____.

Discussion Exercises

Explain how you could shift, flip, and/or stretch the graph of $y = x^2$ (shown in Fig. 8.43) to obtain the graphs of the quadratic functions in Exercises 19 through 34.

19. $y = x^2 - 5$
20. $y = x^2 + 1$
21. $y = (x + 4)^2$
22. $y = (x - 3)^2$
23. $y = (x - 1)^2 + 2$
24. $y = (x + 2)^2 - 5$
25. $y = 3x^2$
26. $y = 4x^2$
27. $y = -\frac{1}{4}x^2$
28. $y = -\frac{1}{3}x^2$
29. $y = \frac{1}{2}x^2 - 4$
30. $y = 2x^2 + 1$
31. $y = -2(x + 1)^2$
32. $y = -\frac{1}{2}(x - 4)^2$
33. $y = -(x - 3)^2 + 1$
34. $y = -(x + 2)^2 - 3$

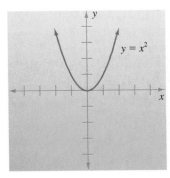

Figure 8.43

Graph each parabola by writing the equation in standard form.

35. $y = x^2 + 2x$
36. $y = x^2 - 6x$
37. $y = x^2 - 4x - 5$
38. $y = x^2 + 2x - 8$
39. $y = -x^2 - 8x - 7$
40. $y = -x^2 + 6x - 5$
41. $y = -2x^2 + 4x + 1$
42. $y = -2x^2 - 4x - 5$

Discussion Exercises

43. If $y = a(x - 4)^2 - 1$ and $a < 0$, how many x-intercepts does the graph have? Explain.

44. How can you tell whether a parabola is horizontal or vertical just by looking at its equation?

Graph each horizontal parabola.

45. $x = 2y^2$
46. $x = \frac{1}{2}y^2$
47. $x = -y^2 + 9$
48. $x = -y^2 + 1$
49. $x = y^2 - 4y$
50. $x = y^2 + 6y$
51. $x = y^2 + 2y + 1$
52. $x = y^2 - 4y + 4$
53. $x = y^2 - 6y + 5$
54. $x = y^2 + 2y - 3$

True-False Exercises

55. Every vertical parabola represents a function.
56. Every horizontal parabola represents a function.

Getting Acquainted with Your Graphing Calculator

Graph all of the equations in each exercise in the same Viewing Rectangle and compare the graphs.

57. $y = x^2$, $y = x^2 - 3$, $y = x^2 + 3$
58. $y = x^2$, $y = (x - 2)^2$, $y = (x + 2)^2$
59. $y = x^2$, $y = \frac{1}{4}x^2$, $y = 4x^2$
60. $y = -x^2$, $y = -2x^2$, $y = -\frac{1}{2}x^2$

Graph each equation.

61. $y = -(x - 3)^2 + 1$
62. $y = (x + 4)^2 - 2$

GRAPHING A RELATION The relation $x = y^2$ is not a function. To graph, solve for y and graph the resulting two functions $y = \sqrt{x}$ and $y = -\sqrt{x}$ in the same Viewing Rectangle.

Graph each relation in Exercises 63 through 68.

63. $x = \frac{1}{3}y^2$
64. $x = -y^2$
65. $x = -y^2 + 4$
66. $x = y^2 - 9$
67. $x = (y - 1)^2$
68. $x = \frac{1}{2}(y + 3)^2$

8.5 The Circle and the Ellipse

In Secs. 8.3 and 8.4 you learned to graph equations of the form $y = ax^2 + bx + c$ and $x = ay^2 + by + c$. If $a \neq 0$, the graph of both of these equations is a parabola. In this section you will learn to graph equations that contain *both* an x^2 term and a y^2 term.

> ***DEFINITION OF A CIRCLE***
>
> A **circle** is the set of all points in the plane that are a fixed distance from a fixed point. The fixed point is called the **center** of the circle, and the fixed distance is called the **radius.**

Consider the circle with center at (h, k) and radius r, as shown in Fig. 8.44.

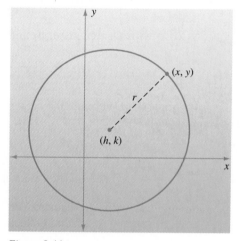

Figure 8.44

If (x, y) is any point on the circle, then the distance between (x, y) and (h, k) is r. Using the distance formula from Sec. 7.1, namely,

$$\sqrt{(x_2 - x_1)^2 + (y_2 - y_1)^2} = d$$

we can write

$$\sqrt{(x - h)^2 + (y - k)^2} = r$$

If we square both sides of this equation, we get the **standard form** for the equation of a circle.

Standard Form of a Circle

An equation for the circle with center (h, k) and radius r is

$$(x - h)^2 + (y - k)^2 = r^2$$

Example 1   Write an equation for the circle with center at $(1, 6)$ and radius 9.

Solution: Use the standard form for the equation of a circle with $(h, k) = (1, 6)$ and $r = 9$.

$$(x - h)^2 + (y - k)^2 = r^2 \qquad \text{Standard form}$$
$$(x - 1)^2 + (y - 6)^2 = 9^2 \qquad \text{Let } h = 1, k = 6, \text{ and } r = 9.$$
$$(x - 1)^2 + (y - 6)^2 = 81 \qquad \text{Simplify.} \qquad \square$$

Try Exercise 9
_____

Write an equation for the circle with center $(3, 5)$ and radius 4.

Example 2   Determine the center and radius of the circle $x^2 + y^2 = 9$.

Solution: This equation can be written as follows:

$$(x - 0)^2 + (y - 0)^2 = 3^2$$

Therefore the center is at $(0, 0)$, and the radius is 3. $\square$

Try Exercise 17
_____

Determine the center and radius of the circle $x^2 + y^2 = 4$.

Generalizing on Example 2, we have the following special case of the standard form of a circle.

Circle with Center at $(0, 0)$ and Radius r

$$x^2 + y^2 = r^2$$

Example 3   Determine the center and radius of the circle $(x - 5)^2 + (y + 2)^2 = 16$.

Solution: The equation can be written as

$$(x - 5)^2 + (y - (-2))^2 = 4^2$$

Therefore the center is at $(5, -2)$, and the radius is 4. $\square$

Try Exercise 23
_____

Determine the center and radius of the circle $(x - 1)^2 + (y + 6)^2 = 144$.

We can write the equation of Example 3 in a different form by multiplying out $(x - 5)^2$ and $(y + 2)^2$ and collecting terms on the left side.

$$(x - 5)^2 + (y + 2)^2 = 16$$
$$x^2 - 10x + 25 + y^2 + 4y + 4 = 16$$
$$x^2 + y^2 - 10x + 4y + 13 = 0$$

This is called the *general form* for the equation of a circle. To graph a circle whose equation is given in general form, we must first convert the equation to standard form by completing the square (refer to Sec. 6.1).

Example 4   Graph: $x^2 + y^2 - 6x + 4y - 12 = 0$.

Solution: Rearrange terms as follows:

$$(x^2 - 6x \quad) + (y^2 + 4y \quad) = 12$$

Then complete the square on $x^2 - 6x$ by adding $(\frac{-6}{2})^2 = 9$ to both sides. Complete the square on $y^2 + 4y$ by adding $(\frac{4}{2})^2 = 4$ to both sides.

$$(x^2 - 6x + 9) + (y^2 + 4y + 4) = 12 + 9 + 4$$

Factor the two perfect-square trinomials on the left side.

Try Exercise 29
_____

Find the center and the radius: $x^2 + y^2 - 4x + 6y - 12 = 0$. Then graph.

$$(x - 3)^2 + (y + 2)^2 = 25$$

The graph is the circle with center at $(3, -2)$ and radius 5 shown in Fig. 8.45. $\square$

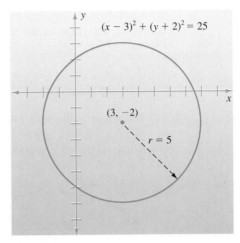

Figure 8.45

The best way to graph the circle in Fig. 8.45 is to use a *compass* (see Fig. 8.46). Open the compass so that the distance between the metal point of the compass and the point of the pencil is 5 units (as measured on the *x*-axis or *y*-axis). Then place the metal point on the center of the circle at $(3, -2)$ and rotate the compass so that the point of the pencil draws the circle.

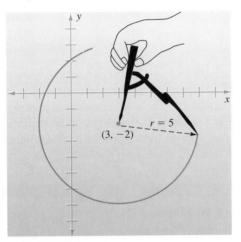

Figure 8.46

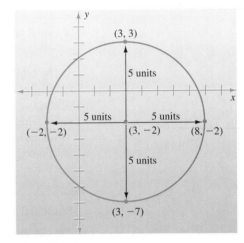

Figure 8.47

If you do not have a compass, start at the center of the circle and draw the four points that are 5 units above, 5 units below, 5 units to the right, and 5 units to the left of the center (see Fig. 8.47). Then draw a circle through these four points.

CAUTION

Equations like

$$(x - 2)^2 + (y - 5)^2 = 0 \quad \text{and} \quad (x - 2)^2 + (y - 5)^2 = -9$$

are special cases. The first equation represents a "circle" with center at (2, 5) and radius 0; that is, the graph is the single point (2, 5). The second equation has no graph, since the sum of two squares cannot be a negative number. ∎

DEFINITION OF AN ELLIPSE

An **ellipse** is the set of all points in the plane, the sum of whose distances from two fixed points is a constant. The two fixed points are called the **foci** (plural of focus) of the ellipse (see Fig. 8.48).

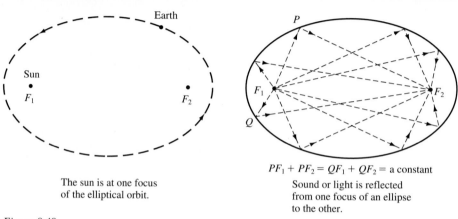

$PF_1 + PF_2 = QF_1 + QF_2 =$ a constant

The sun is at one focus of the elliptical orbit.

Sound or light is reflected from one focus of an ellipse to the other.

Figure 8.48

The orbits of the planets and of artificial satellites are ellipses, as are the cross sections of many camshafts and gears. The Whispering Gallery in the Rotunda of the Capitol Building in Washington, DC, uses its elliptical shape to bounce sound waves from a person whispering at one focus of the ellipse to a person listening at the other focus. In a similar manner, the elliptical reflector in a dentist's light concentrates the rays from a light placed at one focus to the patient's mouth at the other focus.

Standard Form of an Ellipse

An equation of the form

$$\frac{x^2}{a^2} + \frac{y^2}{b^2} = 1 \qquad (a \neq 0, b \neq 0)$$

represents an ellipse with center at the origin.

To find the x-intercepts of an ellipse, let $y = 0$.

$$\frac{x^2}{a^2} + \frac{0^2}{b^2} = 1 \qquad \text{Let } y = 0.$$
$$\frac{x^2}{a^2} = 1 \qquad \text{Since } \frac{0^2}{b^2} = 0$$
$$x^2 = a^2 \qquad \text{Multiply by } a^2.$$
$$x = \pm a \qquad \text{Square root property}$$

Therefore the x-intercepts are a and $-a$ (the square roots of the denominator of x^2/a^2). By letting $x = 0$, we can show that the y-intercepts are b and $-b$ (the square roots of the denominator of y^2/b^2).

Example 5   Graph: $\dfrac{x^2}{9} + \dfrac{y^2}{4} = 1$.

Solution: This represents an ellipse with center at $(0, 0)$. The x-intercepts are 3 and -3, and the y-intercepts are 2 and -2. Draw a smooth, oval-shaped curve through the four intercepts to produce the ellipse shown in Fig. 8.49.   ❏

Try Exercise 39

Graph: $\dfrac{x^2}{25} + \dfrac{y^2}{9} = 1$. Label the intercepts.

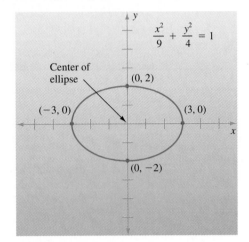

Figure 8.49

We can summarize the procedure for graphing an ellipse as follows.

To Graph an Ellipse Centered at the Origin

1. Write the equation in the **standard form**

$$\frac{x^2}{a^2} + \frac{y^2}{b^2} = 1$$

2. Locate the x-intercepts at a and $-a$, and locate the y-intercepts at b and $-b$.
3. Draw a smooth, oval-shaped curve through the four intercepts.

We illustrate these steps in Example 6.

Example 6   Graph: $25x^2 + 4y^2 = 100$.

Solution: Step 1:   Write the equation in standard form.

$$25x^2 + 4y^2 = 100 \qquad \text{Original equation}$$

$$\frac{25x^2}{100} + \frac{4y^2}{100} = \frac{100}{100} \qquad \text{Divide by 100.}$$

$$\frac{x^2}{4} + \frac{y^2}{25} = 1 \qquad \text{Simplify.}$$

Step 2:   Locate the x-intercepts at 2 and -2, and locate the y-intercepts at 5 and -5 [see Fig. 8.50(a)].

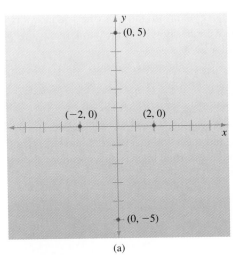

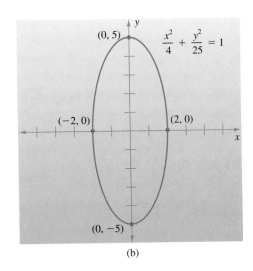

Figure 8.50

Step 3: Draw a smooth, oval-shaped curve through the four intercepts [see Fig. 8.50(b)]. ❏

Try Exercise 43

Graph: $25x^2 + 16y^2 = 400$. Label the intercepts.

If $a^2 = b^2$, the equation $x^2/a^2 + y^2/b^2 = 1$ becomes

$$\frac{x^2}{a^2} + \frac{y^2}{a^2} = 1 \qquad \text{Replace } b^2 \text{ with } a^2.$$

$$x^2 + y^2 = a^2 \qquad \text{Multiply by } a^2.$$

This is a circle with center at the origin and radius a. Therefore a circle is actually a special case of an ellipse whose center and foci are the same point.

Just as some circles are not centered at the origin, some ellipses are not centered at the origin.

Example 7 Graph: $\dfrac{(x-3)^2}{49} + \dfrac{(y+2)^2}{16} = 1$.

Solution: The graph is an ellipse of the same size, shape, and direction as the ellipse $x^2/49 + y^2/16 = 1$ (shown in Fig. 8.51), but the center is at $(3, -2)$, rather than at $(0, 0)$. The graph is shown in Fig. 8.52. Note that the points A and B in both Fig. 8.51 and Fig. 8.52 are located 4 units above and below the center of the ellipse. The points C and D in both figures are located 7 units to the left and right of the center. ❏

Try Exercise 47

Graph:

$\dfrac{(x-4)^2}{36} + \dfrac{(y+1)^2}{16} = 1$.

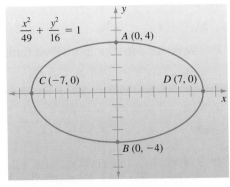

Figure 8.51

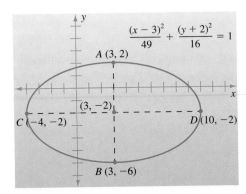

Figure 8.52

Exercises 8.5

Completion Exercises

1. The set of all points in the plane that are a fixed distance from a fixed point is called a(n) _____. The fixed point is called the _____, and the fixed distance is called the _____.

2. The standard form for the equation of a circle with center at (h, k) and radius r is _____.

3. The set of all points in the plane, the sum of whose distances from two fixed points is a constant, is called a(n) _____. The two fixed points are called the _____.

4. The standard form for the equation of an ellipse with center at $(0, 0)$, x-intercepts $\pm a$, and y-intercepts $\pm b$ is _____.

Matching Exercises

Match each equation in Exercises 5 through 8 with its graph in letters A through D.

5. $\dfrac{x}{4} + \dfrac{y}{4} = 1$

6. $\dfrac{x^2}{4} + \dfrac{y}{4} = 1$

7. $\dfrac{x^2}{4} + \dfrac{y^2}{4} = 1$

8. $\dfrac{x^2}{4} + \dfrac{y^2}{9} = 1$

A.
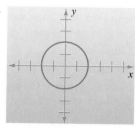

B.
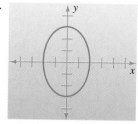

C.
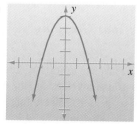

D.
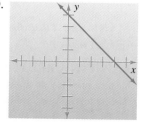

Write an equation for the circle with the given center and radius.

9. Center at $(3, 5)$, radius 4

10. Center at $(1, 4)$, radius 25

11. Center at $(4, -10)$, radius 5

12. Center at $(-9, 5)$, radius 6

13. Center at $(-\frac{1}{2}, -2)$, radius 1

14. Center at $(-3, -\frac{1}{3})$, radius 1

15. Center at $(0, 0)$, radius $\sqrt{2}$

16. Center at $(0, 0)$, radius $\sqrt{3}$

Determine the center and radius of each circle.

17. $x^2 + y^2 = 4$

18. $x^2 + y^2 = 16$

19. $4x^2 + 4y^2 = 25$

20. $9x^2 + 9y^2 = 16$

21. $x^2 + (y - 4)^2 = 16$

22. $x^2 + (y + 2)^2 = 4$

23. $(x - 1)^2 + (y + 6)^2 = 144$

24. $(x + 3)^2 + (y - 4)^2 = 64$

25. $(x + \frac{1}{4})^2 + (y + \frac{3}{4})^2 = \frac{1}{9}$

26. $(x + \frac{1}{5})^2 + (y + \frac{3}{5})^2 = \frac{1}{4}$

27. $(x - 8)^2 + (y - 8)^2 = 8$

28. $(x - 27)^2 + (y - 27)^2 = 27$

Graph each circle. Label the center and the radius.

29. $x^2 + y^2 - 4x + 6y - 12 = 0$

30. $x^2 + y^2 - 6x + 4y - 3 = 0$

31. $x^2 + y^2 - 8x - 10y + 40 = 0$

32. $x^2 + y^2 - 10x - 8y + 40 = 0$

33. $x^2 + y^2 + 2x - 2y - 7 = 0$

34. $x^2 + y^2 + 2x - 2y - 2 = 0$

35. $x^2 + y^2 - 4x = 0$

36. $x^2 + y^2 - 6x = 0$

Discussion Exercises

37. Discuss the graph of each equation.

    (a) $x^2 + y^2 = 0$ (b) $(x - 4)^2 + (y + 1)^2 = 0$

    (c) $x^2 + y^2 + 1 = 0$

38. Tack the ends of a piece of string to a board so that the middle of the string is loose. Then catch the string with the point of a pencil and trace a curve on the board while keeping the string taut (see Fig. 8.53). The curve traced

will be an ellipse. Explain why. How would you modify this procedure to draw a circle?

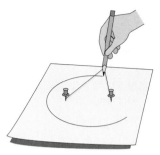

Figure 8.53

Graph each ellipse. Label the intercepts.

39. $\dfrac{x^2}{25} + \dfrac{y^2}{9} = 1$

40. $\dfrac{x^2}{16} + \dfrac{y^2}{9} = 1$

41. $\dfrac{x^2}{81} + \dfrac{y^2}{49} = 1$

42. $\dfrac{x^2}{64} + \dfrac{y^2}{36} = 1$

43. $25x^2 + 16y^2 = 400$

44. $4x^2 + 25y^2 = 100$

45. $x^2 + 4y^2 = 4$

46. $9x^2 + y^2 = 9$

Graph each ellipse.

47. $\dfrac{(x-4)^2}{36} + \dfrac{(y+1)^2}{16} = 1$

48. $\dfrac{(x-3)^2}{81} + \dfrac{(y-1)^2}{9} = 1$

49. $\dfrac{(x+2)^2}{25} + \dfrac{(y+5)^2}{64} = 1$

50. $\dfrac{(x+4)^2}{49} + \dfrac{(y-2)^2}{100} = 1$

Problem Solving

51. An empty cylindrical water glass with a moist rim is inverted and placed so that it covers an equal amount of area in each of the four quadrants. If the diameter of the top of the glass is 3 in., write an equation for the water mark left by the glass.

52. A goat is attached to a stake at the point $(2, -1)$ by a rope that is 9 ft long. If the goat walks clockwise while straining at the rope, write an equation for the goat's path.

53. Write an equation for the circle with center at $(-1, -6)$ that passes through the point $(4, -3)$.

54. Write an equation for the circle with center at $(-2, -4)$ that passes through the point $(5, -1)$.

55. The planet Mars travels around the sun in an elliptical orbit as shown in Fig. 8.54. Draw an x-axis through points A and B on Fig. 8.54. Draw a y-axis through points C and D. Then write an equation for the path of Mars.

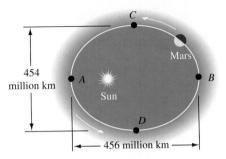

Figure 8.54

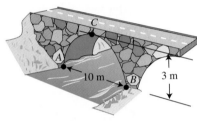

Figure 8.55

56. Stone bridges often have the shape of the top half of an ellipse (see Fig. 8.55). Draw an x-axis through points A and B on Fig. 8.55. Draw a y-axis through point C at the top of the semiellipse. Then write an equation for the semiellipse.

57. Use the definition of an ellipse to sketch, as accurately as you can, the ellipse that passes through the point $(2, 3)$ and has foci at $(2, 0)$ and $(-2, 0)$.

58. Try Exercise 57 by drawing a coordinate system on a board, tacking the ends of a piece of string to the points $(2, 0)$ and $(-2, 0)$, and using the method described in Exercise 38.

Getting Acquainted with Your Graphing Calculator

GRAPHING A RELATION   Circles and ellipses represent relations, but they do not represent functions (they fail the vertical line test). To graph the circle $x^2 + y^2 = 9$, solve for y.

$$x^2 + y^2 = 9 \qquad \text{Equation of circle}$$
$$y^2 = 9 - x^2 \qquad \text{Subtract } x^2.$$
$$y = \pm\sqrt{9 - x^2} \qquad \text{Square root property}$$

The graph of $y = \sqrt{9 - x^2}$ is the upper half of the circle. The graph of $y = -\sqrt{9 - x^2}$ is the lower half. Graph both of these equations in the same Viewing Rectangle to produce the graph of the circle.

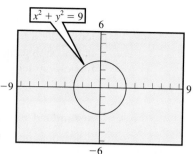

Figure 8.57

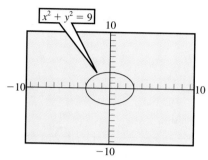

Figure 8.56

Note that the circle in Fig. 8.56 does not appear circular. This is because the tick marks are not spaced equally on both axes. To space the tick marks equally (see Fig. 8.57), either use the SQUARE setting, or set Ymax $-$ Ymin $= \frac{2}{3}$(Xmax $-$ Xmin).

Graph each equation.

59. $x^2 + y^2 = 36$ 60. $16x^2 + 25y^2 = 400$

61. $x^2 + 9y^2 = 9$ 62. $(x - 1)^2 + y^2 = 4$

63. $(x - 2)^2 + (y + 3)^2 = 16$

64. $\dfrac{(x + 4)^2}{9} + \dfrac{(y - 3)^2}{4} = 1$

8.6 The Hyperbola

TAPE AU20

Parabolas, circles, and ellipses fall into a general class of curves known as the *conic sections*. In this section we study the final conic section—the hyperbola. We also summarize all of the conic sections.

DEFINITION OF A HYPERBOLA

A **hyperbola** is the set of all points in the plane such that the positive difference of the distances between each point and two fixed points is a constant (see Fig. 8.58). The two fixed points are called the **foci** of the hyperbola.

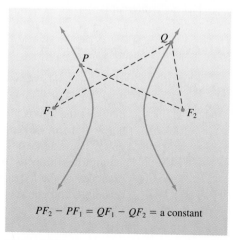

$$PF_2 - PF_1 = QF_1 - QF_2 = \text{a constant}$$

Figure 8.58

Note that a hyperbola consists of two separate curves, called **branches.** Each branch looks like a parabola, but it is not. Hyperbolas are useful in solving problems involving optics, architecture, comets, market-area analysis, and the LORAN (LOng RAnge Navigation) system of navigation for ships and aircraft.

Standard Form of a Hyperbola

An equation of the form

$$\frac{x^2}{a^2} - \frac{y^2}{b^2} = 1 \quad \text{or} \quad \frac{y^2}{b^2} - \frac{x^2}{a^2} = 1 \quad (a \neq 0,\, b \neq 0)$$

represents a hyperbola with center at the origin.

The x-intercepts of the hyperbola whose equation is

$$\frac{x^2}{a^2} - \frac{y^2}{b^2} = 1$$

are a and $-a$. To see this, let $y = 0$.

$$\frac{x^2}{a^2} - \frac{0^2}{b^2} = 1 \qquad \text{Let } y = 0.$$

$$\frac{x^2}{a^2} = 1 \qquad \text{Since } \frac{0^2}{b^2} = 0$$

$$x^2 = a^2 \qquad \text{Multiply by } a^2.$$

$$x = \pm a \qquad \text{Square root property}$$

However, there are no y-intercepts. To see this, let $x = 0$.

$$\frac{0^2}{a^2} - \frac{y^2}{b^2} = 1 \qquad \text{Let } x = 0.$$

$$-\frac{y^2}{b^2} = 1 \qquad \text{Since } \frac{0^2}{a^2} = 0$$

$$y^2 = -b^2 \qquad \text{Multiply by } -b^2.$$

$$y = \pm \sqrt{-b^2} \qquad \text{Square root property}$$

Since $-b^2$ is a negative number, $\sqrt{-b^2}$ is not a real number. Therefore the hyperbola has no y-intercepts.

Example 1 Graph: $\dfrac{x^2}{9} - \dfrac{y^2}{4} = 1.$

Solution: This represents the hyperbola with x-intercepts 3 and -3 shown in Fig. 8.59. To help draw the hyperbola, form the rectangle whose vertical sides intersect the x-axis at 3 and -3 (the square roots of 9 in $x^2/9$) and whose horizontal sides intersect the y-axis at 2 and -2 (the square roots of 4 in $y^2/4$). The extended diagonals of the rectangle are called **asymptotes.** The asymptotes help us to sketch the hyperbola accurately, since the two branches of the hyperbola approach, but never touch, the asymptotes. The rectangle and the asymptotes are dashed, since they are not part of the hyperbola itself. ❐

Try Exercise 11

Graph: $\dfrac{x^2}{16} - \dfrac{y^2}{9} = 1.$ Label the intercepts.

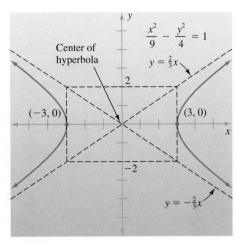

Figure 8.59

The dashed rectangle in Fig. 8.59 is called the **fundamental rectangle** for the hyperbola. The equations of the asymptotes are $y = \frac{2}{3}x$ and $y = -\frac{2}{3}x$. To see why, solve the equation of a hyperbola for y.

$$\frac{x^2}{a^2} - \frac{y^2}{b^2} = 1 \qquad \text{Equation of a hyperbola}$$

$$\frac{x^2}{a^2} - 1 = \frac{y^2}{b^2} \qquad \text{Add } \frac{y^2}{b^2}, \text{ subtract 1.}$$

$$\frac{x^2 - a^2}{a^2} = \frac{y^2}{b^2} \qquad \text{Write the left side as one fraction.}$$

$$\frac{y^2}{b^2} = \frac{1}{a^2}(x^2 - a^2) \qquad \text{Reverse sides, write the quotient as a product.}$$

$$y^2 = \frac{b^2}{a^2}(x^2 - a^2) \qquad \text{Multiply by } b^2.$$

$$y = \pm \frac{b}{a}\sqrt{x^2 - a^2} \qquad \text{Square root property}$$

Now, as the values of x get large, x^2 gets large and $x^2 - a^2$ gets close to x^2. Therefore, the points on the hyperbola $y = \pm \frac{b}{a}\sqrt{x^2 - a^2}$ approach the lines $y = \frac{b}{a}x$ and $y = -\frac{b}{a}x$.

If the equation is of the form

$$\frac{y^2}{b^2} - \frac{x^2}{a^2} = 1 \qquad (a \neq 0,\ b \neq 0)$$

the hyperbola has y-intercepts b and $-b$, and no x-intercepts. However, the fundamental rectangle and the equations of the asymptotes are the same as they were for the hyperbola $x^2/a^2 - y^2/b^2 = 1$.

Example 2   Graph: $\dfrac{y^2}{4} - \dfrac{x^2}{9} = 1$.

Try Exercise 13

Graph: $\dfrac{y^2}{9} - \dfrac{x^2}{16} = 1$. Label the intercepts.

Solution: This equation produces the same fundamental rectangle and asymptotes shown in Fig. 8.59. In this case, however, the y-intercepts of the hyperbola are 2 and -2, and there are no x-intercepts. The graph of this equation is shown in Fig. 8.60.   ❒

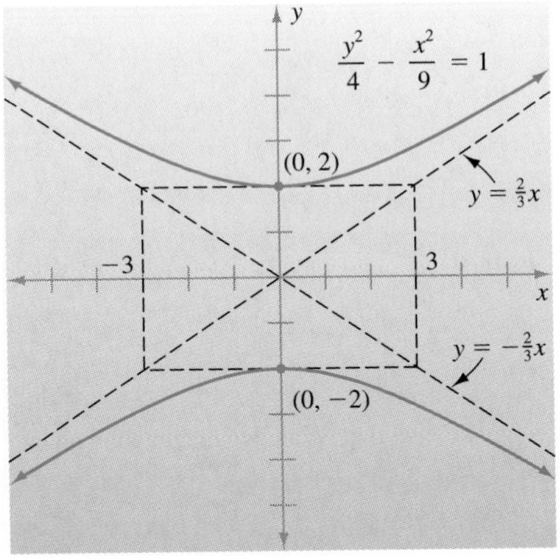

Figure 8.60

We can summarize the procedure for graphing a hyperbola as follows.

To Graph a Hyperbola Centered at the Origin

1. Write the equation in one of the standard forms.

$$\frac{x^2}{a^2} - \frac{y^2}{b^2} = 1 \qquad \text{or} \qquad \frac{y^2}{b^2} - \frac{x^2}{a^2} = 1$$

2. Draw a dashed rectangle whose vertical sides intersect the x-axis at a and $-a$ and whose horizontal sides intersect the y-axis at b and $-b$. This is the fundamental rectangle. Draw two dashed asymptotes by extending the diagonals of the fundamental rectangle.

3. If the x^2-term is positive (the left equation above), locate x-intercepts at a and $-a$. If the y^2-term is positive (the right equation above), locate y-intercepts at b and $-b$. Then draw each branch of the hyperbola so that it passes through an intercept and approaches the asymptotes. The equations of the asymptotes (for either equation above) are $y = \pm\frac{b}{a}x$.

We illustrate these steps in Example 3.

Example 3   Graph: $16x^2 - 25y^2 = 400$.

Solution: Step 1:   Write the equation in standard form.

$$16x^2 - 25y^2 = 400 \qquad \text{Original equation}$$

$$\frac{16x^2}{400} - \frac{25y^2}{400} = \frac{400}{400} \qquad \text{Divide by 400.}$$

$$\frac{x^2}{25} - \frac{y^2}{16} = 1 \qquad \text{Simplify.}$$

Step 2:   Draw a dashed rectangle whose vertical sides intersect the x-axis at 5 and -5, and whose horizontal sides intersect the y-axis at 4 and -4 [see Fig. 8.61(a)]. Draw two dashed asymptotes by extending the diagonals of the fundamental rectangle.

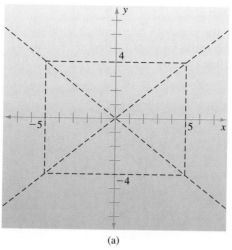
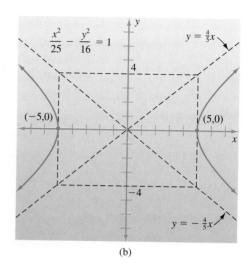

(a)

(b)

Figure 8.61

Step 3:   Since the x^2-term is positive, locate x-intercepts at 5 and -5. Then draw each branch of the hyperbola so that it passes through an intercept and approaches the asymptotes [see Fig. 8.61(b)]. ◻

Try Exercise 17

Graph: $9x^2 - 64y^2 = 576$. Label the intercepts.

Here is an example of how to graph a hyperbola whose center is *not* at the origin.

Example 4   Graph: $\dfrac{(y + 2)^2}{25} - \dfrac{(x - 4)^2}{9} = 1$.

Solution: Shift the graph of $\dfrac{y^2}{25} - \dfrac{x^2}{9} = 1$ (see Fig. 8.62) so that the center is at $(4, -2)$, rather than at $(0, 0)$. This produces the graph shown in Fig. 8.63. Points A and B are located 5 units above and below the center in both figures. Points C and D are located 3 units to the right and left of the center in both figures. ◻

Try Exercise 19

Graph:
$\dfrac{(x - 1)^2}{9} - \dfrac{(y - 3)^2}{36} = 1$.

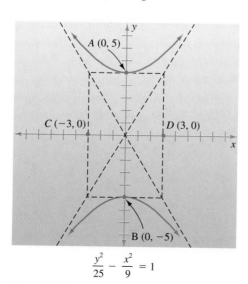

$$\frac{y^2}{25} - \frac{x^2}{9} = 1$$

Figure 8.62

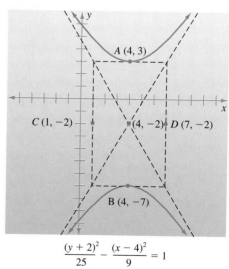

$$\frac{(y + 2)^2}{25} - \frac{(x - 4)^2}{9} = 1$$

Figure 8.63

The Conic Sections

The four curves—the parabola, the circle, the ellipse, and the hyperbola—are collectively known as the **conic sections.*** They are called conic sections because each curve can be formed by intersecting a plane with a (double-napped) cone, as shown in Fig. 8.64.

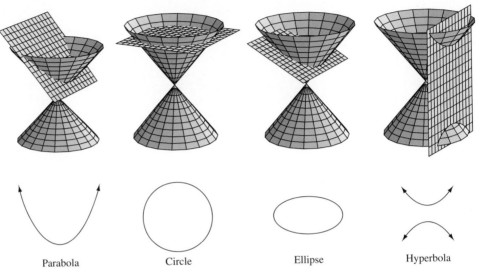

| Parabola | Circle | Ellipse | Hyperbola |

Figure 8.64 The conic sections.

If the plane passes through the *vertex* of the cone, a **degenerate conic** is formed. A degenerate conic may be a point, a line, or two intersecting lines, as shown in Fig. 8.65.

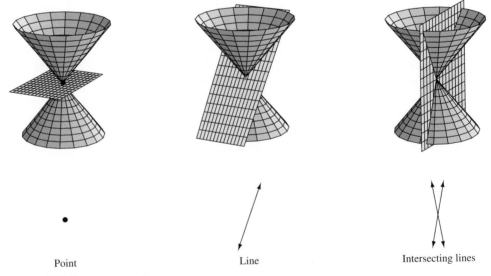

| Point | Line | Intersecting lines |

Figure 8.65 Degenerate conics.

---

*HISTORICAL NOTE: The conic sections were first written about in detail by Appollonius of Perga (ca. 262 B.C.–190 B.C.). It was the work of Appollonius, Archimedes, and Euclid that caused the period from about 300 B.C. to 200 B.C. to be known as the "Golden Age" of Greek mathematics.

Here is a summary of the conic sections.

Vertical Parabola

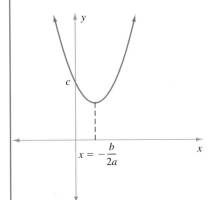

$y = ax^2 + bx + c$

Opens up if $a > 0$.

Opens down if $a < 0$.

y-intercept at c

Let $y = 0$ to find x-intercepts.

Vertex at $x = \dfrac{-b}{2a}$

Horizontal Parabola

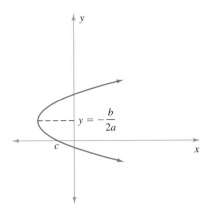

$x = ay^2 + by + c$

Opens right if $a > 0$.

Opens left if $a < 0$.

x-intercept at c

Let $x = 0$ to find y-intercepts.

Vertex at $y = \dfrac{-b}{2a}$

The Circle

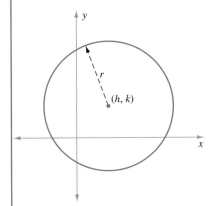

$(x - h)^2 + (y - k)^2 = r^2$

Center at (h, k)

Radius r

The Ellipse

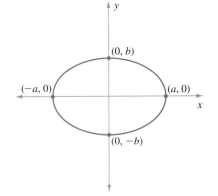

$\dfrac{x^2}{a^2} + \dfrac{y^2}{b^2} = 1$

x-intercepts are a and $-a$.

y-intercepts are b and $-b$.

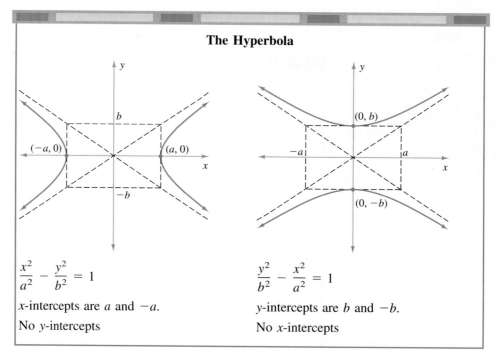

The Hyperbola

$$\frac{x^2}{a^2} - \frac{y^2}{b^2} = 1$$

x-intercepts are a and $-a$.

No y-intercepts

$$\frac{y^2}{b^2} - \frac{x^2}{a^2} = 1$$

y-intercepts are b and $-b$.

No x-intercepts

Example 5 Identify each equation as representing a parabola, a circle, an ellipse, or a hyperbola. Then sketch the graph. (a) $y^2 = 144 - x^2$ (b) $x^2 = y + 8$ (c) $4y^2 = 25x^2 + 100$

Solution: (a) Write the equation as $x^2 + y^2 = 144$. This is a circle with center at $(0, 0)$ and radius 12 (see Fig. 8.66).

(b) Write the equation as $y = x^2 - 8$. Since x is squared but y is not squared, this is a vertical parabola (see Fig. 8.67).

(c) Write the equation in standard form.

$$4y^2 = 25x^2 + 100 \qquad \text{Original equation}$$
$$4y^2 - 25x^2 = 100 \qquad \text{Subtract } 25x^2.$$
$$\frac{4y^2}{100} - \frac{25x^2}{100} = \frac{100}{100} \qquad \text{Divide by 100.}$$
$$\frac{y^2}{25} - \frac{x^2}{4} = 1 \qquad \text{Simplify.}$$

Try Exercise 31

Identify $x^2 = y + 12$ as representing a parabola, a circle, an ellipse, or a hyperbola. Then graph.

This is a hyperbola with y-intercepts 5 and -5 (see Fig. 8.68.) ❐

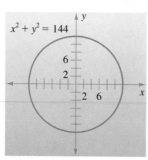

Figure 8.66

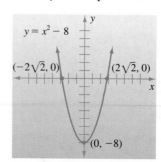

Figure 8.67

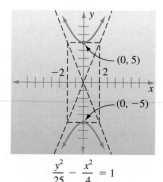

$$\frac{y^2}{25} - \frac{x^2}{4} = 1$$

Figure 8.68

Exercises 8.6

Completion Exercises

1. The set of all points in the plane such that the positive difference of the distances between each point and two fixed points is a constant is called a(n) _____. The two fixed points are called the _____.

2. The standard form for the equation of a hyperbola with center at $(0, 0)$ and fundamental rectangle with vertical sides through $(\pm a, 0)$ and horizontal sides through $(0, \pm b)$ is either _____ or _____.

3. The x-intercepts of $\dfrac{x^2}{a^2} - \dfrac{y^2}{b^2} = 1$ are _____; the y-intercepts are _____.

4. The two separate curves that form a single hyperbola are called _____ of the hyperbola.

5. The extended diagonals of the fundamental rectangle, which are lines that the hyperbola approaches but never touches, are called _____.

6. Parabolas, circles, ellipses, and hyperbolas are collectively known as _____, because they are formed by intersecting a(n) _____ with a(n) _____.

Matching Exercises

Match each equation in Exercises 7 through 10 with its graph in letters A through D.

7. $\dfrac{x^2}{9} - \dfrac{y^2}{4} = 1$

A.
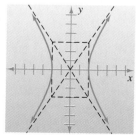

8. $\dfrac{x^2}{4} - \dfrac{y^2}{9} = 1$

B.
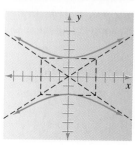

9. $\dfrac{y^2}{9} - \dfrac{x^2}{4} = 1$

C.
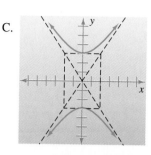

10. $\dfrac{y^2}{4} - \dfrac{x^2}{9} = 1$

D.
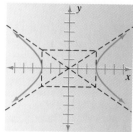

Graph each hyperbola. Label the intercepts.

11. $\dfrac{x^2}{16} - \dfrac{y^2}{9} = 1$

12. $\dfrac{x^2}{25} - \dfrac{y^2}{9} = 1$

13. $\dfrac{y^2}{9} - \dfrac{x^2}{16} = 1$

14. $\dfrac{y^2}{9} - \dfrac{x^2}{25} = 1$

15. $\dfrac{x^2}{4} - \dfrac{y^2}{4} = 1$

16. $\dfrac{x^2}{16} - \dfrac{y^2}{16} = 1$

17. $9x^2 - 64y^2 = 576$

18. $4x^2 - 49y^2 = 196$

Graph each hyperbola.

19. $\dfrac{(x - 1)^2}{9} - \dfrac{(y - 3)^2}{36} = 1$

20. $\dfrac{(x + 4)^2}{25} - \dfrac{(y + 2)^2}{4} = 1$

21. $\dfrac{(y + 5)^2}{49} - \dfrac{x^2}{81} = 1$

22. $\dfrac{y^2}{16} - \dfrac{(x - 3)^2}{64} = 1$

Matching Exercises

Match each equation in Exercises 23 through 26 with its graph in letters A through D.

23. $x^2 - y = 1$

A.

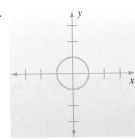

24. $x^2 - y^2 = 1$

B.

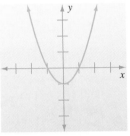

25. $y^2 - x^2 = 1$

C.

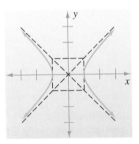

26. $x^2 + y^2 = 1$

D.

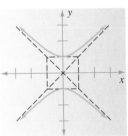

Identify each equation as representing a parabola, a circle, an ellipse, or a hyperbola. Then sketch the graph.

27. $y^2 = 81 - x^2$

28. $x^2 = y - 3$

29. $x^2 = y^2 + 36$

30. $-y^2 = x^2 - 121$

31. $x^2 = y + 12$

32. $x^2 = y^2 + 9$

33. $100y^2 = 900 - 9x^2$

34. $3x^2 - y = 0$

35. $4x^2 + 4y^2 - 64 = 0$

36. $36y^2 = 900 - 25x^2$

37. $2x^2 + y = 0$

38. $36 - 9x^2 - 9y^2 = 0$

39. Graph the degenerate hyperbola $x^2 - y^2 = 0$.

40. Use the definition of a hyperbola to sketch, as accurately as you can, the hyperbola that passes through the point $(2, 3)$ and has foci at $(2, 0)$ and $(-2, 0)$. Can you write an equation for this hyperbola?

Discusson Exercises

41. Explain how to tell from its equation whether a hyperbola opens right and left or up and down.

42. Outline the steps in graphing a hyperbola centered at the origin.

43. A flashlight produces a cone of light. By shining the flashlight against a wall (a plane) at various angles, try to produce a circle, an ellipse, a parabola, and one branch of a hyperbola.

44. Under that conditions would the equation $ax^2 + by^2 = c$ $(a \neq 0, b \neq 0)$ represent (a) a circle? (b) an ellipse? (c) a hyperbola? This equation could not represent a parabola. Why?

Getting Acquainted with Your Graphing Calculator

Graph each equation. (*Hint:* First solve for y.)

45. $y^2 - x^2 = 9$

46. $4x^2 - 25y^2 = 100$

47. $\dfrac{y^2}{16} - \dfrac{(x - 2)^2}{4} = 1$

48. $y - x^2 = 3x + 2$

49. $x^2 + y^2 + 4y - 21 = 0$

50. $16y^2 - 9x^2 + 96y + 72x = 144$

8.7 Special Functions

Three functions that often occur in applications are the *absolute value function,* the *square root function,* and the *greatest integer function.* In this section we discuss the graphs of these functions and their relatives, and we look at an application of one of these functions.

DEFINITION OF THE ABSOLUTE VALUE FUNCTION

The function
$$f(x) = |x|$$

is called the **absolute value function.**

Example 1   Graph: $f(x) = |x|$.

Solution: If $x = 3$, then $f(x) = f(3) = |3| = 3$. By letting x take on the values 3, 2, 1, 0, -1, -2, and -3, we construct the table of values shown below. The graph consists of two half lines joined at their endpoints, as shown in Fig. 8.69. Note that the doman of f is $(-\infty, \infty)$, and the range is $[0, \infty)$.   ❏

Try Exercise 9

Graph: $f(x) = |x| + 1$.

x	$f(x)$
3	3
2	2
1	1
0	0
-1	1
-2	2
-3	3

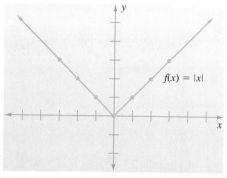

Figure 8.69

A more precise way to graph $y = |x|$ utilizes the definition of absolute value. Recall that

$$y = |x| = \begin{cases} x & \text{if } x \geq 0 \\ -x & \text{if } x < 0 \end{cases}$$

Therefore, the graph of $y = |x|$ consists of the portion of the line $y = x$ where $x \geq 0$, and the portion of the line $y = -x$ where $x < 0$ (see Fig. 8.70).

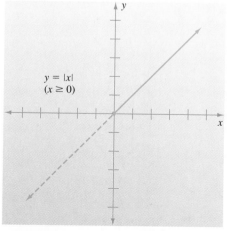

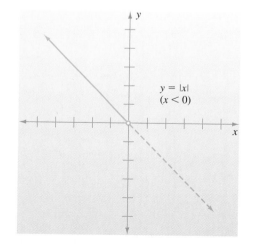

Figure 8.70

Figure 8.71 illustrates how we can use the principles of shifting, flipping, and stretching, discussed in Sec. 8.4, to graph three relatives of the absolute value function $y = |x|$.

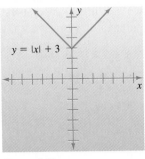
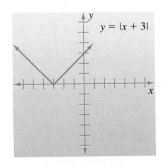
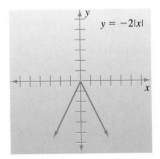

Figure 8.71

Another useful function is the square root function.

DEFINITION OF THE SQUARE ROOT FUNCTION

The function

$$f(x) = \sqrt{x}$$

is called the **square root function.**

Example 2   Graph: $f(x) = \sqrt{x}$.

Solution: Since the square root of a negative number is not a real number, choose only nonnegative x-values when constructing the table of values. From the graph, shown in Fig. 8.72, we note that both the domain and range of f is $[0, \infty)$.   ❑

Try Exercise 21

Graph: $f(x) = \sqrt{x} + 4$.

x	$f(x)$
0	0
1	1
2	$\sqrt{2}$
3	$\sqrt{3}$
4	2
9	3

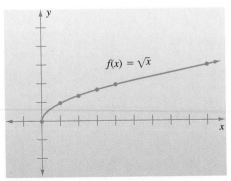

Figure 8.72

Figure 8.73 illustrates the graphs of three relatives of the square root function.

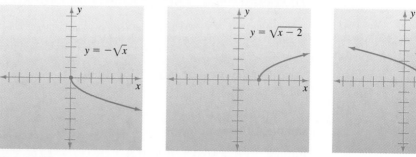

Figure 8.73

If you square both sides of either $y = \sqrt{x}$ or $y = -\sqrt{x}$, you get $y^2 = x$, which is the horizontal parabola shown in Fig. 8.74. The function $y = \sqrt{x}$ represents the top half of this parabola, and the function $y = -\sqrt{x}$ represents the bottom half.

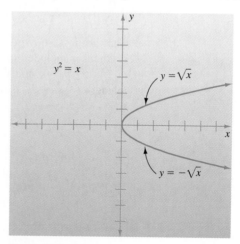

Figure 8.74

Before we define the greatest integer function, we need to define the symbol $[\![x]\!]$.

$$[\![x]\!] = \text{the greatest integer less than or equal to } x$$

Example 3   Evaluate: (a) $[\![4\frac{1}{2}]\!]$, (b) $[\![\pi]\!]$, (c) $[\![10]\!]$, (d) $[\![-1.1]\!]$.

Solution: (a) $[\![4\frac{1}{2}]\!] = 4$ 4 is the greatest integer less than or equal to $4\frac{1}{2}$.

(b) $[\![\pi]\!] = 3$ 3 is the greatest integer less than or equal to π.

(c) $[\![10]\!] = 10$ 10 is the greatest integer less than or equal to 10.

(d) $[\![-1.1]\!] = -2$ -2 is the greatest integer less than or equal to -1.1.   ∎

Try Exercise 29

Evaluate: $[\![7.8]\!]$.

Note that if x is an integer, $[\![x]\!]$ is x. If x is not an integer, $[\![x]\!]$ is the integer immediately to the left of x on the number line.

DEFINITION OF THE GREATEST INTEGER FUNCTION

The function

$$f(x) = [\![x]\!]$$

is called the **greatest integer function.**

Example 4   Graph: $f(x) = [\![x]\!]$.

Solution: If $x = 0.5$, then $f(x) = f(0.5) = [\![0.5]\!] = 0$. In fact, if x is any number in the interval $[0, 1)$, then $f(x) = [\![x]\!] = 0$. If x is in $[1, 2)$, then $f(x) = [\![x]\!] = 1$. In general,

$$f(x) = [\![x]\!] = \begin{cases} 2 & \text{if } 2 \le x < 3 \\ 1 & \text{if } 1 \le x < 2 \\ 0 & \text{if } 0 \le x < 1 \\ -1 & \text{if } -1 \le x < 0 \\ -2 & \text{if } -2 \le x < -1 \end{cases}$$

The graph is shown in Fig. 8.75. Note that the left endpoint of each line segment is included in the graph, but the right endpoint is not. The domain of f is $(-\infty, \infty)$, and the range is $\{\ldots, -2, -1, 0, 1, 2, \ldots\}$, the set of integers.   ❑

Try Exercise 37

Graph: $y = 2[\![x]\!]$.

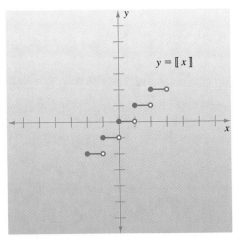

Figure 8.75

The greatest integer function belongs to a class of functions known as **step functions.** They are called step functions because their graphs resemble a series of steps. Step functions appear in many types of applied problems.

Problem Solving

Example 5   A parking garage charges $2 for the first hour (or part of an hour) and $1 for each hour (or part of an hour) after the first (up to a maximum of 4 hr). Write the cost $C(x)$ as a function of the number of hours parked, x. Then graph the function.

Solution: If you park for an hour or less, the cost is $2. That is, if x is in (0, 1], then $C(x) = \$2$. If x is in (1, 2], then $C(x) = \$2 + \$1 = \$3$. Continuing to reason in this fashion, we obtain the function below.

$$C(x) = \begin{cases} 2 & \text{if } 0 < x \le 1 \\ 3 & \text{if } 1 < x \le 2 \\ 4 & \text{if } 2 < x \le 3 \\ 5 & \text{if } 3 < x \le 4 \end{cases}$$

We can write this function using the greatest integer function as follows:

$$C(x) = 1 - [\![-x]\!] \qquad \text{if } 0 < x \le 4$$

The graph is shown in Fig. 8.76. In contrast to the graph shown in Fig. 8.75, this graph includes the right end point of each line segment instead of the left.   ◻

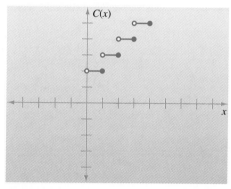

Figure 8.76

When a function is defined by different rules over different parts of its domain, it is called a **piecewise-defined function.**

Example 6   Given the piecewise-defined function

$$f(x) = \begin{cases} 2 & \text{if } x > 1 \\ 2x & \text{if } -1 \le x \le 1 \\ -2 & \text{if } x < -1 \end{cases}$$

find each of the following: (a) $f(5)$, (b) $f(0.7)$, (c) $f(-3)$.

Solution: (a) Since $5 > 1$, use the top rule for f.

$$f(5) = 2$$

(b) Since $-1 \le 0.7 \le 1$, use the middle rule for f.

$$f(0.7) = 2(0.7) = 1.4$$

(c) Since $-3 < -1$, use the bottom rule for f.

$$f(-3) = -2 \qquad ◻$$

Try Exercise 41

To rent a post-hole digger, it costs $3 for the first hour (or part of an hour) and $1 for each hour (or part of an hour) after the first (up to a maximum of 4 hr). Write the cost $C(x)$ of renting a post-hole digger as a function of the number of hours rented, x. Then graph the function.

Try Exercise 45

Given

$$f(x) = \begin{cases} x^2 & \text{if } x \ge 2 \\ x + 1 & \text{if } x < 2 \end{cases}$$

find $f(1)$.

Example 7   Graph the piecewise-defined function of Example 6.

Solution: The graph of $f(x) = 2$ is a horizontal line through $(0, 2)$, but we graph only that part of the line for which $x > 1$ [see Fig. 8.77(a)]. The graph of $f(x) = 2x$ is a line with slope 2 and y-intercept 0, but we graph only that part of the line for which $-1 \le x \le 1$ [see Fig. 8.77(b)]. Finally, we graph that part of the horizontal line $f(x) = -2$ for which $x < -1$ [see Fig. 8.77(c)]. Combining these graphs produces the graph shown in Fig. 8.78. ❏

Try Exercise 55

Graph:

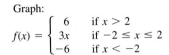

$$f(x) = \begin{cases} 6 & \text{if } x > 2 \\ 3x & \text{if } -2 \le x \le 2 \\ -6 & \text{if } x < -2 \end{cases}$$

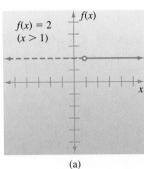

$f(x) = 2$
$(x > 1)$

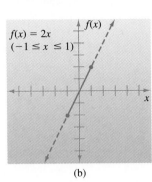

$f(x) = 2x$
$(-1 \le x \le 1)$

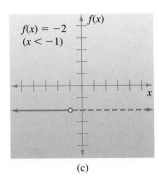

$f(x) = -2$
$(x < -1)$

Figure 8.77   (a)   (b)   (c)

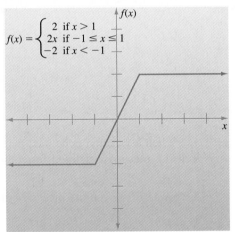

$$f(x) = \begin{cases} 2 & \text{if } x > 1 \\ 2x & \text{if } -1 \le x \le 1 \\ -2 & \text{if } x < -1 \end{cases}$$

Figure 8.78

Exercises 8.7

Completion Exercises

1. The function $f(x) = |x|$ is called the _____ function. The domain is _____. The range is _____.

2. The function $f(x) = \sqrt{x}$ is called the _____ function. The domain is _____. The range is _____.

3. The function $f(x) = [\![x]\!]$ is called the _____ function. The domain is _____. The range is _____.

4. The symbol $[\![x]\!]$ represents the _____ integer _____ or equal to x.

Matching Exercises

Match each function in Exercises 5 through 8 with its graph in letters A through D.

5. $y = |x - 2|$

6. $y = |x| - 2$

7. $y = 2|x|$

8. $y = \frac{1}{2}|x|$

A.

B.

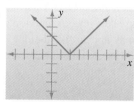

C.

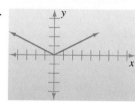

D.

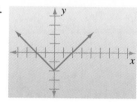

Graph each function.

9. $f(x) = |x| + 1$

10. $f(x) = |x| - 4$

11. $g(x) = |x - 3|$

12. $g(x) = |x + 2|$

13. $y = -\frac{1}{4}|x|$

14. $y = -3|x|$

15. $y = |x + 1| - 2$

16. $y = |x - 1| + 3$

Matching Exercises

Match each function in Exercises 17 through 20 with its graph in letters A through D.

17. $y = \sqrt{x} - 1$

18. $y = \sqrt{x - 1}$

19. $y = \sqrt{1 - x}$

20. $y = -\sqrt{x} - 1$

A.

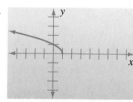

B.

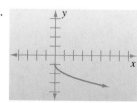

C.

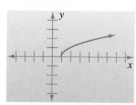

D.

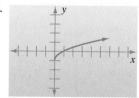

Graph each function.

21. $f(x) = \sqrt{x + 4}$

22. $f(x) = -\sqrt{x + 4}$

23. $h(x) = -\sqrt{x} + 2$

24. $h(x) = \sqrt{x} - 2$

25. $y = \sqrt{x - 5}$

26. $y = \sqrt{3 - x}$

27. $y = \sqrt{-x}$

28. $y = -\sqrt{-x}$

Evaluate.

29. $[\![7.8]\!]$

30. $[\![15]\!]$

31. $[\![0]\!]$

32. $[\![\frac{7}{3}]\!]$

33. $[\![-3\frac{1}{5}]\!]$

34. $[\![-0.2]\!]$

35. $[\![-\pi]\!]$

36. $[\![-\sqrt{2}]\!]$

Matching Exercises

Match each function in Exercises 37 through 40 with its graph in letters A through D.

37. $y = 2[\![x]\!]$

38. $y = [\![2x]\!]$

39. $y = -[\![x]\!]$

40. $y = [\![-x]\!]$

A.

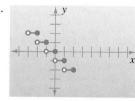

B.

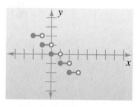

C.

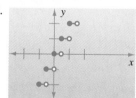

D.

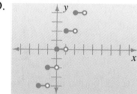

Problem Solving

41. To rent a post-hole digger, it costs $3 for the first hour (or part of an hour) and $1 for each hour (or part of an hour) after the first (up to a maximum of 4 hr). Write the cost $C(x)$ of renting a post-hole digger as a function of the number of hours rented, x. Then graph the function.

42. A parcel delivery service charges $5 for the first pound (or part of a pound) and $3 for each pound (or part of a pound) after the first (up to a maximum of 4 lb). Write the cost $C(x)$ of delivering a parcel as a function of the number of pounds, x. Then graph the function.

43. A long-distance phone call costs $1.50 for the first minute (or part of a minute) and $0.50 for each minute (or part of a minute) after the first (up to a maximum of 5 min). Write the cost $C(x)$ of a call as a function of the number of minutes, x. Then graph the function.

44. A taxi charges $4.50 for the first mile (or part of a mile) and $1.50 for each mile (or part of a mile) after the first (up to a maximum of 5 mi). Write the cost $C(x)$ of a taxi

trip as a function of the number of miles, x. Then graph the function.

Given the piecewise-defined function

$$f(x) = \begin{cases} x^2 & \text{if } x \geq 2 \\ x + 1 & \text{if } x < 2 \end{cases}$$

find each of the following:

45. $f(1)$ 46. $f(3)$

47. $f(2)$ 48. $f(0)$

49. $f(5)$ 50. $f(-2)$

Graph each piecewise-defined function.

51. $f(x) = \begin{cases} 1 & \text{if } x \geq 0 \\ -1 & \text{if } x < 0 \end{cases}$

52. $f(x) = \begin{cases} x^2 & \text{if } x \geq 0 \\ -2 & \text{if } x < 0 \end{cases}$

53. $g(x) = \begin{cases} x^2 & \text{if } x \neq 2 \\ 0 & \text{if } x = 2 \end{cases}$ 54. $g(x) = \begin{cases} 3 & \text{if } x \neq 1 \\ 0 & \text{if } x = 1 \end{cases}$

55. $f(x) = \begin{cases} 6 & \text{if } x > 2 \\ 3x & \text{if } -2 \leq x \leq 2 \\ -6 & \text{if } x < -2 \end{cases}$

56. $f(x) = \begin{cases} 2 & \text{if } x > 4 \\ \frac{1}{2}x & \text{if } -4 \leq x \leq 4 \\ -2 & \text{if } x < -4 \end{cases}$

Each function in Exercises 57 through 60 represents either the upper or the lower half of a parabola, a circle, an ellipse, or a hyperbola. Square both sides of the equation to identify the conic section. Then sketch the graph of the original function.

57. $y = \sqrt{16 - x^2}$ 58. $y = \sqrt{x^2 - 25}$

59. $y = -\sqrt{x + 1}$ 60. $y = -\frac{3}{2}\sqrt{4 - x^2}$

/ **Discussion Exercises**

61. Explain why $y = \sqrt{x}$ gives the top half of the parabola $y^2 = x$, and $y = -\sqrt{x}$ gives the bottom half.

62. Think of a real-world situation that could be described using a step function.

 Getting Acquainted with Your Graphing Calculator

Graph each function.

63. $f(x) = |x| - 3$ 64. $f(x) = |x - 4|$

65. $g(x) = |x + 5| + 2$ 66. $g(x) = -\frac{1}{3}|x|$

67. $f(x) = -\sqrt{x}$ 68. $f(x) = \sqrt{x} + 1$

69. $h(x) = \sqrt{4 - x}$ 70. $h(x) = \sqrt{x - 6}$

Evaluate each expression using your calculator.

71. $[\![5.87]\!]$ 72. $[\![-2.25]\!]$

CONNECTED VERSUS DOT MODE   If you graph the greatest integer function $y = [\![x]\!]$ in connected mode, the graph may appear like the one shown in the figure below. Note that your calculator has incorrectly connected the line segments in the graph.

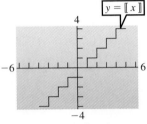

Connected mode

To get the more accurate graph shown in the figure below, change your calculator to dot mode (your calculator may have different names for "connected mode" and "dot mode"). When sketching a graph that has breaks, it is often better to use dot mode.

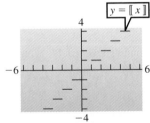

Dot mode

Graph each function in dot mode.

73. $y = [\![x]\!] + 2$ 74. $y = [\![x - 1]\!]$

75. $y = [\![x]\!] - x$ 76. $y = -[\![0.5x]\!]$

GRAPHING PIECEWISE-DEFINED FUNCTIONS
Graph each piecewise-defined function.

77. $f(x) = \begin{cases} 3 & \text{if } x \geq 1 \\ -2 & \text{if } x < 1 \end{cases}$

78. $f(x) = \begin{cases} x^2 & \text{if } x \geq -2 \\ x + 6 & \text{if } x < -2 \end{cases}$

8.8 Increasing/Decreasing Functions; Symmetry

In this section we define what we mean when we say a function is increasing or decreasing. We also discuss how to determine whether a graph is symmetric with respect to the x-axis, the y-axis, and the origin.

Increasing and Decreasing Functions

A function is said to be *increasing* if its graph is rising (from left to right). To describe where a function is increasing, we use x-values only. For example, since the graph of $f(x) = x^2 + 1$ (see Fig. 8.79) is rising to the right of the origin, we say that f is increasing when $x > 0$.

Alternatively, we say that f is increasing on the interval $(0, \infty)$. On the other hand, f is *decreasing* on the interval $(-\infty, 0)$.

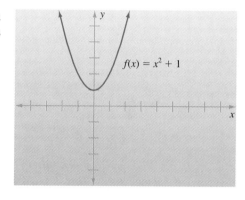

Figure 8.79

Note

Some authors say that $f(x) = x^2 + 1$ is increasing on $[0, \infty)$, but in this text we shall always use *open* intervals when describing where a function is increasing or decreasing.

The following definition of increasing and decreasing functions is illustrated in Fig. 8.80.

> **DEFINITION OF INCREASING AND DECREASING FUNCTIONS**
>
> Let f be a function, and let x_1 and x_2 be two x-values in an open interval I.
>
> 1. f is **increasing** on I if $f(x_1) < f(x_2)$ whenever $x_1 < x_2$.
> 2. f is **decreasing** on I if $f(x_1) > f(x_2)$ whenever $x_1 < x_2$.
> 3. f is **constant** on I if $f(x_1) = f(x_2)$ for all x_1 and x_2 in I.

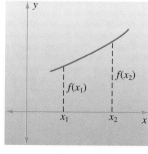
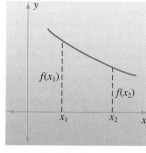
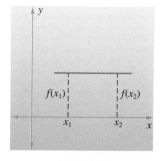

Figure 8.80

$f(x_1) < f(x_2)$
Increasing function

$f(x_1) > f(x_2)$
Decreasing function

$f(x_1) = f(x_2)$
Constant function

Example 1   State the interval(s) on which each function is increasing, decreasing, or constant.

(a) The function f graphed in Fig. 8.81.
(b) The function g graphed in Fig. 8.82.

Solution: (a) f is increasing on $(-\infty, \infty)$.
(b) g is increasing on $(-\infty, -2) \cup (3, \infty)$. g is constant on $(-2, 3)$.   ☐

Try Exercise 15

State the interval(s) on which the function graphed below is increasing, decreasing, or constant.

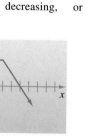

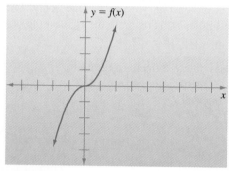

Figure 8.81

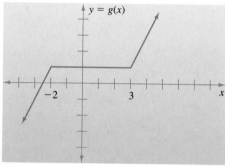

Figure 8.82

Symmetry

The graph of $y = x^2$, shown in Fig. 8.83, is *symmetric* with respect to the y-axis. This means we could "fold" the graph along the y-axis, and the two halves of the graph would coincide. For a graph to be symmetric with respect to the y-axis, the point $(-x, y)$ must be on the graph whenever the point (x, y) is on the graph (see Fig. 8.83). Therefore, we can test the graph of $y = x^2$ for symmetry with respect to the y-axis by replacing x with $-x$ and noting whether the result is equivalent to the original equation.

$$
\begin{aligned}
y &= x^2 && \text{Original equation} \\
y &= (-x)^2 && \text{Replace } x \text{ with } -x. \\
y &= x^2 && \text{Since } (-x)^2 = x^2
\end{aligned}
$$

Since this last equation is equivalent to the original equation, the graph is symmetric with respect to the y-axis.

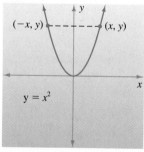

Figure 8.83

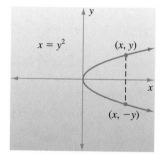

Figure 8.84

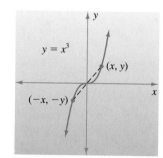

Figure 8.85

The graph of $x = y^2$ in Fig. 8.84 is symmetric with respect to the x-axis. This means that the point $(x, -y)$ is on the graph whenever (x, y) is on the graph. Therefore, when y is replaced by $-y$ in the equation $x = y^2$, the result is equivalent to the original equation.

The graphs in Figs. 8.83 and 8.84 are symmetric with respect to a *line*. The graph in Fig. 8.85 is symmetric with respect to a *point,* namely, the origin. A graph is symmetric with respect to the origin if when it is rotated 180° about the origin, the result coincides with the original graph. For a graph to be symmetric with respect to the origin, the point $(-x, -y)$ must be on the graph whenever (x, y) is on the graph (see Fig. 8.85).

We can summarize the tests for symmetry as follows.

Tests for Symmetry

A graph is **symmetric** with respect to

1. the **x-axis** if replacing y with $-y$ results in an equivalent equation.
2. the **y-axis** if replacing x with $-x$ results in an equivalent equation.
3. the **origin** if replacing both x with $-x$ and y with $-y$ results in an equivalent equation.

Example 2   Test each equation for symmetry with respect to the x-axis, the y-axis, and the origin. (a) $y = 2x$ (b) $y = 3 - |x|$ (c) $x^2 + y^2 = 9$

Solution:
(a) Replacing y with $-y$ gives $-y = 2x$, which is *not* equivalent to the original equation, $y = 2x$. Therefore, the graph is not symmetric with respect to the x-axis. Replacing x with $-x$ gives $y = 2(-x)$, which is *not* equivalent to $y = 2x$. Therefore, the graph is not symmetric with respect to the y-axis. Replacing both x with $-x$ and y with $-y$ gives $-y = 2(-x)$, which *is* equivalent to $y = 2x$. Therefore, the graph is symmetric with respect to the origin. The graph is shown in Fig. 8.86.

(b) *x-axis*

$$-y = 3 - |x|$$
$$y = |x| - 3$$
No

y-axis

$$y = 3 - |-x|$$
$$y = 3 - |x|$$
Yes

Origin

$$-y = 3 - |-x|$$
$$-y = 3 - |x|$$
No

The graph is symmetric with respect to the y-axis (see Fig. 8.87).

(c) *x-axis*

$$x^2 + (-y)^2 = 9$$
$$x^2 + y^2 = 9$$
Yes

y-axis

$$(-x)^2 + y^2 = 9$$
$$x^2 + y^2 = 9$$
Yes

Origin

$$(-x)^2 + (-y)^2 = 9$$
$$x^2 + y^2 = 9$$
Yes

The graph is symmetric with respect to the x-axis, the y-axis, and the origin (see Fig. 8.88). ❏

Try Exercise 35

Test $y = 3x$ for symmetry with respect to the x-axis, the y-axis, and the origin.

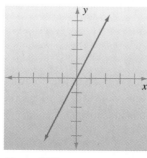

Figure 8.86

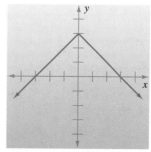

Figure 8.87

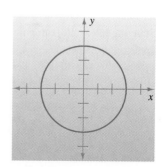

Figure 8.88

<u>CAUTION</u>

A graph may have no, one, or all three of the types of symmetry discussed in this section. **A graph cannot have exactly two of these types of symmetry—x-axis, y-axis, origin.** For example, suppose a graph is symmetric with respect to the *x*-axis but not symmetric with respect to the *y*-axis. Without testing, we know it cannot be symmetric with respect to the origin, because then it would have exactly two of these types of symmetry. ■

Exercises 8.8

Completion Exercises

For Exercises 1 through 3, let *f* be a function, and let x_1 and x_2 be two *x*-values in an open interval *I*.

1. *f* is increasing on *I* if _____ whenever $x_1 < x_2$.

2. *f* is decreasing on *I* if _____ whenever $x_1 < x_2$.

3. *f* is constant on *I* if _____ for all x_1 and x_2 in *I*.

4. A graph is symmetric with respect to the *y*-axis if _____ is on the graph whenever (x, y) is on the graph.

5. A graph is symmetric with respect to the *x*-axis if _____ is on the graph whenever (x, y) is on the graph.

6. A graph is symmetric with respect to the origin if _____ is on the graph whenever (x, y) is on the graph.

True-False Exercises

7. To test for symmetry with respect to the *x*-axis, replace *x* with −*x*.

8. To test for symmetry with respect to the origin, replace *x* with −*y*.

State the interval(s) on which each function is increasing, decreasing, or constant.

9.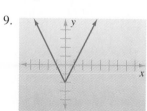

10.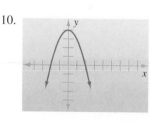

11.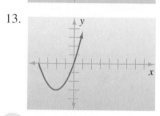

12.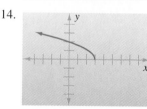

13.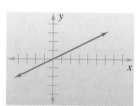

14.

15.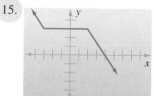

16.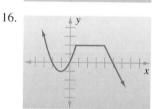

Sketch the graph of each function. Then state the interval(s) on which the function is increasing, decreasing, or constant.

17. $f(x) = x - 3$

18. $f(x) = 5$

19. $g(x) = x^2 + 4x$

20. $g(x) = x^2 - 2x - 3$

21. $h(x) = \begin{cases} 2x & \text{if } x \geq 1 \\ 2 & \text{if } x < 1 \end{cases}$

22. $h(x) = \begin{cases} x^2 & \text{if } x \geq -2 \\ 4 & \text{if } x < -2 \end{cases}$

Plot each point. Then plot the point that is symmetric to the given point with respect to (a) the *x*-axis, (b) the *y*-axis, and (c) the origin.

23. $(4, 2)$

24. $(1, 3)$

25. $(-5, 3)$

26. $(-4, -2)$

27. $(0, -6)$

28. $(-7, 0)$

State whether each graph is symmetric with respect to (a) the x-axis, (b) the y-axis, and (c) the origin.

29.
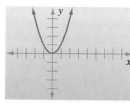

30.
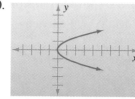

31.
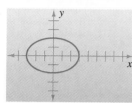

32.
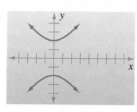

33.
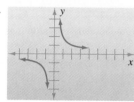

34.
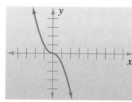

Without graphing, test each equation for symmetry with respect to the x-axis, the y-axis, and the origin.

35. $y = 3x$

36. $y = 5x + 4$

37. $y = |x| - 2$

38. $y = x^3$

39. $x + y = 6$

40. $x + |y| = 3$

41. $x^2 + y^2 = 25$

42. $y = \dfrac{1}{x^2}$

43. $y^2 = x + 1$

44. $x^2 - y^2 = 16$

45. $y = x^3 - x$

46. $y = x^4 + x^2$

Draw a graph that satisfies the conditions of each exercise.

47. Symmetric with respect to the x-axis but not the y-axis

48. Symmetric with respect to the y-axis but not the x-axis

Discussion Exercises

49. Can a graph be symmetric with respect to the x-axis and the y-axis, but not the origin? Explain.

50. Can you think of a function that is symmetric with respect to the x-axis? Can you think of another one? Explain.

Getting Acquainted with Your Graphing Calculator

Graph each function to determine the interval(s) on which the function is increasing, decreasing, or constant.

51. $f(x) = |x - 2|$

52. $f(x) = x^3$

53. $g(x) = x^3 - 3x^2$

54. $g(x) = |x + 1| + |x - 1|$

Graph each equation. Then state whether the graph is symmetric with respect to the x-axis, the y-axis, or the origin.

55. $x = y^2$

56. $y = x^2 - 4$

57. $y = \dfrac{1}{x}$

58. $x^2 + y^2 = 16$

Chapter 8 Key Terms

8.1 Relation A set of ordered pairs

Domain (of a relation) The set of all first coordinates of a relation

Range (of a relation) The set of all second coordinates of a relation

Function A relation in which each first coordinate corresponds to exactly one second coordinate

Dependent variable If y is a function of x, then y is the dependent variable.

Independent variable If y is a function of x, then x is the independent variable.

Vertical line test A graph in the plane represents a function provided that no vertical line intersects the graph at more than one point.

8.3 Constant function $f(x) = k$

Linear function $f(x) = mx + b$

Quadratic function $f(x) = ax^2 + bx + c$ $(a \neq 0)$

Parabola The set of all points in the plane that are equidistant from a fixed line (the **directrix**) and a fixed point (the **focus**) not on the line

Axis of symmetry (of a parabola) The line on which, if the parabola were folded, the two halves of the parabola would coincide

Vertex (of a parabola) The point at which the parabola intersects its axis of symmetry

8.4 **Vertical parabola** A parabola whose axis of symmetry is vertical

Horizontal parabola A parabola whose axis of symmetry is horizontal

8.5 **Circle** The set of all points in the plane that are a fixed distance (the **radius**) from a fixed point (the **center**)

Ellipse The set of all points in the plane, the sum of whose distances from two fixed points (the **foci**) is a constant

8.6 **Hyperbola** The set of all points in the plane such that the positive difference of the distances between each point and two fixed points (the **foci**) is a constant

Branches (of a hyperbola) The two separate curves that form a hyperbola

Asymptotes (of a hyperbola) The two lines that a hyperbola approaches, but never touches

Fundamental rectangle (of a hyperbola) The rectangle whose extended diagonals are the asymptotes of the hyperbola

Conic section A curve formed by intersecting a plane with a cone

Degenerate conic A point, a line, or two intersecting lines formed by passing a plane through the vertex of a cone

8.7 **Absolute-value function** $f(x) = |x|$

Square root function $f(x) = \sqrt{x}$

Greatest integer function $f(x) = [\![x]\!]$

Step function A function whose graph resembles a series of steps

Piecewise-defined function A function defined by different rules over different parts of its domain

8.8 **Increasing function** A function whose graph is rising

Decreasing function A function whose graph is falling

Constant function A function whose graph is neither rising nor falling

Chapter 8 Key Rules/Steps

8.1 Defining Relations and Functions

Identifying Functions

Functions	Not functions
$\{(1, 8), (5, 8)\}$	$\{(3, 4), (3, 2)\}$
$y = x^2$	$y^2 = x$

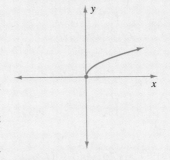 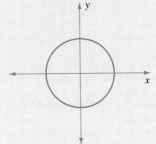

Domain and Range

Example: $\{(-1, 0), (6, 9)\}$, domain $= \{-1, 6\}$, range $= \{0, 9\}$

Example: $y = \dfrac{1}{x^2 - 4}$, domain $= \{x \mid x \neq \pm 2\}$

Example: $y = \sqrt{x - 3}$, domain $= [3, \infty)$, range $= [0, \infty)$

Example: 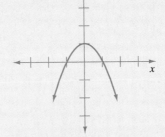 , domain $= (-\infty, \infty)$, range $= (-\infty, 1]$

8.2 Function Notation and Operations on Functions

Function Notation

Range value or value of function

$$y = \widehat{f(x)}$$

Name of function   Domain value

$$f(x) = x^2 + 3x + 1$$
$$f(-5) = (-5)^2 + 3(-5) + 1$$
$$= 25 - 15 + 1$$
$$= 11$$

Operations on Functions

Let $f(x) = x^2$ and $g(x) = 3x + 5$.

Sum:
$$(f + g)(x) = f(x) + g(x) = x^2 + 3x + 5$$

Difference:
$$(f - g)(x) = f(x) - g(x) = x^2 - 3x - 5$$

Product:
$$(fg)(x) = f(x)g(x) = x^2(3x + 5)$$

Quotient:
$$\left(\frac{f}{g}\right)(x) = \frac{f(x)}{g(x)} = \frac{x^2}{3x + 5} \qquad \left(x \neq -\frac{5}{3}\right)$$

Composite:
$$(f \circ g)(x) = f(g(x)) = (3x + 5)^2$$

8.3 Constant, Linear, and Quadratic Functions

Constant function

$f(x) = k$

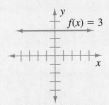

Linear function

$f(x) = mx + b$

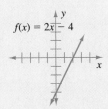

Quadratic function

$f(x) = ax^2 + bx + c \qquad (a \neq 0)$

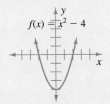

To Graph the Parabola $y = x^2 - 2x - 3$

1. The y-intercept is -3.
2. Let $y = 0$ to find the x-intercepts.

$$0 = x^2 - 2x - 3$$
$$0 = (x + 1)(x - 3)$$
$$x = -1 \qquad x = 3$$

3 & 4. The vertex is at

$$x = \frac{-b}{2a} = \frac{-(-2)}{2(1)} = 1$$
$$y = 1^2 - 2(1) - 3 = -4$$

5. Since $a = 1$ (a positive number), the parabola opens up.
6. Draw a smooth curve through $(-1, 0)$, $(0, -3)$, $(1, -4)$, and $(3, 0)$.

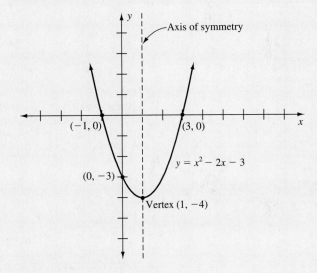

8.4 More about Parabolas

Shifting a Graph

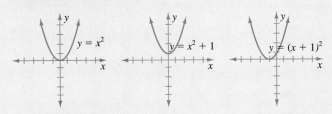

Flipping and Stretching a Graph

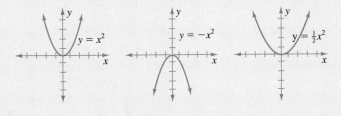

Standard Form of a Vertical Parabola

The graph of

$$y = a(x - h)^2 + k \qquad (a \neq 0)$$

is a vertical parabola with vertex at (h, k). The parabola opens up if $a > 0$ and down if $a < 0$.

To Graph a Parabola by Writing It in Standard Form

$$y = x^2 + 2x - 3$$
$$y + 3 = x^2 + 2x$$
$$y + 3 + 1 = x^2 + 2x + 1 \qquad \text{Complete the square.}$$
$$y + 4 = (x + 1)^2$$
$$y = (x + 1)^2 - 4$$

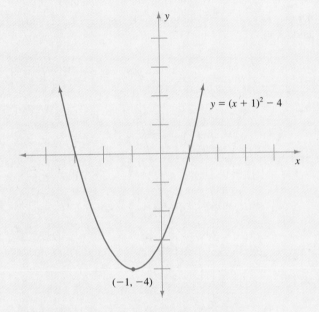

$$y = (x + 1)^2 - 4$$

$(-1, -4)$

Standard Form of a Horizontal Parabola

The graph of

$$x = a(y - k)^2 + h \qquad (a \neq 0)$$

is a horizontal parabola with vertex at (h, k). The parabola opens right if $a > 0$ and left if $a < 0$.

8.5 The Circle and the Ellipse

Standard Form of a Circle

An equation for the circle with center at (h, k) and radius r is

$$(x - h)^2 + (y - k)^2 = r^2$$

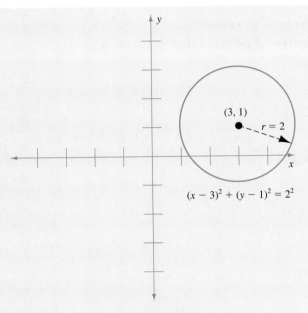

$(3, 1)$

$r = 2$

$$(x - 3)^2 + (y - 1)^2 = 2^2$$

Writing a Circle in Standard Form

$$x^2 + y^2 - 6x + 2y - 15 = 0$$
$$x^2 - 6x + 9 + y^2 + 2y + 1 = 15 + 9 + 1$$
$$(x - 3)^2 + (y + 1)^2 = 25$$

Ellipse Centered at the Origin

$$\frac{x^2}{9} + \frac{y^2}{4} = 1$$

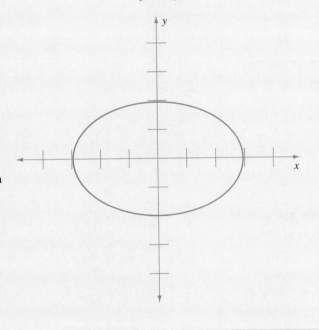

8.6 The Hyperbola

Hyperbola Centered at the Origin

$$\frac{x^2}{9} - \frac{y^2}{4} = 1$$

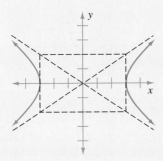

$$\frac{y^2}{4} - \frac{x^2}{9} = 1$$

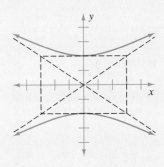

8.7 Special Functions

Absolute-Value Function

$$f(x) = |x|$$

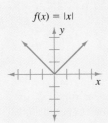

Square Root Function

$$f(x) = \sqrt{x}$$

Greatest Integer Function

$$f(x) = [\![x]\!]$$

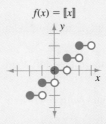

Greatest Integer in x

$[\![x]\!]$ = the greatest integer less than or equal to x

$[\![4.8]\!] = 4$ $[\![7]\!] = 7$ $[\![-2.2]\!] = -3$

Piecewise-Defined Function

$$f(x) = \begin{cases} x^2 & \text{if } x \geq 0 \\ 1 & \text{if } x < 0 \end{cases}$$

$f(3) = 9$ $f(0) = 0$ $f(-2) = 1$

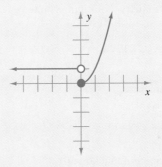

8.8 Increasing/Decreasing Functions; Symmetry

Increasing/Decreasing Functions

Suppose $x_1 < x_2$ in an open interval I.

1. f is increasing on I if $f(x_1) < f(x_2)$.
2. f is decreasing on I if $f(x_1) > f(x_2)$.
3. f is constant on I if $f(x_1) = f(x_2)$.

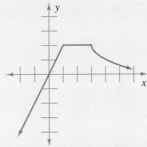

f is increasing on $(-\infty, 1)$, constant on $(1, 3)$, and decreasing on $(3, \infty)$.

Symmetry

A graph is symmetric with respect to

1. the x-axis if replacing y with $-y$ results in an equivalent equation.

2. the y-axis if replacing x with $-x$ results in an equivalent equation.

3. the origin if replacing both x with $-x$ and y with $-y$ results in an equivalent equation.

A graph cannot have exactly two of these types of symmetry.

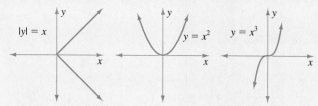

$|y| = x$ $y = x^2$ $y = x^3$

x-Axis symmetry   y-Axis symmetry   Origin symmetry

Chapter 8 Review Exercises

[8.1] State the domain and the range of each relation. Determine whether the relation is a function.

1. $\{(1, 3), (2, 3), (-6, 1), (3, 2)\}$

2. $\{(0, 4), (4, 8), (0, -10)\}$

For each rule, determine whether y is a function of x.

3. $3x - y = 5$ 4. $x = y^2$

5. $y = \sqrt{2 - x}$

Find the domain of each function.

6. $y = 4 - 9x$ 7. $y = \sqrt{7 - x}$

8. $y = \dfrac{x}{x^2 - 9}$

Find the range of each function.

9. $y = x^2 - 5$ 10. $y = 13 - |x|$

Find the domain and the range of each relation graphed in Exercises 11 through 13. Determine whether the relation is a function.

11. 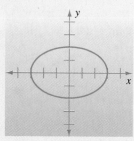   12.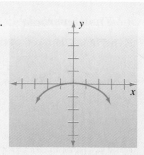

13.

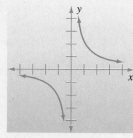

[8.2] For each function, determine (a) $f(2)$, (b) $f(-3)$, (c) $f(t)$, (d) $f(x + h)$.

14. $f(x) = 4x + 9$ 15. $f(x) = 3x^2 - x + 5$

Suppose $f(x) = x^2 - 2x$ and $g(x) = 6 - x$. Determine each of the following. State the domain for Exercises 17, 19, and 21.

16. $(f + g)(4)$ 17. $(f - g)(x)$

18. $(fg)(-1)$ 19. $\left(\dfrac{f}{g}\right)(x)$

20. $(f \circ g)(5)$ 21. $(g \circ f)(x)$

Suppose $f = \{(8, -3), (-4, 5)\}$ and $g = \{(0, -4), (7, 5)\}$. Determine each of the following:

22. $f(8) + g(7)$ 23. $f(g(0))$

24. $g(-4)$

[8.3] Identify as a constant function, a linear function, or a quadratic function. Then graph the function. Label all intercepts and vertices.

25. $f(x) = -1$ 26. $f(x) = -2x - 6$

27. $g(x) = -x^2 + 6x$ 28. $3x - 6y = 18$

29. $y - 7 = 0$ 30. $3x + 4 = y + x^2$

[8.4] Graph each quadratic function by shifting, flipping, and/or stretching the graph of $y = x^2$.

31. $y = x^2 + 3$ 32. $y = -\frac{1}{4}x^2$

33. $y = (x - 2)^2 - 1$

Graph each parabola by writing the equation in standard form.

34. $y = x^2 + 6x$

35. $y = x^2 - 2x - 8$

36. $y = -x^2 - 4x - 4$

Graph each horizontal parabola.

37. $x = y^2$

38. $x = -2y^2 + 8$

39. $x = y^2 - 6y + 8$

40. A farmer plans to use 260 ft of fencing to construct a rectangular pasture. One side of the pasture will lie along a natural barrier and needs no additional fencing. What should the width and length of the pasture be to maximize its area? What is the maximum area?

[8.5]

41. Write an equation for the circle with center at $(2, -3)$ and radius 8.

42. Determine the center and the radius of the circle $(x + 4)^2 + y^2 = 10$.

Graph each circle. Label the center and the radius.

43. $x^2 + y^2 = 16$

44. $x^2 + y^2 + 4x - 6y = 12$

45. $4x^2 + 4y^2 - 8y + 3 = 0$

Graph each ellipse. Label the intercepts in Exercises 46 and 47.

46. $\dfrac{x^2}{49} + \dfrac{y^2}{16} = 1$

47. $25x^2 + y^2 = 25$

48. $\dfrac{(x + 3)^2}{4} + \dfrac{(y - 2)^2}{9} = 1$

49. Write an equation for the circle with center at $(-2, 1)$ that passes through the point $(2, -2)$.

[8.6] Graph each hyperbola. Label the intercepts in Exercises 50, 51, and 52.

50. $\dfrac{x^2}{36} - \dfrac{y^2}{25} = 1$

51. $\dfrac{y^2}{9} - \dfrac{x^2}{9} = 1$

52. $4x^2 - y^2 = 4$

53. $\dfrac{(y - 1)^2}{49} - \dfrac{(x - 4)^2}{16} = 1$

Identify each equation as representing a straight line, a parabola, a circle, an ellipse, or a hyperbola. Do not graph.

54. $9x^2 + 9y^2 = 1$

55. $9x^2 - 16y = 144$

56. $4x - 9y = 1$

57. $y^2 = 64 + x^2$

58. $y^2 = 100 - 2x^2$

59. $x^2 + y^2 + 8x - 4y = 5$

[8.7] Graph each function.

60. $f(x) = |x + 3|$

61. $y = |x - 5| - 2$

62. $g(x) = -\frac{1}{2}|x|$

63. $g(x) = -\sqrt{x} + 1$

64. $y = \sqrt{4 - x}$

65. $f(x) = [\![x]\!]$

Evaluate.

66. $[\![9.9]\!]$

67. $[\![25]\!]$

68. $[\![-7.1]\!]$

69. Given the piecewise-defined function

$$f(x) = \begin{cases} x^2 + 1 & \text{if } x \geq 1 \\ 1 & \text{if } x < 1 \end{cases}$$

find each of the following:

(a) $f(3)$ (b) $f(1)$ (c) $f(-2)$

70. Graph the piecewise-defined function of Exercise 69.

[8.8] State the interval(s) on which each function is increasing, decreasing, or constant.

71.

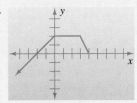

72.

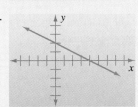

73.

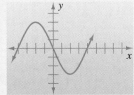

Without graphing, test each equation for symmetry with respect to the x-axis, the y-axis, and the origin.

74. $y = 2x + 5$

75. $y = \dfrac{1}{x}$

76. $x^2 - y^2 = 4$

77. $y = x^2 - 3$

Chapter 8 Test

For Exercises 1 through 3, let $f = \{(4, -6), (2, 3), (3, -9)\}$ and let $g(x) = 5x - 7$.

1. State the domain and range of f. Is f a function?

2. State the domain and range of g. Is g a function?

3. Find $(f + g)(4)$.

If $f(x) = 2x^2 - 3x - 14$, determine the following:

4. $f(-4)$

5. $(f \circ f)(3)$

6. Find the domain of $y = \sqrt{3x - 13}$

7. Evaluate: $[\![8.2]\!]$.

8. State the domain and range of the relation graphed in Fig. 8.89. Does this graph represent a function?

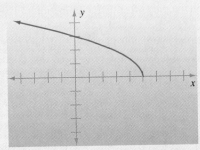

Figure 8.89

9. State the intervals on which the function graphed in Fig. 8.90 is increasing, decreasing, or constant.

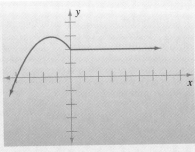

Figure 8.90

10. Without graphing, test $y = x^4 - 3x^2$ for symmetry with respect to the x-axis, the y-axis, and the origin.

11. Write an equation for the circle with center at $(-3, 2)$ that passes through the point $(1, -1)$.

12. The height h of a certain projectile after t seconds is $h = -16t^2 + 80t + 35$. When does the projectile reach maximum height? What is the maximum height?

Graph each equation.

13. $f(x) = 4 - 2x$

14. $y = x^2 + 4x - 5$

15. $x = y^2 - 2y$

16. $g(x) = \sqrt{x + 3}$

17. $\dfrac{x^2}{25} + \dfrac{y^2}{16} = 1$

18. $y = 1 - |x|$

19. $x^2 + y^2 + 4x - 6y = 3$

20. $\dfrac{y^2}{9} - \dfrac{x^2}{4} = 1$

Polynomial and Rational Functions

Many real-life situations can be described using polynomial functions. For example, in Sec. 9.1 we use a polynomial function and its graph to determine the percentage P of alcohol in a person's bloodstream t hours after consuming 4 oz of alcohol.

9.1 Polynomial Functions and Their Graphs

Polynomial Functions

In Sec. 8.3 we studied constant functions, linear functions, and quadratic functions. These functions belong to a broader class of functions known as *polynomial functions*. Some examples of polynomial functions are given below.

Type	Example	Graph
Constant function	$f(x) = 8$	Horizontal line through $(0, 8)$
Linear function	$f(x) = 2x - 6$	Line with slope 2 and y-intercept -6
Quadratic function	$f(x) = x^2 - 4x - 5$	Parabola with vertex at $(2, -9)$
Polynomial function of degree 3	$f(x) = x^3 - x^2 - 4x + 4$	See Example 2

Note that a polynomial function is simply a function of the form

$$f(x) = \text{a polynomial}$$

Here is a complete definition of a polynomial function.

DEFINITION OF A POLYNOMIAL FUNCTION

A **polynomial function of degree n** is a function of the form

$$f(x) = a_n x^n + a_{n-1} x^{n-1} + \cdots + a_2 x^2 + a_1 x + a_0$$

where $a_n, a_{n-1}, \ldots, a_2, a_1,$ and a_0 are real numbers and $a_n \neq 0$.

The number a_n in the definition of a polynomial function is called the **leading coefficient** of $f(x)$, and the number a_0 is called the **constant term.** For example, the leading coefficient of $f(x) = x^3 - x^2 - 4x + 4$ is 1, and the constant term is 4.

The definition of a polynomial function can be extended so that the coefficients $a_n, \ldots, a_1, a_0$ are complex numbers, but in this chapter we shall be interested primarily in polynomial functions with real coefficients.

Graphing $f(x) = x^n$

The simplest polynomial functions are of the form $f(x) = x^n$.

Example 1   Graph each polynomial function: (a) $f(x) = x^3$, (b) $g(x) = x^4$.

Solution: (a) Choose several values for x and compute the corresponding values for $f(x)$. For example, $f(2) = 2^3 = 8$.

x	2	1	$\frac{1}{2}$	0	$-\frac{1}{2}$	-1	-2
$f(x)$	8	1	$\frac{1}{8}$	0	$-\frac{1}{8}$	-1	-8

Plot the points from the table and draw a smooth curve through them to produce the graph shown in Fig. 9.1. Note that the graph is symmetric with respect to the origin.

(b) A table of values for this function is given below:

x	2	1	$\frac{1}{2}$	0	$-\frac{1}{2}$	-1	-2
$g(x)$	16	1	$\frac{1}{16}$	0	$\frac{1}{16}$	1	16

Try Exercise 13

Graph: $y = x^3 - 1$.

The graph is shown in Fig. 9.2. Note that the graph is symmetric with respect to the y-axis. Although the graph of $y = x^4$ is a U-shaped curve, it is not a parabola.

$\square$

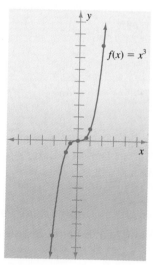

Figure 9.1   Figure 9.2

The graphs of $f(x) = x^n$, where n is a positive *odd* integer, are all similar (see Fig. 9.3). Also, the graphs of $f(x) = x^n$, where n is a positive *even* integer, are all similar (see Fig. 9.4).

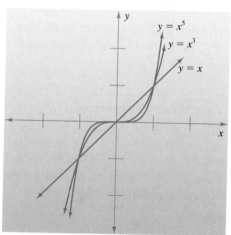

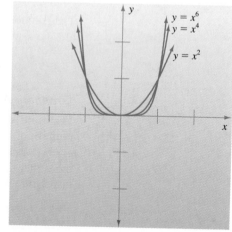

Figure 9.3

Figure 9.4

Note that in both Figs. 9.3 and 9.4, as n gets larger, the graph of $f(x) = x^n$ gets "flatter" inside the interval $[-1, 1]$ on the x-axis and "steeper" outside the interval $[-1, 1]$.

Stretching, Shifting, and Flipping

Compared to the graph of $y = x^3$, the graph of $y = 2x^3$ will be "thinner" (see Fig. 9.5), and the graph of $y = \frac{1}{2}x^3$ will be "fatter" (see Fig. 9.5).

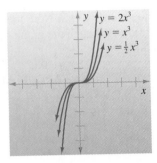

Figure 9.5

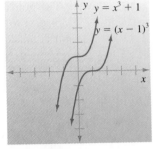

Figure 9.6

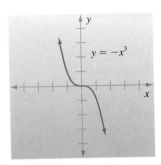

Figure 9.7

The graph of $y = x^3 + 1$ has the same size and shape as the graph of $y = x^3$, but it is shifted up 1 unit (see Fig. 9.6). The graph of $y = (x - 1)^3$ is the graph of $y = x^3$ shifted right 1 unit (see Fig. 9.6). The graph of $y = -x^3$ is the graph of $y = x^3$ flipped over the x-axis (see Fig. 9.7).

Polynomial Equations versus Polynomial Functions

A **polynomial equation of degree n** is an equation that can be written in the form

$$a_n x^n + \cdots + a_1 x + a_0 = 0 \qquad (a_n \neq 0)$$

Note the difference between a polynomial equation and a polynomial function.

Polynomial equation

$$x^2 - 2x - 3 = 0$$
$$(x - 3)(x + 1) = 0$$
$$x = 3 \text{ or } x = -1$$

Polynomial function

$$y = x^2 - 2x - 3$$

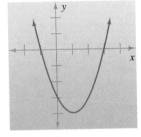

The solutions of the polynomial *equation* $x^2 - 2x - 3 = 0$ are the x-intercepts of the graph of the polynomial *function* $y = x^2 - 2x - 3$. (See Fig. 9.8.)

Figure 9.8

Graphing More Complicated Polynomial Functions

There are some general principles that apply to every polynomial function. First, the **domain** of every polynomial function consists of all real numbers. Second, the graph of every polynomial function is a *smooth, continuous* curve (see Fig. 9.9). By **smooth,** we mean that there are no "corners" in the graph (see Fig. 9.10). By **continuous,** we mean that there are no breaks in the graph (see Fig. 9.11).

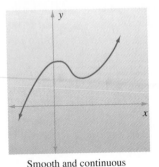

Smooth and continuous

Figure 9.9

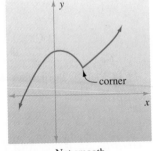

Not smooth

Figure 9.10

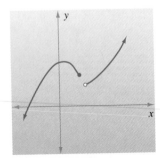

Not continuous

Figure 9.11

Third, the graph of every polynomial function will eventually either rise or fall as x moves to the right and as x moves to the left. To determine whether the graph rises or falls, we note whether the function's degree is even or odd and whether the leading coefficient is positive or negative.

Right and Left Behavior of the Graph of $f(x) = a_n x^n + \cdots + a_1 x + a_0$ $(a_n \neq 0, n \neq 0)$

The dashed portions of the curve indicate that the shape of that part of the curve cannot be determined from the information given.

n is even

The graph either rises on both ends (Fig. 9.12), or falls on both ends (Fig. 9.13), depending on whether a_n is positive or negative.

n is odd

The graph either rises on the right and falls on the left (Fig. 9.14), or rises on the left and falls on the right (Fig. 9.15), depending on whether a_n is positive or negative.

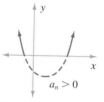

Figure 9.12

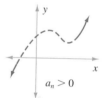

Figure 9.14

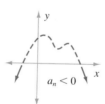

Figure 9.13

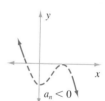

Figure 9.15

The reason that the right and left behavior of $f(x) = a_n x^n + \cdots a_1 x + a_0$ depend only on the highest-degree term $a_n x^n$ is this: The term $a_n x^n$ dominates all of the other terms as x increases without bound (written $x \to \infty$) and as x decreases without bound (written $x \to -\infty$). For example, when $x = 100$, the value of x^3 is 1,000,000, but the value of x^2 is only 10,000. Therefore, the value of $f(x)$ depends more and more on the value of $a_n x^n$ as $x \to \infty$ and as $x \to -\infty$.

To see what makes the graph rise or fall, suppose $a_n x^n = 2x^3$. Since $2x^3 \to \infty$ as $x \to \infty$, the right behavior is $f(x) \to \infty$ (graph rises). Since $2x^3 \to -\infty$ as $x \to -\infty$, the left behavior is $f(x) \to -\infty$ (graph falls).

Recall that to find the y-intercept of a function, we let $x = 0$. If we let $x = 0$ in the polynomial function

$$f(x) = a_n x^n + \cdots + a_1 x + a_0$$

we get

$$f(0) = a_n(0)^n + \cdots + a_1(0) + a_0 = a_0$$

Therefore, the y-intercept of a polynomial function is the constant term.

Here is a summary of the steps we will use to graph more complicated polynomial functions.

To Graph the Polynomial Function $f(x) = a_n x^n + \cdots + a_1 x + a_0$ $(a_n \neq 0, n \neq 0)$

1. Note that the y-intercept is a_0.
2. Let $f(x) = 0$ to find the x-intercepts.
3. Determine the right and left behavior.
4. Choose x-values between the x-intercepts and compute the corresponding $f(x)$-values.
5. Draw a smooth, continuous curve based on the information obtained in steps 1 through 4. Plot additional points if necessary.

We illustrate these steps in Example 2.

Example 2   Graph: $f(x) = x^3 - x^2 - 4x + 4$.

Solution: Step 1:   Note that the y-intercept is a_0.
The y-intercept is 4 (see Fig. 9.16).

Step 2:   Let $f(x) = 0$ to find the x-intercepts.

$$0 = x^3 - x^2 - 4x + 4 \qquad \text{Let } f(x) = 0.$$
$$0 = x^2(x - 1) - 4(x - 1) \qquad \text{Factor by grouping.}$$
$$0 = (x - 1)(x^2 - 4)$$
$$0 = (x - 1)(x - 2)(x + 2) \qquad \text{Factor } x^2 - 4.$$

$$x - 1 = 0 \qquad x - 2 = 0 \qquad x + 2 = 0 \qquad \text{Set each factor equal to 0.}$$
$$x = 1 \qquad \quad x = 2 \qquad \quad x = -2 \qquad \text{Three x-intercepts}$$

Plot the x-intercepts as shown in Fig. 9.16.

Step 3:   Determine the right and left behavior.
Since the degree is odd ($n = 3$) and the leading coefficient is positive ($a_n = 1$), the graph rises to the right and falls to the left (see Fig. 9.16).

Step 4:   Choose x-values between the x-intercepts and compute the corresponding $f(x)$-values.
Choose $x = -1$ and $x = \frac{3}{2}$.

$$f(-1) = (-1)^3 - (-1)^2 - 4(-1) + 4 = -1 - 1 + 4 + 4 = 6$$
$$f\left(\frac{3}{2}\right) = \left(\frac{3}{2}\right)^3 - \left(\frac{3}{2}\right)^2 - 4\left(\frac{3}{2}\right) + 4 = \frac{27}{8} - \frac{9}{4} - 6 + 4 = -\frac{7}{8}$$

Therefore the points $(-1, 6)$ and $(\frac{3}{2}, -\frac{7}{8})$ are on the graph.

Step 5:   Draw a smooth, continuous curve based on the information obtained in steps 1 through 4.
This gives the graph shown in Fig. 9.17.   ❐

Try Exercise 33

Graph:
$f(x) = x^3 + x^2 - 4x - 4$.

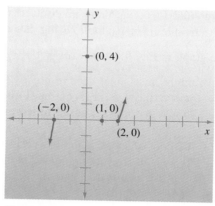

Figure 9.16

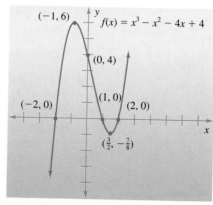

Figure 9.17

Example 3   Graph: $f(x) = -x^4 + 9x^2$.

Solution: Since $f(x) = -x^4 + 9x^2 + 0$, the y-intercept is 0. Let $f(x) = 0$ to find the x-intercepts.

$$0 = -x^4 + 9x^2 \qquad \text{Let } f(x) = 0.$$
$$x^4 - 9x^2 = 0 \qquad \text{Add } x^4, \text{ subtract } 9x^2.$$
$$x^2(x^2 - 9) = 0 \qquad \text{Factor.}$$
$$x^2(x - 3)(x + 3) = 0 \qquad \text{Factor.}$$
$$x^2 = 0 \qquad x - 3 = 0 \qquad x + 3 = 0 \qquad \text{Set each factor equal to 0.}$$
$$x = 0 \qquad x = 3 \qquad x = -3 \qquad \text{Three x-intercepts}$$

The degree is even ($n = 4$) and the leading coefficient is negative ($a_n = -1$), so the graph falls to the right and falls to the left. Letting $x = 2, 1, -1,$ and -2, and computing the corresponding $f(x)$-values, gives the table of values below. For example,

$$f(2) = -2^4 + 9(2)^2 = -16 + 9(4) = -16 + 36 = 20$$

The graph is shown in Fig. 9.18. To make it easier to fit the graph on the page, we changed the scale on the y-axis. Note that the graph is symmetric with respect to the y-axis.   ❒

Try Exercise 35

Graph: $f(x) = -x^4 + 4x^2$.

x	$f(x)$
2	20
1	8
−1	8
−2	20

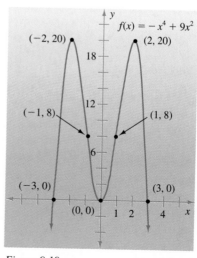

Figure 9.18

To graph polynomial functions more precisely than we have done in this section requires calculus techniques. For example, using calculus we could determine that the high points on the graph in Fig. 9.18 are *exactly* $\left(\dfrac{3\sqrt{2}}{2}, \dfrac{81}{4}\right)$ and $\left(-\dfrac{3\sqrt{2}}{2}, \dfrac{81}{4}\right)$. Also, without calculus we might miss some *turning points* of a graph. A **turning point** is a point at which the graph changes from increasing to decreasing, or from decreasing to increasing. It can be shown that **a polynomial function of degree n has *at most* $n - 1$ turning points.**

Using only the methods discussed in this section, we might mistakenly graph $f(x) = 2x^3 - 9x^2 + 12x$ without any turning points, as shown in Fig. 9.19. A more precise graph, shown in Fig. 9.20, illustrates that the graph has the two turning points $(1, 5)$ and $(2, 4)$.

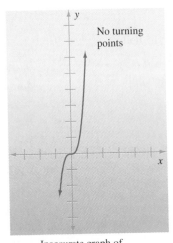

Inaccurate graph of
$f(x) = 2x^3 - 9x^2 + 12x$

Figure 9.19

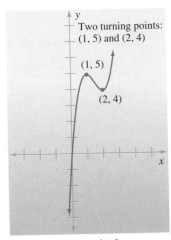

Accurate graph of
$f(x) = 2x^3 - 9x^2 + 12x$

Figure 9.20

Exercises 9.1

Completion Exercises

1. If $a_n \neq 0$, the equation $f(x) = a_n x^n + \cdots + a_1 x + a_0$ is called a _____ function of degree _____. The leading coefficient is _____, and the constant term is _____.

2. When a graph has no corners, we say the graph is _____. When a graph has no breaks, we say the graph is _____.

3. The graph of $f(x) = 6x + x^2 - 2x^3$ _____ (rises, falls) to the right, and _____ (rises, falls) to the left. The graph of $f(x) = x^4 - 3x^3$ _____ (rises, falls) to the right, and _____ (rises, falls) to the left.

4. The equation $x^4 - x^2 - 4x + 4 = 0$ is called a polynomial _____ (equation, function). The equation $y = x^4 - x^2 - 4x + 4$ is called a polynomial _____ (equation, function).

True-False Exercises

5. Every linear function is also a polynomial function.

6. Every quadratic function is also a polynomial function.

7. The graph of $y = x^4$ is "flatter" inside the interval $[-1, 1]$ than the graph of $y = x^2$.

8. The graph of $y = x^4$ is "steeper" outside the interval $[-1, 1]$ than the graph of $y = x^2$.

9. The graph of $y = 2x^4$ is "fatter" than the graph of $y = x^4$.

10. The graph of $y = \frac{1}{2}x^4$ is "thinner" than the graph of $y = x^4$.

11. The graph of every polynomial function is a smooth curve.

12. The graph of every polynomial function is a continuous curve.

By shifting, stretching, and/or flipping the graph of $y = x^3$ (shown in Fig. 9.21), graph each polynomial function in Exercises 13 through 16.

13. $y = x^3 - 1$

14. $y = (x + 2)^3$

15. $y = -\frac{1}{2}x^3$

16. $y = (x - 1)^3 + 3$

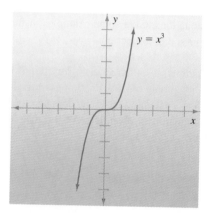

Figure 9.21

By shifting, stretching, and/or flipping the graph of $y = x^4$ (shown in Fig. 9.22), graph each polynomial function in Exercises 17 through 20.

17. $y = (x - 5)^4$

18. $y = x^4 + 4$

19. $y = (x + 1)^4 - 3$

20. $y = -2x^4$

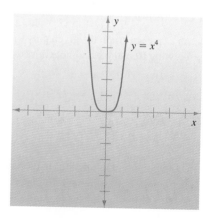

Figure 9.22

Completion Exercises

Suppose $f(x) = a_n x^n + \cdots + a_1 x + a_0$ $(a_n \neq 0, n \neq 0)$ is a polynomial function for Exercises 21 through 26.

21. If n is even and $a_n < 0$, the graph of f _____ (rises, falls) to the right, and _____ (rises, falls) to the left.

22. If n is odd and $a_n > 0$, the graph of f _____ (rises, falls) to the right, and _____ (rises, falls) to the left.

23. The y-intercept of the graph of f is _____.

24. A point at which the graph of f changes from increasing to decreasing, or from decreasing to increasing, is called a(n) _____ of the graph.

25. The graph of f has _____ (how many) turning points.

26. The solutions of the equation $a_n x^n + \cdots + a_1 x + a_0 = 0$ are the _____ of the graph of the polynomial function f.

Graph each polynomial function. As sketching aids, find all intercepts and the right and left behavior.

27. $f(x) = x^3 - 3x^2$

28. $f(x) = x^3 + 2x^2$

29. $f(x) = 2x^3 - 8x$

30. $f(x) = x^3 - 4x$

31. $f(x) = -x^3 - x^2 + 6x$

32. $f(x) = -x^3 + 3x^2 + 10x$

33. $f(x) = x^3 + x^2 - 4x - 4$

34. $f(x) = x^3 - x^2 - 9x + 9$

35. $f(x) = -x^4 + 4x^2$

36. $f(x) = -\frac{1}{2}x^4 + 2x^2$

37. $f(x) = x^4 - 17x^2 + 16$

38. $f(x) = x^4 - 6x^2 + 5$

39. $f(x) = x^5 - 8x^3 + 16x$

40. $f(x) = x^5 - 5x^3 + 4x$

Getting Acquainted with Your Scientific Calculator

41. Show that the graph of $f(x) = x^5$ is "flatter" than the graph of $g(x) = x^3$ inside the interval $[-1, 1]$, but "steeper" outside the interval $[-1, 1]$, by completing the table of values below and then graphing the two functions on the same coordinate system.

x	1.5	1	0.5	0	-0.5	-1	-1.5
$f(x)$							
$g(x)$							

42. Show that the graph of $f(x) = x^4$ is "flatter" than the graph of $g(x) = x^2$ inside the interval $[-1, 1]$, but steeper outside the interval $[-1, 1]$, by completing the table of values on the next page and then graphing the two functions on the same coordinate system.

x	1.5	1	0.5	0	-0.5	-1	-1.5
$f(x)$							
$g(x)$							

Problem Solving

43. The percentage $P(t)$ of alcohol in the average adult's bloodstream t hours after consuming 4 oz of alcohol is approximately given by the polynomial function $P(t) = -0.0015t^3 + 0.106t$. This function is valid over the domain 0 hr $\leq t \leq$ 8 hr. The graph is shown in Fig. 9.23. If a person is legally drunk when the blood alcohol level is greater than 0.10 percent, use Fig. 9.23 to determine the approximate time interval over which a person who consumes 4 oz of alcohol is legally drunk. When is the person's blood alcohol level the highest?

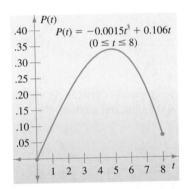

$$P(t) = -0.0015t^3 + 0.106t$$
$$(0 \leq t \leq 8)$$

Figure 9.23

44. A cake pan is to be made by cutting equal size squares from the corners of a piece of metal that is 18 in. by 18 in. and turning up the sides (see Fig. 9.24). Note that the volume of the pan is given by the polynomial function $V(x) = (18 - 2x)^2 x$. The graph of this function is shown in Fig. 9.25. Over what domain is this function valid? What size squares should be cut from the corners to produce maximum volume for the pan?

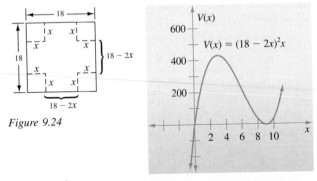

Figure 9.24

$$V(x) = (18 - 2x)^2 x$$

Figure 9.25

Discussion Exercises

45. What is the domain of every polynomial function? Explain your answer.

46. What is the range of a polynomial function of even degree? Does a polynomial function of even degree always intersect the x-axis? Draw graphs to illustrate your answers.

47. What is the range of a polynomial function of odd degree? Why must a polynomial function of odd degree have at least one x-intercept? Draw graphs to illustrate your answers.

48. Outline the steps in graphing the polynomial function $f(x) = a_n x^n + \cdots + a_1 x + a_0$ $(a_n \neq 0, n \neq 0)$.

49. Discuss the difference between a polynomial equation and a polynomial function. Give examples of each.

50. Give some examples of equations that are not polynomial equations. Give some examples of functions that are not polynomial functions.

A function f is called an **even function** if $f(-x) = f(x)$ for all x in the domain of f. A function f is called an **odd function** if $f(-x) = -f(x)$ for all x in the domain of f. Determine whether each function in Exercises 51 through 56 is even, odd, or neither.

51. $f(x) = x^2$

52. $f(x) = 4x^3$

53. $f(x) = x^3 - 6x$

54. $f(x) = x^4 - 3x^2$

55. $f(x) = x^4 + 2x^3 + 8$

56. $f(x) = x^5 + 5x^3 - 7$

Note

The graph of an even function is symmetric with respect to the y-axis, and the graph of an odd function is symmetric with respect to the origin.

Getting Acquainted with Your Graphing Calculator

For each exercise, graph the given polynomial functions in the same Viewing Rectangle and note how the graphs are related.

57. $y = x^3, y = x^3 + 1, y = x^3 - 2$

58. $y = x^4, y = (x + 2)^4, y = (x - 3)^4$

59. $y = x^4, y = \frac{1}{2}x^4, y = 2x^4$

60. $y = x^3, y = -x^3, y = 0.1x^3$

Graph both polynomial functions in the same Viewing Rectangle. Then zoom out far enough to observe that the right and left behavior of both graphs is the same. What seems to determine the right and left behavior of the graph of a polynomial function?

61. $f(x) = 2x^3 - 5x + 1$, $g(x) = 2x^3$

62. $f(x) = \frac{1}{2}x^4 - 2x^2 + 3$, $g(x) = \frac{1}{2}x^4$

Graph each polynomial function. Note that the y-intercept is the constant term of the function. Use your cursor and the zoom-in feature to determine (to the nearest tenth) the coordinates of the x-intercepts and the turning points.

63. $f(x) = x^3 - 5x^2 + 6x + 2$

64. $f(x) = -x^3 + 4x + 1$

65. $f(x) = x^4 - 3x^2 - 1$ 66. $f(x) = 0.1x^4 - x^3 + 2x^2$

9.2 Synthetic Division; Remainder and Factor Theorems

There is a shortcut to the process of long division of polynomials discussed in Sec. 4.7 that can be used when the divisor is of the form $x - k$. To see how this shortcut works, consider the long division shown on the left below:

$$
\begin{array}{r}
2x^2 + 5x + 7 \\
x - 3{\overline{\smash{\big)}\,2x^3 - x^2 - 8x - 10}} \\
\underline{2x^3 - 6x^2} \\
5x^2 - 8x \\
\underline{5x^2 - 15x} \\
7x - 10 \\
\underline{7x - 21} \\
11
\end{array}
\qquad
\begin{array}{r}
2 \quad 5 \quad 7 \\
1 - 3{\overline{\smash{\big)}\,2 \ -1 \ -8 \ -10}} \\
\underline{2 \ -6} \\
5 \ -8 \\
\underline{5 \ -15} \\
7 \ -10 \\
\underline{7 \ -21} \\
11
\end{array}
$$

Now look at this same division with the variables removed, as shown on the right above. The numbers in color are simply repetitions of the numbers directly above them. Omit these numbers and obtain the arrangement shown on the left below:

$$
\begin{array}{r}
2 \quad 5 \quad 7 \\
1 - 3{\overline{\smash{\big)}\,2 \ -1 \ -8 \ -10}} \\
\underline{-6} \\
5 \\
-15 \\
\underline{7} \\
-21 \\
11
\end{array}
\qquad
\begin{array}{r}
1 - 3{\overline{\smash{\big)}\,2 \ -1 \ \ -8 \ \ -10}} \\
\downarrow \ -6 \ -15 \ -21 \\
\hline
2 \ \ 5 \ \ \ 7 \ \ \ 11
\end{array}
$$

Note that the numbers in color in the arrangement on the left above duplicate the numbers on the top row. Therefore omit the top row and condense the remaining numbers as shown on the right above.

The primary cause of the repetitions is the number 1 in $1 - 3$. These repetitions occur because, for any real number a, $a/1 = a$ and $a \cdot 1 = a$. Therefore omit the 1 in $1 - 3$. Then change the sign of the number -3 to obtain its opposite $+3$. Now when each term in the third row is multiplied by $+3$, the resulting term in the second row is the opposite of what it was in the previous arrangement. This means that instead of subtracting we can add, which is easier. The final arrangement is shown in black on the next page:

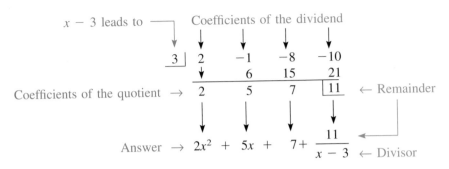

Note that we box the remainder to make it easier to write the answer.

The shortcut procedure outlined above is called **synthetic division**. Remember, **synthetic division can be used only when the divisor can be written in the form $x - k$.**

Example 1   Use synthetic division to divide $3x^2 + x - 14$ by $x + 2$.

Solution: Write $x + 2$ in the form $x - k$ by writing $x + 2 = x - (-2)$. Then write down the coefficients of the dividend and place -2 in a box to the left.

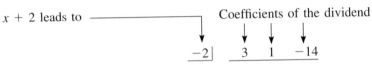

Now follow these steps:

Step 1: Bring down the 3.

Step 2: Multiply -2 and 3.

Step 3: Add 1 and -6.

Step 4: Multiply -2 and -5.

Step 5: Add -14 and 10.

The numbers 3 and -5 are the coefficients of the quotient $3x - 5$. The number -4 is the remainder. Therefore the answer is

$$3x - 5 - \frac{4}{x + 2}$$

Check by multiplication.

Try Exercise 7

Use synthetic division to divide $4x^2 + 10x - 7$ by $x + 3$.

CAUTION

When using synthetic division, make sure you write the dividend in descending order and make sure you insert a 0 for each missing power of the variable in the dividend. ∎

Example 2   Use synthetic division to divide $\dfrac{2x^3 + x^4 - 5x^2 - 58}{x + 4}$.

Solution: Write the dividend in descending order as $x^4 + 2x^3 - 5x^2 - 58$. Then arrange your work as shown below:

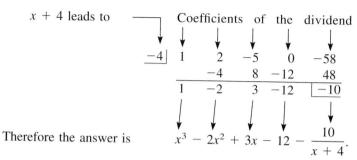

Therefore the answer is $x^3 - 2x^2 + 3x - 12 - \dfrac{10}{x + 4}$.

Try Exercise 17

Use synthetic division to divide:
$\dfrac{3x^3 + x^4 - 2x^2 - 42}{x + 4}$.

The Remainder Theorem

In Example 2, suppose we call the polynomial in the dividend $P(x)$. Consider what happens when we replace x with the number in the box.

$$
\begin{aligned}
P(x) &= x^4 + 2x^3 - 5x^2 - 58 \\
P(-4) &= (-4)^4 + 2(-4)^3 - 5(-4)^2 - 58 \\
&= 256 + 2(-64) - 5(16) - 58 \\
&= 256 - 128 - 80 - 58 \\
&= -10
\end{aligned}
$$

Note that the result -10 is the same as the remainder we obtained in Example 2 when we divided $P(x)$ by $x + 4$. This is more than a coincidence. For example, you can also verify that $P(3)$ equals the remainder obtained when $P(x)$ is divided by $x - 3$. This suggests the following theorem.

The Remainder Theorem

Suppose $P(x)$ is a polynomial. Then $P(k)$ equals the remainder obtained when $P(x)$ is divided by $x - k$.

Sometimes it is easier to find $P(k)$ using the remainder theorem and synthetic division than by substituting k into the polynomial.

Example 3   If $P(x) = 4x^3 - 13x^2 - 30x + 40$, find $P(5)$.

Solution: According to the remainder theorem, $P(5)$ equals the remainder obtained when $P(x)$ is divided by $x - 5$. Use synthetic division to perform the division.

$$
\begin{array}{r|rrrr}
5 & 4 & -13 & -30 & 40 \\
  &   & 20 & 35 & 25 \\
\hline
  & 4 & 7   & 5   & \boxed{65}
\end{array}
$$

Since the remainder is 65, we have $P(5) = 65$. Note that calculating $P(5)$ by substituting 5 for x gives

$$
\begin{aligned}
P(5) &= 4(5)^3 - 13(5)^2 - 30(5) + 40 \\
&= 4(125) - 13(25) - 150 + 40 \\
&= 500 - 325 - 150 + 40 \\
&= 65
\end{aligned}
$$

Try Exercise 29

Use the remainder theorem and synthetic division to find $P(3)$ if
$P(x) = 6x^3 - 7x^2 - 25x + 50$.

Example 4   Determine whether -3 is a solution of $2x^3 - 5x^2 - 39x - 18 = 0$.

Solution: We could determine whether -3 is a solution by substituting $x = -3$ into the equation. However, it is easier to use the remainder theorem and synthetic division.

$$
\begin{array}{r|rrrr}
-3 & 2 & -5 & -39 & -18 \\
   &   & -6 & 33 & 18 \\
\hline
   & 2 & -11 & -6 & \underline{|\,0}
\end{array}
$$

Since the remainder is 0, the value of $2x^3 - 5x^2 - 39x - 18$ is 0 when $x = -3$. This means that -3 is a solution of the equation.   ❏

Try Exercise 37

Use the remainder theorem and synthetic division to determine whether -3 is a solution of $2x^3 + 5x^2 - 13x - 30 = 0$.

The Factor Theorem

In Example 4, we used synthetic division to divide $2x^3 - 5x^2 - 39x - 18$ by $x + 3$. Since the quotient was $2x^2 - 11x - 6$ and the remainder was 0, we can write

$$
\begin{aligned}
2x^3 - 5x^2 - 39x - 18 &= (x + 3)(2x^2 - 11x - 6) + 0 \\
&= (x + 3)(2x^2 - 11x - 6)
\end{aligned}
$$

This means that $x + 3 = x - (-3)$ is a factor of $2x^3 - 5x^2 - 39x - 18$. But Example 4 also showed us that -3 was a solution of the equation $2x^3 - 5x^2 - 39x - 18 = 0$. The next theorem states the relationship between the solutions of a polynomial equation and its factors.

The Factor Theorem

Suppose $P(x)$ is a polynomial. The number k is a solution of the polynomial equation $P(x) = 0$ if and only if $x - k$ is a factor of $P(x)$.

Example 5   Solve: $2x^3 - 5x^2 - 39x - 18 = 0$.

Solution: From Example 4, we know that -3 is a solution. By the factor theorem, $x + 3$ must be one of the factors of the polynomial. The other factor is determined from the bottom row in the synthetic division process of Example 4.

$$
\begin{aligned}
2x^3 - 5x^2 - 39x - 18 &= 0 \qquad \text{Original equation} \\
(x + 3)(2x^2 - 11x - 6) &= 0 \qquad \text{Factor the left side.}
\end{aligned}
$$

Now factor the trinomial and set each factor equal to 0.

$$(x + 3)(2x + 1)(x - 6) = 0$$

$$
\begin{array}{ccccc}
x + 3 = 0 & \text{or} & 2x + 1 = 0 & \text{or} & x - 6 = 0 \\
x = -3 & & x = -\frac{1}{2} & & x = 6
\end{array}
$$

The solution set is $\{-3, -\frac{1}{2}, 6\}$. You can check these solutions by substituting them into the original equation, or by using synthetic division and the remainder theorem.   ❏

Try Exercise 57

Show that $x = -5$ is a solution of $2x^3 + 5x^2 - 28x - 15 = 0$. Then solve the equation.

Example 6   Write a polynomial equation with integer coefficients that has the given solution set. (a) $\{3, -5\}$ (b) $\{\frac{1}{2}, \sqrt{7}, -\sqrt{7}\}$

Solution: (a) Since 3 is a solution of the equation, $x - 3$ is a factor. Since -5 is a solution, $x - (-5) = x + 5$ is a factor. Therefore, an equation with the given solutions is

$$(x - 3)(x + 5) = 0$$
$$x^2 + 2x - 15 = 0$$

Note that we could multiply both sides of this equation by any nonzero number and obtain an equivalent equation (an equation with the same solution set). For example,

$$2(x^2 + 2x - 15) = 2(0)$$
$$2x^2 + 4x - 30 = 0$$

also has the solution set $\{3, -5\}$.

(b) The solution $\frac{1}{2}$ corresponds to either of the factors $x - \frac{1}{2}$ or $2x - 1$. To avoid fractions, choose the factor $2x - 1$.

$$(2x - 1)(x - \sqrt{7})(x + \sqrt{7}) = 0$$
$$(2x - 1)(x^2 - 7) = 0$$
$$2x^3 - x^2 - 14x + 7 = 0 \qquad \qquad \square$$

Try Exercise 61

Write a polynomial equation with integer coefficients that has the solution set $\{4, -7\}$.

Sometimes a solution of a polynomial equation appears twice. Note what happens when we solve the equation $x^3 - 8x^2 + 16x = 0$.

$$x^3 - 8x^2 + 16x = 0 \qquad \text{Original equation}$$
$$x(x^2 - 8x + 16) = 0 \qquad \text{Factor out } x.$$
$$x(x - 4)(x - 4) = 0 \qquad \text{Factor the trinomial.}$$
$$x = 0 \qquad x - 4 = 0 \qquad x - 4 = 0 \qquad \text{Set each factor equal to 0.}$$
$$x = 4 \qquad x = 4 \qquad \text{Solve each linear equation.}$$

The solution 4 appears twice because the factor $x - 4$ appears twice. We say that 4 is a solution of **multiplicity** 2. In the equation

$$x^2(x - 5)^3(x + 8) = 0$$

the number 0 is a solution of multiplicity 2, the number 5 is a solution of multiplicity 3, and the number -8 is a solution of multiplicity 1. **If a solution of multiplicity k is counted k times, a polynomial equation of positive degree n has exactly n solutions.** For example, counting multiplicity the polynomial equation $x^3 + x^2 - 16x + 20 = 0$ has exactly three solutions.

Exercises 9.2

◆◆

True-False Exercises

1. We can use synthetic division to divide $x^3 + x^2 + x - 1$ by $x^2 - 1$.

2. Synthetic division is a shortcut to long division of polynomials that can only be used if the divisor can be written in the form $x - k$.

3. In synthetic division, if the divisor is $x + 6$, the number we place in the box in the upper left corner is 6.

4. When using synthetic division, we must write the dividend in descending order, and we must insert a 0 for each missing power of the variable.

5. Given the synthetic division below, state the divisor, the dividend, the quotient, and the remainder.

$$\begin{array}{r|rrr} -2 & 1 & 2 & -6 & -3 \\ & & -2 & 0 & 12 \\ \hline & 1 & 0 & -6 & \boxed{9} \end{array}$$

6. Use the result of the synthetic division below to factor the polynomial $x^3 - 7x^2 + 7x + 15$.

$$\begin{array}{r|rrrr} 3 & 1 & -7 & 7 & 15 \\ & & 3 & -12 & -15 \\ \hline & 1 & -4 & -5 & \boxed{0} \end{array}$$

Use synthetic division to divide.

7. $(4x^2 + 10x - 7) \div (x + 3)$

8. $(3x^2 + 10x - 13) \div (x + 4)$

9. $(x^2 + 2x - 15) \div (x - 3)$

10. $(x^2 + 3x - 10) \div (x - 2)$

11. $(2x^3 - 5x^2 - 6x - 10) \div (x - 4)$

12. $(2x^3 - 7x^2 - 6x - 25) \div (x - 5)$

13. $\dfrac{y^3 + y^2 - 4}{y - 1}$

14. $\dfrac{y^3 + y^2 - 5}{y - 1}$

15. $\dfrac{-4m^3 + 2m^2 - 7m}{m + 1}$

16. $\dfrac{-5m^3 + 3m^2 - 9m}{m + 1}$

17. $\dfrac{3x^3 + x^4 - 2x^2 - 42}{x + 4}$

18. $\dfrac{6x^3 + 11x^2 + x^4 - 59}{x + 4}$

19. $\dfrac{2k^4 + 2k^3 - 5k + 9}{k + 1}$

20. $\dfrac{3k^4 + 3k^3 - 7k + 4}{k + 1}$

21. $(y^5 + 32) \div (y + 2)$

22. $(y^5 + 1) \div (y + 1)$

23. $(4p^3 + 5p - 3) \div (p - \frac{1}{2})$

24. $(8p^3 + 2p - 2) \div (p - \frac{1}{2})$

Completion Exercises

25. The remainder theorem states that if the remainder obtained when the polynomial $P(x) = 3x^3 - 5x^2 + 4x - 4$ is divided by $x - 2$ is 8, then $P(2) = $ _____.

26. The remainder theorem states that if $P(x)$ is a polynomial, then the remainder obtained if $P(x)$ is divided by $x - k$ is equal to _____.

✏ Discussion Exercises

27. If $P(x) = x^4 + 9x^3 + 17x^2 + 26x + 43$, find $P(-7)$ by (a) substituting -7 for x, and (b) using the remainder theorem and synthetic division. Explain why the method used in part (b) was easier.

28. Given the two synthetic divisions below, which number, 1 or 4, is a solution of $x^3 - 6x^2 + 10x - 8 = 0$? Explain why.

$$\begin{array}{r|rrrr} 1 & 1 & -6 & 10 & -8 \\ & & 1 & -5 & 5 \\ \hline & 1 & -5 & 5 & \boxed{-3} \end{array} \qquad \begin{array}{r|rrrr} 4 & 1 & -6 & 10 & -8 \\ & & 4 & -8 & 8 \\ \hline & 1 & -2 & 2 & \boxed{0} \end{array}$$

Use the remainder theorem and synthetic division to find $P(k)$ for each exercise.

29. $P(x) = 6x^3 - 7x^2 - 25x + 50; k = 3$

30. $P(x) = 5x^3 - 9x^2 - 34x + 45; k = 4$

31. $P(y) = y^3 - y^2 - 10y; k = 5$

32. $P(y) = y^3 - y^2 - 15y; k = 6$

33. $P(z) = -3z^4 + 6z^3 - 14z^2 + 4z + 9; k = -2$

34. $P(z) = -4z^4 - 8z^3 + 17z^2 - 19z + 16; k = -1$

35. $P(r) = r^5 - 10r^3 + 13r^2 + 31r - 15; k = -4$

36. $P(r) = r^5 + 12r^4 - 4r^3 + 29r - 20; k = -3$

Use the remainder theorem and synthetic division to determine whether the given number is a solution of the given equation.

37. $2x^3 + 5x^2 - 13x - 30 = 0; x = -3$

38. $3x^3 - 7x^2 - 43x + 15 = 0; x = 5$

39. $x^4 - 7x^3 + 14x^2 - 13x + 21 = 0; x = 4$

40. $x^4 + 6x^3 + 6x^2 - 11x - 13 = 0; x = -2$

41. $-4x^5 - 4x^4 - 9x^3 - 6x^2 + 20x - 43 = 0; x = -1$

42. $-5x^5 - 18x^4 + 8x^3 - 6x^2 - 5x - 28 = 0; x = -4$

43. $y^6 - 6y^5 - 49y^3 + 4y^2 - 26y - 14 = 0; y = 7$

44. $y^6 - 42y^4 + 39y^3 - 18y^2 - 2y + 12 = 0; y = 6$

Completion Exercises

45. The factor theorem states that if 6 is a solution of the polynomial equation $x^3 - 4x^2 - 12x = 0$, then _____ is a factor of $x^3 - 4x^2 - 12x$.

46. The factor theorem states that if $P(x)$ is a polynomial, the number k is a solution of the equation _____ if and only if _____ is a factor of $P(x)$.

47. If multiplicity is counted, the polynomial equation $x^5 - 8x^4 + 25x^3 - 38x^2 + 28x - 8 = 0$ has exactly _____ (how many) solutions.

48. Counting multiplicity, a polynomial equation of degree n has exactly _____ (how many) solutions?

State the solutions of each equation and give the multiplicity of each solution.

49. $x - 5 = 0$

50. $x + 7 = 0$

51. $x^3(x + 4)(x - 8)^6 = 0$

52. $(2x - 5)^3(x - 9) = 0$

53. $(x - (2 + i))^4(x - (2 - i))^4 = 0$

54. $x^2(x + 6i)^5(x - 6i)^5 = 0$

Use the remainder theorem and synthetic division to show that the given number is a solution of the given equation. Then use the factor theorem to completely solve the equation. Note that, counting multiplicity, the number of solutions equals the degree of the polynomial equation.

55. $x^3 + x^2 - 10x + 8 = 0$; $x = 2$

56. $x^3 + 3x^2 - 22x - 24 = 0$; $x = -1$

57. $2x^3 + 5x^2 - 28x - 15 = 0$; $x = -5$

58. $4x^3 - 5x^2 - 23x + 6 = 0$; $x = 3$

59. $z^4 + 5z^3 - 22z^2 - 56z = 0$; $z = 4$

60. $z^4 + 10z^3 + 7z^2 - 18z = 0$; $z = -2$

Write a polynomial equation with integer coefficients that has the given solution set.

61. $\{4, -7\}$

62. $\{6, -1\}$

63. $\{0, 3\}$; 3 of multiplicity 2

64. $\{0, -5\}$; -5 of multiplicity 2

65. $\{\frac{1}{3}, \sqrt{5}, -\sqrt{5}\}$

66. $\{-\frac{1}{4}, \sqrt{10}, -\sqrt{10}\}$

67. $\{-\frac{1}{2}, 3i, -3i\}$

68. $\{\frac{1}{5}, 1 + i, 1 - i\}$

 Getting Acquainted with Your Graphing Calculator

69. If $P(x) = 3x^4 - 5x^3 + 7x^2 + 4x - 11$, find each of the following without retyping the polynomial for each part: (a) $P(3)$, (b) $P(-6)$, (c) $P(2.7)$.

9.3 The Rational Root Test

In Secs. 3.10 and 6.4, we learned to solve certain polynomial equations of degree 3 or higher, such as

$$x^4 - 10x^2 + 9 = 0$$

by factoring. In Sec. 9.2, we learned to solve polynomial equations of degree 3 or higher that were not factorable by inspection, such as

$$2x^3 - 5x^2 - 39x - 18 = 0$$

However, to solve these equations, we had to know one of the solutions ahead of time.

In this section, we will solve certain types of polynomial equations of degree 3 and higher that are not factorable by inspection, and we will solve them without knowing a solution ahead of time. To do this, we use a theorem called the **rational root test.** The word "root" in this context does *not* mean "square root" or "cube root." A **root** of an equation is another word for a *solution* of an equation. For example, $x^2 - x - 6 = 0$ has the two roots (or solutions) 3 and -2.

The Rational Root Test

Let $a_n x^n + a_{n-1}x^{n-1} + \cdots + a_1 x + a_0 = 0$ $(a_n \neq 0)$ be a polynomial equation with *integer* coefficients. If $\frac{p}{q}$ is a rational number in lowest terms, and $\frac{p}{q}$ is a root (solution) of the equation, then p is a factor of the constant term a_0, and q is a factor of the leading coefficient a_n.

To see why the rational root test works, note that if $\frac{p}{q}$ is a solution of the equation

$$a_n x^n + \cdots + a_1 x + a_0 = 0$$

where the coefficients $a_n, \ldots, a_1, a_0$ are integers, then

$$a_n\left(\frac{p}{q}\right)^n + \cdots + a_1\left(\frac{p}{q}\right) + a_0 = 0 \qquad \text{Substitute } \frac{p}{q} \text{ for } x.$$

$$a_n\frac{p^n}{q^n} + \cdots + a_1\frac{p}{q} + a_0 = 0 \qquad \text{Since } \left(\frac{p}{q}\right)^n = \frac{p^n}{q^n}.$$

$$a_np^n + \cdots + a_1pq^{n-1} + a_0q^n = 0 \qquad \text{Multiply by } q^n.$$

$$a_np^n + \cdots + a_1pq^{n-1} = -a_0q^n \qquad \text{Subtract } a_0q^n.$$

$$p(a_np^{n-1} + \cdots + a_1q^{n-1}) = -a_0q^n \qquad \text{Factor out } p.$$

Since p is a factor of the left side, and since the left side and the right side are equal, p must be a factor of the right side. That is, p must be a factor of $-a_0q^n$. But $\frac{p}{q}$ is in lowest terms, so p and q have no common factor. Therefore, p must be a factor of a_0. In a similar manner, we can show that q must be a factor of a_n.

Example 1 illustrates how to use the rational root test.

Example 1   List all possible rational roots of the equation $2x^3 - x^2 - 12x + 6 = 0$.

Solution: Since this is a polynomial equation with integer coefficients, we can use the rational root test. This means that if $\frac{p}{q}$ is a rational number in lowest terms, and $\frac{p}{q}$ is a root of the equation, then p must be a factor of the constant term 6, and q must be a factor of the leading coefficient 2. Therefore, the possible values of p are 1, -1, 2, -2, 3, -3, 6, and -6, and the possible values of q are 1, -1, 2, and -2. We usually write this as follows:

$$p: \pm 1, \pm 2, \pm 3, \pm 6$$
$$q: \pm 1, \pm 2$$

The possible rational roots consist of all possible quotients, $\frac{p}{q}$.

$$\frac{p}{q}: \pm 1, \pm 2, \pm 3, \pm 6, \pm \tfrac{1}{2}, \pm \tfrac{3}{2} \qquad \square$$

Try Exercise 9

List all possible rational roots of $2x^3 + x^2 - 16x - 8 = 0$.

In Example 1 we determined that there were 12 possible rational roots of the equation $2x^3 - x^2 - 12x + 6 = 0$. Counting multiplicity, a polynomial equation of degree 3 has only three roots. Therefore, not all of the possible roots can be actual roots. In fact, the only rational root of this equation is $\frac{1}{2}$ (note that $\frac{1}{2}$ is listed as a possible rational root). Solving the equation reveals that the other two roots are *irrational* numbers, and hence would not appear as possible rational roots.

$$2x^3 - x^2 - 12x + 6 = 0 \qquad \text{Original equation}$$
$$x^2(2x - 1) - 6(2x - 1) = 0 \qquad \text{Factor by grouping.}$$
$$(2x - 1)(x^2 - 6) = 0$$

$$2x - 1 = 0 \quad \text{or} \quad x^2 - 6 = 0 \qquad \text{Set each factor equal to 0.}$$
$$2x = 1 \qquad\qquad x^2 = 6 \qquad \text{Solve each equation.}$$
$$x = \tfrac{1}{2} \qquad\qquad x = \pm\sqrt{6} \qquad \text{One rational root, two irrational roots}$$

Since the equation of Example 1 could be factored by inspection, we did not have to use the rational root test to solve the equation. This is not the case in Example 2.

Example 2   Solve: $x^3 - 19x + 30 = 0$.

Solution: We cannot factor the equation by inspection, so we use the rational root test. Since p must be a factor of the constant term 30, p can be ± 1, ± 2, ± 3, ± 5, ± 6, ± 10, ± 15, or ± 30. Since q must be a factor of the leading coefficient 1, q can be ± 1. Therefore, the possible rational roots $\frac{p}{q}$ are the same as the possible values of p.

$$\frac{p}{q}: \pm 1, \pm 2, \pm 3, \pm 5, \pm 6, \pm 10, \pm 15, \pm 30$$

Use synthetic division to determine whether any of these possible rational roots are actual roots.

$$\begin{array}{r|rrrr} 1 & 1 & 0 & -19 & 30 \\ & & 1 & 1 & -18 \\ \hline & 1 & 1 & -18 & \boxed{12} \end{array} \qquad \begin{array}{r|rrrr} -1 & 1 & 0 & -19 & 30 \\ & & -1 & 1 & 18 \\ \hline & 1 & -1 & -18 & \boxed{48} \end{array} \qquad \begin{array}{r|rrrr} 2 & 1 & 0 & -19 & 30 \\ & & 2 & 4 & -30 \\ \hline & 1 & 2 & -15 & \boxed{0} \end{array}$$

Remainder 12 is not 0, so $x - 1$ is not a factor.

Remainder 48 is not 0, so $x + 1$ is not a factor.

Remainder is 0, so $x - 2$ is a factor, and 2 is a root.

Not only does the third synthetic division tell us that $x - 2$ is factor of the equation, but the bottom row gives us the coefficients of the other factor as well.

$$\begin{aligned} x^3 - 19x + 30 &= 0 & &\text{Original equation} \\ (x - 2)(x^2 + 2x - 15) &= 0 & &\text{Factor, using the synthetic} \\ & & &\text{division above.} \end{aligned}$$

$$\begin{array}{ll} x - 2 = 0 \quad \text{or} & x^2 + 2x - 15 = 0 \qquad \text{Set each factor equal to 0.} \\ \qquad x = 2 & (x - 3)(x + 5) = 0 \qquad \text{Solve each equation.} \\ & \qquad x = 3 \quad \text{or} \quad x = -5 \end{array}$$

The solution set is $\{2, 3, -5\}$.   ❏

Try Exercise 25

Solve: $x^3 - 19x - 30 = 0$.

From Example 2, we see that **whenever the leading coefficient of a polynomial equation with integer coefficients is 1 or −1, the possible rational roots are simply the factors of the constant term.**

Example 3   Solve: $x^4 - 6x^3 + 22x^2 - 30x + 13 = 0$.

Solution: Since the leading coefficient is 1, the possible rational roots are simply the factors of the constant term, 13.

$$\frac{p}{q}: \pm 1, \pm 13$$

Use synthetic division to determine whether any of these possible roots are actual roots.

$$\begin{array}{r|rrrrr} 1 & 1 & -6 & 22 & -30 & 13 \\ & & 1 & -5 & 17 & -13 \\ \hline & 1 & -5 & 17 & -13 & \boxed{0} \end{array}$$

Since the remainder is 0, the equation can be factored as follows:

$$\begin{aligned} x^4 - 6x^3 + 22x^2 - 30x + 13 &= 0 \\ (x - 1)(x^3 - 5x^2 + 17x - 13) &= 0 \end{aligned}$$

$$x - 1 = 0 \quad \text{or} \quad x^3 - 5x^2 + 17x - 13 = 0$$
$$\qquad x = 1$$

Use the rational root test again to solve the second equation.

$$\frac{p}{q}: \pm 1, \pm 13 \qquad \begin{array}{r|rrrr} 1 & 1 & -5 & 17 & -13 \\ & & 1 & -4 & 13 \\ \hline & 1 & -4 & 13 & \boxed{0} \end{array}$$

Note that $x - 1$ is a factor of the second equation as well.

$$(x - 1)(x^2 - 4x + 13) = 0$$
$$x - 1 = 0 \quad \text{or} \quad x^2 - 4x + 13 = 0$$
$$x = 1$$

Use the quadratic formula to solve $x^2 - 4x + 13 = 0$.

$$x = \frac{-(-4) \pm \sqrt{(-4)^2 - 4(1)(13)}}{2(1)} = \frac{4 \pm \sqrt{16 - 52}}{2}$$

$$= \frac{4 \pm \sqrt{-36}}{2} = \frac{4 \pm 6i}{2} = 2 \pm 3i$$

Therefore, the solution set is $\{1, 2 + 3i, 2 - 3i\}$. ❑

Try Exercise 35

Solve:
$x^4 - 4x^3 + 10x^2 - 12x + 5 = 0.$

In Example 3, the number 1 was a root of multiplicity 2, because the original equation had two factors of $x - 1$. Also, it was no coincidence that the two imaginary roots, $2 + 3i$ and $2 - 3i$, were complex conjugates. Imaginary roots of a polynomial equation with real coefficients, if they exist at all, must occur in conjugate pairs.

Conjugate Pairs Theorem

If the imaginary number $a + bi$ is a root of a polynomial equation with *real* coefficients, then $a - bi$ must also be a root.

Here is an example of an applied problem that we can solve using the rational root test.

Problem Solving

Example 4   A company wants to send its product in rectangular boxes having square cross sections (see Fig. 9.26). The product requires a volume of 1984 in.^{3}. The U. S. Post Office will not mail a package if the sum of its length and girth (perimeter of a cross section) is more than 108 in. If the company sends the largest package permitted by the Post Office, what will the dimensions of the package be?

Solution: The sum of the length y and the girth $4x$ is 108 in.

$$y + 4x = 108$$
$$y = 108 - 4x \qquad \text{Solve for } y.$$

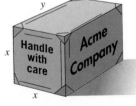

Figure 9.26

The volume is

$$V = lwh$$
$$1984 = y \cdot x \cdot x \qquad \text{From Fig. 9.26}$$
$$1984 = (108 - 4x)x^2 \qquad \text{Substitute } 108 - 4x \text{ for } y.$$
$$1984 = 108x^2 - 4x^3 \qquad \text{Multiply.}$$
$$4x^3 - 108x^2 + 1984 = 0 \qquad \text{Add } 4x^3, \text{ subtract } 108x^2.$$
$$x^3 - 27x^2 + 496 = 0 \qquad \text{Divide by 4.}$$

Since the leading coefficient is 1, the possible rational roots are the factors of the constant term 496.

$$\frac{p}{q}: \pm 1, \ \pm 2, \ \pm 4, \ \pm 8, \ \pm 16, \ \pm 31, \ \pm 62, \ \pm 124, \ \pm 248, \ \pm 496$$

Synthetic division reveals that -4 is a root.

$$
\begin{array}{r|rrrr}
-4 & 1 & -27 & 0 & 496 \\
   &   & -4 & 124 & -496 \\
\hline
   & 1 & -31 & 124 & \enclose{updiagonalstrike}{} \ 0
\end{array}
$$

Therefore, we can factor and solve the equation as follows:

$$x^3 - 27x^2 + 496 = 0$$
$$(x + 4)(x^2 - 31x + 124) = 0$$

$$x + 4 = 0 \quad \text{or} \quad x^2 - 31x + 124 = 0$$
$$x = -4$$

$$x = \frac{-(-31) \pm \sqrt{(-31)^2 - 4(1)(124)}}{2(1)}$$

$$= \frac{31 \pm \sqrt{961 - 496}}{2}$$

$$= \frac{31 \pm \sqrt{465}}{2}$$

Since x cannot be negative, discard $x = -4$. That leaves two possible solutions for the box.

Solution 1: The width and height are both $x = \dfrac{31 + \sqrt{465}}{2} \approx 26.28$ in., and the length is $y = 108 - 4x \approx 2.87$ in. Solution 2: The width and height are both $x = \dfrac{31 - \sqrt{465}}{2} \approx 4.72$ in., and the length is $y = 108 - 4x \approx$ 89.13 in. ◻

Try Exercise 51

Repeat Example 4 if the product requires a volume of 1080 in.^{3}.

In Example 4, it was the irrational roots $\dfrac{31 \pm \sqrt{465}}{2}$ that were used to determine the dimensions of the package. However, we were able to find these irrational roots only because we first found the rational root, -4. This brings us to an important point: If you can find enough rational roots to get the remaining factor down to a quadratic polynomial, you are home free. You can use the quadratic formula to find the remaining two roots, even if they are irrational or imaginary.

CAUTION

Remember, the rational root test can be used only if the coefficients of the polynomial equation are integers. We could not use the rational root test to solve either of the equations

$$x^3 + \sqrt{7}x^2 - 6x + 9 = 0 \quad \text{or} \quad 2x^3 - 5x^2 - ix + 10 = 0$$

The first equation has the irrational coefficient $\sqrt{7}$, and the second equation has the imaginary coefficient $-i$. We can, however, use the rational root test to solve the equation

$$\tfrac{1}{2}x^3 - \tfrac{3}{2}x^2 + 2x - 6 = 0$$

even though the coefficients $\tfrac{1}{2}$ and $-\tfrac{3}{2}$ are not integers. Simply multiply both sides of the equation by the LCD, 6, to get the equivalent equation

$$3x^3 - 9x^2 + 12x - 36 = 0 \qquad\qquad \blacksquare$$

Exercises 9.3

Completion Exercises

1. If $a_n x^n + \cdots a_1 x + a_0\ (a_n \neq 0)$ is a polynomial equation with integer coefficients, and $\tfrac{p}{q}$ is a rational root in lowest terms, the rational root test states that p is a factor of _____ and q is a factor of _____.

2. The roots of an equation are the _____ of the equation.

3. We cannot use the rational root test to solve the equation $x^3 + 3x^2 - \sqrt{5}x + 6 = 0$ because _____.

4. If $a - bi$ is a root of a polynomial equation with real coefficients, then _____ must also be a root.

True-False Exercises

5. If all of the roots of a polynomial equation are irrational numbers and/or imaginary numbers, the rational root test will be of no help in solving the equation.

6. If the leading coefficient of a polynomial equation with integer coefficients is 1 or -1, the possible rational roots are simply the factors of the constant term.

7. Imaginary roots of a polynomial equation with real coefficients, if they exist, must occur in conjugate pairs.

8. We usually try the rational root test only if we cannot factor the equation by inspection.

List all possible rational roots of each equation. Do not solve the equation.

9. $2x^3 + x^2 - 16x - 8 = 0$

10. $3x^3 - x^2 - 18x + 6 = 0$

11. $x^3 - 6x^2 - 11x + 12 = 0$

12. $x^3 - 6x^2 + 3x + 10 = 0$

13. $6x^3 + 5x^2 - 19x + 3 = 0$

14. $12x^3 - 47x^2 - 28x - 2 = 0$

15. $x^4 + 2x^3 - 2x - 1 = 0$ 16. $x^4 - 2x^3 + 2x - 1 = 0$

17. $\tfrac{1}{2}x^3 - \tfrac{3}{2}x^2 + 2x - 6 = 0$

18. $\tfrac{1}{3}x^3 - \tfrac{2}{3}x^2 + 3x - 6 = 0$

Solve each equation. Try to factor the equation before trying the rational root test.

19. $x^3 - 2x^2 - 9x + 18 = 0$

20. $x^3 + 3x^2 - 4x - 12 = 0$

21. $x^4 - 3x^2 + 2 = 0$ 22. $x^4 - 6x^2 + 5 = 0$

23. $2x^3 - 9x^2 + 3x + 4 = 0$

24. $2x^3 - 9x^2 + 10x - 3 = 0$

25. $x^3 - 19x - 30 = 0$ 26. $x^3 - 39x + 70 = 0$

27. $3x^3 - 17x^2 + 16x - 4 = 0$

28. $2x^3 - 17x^2 + 31x - 15 = 0$

29. $x^4 + x^3 - 10x^2 - 4x + 24 = 0$

30. $x^4 - 4x^3 - 18x^2 + 108x - 135 = 0$

31. $2x^4 - 3x^3 + 32x^2 - 48x = 0$

32. $3x^5 - 5x^4 + 27x^3 - 45x^2 = 0$

33. $4x^4 - 22x^3 + 5x^2 + 26x - 5 = 0$

34. $6x^4 - 44x^3 + 7x^2 + 50x - 7 = 0$

35. $x^4 - 4x^3 + 10x^2 - 12x + 5 = 0$

36. $x^4 - 8x^3 + 33x^2 - 68x + 52 = 0$

37. $x^4 - 6x^2 - 7 = 0$ 38. $x^4 - 9x^2 - 10 = 0$

39. $\tfrac{2}{3}x^4 - \tfrac{14}{3}x^3 + \tfrac{61}{6}x^2 - 7x + \tfrac{3}{2} = 0$

40. $\tfrac{3}{2}x^4 - 13x^3 + \tfrac{193}{6}x^2 - \tfrac{52}{3}x + \tfrac{8}{3} = 0$

41. $x^5 + 3x^4 - 5x^3 - 15x^2 + 4x + 12 = 0$

42. $x^5 - 2x^4 - 10x^3 + 20x^2 + 9x - 18 = 0$

43. Verify that $x^3 + 5x - 3 = 0$ has no rational roots.

44. The number $\sqrt{2}$ is a real root (solution) of the equation $x^2 - 2 = 0$. Verify that $\sqrt{2}$ is not a rational number (and hence must be an irrational number) by showing that the equation $x^2 - 2 = 0$ has no rational roots.

Discussion
Exercises

45. Suppose that a polynomial equation of degree 3 has three distinct roots, one rational and two irrational. How does the rational root test enable us to find the two irrational roots?

46. Discuss the disadvantages of the rational root test.

47. The solutions of the equation $ix^2 + 3x + 4i = 0$ are $4i$ and $-i$. Does this contradict the conjugate pairs theorem?

48. The equation $x^3 - 5x^2 + 6x = 0$ has the three rational roots 0, 2, and 3. Since the constant term is 0, does this contradict the rational root test?

Problem Solving

49. The cost C in thousands of dollars of producing x liters of a drug is $C = x^3 - 5x^2 + 8x + 3$. The revenue R in thousands of dollars is $R = -x^2 + 10x$. Find the break-even points. Breakeven occurs when cost = revenue.

50. Find the dimensions of the box in Fig. 9.27 if the volume of the box is 4 m³.

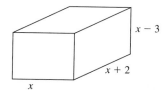

Figure 9.27

51. A company wants to send its product in rectangular boxes having square cross sections (see Fig. 9.28). The product requires a volume of 1080 in.³. The U. S. Post Office will not mail a package if the sum of its length and girth (perimeter of a cross section) is more than 108 in. If the company sends the largest package permitted by the Post Office, what will the dimensions of the package be?

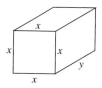

Figure 9.28

52. A box is to be made by cutting equal size squares from the corners of a piece of cardboard that is 12 in. by 12 in. and then turning up the sides (see Fig. 9.29). If the volume of the box is to be 128 in.³, determine the size of the squares that should be cut.

Figure 9.29

9.4 Zeros of a Function

From the first three sections of this chapter, it should be clear that there is a strong relationship between the *graph* of a polynomial function, the *roots* of a polynomial equation, and the *factors* of a polynomial. Before discussing this relationship in detail, we define the term *zero* of a function.

DEFINITION OF A ZERO OF A FUNCTION

The number k is a **zero** of the function f if $f(k) = 0$.

Example 1   Find the zeros of the function $f(x) = x^2 - 4$.

Try Exercise 9

Find the zeros of $f(x) = x^2 - 9$.

Solution: Since $f(2) = 2^2 - 4 = 0$ and $f(-2) = (-2)^2 - 4 = 0$, the zeros of f are 2 and -2.   ❏

Since the zeros of the function $y = f(x)$ are those x-values that make $f(x)$ zero, we can find the zeros by letting $f(x) = 0$ and solving for x. Therefore, we could have found the zeros in Example 1 as follows:

$$f(x) = x^2 - 4 \qquad \text{Original function}$$
$$0 = x^2 - 4 \qquad \text{Let } f(x) = 0.$$
$$4 = x^2 \qquad \text{Add 4.}$$
$$x = \pm 2 \qquad \text{Square root property}$$

Note that we find the zeros of a function in exactly the same way that we find the x-intercepts of the graph of the function.

Relationship between Zeros, Roots, x-Intercepts, and Factors

1. 2 and -2 are the **zeros** of the function $f(x) = x^2 - 4$.
2. 2 and -2 are the **roots** (solutions) of the equation $x^2 - 4 = 0$.
3. 2 and -2 are the **x-intercepts** of the graph of $f(x) = x^2 - 4$.
4. $x - 2$ and $x + 2$ are the **factors** of the polynomial $f(x)$.

(See Fig. 9.30)

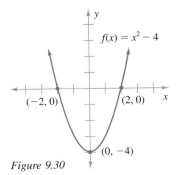

Figure 9.30

The precise relationship between the zeros of a polynomial and its factors is given by the **linear factorization theorem.** The factors are linear because the variable x appears to the first power only.

Linear Factorization Theorem

If $f(x)$ is a polynomial of positive degree n, then

$$f(x) = a(x - k_1)(x - k_2) \cdots (x - k_n)$$

where $k_1, k_2, \ldots, k_n$ are the n zeros of f and a is the leading coefficient of $f(x)$.

The linear factorization theorem can be proved using a theorem which occupies such a central position in the study of algebra that it is called the **fundamental theorem of algebra.**

The Fundamental Theorem of Algebra*

A polynomial function of positive degree with complex coefficients (which includes real coefficients) has at least one complex zero.

Sometimes we factor a polynomial to find the zeros, and sometimes we find the zeros first and then use the zeros to factor the polynomial.

---

*HISTORICAL NOTE: The fundamental theorem of algebra was proved by the German mathematician Carl Friedrich Gauss (1777–1855) when Gauss was only 22 years old. Stories abound attesting to Gauss's status as an infant prodigy, such as the one in which he detected an error in his father's bookkeeping when he was only 2 years old. At 19 years of age, he had constructed a regular polygon of 17 sides using only straightedge and compass, a problem that had eluded mathematicians for 2000 years.

Example 2 Find the zeros of $f(x) = 5x^3 - 3x^2 + 80x - 48$. Then use the zeros to factor $f(x)$ into linear factors.

Solution: To find the zeros, let $f(x) = 0$ and solve for x.

$$5x^3 - 3x^2 + 80x - 48 = 0 \qquad \text{Let } f(x) = 0.$$

$$\left.\begin{array}{l} x^2(5x - 3) + 16(5x - 3) = 0 \\ (5x - 3)(x^2 + 16) = 0 \end{array}\right\} \quad \text{Factor by grouping.}$$

$$\left.\begin{array}{lll} 5x - 3 = 0 & \text{or} & x^2 + 16 = 0 \\ 5x = 3 & & x^2 = -16 \\ x = \frac{3}{5} & & x = \pm 4i \end{array}\right\} \quad \text{Solve each equation.}$$

The zeros of f are $\frac{3}{5}$, $4i$, and $-4i$. Therefore the factors of $f(x)$ are $x - \frac{3}{5}$, $x - 4i$, and $x - (-4i) = x + 4i$. Since the leading coefficient is 5, we can factor $f(x)$ using the linear factorization theorem as follows:

$$f(x) = 5\left(x - \frac{3}{5}\right)(x - 4i)(x + 4i)$$

Note that we could remove the fraction $\frac{3}{5}$ by distributing the leading coefficient, 5, over the first factor, $x - \frac{3}{5}$.

$$f(x) = (5x - 3)(x - 4i)(x + 4i) \qquad \qquad \square$$

We can check the answer to Example 2 by multiplying the factors.

$$\begin{aligned} (5x - 3)(x - 4i)(x + 4i) &= (5x - 3)(x^2 - 16i^2) \\ &= (5x - 3)(x^2 + 16) \\ &= 5x^3 - 3x^2 + 80x - 48 \end{aligned}$$

Try Exercise 21

Find the zeros of the function $f(x) = 5x^3 - 2x^2 + 125x - 50$. Then use the zeros to factor $f(x)$ into linear factors.

Example 3 Find the zeros of $f(x) = x^4 - 6x^3 + 12x^2 - 8x$. Then use the zeros to factor $f(x)$ into linear factors.

Solution: Let $f(x) = 0$ and solve for x.

$$\begin{array}{ll} x^4 - 6x^3 + 12x^2 - 8x = 0 & \text{Let } f(x) = 0. \\ x(x^3 - 6x^2 + 12x - 8) = 0 & \text{Factor out } x. \end{array}$$

$$x = 0 \quad \text{or} \quad x^3 - 6x^2 + 12x - 8 = 0 \qquad \text{Set each factor equal to 0.}$$

Using the rational root test, we find that 2 is a root of the second equation.

$$\begin{array}{r|rrrr} 2 & 1 & -6 & 12 & -8 \\ & & 2 & -8 & 8 \\ \hline & 1 & -4 & 4 & \underline{|\,0} \end{array}$$

Factor the second equation and solve as follows:

$$\begin{array}{lll} (x - 2)(x^2 - 4x + 4) = 0 & \text{Result of synthetic division} \\ x - 2 = 0 \quad \text{or} \quad x^2 - 4x + 4 = 0 & \text{Set each factor equal to 0 and} \\ x = 2 \qquad \qquad (x - 2)(x - 2) = 0 & \text{solve.} \\ \qquad \qquad x = 2 \quad \text{or} \quad x = 2 \end{array}$$

The zeros of f are 0 and 2 (multiplicity 3). Since the leading coefficient is 1, we factor $f(x)$ into linear factors as follows:

$$f(x) = 1(x - 0)(x - 2)(x - 2)(x - 2) = x(x - 2)^3 \qquad \qquad \square$$

Try Exercise 25

Find the zeros of the function $f(x) = x^4 - 3x^3 + 3x^2 - x$. Then use the zeros to factor $f(x)$ into linear factors.

Example 4   Find the zeros of $f(x) = x^2 - 7x + 9$. Then use the zeros to factor $f(x)$ into linear factors.

Solution: Set $f(x) = 0$ and use the quadratic formula to solve the equation.
$$x^2 - 7x + 9 = 0$$
$$x = \frac{-(-7) \pm \sqrt{(-7)^2 - 4(1)(9)}}{2(1)} = \frac{7 \pm \sqrt{49 - 36}}{2} = \frac{7 \pm \sqrt{13}}{2}$$

Since $\dfrac{7 + \sqrt{13}}{2}$ is a zero, $x - \dfrac{7 + \sqrt{13}}{2}$ is a factor. Since $\dfrac{7 - \sqrt{13}}{2}$ is a

zero, $x - \dfrac{7 - \sqrt{13}}{2}$ is a factor. The leading coefficient is 1, so

$$f(x) = \left(x - \frac{7 + \sqrt{13}}{2} \right)\left(x - \frac{7 - \sqrt{13}}{2} \right)$$ ❑

Try Exercise 29

Find the zeros of the function $f(x) = x^2 - 9x + 16$. Then use the zeros to factor $f(x)$ into linear factors.

Sometimes we find the zeros (x-intercepts) of a polynomial function to help in drawing its graph, and sometimes we draw the graph of a polynomial function to help in finding its zeros.

Example 5   Graph: $f(x) = x^3 - 2x^2 - 5x + 6$.

Solution: The y-intercept is 6. Using the rational root test we find the zeros (x-intercepts) to be 1, 3, and -2. Since the degree is odd and the leading coefficient is positive, the graph rises to the right and falls to the left. Choosing x-values between the x-intercepts, and computing the corresponding $f(x)$-values produces the table of values below. Drawing a smooth, continuous curve through the intercepts and the points from the table gives the graph shown in Fig. 9.31.   ❑

Try Exercise 33

Graph:
$f(x) = x^3 - 5x^2 + 2x + 8$.

x	y
2	-4
-1	8

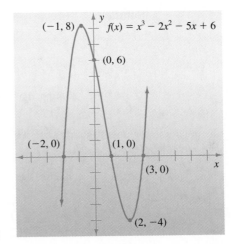

Figure 9.31

Example 6   Solve: $2x^3 - 7x^2 + 11x - 4 = 0$.

Solution: By the rational root test, the possible rational roots are
$$\frac{p}{q}: \pm 1, \pm 2, \pm 4, \pm \tfrac{1}{2}$$

Now, the roots of $2x^3 - 7x^2 + 11x - 4 = 0$ are the x-intercepts of the graph of $f(x) = 2x^3 - 7x^2 + 11x - 4$, shown in Fig. 9.32.

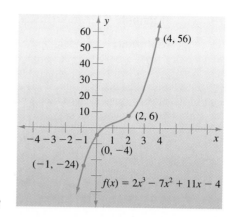

Figure 9.32

Noting where the graph appears to intersect the *x*-axis, we determine that the only possible roots that have a chance of being an actual root are $\frac{1}{2}$ and 1. Synthetic division verifies that $\frac{1}{2}$ is a root.

$$
\begin{array}{r|rrrr}
\frac{1}{2} & 2 & -7 & 11 & -4 \\
& & 1 & -3 & 4 \\
\hline
& 2 & -6 & 8 & \boxed{0}
\end{array}
$$

Factor and solve as follows:

$$(x - \tfrac{1}{2})(2x^2 - 6x + 8) = 0$$

$$
\begin{aligned}
x - \tfrac{1}{2} &= 0 \quad \text{or} \quad & 2x^2 - 6x + 8 &= 0 \\
x &= \tfrac{1}{2} & x^2 - 3x + 4 &= 0 \quad \text{\small Divide by 2.} \\
& & x &= \frac{3 \pm \sqrt{9 - 16}}{2} \\
& & &= \frac{3 \pm \sqrt{-7}}{2} = \frac{3 \pm i\sqrt{7}}{2}
\end{aligned}
$$

The solution set is $\left\{ \dfrac{1}{2}, \dfrac{3 + i\sqrt{7}}{2}, \dfrac{3 - i\sqrt{7}}{2} \right\}$. Since only real values are represented on the *x*-axis, the imaginary roots do not appear as *x*-intercepts in Fig. 9.32.   ❏

Approximating Real Roots

Suppose we wanted to solve the equation $x^3 - x - 1 = 0$. We cannot factor the left side by inspection, so we try the rational root test. But neither of the possible rational roots ± 1 are actual roots. So how do we solve the equation?

There is a formula that will solve any polynomial equation of degree 3, but it is complicated.* Therefore, to solve a polynomial equation of degree 3 or higher, we often use approximation methods.

Try Exercise 39

Use the graph below to assist in solving $2x^3 + 3x^2 - x - 12 = 0$.

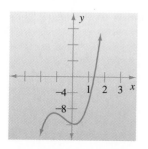

---

*HISTORICAL NOTE: The discovery by Italian mathematicians of formulas for solving cubic (third-degree) and quartic (fourth-degree) equations were probably the most spectacular mathematical achievements of the 16th century. The formulas were first published in 1545 in a book entitled *Ars Magna* by Geronimo Cardan (1501–1576), who had already gained world reknown as a physician. In his book, Cardan attributed the cubic solution to Niccolo Tartaglia (ca. 1499–1557) and the quartic soluton to Ludovico Ferrari (1522–1565). Tartaglia's birthname was Niccolo Fontana, but as a child he suffered a knife wound that impaired his speech and thereafter went by Tartaglia (meaning "stammerer"). Tartaglia had extracted a solemn oath from Cardan not to reveal the secret of the cubic—Tartaglia meant to do this himself—but Cardan published it anyway, and a bitter dispute resulted. Tartaglia himself probably got hint of the solution from Scipione del Ferro (ca. 1465–1526), a little-known mathematics professor who had disclosed it to one of his students. Many great mathematicians tried to find a formula for the quintic (fifth-degree) equation, but the Italian physician Paolo Ruffin (1765–1822) and the Norwegian mathematician Niels Henrik Abel (1802–1829) finally proved that no such formula was possible.

Example 7   Approximate the real root of $x^3 - x - 1 = 0$ to the nearest hundredth.

Solution: Consider the graph of $f(x) = x^3 - x - 1$, shown in Fig. 9.33. Note that f has a zero (x-intercept) between $x = 1$ and $x = 2$. This means that the equation $x^3 - x - 1 = 0$ has a root (solution) between 1 and 2. Since $f(1)$ is closer to 0 than $f(2)$, the solution is probably closer to 1 than 2.

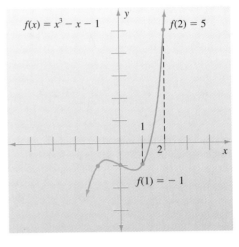

Figure 9.33

Now, calculate $f(1.1)$, $f(1.2)$, etc., until you get a change in sign. You can use direct substitution or synthetic division to calculate these values.

$$f(1.1) = (1.1)^3 - (1.1) - 1 = -0.769$$
$$f(1.2) = (1.2)^3 - (1.2) - 1 = -0.472$$
$$f(1.3) = (1.3)^3 - (1.3) - 1 = -0.103$$
$$f(1.4) = (1.4)^3 - (1.4) - 1 = +0.344$$
Sign change

Since $f(1.3)$ is negative and $f(1.4)$ is positive, the solution must lie between 1.3 and 1.4 (see Fig. 9.34). Since $f(1.3)$ is closer to 0 than $f(1.4)$, the solution is probably closer to 1.3 than 1.4. To approximate the solution to the nearest hundredth, calculate $f(1.31)$, $f(1.32)$, etc., until you get a change in sign.

$$f(1.31) = (1.31)^3 - (1.31) - 1 = -0.061909$$
$$f(1.32) = (1.32)^3 - (1.32) - 1 = -0.020032$$
$$f(1.33) = (1.33)^3 - (1.33) - 1 = +0.022637$$
Sign change

Since $f(1.32)$ is closer to 0 than $f(1.33)$, the solution to the nearest hundredth is probably 1.32 (see Fig. 9.35). ❏

Try Exercise 45

Use the graph below to assist in approximating the real root of $x^3 - 2x^2 - 2 = 0$ to the nearest hundredth.

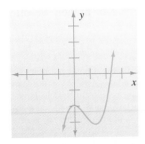

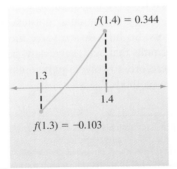

Figure 9.34

Figure 9.35

The approximation method presented in Example 7 is somewhat inefficient. The purpose in discussing this method was simply to illustrate how these methods work. More efficient approximation methods, such as *Newton's method,** exist, but they require calculus techniques. Approximation methods are tedious to perform by hand, but a computer can use them to generate answers to any desired accuracy in a very short time.

---

*HISTORICAL NOTE: The Englishman Sir Isaac Newton (1642–1727) discovered Newton's method in 1665 while confined at home during the plague year 1665–1666. He had returned home when Trinity College, where he was a student, was forced to close because of the plague. His discoveries during that year (the law of gravitation, the calculus, the nature of colors, and the binomial theorem for fractional exponents) comprise the most productive period of mathematical discovery in history.

Exercises 9.4

Completion Exercises

1. The number k is a zero of the function f if _____.

2. To find the zeros of the function f, let _____ and solve for x.

3. If 5 is a zero of the polynomial function f, then _____ is a factor of $f(x)$.

4. The linear factorization theorem states that if $f(x)$ is a polynomial of positive degree n, then $f(x) =$ _____ where $k_1, k_2, \ldots, k_n$ are the zeros of f and a is the leading coefficient of $f(x)$.

True-False Exercises

5. If 3 is a zero of the function f, then 3 is a y-intercept of the graph of f.

6. The fundamental theorem of algebra states that a polynomial function of positive degree with complex coefficients has at least one complex zero.

Find the zeros of each polynomial function.

7. $f(x) = x - 8$
8. $f(x) = x + 6$
9. $f(x) = x^2 - 9$
10. $f(x) = x^2 - 5x$
11. $g(x) = 5$
12. $g(x) = -3$
13. $h(x) = 2x^3 + 7x^2 - 4x$
14. $h(x) = x^4 - 10x^2 + 9$

Use Fig. 9.36 to answer Exercises 15 through 18.

15. What are the zeros of f?

16. What are the x-intercepts of the graph of f?

17. Factor $x^4 - x^3 - 7x^2 + x + 6$ into linear factors.

18. What are the solutions of $x^4 - x^3 - 7x^2 + x + 6 = 0$?

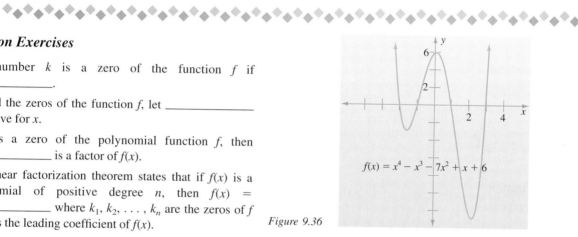

Figure 9.36

$$f(x) = x^4 - x^3 - 7x^2 + x + 6$$

Find the zeros of each polynomial function. Then use the zeros to factor $f(x)$ into linear factors.

19. $f(x) = 2x^4 - 162$
20. $f(x) = 2x^4 - 1250$
21. $f(x) = 5x^3 - 2x^2 + 125x - 50$
22. $f(x) = 5x^3 - 4x^2 + 80x - 64$
23. $f(x) = x^4 - 13x^2 - 48$
24. $f(x) = x^4 - 6x^2 - 7$
25. $f(x) = x^4 - 3x^3 + 3x^2 - x$
26. $f(x) = x^4 - 9x^3 + 27x^2 - 27x$
27. $f(x) = x^3 + 4x^2 - 7x - 10$
28. $f(x) = x^3 + 3x^2 - 13x - 15$
29. $f(x) = x^2 - 9x + 16$
30. $f(x) = x^2 + 7x + 11$
31. $f(x) = 4x^2 - 8x + 5$
32. $f(x) = 4x^2 - 12x + 10$

Find the zeros and the right and left behavior for each function. Then use this information to graph the function.

33. $f(x) = x^3 - 5x^2 + 2x + 8$
34. $f(x) = x^3 - 3x^2 - 6x + 8$
35. $f(x) = x^4 - x^3 - 7x^2 + x + 6$
36. $f(x) = x^4 - 4x^3 - x^2 + 16x - 12$

Use the graphs in Exercises 37 through 40 to assist in solving the given equations.

37. $x^3 - 4x^2 - 2x + 12 = 0$ 38. $x^3 - 4x^2 + 8 = 0$

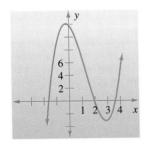

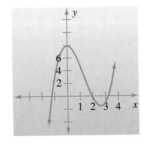

39. $2x^3 + 3x^2 - x - 12 = 0$ 40. $2x^3 - 3x^2 - 8x + 30 = 0$

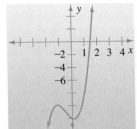

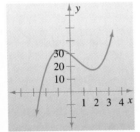

Discussion Exercises

41. Can a polynomial function of degree 3 with real coefficients have (a) one real zero? (b) two distinct real zeros? (c) three distinct real zeros? (d) no real zeros? Draw graphs to illustrate your answers.

42. Can a polynomial function of degree 4 with real coefficients have (a) one real zero? (b) two distinct real zeros? (c) three distinct real zeros? (d) four distinct real zeros? (e) no real zeros? Draw graphs to illustrate your answers.

43. Let f be a polynomial function. If $f(2) = -1$ and $f(3) = 4$, then f must have a real zero between 2 and 3. However, if $f(2) = 1$ and $f(3) = 4$, then f may or may not have a real zero between 2 and 3. Explain both statements. Draw graphs to illustrate your answer.

44. Graph $f(x) = (x + 1)(x - 2)^2$ and note that the graph *crosses* the x-axis at $x = -1$, but only *touches* the x-axis at $x = 2$. State a rule for determining whether the graph of a polynomial function will cross the x-axis or touch the x-axis at a particular zero.

 Getting Acquainted with Your Scientific Calculator

Each of the following equations has one real root. Approximate this root to the nearest hundredth.

45. $x^3 - 2x^2 - 2 = 0$ 46. $x^3 + 2x + 1 = 0$

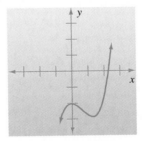

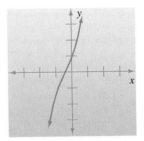

 Getting Acquainted with Your Graphing Calculator

Solve each equation by graphing the appropriate function and zooming in on the x-intercepts. Write approximate answers to the nearest hundredth.

47. $2x^3 + 5x - 3 = 0$ 48. $3x^3 - x + 5 = 0$

49. $x^3 - 4x = -2$ 50. $x^3 - 6x^2 + 7x = -4$

9.5 Rational Functions and Their Graphs

In Sec. 9.1 we noted that the graph of a polynomial function is always a smooth, continuous curve. The graph of a *rational function* is also a smooth curve on its domain, but it may not be a continuous curve. That is, the graph may have "gaps" or "holes."

> **DEFINITION OF A RATIONAL FUNCTION**
>
> A **rational function** is a function of the form
>
> $$f(x) = \frac{p(x)}{q(x)}$$
>
> where $p(x)$ and $q(x)$ are polynomials and $q(x) \neq 0$.

Note that a rational function is simply a function of the form

$$f(x) = \text{a rational expression}$$

Examples of rational functions are

$$f(x) = \frac{1}{x} \qquad g(x) = \frac{3x}{x^2 - 4} \qquad h(x) = \frac{x - 5}{x^2 + 1}$$

The **domain** of a rational function consists of all real numbers except those that make the denominator 0. Therefore, the domain of the function f above is all real numbers except 0. The domain of g is all real numbers except 2 and -2. The domain of h is all real numbers, since no real number makes the denominator 0.

Horizontal and Vertical Asymptotes

Example 1   Graph the rational function $f(x) = \dfrac{1}{x}$.

Solution: Construct the table of values below.

x	-3	-2	-1	1	2	3
$f(x)$	$-\frac{1}{3}$	$-\frac{1}{2}$	-1	1	$\frac{1}{2}$	$\frac{1}{3}$

We avoided choosing $x = 0$ for our table, since 0 is not in the domain of f. Joining the points from the table with smooth curves gives the part of the graph shown in Fig. 9.37. Note that $f(x)$ approaches 0 as x increases without bound, written

$$f(x) \to 0 \text{ as } x \to \infty$$

Also, $f(x)$ approaches 0 as x decreases without bound, written

$$f(x) \to 0 \text{ as } x \to -\infty$$

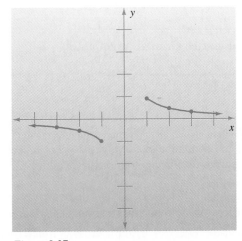

Figure 9.37

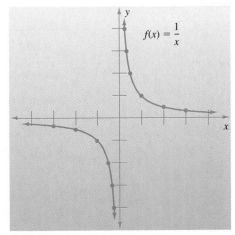

Figure 9.38

To determine what happens between -1 and 1, construct the table of values below:

x	$-\frac{1}{2}$	$-\frac{1}{3}$	$-\frac{1}{4}$	$\frac{1}{4}$	$\frac{1}{3}$	$\frac{1}{2}$
$f(x)$	-2	-3	-4	4	3	2

Draw smooth curves through these points, as well as the points already plotted in Fig. 9.37 to produce the graph shown in Fig. 9.38. Note that $f(x)$ increases without bound as x approaches 0 from the right, written

$$f(x) \to \infty \text{ as } x \to 0 \text{ from the right}$$

Also, $f(x)$ decreases without bound as x approaches 0 from the left, written

$$f(x) \to -\infty \text{ as } x \to 0 \text{ from the left}$$

Try Exercise 15

Graph: $f(x) = -\dfrac{1}{x}$.

Finally, note that the graph is symmetric with respect to the origin. ❏

The y-axis (the vertical line $x = 0$) is called a *vertical asymptote* of the graph of $f(x) = \frac{1}{x}$. The x-axis (the horizontal line $y = 0$) is called a *horizontal asymptote* of the graph.

DEFINITION OF VERTICAL AND HORIZONTAL ASYMPTOTES

Let f be a function.

1. The line $x = k$ is a **vertical asymptote** of the graph of f if

$$f(x) \to \infty \qquad \text{or} \qquad f(x) \to -\infty$$

as $x \to k$ either from the right or from the left.
2. The line $y = k$ is a **horizontal asymptote** of the graph of f if

$$f(x) \to k$$

as $x \to \infty$ or as $x \to -\infty$.

In other words, an **asymptote** of a graph is a line that a portion of the graph gets closer and closer to as that portion of the graph moves away from the origin.

The following rules summarize how to find the vertical and horizontal asymptotes of the graph of a rational function.

Finding Vertical and Horizontal Asymptotes

Assume $f(x) = \dfrac{a_n x^n + \cdots + a_1 x + a_0}{b_m x^m + \cdots + b_1 x + b_0}$ $(a_n \neq 0, b_m \neq 0)$ is a rational function in lowest terms.

Vertical asymptotes
If k is a zero of the denominator (a value of x that makes the denominator 0), then the line $x = k$ is a vertical asymptote of the graph of f.

Horizontal asymptote
Note that the degree of the numerator is n and the degree of the denominator is m.

1. If $n < m$, then the line $y = 0$ (the x-axis) is a horizontal asymptote of the graph of f.

2. If $n = m$, then the line $y = \dfrac{a_n}{b_m}$ is a horizontal asymptote of the graph of f (assuming n and m are not 0).

3. If $n > m$, then there is no horizontal asymptote of the graph of f.

We will discuss the rationale behind these rules momentarily, but first let us look at a few examples.

Example 2   Find all vertical and horizontal asymptotes for each function. Do not graph the function.

$$\text{(a) } f(x) = \frac{x - 3}{x + 2} \qquad \text{(b) } f(x) = \frac{2x^2}{x^2 + 1} \qquad \text{(c) } f(x) = \frac{x^3}{x^2 - 1}$$

Solution: (a) To find the vertical asymptotes, set the denominator equal to 0 and solve for x.

$$x + 2 = 0$$
$$x = -2$$

Therefore, the vertical line $x = -2$ is a vertical asymptote.

To find the horizontal asymptote, note that $n = m$ (that is, the degree of the numerator equals the degree of the denominator). Therefore, the line

$$y = \frac{a_n}{b_m} = \frac{\text{leading coefficient of the numerator}}{\text{leading coefficient of the denominator}} = \frac{1}{1} = 1$$

or $y = 1$, is a horizontal asymptote.

(b) Since the solutions of $x^2 + 1 = 0$ are not real numbers (the solutions are $x = \pm i$), there are no vertical asymptotes. To find the horizontal asymptote, note that $n = m$ (the degree of the numerator and the denominator are both 2). Therefore, the line

$$y = \frac{a_n}{b_m} = \frac{\text{leading coefficient of the numerator}}{\text{leading coefficient of the denominator}} = \frac{2}{1} = 2$$

or $y = 2$, is a horizontal asymptote.

(c) Since the solutions of $x^2 - 1 = 0$ are $x = \pm 1$, the vertical asymptotes are $x = 1$ and $x = -1$. Since $n > m$ (the degree of the numerator is 3 and the degree of the denominator is 2), there is no horizontal asymptote.   ❏

Try Exercise 23

Find all vertical and horizontal asymptotes for $f(x) = \dfrac{5}{x - 4}$. (Ignore slant asymptotes.)

To illustrate the rationale behind the rule for finding the vertical asymptotes of a rational function, consider

$$f(x) = \frac{1}{x - 2}$$

As $x \to 2$, the denominator $x - 2$ approaches 0, and the fraction $\dfrac{1}{x - 2}$ gets larger in absolute value. Therefore, $f(x) \to \infty$ or $f(x) \to -\infty$ as $x \to 2$, which means $x = 2$ (the zero of the denominator) is a vertical asymptote.

To illustrate the rationale behind the rules for finding the horizontal asymptote, consider the three cases below.

Case 1 ($n < m$):

$$f(x) = \frac{5}{x + 3} = \frac{\dfrac{5}{x}}{\dfrac{x}{x} + \dfrac{3}{x}} = \frac{\dfrac{5}{x}}{1 + \dfrac{3}{x}}$$

As $x \to \infty$ (or $x \to -\infty$), the fractions $\frac{5}{x}$ and $\frac{3}{x}$ approach 0, so $f(x) \to \dfrac{0}{1+0} = 0$. There-fore, $y = 0$ is a horizontal asymptote.

Case 2 $(n = m)$:

$$f(x) = \frac{2x-1}{3x+7} = \frac{\dfrac{2x}{x} - \dfrac{1}{x}}{\dfrac{3x}{x} + \dfrac{7}{x}} = \frac{2 - \dfrac{1}{x}}{3 + \dfrac{7}{x}}$$

As $x \to \infty$ (or $x \to -\infty$), the fractions $\frac{1}{x}$ and $\frac{7}{x}$ approach 0, so $f(x) \to \dfrac{2-0}{3+0} = \dfrac{2}{3}$. Therefore, $y = \frac{2}{3}$ is a horizontal asymptote.

Case 3 $(n > m)$:

$$f(x) = \frac{x^2}{x+4} = \frac{\dfrac{x^2}{x^2}}{\dfrac{x}{x^2} + \dfrac{4}{x^2}} = \frac{1}{\dfrac{1}{x} + \dfrac{4}{x^2}}$$

As $x \to \infty$ (or $x \to -\infty$), the fractions $\frac{1}{x}$ and $\frac{4}{x^2}$ approach 0, so $f(x) \to \dfrac{1}{0+0} = \dfrac{1}{0}$, which is undefined. Therefore, there is no horizontal asymptote.

In Case 3, if we divide the numerator by the denominator, we get

$$
\begin{array}{r}
x - 4 \\
x+4 \overline{\smash{)}\,x^2 } \\
\underline{x^2 + 4x } \\
-4x \\
\underline{-4x - 16} \\
16
\end{array}
$$

Therefore, we can write the original function as follows:

$$f(x) = \frac{x^2}{x+4} = x - 4 + \frac{16}{x+4}$$

Now as $x \to \infty$ (or $x \to -\infty$), the fraction $\dfrac{16}{x+4}$ approaches 0, so $f(x) \to (x-4)$. This means that the graph of f approaches the line $y = x - 4$. Since the line $y = x - 4$ is neither vertical nor horizontal, we call the line $y = x - 4$ a **slant asymptote** of the graph of f. Slant asymptotes occur when the degree of the numerator of a rational function is exactly 1 more than the degree of the denominator (assuming the function is in lowest terms and the denominator is not a constant).

Examples 3 and 4 illustrate how to use asymptotes as an aid in graphing rational functions.

Example 3   Graph: $f(x) = \dfrac{x-3}{x+2}$.

Solution: From Example 2(a), we know that $x = -2$ is a vertical asymptote and $y = 1$ is a horizontal asymptote. Find the intercepts.

y-Intercept (let *x* = 0)

$$f(0) = \frac{0 - 3}{0 + 2} = -\frac{3}{2}$$

x-Intercepts [let *f*(*x*) = 0]

$$\frac{x - 3}{x + 2} = 0$$

$$x - 3 = 0 \qquad \text{Multiply by } x + 2.$$

$$x = 3$$

The *y*-intercept is $-\frac{3}{2}$, and the *x*-intercept is 3. Choose *x*-values between and beyond the vertical asymptote and the *x*-intercept and construct the table of values below. Then draw smooth curves through the intercepts and the points from the table so that each curve approaches the asymptotes as the curve moves away from the origin (see Fig. 9.39). We draw the asymptotes as dashed lines, because the asymptotes are not actually part of the graph of *f*. ☐

Try Exercise 37

Graph: $f(x) = \dfrac{x - 4}{x + 1}$.

x	*f*(*x*)
4	$\frac{1}{6}$
2	$-\frac{1}{4}$
1	$-\frac{2}{3}$
−1	−4
−3	6

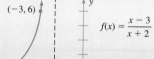
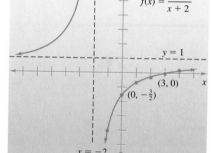

Figure 9.39

Example 4 Graph: $f(x) = \dfrac{2x^2}{x^2 + 1}$.

Solution: From Example 2(b), we know that there are no vertical asymptotes, and that *y* = 2 is a horizontal asymptote. Find the intercepts.

y-Intercept (let *x* = 0)

$$f(0) = \frac{2(0)^2}{0^2 + 1} = \frac{0}{1} = 0$$

x-Intercepts [let *f*(*x*) = 0]

$$\frac{2x^2}{x^2 + 1} = 0$$

$$2x^2 = 0 \qquad \text{Multiply by } x^2 + 1.$$

$$x^2 = 0$$

$$x = 0$$

The *y*-intercept and the *x*-intercept are both 0 (the graph intersects both axes at the origin). Choose *x*-values beyond the *x*-intercept and construct the table of values below. The graph is shown in Fig. 9.40. Note that the graph is symmetric with respect to the *y*-axis. ☐

Try Exercise 45

Graph: $f(x) = \dfrac{3x^2}{x^2 + 1}$.

x	*f*(*x*)
2	$\frac{8}{5}$
1	1
−1	1
−2	$\frac{8}{5}$

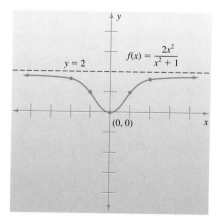

Figure 9.40

Here is a summary of the steps in graphing a rational function.

To Graph the Rational Function $f(x) = \dfrac{a_n x^n + \cdots + a_1 x + a_0}{b_m x^m + \cdots + b_1 x + b_0}$

$(a_n \neq 0, \, b_m \neq 0)$

First make sure the function is in lowest terms.

1. Set the denominator equal to 0 to find the vertical asymptotes.
2. Find the horizontal asymptote.

    (a) If $n < m$, the horizontal asymptote is $y = 0$.

    (b) If $n = m$, the horizontal asymptote is $y = \dfrac{a_n}{b_m}$ (assuming n and m are not 0).

    (c) If $n > m$, there is no horizontal asymptote. If $n = m + 1$ ($m \neq 0$), there is a slant asymptote.

3. Let $x = 0$ to find the y-intercept.
4. Let $f(x) = 0$ to find the x-intercepts.
5. Choose x-values between and beyond the vertical asymptotes and the x-intercepts and construct a table of values.
6. Draw smooth curves through the intercepts and the points from the table so that a portion of each curve approaches an asymptote as that portion of the curve moves away from the origin.

Example 5 Graph: $f(x) = \dfrac{6}{x^2 - x - 6}$.

Solution: Step 1:   Set the denominator equal to 0 to find the vertical asymptotes.

$$x^2 - x - 6 = 0$$
$$(x + 2)(x - 3) = 0$$
$$x + 2 = 0 \quad \text{or} \quad x - 3 = 0$$
$$x = -2 \qquad\qquad x = 3$$

Step 2:   Find the horizontal asymptote.
Since n (the degree of the numerator) is 0, and m (the degree of the denominator) is 2, we have $n < m$. Therefore, $y = 0$ is the horizontal asymptote.

Step 3:   Let $x = 0$ to find the y-intercept.

$$f(0) = \frac{6}{0^2 - 0 - 6} = \frac{6}{-6} = -1$$

The y-intercept is -1.

Step 4:   Let $f(x) = 0$ to find the x-intercepts.

$$\frac{6}{x^2 - x - 6} = 0$$
$$6 = 0 \qquad \text{Multiply by } x^2 - x - 6.$$

Since $6 = 0$ is a contradiction, the equation has no solution and there are no x-intercepts.

Step 5: Choose x-values between and beyond the vertical asymptotes and construct a table of values.

x	-3	-1	1	2	4
$f(x)$	1	$-\frac{3}{2}$	-1	$-\frac{3}{2}$	1

Step 6: Draw smooth curves through the intercepts and the points from the table so that a portion of each curve approaches an asymptote as that portion of the curve moves away from the origin. The graph is shown in Fig. 9.41. ❏

Try Exercise 49

Graph: $f(x) = \dfrac{6}{x^2 + x - 6}$.

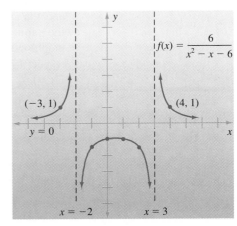

Figure 9.41

Slant Asymptotes

If a rational function with a nonconstant denominator is in lowest terms, and the degree of the numerator is exactly 1 more than the degree of the denominator, then the graph of the function will have a slant asymptote. Since there is no horizontal asymptote when the degree of the numerator is greater than the degree of the denominator, **the graph of a rational function cannot have both a horizontal asymptote and a slant asymptote.**

To find the equation of the slant asymptote, divide the numerator by the denominator. The equation $y = $ the quotient (ignore the remainder) is the equation of the slant asymptote.

Example 6 Graph: $f(x) = \dfrac{x^3}{x^2 - 1}$.

Solution: From Example 2(c), we know that the vertical asymptotes are $x = 1$ and $x = -1$ and there is no horizontal asymptote. Since the denominator is not a constant and the function is in lowest terms, and since $n = m + 1$, the graph has a slant asymptote. To find the equation of the slant asymptote, divide.

$y = $ this quotient is the equation of the slant asymptote

$$
\begin{array}{r}
x \phantom{{}^3} \\
x^2 - 1 \overline{\smash{)}\, x^3 \phantom{{}- x}} \\
\underline{x^3 - x} \\
x
\end{array}
$$

Therefore, the equation of the slant asymptote is $y = x$. Find the intercepts.

y-Intercept (let $x = 0$)

$$f(0) = \frac{0^3}{0^2 - 1} = \frac{0}{-1} = 0$$

x-Intercepts [let $f(x) = 0$]

$$\frac{x^3}{x^2 - 1} = 0$$

$$x^3 = 0 \qquad \text{Multiply by } x^2 - 1.$$

$$x = 0$$

The x- and the y-intercept are both 0.

Choose x-values between and beyond the vertical asymptotes and the x-intercept and construct the table of values below. The graph, which is symmetric with respect to the origin, is shown in Fig. 9.42.   ◻

Try Exercise 55

Graph: $f(x) = \dfrac{2x^3}{2x^2 - 8}$.

x	$f(x)$
2	$\frac{8}{3}$
$\frac{1}{2}$	$-\frac{1}{6}$
$-\frac{1}{2}$	$\frac{1}{6}$
-2	$-\frac{8}{3}$

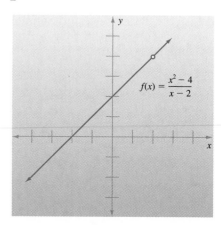

Figure 9.42

CAUTION

The graph of a rational function will never intersect a vertical asymptote. However, the graph of a rational function may intersect a horizontal asymptote or a slant asymptote (see Fig. 9.42). Also, the graph of a rational function may have any number of vertical asymptotes (see Figs. 9.39, 9.40, and 9.41), but it will have at most one nonvertical asymptote (horizontal or slant).   ∎

Graphing Rational Functions That Are Not in Lowest Terms

The graph of a rational function that is not in lowest terms may contain a "hole."

Example 7   Graph: $f(x) = \dfrac{x^2 - 4}{x - 2}$.

Solution: Reduce the function to lowest terms, noting that x cannot equal 2.

$$f(x) = \frac{x^2 - 4}{x - 2} = \frac{(x + 2)(x - 2)}{x - 2} = x + 2 \qquad \text{if } x \neq 2$$

The graph, shown in Fig. 9.43, is the line $f(x) = x + 2$, with a "hole" in the line at the point $(2, 4)$.   ◻

Try Exercise 59

Graph: $f(x) = \dfrac{x^2 - 1}{x - 1}$.

Figure 9.43

Exercises 9.5

Completion Exercises

1. A rational function is of the form $f(x) = p(x)/q(x)$, where $p(x)$ and $q(x)$ are _____ and $q(x) \neq 0$.

2. The domain of a rational function consists of all real numbers except those that _____.

3. If $f(x) \to 3$ as $x \to \infty$, then the line _____ is a _____ (vertical, horizontal) asymptote.

4. If $f(x) \to \infty$ as $x \to -4$, then the line _____ is a _____ (vertical, horizontal) asymptote.

Assume $f(x) = \dfrac{a_n x^n + \cdots + a_1 x + a_0}{b_m x^m + \cdots + b_1 x + b_0}$ $(a_n \neq 0, b_m \neq 0)$ is a rational function in lowest terms for Exercises 5 and 6.

5. If k is a zero of the denominator, then the line _____ is a vertical asymptote.

6. If $n < m$, then the line _____ is a horizontal asymptote. If $n = m$ (neither 0), then the line _____ is a horizontal asymptote. If $n > m$, then _____.

True-False Exercises

7. The graph of a rational function is a smooth curve on its domain.

8. The graph of a rational function is always a continuous curve.

9. The graph of a rational function cannot intersect a vertical asymptote.

10. The graph of a rational function cannot intersect a horizontal asymptote or a slant asymptote.

11. The graph of a rational function can have any number of vertical asymptotes.

12. The graph of a rational function can have any number of horizontal asymptotes.

13. The graph of a rational function can have any number of slant asymptotes.

14. The graph of a rational function that is not in lowest terms may contain a "hole."

Matching Exercises

Match each rational function in Exercises 15 through 18 with its graph in letters A through D.

15. $f(x) = -\dfrac{1}{x}$

16. $f(x) = \dfrac{1}{x}$

17. $f(x) = \dfrac{1}{x - 2}$

18. $f(x) = \dfrac{1}{x + 2}$

A. 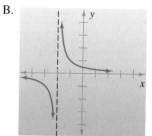   B.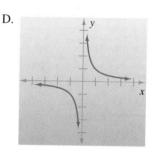

C. 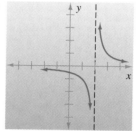   D.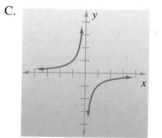

Match each rational function in Exercises 19 through 22 with its graph in letters A through D.

19. $f(x) = \dfrac{1}{x^2}$

20. $f(x) = -\dfrac{1}{x^2}$

21. $f(x) = \dfrac{1}{(x + 1)^2}$

22. $f(x) = \dfrac{1}{(x - 1)^2}$

A. 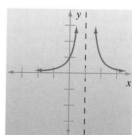   B.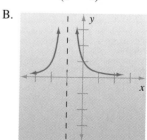

C.

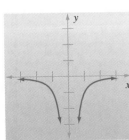

D.

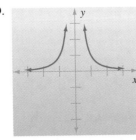

Find all vertical, horizontal, and slant asymptotes for each function. Do not graph the function.

23. $f(x) = \dfrac{5}{x - 4}$

24. $f(x) = \dfrac{1 - 6x}{2x}$

25. $f(x) = \dfrac{-4x^2}{2x^2 - 50}$

26. $f(x) = \dfrac{10x}{2x^2 - 18}$

27. $f(x) = \dfrac{2x^2 + x + 1}{x + 1}$

28. $f(x) = \dfrac{4x^2 - x + 2}{x - 1}$

29. $f(x) = \dfrac{x^4}{x^2 + 1}$

30. $f(x) = \dfrac{x^5 + 1}{x^2 + 1}$

Discussion Exercises

31. In your own words, explain what is meant by an asymptote of a graph.

32. How can you tell whether the graph of a rational function has a slant asymptote? If it has a slant asymptote, how can you find the equation of the slant asymptote?

33. The graph of a rational function cannot have both a horizontal asymptote and a slant asymptote. Why?

34. Outline the steps in graphing a rational function.

Graph each rational function. As sketching aids, find all vertical and horizontal asymptotes and all intercepts.

35. $f(x) = \dfrac{6}{x}$

36. $f(x) = \dfrac{6}{x - 3}$

37. $f(x) = \dfrac{x - 4}{x + 1}$

38. $f(x) = \dfrac{2 - 2x}{x}$

39. $f(x) = \dfrac{4}{x^2 - 4}$

40. $f(x) = \dfrac{-8x}{x^2 - 9}$

41. $f(x) = \dfrac{5x}{x^2 - 9}$

42. $f(x) = \dfrac{x^2}{x^2 - 1}$

43. $f(x) = \dfrac{-2x^2}{x^2 - 16}$

44. $f(x) = \dfrac{x^2 - 9}{x^2 - 4}$

45. $f(x) = \dfrac{3x^2}{x^2 + 1}$

46. $f(x) = \dfrac{12}{x^2 + 3}$

47. $f(x) = \dfrac{3x^2 - 12}{x^2 + 4}$

48. $f(x) = \dfrac{5x}{x^2 + 1}$

49. $f(x) = \dfrac{6}{x^2 + x - 6}$

50. $f(x) = \dfrac{4x}{x^2 - 3x - 4}$

51. $f(x) = \dfrac{x^2 - x}{x^2 + 2x + 1}$

52. $f(x) = \dfrac{x^2 + x}{x^2 - 2x + 1}$

Graph each rational function. As sketching aids, find all vertical and slant asymptotes and all intercepts.

53. $f(x) = \dfrac{x^2 + 1}{x}$

54. $f(x) = \dfrac{x^2}{x - 1}$

55. $f(x) = \dfrac{2x^3}{2x^2 - 8}$

56. $f(x) = \dfrac{x^3 - 8}{2x^2}$

57. $f(x) = \dfrac{x^2 - 4}{x + 1}$

58. $f(x) = \dfrac{x^2 + x - 2}{x + 1}$

Reduce each rational function to lowest terms. Then graph.

59. $f(x) = \dfrac{x^2 - 1}{x - 1}$

60. $f(x) = \dfrac{x^2 - 2x}{x}$

61. $f(x) = \dfrac{x^3 + 2x^2}{x + 2}$

62. $f(x) = \dfrac{x - 1}{x^2 - x}$

63. $f(x) = \dfrac{x^2 - 3x}{x^2}$

64. $f(x) = \dfrac{6x + 18}{(x + 3)^2}$

Problem Solving

65. The graph in Fig. 9.44 shows the number of people, $P(t)$, in a community of 100,000 people who have heard a rumor t days after it has started. At first, only a few people have heard the rumor. Later, the rumor spreads more rapidly since there are more persons to spread it. As time passes, the rumor spreads less rapidly since there are fewer people to hear it. (a) What does $P(t)$ approach as $t \to \infty$? (b) When is the rumor spreading most rapidly?

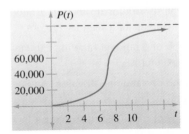

Figure 9.44

66. Newton's law of cooling states that the rate at which the temperature of a body changes to meet that of the surrounding medium is directly proportional to the difference in temperature between the body and the medium. The graph in Fig. 9.45 depicts the temperature $T(t)$ of a

cup of hot coffee t minutes after it is placed in a room whose temperature is 70°. What does $T(t)$ approach as $t \to \infty$?

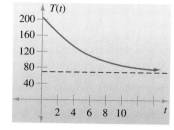

Figure 9.45

67. The average cost per unit, $AC(x)$, of producing x units of a product is shown in Fig. 9.46. How many units should be produced to minimize the average cost per unit?

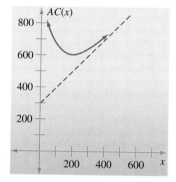

Figure 9.46

Getting Acquainted with Your Graphing Calculator

Graph each rational function. (*Note:* Your calculator may or may not show the vertical asymptotes and/or the "holes" when graphing a rational function. It will depend on the RANGE settings and on whether your calculator is in connected mode or dot mode.)

68. $y = \dfrac{1}{x - 3}$

69. $y = \dfrac{x - 2}{x + 2}$

70. $y = \dfrac{4x}{x^2 - 4}$

71. $y = \dfrac{3x^2 - 3}{x^2 + 1}$

72. $y = \dfrac{x^2 - 1}{x + 1}$

73. $y = \dfrac{x^2 + 1}{x}$

Chapter 9 Key Terms

9.1 Polynomial function (of degree n) A function of the form $f(x) = a_n x^n + \cdots + a_1 x + a_0$, where $a_n, \ldots, a_1, a_0$ are real numbers and $a_n \neq 0$

Leading coefficient (of a polynomial) The number a_n in the definition of a polynomial function

Constant term (of a polynomial) The number a_0 in the definition of a polynomial function

Polynomial equation (of degree n) An equation of the form $a_n x^n + \cdots + a_1 x + a_0 = 0$ $(a_n \neq 0)$

Domain (of a polynomial function) All real numbers

Smooth curve A curve with no corners

Continuous curve A curve with no breaks

Turning point A point at which a graph changes from increasing to decreasing, or from decreasing to increasing

Even function A function such that $f(-x) = f(x)$ for all x in the domain

Odd function A function such that $f(-x) = -f(x)$ for all x in the domain

9.2 Synthetic division A shortcut to long division of polynomials that can be used when the divisor is of the form $x - k$

Multiplicity (of a solution) For $(x - 5)^2 \cdot (x + 4)^3 = 0$, the number 5 is a solution of multiplicity 2, and the number -4 is a solution of multiplicity 3.

9.3 Root (of an equation) A solution of the equation

9.4 Zero (of a function) The number k is a zero of the function f if $f(k) = 0$.

9.5 Rational function A function that can be written in the form $f(x) = p(x)/q(x)$ where $p(x)$ and $q(x)$ are polynomials and $q(x) \neq 0$

Domain (of a rational function) All real numbers except those that make the denominator 0

Vertical asymptote The line $x = k$ is a vertical asymptote if $f(x) \to \infty$ or $f(x) \to -\infty$ as $x \to k$ either from the right or from the left.

Horizontal asymptote The line $y = k$ is a horizontal asymptote if $f(x) \to k$ as $x \to \infty$ or $x \to -\infty$.

Asymptote A line that a portion of a graph gets closer and closer to as that portion of the graph moves away from the origin

Slant asymptote An asymptote that is neither vertical nor horizontal

Chapter 9 Key Rules/Steps

9.1 Polynomial Functions and Their Graphs

Graphing $f(x) = x^n$

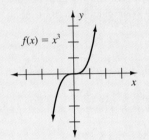

$f(x) = x^3$

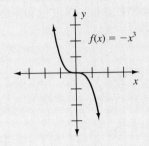

$f(x) = -x^3$

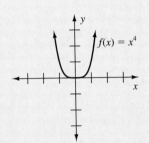

$f(x) = x^4$

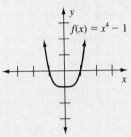

$f(x) = x^4 - 1$

Right and Left Behavior of $f(x) = a_n x^n + \cdots + a_1 x + a_0$
($a_n \neq 0, n \neq 0$)

n even

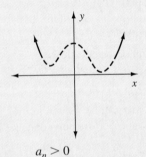

$a_n > 0$

n odd

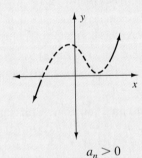

$a_n > 0$

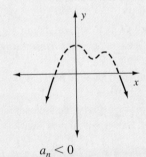

$a_n < 0$

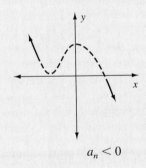

$a_n < 0$

To Graph $f(x) = x^4 - 5x^2 + 4$

1. The y-intercept is 4.
2. Let $f(x) = 0$ to find the x-intercepts.

$$x^4 - 5x^2 + 4 = 0$$
$$(x^2 - 1)(x^2 - 4) = 0$$
$$x^2 - 1 = 0 \qquad x^2 - 4 = 0$$
$$x^2 = \pm 1 \qquad x^2 = \pm 2$$

3. Determine the right and left behavior. The degree is even and the leading coefficient is positive, so the graph rises on both ends.

4. Choose x-values between the x-intercepts.

x	-1.5	1.5
$f(x)$	-2.1875	-2.1875

5. Draw a smooth, continuous curve based on the information obtained in steps 1 through 4.

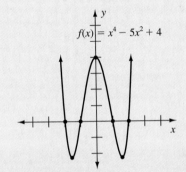

$f(x) = x^4 - 5x^2 + 4$

Turning Points

The graph of a polynomial function of degree n has *at most* $n - 1$ turning points.

Example: The graph of $f(x) = x^3 + 2x^2 + 1$ has *at most* two turning points.

9.2 Synthetic Division; Remainder and Factor Theorems

Synthetic Division

Problem	Synthetic division	Answer

$\dfrac{x^2 - 5x + 10}{x - 2}$

$$\begin{array}{r|rrr} 2 & 1 & -5 & 10 \\ & & 2 & -6 \\ \hline & 1 & -3 & \boxed{4} \end{array}$$

$x - 3 + \dfrac{4}{x - 2}$

The Remainder Theorem

Suppose $P(x)$ is a polynomial. Then $P(k)$ equals the remainder when $P(x)$ is divided by $x - k$.

Example: $\quad P(x) = x^2 + 7$
$\qquad\qquad P(3) = 16$

$$\begin{array}{r|rrr} 3 & 1 & 0 & 7 \\ & & 3 & 9 \\ \hline & 1 & 3 & \boxed{16} \end{array}$$

The Factor Theorem

Suppose $P(x)$ is a polynomial. The number k is a solution of $P(x) = 0$ if and only if $x - k$ is a factor of $P(x)$.

Example: The number 4 is a solution of $x^2 - 3x - 4 = 0$ if and only if $x - 4$ is a factor of $x^2 - 3x - 4$.

Write an Equation with the Solution Set {0, 2, −1, −1}

$$(x - 0)(x - 2)(x + 1)^2 = 0$$
$$(x^2 - 2x)(x^2 + 2x + 1) = 0$$
$$x^4 - 3x^2 - 2x = 0$$

Number of Solutions of a Polynomial Equation

If a solution of multiplicity k is counted k times, a polynomial equation of degree n has exactly n solutions.

Example: Counting multiplicity, the equation $x^5 + x^3 - 2 = 0$ has exactly five solutions.

9.3 The Rational Root Test

The Rational Root Test

Let $a_n x^n + \cdots + a_1 x + a_0 = 0$ $(a_n \neq 0)$ be a polynomial equation with *integer* coefficients. If p/q is a rational root in lowest terms, then p is a factor of a_0 and q is a factor of a_n.

Example:

$$2x^3 - 5x^2 - x + 6 = 0$$

p: $\pm 1, \pm 2, \pm 3, \pm 6$

q: $\pm 1, \pm 2$

$\dfrac{p}{q}$: $\pm 1, \pm 2, \pm 3, \pm 6, \pm\frac{1}{2}, \pm\frac{3}{2}$

$$\begin{array}{r|rrrr} -1 & 2 & -5 & -1 & 6 \\ & & -2 & 7 & -6 \\ \hline & 2 & -7 & 6 & \boxed{0} \end{array}$$

$$(x + 1)(2x^2 - 7x + 6) = 0$$
$$(x + 1)(2x - 3)(x - 2) = 0$$
$$x = -1 \qquad x = \tfrac{3}{2} \qquad x = 2$$

Conjugate Pairs Theorem

If the imaginary number $a + bi$ is a root of a polynomial equation with *real* coefficients, then $a - bi$ must also be a root.

Example: Since $2 + i$ is a root of $x^2 - 4x + 5 = 0$, then $2 - i$ must also be a root.

9.4 Zeros of a Function

Relationship between Zeros, Roots, x-Intercepts, and Factors

1. The numbers -1 and 3 are zeros of $f(x) = x^2 - 2x - 3$.

2. The numbers -1 and 3 are roots of $x^2 - 2x - 3 = 0$.

3. The numbers -1 and 3 are x-intercepts of the graph of $f(x) = x^2 - 2x - 3$.

4. The binomials $x + 1$ and $x - 3$ are factors of $f(x)$.

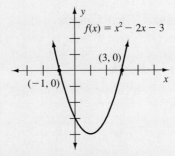

Linear Factorization Theorem

If $f(x)$ is a polynomial of positive degree, then

$$f(x) = a(x - k_1)(x - k_2) \cdots (x - k_n)$$

where $k_1, k_2, \ldots, k_n$ are the zeros of f and a is the leading coefficient of $f(x)$.

Example: To factor $f(x) = 3x^2 - 5x + 1$ into linear factors, let $f(x) = 0$ to find the zeros.

$$3x^2 - 5x + 1 = 0$$

$$x = \frac{5 \pm \sqrt{25 - 12}}{2} = \frac{5 \pm \sqrt{13}}{2}$$

$$f(x) = 3\left(x - \frac{5 + \sqrt{13}}{2}\right)\left(x - \frac{5 - \sqrt{13}}{2}\right)$$

Fundamental Theorem of Algebra

A polynomial function of positive degree with complex coefficients has at least one complex zero.

Example: $f(x) = x^5 - 4x^3 - x + 2$ has at least one complex zero.

Approximating the Real Root of $x^3 + x + 1 = 0$

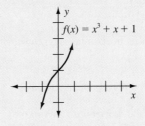

$$f(-1) = (-1)^3 + (-1) + 1 = -1$$
$$f(-0.9) = (-0.9)^3 + (-0.9) + 1 = -0.629$$
$$f(-0.8) = (-0.8)^3 + (-0.8) + 1 = -0.312$$
$$f(-0.7) = (-0.7)^3 + (-0.7) + 1 = -0.043 \quad \Big\}\ \text{Sign change}$$
$$f(-0.6) = (-0.6)^3 + (-0.6) + 1 = +0.184$$

To the nearest tenth, the root (solution) is probably $x = -0.7$.

9.5 Rational Functions and Their Graphs

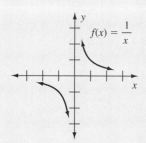

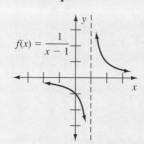

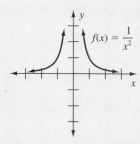

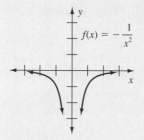

Finding Vertical and Horizontal Asymptotes

Assume $f(x) = \dfrac{a_n x^n + \cdots + a_1 x + a_0}{b_m x^m + \cdots + b_1 x + b_0}$ $(a_n \neq 0, b_m \neq 0)$ is a rational function in lowest terms.

Vertical asymptotes

If k is a zero of the denominator, the line $x = k$ is a vertical asymptote.

Example: $f(x) = \dfrac{1}{x^2 - 4x}$ has the two vertical asymptotes $x = 0$ and $x = 4$.

Horizontal asymptotes

1. If $n < m$, $y = 0$ is a horizontal asymptote.

   Example: $f(x) = \dfrac{6}{x - 3}$ has the horizontal asymptote $y = 0$.

2. If $n = m$ (neither 0), $y = a_n/b_m$ is a horizontal asymptote.

   Example: $f(x) = \dfrac{2x^2}{x^2 - 9}$ has the horizontal asymptote $y = 2$.

3. If $n > m$, there is no horizontal asymptote. If $n = m + 1$ ($m \neq 0$), there is a slant asymptote.

   Example: $f(x) = \dfrac{x^2}{x + 2}$ has no horizontal asymptote, but it has a slant asymptote.

To Graph the Rational Function $f(x) = \dfrac{x^2 - 4}{x^2 - 9}$

1. Set the denominator equal to 0 to find the vertical asymptotes.

$$x^2 - 9 = 0$$
$$x^2 = 9$$
$$x = \pm 3$$

2. Find the horizontal asymptote.

   Since $n = m$, $y = 1$ is the horizontal asymptote.

3. Let $x = 0$ to find the y-intercept.

$$f(0) = \frac{0^2 - 4}{0^2 - 9} = \frac{4}{9}$$

4. Let $f(x) = 0$ to find the x-intercepts.

$$\frac{x^2 - 4}{x^2 - 9} = 0$$
$$x^2 - 4 = 0 \qquad \text{Multiply by } x^2 - 9.$$
$$x = \pm 2$$

5. Choose x-values between and beyond the vertical asymptotes and the x-intercepts.

x	-4	$-\frac{5}{2}$	$\frac{5}{2}$	4
$f(x)$	$\frac{12}{7}$	$-\frac{9}{11}$	$-\frac{9}{11}$	$\frac{12}{7}$

6. Draw smooth curves through the intercepts and the points from the table so that a portion of each curve approaches an asymptote.

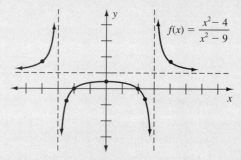

Finding a Slant Asymptote

A rational function has a slant asymptote when the degree of the numerator is exactly 1 more than the degree of the denominator and the function is in lowest terms and the denominator is not a constant.

Example: Find the slant asymptote for $f(x) = \dfrac{2x^2}{x + 2}$.

$$
\begin{array}{r}
2x - 4 \\
x + 2 \overline{)\, 2x^2 } \\
\underline{2x^2 + 4x } \\
-4x \\
\underline{-4x - 8} \\
8
\end{array}
$$

The slant asymptote is $y = 2x - 4$.

To Graph a Rational Function Not in Lowest Terms

$$f(x) = \frac{x^3 - x^2}{x - 1}$$

$$f(x) = \frac{x^2(x - 1)}{x - 1}$$

$$f(x) = x^2 \quad \text{if } x \neq 1$$

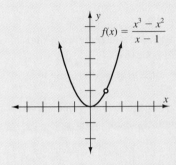

Chapter 9 Review Exercises

[9.1] Graph each polynomial function.

1. $y = (x - 1)^3$
2. $y = x^4 - 2$

Graph each polynomial function. As sketching aids, find all intercepts and the right and left behavior.

3. $f(x) = x^3 - x$
4. $f(x) = -x^3 + x^2 + 6x$
5. $f(x) = x^3 + 2x^2 - 4x - 8$
6. $f(x) = x^4 - 9x^2$

7. Explain what a turning point of a graph is.

8. Is the domain of every polynomial function all real numbers? Is the range of every polynomial function all real numbers? Draw graphs to illustrate your answers.

9. Show that $f(x) = 2x^3 - 5x$ is an odd function.

[9.2] Use synthetic division to divide.

10. $(x^2 - x - 6) \div (x - 3)$
11. $(2x^4 + 13x - 3) \div (x - 2)$
12. $\dfrac{t^3 - 6t^2 + 11t - 6}{t - 2}$ 13. $\dfrac{p^3 + 2p^2 - 8p - 24}{p - 3}$

Use the remainder theorem and synthetic division to find $P(k)$ for each exercise.

14. $P(x) = 4x^3 - 8x^2 - 13x + 31; k = 2$
15. $P(y) = y^5 + 5y^4 - 7y^2 + 29y - 20; k = -3$

Use the remainder theorem and synthetic division to deter-mine whether the given number is a solution of the given equation.

16. $3x^3 - 6x^2 + x + 10 = 0; x = -1$
17. $x^4 - 28x^2 + 9x + 58 = 0; x = 5$

18. Counting multiplicity, how many solutions does a polynomial equation of degree n have?

19. Use the remainder theorem and synthetic division to show that 4 is a solution of $x^3 - x^2 - 30x + 72 = 0$. Then use the factor theorem to completely solve the equation.

20. Write a polynomial equation with integer coefficients that has the solution set $\{0, \frac{2}{5}, -6\}$.

[9.3]

21. List all possible rational roots of the polynomial equation $3x^3 + 23x^2 + 28x - 12 = 0$. Do not solve the equation.

Solve each equation.

22. $2x^3 + 5x^2 - 32x - 80 = 0$
23. $x^4 - 12x^2 - 64 = 0$
24. $\frac{1}{2}x^3 - \frac{3}{2}x^2 - x + 1 = 0$
25. $x^4 + 9x^3 + 18x^2 - 4x - 24 = 0$

26. Show that the equation $x^3 + x^2 - 3 = 0$ has no rational roots.

27. What does the conjugate pairs theorem say?

28. Find the dimensions of the box in Fig. 9.47, if the volume of the box is 24 cubic feet.

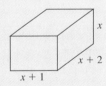

Figure 9.47

[9.4] Find the zeros of each polynomial function. Then use the zeros to factor $f(x)$ into linear factors.

29. $f(x) = x^2 + 9$
30. $f(x) = x^3 + 3x^2 - 5x - 15$
31. $f(x) = x^2 - 8x + 3$
32. $f(x) = 2x^3 + x^2 - 25x + 12$

33. Find the zeros and the right and left behavior for the function $f(x) = x^3 + 2x^2 - 5x - 6$. Then use this information to graph the function.

34. Approximate the real root of $x^3 + 2x - 1 = 0$ to the nearest hundredth (see Fig. 9.48).

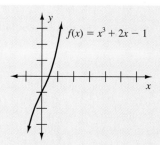

Figure 9.48

35. Draw a graph of a polynomial function of degree 4 that has three distinct real zeros.

36. State the fundamental theorem of algebra.

[9.5] Find all vertical, horizontal, and slant asymptotes for each function. Do not graph the function.

37. $f(x) = \dfrac{4}{x^2}$ 38. $f(x) = \dfrac{x^2 + 1}{2x^2 - 6x}$

39. $f(x) = \dfrac{3x^2}{x + 2}$

Graph each rational function. As sketching aids, find all vertical, horizontal, and slant asymptotes and all intercepts.

40. $f(x) = \dfrac{6}{x + 2}$ 41. $f(x) = \dfrac{-2x^2}{x^2 - 4}$

42. $f(x) = \dfrac{3x}{x^2 - 2x - 3}$ 43. $f(x) = \dfrac{5}{x^2 + 1}$

44. $f(x) = \dfrac{x^3}{x^2 - 1}$ 45. $f(x) = \dfrac{x^2 - 9}{x - 3}$

46. Can the graph of a rational function intersect a vertical asymptote? A horizontal asymptote? A slant asymptote?

Chapter 9 Test

1. If $5 - 3i$ is a solution of a polynomial equation with real coefficients, name another solution.

2. What is the most number of turning points the graph of $f(x) = x^3 - 4x^2 - 5x - 3$ could have?

3. Discuss the number of vertical, horizontal, and slant asymptotes the graph of a rational function may have.

4. Explain what is meant by a smooth, continuous curve. Draw graphs to illustrate your answer.

5. Show that $f(x) = x^4 + 3x^2$ is an even function.

6. Write a polynomial equation with integer coefficients that has the solution set $\{1, 2, -4\}$.

7. State the solutions of $x^2(x - 6)^3 = 0$, and give the multi-plicity of each solution.

8. Use synthetic division to divide $x^3 - 2x^2 - x + 2$ by $x - 3$.

9. List all possible rational roots of the polynomial equation $4x^3 - x^2 - 17x + 15 = 0$. Do not solve the equation.

10. Find the zeros of $f(x) = x^2 - 4x + 2$. Then use the zeros to factor $f(x)$ into linear factors.

11. Solve $x^3 + 5x^2 + 12x + 8 = 0$.

Graph each polynomial function. As sketching aids, find all intercepts and the right and left behavior.

12. $f(x) = x^3 + 3x^2$ 13. $f(x) = x^4 - 10x^2 + 9$

Graph each rational function. As sketching aids, find all vertical, horizontal, and slant asymptotes and all intercepts.

14. $f(x) = \dfrac{4x}{x^2 - 1}$ 15. $f(x) = \dfrac{x^2 + 1}{x}$

16. Approximate the real root of $x^3 - x - 2 = 0$ to the nearest tenth (see Fig. 9.49).

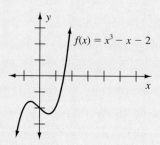

Figure 9.49

Cumulative Test for Chapters 7, 8, and 9

1. Evaluate: $[\![7.9]\!]$.

2. Find the domain of $f(x) = \dfrac{3x}{x^2 - 16}$.

3. Find the distance between $(-4, 5)$ and $(-1, 8)$.

4. Find the midpoint of the line segment joining $(1, -2)$ and $(-9, 7)$.

5. Write a polynomial equation with integer coefficients that has the solution set $\{0, 2, -5\}$.

6. Without graphing, test $y = x^3 - 4x$ for symmetry with respect to the x-axis, the y-axis, and the origin.

7. Find the slope and the y-intercept of the line $5x - 3y = 6$. Then use this information to graph the line.

8. Find the zeros of $f(x) = x^2 - x - 5$. Then use the zeros to factor $f(x)$ into linear factors.

9. List all possible rational roots of the polynomial equation $3x^3 - 4x^2 - 17x + 6 = 0$. Do not solve the equation.

10. Use synthetic division to divide $2x^3 - 7x + 13$ by $x + 3$.

11. Solve $2x^3 - 3x^2 - x + 1 = 0$.

12. The distance y that a spring is stretched varies directly as the weight x on the spring. If a weight of 16 kg stretches the spring 10 cm, how far will a weight of 20 kg stretch the spring?

13. If $f(x) = x^2 - 9$ and $g(x) = x + 3$, find each of the following: (a) $f(-5)$, (b) $(f - g)(4)$, (c) $(f \circ g)(x)$.

Find an equation of the line satisfying the given conditions. Then write your answer in standard form.

14. Through $(-4, 7)$ and $(-3, -2)$

15. Perpendicular to $y = 4x - 1$ with y-intercept 6

Graph each equation.

16. $f(x) = 3x - 6$

17. $f(x) = x^2 - 4x - 5$

18. $\dfrac{x^2}{16} + \dfrac{y^2}{36} = 1$

19. $\dfrac{x^2}{25} - \dfrac{y^2}{9} = 1$

20. $x^2 + y^2 - 6x - 4y - 3 = 0$

21. $y = \sqrt{1 - x}$

22. $f(x) = x^3 + 3x^2 - x - 3$

23. $f(x) = \dfrac{8x}{x^2 - 9}$

Graph each inequality.

24. $x + 2y < 4$

25. $|y| \geq 3$

Exponential and Logarithmic Functions

Above-ground nuclear explosions produce various types of radioactive fallout. Strontium 90 is one of the most dangerous of these because it has a long half-life and is easily absorbed into the bone structure. In Sec. 10.1 we use exponential functions to describe the decay of a radioactive substance.

10.1 Exponential Functions

TAPE AU21

In this chapter we study two new functions—*exponential functions* and *logarithmic functions*. We will use exponential functions to calculate the decay of a radioactive substance and the value of an account that compounds interest continuously. We will use logarithmic functions to measure the pH of a solution and the magnitude of an earthquake.

Defining the Exponential Function

In Sec. 5.2 we defined rational exponents. For example,

$$2^{1/3} = \sqrt[3]{2} \quad \text{and} \quad 2^{1.5} = 2^{3/2} = \sqrt{8}$$

We can also define *irrational* powers, like 2^π. Since $\pi = 3.1416 \ldots$, we can think of 2^π as that number which has the successively closer approximations

$$2^3, \ 2^{3.1}, \ 2^{3.14}, \ 2^{3.141}, \ 2^{3.1416}, \ \ldots$$

Therefore we assume that 2^x is defined for all real numbers x, and that the rules for exponents stated in Sec. 5.2 hold for all real exponents.

Let a, b, x, and y be any real numbers such that the expressions below are defined:

(a) $a^x a^y = a^{x+y}$ (b) $\dfrac{a^x}{a^y} = a^{x-y}$

(c) $(a^x)^y = a^{xy}$ (d) $(ab)^x = a^x b^x$

(e) $\left(\dfrac{a}{b}\right)^x = \dfrac{a^x}{b^x}$

We are now ready to define the exponential function.

DEFINITION OF THE EXPONENTIAL FUNCTION

Suppose $a > 0$ and $a \neq 1$. The **exponential function with base a** is a function of the form

$$f(x) = a^x$$

Note that if $a = 1$, the function $f(x) = a^x$ becomes $f(x) = 1^x = 1$, which is a constant function. That is why we eliminate 1 as a base. If $a = 0$, then $f(x) = a^x = 0^x = 0$ (assuming $x > 0$), which is also a constant function. If $a = -4$, then $f(x) = (-4)^x$, which is not a real number when $x = \frac{1}{2}$. For these reasons, we also eliminate 0 and negative numbers as bases for the exponential function.

Graphing Exponential Functions

Example 1 Graph: $y = 2^x$.

Solution: This is the exponential function with base 2. If we substitute values for x and compute the corresponding values of y, we obtain the table of values below.

x	-3	-2	-1	0	1	2	3
y	$\frac{1}{8}$	$\frac{1}{4}$	$\frac{1}{2}$	1	2	4	8

Try Exercise 15

Graph: $y = 3^x$.

Plot these points and draw a smooth curve through them to produce the graph shown in Fig. 10.1.   ❐

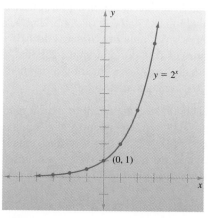

Figure 10.1   Figure 10.2

Since a portion of the graph of $y = 2^x$ gets closer and closer to the x-axis as x decreases without bound (written $y \to 0$ as $x \to -\infty$), the x-axis is called a **horizontal asymptote** of the graph of $y = 2^x$.

Example 2   Graph: $y = (\frac{1}{2})^x$.

Solution: This is the exponential function with base $\frac{1}{2}$. Using this function, we construct the table of values below.

x	-3	-2	-1	0	1	2	3
y	8	4	2	1	$\frac{1}{2}$	$\frac{1}{4}$	$\frac{1}{8}$

Plot these points and draw a smooth curve through them to produce the graph shown in Fig. 10.2. Note that the x-axis is a horizontal asymptote of the graph.   ❐

Try Exercise 17

Graph: $y = (\frac{1}{3})^x$.

Since the y-values in $y = 2^x$ increase as the x-values increase (see Fig. 10.1), we say that $y = 2^x$ is an **increasing function.** Since the y-values in $y = (\frac{1}{2})^x$ decrease as the x-values increase (see Fig. 10.2), we say that $y = (\frac{1}{2})^x$ is a **decreasing function.**

Generalizing on these last two statements, we have the following rule: The graph of $y = a^x$ is increasing if $a > 1$ and decreasing if $0 < a < 1$.

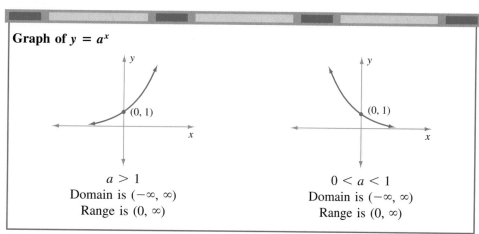

Graph of $y = a^x$

$a > 1$
Domain is $(-\infty, \infty)$
Range is $(0, \infty)$

$0 < a < 1$
Domain is $(-\infty, \infty)$
Range is $(0, \infty)$

Problem Solving

Example 3   The radioactive substance plutonium-239 is present in nuclear reactor wastes. It has a half-life of about 25,000 yr. This means that it takes 25,000 yr for a given quantity to decay to half of its original quantity. If Q is the quantity of 100 g of plutonium-239 that remains after t years, write an equation that gives Q as a function of t. Then use this equation to find the quantity that remains after 100,000 yr.

Solution: After 25,000 yr, the quantity that remains is $100 \cdot \frac{1}{2} = 100(2)^{-1}$.
After 50,000 yr, the quantity that remains is $100 \cdot \frac{1}{4} = 100(2)^{-2}$.
After 75,000 yr, the quantity that remains is $100 \cdot \frac{1}{8} = 100(2)^{-3}$.
After t years, the quantity that remains is $100(2)^{-t/25,000}$.
Therefore an equation that gives Q as a function of t is

$$Q = 100(2)^{-t/25,000}$$

To find the quantity that remains after 100,000 years, replace t with 100,000.

$$Q = 100(2)^{-100,000/25,000}$$
$$Q = 100(2)^{-4}$$
$$Q = 100\left(\frac{1}{16}\right)$$
$$Q = 6.25$$

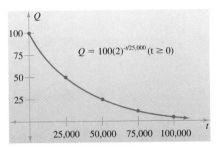

Figure 10.3

Therefore, 6.25 g will remain after 100,000 yr. A graph of this function for $t \geq 0$ is shown in Fig. 10.3.   ❐

Try Exercise 33

The radioactive substance iodine-131 has a half-life of 8 days. If Q is the quantity of 40 g of iodine-131 that remains after t days, write an equation that gives Q as a function of t. Then graph the function for $t \geq 0$.

The Number e

Although we can use any positive number except 1 as the base for the exponential function, the base that is used most often in applied problems is the number e. The number e is irrational, like π, so its decimal value neither terminates nor repeats. The value of e correct to nine decimal places is

$$e \approx 2.718281828$$

The number e comes from the expression $\left(1 + \frac{1}{n}\right)^n$, where n is a number that grows large without bound. Because this expression kept appearing in many different types of applied problems, the Swiss mathematician Leonhard Euler (1701–1783) decided to give this number a name, and the name he chose was e.*

The table below illustrates that $\left(1 + \frac{1}{n}\right)^n$ approaches $e = 2.71828\ldots$ as n grows larger:

n	1	10	100	1000	10,000 $\to \infty$
$\left(1 + \frac{1}{n}\right)^n$	2	2.59374	2.70481	2.71692	2.71815 $\to e$

*HISTORICAL NOTE: Euler, who also popularized the symbols π, i, Σ, and $f(x)$, was probably the greatest notation builder of all time. Fluent in several languages, he was also trained in physics, astronomy, medicine, and theology.

The exponential function with base e, namely,

$$y = e^x$$

is used so often that we usually refer to $y = e^x$ as *the* exponential function. Figure 10.4 illustrates how the graph of $y = e^x$ compares with the graphs of $y = 2^x$ and $y = 3^x$.

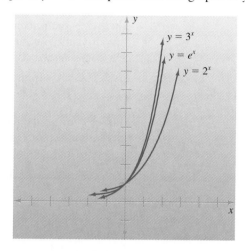

Figure 10.4

To illustrate an applied problem that uses the number e, consider the formula below:

$$V = P\left(1 + \frac{r}{n}\right)^{nt}$$

This formula gives the value V of a savings account that compounds interest n times per year. P is the principal (original amount), r is the interest rate in decimal form, and t is the time in years. For example, suppose you decide to deposit \$3500 into an account that pays 9 percent interest compounded monthly. The value V of the account after 5 yr is

$$V = 3500\left(1 + \frac{0.09}{12}\right)^{12(5)}$$

Using a calculator, we find that

$$V = 3500(1.0075)^{60} \approx 3500(1.56568) = \$5479.88$$

Now suppose that the interest is compounded daily (365 times per year), or hourly (8760 times per year), or every minute (525,600 times per year). If we allow n, the number of times that the interest is compounded each year, to grow large without bound, we say that the interest is **compounded continuously.** To see what happens in that case, let $m = \frac{n}{r}$. Then the formula

$$V = P\left(1 + \frac{r}{n}\right)^{nt}$$

becomes

$$V = P\left(1 + \frac{1}{m}\right)^{mrt} = P\left[\left(1 + \frac{1}{m}\right)^{m}\right]^{rt}$$

Now, if $n \to \infty$, then $m \to \infty$, and $(1 + \frac{1}{m})^{m} \to e$. Therefore, the formula for continuously compounded interest is

$$V = Pe^{rt}$$

Problem Solving

Example 4 If interest is compounded continuously, the value V of a savings account is given by the formula

$$V = Pe^{rt}$$

where P is the principal, r is the interest rate in decimal form, and t is the time in years. Determine the value of $1000 invested for 2 years at 8 percent compounded continuously.

Solution: Substitute $P = 1000$, $r = 0.08$, and $t = 2$ into the formula $V = Pe^{rt}$.
$$V = 1000e^{0.08(2)} = 1000e^{0.16}$$

Find $e^{0.16}$ using the table in Appendix 3 or by pressing the following keys on your calculator. (In Sec. 10.5 we will see that the function $y = e^x$ is the *inverse* of the natural logarithmic function $y = \ln x$. That is why we can find values for e^x by pressing $\boxed{\text{INV}}\boxed{\ln x}$.)

$$\boxed{\text{Clear}}\ .16\ \boxed{\text{INV}}\ \boxed{\ln x}\ \boxed{1.173510871}$$

$$e^x$$

Therefore $e^{0.16} \approx 1.17351$, and

$$V \approx 1000(1.17351) = 1173.51$$

The value is $1173.51.

Exercises 10.1

Completion Exercises

1. Suppose $a > 0$ and $a \neq 1$. The exponential function with base a is a function of the form _____.

2. The domain of the exponential function $f(x) = a^x$ is _____, and the range is _____.

3. The graph of $y = a^x$ is increasing if _____, and decreasing if _____.

4. The number e is _____ (rational, irrational). The value of e accurate to three decimal places is _____. The expression $(1 + \frac{1}{n})^n$ approaches e as n _____.

True-False Exercises

5. The function $f(x) = x^2$ is an exponential function.

6. The function $f(x) = x^{-1}$ is an exponential function.

Matching Exercises

Match each function in Exercises 7 through 10 with its graph in letters A through D.

7. $y = 2^x$

8. $y = 2^{-x}$

9. $y = -2^x$

10. $y = -2^{-x}$

A.

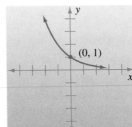

B.

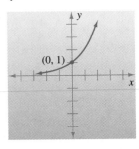

C.

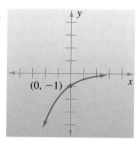

D.

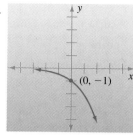

Match each function in Exercises 11 through 14 with its graph in letters A through D.

11. $y = 2^x + 1$

12. $y = 2^{x+1}$

13. $y = 2^{x-1}$

14. $y = -2^x + 1$

A.

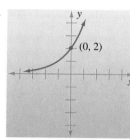

B.

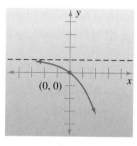

C.

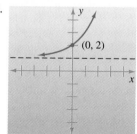

D.

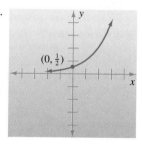

Graph each function.

15. $y = 3^x$

16. $y = 4^x$

17. $y = (\frac{1}{3})^x$

18. $y = 4^{-x}$

19. $y = 4^{x/2}$

20. $y = 8^{x/3}$

21. $y = 2^{|x|}$

22. $y = (1.5)^{x^2}$

23. $y = 8 \cdot 2^{-x^2}$

24. $y = 6 \cdot 3^{-|x|}$

✏ Discussion Exercises

25. Why do we eliminate 0 and 1 as possible values of the base a for the exponential function $f(x) = a^x$?

26. Why do we eliminate negative numbers as possible values of the base a for the exponential function $f(x) = a^x$?

27. Compare the graphs of $y = 2^{-x}$ and $y = (\frac{1}{2})^x$.

28. Discuss the difference between the quadratic function $y = x^2$ and the exponential function $y = 2^x$.

29. What can be said about the base a in Fig. 10.5? Explain.

30. What can be said about the base a in Fig. 10.6? Explain.

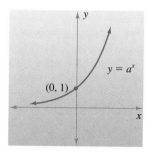

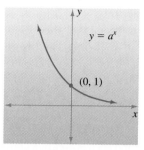

Figure 10.5   *Figure 10.6*

Problem Solving

31. The value of a comic book that was purchased for $3 doubles each year. This means that its value V after t years is given by the function

$$V = 3(2)^t$$

Find the value of the comic book at each time below. Then graph the function for $t \geq 0$.

(a) $t = 0$ (b) $t = 1$ (c) $t = 2$ (d) $t = 3$

32. In a petri dish, a bacteria culture that contains five cells quadruples its number each hour. This means that the number N of bacteria present after t hours is given by the function

$$N = 5(4)^t$$

Find the number of bacteria present at each time below. Then graph the function for $t \geq 0$.

(a) $t = 0$ (b) $t = 1$ (c) $t = 2$ (d) $t = 3$

33. The radioactive substance iodine-131 has a half-life of 8 days. If Q is the quantity of 40 g of iodine-131 that remains after t days, write an equation that gives Q as a function of t. Then graph the function for $t \geq 0$.

34. The radioactive substance radium-226 has a half-life of 1600 yr. If Q is the quantity of 56 g of radium-226 that remains after t years, write an equation that gives Q as a function of t. Then graph the function for $t \geq 0$.

Getting Acquainted with Your Scientific Calculator

To find 2^π on your calculator, press

[Clear] 2 [y^x] [π] [=]   [8.824977827]

This answer is correct to nine decimal places.

Find each power in Exercises 35 through 37 correct to three decimal places.

35. 3^{π}

36. $5^{-\pi}$

37. $6^{\sqrt{2}}$

38. Find $2^{3.1}$, $2^{3.14}$, $2^{3.141}$, $2^{3.1416}$, and compare these values to the value of 2^{π} given above.

To find e^3 on your calculator, press

$$\boxed{\text{Clear}}\ 3\ \boxed{\text{INV}}\ \boxed{\ln x}\ \boxed{20.08553692}$$

This answer is correct to eight decimal places.

Find each power of e correct to five decimal places.

39. e^2

40. e^4

41. $e^{3.8}$

42. $e^{3.9}$

43. $e^{1.52}$

44. $e^{1.26}$

45. $e^{-0.3}$

46. $e^{-0.8}$

Graph each function.

47. $y = e^{-x}$

48. $y = -e^x$

Problem Solving

49. If you invest \$10,000 in a fund that pays $10\frac{3}{4}$ percent interest compounded quarterly, what is the value of the fund after 11 yr?

50. Suppose you deposit \$2800 into an account that compounds interest semiannually. Find the value of the account after 3 yr if the interest rate is $7\frac{1}{2}$ percent.

51. Determine the value of \$2000 invested for 5 yr at 8 percent compounded continuously.

52. Determine the value of \$3000 invested for 10 yr at 12 percent compounded continuously.

53. Should you invest your money at 11.4 percent compounded annually or at 11 percent compounded continuously?

54. Should you invest your money at 9.1 percent compounded monthly or at 9 percent compounded continuously?

The atmospheric pressure P in pounds per square inch at an altitude of a miles is given by the formula

$$P = 14.7e^{-0.21a}$$

55. Find the atmospheric pressure at sea level (see Fig. 10.7). Find the atmospheric pressure at an altitude of 10 mi.

56. Find the atmospheric pressure on Mt. Everest, about 5.5 mi above sea level (see Fig. 10.7). Find the atmospheric pressure in Death Valley, about 0.05 mi below sea level (see Fig. 10.7).

Figure 10.7

Getting Acquainted with Your Graphing Calculator

Graph the three functions in each exercise in the same Viewing Rectangle and compare the graphs.

57. $y = 2^x$, $y = e^x$, $y = 3^x$

58. $y = 4^x$, $y = 4^{-x}$, $y = -4^x$

Graph each function.

59. $y = 2.3^{|x|}$

60. $y = 4(1.1)^{-x^2}$

10.2 Inverse of a Function

TAPE AU19

There are several types of *inverses* in mathematics. For example, the *additive* inverse of the number 5 is -5. The *multiplicative* inverse of 5 is $\frac{1}{5}$. We can also find the inverse of certain kinds of functions. For example, the inverse of the function $f(x) = x^2$ $(x \ge 0)$ is the function $g(x) = \sqrt{x}$. Each of the functions f and g "undoes" what the other function "does" (see Fig. 10.8).

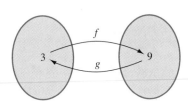

Figure 10.8

In this section we learn how to find the inverse of a function. In Sec. 10.3 we see that a logarithmic function is simply the inverse of an exponential function.

One-to-One Functions

A function f will have an inverse function if and only if f is a *one-to-one* function.

DEFINITION OF A ONE-TO-ONE FUNCTION

A function f is **one-to-one** if for any x_1 and x_2 in its domain,

$$x_1 \neq x_2 \text{ implies } f(x_1) \neq f(x_2)$$

Note that f is one-to-one if different x-values imply different y-values. Therefore, **a function f is one-to-one if each y-value corresponds to exactly one x-value.**

Example 1   Decide whether each function is one-to-one.
    (a) $f = \{(3, 5), (7, -2)\}$ (b) $f = \{(0, 1), (4, 1)\}$
    (c) $f(x) = 2x - 6$ (d) $f(x) = x^2$

Solution:
    (a) f is one-to-one, since $3 \neq 7$ implies $f(3) \neq f(7)$. Note that $f(3) = 5$ and $f(7) = -2$. f is illustrated in Fig. 10.9.
    (b) f is not one-to-one, since $0 \neq 4$, but $f(0) = f(4)$. In fact, $f(0)$ and $f(4)$ are both 1. f is illustrated in Fig. 10.10.
    (c) Suppose $x_1 \neq x_2$. Then $2x_1 \neq 2x_2$ and $2x_1 - 6 \neq 2x_2 - 6$. Therefore $f(x_1) \neq f(x_2)$, so f is one-to-one. f is illustrated in Fig. 10.11.
    (d) Let $x_1 = 2$ and $x_2 = -2$. Then $x_1 \neq x_2$, but $f(x_1) = f(2) = 2^2 = 4$ and $f(x_2) = f(-2) = (-2)^2 = 4$. Therefore, f is not one-to-one. f is illustrated in Fig. 10.12. ❏

Try Exercise 9(a)

Determine whether the function $f = \{(2, 5), (-1, 2), (-3, 0)\}$ is one-to-one.

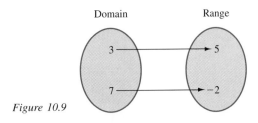

Figure 10.9

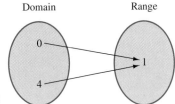

Figure 10.10

We can also determine whether a function is one-to-one from its graph.

Horizontal Line Test

A function is one-to-one provided that no horizontal line intersects its graph at more than one point.

The graph in Fig. 10.11 [part (c) of Example 1] represents a one-to-one function; the graph in Fig. 10.12 [part (d) of Example 1] does not. Note that the y-value 4 in Fig. 10.12 corresponds to the two x-values 2 and -2.

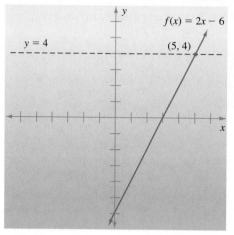

Figure 10.11

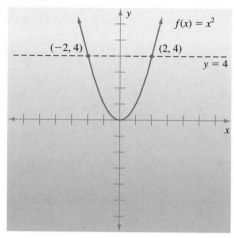

Figure 10.12

The Inverse of a Function

Consider the one-to-one function f below:

$$f = \{(3, 5), (-1, 6), (-4, 0)\}$$

If we interchange the coordinates of each ordered pair, the resulting function is called the *inverse* of f, written f^{-1}. The symbol f^{-1} is read "f inverse."

$$f^{-1} = \{(5, 3), (6, -1), (0, -4)\}$$

Notice that the range of f becomes the domain of f^{-1}, and the domain of f becomes the range of f^{-1} (see Fig. 10.13). The function f^{-1} "undoes" what the function f "does."

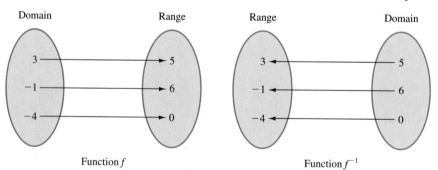

Figure 10.13

CAUTION

The symbol -1 in the expression f^{-1} is not an exponent, and f^{-1} does *not* mean $\frac{1}{f}$. The expression $f^{-1}(x)$ represents the range value of the function f^{-1}, and $f^{-1}(x)$ is read "f inverse of x." ∎

Now consider the function

$$f = \{(4, 6), (-2, -1), (-5, 6)\}$$

Note that f is not a one-to-one function, since $4 \neq -5$, but $f(4) = f(-5)$. If we try to form the inverse of f, we get

$$\{(6, 4), (-1, -2), (6, -5)\}$$

But this last set of ordered pairs is not a function since the first coordinate 6 corresponds to two different second coordinates, namely, 4 and -5. Can you see why only one-to-one functions have inverse functions?

DEFINITION OF THE INVERSE OF A FUNCTION

Let f be a one-to-one function. The **inverse of f,** denoted f^{-1}, is the set of all ordered pairs of the form (b, a), where (a, b) belongs to f.

When a function is defined by an equation, we find its inverse by interchanging the independent and the dependent variables.

Example 2 Given $f(x) = 2x - 4$, find f^{-1}.

Solution: The graph of f is a line with slope 2 and y-intercept -4 (see Fig. 10.14). Since no horizontal line intersects the graph at more than one point, f is one-to-one and therefore f has an inverse function. To find f^{-1}, begin by replacing $f(x)$ with y.

$$y = 2x - 4$$

Interchange x and y to get the equation that defines f^{-1}.

$$x = 2y - 4$$

Solve for the new y.

$$2y = x + 4 \qquad \text{Add 4.}$$
$$y = \tfrac{1}{2}x + 2 \qquad \text{Divide by 2.}$$

Replace the new y with $f^{-1}(x)$.

$$f^{-1}(x) = \tfrac{1}{2}x + 2 \qquad \qquad \square$$

Try Exercise 17

Show $f(x) = 3x - 6$ is one-to-one. Then find f^{-1}.

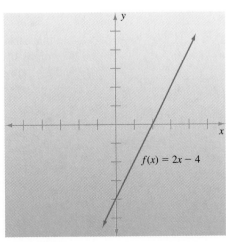

Figure 10.14

The graphs of $f(x) = 2x - 4$ and $f^{-1}(x) = \frac{1}{2}x + 2$ are shown in Fig. 10.15. If we "fold" Fig. 10.15 along the line $y = x$, the graphs of f and f^{-1} will coincide. That is, the graphs of f and f^{-1} are symmetric with respect to the line $y = x$. In fact, **the graphs of *any* one-to-one function f and its inverse f^{-1} are symmetric with respect to the line $y = x$.** This occurs because when (a, b) belongs to f, then (b, a) belongs to f^{-1} (see Fig. 10.16).

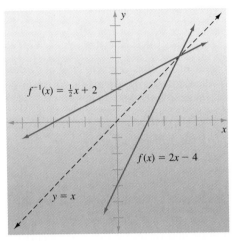

Figure 10.15   *Figure 10.16*

Example 3   Given $f(x) = x^2 + 2$ $(x \geq 0)$, find f^{-1}. Then graph f and f^{-1} on the same axes.

Soluton: The function $f(x) = x^2 + 2$ is *not* one-to-one (see Fig. 10.17). However, the function $f(x) = x^2 + 2$ $(x \geq 0)$ *is* one-to-one (see Fig. 10.18). Therefore, replace $f(x)$ with y and find f^{-1} as follows:

$$y = x^2 + 2 \qquad x \geq 0$$

Interchange x and y.

$$x = y^2 + 2 \qquad y \geq 0$$

Solve for the new y.

$$y^2 = x - 2 \qquad y \geq 0$$
$$f^{-1}(x) = y = \sqrt{x - 2} \qquad y \geq 0$$

Note that since $y \geq 0$, we write $y = \sqrt{x - 2}$, rather than $y = \pm\sqrt{x - 2}$. The graphs of f and f^{-1} are shown in Fig. 10.19. Note that the graphs of f and f^{-1} are symmetric with respect to the line $y = x$.   ❑

Try Exercise 23

Show $f(x) = x^2 + 1$ $(x \geq 0)$ is one-to-one, and find f^{-1}. Then graph f and f^{-1} on the same axes.

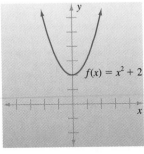   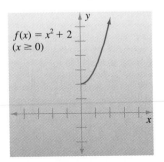   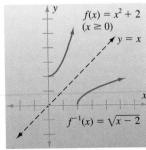

Figure 10.17   *Figure 10.18*   *Figure 10.19*

Here is a summary of the steps in finding the inverse of a function.

To Find the Inverse of the Function $y = f(x)$

1. Verify that f is one-to-one.
2. Replace $f(x)$ with y.
3. Interchange x and y.
4. Solve the equation resulting from step 3 for the new y.
5. Replace the new y with $f^{-1}(x)$.

We illustrate these steps in Example 4.

Example 4   Find the inverse of $f(x) = x^3 - 1$.

Solution: Step 1:   Verify that f is one-to-one.
Suppose $x_1 \neq x_2$. Then $x_1^3 \neq x_2^3$ and $x_1^3 - 1 \neq x_2^3 - 1$. Therefore $f(x_1) \neq f(x_2)$ and f is one-to-one.

Step 2:   Replace $f(x)$ with y.

$$y = x^3 - 1$$

Step 3:   Interchange x and y.

$$x = y^3 - 1$$

Step 4:   Solve for the new y.

$$x + 1 = y^3 \qquad \text{Add 1.}$$
$$y = \sqrt[3]{x + 1} \qquad \text{Cube root of both sides}$$

Step 5:   Replace the new y with $f^{-1}(x)$.

$$f^{-1}(x) = \sqrt[3]{x + 1}$$

The graphs of f and f^{-1} are shown in Fig. 10.20.   ❏

Try Exercise 27

Find the inverse of the function
$f(x) = x^3 - 2$.

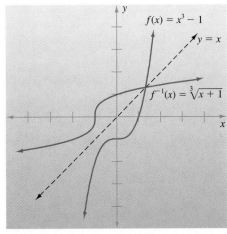

Figure 10.20

Exercises 10.2

Completion Exercises

1. The inverse of a one-to-one function f is the set of all ordered pairs of the form _____, where (a, b) belongs to f.

2. The additive inverse of -3 is _____. The multiplicative inverse of 2 is _____. The inverse of the function $f(x) = \sqrt[3]{x}$ is _____.

3. The graphs of a one-to-one function f and its inverse f^{-1} are symmetric with respect to _____.

4. The domain of a one-to-one function f is the _____ of f^{-1}, and the range of f is the _____ of f^{-1}.

True-False Exercises

5. If f is a one-to-one function, the symbol f^{-1} means $1/f$.

6. The expression $f^{-1}(x)$ represents the range value of the function f^{-1}.

7. Every function has an inverse function.

8. A function is one-to-one provided that no horizontal line intersects its graph more than once.

For Exercises 9 through 14, (a) determine whether the function f is one-to-one, (b) if f is one-to-one, find the inverse function.

9. $f = \{(2, 5), (-1, 2), (-3, 0)\}$

10. $f = \{(3, 4), (-2, 3), (-1, 0)\}$

11. $f = \{(0, 0), (2, 2), (4, -4)\}$

12. $f = \{(3, 3), (-1, -1), (-5, 5)\}$

13. $f = \{(3, 6), (-4, -1), (5, -2), (-7, 6)\}$

14. $f = \{(0, -3), (2, 7), (-5, -1), (-4, 7)\}$

Determine whether the given function is one-to-one. If it is, find f^{-1} and then graph f and f^{-1} on the same axes.

15. $f(x) = x + 2$

16. $f(x) = 2x$

17. $f(x) = 3x - 6$

18. $f(x) = 2x - 6$

19. $f(x) = -x + 3$

20. $f(x) = -x + 5$

21. $f(x) = x^2 + 1$

22. $f(x) = x^2 - 3$

23. $f(x) = x^2 + 1 \ (x \ge 0)$

24. $f(x) = x^2 - 3 \ (x \ge 0)$

25. $f(x) = 2$

26. $f(x) = -4$

Each function below is one-to-one. Find its inverse.

27. $f(x) = x^3 - 2$

28. $f(x) = x^3 + 1$

29. $2x - 3y = 6$

30. $2x - 5y = 10$

31. $y = \sqrt{x}$

32. $y = \sqrt{x + 1}$

33. $y = \dfrac{1}{x}$

34. $y = -\dfrac{8}{x}$

Given $f = \{(-1, 8), (5, 2), (3, 7), (-4, 6)\}$, determine each of the following.

35. $f(5)$

36. $f(3)$

37. $f^{-1}(2)$

38. $f^{-1}(7)$

39. $f^{-1}(f(-1))$

40. $f^{-1}(f(-4))$

41. $f(f^{-1}(6))$

42. $f(f^{-1}(8))$

43. Given $f(x) = 5x$, find $f^{-1}(x)$. Then determine each of the following: (a) $f(4)$, (b) $f^{-1}(20)$, (c) $f^{-1}(f(a))$, (d) $f(f^{-1}(a))$.

44. Given $f(x) = x + 1$, find $f^{-1}(x)$. Then determine each of the following: (a) $f(7)$, (b) $f^{-1}(8)$, (c) $f^{-1}(f(a))$, (d) $f(f^{-1}(f(a)))$.

Discussion Exercises

45. The inverse of the function $f(x) = 4x + 5$ is the function $f^{-1}(x) = \dfrac{x - 5}{4}$. Find $f(f^{-1}(x))$ and $f^{-1}(f(x))$. Can you make a general conclusion about inverse functions? Explain what this says about inverse functions.

46. Why do only one-to-one functions have inverse functions?

47. Describe two ways to determine whether a function is one-to-one.

48. Outline the steps in finding the inverse of the function $y = f(x)$.

For Exercises 49 through 58, state whether f is one-to-one. If f is one-to-one, sketch f^{-1} on the same axes.

49.
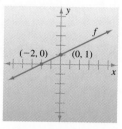

50.
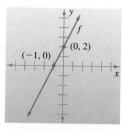

51.
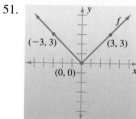

52.
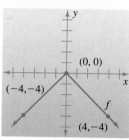

53.
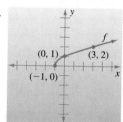

54.
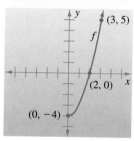

55.
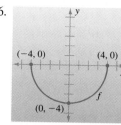

56.
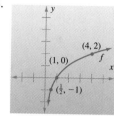

57.
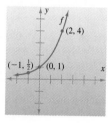

58.

Getting Acquainted with Your Graphing Calculator

Graph f, f^{-1}, and the line $y = x$ in the same Viewing Rectangle and note that the graphs of f and f^{-1} are symmetric with respect to $y = x$.

59. $f(x) = 2x - 2$, $f^{-1}(x) = \dfrac{x + 2}{2}$

60. $f(x) = x^3 + 3$, $f^{-1}(x) = \sqrt[3]{x - 3}$

◆ 10.3 Logarithmic Functions

TAPE AU21

The inverse of an exponential function is called a *logarithmic function*. In this section we define logarithmic functions and study their graphs.*

Defining the Logarithmic Function

The exponential function $y = 2^x$ is a one-to-one function, since its graph passes the horizontal line test (see Fig. 10.1). This means that its inverse $x = 2^y$ is a function. However, we currently have no way of solving $x = 2^y$ for y. To solve $x = 2^y$ for y, we introduce the following definition.

DEFINITION OF THE LOGARITHMIC FUNCTION

Suppose $a > 0$ and $a \neq 1$. The **logarithmic function with base a** is a function of the form $y = \log_a x$, where

$$y = \log_a x \text{ means } x = a^y$$

---

*HISTORICAL NOTE: Logarithms were originally invented by the Scottish baron, John Napier (1550–1617), to simplify the tedious computations involved in astronomy. Today these computations can be done on a calculator. Even so, logarithms are more important than ever. They are used to describe many natural phenomena and are invaluable in solving certain types of exponential equations.

The symbol **log** is an abbreviation for the word logarithm. The symbol $\log_a x$ is read "the logarithm of x to the base a," or simply "log base a of x."

The definition of the logarithmic function allows us to write the equation $x = a^y$ in two ways—in exponential form or in logarithmic form:

Exponential form   *Logarithmic form*

Exponent

$$x = a^y \qquad\qquad y = \log_a x$$

Base

Note that $\log_a x$ equals y, and y is the exponent on a that gives x. Therefore if $a > 0$ and $a \neq 1$,

$$\log_a x = \text{ the exponent on } a \text{ that gives } x$$

In other words, a logarithm is an exponent.

Example 1   Evaluate each logarithm: (a) $\log_2 8$, (b) $\log_7 7$, (c) $\log_3 1$.

Solution: (a) $\log_2 8 = $ the exponent on 2 that gives 8
$\qquad = 3$
(b) $\log_7 7 = $ the exponent on 7 that gives 7
$\qquad = 1$
(c) $\log_3 1 = $ the exponent on 3 that gives 1
$\qquad = 0$

Try Exercise 11

Evaluate: $\log_2 16$.

Parts (b) and (c) of Example 1 suggest the following rules.

Suppose $a > 0$ and $a \neq 1$. Then

$$\log_a a = 1 \qquad \text{and} \qquad \log_a 1 = 0$$

Example 2   Convert to logarithmic form: (a) $9 = 3^2$, (b) $10^3 = 1000$, (c) $6^{-1} = \frac{1}{6}$.

Solution: (a) $9 = 3^2$ is equivalent to $\log_3 9 = 2$.
(b) $10^3 = 1000$ is equivalent to $\log_{10} 1000 = 3$.
(c) $6^{-1} = \frac{1}{6}$ is equivalent to $\log_6 \frac{1}{6} = -1$.

Try Exercise 23

Convert $81 = 3^4$ to logarithmic form.

Example 3   Convert to exponential form: (a) $\log_4 64 = 3$, (b) $\log_{1/5} 25 = -2$, (c) $\log_9 3 = \frac{1}{2}$.

Solution: (a) $\log_4 64 = 3$ is equivalent to $64 = 4^3$.
(b) $\log_{1/5} 25 = -2$ is equivalent to $25 = (\frac{1}{5})^{-2}$.
(c) $\log_9 3 = \frac{1}{2}$ is equivalent to $3 = 9^{1/2}$.

Try Exercise 31

Convert $\log_5 125 = 3$ to exponential form.

Solving $y = \log_a x$ for y, a, or x

If we know any two of the numbers y, a, or x in the equation $y = \log_a x$, we can find the third number. We do this by writing the equation in exponential form.

Example 4   Find x if $\log_{64} x = \frac{2}{3}$.

Try Exercise 41

Find x if $\log_8 x = \frac{2}{3}$.

Solution: $\qquad x = 64^{2/3}$ Convert to exponential form.
$\qquad x = (\sqrt[3]{64})^2 = (4)^2 = 16$ Simplify.

Example 5   Find a if $\log_a 49 = 2$.

Solution:

$$a^2 = 49 \qquad \text{Write in exponential form.}$$
$$a = \pm 7 \qquad \text{Square root property}$$

Try Exercise 43

Find a if $\log_a 16 = 2$.

Eliminate $a = -7$ since a must be positive. Therefore $a = 7$. ❏

Since the exponential function $y = a^x$ is a one-to-one function (its graph passes the horizontal line test), the two y-values $y_1 = a^{x_1}$ and $y_2 = a^{x_2}$ are equal only if their corresponding x-values are equal.

One-to-One Property of the Exponential Function

Suppose $a > 0$ and $a \neq 1$.

$$\text{If } a^{x_1} = a^{x_2}, \text{ then } x_1 = x_2.$$

We use this property in Examples 6, 7, and 8.

Example 6   Find y if $y = \log_2 16$.

Solution: Convert to exponential form.

$$2^y = 16$$

Write each side with the same base.

$$2^y = 2^4$$

Use the one-to-one property of the exponential function to equate exponents.

Try Exercise 47

Find y if $y = \log_4 64$.

$$y = 4 \qquad ❏$$

Example 7   Find y if $y = \log_{1/2} 8$.

Solution:

$$\left(\tfrac{1}{2}\right)^y = 8 \qquad \text{Convert to exponential form.}$$
$$\left.\begin{array}{l} (2^{-1})^y = 2^3 \\ 2^{-y} = 2^3 \end{array}\right\} \quad \begin{array}{l}\text{Write each side with the same} \\ \text{base.}\end{array}$$
$$-y = 3 \qquad \text{Equate exponents.}$$
$$y = -3 \qquad \text{Solve for } y.$$

Try Exercise 57

Find y if $y = \log_{1/3} 27$.

❏

Example 8   Find y if $y = \log_3 \sqrt{3}$.

Solution:

$$3^y = \sqrt{3} \qquad \text{Exponential form}$$
$$3^y = 3^{1/2} \qquad \text{Write with the same base.}$$
$$y = \tfrac{1}{2} \qquad \text{Equate exponents.}$$

Try Exercise 61

Find y if $y = \log_{10} 10$.

❏

Graphing Logarithmic Functions

The best way to graph the logarithmic function $y = \log_a x$ is to convert to the exponential form $x = a^y$. Then substitute values for y (not x), and compute the corresponding values of x.

Example 9   Graph: $y = \log_2 x$.

Solution: This is the logarithmic function with base 2. Convert to the exponential form $x = 2^y$. Substitute values for y and compute the corresponding values of x to obtain the table of values below:

x	8	4	2	1	$\frac{1}{2}$	$\frac{1}{4}$	$\frac{1}{8}$
y	3	2	1	0	-1	-2	-3

Plot these points and draw a smooth curve through them to produce the graph shown in Fig. 10.21. Since the functions $y = \log_2 x$ and $y = 2^x$ are inverse functions, their graphs are symmetric about the line $y = x$ (see Fig. 10.21). ❏

Try Exercise 75

Graph: $y = \log_3 x$.

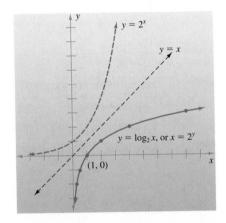

Figure 10.21

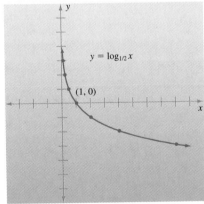

Figure 10.22

Example 10   Graph: $y = \log_{1/2} x$.

Solution: This is the logarithmic function with base $\frac{1}{2}$. Convert to the exponential form $x = (\frac{1}{2})^y$. Then assign values to y to get the following table of values.

x	$\frac{1}{8}$	$\frac{1}{4}$	$\frac{1}{2}$	1	2	4	8
y	3	2	1	0	-1	-2	-3

Plot these points and draw a smooth curve through them to produce the graph shown in Fig. 10.22. ❏

Try Exercise 77

Graph: $y = \log_{1/3} x$.

Note that the graph of $y = \log_a x$ is increasing if $a = 2$ (see Fig. 10.21) and decreasing if $a = \frac{1}{2}$ (see Fig. 10.22).

We summarize our observations about the graph of $y = \log_a x$ in Figs. 10.23 and 10.24.

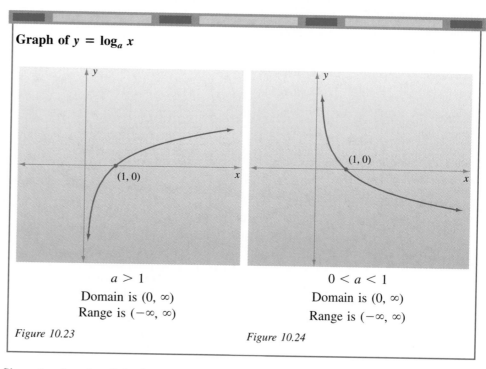

Graph of $y = \log_a x$

$a > 1$
Domain is $(0, \infty)$
Range is $(-\infty, \infty)$

Figure 10.23

$0 < a < 1$
Domain is $(0, \infty)$
Range is $(-\infty, \infty)$

Figure 10.24

Since the domain of the logarithmic function $y = \log_a x$ consists only of positive real numbers, there is no real value for $\log_2 0$ or $\log_2 (-4)$.

Exercises 10.3

Completion Exercises

1. Suppose $a > 0$ and $a \neq 1$. The logarithmic function with base a is a function of the form $y = \log_a x$, where $y = \log_a x$ means _____.

2. The domain of the logarithmic function $y = \log_a x$ is _____, and the range is _____.

3. If $a > 0$ and $a \neq 1$, then $\log_a a =$ _____, and $\log_a 1 =$ _____.

4. The logarithmic function $t = \log_b u$ in exponential form is _____.

5. The exponential function $z = b^w$ in logarithmic form is _____.

6. The graphs of $y = \log_3 x$ and $y = 3^x$ are symmetric with respect to _____.

True-False Exercises

7. The function $y = \log_4 x$ is the inverse of the function $y = 4^x$.

8. The symbol $\log_a x$ is read "log base a of x."

9. Suppose $a > 0$ and $a \neq 1$. Then $\log_a x$ is the exponent on a that gives x.

10. If $5^{2x} = 5^{y+1}$, then $2x = y + 1$.

Evaluate each logarithm.

11. $\log_2 16$

12. $\log_3 27$

13. $\log_5 25$

14. $\log_7 49$

15. $\log_8 8$

16. $\log_{12} 12$

17. $\log_{11} 1$

18. $\log_6 1$

19. $\log_3 \frac{1}{3}$

20. $\log_2 \frac{1}{4}$

21. $\log_{10} \sqrt{10}$

22. $\log_5 \sqrt{5}$

Convert each equation to logarithmic form. Assume all variables are positive and $b \neq 1$.

23. $81 = 3^4$

24. $32 = 2^5$

25. $10^{-4} = 0.0001$

26. $10^{-2} = 0.01$

27. $4^{1/2} = 2$

28. $100^{1/2} = 10$

29. $t = b^s$

30. $b^u = v$

Convert each equation to exponential form.

31. $\log_5 125 = 3$

32. $\log_5 625 = 4$

33. $\log_{1/7} 49 = -2$

34. $\log_{1/9} 9 = -1$

35. $\log_{27} 3 = \frac{1}{3}$

36. $\log_{16} 2 = \frac{1}{4}$

37. $\log_{\sqrt{3}} 3 = 2$

38. $\log_{\sqrt{5}} 5 = 2$

Find the unknown in each equation.

39. $\log_9 x = 2$

40. $\log_4 x = 3$

41. $\log_8 x = \frac{2}{3}$

42. $\log_{27} x = \frac{4}{3}$

43. $\log_a 16 = 2$

44. $\log_a 64 = 2$

45. $y = \log_{10} 10{,}000$

46. $y = \log_{10} 100{,}000$

47. $y = \log_4 64$

48. $y = \log_2 \frac{1}{8}$

49. $\log_a \frac{1}{125} = 3$

50. $\log_a 16 = 4$

51. $y = \log_7 1$

52. $y = \log_3 1$

53. $\log_2 x = -5$

54. $\log_3 x = -4$

55. $y = \log_{10} 0.001$

56. $y = \log_{10} 0.1$

57. $y = \log_{1/3} 27$

58. $y = \log_{1/5} 25$

59. $\log_a 4 = \frac{1}{2}$

60. $\log_a 9 = \frac{1}{2}$

61. $y = \log_{10} 10$

62. $y = \log_{13} 13$

Discussion Exercises

63. Show that there is no real value for $\log_2 0$ or for $\log_2 (-4)$ by trying to find y in the equations $y = \log_2 0$ and $y = \log_2 (-4)$.

64. Why do we eliminate 0, 1, and negative numbers as possible values of the base a for the logarithmic function $y = \log_a x$? (*Hint:* Try to find a value for y in the equations $y = \log_0 8$, $y = \log_1 8$, and $y = \log_{-2} 8$.)

65. What can be said about the base a in Fig. 10.25?

66. What can be said about the base a in Fig. 10.26?

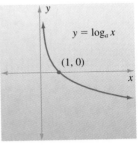
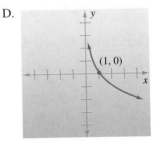

Figure 10.25

Figure 10.26

Matching Exercises

Match each function in Exercises 67 through 70 with its graph in letters A through D.

67. $y = \log_2 x$

68. $y = -\log_2 x$

69. $y = \log_2 (-x)$

70. $y = \log_2 |x|$

A.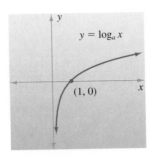

B.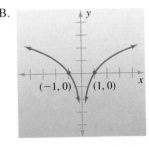

C.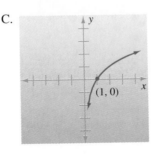

D.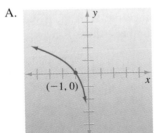

Match each function in Exercises 71 through 74 with its graph in letters A through D.

71. $y = \log_2 (x - 1)$

72. $y = \log_2 (x + 1)$

73. $y = \log_2 x + 1$

74. $y = \log_2 x - 1$

A.
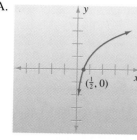

B.
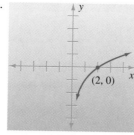

C.
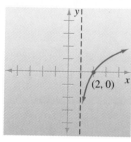

D.
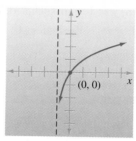

Graph each function.

 75. $y = \log_3 x$

76. $y = \log_4 x$

 77. $y = \log_{1/3} x$

78. $y = \log_{1/4} x$

79. $y = \log_2 2x$

80. $y = 2 \log_2 x$

81. $y = -\log_4 x$

82. $y = 1 + \log_3 x$

Problem Solving

83. The magnitude M of an earthquake measured on the *Richter scale* is given by the formula

$$M = \log_{10} \frac{I}{I_0}$$

where I is the intensity of the earthquake, and I_0 is the intensity of a "minimum" earthquake. Find the magnitude of an earthquake with the given intensity I.*

(a) $I = I_0$ — Minimum earthquake

(b) $I = 10^{6.7} I_0$ — Los Angeles earthquake of 1971

(c) $I = 10^{8.3} I_0$ — San Francisco earthquake of 1906

84. The loudness L of a sound wave in decibels is given by the formula

$$L = 10 \log_{10} \frac{I}{I_0}$$

where I is the intensity of the wave, and I_0 is the intensity of a wave at the threshold of audibility. Find the loudness of a sound wave with the given intensity I.*

(a) $I = I_0$ — Threshold of audibility

(b) $I = 10^{6.5} I_0$ — Normal conversation

(c) $I = 10^{12} I_0$ — Threshold of pain

Getting Acquainted with Your Graphing Calculator

85. Graph $f(x) = 10^x$, $g(x) = \log_{10} x$, and the line $y = x$ in the same Viewing Rectangle and note that the graphs of f and g are symmetric with respect to $y = x$.

Graph the two functions in each exercise in the same Viewing Rectangle and compare the graphs.

86. $y = -\log_{10} x$, $y = \log_{10} (-x)$

87. $y = \log_{10} x + 2$, $y = \log_{10} (x + 2)$

10.4 Properties of Logarithms

 TAPE AU21

Since logarithms are exponents, we can use the laws of exponents to prove several important properties of logarithms. It is these properties that make logarithms so useful. We will use the properties of logarithms to solve exponential and logarithmic equations in Sec. 10.6. The purpose of this section is to familiarize you with these properties.

Product Rule for Logarithms

The first property states that **the logarithm of a product is the sum of the logarithms of the factors.**

---

*HISTORICAL NOTE: Both the magnitude of an earthquake and the loudness of a soundwave are measured on what is called a *"logarithmic scale."* The Richter scale is named after the American physicist Charles Richter (1900–1985). The term *"decibel"* was coined in honor of Alexander Graham Bell (1847–1922), the inventor of the telephone.

Product Rule for Logarithms

If a, M, and N are positive numbers and $a \neq 1$, then

$$\log_a M \cdot N = \log_a M + \log_a N$$

To prove the product rule, let $x = \log_a M$ and $y = \log_a N$. Then write both equations in exponential form.

$$x = \log_a M \text{ is equivalent to } M = a^x$$
$$y = \log_a N \text{ is equivalent to } N = a^y$$

Therefore,

$$M \cdot N = a^x \cdot a^y = a^{x+y}$$

Now convert back to logarithmic form.

$$M \cdot N = a^{x+y} \text{ is equivalent to } \log_a M \cdot N = x + y$$

Finally, replacing x with $\log_a M$ and y with $\log_a N$ gives

$$\log_a M \cdot N = \log_a M + \log_a N$$

which is the desired result.

Example 1   Express as a sum of two logarithms. Assume $x > 0$.
(a) $\log_2 5 \cdot 7$ (b) $\log_3 8x$ (c) $\log_{10} x(x + 2)$

Solution: (a) $\log_2 5 \cdot 7 = \log_2 5 + \log_2 7$
(b) $\log_3 8x = \log_3 8 + \log_3 x$
(c) $\log_{10} x(x + 2) = \log_{10} x + \log_{10} (x + 2)$ ❐

Try Exercise 7

Express $\log_2 3 \cdot 7$ as a sum of two logarithms.

Quotient Rule for Logarithms

The next property states that **the logarithm of a fraction is the logarithm of the numerator minus the logarithm of the denominator.**

Quotient Rule for Logarithms

If a, M, and N are positive numbers and $a \neq 1$, then

$$\log_a \frac{M}{N} = \log_a M - \log_a N$$

The proof of the quotient rule is similar to the proof of the product rule.

Example 2   Express as a difference of two logarithms. Assume $x > 1$.
(a) $\log_5 \dfrac{2}{7}$ (b) $\log_2 \dfrac{3}{x}$ (c) $\log_{10} \dfrac{x - 1}{x}$

Solution: (a) $\log_5 \dfrac{2}{7} = \log_5 2 - \log_5 7$

(b) $\log_2 \dfrac{3}{x} = \log_2 3 - \log_2 x$

(c) $\log_{10} \dfrac{x - 1}{x} = \log_{10} (x - 1) - \log_{10} x$ ❐

Try Exercise 9

Express $\log_3 \dfrac{2}{y}$ as a difference of two logarithms. Assume $y > 0$.

Power Rule for Logarithms

The third property states that **the logarithm of a number to a power is the power times the logarithm of the number.** Its proof is similar to the proof of the product rule.

Power Rule for Logarithms

If a and M are positive numbers and $a \neq 1$, then

$$\log_a M^r = r \log_a M$$

for any real number r.

Example 3   Express as a product of a number and a logarithm. Assume $x > 0$, $a > 0$, and $a \neq 1$.

$$\text{(a) } \log_4 x^3 \qquad \text{(b) } \log_7 x^{-2} \qquad \text{(c) } \log_a \sqrt{x}$$

Solution: (a) $\log_4 x^3 = 3 \log_4 x$
(b) $\log_7 x^{-2} = -2 \log_7 x$
(c) $\log_a \sqrt{x} = \log_a x^{1/2} = \frac{1}{2} \log_a x$ ❏

Try Exercise 11

Express $\log_6 x^2$ as a product of a number and a logarithm. Assume $x > 0$.

Additional Properties of Logarithms

Here are two more useful properties of logarithms.

Suppose $a > 0$ and $a \neq 1$. Then

$$\log_a a^t = t \qquad \text{and} \qquad a^{\log_a t} = t \qquad (t > 0)$$

To prove the first property, apply the power rule.

$$\log_a a^t = t \log_a a = t \cdot 1 = t$$

To prove the second property, let $x = \log_a t$. Then convert to exponential form.

$$a^x = t$$

Replacing x with $\log_a t$ gives

$$a^{\log_a t} = t$$

which is the desired result.

Example 4   Simplify. Assume $x > 0$.

$$\text{(a) } \log_3 81 \qquad \text{(b) } \log_5 \sqrt[3]{25} \qquad \text{(c) } 4^{\log_4 7x}$$

Solution: (a) $\log_3 81 = \log_3 3^4 = 4$
(b) $\log_5 \sqrt[3]{25} = \log_5 \sqrt[3]{5^2} = \log_5 5^{2/3} = \frac{2}{3}$
(c) $4^{\log_4 7x} = 7x$ ❏

Try Exercise 19

Simplify: $\log_4 64$.

Expanding a Logarithm

To **expand** a logarithm means to apply the product, quotient, and power rules to the logarithm wherever possible.

Example 5   Expand and simplify. Assume $x > 0$.

$$\text{(a) } \log_2 8x \qquad \text{(b) } \log_5 \frac{1}{x}$$

Solution: (a) $\log_2 8x = \log_2 8 + \log_2 x$
$$= \log_2 2^3 + \log_2 x$$
$$= 3 + \log_2 x$$

(b) $\log_5 \dfrac{1}{x} = \log_5 1 - \log_5 x$
$$= 0 - \log_5 x$$
$$= -\log_5 x$$

You can also simplify Example 5(b) as follows:

$$\log_5 \frac{1}{x} = \log_5 x^{-1} = (-1)\log_5 x = -\log_5 x$$

Example 6 Expand $\log_3 \dfrac{x\sqrt{y}}{z^3}$. Assume all variables are positive.

Solution: $\log_3 \dfrac{x\sqrt{y}}{z^3} = \log_3 x\sqrt{y} - \log_3 z^3$ Quotient rule

$$= \log_3 x + \log_3 y^{1/2} - \log_3 z^3 \quad \text{Product rule}$$

$$= \log_3 x + \tfrac{1}{2}\log_3 y - 3\log_3 z \quad \text{Power rule}$$

We can combine a sum or difference of logarithms having the same base into a single logarithm by using the properties of logarithms in reverse.

Example 7 Express as a single logarithm with a coefficient of 1. Assume all variables are positive.

      (a) $6\log_5 x$ (b) $\log_7 r + \log_7 s$ (c) $\log_{10} r - \log_{10} s$

Solution: (a) $6\log_5 x = \log_5 x^6$ Power rule

(b) $\log_7 r + \log_7 s = \log_7 rs$ Product rule

(c) $\log_{10} r - \log_{10} s = \log_{10} \dfrac{r}{s}$ Quotient rule

Example 8 Express $\tfrac{1}{3}\log_a x - \log_a y - \log_a z$ as a single logarithm with a coefficient of 1. Assume all variables are positive and $a \neq 1$.

Solution: $\tfrac{1}{3}\log_a x - \log_a y - \log_a z = \log_a x^{1/3} - (\log_a y + \log_a z)$

$$= \log_a \sqrt[3]{x} - \log_a yz$$

$$= \log_a \frac{\sqrt[3]{x}}{yz}$$

CAUTION

There are many pitfalls to avoid when working with logarithms. Here are several examples of correct and incorrect ways to apply the properties of logarithms.

Correct

$\log_3 5x = \log_3 5 + \log_3 x$

$\log_2 \tfrac{9}{10} = \log_2 9 - \log_2 10$

$\log_8 y^2 = 2\log_8 y$

Wrong

$\log_3 5x = (\log_3 5)(\log_3 x)$

$\log_4 (x + 6) = \log_4 x + \log_4 6$

$\log_2 \tfrac{9}{10} = \dfrac{\log_2 9}{\log_2 10}$

$\dfrac{\log_2 9}{\log_2 10} = \log_2 9 - \log_2 10$

$(\log_8 y)^2 = 2\log_8 y$

Try Exercise 37

Expand $\log_6 \dfrac{1}{x^2}$ and simplify. Assume $x > 0$.

Try Exercise 43

Expand $\log_5 \dfrac{x\sqrt{y}}{z^2}$. Assume all variables are positive.

Try Exercise 51

Express $7\log_5 x$ as a single logarithm with coefficient 1. Assume $x > 0$.

Try Exercise 61

Express as a single logarithm with a coefficient of 1. Assume all variables are positive and $a \neq 1$: $\tfrac{1}{2}\log_a x - \log_a y - \log_a z$.

Exercises 10.4

Completion Exercises

Assume a, M, and N are positive numbers and $a \neq 1$.

1. $\log_a M \cdot N =$ _____ , and $\log_a \dfrac{M}{N} =$ _____ .

2. $\log_a M^r =$ _____ , $\log_a a^t =$ _____ , and (if $t > 0$) $a^{\log_a t} =$ _____ .

True-False Exercises

3. $\log_2 (8 + 8) = \log_2 8 + \log_2 8$

4. $\log_3 \dfrac{27}{9} = \dfrac{\log_3 27}{\log_3 9}$

5. $(\log_5 25)^3 = 3 \log_5 25$

6. $\log_4 4 \cdot 16 = (\log_4 4)(\log_4 16)$

Expand each logarithm. Assume all variables are such that all expressions are defined.

7. $\log_2 3 \cdot 7$ 8. $\log_5 4x$

9. $\log_3 \dfrac{2}{y}$ 10. $\log_5 \dfrac{x}{8}$

11. $\log_6 x^2$ 12. $\log_9 y^3$

13. $\log_8 z^{-3}$ 14. $\log_4 p^{-2}$

15. $\log_{10} x(x + 1)$ 16. $\log_6 x(x - 3)$

17. $\log_b \dfrac{x - 4}{x + 2}$ 18. $\log_a \dfrac{x + 6}{x + 5}$

Simplify each expression. Assume all variables are such that all expressions are defined.

19. $\log_4 64$ 20. $\log_2 64$

21. $\log_3 \sqrt[5]{81}$ 22. $\log_5 \sqrt[4]{125}$

23. $\log_a a^{2x}$ 24. $\log_a a^{x+1}$

25. $\log_3 (\log_2 8)$ 26. $\log_4 (\log_2 16)$

27. $7^{\log_7 13}$ 28. $9^{\log_9 21}$

29. $b^{\log_b (y-1)}$ 30. $b^{\log_b 6y}$

✏ *Discussion Exercises*

31. Assuming all variables have appropriate values, what is the logarithm of a product? the logarithm of a fraction? the logarithm of a number to a power?

32. The condition that M and N are positive numbers is important when stating the properties of logarithms. Show that each of the following properties is false if $M = -8$ and $N = -4$.

Product Rule: $\log_2 M \cdot N = \log_2 M + \log_2 N$

Quotient rule: $\log_2 \dfrac{M}{N} = \log_2 M - \log_2 N$

Power rule: $\log_2 M^2 = 2 \log_2 M$

Expand each logarithm. Simplify where possible. Assume all variables are such that all expressions are defined.

33. $\log_{10} 3 \cdot 100$ 34. $\log_3 27x$

35. $\log_5 \dfrac{x}{25}$ 36. $\log_2 \dfrac{4}{3}$

37. $\log_6 \dfrac{1}{x^2}$ 38. $\log_8 \dfrac{1}{x^3}$

39. $\log_2 2x^4$ 40. $\log_6 6x^2$

41. $\log_3 \dfrac{x^5}{9}$ 42. $\log_3 \dfrac{81}{x^4}$

43. $\log_5 \dfrac{x\sqrt{y}}{z^2}$ 44. $\log_7 \dfrac{x\sqrt{y}}{z^4}$

45. $\log_b \sqrt[4]{xy^3}$ 46. $\log_b \sqrt[3]{x^2 y}$

47. $\log_2 \sqrt[3]{\dfrac{x^2}{32y}}$ 48. $\log_5 \sqrt[4]{\dfrac{x^3}{625y}}$

49. $\log_{10} (x + 8)^4 (x - 5)^6$ 50. $\log_{10} (x - 6)^7 (x + 1)^5$

Express as a single logarithm with a coefficient of 1. Assume all variables are such that all expressions are defined.

51. $7 \log_5 x$ 52. $\frac{1}{2} \log_3 x$

53. $-3 \log_{10} x$ 54. $-\log_{10} x$

55. $-\log_3 8y$ 56. $-2 \log_2 7y$

57. $\log_2 r + \log_2 s$ 58. $\log_5 r + \log_5 2$

59. $\log_4 15 - \log_4 s$ 60. $\log_6 r - \log_6 s$

61. $\frac{1}{2} \log_a x - \log_a y - \log_a z$

62. $\frac{1}{4} \log_a x - \log_a y - \log_a z$

63. $\log_a x - 2 \log_a y - \log_a z$

64. $\log_a x - \log_a y - 3 \log_a z$

65. $\log_{10} (x + 1) + \log_{10} (x - 1)$

66. $\log_{10}(x + 2) + \log_{10}(x - 2)$

67. $\log_{10}(x^2 + 7x + 12) - \log_{10}(x + 4)$

68. $\log_{10}(x^2 + 8x + 15) - \log_{10}(x + 5)$

 Getting Acquainted with Your Graphing Calculator

Graph each pair of functions in Exercises 69 through 72 and note that the graphs are identical.

69. $y = \log_{10} 2x$, $y = \log_{10} 2 + \log_{10} x$

70. $y = \log_{10}\left(\dfrac{x}{5}\right)$, $y = \log_{10} x - \log_{10} 5$

71. $y = \log_{10} 10^x$, $y = x$

72. $y = 10^{\log_{10} x}$, $y = x$ \qquad $(x > 0)$

73. Graph $y = 2 \log_{10} x$ and $y = \log_{10} x^2$. Why are the graphs different?

 # 10.5 Common Logarithms; Natural Logarithms; Change of Base

TAPE AU21

Any positive real number except 1 can be used as the base for a logarithm. However, the two numbers that are used most often as bases are 10 and e. In this section we learn how to find $\log_{10}$ and $\log_e$ on a calculator, and we investigate several applications of these logarithms. We also learn how to change from one logarithmic base to another.

Common Logarithms

Since our number system is based on 10, logarithms with base 10 are more efficient in many situations than logarithms with other bases. A logarithm with base 10 is called a **common logarithm.** For simplicity, $\log_{10} x$ is written **log x.**

$$\log x = \log_{10} x$$

We can determine common logarithms of powers of 10 using the rule $\log_a a^t = t$.

$$\begin{aligned}
\log 1000 &= \log 10^3 &= 3 \\
\log 100 &= \log 10^2 &= 2 \\
\log 10 &= \log 10^1 &= 1 \\
\log 1 &= \log 10^0 &= 0 \\
\log 0.1 &= \log 10^{-1} &= -1 \\
\log 0.01 &= \log 10^{-2} &= -2 \\
\log 0.001 &= \log 10^{-3} &= -3
\end{aligned}$$

Note that numbers greater than 1 have common logarithms that are positive, and numbers between 0 and 1 have common logarithms that are negative.

To determine common logarithms of other positive numbers, we use either a table or a calculator. Appendix 4 contains a table of common logarithms.* A calculator is faster, however, so we will use a calculator throughout this section.

Let us begin by trying to get a feel for the value of a common logarithm. Suppose we want the common logarithm of 55. First, we note that

$$10 < 55 < 100$$

Therefore we expect

$$\log 10 < \log 55 < \log 100$$

That is, we expect

$$1 < \log 55 < 2$$

---

*HISTORICAL NOTE: The first table of common logarithms was published by the Englishman Henry Briggs (1561–1631), a professor of mathematics at Oxford. As a result, sometimes common logarithms are called *Briggsian* logarithms.

Since 55 is halfway between 10 and 100, we might also expect that log 55 is halfway between 1 and 2. This would be the case if the logarithmic function were a linear function, but it is not.

Example 1   Find: log 55.

Solution: Press the following keys on your calculator:

$$\boxed{\text{Clear}}\ 55\ \boxed{\text{log}}\ \boxed{1.740362689}$$

Try Exercise 11

Find log 65 to four decimal places.

Rounding off to four decimal places, we have log 55 ≈ 1.7404.   ❐

You can use your calculator to check the answer to Example 1 by showing that the equivalent exponential equation below is true:

$$10^{1.7404} \approx 55$$

To do this, press

$$\boxed{\text{Clear}}\ 10\ \boxed{y^x}\ 1.7404\ \boxed{=}\ \boxed{55.00472529}$$

If a number is too large or too small to be entered into your calculator in standard form, you must first write the number in scientific notation.

Example 2   Find each common logarithm to four decimal places.
    (a) log 437,000,000,000   (b) log 0.00000000437

Solution: (a)
$$
\begin{aligned}
\log 437{,}000{,}000{,}000 &= \log (4.37 \times 10^{11}) && \text{Scientific notation}\\
&= \log 4.37 + \log 10^{11} && \text{Product rule for logarithms}\\
&\approx 0.6405 + 11 && \text{Find each logarithm.}\\
&= 11.6405 && \text{Add.}
\end{aligned}
$$

(b)
$$
\begin{aligned}
\log 0.00000000437 &= \log (4.37 \times 10^{-9}) && \text{Scientific notation}\\
&= \log 4.37 + \log 10^{-9} && \text{Product rule for logarithms}\\
&\approx 0.6405 + (-9) && \text{Find each logarithm.}\\
&= -8.3595 && \text{Add.} \quad ❐
\end{aligned}
$$

Try Exercise 17

Find log 37,500,000,000 to four decimal places.

You can avoid the product rule in Example 2(a) by entering 437,000,000,000 into your calculator in scientific notation as follows:

$$\boxed{\text{Clear}}\ 4.37\ \boxed{\text{EXP}}\ 11\ \boxed{\text{log}}\ \boxed{11.64048144}$$

This method also works for Example 2(b).

Sometimes we want to reverse the procedure of finding the logarithm of a number. For example, since

$$\log 100 = 2$$

we say that 2 is the logarithm of 100. On the other hand, we say that 100 is the **anti-logarithm** (abbreviated **antilog**) of 2.

DEFINITION OF ANTILOGARITHM

If log $x = y$ then $x =$ antilog y

Example 3   Find x, if log $x = 2.49$.

Solution: There are two ways to do this on most calculators

Method 1   Use the inverse of the log function.

$$\boxed{\text{Clear}}\ 2.49\ \boxed{\text{INV}}\ \boxed{\text{log}}\ \boxed{309.0295433}$$

Method 2   Use the fact that $x = 10^{2.49}$.

$$\boxed{\text{Clear}}\ 10\ \boxed{y^x}\ 2.49\ \boxed{=}\ \boxed{309.0295433}$$

The answer to three significant figures is x = antilog $2.49 \approx 309$.   ❐

Try Exercise 25

Find x to three significant figures if log $x = 2.42$.

You can check the solution to Example 3 by substituting it into the original equation. That is,

$$\log 309 \approx 2.49$$

CAUTION

Do not confuse a logarithm with an antilogarithm.

$$\underset{\underset{\log 100}{\downarrow}}{\text{antilog}}\ \underset{\underset{=\ 2}{\downarrow}}{\text{log}}\ \text{is equivalent to}\ \underset{\underset{10^2}{\downarrow}}{\text{log}}\ \underset{\underset{=\ 100}{\downarrow}}{\text{antilog}}$$ ■

Example 4   Find x to three significant figures if log $x = 37.98$.

Solution: Press

$$\boxed{\text{Clear}}\ 37.98\ \boxed{\text{INV}}\ \boxed{\log}\ \boxed{9.54992586\ 37}$$

The antilogarithm of 37.98 is very large, so your calculator expressed the answer in scientific notation. That is, x = antilog $37.98 \approx 9.55 \times 10^{37}$.   ❐

Try Exercise 27

Find x to three significant figures if log $x = 17.6$.

Example 5   Find x to three significant figures if log $x = -0.158$.

Solution: Press

$$\boxed{\text{Clear}}\ .158\ \boxed{+/-}\ \boxed{\text{INV}}\ \boxed{\log}\ \boxed{0.695024317}$$

Therefore, x = antilog $(-0.158) \approx 0.695$.   ❐

Try Exercise 29

Find x to three significant figures if log $x = -0.5717$.

Here is an example of an applied problem involving common logarithms.

Problem Solving

Example 6   Chemists define the pH (power of hydrogen) of a solution as

$$\text{pH} = -\log\ [\text{H}^+]$$

The symbol $[\text{H}^+]$ stands for the hydrogen ion concentration of the solution in moles per liter. On a scale of 0 to 14, pure distilled water has a pH of 7, acidic solutions have a pH of less than 7, and alkaline solutions have a pH of greater than 7 (see Fig. 10.27).

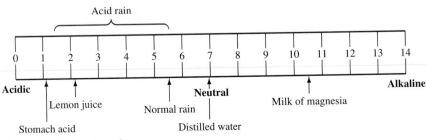

Figure 10.27 The pH scale.

Acid rains have been measured that have hydrogen ion concentrations of 3.16×10^{-2} moles per liter. Determine the pH of this acid rain.

Solution:

$$\begin{aligned} \text{pH} = -\log [\text{H}^+] &= -\log (3.16 \times 10^{-2}) && \text{Substitute for } [\text{H}^+]. \\ &= -(\log 3.16 + \log 10^{-2}) && \text{Product rule for logarithms} \\ &\approx -(0.4997 + (-2)) && \text{Find each logarithm.} \\ &= -(-1.5003) && \text{Add.} \\ &= 1.5003 && \text{Simplify.} \end{aligned}$$

Generally pH is expressed only to be the nearest tenth, so the pH is 1.5.

Try Exercise 33

Determine the pH of milk of magnesia, given that it has a hydrogen ion concentration of 3.16×10^{-11} moles per liter.

Natural Logarithms

Just as the number e is the most important base for the exponential function, so is e the most important base for the logarithmic function. A logarithm with base e is called a **natural logarithm.*** For simplicity, $\log_e x$ is written **ln x.**

$$\ln x = \log_e x$$

We can determine the natural logarithm of any positive number using either the table in Appendix 5 or a calculator.

Example 7 Find ln 1350 to four decimal places.

Solution: Press

$$\boxed{\text{Clear}} \; 1350 \; \boxed{\text{ln } x} \; \boxed{7.207859871}$$

Therefore, $\ln 1350 \approx 7.2079$.

Try Exercise 39

Find ln 1260 to four decimal places.

Example 8 Find x to three significant figures if $\ln x = 3.18$.

Solution: Press

$$\boxed{\text{Clear}} \; 3.18 \; \boxed{\text{INV}} \; \boxed{\text{ln } x} \; \boxed{24.04675355}$$

Therefore, $x \approx 24.0$.

Try Exercise 47

Find x to three significant figures if $\ln x = 1.91$.

In Sec. 10.4 we learned that $\log_a a^t = t$, assuming $a > 0$ and $a \neq 1$. If we replace the base a with base e, then for any real number t

$$\ln e^t = t$$

Example 9 Find each natural logarithm without using a calculator.

(a) $\ln e^5$ (b) $\ln \dfrac{1}{\sqrt{e}}$

Solution: Use the property $\ln e^t = t$.

(a) $\ln e^5 = 5$

(b) $\ln \dfrac{1}{\sqrt{e}} = \ln \dfrac{1}{e^{1/2}} = \ln e^{-1/2} = -\dfrac{1}{2}$

Try Exercise 59

Find $\ln e^3$ without using a calculator.

---

*HISTORICAL NOTE: Natural logarithms are also called *Napierian* logarithms in honor of the inventor of logarithms, John Napier (1550–1617). Napier was not a mathematician by profession, but rather a Scottish nobleman who dabbled in mathematics. He also invented *Napier's rods,* which are sticks on which multiplication tables are carved in a manner that allows multiplication to be done mechanically.

The functions $y = \ln x$ and $y = e^x$ are inverse functions, so their graphs are symmetric about the line $y = x$ (see Fig. 10.28).

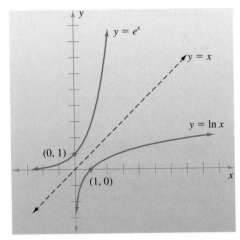

Figure 10.28

In Sec. 10.1 we learned that if interest is compounded continuously, the value V of an account after t years is

$$V = Pe^{rt}$$

where P is the principal, and r is the interest rate in decimal form. We can use this formula to develop a formula that tells us how long it will take a given sum of money to double in value. Since we want the value V to be twice the original amount P, we proceed as follows:

$2P = Pe^{rt}$	Replace V with $2P$.
$2 = e^{rt}$	Divide by P.
$rt = \ln 2$	Convert to logarithmic form.
$t = \dfrac{\ln 2}{r}$	Divide by r.

Problem Solving

Example 10   The doubling time t in years of a sum of money is given by the formula

$$t = \frac{\ln 2}{r}$$

where r is the compounded continuously interest rate in decimal form. Find the doubling time of a sum of money if the interest rate is 7.46 percent compounded continuously.

Solution: Substitute $r = 0.0746$ in the formula $t = \dfrac{\ln 2}{r}$.

$$t = \frac{\ln 2}{0.0746} \approx \frac{0.6931}{0.0746} \approx 9.3$$

The doubling time is approximately 9.3 yr.   ❑

Try Exercise 67

Find the doubling time of a sum of money if the interest rate is 12.57 percent compounded continuously.

Change of Base

Since there are an infinite number of possible bases for the logarithmic function, it would be impossible to construct a table or a calculator key for every base. However, we can find logarithms to any base using the **change-of-base formula** below.

Change-of-Base Formula

Suppose a, b, and x are positive numbers and neither a nor b is 1. Then

$$\log_b x = \frac{\log_a x}{\log_a b}$$

To prove the change-of-base formula, let $y = \log_b x$. Then write this equation in exponential form.

$$b^y = x \qquad \text{Exponential form}$$
$$\log_a b^y = \log_a x \qquad \text{Take } \log_a \text{ of each side.}$$
$$y \log_a b = \log_a x \qquad \text{Power rule for logarithms}$$
$$y = \frac{\log_a x}{\log_a b} \qquad \text{Divide by } \log_a b.$$
$$\log_b x = \frac{\log_a x}{\log_a b} \qquad \text{Replace } y \text{ with } \log_b x.$$

We can now find logarithms to any base by using the change-of-base formula to convert the logarithm to a common logarithm or to a natural logarithm.

Example 11 Find $\log_2 5$ to four decimal places.

Solution: Use the change-of-base formula with $b = 2$, $x = 5$, and $a = 10$.

$$\log_2 5 = \frac{\log_{10} 5}{\log_{10} 2} \approx \frac{0.69897}{0.30103} \approx 2.3219$$
❏

Try Exercise 71

Find $\log_2 3$ to four decimal places.

You could also do Example 11 by converting to natural logarithms instead of to common logarithms.

$$\log_2 5 = \frac{\ln 5}{\ln 2} \approx \frac{1.60944}{0.69315} \approx 2.3219$$

CAUTION

In Example 11, suppose we performed the intermediate calculations to just four decimal places:

$$\log_2 5 = \frac{\log_{10} 5}{\log_{10} 2} \approx \frac{0.6990}{0.3010} \approx 2.3223$$

This answer is not correct to four decimal places. Rounding off before the final calculation is performed may cause the roundoff error to accumulate and produce a less accurate answer. That is why we wrote the intermediate calculations to *five* decimal places. Even better, use the values stored in your calculator to do the intermediate calculations.

■

Exercises 10.5

Completion Exercises

1. The expression log x means _____ (write with the correct base); the expression ln x means _____ (write with the correct base).

2. If $x > 1$, then log x is a _____ (positive, negative) number; if $0 < x < 1$, then log x is a _____ (positive, negative) number.

3. Supply a reason for each step.

$$\log 3,270,000,000 = \log 3.27 \times 10^9 \qquad \text{_____}$$
$$= \log 3.27 + \log 10^9 \qquad \text{_____}$$
$$\approx 0.5145 + 9 \qquad \text{_____}$$
$$= 9.5145 \qquad \text{_____}$$

4. Since log $68 \approx 1.8325$, the number 68 is called the _____ of 1.8325.

5. For any real number t, $\ln e^t =$ _____.

6. If a, b, and x are positive numbers ($a \neq 1$, $b \neq 1$), then the change-of-base formula states that $\log_b x =$ _____.

7. A logarithm to the base 10 is called a _____ logarithm; a logarithm to the base e is called a _____ logarithm.

8. The functions $y = \ln x$ and _____ are inverse functions.

Find each common logarithm to four decimal places.

9. log 7.52
10. log 3.49
11. log 65
12. log 75
13. log 6500
14. log 75,000
15. log 0.514
16. log 0.0371
17. log 37,500,000,000
18. log 8,650,000,000
19. log 0.000000000375
20. log 0.0000000865

Find each antilogarithm x to three significant figures.

21. log $x = 0.7443$
22. log $x = 0.8293$
23. log $x = 2.7443$
24. log $x = 3.8293$
25. log $x = 2.42$
26. log $x = 2.73$
27. log $x = 17.6$
28. log $x = 54.29$
29. log $x = -0.5717$
30. log $x = -1.7126$
31. log $x = -15.384$
32. log $x = -17.684$

Problem Solving

33. Determine the pH of milk of magnesia, given that it has a hydrogen ion concentration of 3.16×10^{-11} moles per liter.

34. Determine the pH of stomach acid, given that it has a hydrogen ion concentration of 6.31×10^{-2} moles per liter.

35. Find the hydrogen ion concentration of normal rainwater, given that it has a pH of 5.6.

36. Find the hydrogen ion concentration of lemon juice, given that it has a pH of 2.1.

37. How many times greater is the hydrogen ion concentration of solution A than solution B, given that A has a pH of 5 and B has a pH of 6?

38. How many times greater is the hydrogen ion concentration of solution A than solution B, given that A has a pH of 7 and B has a pH of 9?

Find each natural logarithm to four decimal places.

39. ln 1260
40. ln 1480
41. ln 100
42. ln 1000
43. ln 5.13
44. ln 24.2
45. ln 0.0707
46. ln 0.0033

Find x in each equation to three significant figures.

47. ln $x = 1.91$
48. ln $x = 2.57$
49. ln $x = 0.4386$
50. ln $x = 0.9644$
51. ln $x = 20.5$
52. ln $x = 8.02$
53. ln $x = -0.0379$
54. ln $x = -13.6$

Discussion Exercises

55. Why does your calculator register an error when you try to find log (-5) or ln 0?

56. When you are doing a problem on your calculator that requires several steps, it is generally best to wait until the final calculation has been performed before rounding off to the required number of digits. Why? Give examples.

Find each natural logarithm without using a calculator.

57. ln e
58. ln e^2

59. $\ln e^3$

60. $\ln e^4$

61. $\ln \dfrac{1}{e}$

62. $\ln \dfrac{1}{e^2}$

63. $\ln \sqrt{e}$

64. $\ln \sqrt[3]{e}$

Graph each function.

65. $y = -\ln x$

66. $y = \ln(-x)$

Problem Solving

67. Find the doubling time of a sum of money if the interest rate is 12.57 percent compounded continuously.

68. Find the doubling time of a sum of money if the interest rate is 9.13 percent compounded continuously.

69. Show that the formula for tripling time is $t = \dfrac{\ln 3}{r}$. Then find the tripling time if $r = 7.64$ percent compounded continuously.

70. Show that the formula for quadrupling time is $t = \dfrac{\ln 4}{r}$. Then find the quadrupling time if $r = 12.5$ percent compounded continuously.

Use the change-of-base formula to find each logarithm to four decimal places.

71. $\log_2 3$

72. $\log_3 7$

73. $\log_5 632$

74. $\log_6 588$

75. $\log_{11} 0.514$

76. $\log_{12} 3.16$

77. $\log_{1/2} 9.1$

78. $\log_{1/4} 9.7$

Use the change-of-base formula in Exercises 79 through 82.

79. Express $\log_2 15$ in terms of $\log_9$.

80. Express $\log_3 25$ in terms of $\log_8$.

81. Express $\log_7 5$ in terms of $\log_5$.

82. Express $\log_5 11$ in terms of $\log_{11}$.

 ### Getting Acquainted with Your Graphing Calculator

83. You can graph $y = \log_2 x$ by using the change-of-base formula to write this equation as $y = \dfrac{\log x}{\log 2}$ or as $y = \dfrac{\ln x}{\ln 2}$. Graph $y = \log_2 x$, $y = \log_3 x$, and $y = \log_4 x$ in the same Viewing Rectangle.

10.6 Exponential and Logarithmic Equations

TAPE AU21

We have already learned how to solve several different types of equations—linear equations, quadratic equations, radical equations, absolute-value equations, and equations with rational expressions. In this section we learn how to solve two new types of equations—*exponential equations* and *logarithmic equations*.

Exponential Equations

An **exponential equation** is an equation that contains a variable in an exponent. Examples of exponential equations are

$$2^x = 8 \qquad 3^{y+1} = 10 \qquad \text{and} \qquad 5^{t^2} = 5^{2t-1}$$

We can solve some exponential equations using the one-to-one property of the exponential function, which was first stated in Sec. 10.3.

One-to-One Property of the Exponential Function

Suppose $a > 0$ and $a \neq 1$.

$$\text{If } a^{x_1} = a^{x_2}, \text{ then } x_1 = x_2$$

Example 1 Solve: $5^x = 25$.

Solution: Write both sides with the same base, 5.

$$5^x = 5^2$$

Use the one-to-one property of the exponential function to equate exponents.

$$x = 2$$

Try Exercise 9

Solve: $2^x = 4$.

The solution set is $\{2\}$. Check in the original equation.   ◻

If we cannot easily write both sides of an exponential equation with the same base, we solve by taking the logarithm of both sides. Since $y = \log_a x$ is a function, equal x-values produce equal y-values. That is,

$$\text{If } x_1 = x_2, \text{ then } \log_a x_1 = \log_a x_2$$

assuming x_1, x_2, and a are positive numbers and $a \neq 1$. We can use a logarithm to any legitimate base, but we shall use common logarithms since their values are easily found using a calculator or a table.

Example 2   Solve: $5^x = 7$.

Solution: Take the log of each side.
$$\log 5^x = \log 7$$

Use the power rule for logarithms to bring the exponent down as a multiplier.
$$x \log 5 = \log 7$$

Divide each side by $\log 5$.

$$x = \frac{\log 7}{\log 5}$$

This is the *exact* value of x. You can approximate this answer to the nearest hundredth using a table or a calculator as follows:

$$x = \frac{\log 7}{\log 5} \approx \frac{0.8451}{0.6990} \approx 1.21$$

Try Exercise 11

Solve: $2^x = 5$ to the nearest hundredth.

The solution set is $\{1.21\}$. Check in the original equation.   ◻

We can summarize the procedure for solving an exponential equation as follows:

To Solve an Exponential Equation

1. Write both sides with the same base, then equate exponents.
2. If step 1 fails, take the common logarithm of both sides. Then use the power rule for logarithms to bring the variable exponent down as a multiplier.
3. Solve the equation resulting from step 1 or step 2.
4. Check your solution in the original equation.

We illustrate these steps in Examples 3 and 4.

Example 3   Solve: $9^{x-1} = 27$.

Solution: Step 1:   Write both sides with the same base, then equate exponents.

$$9^{x-1} = 27 \qquad \text{Original equation}$$
$$\left.\begin{array}{l} (3^2)^{x-1} = 3^3 \\ 3^{2x-2} = 3^3 \end{array}\right\} \qquad \text{Write with the same base, 3.}$$
$$2x - 2 = 3 \qquad \text{Equate exponents.}$$

Step 2:   Since step 1 succeeded, bypass step 2.

Step 3:   Solve the equation resulting from step 1.

$$2x = 5 \qquad \text{Add 2.}$$
$$x = \tfrac{5}{2} \qquad \text{Divide by 2.}$$

Step 4:   Check your solution in the original equation.

$$9^{x-1} = 27 \qquad \text{Original equation}$$
$$9^{(5/2)-1} = 27 \qquad \text{Let } x = \tfrac{5}{2}.$$
$$9^{3/2} = 27 \qquad \text{True}$$

Try Exercise 17

Solve: $4^{x+1} = 8$.

Therefore, the solution set is $\{\tfrac{5}{2}\}$.   ❏

Example 4   Solve: $9^{x-1} = 15$ to the nearest hundredth.

Solution: Step 1:   We cannot easily write both sides with the same base, so go to step 2.

Step 2:   Take the common logarithm of both sides. Then use the power rule to bring the variable exponent down as a multiplier.

$$\log 9^{x-1} = \log 15 \qquad \text{Take the log of both sides.}$$
$$(x - 1) \log 9 = \log 15 \qquad \text{Power rule for logarithms}$$

Step 3:   Solve the equation resulting from step 2.

$$x - 1 = \frac{\log 15}{\log 9} \qquad \text{Divide by log 9.}$$
$$x = \frac{\log 15}{\log 9} + 1 \qquad \text{Add 1.}$$
$$\left.\begin{array}{l} x = \dfrac{1.1761}{0.9542} + 1 \\ x = 2.23 \end{array}\right\} \qquad \text{Approximate.}$$

Step 4:   Check your solution in the original equation.

$$9^{x-1} = 15 \qquad \text{Original equation}$$
$$9^{2.23-1} = 15 \qquad \text{Replace } x \text{ with 2.23.}$$
$$9^{1.23} \approx 15 \qquad \text{True}$$

Try Exercise 19

Solve: $4^{x+1} = 15$ to the nearest hundredth.

Therefore, the solution set is $\{2.23\}$.   ❏

Logarithmic Equations

A **logarithmic equation** is an equation that contains a logarithm of a variable quantity. Examples of logarithmic equations are

$$\log_2 x = 5 \qquad \log_6 (x + 1) = \log_6 7 \qquad \text{and} \qquad \log y - \log 4 = 3$$

We can solve some logarithmic equations using the fact that the logarithmic function $y = \log_a x$ is one-to-one (its graph passes the horizontal line test).

One-to-One Property of the Logarithmic Function

Suppose a, x_1, and x_2 are positive numbers and $a \neq 1$.

$$\text{If } \log_a x_1 = \log_a x_2, \text{ then } x_1 = x_2$$

For simplicity, we call the quantity that forms the input of a function the **argument** of the function. For example, the argument of $\log_7 (2x + 5)$ is $2x + 5$, the argument of $\log_7 11$ is 11, and the argument of $\log (x^2 - 2x - 8)$ is $x^2 - 2x - 8$. Therefore the one-to-one property of the logarithmic function states that, **if two logarithms with the same base are equal, then their arguments are equal.**

Example 5   Solve: $\log_7 (2x + 5) = \log_7 11$.

Solution: Both logarithms have the same base. Therefore apply the one-to-one property of the logarithmic function.

$$2x + 5 = 11 \qquad \text{Equate arguments.}$$

Then solve this equation.

$$2x = 6 \qquad \text{Subtract 5.}$$
$$x = 3 \qquad \text{Divide by 2.}$$

The solution set is $\{3\}$. Check in the original equation.   ❒

Try Exercise 31

Solve: $\log_7 (3x + 5) = \log_7 8$.

If we cannot easily isolate a logarithm on both sides of a logarithmic equation, we solve by converting to exponential form.

Example 6   Solve: $\log (x + 1) - \log (x - 2) = 1$.

Solution: Use the quotient rule for logarithms to write the left side as a single logarithm.

$$\log \frac{x + 1}{x - 2} = 1 \qquad \text{Quotient rule for logarithms}$$

Convert to exponential form. Remember, the base of this logarithm is 10.

$$\frac{x + 1}{x - 2} = 10^1 \qquad \text{Exponential form}$$

Solve for x.

$$x + 1 = 10x - 20 \qquad \text{Multiply by } x - 2.$$
$$21 = 9x \qquad \text{Subtract } x, \text{ add 20.}$$
$$x = \frac{21}{9} = \frac{7}{3} \qquad \text{Divide by 9 and simplify.}$$

The solution set is $\{\frac{7}{3}\}$. Check in the original equation.   ❒

Try Exercise 45

Solve:
$\log (x + 2) - \log (x - 1) = 1$.

We can summarize the procedure for solving a logarithmic equation as follows.

To Solve a Logarithmic Equation
1. Write both sides as a single logarithm with the same base. Then equate arguments.
2. If step 1 fails, write one side as a single logarithm and the other side as a number. Then convert to exponential form.
3. Solve the equation resulting from step 1 or step 2.
4. Check your solution in the original equation. Discard any solution that causes a logarithm to be undefined.

We illustrate these steps in Example 7.

Example 7   Solve: $\log (x + 2) + \log (x - 4) = \log 5x$.

Solution: Step 1:   Write both sides as a single logarithm with the same base. Then equate arguments.

$$\log (x + 2) + \log (x - 4) = \log 5x \quad \text{Original equation}$$
$$\log (x + 2)(x - 4) = \log 5x \quad \text{Product rule for logarithms}$$
$$(x + 2)(x - 4) = 5x \quad \text{Equate arguments.}$$

Step 2:   Since step 1 succeeded, bypass step 2.

Step 3:   Solve the equation resulting from step 1.

$$x^2 - 2x - 8 = 5x \quad \text{Multiply out.}$$
$$x^2 - 7x - 8 = 0 \quad \text{Subtract } 5x.$$
$$(x - 8)(x + 1) = 0 \quad \text{Factor.}$$
$$x - 8 = 0 \quad \text{or} \quad x + 1 = 0 \quad \text{Set each factor equal to 0.}$$
$$x = 8 \qquad\qquad x = -1 \quad \text{Solve each linear equation.}$$

Step 4:   Check your solution in the original equation.

Check x = 8:

$$\log (x + 2) + \log (x - 4) = \log 5x \quad \text{Original equation}$$
$$\log (8 + 2) + \log (8 - 4) = \log 5(8) \quad \text{Let } x = 8.$$
$$\log 10 + \log 4 = \log 40 \quad \text{Simplify.}$$
$$\log 40 = \log 40 \quad \text{Product rule for logarithms}$$

Check x = -1:

$$\log (-1 + 2) + \log (-1 - 4) = \log 5(-1) \quad \text{Let } x = -1.$$
$$\log 1 + \log (-5) = \log (-5) \quad \text{Simplify.}$$

Since the domain of the logarithmic function is $(0, \infty)$, the expression $\log (-5)$ is undefined. Therefore $x = -1$ is an extraneous solution and must be discarded. The solution set is $\{8\}$.   ❑

Try Exercise 47

Solve:
$\log (x + 3) + \log (x - 2) = \log 6x$.

CAUTION

When solving a logarithmic equation, always check each solution to make sure the solution makes the argument of every logarithm a positive number. If it does not, the solution is extraneous and must be discarded.   ∎

To illustrate an applied problem, recall that the value V of an account that compounds interest annually is given by the formula

$$V = P(1 + r)^t$$

where P is the principal, r is the interest rate in decimal form, and t is the time in years.

Problem Solving

Example 8   How long will it take $1000 to grow to $1800 at 9.32 percent interest compounded annually?

Solution: Write down the formula.

$$V = P(1 + r)^t$$

Substitute $V = 1800$, $P = 1000$, and $r = 0.0932$.

$$1800 = 1000(1 + 0.0932)^t$$

Simplify.

$$1800 = 1000(1.0932)^t \qquad \text{Add in parentheses.}$$
$$1.8 = (1.0932)^t \qquad \text{Divide by 1000.}$$

This is an exponential equation. Since we cannot easily write both sides with the same base, take the logarithm of both sides.

$$\log 1.8 = \log (1.0932)^t$$
$$\log 1.8 = t \log (1.0932) \qquad \text{Power rule for logarithms}$$
$$t = \frac{\log 1.8}{\log 1.0932} \qquad \text{Divide by log 1.0932.}$$
$$t \approx 6.6 \qquad \text{Find each log and divide.}$$

It will take approximately 6.6 yr.   ◻

Try Exercise 51

How long will it take $800 to grow to $1200 at 7.94 percent compounded annually?

Exercises 10.6

Completion Exercises

1. An equation that contains a variable in an exponent is called a(n) _____ equation.

2. An equation that contains a logarithm of a variable quantity is called a(n) _____ equation.

3. Suppose $a > 0$ and $a \neq 1$. The one-to-one property of the exponential function states that if $a^{x_1} = a^{x_2}$, then _____.

4. Suppose a, x_1, and x_2 are positive numbers and $a \neq 1$. The one-to-one property of the logarithmic function states that if $\log_a x_1 = \log_a x_2$, then _____.

True-False Exercises

5. Suppose a, x_1, and x_2 are positive numbers and $a \neq 1$. If $x_1 = x_2$, then $\log_a x_1 = \log_a x_2$.

6. The quantity that forms the input of a function is called the argument of the function.

7. The solution of a logarithmic equation cannot be a negative number or 0.

8. The solution of a logarithmic equation cannot be a number that makes an argument of a logarithmic function a negative number or 0.

Solve each exponential equation. Round approximate answers to the nearest hundredth.

9. $2^x = 4$

10. $3^x = 9$

11. $2^x = 5$

12. $3^x = 11$

13. $3^{2x} = 81$

14. $2^{2x} = 16$

15. $3^{2x} = 13$

16. $2^{2x} = 7$

17. $4^{x+1} = 8$

18. $9^{x+1} = 27$

19. $4^{x+1} = 15$

20. $9^{x+1} = 14$

21. $5^{-x} = 125$

22. $5^{-x} = 625$

23. $5^{-x} = 100$

24. $5^{-x} = 1000$

25. $6^{3x-1} = 1$

26. $7^{3x+1} = 1$

27. $2^{x^2+3x} = \frac{1}{4}$

28. $3^{x^2-3x} = \frac{1}{9}$

Solve each logarithmic equation.

29. $\log_4 5x = \log_4 15$

30. $\log_4 7x = \log_4 14$

31. $\log_7 (3x + 5) = \log_7 8$

32. $\log_5 (2x - 1) = \log_5 9$

33. $\log (2x - 4) = \log (x + 1)$

34. $\log (3x + 2) = \log (2x + 5)$

35. $\log_2 x = 5$

36. $\log_3 x = 4$

37. $\log (x - 1) - 3 = 0$

38. $\log (x + 1) - 2 = 0$

39. $\log_3 (x^2 - 8x) = 2$

40. $\log_2 (x^2 - 7x) = 3$

41. $\log_2 x = 1 + \log_2 3$

42. $\log_3 x = 1 + \log_3 2$

43. $\log x + \log (x - 15) = 2$

44. $\log x + \log (x - 21) = 2$

45. $\log (x + 2) - \log (x - 1) = 1$

46. $\log (x + 5) - \log (x - 1) = 1$

47. $\log (x + 3) + \log (x - 2) = \log 6x$

48. $\log (x + 4) + \log (x - 3) = \log 5x$

Discussion Exercises

49. Outline the steps in solving an exponential equation.

50. Outline the steps in solving a logarithmic equation.

Problem Solving

51. How long will it take $800 to grow to $1200 at 7.94 percent compounded annually?

52. If $3200 is invested at 11.6 percent compounded annually, how long will it take to become $5000?

53. The population of a town increases 2 percent each year. How long will it take the population to double?

54. If you get an 8 percent raise each year, how long will it take your salary to double?

55. A person is murdered in a room where the temperature is 70°. According to Newton's law of cooling, the temperature T of the body t hours after death is given by the formula

$$T = 70 + 28.6(0.97)^t$$

The temperature of the body when it is discovered at 9 a.m. is 95°. At what time did the murder take place?

56. The half-life of carbon-14 is 5700 yr. This means that the quantity Q of carbon-14 that remains t years after an organism dies is given by the formula

$$Q = Q_0(2)^{-t/5700}$$

where Q_0 is the original quantity. An old bone is unearthed and determined to contain only 30 percent of its original carbon-14 content. Determine the age of the bone.

Getting Acquainted with Your Graphing Calculator

The solution of the exponential equation $3^x = 10$ is the x-intercept of the graph of $y = 3^x - 10$. Use this fact to solve each of the following equations to the nearest hundredth by graphing the appropriate function and zooming in on the x-intercept(s).

57. $3^x = 10$

58. $\ln x = 1.7$

59. $\ln 3x = x^2 - 1$

60. $e^x = x + 2$

Chapter 10 Key Terms

10.1 **Exponential function** (with base a) A function of the form $f(x) = a^x$ $(a > 0, a \neq 1)$

Horizontal asymptote A horizontal line that a portion of a graph approaches as that portion of the graph moves away from the origin

Increasing function A function whose y-values increase as x increases

Decreasing function A function whose y-values decrease as x increases

e An irrational number, approximately 2.718, that the expression $\left(1 + \frac{1}{n}\right)^n$ approaches as $n \to \infty$

Continuously compounded interest Interest that occurs when the number of compounding periods per year grows large without bound

10.2 **One-to-one function** A function f is one-to-one if, for any x_1 and x_2 in its domain, $x_1 \neq x_2$ implies $f(x_1) \neq f(x_2)$; that is, if each y-value corresponds to exactly one x-value.

Horizontal line test A function is one-to-one provided that no horizontal line intersects the graph at more than one point

Inverse of a one-to-one function f The set of all ordered pairs of the form (b, a), where (a, b) belongs to f

10.3 **Logarithmic function** (with base a) A function of the form $y = \log_a x$, which means $x = a^y$ $(a > 0, a \neq 1)$

10.4 **Expand** (a logarithm) Apply the product, quotient, and power rules to the logarithm wherever possible

10.5 **Common logarithm** A logarithm with base 10

Antilogarithm (of a number) The number whose logarithm is the given number

Natural logarithm A logarithm with base e

10.6 **Exponential equation** An equation that contains a variable in an exponent

Logarithmic equation An equation that contains a logarithm of a variable quantity

Argument (of a function) The quantity that forms the input of the function

Chapter 10 Key Rules/Steps

10.1 Exponential Functions

Graphing Exponential Functions

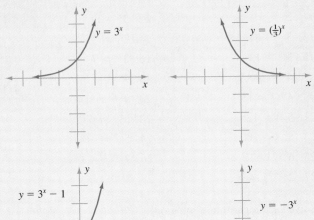

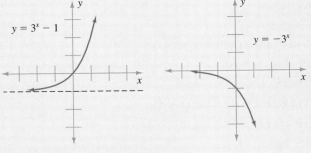

10.2 Inverse of a Function

Finding f^{-1}

Since $f = \{(0, 7), (-2, 4), (3, 8)\}$ is a one-to-one function, then $f^{-1} = \{(7, 0), (4, -2), (8, 3)\}$.

One-to-One Functions

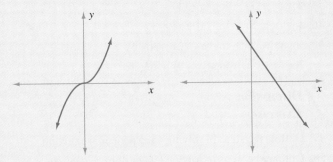

Functions That Are Not One-to-One

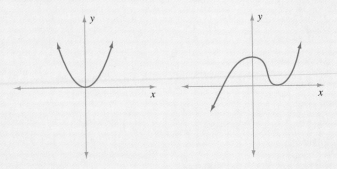

To Find the Inverse of $f(x) = 2x - 4$

1. Verify that f is one-to-one.
   Suppose $x_1 \neq x_2$. Then $2x_1 \neq 2x_2$ and $2x_1 - 4 \neq 2x_2 - 4$.
   Therefore $f(x_1) \neq f(x_2)$ and f is one-to-one.

2. Replace $f(x)$ with y.

   $$y = 2x - 4$$

3. Interchange x and y.

   $$x = 2y - 4$$

4. Solve for y.

   $$x + 4 = 2y$$
   $$y = \frac{x + 4}{2}$$

5. Replace y with $f^{-1}(x)$.

   $$f^{-1}(x) = \frac{x + 4}{2}$$

The graphs of f and f^{-1} are symmetric with respect to the line $y = x$.

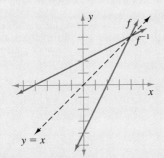

10.3 Logarithmic Functions

Definition of a Logarithm

Suppose $a > 0$ and $a \neq 1$.

$\log_a x$ = the exponent on a that gives x

Example: $\log_2 8 = 3$, $\log_5 5 = 1$, $\log_7 1 = 0$

Logarithmic versus Exponential Form

Suppose $a > 0$ and $a \neq 1$.

$y = \log_a x$ is equivalent to $x = a^y$.

Example: $\log_3 9 = 2$ is equivalent to $9 = 3^2$.

One-to-One Property of the Exponential Function

Suppose $a > 0$ and $a \neq 1$.

If $a^{x_1} = a^{x_2}$, then $x_1 = x_2$.

Example: If $4^{2x} = 4^{10}$, then $2x = 10$.

Graphing Logarithmic Functions

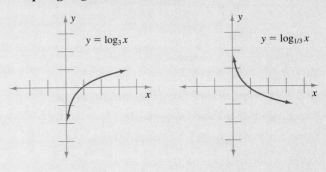

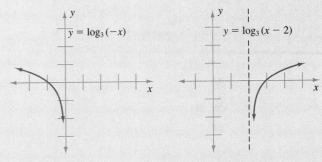

10.4 Properties of Logarithms

Suppose a, M, and N are positive numbers and $a \neq 1$.

Product rule: $\log_a MN = \log_a M + \log_a N$

Example: $\log_6 7x = \log_6 7 + \log_6 x$ $(x > 0)$

Quotient rule: $\log_a \dfrac{M}{N} = \log_a M - \log_a N$

Example: $\log_9 \dfrac{y}{5} = \log_9 y - \log_9 5$ $(y > 0)$

Power rule: $\log_a M^r = r \log_a M$

Example: $\log_4 (x + 3)^3 = 3 \log_4 (x + 3)$ $(x > -3)$

$\log_a a^t = t$ Example: $\log_5 5^x = x$

$a^{\log_a t} = t \ (t > 0)$ Example: $2^{\log_2 10p} = 10p$ $(p > 0)$

10.5 Common Logarithms; Natural Logarithms; Change of Base

Common Logarithm

$\log x = \log_{10} x$ Example: $\log 1000 = \log_{10} 1000 = 3$

Antilogarithm

If $\log x = y$, then $x = $ antilog y.

Example: Since $\log 1000 = 3$, antilog $3 = 1000$.

Natural Logarithm

$\ln x = \log_e x$ Example: $\ln e^5 = \log_e e^5 = 5$

Change-of-Base Formula

Suppose a, b, and x are positive numbers and neither a nor b is 1.

$$\log_b x = \frac{\log_a x}{\log_a b}$$ Example: $\log_5 72 = \dfrac{\ln 72}{\ln 5} = 2.6572$

10.6 Exponential and Logarithmic Equations

A Logarithm-Is-a-Function Property

Suppose a, x_1, and x_2 are positive numbers and $a \neq 1$.

If $x_1 = x_2$, then $\log_a x_1 = \log_a x_2$.

Example: If $3^{z+1} = 15$, then $\log 3^{z+1} = \log 15$.

To Solve the Exponential Equation $5^{2x} = 20$

1. We cannot write both sides with the same base, so go to step 2.

2. Take the log of both sides.

$$\log 5^{2x} = \log 20$$
$$2x \log 5 = \log 20 \qquad \text{Power rule}$$

3. Solve the equation.

$$x = \frac{\log 20}{2 \log 5} \qquad \text{Divide by 2 log 5.}$$
$$x = 0.93 \qquad \text{Approximate.}$$

4. Check.

$$5^{2(0.93)} = 5^{1.86} \approx 20$$

One-to-One Property of the Logarithmic Function

Suppose a, x_1, and x_2 are positive numbers and $a \neq 1$.

If $\log_a x_1 = \log_a x_2$, then $x_1 = x_2$.

Example: If $\log_8 (x + 1) = \log_8 12$, then $x + 1 = 12$.

To Solve the Logarithmic Equation
$\log_2 x + \log_2 (x - 2) = 3$

1. We cannot write both sides as a single logarithm with the same base, so go to step 2.

2. Write one side as a single logarithm and the other side as a number. Then convert to exponential form.

$$\log_2 x(x - 2) = 3 \qquad \text{Product rule}$$
$$x(x - 2) = 2^3 \qquad \text{Convert to exponential form.}$$

3. Solve the equation.

$$x^2 - 2x = 8 \qquad \text{Multiply out.}$$
$$x^2 - 2x - 8 = 0 \qquad \text{Subtract 8.}$$
$$(x - 4)(x + 2) = 0 \qquad \text{Factor.}$$
$$x = 4 \quad \text{or} \quad x = -2$$

4. Check. Discard any solution that causes a logarithm to be undefined.

Check $x = 4$:

$$\log_2 4 + \log_2 (4 - 2) = 3$$
$$\log_2 4 + \log_2 2 = 3$$
$$2 + 1 = 3 \qquad \text{True}$$

Check $x = -2$:

$$\log_2 (-2) + \log_2 (-2 - 2) = 3$$

Since $\log_2 (-2)$ and $\log_2 (-4)$ are undefined, discard the solution $x = -2$. Therefore the solution set is $\{4\}$.

Chapter 10 Review Exercises

[10.1] Graph each function.

1. $y = 5^x$

2. $y = (\frac{1}{3})^x$

3. $y = e^x + 1$

4. $y = 2^{-|x|}$

Find each power of e correct to five decimal places.

5. $e^{2.4}$

6. $e^{-0.5}$

7. What determines whether the exponential function $y = a^x$ is increasing or decreasing?

8. Determine the value of \$5000 invested at 9 percent for 8 yr if interest is compounded (a) annually, (b) quarterly, (c) continuously.

[10.2] Determine whether f is one-to-one. If it is, find f^{-1} and then graph f and f^{-1} on the same axes.

9. $f = \{(2, 2), (3, 4), (-1, 3)\}$

10. $f = \{(-7, 9), (0, 5), (6, 9), (8, 8)\}$

11. $f(x) = -3x$

12. $f(x) = 2x + 4$

13. $f(x) = x^2 + 4$

14. $f(x) = x^2 + 4 \ (x \geq 0)$

15. Given the one-to-one function $f(x) = 7x - 9$, find $f^{-1}(x)$. Then determine each of the following.
   (a) $f^{-1}(-5)$ (b) $f^{-1}(-9)$
   (c) $f(f^{-1}(a))$ (d) $f^{-1}(f(a))$

For Exercises 16 and 17 state whether f is one-to-one. If it is, sketch f^{-1} on the same axes.

16.
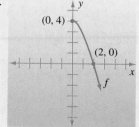

17.
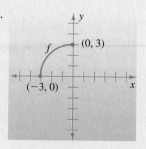

[10.3] Evaluate each logarithm.

18. $\log_2 32$

19. $\log_3 \frac{1}{81}$

20. $\log_7 1$

Convert each equation to logarithmic form.

21. $3^5 = 243$

22. $4 = 8^{2/3}$

Convert each equation to exponential form.

23. $\log_4 1024 = 5$

24. $\log_{1/2} 16 = -4$

Find the unknown in each equation.

25. $y = \log_{10} 10{,}000$

26. $\log_5 x = 4$

27. $y = \log_{10} 0.000001$

28. $\log_a 144 = 2$

29. $\log_9 x = \frac{3}{2}$

30. $\log_a 27 = -3$

Graph each function.

31. $y = \log_{1/2} x$

32. $y = \log_2 (x + 2)$

33. What determines whether the logarithmic function $y = \log_a x$ is increasing or decreasing?

[10.4] Expand each logarithm. Assume all variables represent positive numbers and $a \neq 1$.

34. $\log_4 5 \cdot 7$

35. $\log_2 \dfrac{x}{x + 1}$

36. $\log_6 y^{-1}$

37. $\log_a \sqrt{z}$

Simplify each expression. Assume $t > 0$, $a > 0$, and $a \neq 1$.

38. $\log_4 \sqrt[3]{16}$

39. $\log_7 7^t$

40. $a^{\log_a 3t}$

Expand and simplify. Assume all variables represent positive numbers.

41. $\log_7 \dfrac{1}{x^4}$

42. $\log_5 \dfrac{7x}{25y^2}$

43. $\log_3 \sqrt[4]{\dfrac{x^2 y^3}{z}}$

Express as a single logarithm with a coefficient of 1. Assume all variables represent positive numbers and $a \neq 1$.

44. $-\log_9 x$

45. $\log_3 2x + 4 \log_3 y$

46. $\frac{1}{3} \log_a x - 2 \log_a y - 5 \log_a z$

47. $\log_{10} (x^2 + 8x + 7) - \log_{10} (x + 7)$

48. Give an example to show that $\log_3 x^2$ is *not* equal to $2 \log_3 x$ when x is a negative number.

[10.5] Find each common logarithm to four decimal places.

49. $\log 34{,}100$

50. $\log 0.000425$

Find each antilogarithm x to three significant figures.

51. $\log x = 1.441$

52. $\log x = 15.72$

53. $\log x = -0.3958$

Find each natural logarithm to four decimal places.

54. $\ln 10$

55. $\ln 0.855$

Find x in each equation to three significant figures.

56. $\ln x = 2.46$

57. $\ln x = -0.909$

Find each natural logarithm without using a calculator.

58. $\ln \dfrac{1}{e^3}$

59. $\ln \sqrt[4]{e}$

Find each logarithm to the nearest hundredth.

60. $\log_7 4$

61. $\log_{1/2} 6.2$

62. Express $\log_5 93$ in terms of $\log_8$.

63. Graph: $y = \ln x$.

64. How are the graphs of $y = \ln x$ and $y = e^x$ related?

65. Find the doubling time of a sum of money if the interest rate is 10.4 percent compounded continuously.

66. At what point in a problem should you round off? Explain.

[10.6] Solve each equation. Round approximate answers to the nearest hundredth.

67. $5^x = 125$

68. $5^x = 100$

69. $3^{x^2 - 5} = 9^{2x}$

70. $3^{x-1} = 7$

71. $\log_2 (2x + 1) = \log_2 9$

72. $\log (x + 10) - \log x = \log 6$

73. $\log_6 (x + 3) + \log_6 (x + 2) = 1$

74. $\log_3 x + \log_3 (x - 6) = 3$

75. The population of sea otters in a certain area is 500. If the population increases at the rate of 4 percent each year, how long will it take the population to become 700?

Chapter 10 Test

1. If $f = \{(8, -6), (3, 5), (2, 9)\}$, find f^{-1}.

Find the unknown in each equation.

2. $\log_{27} x = \frac{2}{3}$ 3. $\log_a 8 = 3$

4. $y = \log_8 1$

Simplify each expression. Assume $t > 0$, $b > 0$, and $b \neq 1$.

5. $\log_b b^{5t}$ 6. $6^{\log_6 12}$

Expand and simplify. Assume $x > 0$ and $y > 0$.

7. $\log_2 x^3 \sqrt{y}$ 8. $\log_9 \dfrac{9}{xy}$

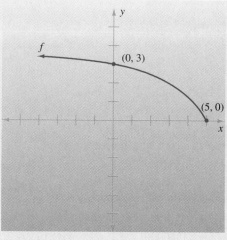

Figure 10.29

Find each logarithm to four decimal places.

9. $\ln 247$ 10. $\log_7 0.0018$

Find x to three significant figures.

11. $\log x = -0.6996$ 12. $\ln x = 73.5$

13. Determine the value of $3200 invested for 12 years at $11\frac{1}{2}$ percent compounded continuously. Use $V = Pe^{rt}$.

14. Why does the graph of f in Fig. 10.29 represent a one-to-one function? Sketch the graph of f^{-1} on the same axes.

15. Explain why it is important to check your solutions when solving a logarithmic equation.

16. Show that $f(x) = 5x + 3$ is one-to-one. Then find f^{-1}.

Solve each equation. Round approximate answers to the nearest hundredth.

17. $2^{2x-1} = 8$ 18. $4^x = 9$

19. $\log_4 x + \log_4 (x - 6) = 2$

Graph each function.

20. $y = (\frac{1}{2})^x$ 21. $y = \log_4 x$

Systems of Equations and Inequalities

Suppose a plane travels 320 mi against the wind in 4 hr, then turns and travels 220 mi with the wind in 2 hr. What is the speed of the wind? In Sec. 11.3 we show how to solve this problem by setting up a system of equations.

11.1 The Graphing Method

TAPE AU22

In this section we define the phrase *system of equations*, and we learn a method for solving a system.

Definition of a System

A demand equation for a product gives the quantity y, demanded at a given price x. For a particular product, the demand equation might be $y = 750 - 10x$. A supply equation gives the quantity y that a firm is willing to supply at a given price x. A supply equation for this product might be $y = 15x$. Together, these two equations form the system of equations below.

$$y = 750 - 10x$$
$$y = 15x$$

When two or more equations are true simultaneously, the equations are called a **system of equations.*** If all of the equations are linear, the equations are called a **linear system.**

Solution of a System

A **solution** of a system of two equations, each in two variables, is an ordered pair that satisfies both equations in the system.

Example 1   Determine whether the ordered pair (30, 450) is a solution of the system stated above.

Solution: Substitute $x = 30$ and $y = 450$ into both equations of the system.

$y = 750 - 10x$	$y = 15x$
$450 = 750 - 10(30)$	$450 = 15(30)$
$450 = 750 - 300$ True	$450 = 450$ True

Since (30, 450) satisfies both equations, (30, 450) is a solution of the system. (This means that a transaction will take place when the price is 30. The quantity changing hands will be 450.)   ❐

Try Exercise 13

Is (9, 7) a solution of the following system?

$$x + y = 16$$
$$x - y = 2$$

Example 2   Determine whether (4, −2) is a solution of the following system:

$$5x - 4y = 28$$
$$x = 8 - 2y$$

Solution: Substitute $(x, y) = (4, -2)$ into both equations.

$5x - 4y = 28$	$x = 8 - 2y$
$5(4) - 4(-2) = 28$	$4 = 8 - 2(-2)$
$20 + 8 = 28$ True	$4 = 8 + 4$ False

Since (4, −2) does not satisfy *both* equations, (4, −2) is not a solution of the system.   ❐

Try Exercise 17

Is (5, −1) a solution of the following system?

$$3x - 7y = 22$$
$$x = 7 - 2y$$

*HISTORICAL NOTE: Systems of equations were studied by the Greek mathematician Diophantus (ca. 250), among others. A particular type of system of equations is still referred to as a *Diophantine system*.

Solving by Graphing

Examples 1 and 2 illustrate how we determine whether a given ordered pair is a solution of a given linear system. But how do we find the ordered pair in the first place? One method is to graph both equations on the same coordinate axes. Then the intersection point of the two lines represents the solution of the system. This method of solution is called the **graphing method.**

Example 3 Solve the following system by the graphing method:

$$x + y = 4$$
$$x - y = 6$$

Solution: Graph both equations on the same axes, as shown in Fig. 11.1. Note that the intersection point of the two lines is $(5, -1)$. Since $(5, -1)$ is the only ordered pair that satisfies both equations, the solution set of the system has just one element, namely, $(5, -1)$. Therefore, the solution set is $\{(5, -1)\}$. Check in both equations. ❐

Try Exercise 19

Solve by graphing:

$$x + y = 2$$
$$x - y = 4$$

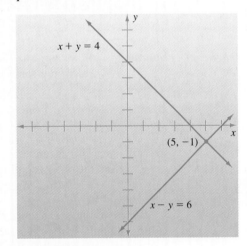

Figure 11.1

Example 4 Solve the following system by the graphing method:

$$2x - y = -1$$
$$4x - 2y = 6$$

Solution: If we write both equations in slope-intercept form, we see that the lines have the same slope but different y-intercepts.

$$
\begin{array}{ll}
2x - y = -1 & \qquad 4x - 2y = 6 \\
-y = -2x - 1 & \qquad -2y = -4x + 6 \\
y = 2x + 1 & \qquad y = 2x - 3 \\
m = 2,\ b = 1 & \qquad m = 2,\ b = -3
\end{array}
$$

Therefore the graphs of the two equations are parallel lines, as shown in Fig. 11.2. Since parallel lines have no points in common, the solution set is the empty set, $\varnothing$. ❐

Try Exercise 27

Solve by graphing:

$$2x - y = -2$$
$$6x - 3y = 12$$

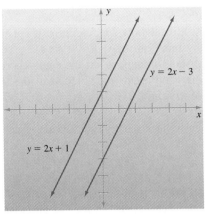

Figure 11.2

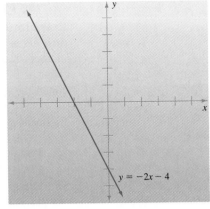

Figure 11.3

Example 5   Solve the following system by the graphing method:

$$-2x - y = 4$$
$$6x + 3y = -12$$

Solution: If we write both equations in slope-intercept form, we see that they represent the same lines.

$$-2x - y = 4 \qquad\qquad 6x + 3y = -12$$
$$-y = 2x + 4 \qquad\qquad 3y = -6x - 12$$
$$y = -2x - 4 \qquad\qquad y = -2x - 4$$

This means that any ordered pair that satisfies one equation also satisfies the other equation. Therefore any point on the line (see Fig. 11.3) represents a solution. We can write the solution set as $\{(x, y)\,|\,y = -2x - 4\}$.   ❏

Try Exercise 29

Solve by graphing:

$$-5x - y = 4$$
$$10x + 2y = -8$$

If the two linear equations that form a system represent two different lines, the system is **independent.** If they represent the same line, the system is **dependent.** If the system has at least one solution, the system is **consistent.** If the system has no solution, the system is **inconsistent.** Therefore the system of Example 3 is independent and consistent. The system of Example 4 is independent and inconsistent. The system of Example 5 is dependent and consistent. These terms are also illustrated in Figs. 11.4, 11.5, and 11.6.

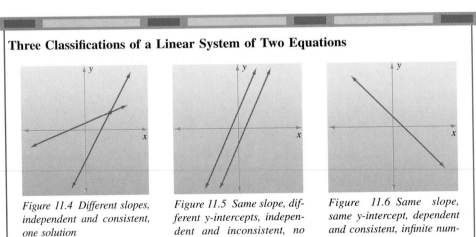

Three Classifications of a Linear System of Two Equations

Figure 11.4 Different slopes, independent and consistent, one solution

Figure 11.5 Same slope, different y-intercepts, independent and inconsistent, no solution

Figure 11.6 Same slope, same y-intercept, dependent and consistent, infinite number of solutions

Exercises 11.1

◆◆◆

Completion Exercises

1. When two or more equations are true simultaneously, the equations are called a(n) _____; if all of the equations are linear, the equations are called a(n) _____.

2. A solution of a system of two equations, each in two variables, is a(n) _____ that satisfies _____ (both, one) of the equations.

3. If the two equations that form a system of linear equations represent two different lines, the system is _____; if they represent the same line, the system is _____.

4. If a system of two linear equations has at least one solution, the system is _____; if it has no solution, the system is _____.

Matching Exercises

Match each description in Exercises 5 through 12 with an appropriate graph in letters A through C.

5. Independent system

6. Dependent system

7. Inconsistent system

8. Consistent system

9. Infinite number of solutions

10. Two solutions

11. One solution

12. No solution

A.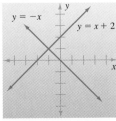

B.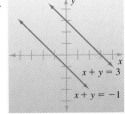

C.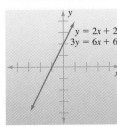

Determine whether the given ordered pair is a solution of the given system.

13. $(9, 7)$
    $x + y = 16$
    $x - y = 2$

14. $(3, 5)$
    $y = 3x - 4$
    $y = x + 2$

15. $(-2, -3)$
    $4x - y = -11$
    $2x - 5y = 19$

16. $(-1, -4)$
    $5x - y = -9$
    $3x - 6y = 27$

17. $(5, -1)$
    $3x - 7y = 22$
    $x = 7 - 2y$

18. $(6, -2)$
    $2x - 7y = 26$
    $x + 4y = 14$

Solve each system by the graphing method. Classify each system as dependent or independent. Also classify each system as consistent or inconsistent.

19. $x + y = 2$
    $x - y = 4$

20. $x + y = 3$
    $x - y = 5$

21. $y = 2x - 2$
    $y = -4$

22. $y = -x + 1$
    $x = -3$

23. $2x - y = -6$
    $4x + 3y = -12$

24. $x + 2y = -10$
    $5x - 4y = 20$

25. $9x + 7y = 21$
    $x - y = 5$

26. $7x - 5y = 10$
    $x + y = 4$

27. $2x - y = -2$
    $6x - 3y = 12$

28. $x - 2y = -2$
    $5x - 10y = 20$

29. $-5x - y = 4$
    $10x + 2y = -8$

30. $-3x + y = 4$
    $9x - 3y = -12$

Discussion Exercises

31. Do you see any disadvantages with the graphing method for solving a system?

32. If the solution set of a system is $\{(6, -3)\}$, does the system have one or two solutions? Explain.

33. Discuss the three things that can happen when you graph two lines.

34. How can you use the slope-intercept form of a line to determine whether a system of two linear equations has no solution, one solution, or an infinite number of solutions?

Without graphing the system, (a) classify as dependent or independent, (b) classify as consistent or inconsistent, and (c) determine the number of solutions.

35. $y = 5x + 2$
    $y = -7x + 3$

36. $y = -x - 4$
    $y = 6x - 9$

37. $3x = 13 - 4y$
    $6x + 8y = 18$

38. $2y = 11 + 5x$
    $9x - 6y = -15$

39. $\frac{1}{2}x - \frac{3}{4}y = -1$
    $y = \dfrac{2x + 4}{3}$

40. $\frac{2}{3}x - \frac{5}{6}y = 2$
    $x = \dfrac{5y + 12}{4}$

41. $\dfrac{x}{4} + \dfrac{y}{10} = \dfrac{2}{5}$
    $\dfrac{x}{6} + \dfrac{y}{2} = \dfrac{4}{3}$

42. $\dfrac{x}{5} - \dfrac{y}{2} = \dfrac{3}{5}$
    $\dfrac{x}{6} - \dfrac{y}{9} = \dfrac{1}{3}$

Problem Solving

Use Fig. 11.7 to answer Exercise 43. Use Fig. 11.8 to answer Exercise 44.

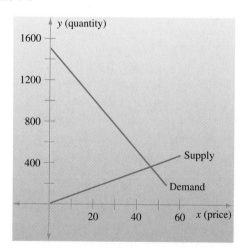

Figure 11.7

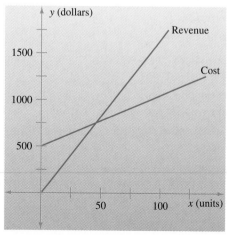

Figure 11.8

43. A demand equation for a product gives the quantity demanded at a given price. A supply equation gives the quantity that firms are willing to supply at a given price.

(a) The price at which supply equals demand is called the equilibrium price. Estimate the equilibrium price.

(b) Estimate the quantity when supply equals demand.

(c) Estimate both the supply and the demand when the price is 30.

(d) What is the demand if the product is given away?

44. The money that a company spends to produce a given number of units of a product is its cost. The money that a company receives from the sale of the product is its revenue.

(a) The breakeven point occurs when revenue equals cost. Estimate the number of units needed to break even.

(b) Estimate the cost at the breakeven point.

(c) Profit equals revenue minus cost. Estimate the profit if 100 units are produced and sold.

(d) Fixed cost is the cost to produce 0 units. What is the fixed cost?

 Getting Acquainted with Your Graphing Calculator

Solve each system by graphing the two lines and zooming in on the intersection point. If the answer is approximate, write the answer correct to two decimal places.

45. $y = 1.5x - 4$
    $y = -x + 6$

46. $5.8x + 2.5y = 14$
    $1.8x - 3.2y = 9.6$

47. $\frac{1}{2}y - x = 2$
    $2x = y + 3$

You can determine whether $(4, -3)$ is a solution of the system in Exercise 48 by storing 4 in x and -3 in y and then calculating the value of the left side of each equation. Use this method to determine whether the given ordered pair is a solution of the given system.

48. $(4, -3)$
    $2x - 5y = 23$
    $x + 7y = -17$

49. $(-2, -5)$
    $3x - 4y = 14$
    $-6x + y = 17$

11.2 The Substitution Method and the Addition Method

TAPE AU22

The main reason we discuss the graphing method for solving a linear system of two equations is to illustrate the three different possibilities that can occur when solving such a system. The graphing method is seldom used to actually solve a linear system because it is often difficult to read the exact coordinates of the intersection point.

In this section we present two algebraic methods for solving a linear system—the **substitution method** and the **addition method.**

The Substitution Method

Here is a summary of the steps in the substitution method.

To Solve a Linear System of Two Equations by Substitution

1. Solve one of the equations for one of its variables.
2. Substitute the result of step 1 into the *other* equation.
3. Solve the equation in one variable that results from step 2. If a contradiction results, the lines are parallel and there is no solution. If an identity results, the lines coincide and any point on the line is a solution.
4. Substitute the value found in step 3 into the equation resulting from step 1 to find the value of the other variable.
5. Check your solution in both of the original equations.

We illustrate these steps in Example 1.

Example 1   Solve the following system by the substitution method:

$$x + 5y = 17$$
$$2x - 3y = -18$$

Solution: Step 1:   Solve one of the equations for one of its variables.
To avoid fractions, we solve the first equation for x.

$$x = 17 - 5y \qquad \text{Subtract } 5y.$$

Step 2:   Substitute the result of step 1 into the other equation.

$$2x - 3y = -18 \qquad \text{Second equation}$$
$$2(17 - 5y) - 3y = -18 \qquad \text{Replace } x \text{ with } 17 - 5y.$$

Step 3:   Solve the equation in one variable that results from step 2.

$$34 - 10y - 3y = -18 \qquad \text{Distributive property}$$
$$34 - 13y = -18 \qquad \text{Simplify.}$$
$$-13y = -52 \qquad \text{Subtract 34.}$$
$$y = 4 \qquad \text{Divide by } -13.$$

Step 4:   Substitute the value found in step 3 into the equation resulting from step 1 to find the value of the other variable. (We could also substitute into either of the original equations.)

$$x = 17 - 5y \qquad \text{Equation from step 1}$$
$$= 17 - 5(4) \qquad \text{Let } y = 4.$$
$$= -3 \qquad \text{Simplify.}$$

Step 5: Check your solution in both of the original equations.
The ordered pair $(-3, 4)$ checks in both equations, so the solution set is $\{(-3, 4)\}$. ❐

Try Exercise 3

Solve by substitution:

$$x + 4y = 16$$
$$3x - 2y = -22$$

Example 2 Solve the following system by the substitution method:

$$6x - 3y = 0$$
$$y = 2x$$

Solution: The second equation is already solved for y, so replace y in the first equation with $2x$.

$$6x - 3(2x) = 0$$
$$6x - 6x = 0$$
$$0 = 0$$

The identity $0 = 0$ means that the two lines coincide, and any point on the line is a solution. Therefore the solution set is $\{(x, y) | y = 2x\}$. ❐

Try Exercise 9

Solve by substitution:

$$8x - 2y = 0$$
$$y = 4x$$

Example 3 Solve the following system by substitution:

$$12x - 4y = 11$$
$$-6x + 2y = 7$$

Solution: In this case we cannot avoid fractions. We decide to solve the second equation for y.

$$2y = 7 + 6x \qquad \text{Add } 6x.$$
$$y = \frac{7 + 6x}{2} \qquad \text{Divide by 2.}$$

Replace y with $\dfrac{7 + 6x}{2}$ in the first equation.

$$12x - 4\left(\frac{7 + 6x}{2}\right) = 11$$

Multiply both sides by 2 to remove the fraction.

$$24x - 4(7 + 6x) = 22 \qquad \text{Multiply by 2.}$$
$$24x - 28 - 24x = 22 \qquad \text{Distributive property}$$
$$-28 = 22 \qquad \text{Simplify.}$$

The contradiction $-28 = 22$ means that the two lines are parallel. Therefore the solution set is $\varnothing$. ❐

Try Exercise 11

Solve by substitution:

$$24x - 6y = 13$$
$$-8x + 2y = 9$$

The Addition Method

We can avoid fractions when solving a linear system by using the addition method. Example 4 illustrates how the addition method works.

Example 4 Solve the following system by the addition method:

$$2x + y = 13$$
$$x - y = 5$$

Solution: Add the corresponding sides of the two equations to eliminate y.

$$2x + y = 13$$
$$\underline{x - y = 5}$$
$$3x \quad\quad = 18$$

Divide each side of $3x = 18$ by 3.

$$x = 6$$

To find y, replace x with 6 in either of the original equations. If we use the second equation, we get

$x - y = 5$	Second equation
$6 - y = 5$	Let $x = 6$.
$-y = -1$	Subtract 6.
$y = 1$	Multiply by -1.

The solution set is $\{(6, 1)\}$. Check in both of the original equations. ❐

Try Exercise 17

Solve by addition:

$$3x + y = 7$$
$$x - y = 1$$

Example 5   Solve the following system by the addition method:

$$-x + 4y = 5$$
$$2x - 3y = -10$$

Solution: In this case, adding the two equations will not eliminate either variable. Therefore multiply the first equation by 2. This will make the coefficient of x in the first equation the opposite of the coefficient of x in the second equation.

$-x + 4y = 5$	First equation
$2(-x + 4y) = 2(5)$	Multiply by 2.
$-2x + 8y = 10$	Simplify.

Add the new first equation to the original second equation to eliminate x.

$-2x + 8y = 10$	New first equation
$\underline{2x - 3y = -10}$	Original second equation
$5y = 0$	Add.
$y = 0$	Divide by 5.

To find x, replace y with 0 in either of the original equations. Using the first equation, we have

$-x + 4(0) = 5$	Let $y = 0$ in the first equation.
$-x = 5$	Simplify.
$x = -5$	Multiply by -1.

Try Exercise 21

Solve by addition:

$$-x + 4y = 4$$
$$2x - 5y = -8$$

The solution set is $\{(-5, 0)\}$. Check in both of the original equations. ❐

Here is a summary of the steps in the addition method.

To Solve a Linear System of Two Equations by Addition

1. Write both equations in the standard form $ax + by = c$ with integer coefficients.
2. If necessary, multiply one or both equations by a suitable number so that the coefficients of x (or y) are opposites.
3. Add the new equations and solve the resulting equation in one variable. If a contradiction results, the lines are parallel and there is no solution. If an identity results, the lines coincide and any point on the line is a solution.
4. Substitute the value found in step 3 into either of the original equations to find the value of the other variable.
5. Check your solution in both of the original equations.

We illustrate these steps in Example 6.

Example 6 Solve the following system by the addition method:

$$\frac{x}{4} - \frac{y}{6} = -\frac{1}{12}$$
$$4x = 5y - 6$$

Solution: **Step 1:** Write both equations in standard form with integer coefficients. Multiply both sides of the first equation by the LCD 12 to clear fractions. Subtract $5y$ from both sides of the second equation.

$$12\left(\frac{x}{4} - \frac{y}{6}\right) = 12\left(-\frac{1}{12}\right) \qquad 4x = 5y - 6$$
$$3x - 2y = -1 \qquad\qquad 4x - 5y = -6$$

Step 2: Multiply $3x - 2y = -1$ by 4, and multiply $4x - 5y = -6$ by -3, so that the coefficients of x are opposites.

$$4(3x - 2y) = 4(-1) \qquad -3(4x - 5y) = -3(-6)$$
$$12x - 8y = -4 \qquad\qquad -12x + 15y = 18$$

Step 3: Add the new equations and solve the resulting equation in one variable.

$12x - 8y = -4$	New first equation
$-12x + 15y = 18$	New second equation
$7y = 14$	Add.
$y = 2$	Divide by 7.

Step 4: Substitute the value found in step 3 into either of the original equations (or either equation resulting from step 1).

$4x = 5y - 6$	Original second equation
$4x = 5(2) - 6$	Let $y = 2$.
$4x = 4$	Simplify.
$x = 1$	Divide by 4.

Step 5: Check your solution in both of the original equations. The ordered pair $(1, 2)$ checks, so the solution set is $\{(1, 2)\}$.

Try Exercise 31

Solve by addition:

$$\frac{x}{4} - \frac{y}{6} = \frac{1}{4}$$
$$4x = 5y - 10$$

Choosing a Method

The chart below summarizes the advantages and disadvantages of the three methods for solving a linear system.

Solving a Linear System of Two Equations

Method	Advantages	Disadvantages
1. Graphing	Illustrates the three types of solution	Time-consuming to draw graphs accurately; hard to read exact coordinates of intersection point
2. Substitution	Best method if one equation is already solved for a variable	Sometimes involves working with fractions
3. Addition	Best method overall; avoids fractions	Must write equations in standard form

Exercises 11.2

Solve each system by the substitution method.

1. $x = y + 1$
   $x + 3y = 9$

2. $x = y + 2$
   $x + 2y = 8$

3. $x + 4y = 16$
   $3x - 2y = -22$

4. $x + 3y = 16$
   $4x - 5y = -38$

5. $2x - y = 5$
   $4x + y = 7$

6. $2x - y = 7$
   $6x + y = 17$

7. $2x + 6y = 5$
   $4x = 3y$

8. $3x + 10y = 5$
   $6x = 5y$

9. $8x - 2y = 0$
   $y = 4x$

10. $12x - 4y = 0$
    $y = 3x$

11. $24x - 6y = 13$
    $-8x + 2y = 9$

12. $24x - 8y = 17$
    $-6x + 2y = 5$

13. $5x + 4y = 2$
    $2x + 3y = 3$

14. $5x + 4y = 1$
    $6x + 3y = 2$

15. $3x - 2y = 18$
    $4x - 3y = 22$

16. $4x - 5y = -1$
    $3x - 2y = 15$

Solve each system by the addition method.

17. $3x + y = 7$
    $x - y = 1$

18. $x + y = 9$
    $4x - y = 6$

19. $3x - 6y = 1$
    $5x + 6y = -9$

20. $2x - 8y = 2$
    $9x + 8y = -13$

21. $-x + 4y = 4$
    $2x - 5y = -8$

22. $-x + 4y = -3$
    $2x - 3y = 6$

23. $2x + 3y = 11$
    $4x + 5y = 19$

24. $3x + 4y = 19$
    $6x + 2y = 32$

25. $7x - 5y = 2$
    $6x + 2y = -3$

26. $3x - 5y = 6$
    $7x + 3y = -8$

27. $-x + 2y = 3$
    $3x - 6y = -9$

28. $x - 3y = 2$
    $-4x + 12y = -8$

29. $-x + \frac{5}{3}y = \frac{1}{3}$
    $\frac{3}{2}x - \frac{5}{2}y = -\frac{1}{4}$

30. $\frac{4}{3}x - \frac{5}{3}y = \frac{1}{3}$
    $-x + \frac{5}{4}y = -\frac{1}{12}$

31. $\frac{x}{4} - \frac{y}{6} = \frac{1}{4}$
    $4x = 5y - 10$

32. $\frac{x}{3} - \frac{y}{4} = \frac{1}{3}$
    $5x = 2y + 19$

/ ***Discussion***
Exercises

33. Suppose you are solving a linear system of two equations, and you arrive at the statement $5 = 2$. What conclusion can you make? If you arrive at the statement $-4 = -4$, what conclusion can you make?

34. Discuss the advantages and disadvantages of the three methods for solving a linear system of two equations.

35. Outline the steps in solving a linear system of two equations by substitution.

36. Outline the steps in solving a linear system of two equations by addition.

Solve each system using any method.

37. $x - 9y = -4$
    $x = \dfrac{5 - 9y}{4}$

38. $-8x + y = -5$
    $y = \dfrac{10 - 4x}{6}$

39. $3x - \frac{7}{2}y = 1$
    $\frac{1}{3}x + \frac{1}{4}y = \frac{2}{3}$

40. $\frac{5}{2}x + 2y = 1$
    $\frac{1}{9}x - \frac{1}{3}y = \frac{1}{2}$

Solve each system by first making the substitutions $u = \dfrac{1}{x}$ and $v = \dfrac{1}{y}$.

41. $\dfrac{1}{x} + \dfrac{1}{y} = 3$
    $\dfrac{2}{x} - \dfrac{1}{y} = 3$

42. $-\dfrac{1}{x} + \dfrac{3}{y} = 5$
    $\dfrac{1}{x} - \dfrac{1}{y} = 1$

43. $\dfrac{3}{x} + \dfrac{4}{y} = \dfrac{5}{2}$
    $\dfrac{5}{x} - \dfrac{3}{y} = \dfrac{7}{4}$

44. $\dfrac{5}{x} - \dfrac{3}{y} = \dfrac{1}{2}$
    $\dfrac{1}{x} + \dfrac{4}{y} = \dfrac{12}{5}$

Solve each system for the variables x and y in terms of the constants a, b, and c.

45. $5ax - y = 6$
    $y = 2ax$

46. $3bx - 7by = 9$
    $x = 2y$

47. $ax + by = c$
    $x - y = 0$

48. $ax - by = c$
    $x + y = 1$

TAPE AU23

11.3 Problem Solving with Linear Systems

In Chap. 2 we solved word problems involving two unknown quantities by representing one unknown by x, and then writing the other unknown in terms of x. Then we wrote an equation involving x.

Sometimes it is easier to solve such problems by representing one unknown by x and the other unknown by y. Then we write a system of two equations involving x and y. In this section we learn how to solve word problems using a system of equations.

To Solve a Word Problem Using a System of Equations

1. Write down the two unknown quantities and represent each by a different variable.
2. Write a system of two equations involving the variables. Sometimes a chart or a diagram is helpful here.
3. Solve the system.
4. Check your solution in the words of the original problem.

We illustrate these steps in Example 1.

Example 1   A chemist wants to make 60 kg of a solution that is 35% alcohol. She has in stock a 15% alcohol solution and a 40% alcohol solution. How many kilograms of each solution should she use?

Solution: Step 1:   Write down the two unknown quantities and represent each by a different variable.

$$x = \text{no. of kg of 15\% solution}$$
$$y = \text{no. of kg of 40\% solution}$$

Step 2:   Write a system of two equations involving the variables.
Since the final mixture contains 60 kg, the first equation is

$$x + y = 60$$

To write the second equation, construct the chart below:

	Kilograms of solution	*Kilograms of alcohol*
15% solution	x	$0.15x$
40% solution	y	$0.40y$
35% solution	60	$0.35(60)$

Sum equals

Since no alcohol is gained or lost during mixing, the second equation is

$$0.15x + 0.40y = 0.35(60)$$

Therefore the system that describes the original problem is

$$x + y = 60$$
$$0.15x + 0.40y = 21$$

Step 3:   Solve the system.

Multiply the first equation by -15 and the second equation by 100. Then add the resulting equations.

$$
\begin{array}{ll}
-15x - 15y = -900 & \text{First equation multiplied by } -15 \\
\underline{15x + 40y = 2100} & \text{Second equation multiplied by } 100 \\
25y = 1200 & \text{Add.} \\
y = 48 & \text{Divide by 25.}
\end{array}
$$

Substitute $y = 48$ into $x + y = 60$, and get $x = 12$. Therefore the chemist should use 12 kg of 15% solution and 48 kg of 40% solution.

Step 4:   Check your solution in the words of the original problem.

There are 1.8 kg of alcohol in 12 kg of 15% solution. There are 19.2 kg of alcohol in 48 kg of 40% solution. Therefore the final mixture contains $1.8 + 19.2 = 21$ kg of alcohol in $12 + 48 = 60$ kg of solution, which is a 35% solution.   ❐

Try Exercise 5

A chemist has on hand a solution that is 5% acid and a solution that is 25% acid. How many kilograms of each solution are needed to make 40 kg of a solution that is 20% acid?

Example 2   A slice of low-fat cheese contains 5 g of protein and 3 g of fat. A slice of wheat bread contains 6 g of protein and 2 g of fat. How many slices of each food should be used to prepare a lunch containing 32 g of protein and 16 g of fat?

Solution: Let $x = $ the number of slices of cheese, and $y = $ the number of slices of bread. Then construct the chart below:

	Number of slices	Grams of protein	Grams of fat
Cheese	x	$5x$	$3x$
Bread	y	$6y$	$2y$

Since the total protein is 32 g, the first equation is

$$5x + 6y = 32$$

Since the total fat is 16 g, the second equation is

$$3x + 2y = 16$$

Multiply the second equation by -3 and add the result to the first equation.

$$
\begin{array}{ll}
5x + 6y = 32 & \text{First equation} \\
\underline{-9x - 6y = -48} & \text{Second equation multiplied by } -3 \\
-4x = -16 & \text{Add.} \\
x = 4 & \text{Divide by } -4.
\end{array}
$$

Substitute $x = 4$ into either of the original equations and get $y = 2$. Therefore four slices of cheese and two slices of bread should be used to prepare the lunch.   ❐

Try Exercise 11

An ounce of macaroni contains 4 g of protein and 3 g of fat. An ounce of tuna contains 6 g of protein and 2 g of fat. How many ounces of each food should be used to prepare a lunch containing 46 g of protein and 27 g of fat?

Example 3   A square and an equilateral triangle (a triangle with three equal sides) have the same perimeter. The sum of one side of the square and one side of the triangle is 21 cm. Find the length of one side of each figure.

Solution: Let $x = $ the length of one side of the square, and $y = $ the length of one side of the triangle.

Figure 11.9

Since the perimeters are the same (see Fig. 11.9), the first equation is

$$\begin{array}{ccc} \text{Perimeter} & & \text{perimeter} \\ \text{of square} & = & \text{of triangle} \\ \downarrow & \downarrow & \downarrow \\ 4x & = & 3y \end{array}$$

Since the sum of one side of the square and one side of the triangle is 21, the second equation is

$$x + y = 21$$

Solve the first equation for x.

$$x = \frac{3y}{4} \qquad \text{Divide the first equation by 4.}$$

Replace x with $\frac{3y}{4}$ in the second equation and solve.

$$\begin{aligned} \frac{3y}{4} + y &= 21 & &\text{Let } x = \frac{3y}{4} \text{ in } x + y = 21. \\ 3y + 4y &= 84 & &\text{Multiply by 4.} \\ 7y &= 84 & &\text{Simplify.} \\ y &= 12 & &\text{Divide by 7.} \end{aligned}$$

Substitute $y = 12$ into $x + y = 21$ and get $x = 9$. Therefore the square has sides of length 9 cm, and the triangle has sides of length 12 cm. ❑

Try Exercise 15

A square and an equilateral triangle have the same perimeter. The sum of one side of the square and one side of the triangle is 28 cm. Find the length of one side of each figure.

Example 4 A plane travels 320 mi against the wind in 4 hr, then turns and travels 220 mi with the wind in half that time. Find the speed of the plane and the speed of the wind.

Solution: Let $p = $ the speed of the plane, and $w = $ the speed of the wind. When the plane travels against the wind, the ground speed is the speed of the plane minus the speed of the wind. When the plane travels with the wind, the ground speed is the speed of the plane plus the speed of the wind. Organize the information in the chart below. Note that r is rate, t is time, and d is distance:

	r	t	d
Against wind	$p - w$	4	$(p - w)4$
With wind	$p + w$	2	$(p + w)2$

Since $d = rt$

Since the distance against the wind is 320 mi, the first equation is

$$(p - w)4 = 320$$

Since the distance with the wind is 220 mi, the second equation is

$$(p + w)2 = 220$$

Dividing the first equation by 4 and the second equation by 2 gives the simpler system

$$p - w = 80$$
$$p + w = 110$$

Add and solve for p.

$$2p = 190$$
$$p = 95$$

Substitute $p = 95$ into $p + w = 110$ and get $w = 15$. Therefore the speed of the plane is 95 mi/hr and the speed of the wind is 15 mi/hr. ❑

To solve Example 5, note that the number 37 can be written as $3 \cdot 10 + 7$. Similarly the two-digit number tu can be written as $t \cdot 10 + u$, or $10t + u$.

Example 5 The sum of the digits of a two-digit number is 10. If the digits are reversed, the new number is 18 less than the original number. Find the original number.

Solution: Let t = tens digit of original number, and u = units digit of original number. Since the sum of the digits is 10, the first equation is

$$t + u = 10$$

The second equation is

New number	=	original number	−	18
↓	↓	↓	↓	↓
$10u + t$	=	$10t + u$	−	18

Simplify the second equation.

$$-9t + 9u = -18 \qquad \text{Subtract } 10t, \text{ subtract } u.$$
$$t - u = 2 \qquad \text{Divide by } -9.$$

Therefore the system that describes the problem is

$$t + u = 10$$
$$t - u = 2$$

Add the two equations and get

$$2t = 12$$
$$t = 6$$

Substitute $t = 6$ into $t + u = 10$ and get $u = 4$. Therefore the original number is 64. ❑

Try Exercise 19

A motorboat travels 21 mi upstream in 3 hr, then turns and travels 22 mi downstream in 2 hr. Find the speed of the boat and the speed of the current.

Try Exercise 29

The sum of the digits of a two-digit number is 11. If the digits are reversed, the new number is 45 less than the original number. Find the original number.

Exercises 11.3

Write a system of equations for each exercise. Then solve the system.

1. Two consecutive odd integers x and y have a sum of 76. Find the integers.

2. One number added to twice another number is 18. Find the numbers if the first is 3 more than the second.

3. One-third of a number is 1 less than one-quarter of another number. Two-thirds of the second number is 5 more than one-half of the first number. Find the numbers.

4. One-fourth of a number is 2 less than one-third of another number. Three-fourths of the second number is 5 more than one-half of the first number. Find the numbers.

5. A chemist has on hand a solution that is 5% acid and a solution that is 25% acid. How many kilograms of each solution are needed to make 40 kg of a solution that is 20% acid?

6. A candy store owner wants to mix candy worth $0.95 per pound with candy worth $1.45 per pound to make 20 lb of candy worth $1.10 per pound. How many pounds of each should she use?

7. How many grams of pure gold and how many grams of an alloy that is 55% gold should be melted together to produce 72 g of an alloy that is 65% gold?

8. How many pounds of pure salt and how many pounds of a solution that is 4% salt should be mixed to produce 84 lb of a solution that is 28% salt?

9. One marigold plant and four geranium plants cost $19. Seven marigolds and two geraniums cost $29. Find the cost of each marigold and each geranium.

10. One ballpoint pen and three fountain pens cost $29. Five ballpoints and two fountain pens cost $28. Find the cost of each ballpoint and each fountain pen.

11. An ounce of macaroni contains 4 g of protein and 3 g of fat. An ounce of tuna contains 6 g of protein and 2 g of fat. How many ounces of each food should be used to prepare a lunch containing 46 g of protein and 27 g of fat?

12. To manufacture one baseball requires 3 units of capital and 4 units of labor. One football requires 5 units of capital and 7 units of labor. How many baseballs and how many footballs can be manufactured with 85 units of capital and 115 units of labor?

13. The total receipts for a basketball game were $1490. Adult tickets were $7 and student tickets were $4. If a total of 305 tickets were sold, how many adults and how many students attended the game?

14. The total receipts for a musical were $2115. Adult tickets were $9 and student tickets were $5. If a total of 315 tickets were sold, how many adults and how many students attended the musical?

15. A square and an equilateral triangle (three equal sides) have the same perimeter. The sum of one side of the square and one side of the triangle is 28 cm. Find the length of one side of each figure.

16. The length of a rectangle is 7 ft less than 3 times the width. Find the width and length if the perimeter is 154 ft.

17. A collection of 51 nickels and dimes is worth $3.45. How many nickels and how many dimes are in the collection?

18. A cash register contains $7.30 in nickels and quarters. If there are eight more nickels than quarters, how many nickels and how many quarters are in the register?

19. A motorboat travels 21 mi upstream in 3 hr, then turns and travels 22 mi downstream in 2 hr. Find the speed of the boat and the speed of the current.

20. A plane travels 300 mi against the wind in 5 hr, then turns and travels 330 mi with the wind in 3 hr. Find the speed of the plane and the speed of the wind.

21. A jogger and a cyclist, 36 mi apart, travel toward each other. If the cyclist travels twice as fast as the jogger and they meet in 1 hr and 20 min, find the speed of each.

22. A car and a train, 161 mi apart, travel toward each other. If the car travels 12 mi/hr faster than the train and they meet in 1 hr and 45 min, find the speed of each.

23. A motorboat can travel 14.5 mi/hr with the current but only 9.5 mi/hr against the current. Find the speed of the boat and the speed of the current.

24. A plane can travel 162.5 mi/hr with the wind but only 107.5 mi/hr against the wind. Find the speed of the plane and the speed of the wind.

25. A financial planner invests twice as much of a client's money at 8 percent than at 12 percent. If the total first year's interest is $1078, how much was invested at each rate?

	Amount	Interest
8% investment	x	$0.08x$
12% investment	y	

26. A stockbroker invests 3 times as much of a client's money at 9 percent than at 14 percent. How much is invested at each rate if the total first year's interest is $984?

	Amount	Interest
9% investment	x	
14% investment	y	$0.14y$

27. A total of $42,000 is invested, part at 6 percent and the rest at 10 percent. How much is invested at each rate if the interest is the same on the two investments?

28. A total of $33,000 is invested, part at 7 percent and the rest at 15 percent. If the interest on the two investments is the same, how much is invested at each rate?

29. The sum of the digits of a two-digit number is 11. If the digits are reversed, the new number is 45 less than the original number. Find the original number.

30. The sum of the digits of a two-digit number is 12. If the digits are reversed, the new number is 54 less than the original number. Find the original number.

31. How old are Brett and Kara if 4 yr ago Brett was 6 times as old as Kara, but 4 yr from now he will only be twice as old?

32. Erika and Jon have just started collecting baseball cards. If Erika gives Jon one card, Jon will have twice as many cards as Erika. If Jon gives Erika one card, each will have the same number of cards. How many cards does each have?

 Discussion Exercise

33. Outline the steps in solving a word problem using a system of equations.

11.4 Linear Systems of Three Equations

TAPE AU23

To solve an applied problem with three unknown quantities, we often use three variables. In this section we discuss how to solve a linear system of three equations, each in three variables.

First, consider the following equation in three variables:

$$2x + 3y + z = 6$$

One solution of this equation is $x = 1$, $y = -2$, and $z = 10$. This solution is usually written as the **ordered triple** $(1, -2, 10)$. We can check this solution as follows:

$2x + 3y + z = 6$	Original equation
$2(1) + 3(-2) + 10 = 6$	Let $(x, y, z) = (1, -2, 10)$.
$2 - 6 + 10 = 6$	Simplify.
$6 = 6$	True

You can verify that the ordered triples $(3, 0, 0)$, $(0, 2, 0)$, and $(0, 0, 6)$ also satisfy the equation.

If every ordered triple that satisfies the equation $2x + 3y + z = 6$ is graphed in a *three-dimensional* coordinate system, the result is a plane. It is difficult to think about graphing in three dimensions. It may be helpful to think of the origin as a corner of a room, where the x-axis is the intersection of one wall with the floor, the y-axis is the intersection of the adjacent wall with the floor, and the z-axis is the intersection of the two walls. The graph of $2x + 3y + z = 6$ in a three-dimensional coordinate system is the plane shown in Fig. 11.12.

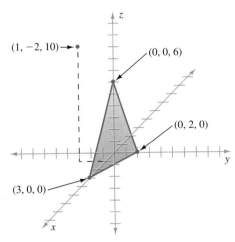

Figure 11.12

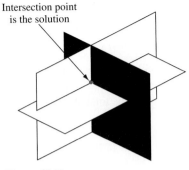

Figure 11.13

To solve a linear system of three equations by graphing would mean graphing three planes and then estimating the coordinates of the intersection point (see Fig. 11.13). Clearly this is impractical. Instead, we extend the algebraic methods discussed in Sec. 11.2 to solve linear systems of three equations in three variables.

Example 1 Solve the following system:

$$
\begin{array}{ll}
x + y + z = 4 & \text{First equation} \\
x - y + 3z = 6 & \text{Second equation} \\
5x + 4y + 2z = 9 & \text{Third equation}
\end{array}
$$

Solution: Add the first two equations to eliminate the variable y.

$$
\begin{array}{ll}
x + y + z = 4 & \text{First equation} \\
\underline{x - y + 3z = 6} & \text{Second equation} \\
2x + 4z = 10 & \text{Add.} \\
x + 2z = 5 & \text{Divide by 2.}
\end{array}
$$

To get another equation in x and z, multiply the second equation by 4 and add the result to the third equation.

$$
\begin{array}{ll}
4x - 4y + 12z = 24 & \text{Second equation multiplied by 4} \\
\underline{5x + 4y + 2z = 9} & \text{Third equation} \\
9x + 14z = 33 & \text{Add.}
\end{array}
$$

Use the two equations in color to form the following system of two equations in two variables.

$$
\begin{array}{r}
x + 2z = 5 \\
9x + 14z = 33
\end{array}
$$

Solve this system using any of the methods discussed in Sec. 11.2. For example, multiplying the first equation by -7 and adding the result to the second equation gives

$$
\begin{array}{r}
-7x - 14z = -35 \\
\underline{9x + 14z = 33} \\
2x = -2 \\
x = -1
\end{array}
$$

To get z, substitute $x = -1$ into the equation $x + 2z = 5$ (you could use $9x + 14z = 33$ instead).

$$-1 + 2z = 5$$
$$2z = 6$$
$$z = 3$$

To get y, substitute $x = -1$ and $z = 3$ into the equation $x + y + z = 4$ (you could use $x - y + 3z = 6$ or $5x + 4y + 2z = 9$ instead).

$$-1 + y + 3 = 4$$
$$y = 2$$

To check the solution, substitute $(x, y, z) = (-1, 2, 3)$ into all three original equations.

$$x + y + z = 4 \qquad\qquad x - y + 3z = 6$$
$$-1 + 2 + 3 = 4 \qquad\qquad -1 - 2 + 3(3) = 6$$
$$4 = 4 \quad \text{True} \qquad\qquad 6 = 6 \quad \text{True}$$

$$5x + 4y + 2z = 9$$
$$5(-1) + 4(2) + 2(3) = 9$$
$$9 = 9 \quad \text{True}$$

Since $(-1, 2, 3)$ satisfies all three equations, the solution set of the system is $\{(-1, 2, 3)\}$.   ◻

Try Exercise 13

Solve:
$$x + \ y + \ z = 6$$
$$x - \ y + 3z = 4$$
$$5x + 4y + 2z = 13$$

Example 2   Solve the following system:

$$2x + \ y - 3z = \ \ 4 \qquad \text{First equation}$$
$$5y - 4z = \ 12 \qquad \text{Second equation}$$
$$3x - \ y + \ z = -7 \qquad \text{Third equation}$$

Solution: The second equation contains only the variables y and z. To produce another equation in y and z, multiply the first equation by 3 and the third equation by -2 and add the resulting equations.

$$6x + 3y - \ 9z = 12 \qquad \text{First equation multiplied by 3}$$
$$\underline{-6x + 2y - \ 2z = 14} \qquad \text{Third equation multiplied by } -2$$
$$5y - 11z = 26 \qquad \text{Add.}$$

Use the two equations in color to produce the following system:

$$5y - \ 4z = 12$$
$$5y - 11z = 26$$

To solve this system, multiply the second equation by -1 and add.

$$5y - \ 4z = \ \ \ 12$$
$$\underline{-5y + 11z = -26}$$
$$7z = -14$$
$$z = \ -2$$

Substitute $z = -2$ into $5y - 4z = 12$.

$$5y - 4(-2) = 12$$
$$5y + 8 = 12$$
$$5y = 4$$
$$y = \tfrac{4}{5}$$

Substitute $y = \frac{4}{5}$ and $z = -2$ into $2x + y - 3z = 4$.

$$2x + \tfrac{4}{5} - 3(-2) = 4$$

$$2x + \tfrac{4}{5} + 6 = 4$$

$$2x + \tfrac{4}{5} = -2 \qquad \text{Subtract 6.}$$

$$10x + 4 = -10 \qquad \text{Multiply by 5.}$$

$$10x = -14$$

$$x = -\tfrac{7}{5}$$

Try Exercise 17

Solve:

$$
\begin{aligned}
2x + y - 3z &= 3 \\
5y - 2z &= 7 \\
3x - y + z &= -8
\end{aligned}
$$

The solution set is $\{(-\tfrac{7}{5}, \tfrac{4}{5}, -2)\}$. Check in the original system. ❏

A linear system of three equations is **inconsistent** when there is no point that is common to all three planes. This can occur in a variety of ways, as illustrated in Fig. 11.14.

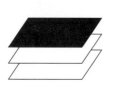 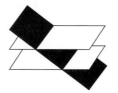  

(a) Three parallel planes   (b) Two parallel planes   (c) No parallel planes   (d) Two planes coincide, with third plane parallel

Figure 11.14 Inconsistent systems, no solution.

Example 3   Solve the following system:

$$
\begin{aligned}
x - 2y + 3z &= 1 \qquad \text{First equation} \\
-2x + 4y - 6z &= 3 \qquad \text{Second equation} \\
5x + y + z &= 8 \qquad \text{Third equation}
\end{aligned}
$$

Solution: Eliminate x by multiplying the first equation by 2 and adding the result to the second equation.

$$
\begin{aligned}
2x - 4y + 6z &= 2 \\
\underline{-2x + 4y - 6z} &= \underline{3} \\
0 &= 5
\end{aligned}
$$

Try Exercise 23

Solve:

$$
\begin{aligned}
x - 3y + 2z &= 1 \\
-2x + 6y - 4z &= 5 \\
5x + y + z &= 9
\end{aligned}
$$

The fact that we arrived at a contradiction means that the original system is inconsistent, and the solution set is $\varnothing$. ❏

Figure 11.14(b) illustrates what happened in Example 3. The first two equations of the original system graph into parallel planes. The third plane intersects each of the other two.

A linear system of three equations is **dependent** when there are an infinite number of points that are common to all three planes. This can occur in a variety of ways, as illustrated in Fig. 11.15.

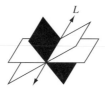  

(a) Three planes intersect   (b) Three planes coincide   (c) Two planes coincide, with
     along line L third plane intersecting

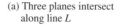 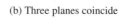

Figure 11.15 Dependent systems, infinite number of solutions.

Example 4   Solve the following system:

$$
\begin{array}{ll}
x + 4y - z = 0 & \text{First equation} \\
-x + 2y - z = 0 & \text{Second equation} \\
2x - y + z = 0 & \text{Third equation}
\end{array}
$$

Solution: Eliminate z by adding the first equation to the third.

$$
\begin{array}{l}
x + 4y - z = 0 \\
\underline{2x - y + z = 0} \\
3x + 3y \qquad = 0 \\
x + y \qquad = 0 \qquad \text{Divide by 3.}
\end{array}
$$

Eliminate z again by adding the second equation to the third.

$$
\begin{array}{l}
-x + 2y - z = 0 \\
\underline{2x - y + z = 0} \\
x + y \qquad = 0
\end{array}
$$

The two equations in color form the following system:

$$
\begin{array}{l}
x + y = 0 \\
x + y = 0
\end{array}
$$

If we multiply the second equation by -1 and add the result to the first equation, we obtain the identity $0 = 0$. This system is dependent, and there are an infinite number of solutions.   ❏

Try Exercise 27

Solve:

$$
\begin{array}{l}
x + 5y - z = 0 \\
-2x + 2y - z = 0 \\
3x - y + z = 0
\end{array}
$$

Figure 11.15(a) illustrates what happened in Example 4. The solutions of the system are represented by the points on line L. If we solve the equation $x + y = 0$ from Example 4 for y, we get $y = -x$. Substituting $-x$ for y in the third equation, $2x - y + z = 0$, and solving for z, gives $z = -3x$. Therefore the points on line L can be represented by ordered triples of the form $(x, -x, -3x)$. For example, three possible solutions of the system are $(1, -1, -3)$, $(4, -4, -12)$, and $(-2, 2, 6)$.

We can summarize our discussion of linear systems of three equations as follows:

To Solve a Linear System of Three Equations

1. Use any two equations to get an equation in two variables.
2. Use a different pair of equations to get another equation in the same two variables as the equation resulting from step 1.
3. Solve the system of two equations that results from steps 1 and 2.
4. Substitute the values found in step 3 into one of the original equations to find the value of the third variable.
5. Check your solution in all three original equations.

Note:

If at any time a contradiction is obtained, the system is inconsistent. If at any time an identity is obtained, the system is either dependent, or inconsistent in the manner of Fig. 11.14(d). If a system is inconsistent or dependent, we shall simply state that there is no unique solution.

The method for solving a linear system of three equations outlined above can be extended to solving a system of four or more equations.

We now illustrate an applied problem that can be solved using a system of three equations.

Problem Solving

Example 5 Tickets for a school concert were priced at $1, $3, and $5. A total of 263 tickets were sold. Twice as many $5 tickets were sold as $3 tickets. If the total receipts were $1023, how many of each kind of ticket were sold?

Solution: Let x = the number of $1 tickets sold, y = the number of $3 tickets sold, and z = the number of $5 tickets sold. Then set up the chart below:

	Number	Value
$1 tickets	x	x
$3 tickets	y	$3y$
$5 tickets	z	$5z$

Since a total of 263 tickets were sold, the first equation is

$$x + y + z = 263$$

Twice as many $5 tickets were sold as $3 tickets, so the second equation is

$$z = 2y$$

Since the total receipts were $1023, the third equation is

$$x + 3y + 5z = 1023$$

Solving this system gives $x = 35$, $y = 76$, and $z = 152$. Therefore 35 $1 tickets, 76 $3 tickets, and 152 $5 tickets were sold. ❏

Try Exercise 39

Tickets for a school play were priced at $1, $2, and $5. A total of 434 tickets were sold. Three times as many $5 tickets were sold as $2 tickets. If the total receipts were $1656, how many of each kind of ticket were sold?

Exercises 11.4

Completion Exercises

1. Three numbers written in a specified order, such as $(5, -2, 0)$, form a(n) _____ of numbers.

2. The graph of a linear equation in three variables, such as $x + 3y - 2z = 6$, in a three-dimensional coordinate system is a(n) _____.

3. Suppose you are solving a linear system of three equations and you arrive at the statement $0 = -2$. You can conclude that the system _____.

4. A linear system of three equations is _____ when there is no point common to all three planes; the system is _____ when there are an infinite number of points common to the three planes.

Determine whether the given ordered triple is a solution of the given system.

5. $(2, -1, 4)$
$$x + y - z = -3$$
$$2x - 3y + 4z = 23$$
$$5x - 2y - 6z = -16$$

6. $(3, 1, -5)$
$$x - y + z = -3$$
$$3x + 5y - 2z = 24$$
$$4x - 7y - 3z = -10$$

7. $(-5, \frac{2}{3}, 0)$
$$x - 3y + 8z = -7$$
$$2x + 15y - z = 0$$
$$-3x + 5z = 15$$

8. $(\frac{3}{4}, 0, -2)$
$$-4x + 9y + z = -5$$
$$16x - y + 6z = 0$$
$$8y - 5z = 10$$

Solve each system.

9. $2x - 3y - z = 2$
   $x + y + z = 3$
   $x + 2y \quad\;\; = 7$

10. $3x - y + z = 1$
    $x + 2y - z = 3$
    $x - y \quad\;\; = 6$

11. $-x + 4y + z = 3$
    $2x - 3y - z = 0$
    $x + 2y + z = 5$

12. $-x + 3y + z = 3$
    $2x - 2y - z = 0$
    $x + y + z = 5$

13. $x + y + z = 6$
    $x - y + 3z = 4$
    $5x + 4y + 2z = 13$

14. $x + y + z = 8$
    $x - y + 3z = 2$
    $5x + 4y + 2z = 21$

15. $2x - 3y + 4z = -2$
    $3x + y - 4z = -8$
    $x + 4y - 2z = -3$

16. $2x - y + 6z = -2$
    $3x - 2y - 6z = -8$
    $x + 3y - 3z = -3$

17. $2x + y - 3z = 3$
    $5y - 2z = 7$
    $3x - y + z = -8$

18. $3x + y - 2z = 2$
    $7y - 4z = 17$
    $4x - y + z = -10$

19. $4x - 3y = -13$
    $2y + 5z = -12$
    $6x - 7z = -10$

20. $3x - 5y = -5$
    $4y + 2z = -10$
    $7x - 9z = -26$

21. $3x - 4y - 6z = 12$
    $-\frac{1}{4}x + \frac{1}{3}y + \frac{1}{2}z = -1$
    $\frac{1}{2}x - \frac{2}{3}y - z = 2$

22. $4x + 5y - 10z = 20$
    $-\frac{1}{5}x - \frac{1}{4}y + \frac{1}{2}z = -1$
    $\frac{2}{5}x + \frac{1}{2}y - z = 2$

23. $x - 3y + 2z = 1$
    $-2x + 6y - 4z = 5$
    $5x + y + z = 9$

24. $x + 2y - 3z = 1$
    $-4x - 8y + 12z = 5$
    $5x + y + z = 8$

25. $2x + 2y - 3z = 0$
    $x - y + z = 0$
    $3x + 5y - 6z = 0$

26. $x + 3y - 2z = 0$
    $3x - y + z = 0$
    $2x + 4y - 3z = 0$

27. $x + 5y - z = 0$
    $-2x + 2y - z = 0$
    $3x - y + z = 0$

28. $x + 6y - z = 0$
    $-3x + 2y - z = 0$
    $4x - y + z = 0$

29. $z = 1 - x + 3y$
    $x + 2z = 8y - 1$
    $2x - 11y + 3z = 2$

30. $z = 2 - x + 2y$
    $x + 2z = 11y - 2$
    $2x - 13y + 3z = 4$

31. $x - y = 3$
    $x + z = 7$
    $y = 3z$

32. $y + z = 5$
    $x + z = 2$
    $y = 4x$

Discussion Exercises

33. Is it possible for three planes to have no points in common? Explain. Draw sketches to support your answer.

34. Is it possible for three planes to have one and only one point in common? An infinite number of points in common? Explain. Draw sketches to support your answer.

35. Suppose you are solving a linear system of three equations and you arrive at the statement $6 = 6$. What conclusion can you make?

36. Outline the steps in solving a linear system of three equations.

Problem Solving

Write a system of three equations for each exercise. Then solve the system.

37. The sum of three numbers is 4. The first plus the second minus the third is -10. Twice the first plus the sum of the other two is 0. Find the numbers.

38. The sum of three numbers is 9. Three times the first plus the difference of the other two is 7. The third is twice the sum of the other two. Find the numbers.

39. Tickets for a school play were priced at $1, $2, and $5. A total of 434 tickets were sold. Three times as many $5 tickets were sold as $2 tickets. If the total receipts were $1656, how many of each kind of ticket were sold?

40. One muffin, two pies, and three cakes cost $23. One muffin, three pies, and two cakes cost $21. One muffin, four pies, and five cakes cost $39. Find the cost of each.

41. The perimeter of a rectangle is twice the perimeter of an equilateral triangle. The rectangle's width plus its length plus a side of the triangle is 100 cm. Four times the rectangle's width equals its length minus one side of the triangle. Find each dimension.

42. The sum of the angles of a triangle is 180°. The largest angle is 12° more than the sum of the other two. The smallest angle is 26° less than the middle angle. Find the angles.

43. The average of Jennifer's three test scores is 76. She scored 4 points better on the second test than the first, and 7 points better on the third test than the second. What was her score on each test?

44. A collection of 129 nickels, dimes, and quarters has a total value of $13.35. If there are 6 times as many nickels as dimes, how many of each kind of coin is in the collection?

45. Larry, Curly, and Moe each have some bananas. If Larry gives Curly a banana, Larry and Curly will have the same number. If Larry gives Moe a banana, Moe will have twice as many as Curly. If Larry eats four of Moe's bananas, Moe will still have three more bananas than Larry. How many bananas does each have?

46. The sum of the digits of a three-digit number is 16. The tens digit is 3 more than the units digit. If the order of the digits is reversed and the new number is subtracted from the original number, the result is 99. Find the original number.

Solve each system.

47.
$$\begin{aligned}
x + y - z - w &= 5 \\
x - 2y + 3z + w &= -12 \\
-x + y - 4z + w &= 15 \\
5x - y + 6z - 2w &= -18
\end{aligned}$$

48.
$$\begin{aligned}
2x - y + z - 3w &= -7 \\
-x + 8y - z + 5w &= 7 \\
x - 4y - 2z - w &= -20 \\
-6x + 3y - z - 4w &= 9
\end{aligned}$$

Getting Acquainted with Your Graphing Calculator

Determine whether the given ordered triple is a solution of the given system by storing the appropriate value in each of the variables x, y, and z, and then calculating the value of the left side of each equation.

49. $(3, -6, -2)$
$$\begin{aligned}
2x - 5y + 4z &= 28 \\
x + 2y - 9z &= 9 \\
-3x - y + 7z &= -17
\end{aligned}$$

50. $(5, 4, -1)$
$$\begin{aligned}
x + 2y - z &= 14 \\
-x - 3y + 4z &= -21 \\
-2x + y - 6z &= 10
\end{aligned}$$

11.5 Determinants

Another method for solving a linear system of equations utilizes a mathematical device known as a *determinant*. In this section we define a determinant and learn how to calculate its value.*

Second-Order Determinants

A **determinant** is a square array of numbers enclosed between two vertical lines. The numbers are called the **elements** of the determinant. For example, in the determinant

Two columns

$$\begin{vmatrix} 7 & 2 \\ 5 & 3 \end{vmatrix}$$ Two rows

the numbers 7, 2, 5, and 3 are the elements. Since this determinant has two rows and two columns, it is called a **2 × 2** determinant (read "2 by 2 determinant"). It is also called a **second-order** determinant.

We define the *value* of a 2 × 2 determinant as follows.

> **DEFINITION**
>
> The **value** of a 2 × 2 determinant is calculated as follows:
> $$\begin{vmatrix} a_1 & b_1 \\ a_2 & b_2 \end{vmatrix} = a_1 b_2 - a_2 b_1$$

---

*HISTORICAL NOTE: The theory of determinants was introduced by the German mathematician Gottfried Leibniz (1646–1716). Leibniz, who also studied law, science, and philosophy, is regarded by many as the last scholar to achieve universal knowledge. His greatest achievement was the discovery of calculus, independent of Newton. Leibniz coined the word "transcendental," and was well known for his mathematical notation. He was the first to use a dot for multiplication, he was the first to use ≅ for "is congruent to," and he was the first to use the calculus notation dx and dy.

Example 1   Evaluate each determinant.

$$\text{(a)} \begin{vmatrix} 7 & 2 \\ 5 & 3 \end{vmatrix} \qquad \text{(b)} \begin{vmatrix} 6 & -1 \\ -4 & -2 \end{vmatrix}$$

Solution: Multiply and subtract as shown below:

(a) $\begin{vmatrix} 7 & 2 \\ 5 & 3 \end{vmatrix} = 7 \cdot 3 - 5 \cdot 2 = 21 - 10 = 11$

(b) $\begin{vmatrix} 6 & -1 \\ -4 & -2 \end{vmatrix} = 6(-2) - (-4)(-1) = -12 - 4 = -16$

Try Exercise 5

Evaluate:

$\begin{vmatrix} 5 & 4 \\ 2 & 3 \end{vmatrix}$

Third-Order Determinants

To calculate the value of a 3×3 determinant, we use what are called *minors*. The **minor** of an element in a 3×3 determinant is the 2×2 determinant that results when both the row and the column that contain that element are deleted. Applying this definition, we obtain the following minors:

$$\text{Minor of } a_1 = \begin{vmatrix} a_1 & b_1 & c_1 \\ a_2 & b_2 & c_2 \\ a_3 & b_3 & c_3 \end{vmatrix} = \begin{vmatrix} b_2 & c_2 \\ b_3 & c_3 \end{vmatrix}$$

$$\text{Minor of } a_2 = \begin{vmatrix} a_1 & b_1 & c_1 \\ a_2 & b_2 & c_2 \\ a_3 & b_3 & c_3 \end{vmatrix} = \begin{vmatrix} b_1 & c_1 \\ b_3 & c_3 \end{vmatrix}$$

$$\text{Minor of } b_2 = \begin{vmatrix} a_1 & b_1 & c_1 \\ a_2 & b_2 & c_2 \\ a_3 & b_3 & c_3 \end{vmatrix} = \begin{vmatrix} a_1 & c_1 \\ a_3 & c_3 \end{vmatrix}$$

Example 2   Determine the minor of 5 in the 3×3 determinant below:

$$\begin{vmatrix} 1 & 2 & 3 \\ 4 & 5 & 6 \\ 7 & 8 & 9 \end{vmatrix}$$

Solution:

$$\text{Minor of } 5 = \begin{vmatrix} 1 & 2 & 3 \\ 4 & 5 & 6 \\ 7 & 8 & 9 \end{vmatrix} = \begin{vmatrix} 1 & 3 \\ 7 & 9 \end{vmatrix}$$

Try Exercise 25

Determine the minor of 1 in

$\begin{vmatrix} 1 & 2 & 3 \\ 4 & 5 & 6 \\ 7 & 8 & 9 \end{vmatrix}$

We are now ready to define the *value* of a 3×3 determinant.

DEFINITION

The **value** of a 3×3 determinant can be calculated as follows:

$$\begin{vmatrix} a_1 & b_1 & c_1 \\ a_2 & b_2 & c_2 \\ a_3 & b_3 & c_3 \end{vmatrix} = a_1(\text{minor of } a_1) - a_2(\text{minor of } a_2) + a_3(\text{minor of } a_3)$$

$$= a_1 \begin{vmatrix} b_2 & c_2 \\ b_3 & c_3 \end{vmatrix} - a_2 \begin{vmatrix} b_1 & c_1 \\ b_3 & c_3 \end{vmatrix} + a_3 \begin{vmatrix} b_1 & c_1 \\ b_2 & c_2 \end{vmatrix}$$

The method of calculating the value of a third-order determinant outlined above is called **expansion by minors** about the first column.

Example 3   Evaluate:

$$\begin{vmatrix} 3 & -1 & -2 \\ 5 & 0 & -4 \\ 2 & -3 & 6 \end{vmatrix}$$

Solution: Expand by minors about the first column.

$$\begin{vmatrix} 3 & -1 & -2 \\ 5 & 0 & -4 \\ 2 & -3 & 6 \end{vmatrix}$$

$$= 3\begin{vmatrix} 0 & -4 \\ -3 & 6 \end{vmatrix} - 5\begin{vmatrix} -1 & -2 \\ -3 & 6 \end{vmatrix} + 2\begin{vmatrix} -1 & -2 \\ 0 & -4 \end{vmatrix}$$

$$= 3[0 \cdot 6 - (-3)(-4)] - 5[(-1)6 - (-3)(-2)] + 2[(-1)(-4) - 0(-2)]$$

$$= 3[-12] - 5[-12] + 2[4]$$

$$= -36 + 60 + 8$$

$$= 32$$ ❒

Try Exercise 35

Evaluate by expanding by minors about the first column.

$$\begin{vmatrix} 4 & -6 & -1 \\ 3 & 0 & -2 \\ 1 & -3 & 2 \end{vmatrix}$$

Actually, we can calculate the value of 3 × 3 determinant using expansion by minors about *any* column or *any* row. When expanding about a particular row or column, prefix the terms of the expansion with the signs from the corresponding row or column in the **sign array** below:

$$\begin{vmatrix} + & - & + \\ - & + & - \\ + & - & + \end{vmatrix}$$

This sign array is easy to remember. Simply start with a + in the first-row, first-column position (upper left corner), and alternate signs as you travel along any row or column.

Example 4   Evaluate the determinant of Example 3 using expansion by minors about the second row.

Solution: Since we are expanding about the second row, use the signs given in the second row of the sign array.

$$\begin{vmatrix} 3 & -1 & -2 \\ 5 & 0 & -4 \\ 2 & -3 & 6 \end{vmatrix} = -5\begin{vmatrix} -1 & -2 \\ -3 & 6 \end{vmatrix} + 0\begin{vmatrix} 3 & -2 \\ 2 & 6 \end{vmatrix} - (-4)\begin{vmatrix} 3 & -1 \\ 2 & -3 \end{vmatrix}$$

Signs from sign array

$$= -5(-6 - 6) + 0 + 4(-9 + 2)$$

$$= 60 + 0 - 28$$

$$= 32$$

This is the same value we obtained in Example 3. ❒

Try Exercise 37

Evaluate by expanding by minors about the second row.

$$\begin{vmatrix} 4 & -6 & -1 \\ 3 & 0 & -2 \\ 1 & -3 & 2 \end{vmatrix}$$

In summary, we evaluate a 3 × 3 determinant as follows.

To Evaluate a 3 × 3 Determinant Using Expansion by Minors

1. Choose a row or column to expand about. Choose the row or column with the most zeros to simplify your calculations.
2. Multiply each element in the row or column chosen in step 1 by its minor.
3. Prefix the terms in step 2 with the signs from the corresponding row or column in the sign array.

The method of expansion by minors can be extended to find the value of a determinant of any order. For example, the sign array for a 4 × 4 determinant would be

$$\begin{vmatrix} + & - & + & - \\ - & + & - & + \\ + & - & + & - \\ - & + & - & + \end{vmatrix}$$

Exercises 11.5

Completion Exercises

1. A square array of numbers enclosed between two vertical lines is called a(n) _____.

2. The individual numbers in a determinant are called the _____ of the determinant.

3. A determinant that has three rows and three columns is called a(n) _____ determinant, or a(n) _____ order determinant.

4. The sign array for a 3 × 3 determinant is _____.

Evaluate each 2 × 2 determinant.

5. $\begin{vmatrix} 5 & 4 \\ 2 & 3 \end{vmatrix}$

6. $\begin{vmatrix} 8 & 2 \\ 5 & 3 \end{vmatrix}$

7. $\begin{vmatrix} 2 & 9 \\ 3 & 6 \end{vmatrix}$

8. $\begin{vmatrix} 4 & 6 \\ 3 & 2 \end{vmatrix}$

9. $\begin{vmatrix} 9 & 6 \\ 3 & 2 \end{vmatrix}$

10. $\begin{vmatrix} 6 & 3 \\ 4 & 2 \end{vmatrix}$

11. $\begin{vmatrix} -10 & \frac{1}{2} \\ 8 & -2 \end{vmatrix}$

12. $\begin{vmatrix} -3 & \frac{1}{3} \\ 9 & -10 \end{vmatrix}$

13. $\begin{vmatrix} 7 & -4 \\ -1 & -3 \end{vmatrix}$

14. $\begin{vmatrix} 8 & -1 \\ -4 & -2 \end{vmatrix}$

15. $\begin{vmatrix} -2 & -2 \\ 5 & 6 \end{vmatrix}$

16. $\begin{vmatrix} -4 & -4 \\ 3 & 5 \end{vmatrix}$

17. $\begin{vmatrix} \frac{1}{5} & \frac{2}{5} \\ -\frac{1}{5} & \frac{3}{5} \end{vmatrix}$

18. $\begin{vmatrix} \frac{4}{5} & \frac{2}{5} \\ -\frac{4}{5} & \frac{3}{5} \end{vmatrix}$

19. $\begin{vmatrix} 1 & 0 \\ 0 & 1 \end{vmatrix}$

20. $\begin{vmatrix} 0 & 1 \\ 1 & 0 \end{vmatrix}$

21. $\begin{vmatrix} 0 & 1 \\ 0 & 1 \end{vmatrix}$

22. $\begin{vmatrix} 0 & 0 \\ 1 & 1 \end{vmatrix}$

23. $\begin{vmatrix} x & y \\ y & x \end{vmatrix}$

24. $\begin{vmatrix} y & x \\ x & y \end{vmatrix}$

Using the determinant below, state the minor of each element given in Exercises 25 through 32.

$$\begin{vmatrix} 1 & 2 & 3 \\ 4 & 5 & 6 \\ 7 & 8 & 9 \end{vmatrix}$$

25. 1

26. 2

27. 3

28. 6

29. 4

30. 7

31. 8

32. 9

Discussion Exercises

33. Explain how to calculate the value of a 2 × 2 determinant.

34. Explain how to find the minor of an element in a 3 × 3 determinant.

Evaluate each 3 × 3 determinant using expansion by minors about the first column.

35. $\begin{vmatrix} 4 & -6 & -1 \\ 3 & 0 & -2 \\ 1 & -3 & 2 \end{vmatrix}$

36. $\begin{vmatrix} 3 & -1 & -2 \\ 5 & 0 & -4 \\ 2 & -3 & 6 \end{vmatrix}$

37. Evaluate the determinant of Exercise 35 using expansion by minors about the second row.

38. Evaluate the determinant of Exercise 36 using expansion by minors about the second row.

Evaluate each 3 × 3 determinant. Expand about the row or column of your choice.

39. $\begin{vmatrix} 0 & 0 & 4 \\ 4 & 7 & 8 \\ 2 & 6 & -5 \end{vmatrix}$

40. $\begin{vmatrix} 0 & 0 & 5 \\ 5 & 9 & 4 \\ 4 & 8 & -6 \end{vmatrix}$

41. $\begin{vmatrix} -3 & 2 & 1 \\ -1 & 4 & 0 \\ 5 & -2 & 1 \end{vmatrix}$

42. $\begin{vmatrix} -3 & 2 & 1 \\ -2 & 8 & 0 \\ 4 & -1 & 1 \end{vmatrix}$

43. $\begin{vmatrix} -1 & 2 & -1 \\ -3 & 1 & -3 \\ -6 & -2 & 5 \end{vmatrix}$

44. $\begin{vmatrix} -5 & 1 & 2 \\ -3 & 1 & -1 \\ 6 & -2 & -2 \end{vmatrix}$

45. $\begin{vmatrix} 3 & 1 & -1 \\ 2 & -1 & -3 \\ 6 & 2 & -2 \end{vmatrix}$

46. $\begin{vmatrix} 2 & 1 & -1 \\ 3 & -1 & -4 \\ 6 & 3 & -3 \end{vmatrix}$

47. $\begin{vmatrix} 1 & 0 & 0 \\ 0 & 1 & 0 \\ 0 & 0 & 1 \end{vmatrix}$

48. $\begin{vmatrix} 0 & 0 & 1 \\ 0 & 1 & 0 \\ 1 & 0 & 0 \end{vmatrix}$

49. $\begin{vmatrix} a & 1 & 1 \\ b & 2 & 1 \\ c & 3 & 0 \end{vmatrix}$

50. $\begin{vmatrix} a & b & c \\ 1 & 2 & 3 \\ 1 & 1 & 0 \end{vmatrix}$

Discussion Exercises

51. Explain how to calculate the value of a 3×3 determinant using expansion by minors.

52. How would you construct a sign array for a 5×5 determinant?

Solve each determinant equation for x.

53. $\begin{vmatrix} x & 1 \\ 2x & 3 \end{vmatrix} = 6$

54. $\begin{vmatrix} x & 1 \\ 3x & 4 \end{vmatrix} = 8$

55. $\begin{vmatrix} x+1 & -4 \\ x-2 & -5 \end{vmatrix} = x - 3$

56. $\begin{vmatrix} x-1 & -7 \\ x+2 & -4 \end{vmatrix} = x + 4$

57. $\begin{vmatrix} x & 2 & -7 \\ 1 & 0 & 2 \\ 3 & x & -1 \end{vmatrix} = 10$

58. $\begin{vmatrix} x & 3 & -5 \\ 1 & 0 & 2 \\ 4 & x & -1 \end{vmatrix} = 24$

59. Evaluate each determinant. Can you draw a general conclusion?

(a) $\begin{vmatrix} 0 & 0 & 0 \\ a & b & c \\ x & y & z \end{vmatrix}$ (b) $\begin{vmatrix} a & b & c \\ 0 & 0 & 0 \\ x & y & z \end{vmatrix}$ (c) $\begin{vmatrix} a & b & c \\ x & y & z \\ 0 & 0 & 0 \end{vmatrix}$

60. Evaluate each determinant. Can you draw a general conclusion?

(a) $\begin{vmatrix} 0 & a & x \\ 0 & b & y \\ 0 & c & z \end{vmatrix}$ (b) $\begin{vmatrix} a & 0 & x \\ b & 0 & y \\ c & 0 & z \end{vmatrix}$ (c) $\begin{vmatrix} a & x & 0 \\ b & y & 0 \\ c & z & 0 \end{vmatrix}$

Evaluate each 4×4 determinant.

61. $\begin{vmatrix} 3 & -1 & 1 & 2 \\ 1 & 0 & 4 & 3 \\ -2 & 1 & -5 & 0 \\ -1 & 4 & 2 & 1 \end{vmatrix}$

62. $\begin{vmatrix} -1 & 3 & 0 & 1 \\ 2 & 1 & 4 & -2 \\ -3 & 1 & 2 & -1 \\ 1 & 0 & 5 & 3 \end{vmatrix}$

Getting Acquainted with Your Graphing Calculator

Use your calculator to evaluate each determinant.

63. $\begin{vmatrix} 5 & -3 \\ 2 & 4 \end{vmatrix}$

64. $\begin{vmatrix} 6 & -1 & 7 \\ 0 & 4 & -8 \\ -2 & 9 & 5 \end{vmatrix}$

65. $\begin{vmatrix} -2 & -4 & 6 & 11 \\ 10 & 5 & -1 & 9 \\ -8 & 23 & 7 & -3 \\ 31 & -6 & 4 & 13 \end{vmatrix}$

11.6 Cramer's Rule

In this section we demonstrate how to use determinants to solve a linear system of equations.

Systems of Two Equations

Consider the general system below:

$$a_1 x + b_1 y = c_1 \qquad \text{First equation}$$
$$a_2 x + b_2 y = c_2 \qquad \text{Second equation}$$

The numbers a_1, b_1, c_1 and a_2, b_2, c_2 are constants. We use the addition method to solve this system for x and y. To eliminate y, multiply the first equation by b_2 and the second equation by $-b_1$. Then add the resulting equations and solve for x.

$$a_1 b_2 x + b_1 b_2 y = c_1 b_2 \qquad \text{First equation multiplied by } b_2$$
$$\underline{-a_2 b_1 x - b_1 b_2 y = -c_2 b_1} \qquad \text{Second equation multiplied by } -b_1$$
$$(a_1 b_2 - a_2 b_1) x = c_1 b_2 - c_2 b_1 \qquad \text{Add.}$$
$$x = \frac{c_1 b_2 - c_2 b_1}{a_1 b_2 - a_2 b_1} \qquad \text{Divide by } a_1 b_2 - a_2 b_1.$$

To eliminate x, multiply the first equation by $-a_2$ and the second equation by a_1. Then add the resulting equations and solve for y.

$$-a_1a_2x - a_2b_1y = -a_2c_1 \qquad \text{First equation multiplied by } -a_2$$
$$\underline{a_1a_2x + a_1b_2y = \quad a_1c_2} \qquad \text{Second equation multiplied by } a_1$$
$$(a_1b_2 - a_2b_1)y = a_1c_2 - a_2c_1 \qquad \text{Add.}$$
$$y = \frac{a_1c_2 - a_2c_1}{a_1b_2 - a_2b_1} \qquad \text{Divide by } a_1b_2 - a_2b_1.$$

Note that the values of x and y are fractions with the same denominator. We can express this common denominator and the two numerators using determinants.

$$x = \frac{\begin{vmatrix} c_1 & b_1 \\ c_2 & b_2 \end{vmatrix}}{\begin{vmatrix} a_1 & b_1 \\ a_2 & b_2 \end{vmatrix}} \qquad y = \frac{\begin{vmatrix} a_1 & c_1 \\ a_2 & c_2 \end{vmatrix}}{\begin{vmatrix} a_1 & b_1 \\ a_2 & b_2 \end{vmatrix}}$$

For convenience, we denote the three determinants as shown below:

$$\begin{vmatrix} a_1 & b_1 \\ a_2 & b_2 \end{vmatrix} = D \qquad \begin{vmatrix} c_1 & b_1 \\ c_2 & b_2 \end{vmatrix} = D_x \qquad \begin{vmatrix} a_1 & c_1 \\ a_2 & c_2 \end{vmatrix} = D_y$$

The elements of D consist of the coefficients of x and y in the original system. To form D_x from D, replace the coefficients of x (a_1 and a_2) with the constant terms (c_1 and c_2). To form D_y from D, replace the coefficients of y with the constant terms.

This method for solving a linear system is called **Cramer's rule*** and is summarized below.

Cramer's Rule for a Linear System of Two Equations

The solution of the system

$$a_1x + b_1y = c_1$$
$$a_2x + b_2y = c_2$$

is given by

$$x = \frac{D_x}{D} = \frac{\begin{vmatrix} c_1 & b_1 \\ c_2 & b_2 \end{vmatrix}}{\begin{vmatrix} a_1 & b_1 \\ a_2 & b_2 \end{vmatrix}} \qquad \text{and} \qquad y = \frac{D_y}{D} = \frac{\begin{vmatrix} a_1 & c_1 \\ a_2 & c_2 \end{vmatrix}}{\begin{vmatrix} a_1 & b_1 \\ a_2 & b_2 \end{vmatrix}}$$

so long as $D \neq 0$.

Example 1 Use Cramer's rule to solve the following system:

$$6x - 2y = \quad 7$$
$$3x + 4y = -9$$

---

*HISTORICAL NOTE: Cramer's rule was known to the British mathematician Colin Maclaurin (1698–1746) as early as 1729, but it was first published by the Swiss mathematician Gabriel Cramer (1704–1752) in 1750, so the rule bears Cramer's name.

Solution: Begin by finding D, D_x, and D_y.

$$D = \begin{vmatrix} 6 & -2 \\ 3 & 4 \end{vmatrix} = 6 \cdot 4 - 3(-2) = 30$$

$$D_x = \begin{vmatrix} 7 & -2 \\ -9 & 4 \end{vmatrix} = 7 \cdot 4 - (-9)(-2) = 10$$

$$D_y = \begin{vmatrix} 6 & 7 \\ 3 & -9 \end{vmatrix} = 6(-9) - 3 \cdot 7 = -75$$

Then apply Cramer's rule.

$$x = \frac{D_x}{D} = \frac{10}{30} = \frac{1}{3} \quad \text{and} \quad y = \frac{D_y}{D} = \frac{-75}{30} = -\frac{5}{2}$$

Therefore the solution set is $\{(\frac{1}{3}, -\frac{5}{2})\}$. Check in the original system.   ❏

Try Exercise 7

Use Cramer's rule to solve:

$$6x - 2y = 9$$
$$3x + 4y = -13$$

<u>CAUTION</u>

Always calculate D before calculating D_x or D_y. If $D = 0$, then Cramer's rule does not apply and the system is inconsistent (no solution), or dependent (infinite number of solutions). In either case, there is no unique solution. Use either the addition method or the substitution method to solve the system and determine whether there is no solution or an infinite number of solutions.   ■

Example 2   Use Cramer's rule to solve the following system:

$$x = \frac{1}{3}y - \frac{2}{3}$$
$$y = 3x + \frac{5}{2}$$

Solution: Multiply the first equation by 3 and the second equation by 2 to clear fractions.

$$3x = y - 2$$
$$2y = 6x + 5$$

Then write the system in standard form.

$$3x - y = -2$$
$$-6x + 2y = 5$$

Calculate D before calculating D_x or D_y.

$$D = \begin{vmatrix} 3 & -1 \\ -6 & 2 \end{vmatrix} = 3 \cdot 2 - (-6)(-1) = 0$$

Since $D = 0$, Cramer's rule does not apply. Solving by addition or substitution, we find that the lines are parallel. Therefore, the solution set is $\varnothing$.   ❏

Try Exercise 11

Use Cramer's rule to solve:

$$x = \frac{1}{2}y - 3$$
$$y = 2x + \frac{4}{3}$$

Systems of Three Equations

Cramer's rule can be extended to a linear system of three equations as follows.

Cramer's Rule for a Linear System of Three Equations
The solution of the system

$$a_1x + b_1y + c_1z = d_1$$
$$a_2x + b_2y + c_2z = d_2$$
$$a_3x + b_3y + c_3z = d_3$$

is given by

$$x = \frac{D_x}{D} \qquad y = \frac{D_y}{D} \qquad z = \frac{D_z}{D}$$

where

$$D = \begin{vmatrix} a_1 & b_1 & c_1 \\ a_2 & b_2 & c_2 \\ a_3 & b_3 & c_3 \end{vmatrix} \qquad D_x = \begin{vmatrix} d_1 & b_1 & c_1 \\ d_2 & b_2 & c_2 \\ d_3 & b_3 & c_3 \end{vmatrix}$$

$$D_y = \begin{vmatrix} a_1 & d_1 & c_1 \\ a_2 & d_2 & c_2 \\ a_3 & d_3 & c_3 \end{vmatrix} \qquad D_z = \begin{vmatrix} a_1 & b_1 & d_1 \\ a_2 & b_2 & d_2 \\ a_3 & b_3 & d_3 \end{vmatrix}$$

so long as $D \neq 0$.

As before, if $D = 0$, then Cramer's rule does not apply and the system does not have a unique solution.

Example 3   Use Cramer's rule to solve the following system:

$$x + y + z = 2$$
$$x + y - z = 0$$
$$2x - 3y - 5z = 7$$

Solution: Calculate D, D_x, D_y, and D_z and obtain the following results:

$$D = \begin{vmatrix} 1 & 1 & 1 \\ 1 & 1 & -1 \\ 2 & -3 & -5 \end{vmatrix} = -10 \qquad D_x = \begin{vmatrix} 2 & 1 & 1 \\ 0 & 1 & -1 \\ 7 & -3 & -5 \end{vmatrix} = -30$$

$$D_y = \begin{vmatrix} 1 & 2 & 1 \\ 1 & 0 & -1 \\ 2 & 7 & -5 \end{vmatrix} = 20 \qquad D_z = \begin{vmatrix} 1 & 1 & 2 \\ 1 & 1 & 0 \\ 2 & -3 & 7 \end{vmatrix} = -10$$

Apply Cramer's rule.

$$x = \frac{D_x}{D} = \frac{-30}{-10} = 3$$

$$y = \frac{D_y}{D} = \frac{20}{-10} = -2$$

$$z = \frac{D_z}{D} = \frac{-10}{-10} = 1$$

Try Exercise 15

Use Cramer's rule to solve:

$$x + y + z = 4$$
$$x + y - z = 0$$
$$3x - 4y - 5z = 3$$

The solution set is $\{(3, -2, 1)\}$. Check in the original system.   ❐

Cramer's rule can be extended to solve linear systems of any number of equations.

Exercises 11.6

Completion Exercises

1. By Cramer's rule, the solution of the system

$$a_1x + b_1y = c_1$$
$$a_2x + b_2y = c_2$$

is given by $x = \dfrac{D_x}{D}$ and $y = \dfrac{D_y}{D}$, where $D = $ _____,

$D_x = $ _____, and $D_y = $ _____, so long as $D \neq 0$.

2. By Cramer's rule, the solution of the system

$$a_1x + b_1y + c_1z = d_1$$
$$a_2x + b_2y + c_2z = d_2$$
$$a_3x + b_3y + c_3z = d_3$$

is given by $x = \dfrac{D_x}{D}$, $y = \dfrac{D_y}{D}$, and $z = \dfrac{D_z}{D}$, where

$D = $ _____, $D_x = $ _____, $D_y = $ _____, and

$D_z = $ _____, so long as $D \neq 0$.

Use Cramer's rule to solve each system.

3. $2x + 3y = 8$
    $5x + 4y = 13$

4. $4x + 3y = 11$
    $2x + 5y = 9$

5. $x + y = 1$
    $3x + 4y = -2$

6. $x + y = 1$
    $5x + 6y = -2$

7. $6x - 2y = 9$
    $3x + 4y = -13$

8. $8x - 3y = 7$
    $4x + 6y = -9$

9. $7x - y = 21$
    $-2x + 5y = -6$

10. $7x - y = 28$
     $-3x + 4y = -12$

11. $x = \frac{1}{2}y - 3$
     $y = 2x + \frac{4}{3}$

12. $x = \frac{1}{4}y - 2$
     $y = 4x + \frac{5}{2}$

13. $4x + 7y = 16$
     $x = -3$

14. $y = -2$
     $8x + 5y = 14$

Use Cramer's rule to solve each system.

15. $x + y + z = 4$
     $x + y - z = 0$
     $3x - 4y - 5z = 3$

16. $x + y + z = 6$
     $x + y - z = 0$
     $2x - 4y - 3z = 3$

17. $x - 3y + z = 2$
     $3x - y + 2z = 2$
     $2x + y + z = 0$

18. $x - 2y + z = 4$
     $4x - y + 2z = 4$
     $3x + y + z = 0$

19. $x + y = -1$
     $2y + 8z = -6$
     $2x - 4z = 7$

20. $x + y = -1$
     $2y + 5z = -8$
     $2x - 10z = 4$

21. $5x - 2z + 2 = 0$
     $x - 2y + 3z = 3$
     $3y + z - 1 = 0$

22. $4x - 3z + 3 = 0$
     $x - 2y + 2z = 2$
     $5y + z - 1 = 0$

23. $x - 2y + z = 1$
     $2x - 4y + 3z = 2$
     $-x + 2y - 3z = 3$

24. $x - 3y + z = 1$
     $2x - 6y + 2z = 3$
     $-x + 3y - 2z = 4$

25. $-x + y - 2z = 0$
     $2x - 3y + z = 0$
     $4x - y - 3z = 0$

26. $-x + 2y - z = 0$
     $3x - y + 4z = 0$
     $5x - 3y - z = 0$

Discussion Exercises

27. Describe how to write D, D_x, D_y, and D_z for a linear system of three equations.

28. Why should you calculate D before you calculate D_x or D_y or D_z?

29. If you are using Cramer's rule to solve a linear system of two equations and find that $D = 0$, what should you do?

30. How do you think Cramer's rule would be stated for a linear system of four equations?

Use Cramer's rule to solve each system for x and y. Assume a and b represent nonzero real constants.

31. $ax - y = b$
     $x + by = a$

32. $bx - ay = 1$
     $ax + by = 1$

Use Cramer's rule to solve each system.

33. $x + 3y - z + w = 4$
     $2x - y + 5z - w = 0$
     $-x + y + 4z + 2w = 7$
     $3x + 2y + z - w = -1$

34. $3x - y + z - 2w = -6$
     $-y - 4z - w = 3$
     $-x + y + 3w = 8$
     $2x - 2y + z + w = -4$

Getting Acquainted with Your Graphing Calculator

Use Cramer's rule to solve each system. Use your calculator to evaluate the determinants involved.

35. $x + 2y = 3$
    $4x + 5y = 6$

36. $2x - 3y + z = 8$
    $4x + y - 2z = 9$
    $-3x - 5y + 6z = 12$

37. $x + y - 2z + 3w = 18$
    $5x - y + 4z - w = 6$
    $y - z + w = 10$
    $3x + 6y + 9z = 9$

11.7 Solving Linear Systems by Matrices

TAPE AU24

In this section we learn to solve a linear system using *matrices*.* The advantages of this method are that it is systematic, and that it is easily adapted to a computer.

Definition of a Matrix

A **matrix** is a rectangular array of numbers. We usually enclose a matrix in square brackets. Three examples of matrices are shown below:

Three columns

$$\text{Two rows} \begin{bmatrix} 5 & -3 & 6 \\ 0 & -2 & 1 \end{bmatrix} \quad \begin{bmatrix} 4 & 7 \\ 5 & 2 \end{bmatrix} \quad \begin{bmatrix} -1 & 3 & 8 & 11 \\ 6 & 0 & 9 & -5 \\ 1 & 1 & -4 & 7 \end{bmatrix}$$

Since the first matrix has two rows and three columns, it is called a **2 × 3** (read "2 by 3") matrix. The numbers 5, −3, 6, 0, −2, and 1 are called the **elements** of the matrix. The second matrix is a 2 × 2 matrix, and the third matrix is a 3 × 4 matrix. Since the matrix in the middle has the same number of rows as columns, it is called a **square matrix.** If a matrix is square, the diagonal from its upper left corner to its lower right corner (formed by the numbers 4 and 2 in the square matrix above) is called the **main diagonal** of the matrix.

Solving a Linear System of Two Equations by Matrices

We can represent the system of equations

$$3x - y = 17$$
$$x + 2y = -6$$

by the matrix

$$\begin{bmatrix} 3 & -1 & \vdots & 17 \\ 1 & 2 & \vdots & -6 \end{bmatrix}$$

This last matrix is called the **augmented matrix** of the system. The dashed vertical line (which is optional) separates the coefficients of the variables from the constant terms of the system. The rectangular array of numbers to the left of this vertical line is called the

---

*HISTORICAL NOTE: Matrices were developed in 1857 by the English mathematician Arthur Cayley (1821–1895). Cayley was probably the third most prolific writer of mathematics in history, behind only the Swiss mathematician Leonhard Euler (1707–1783) and the French mathematician Augustin-Louis Cauchy (1789–1857). Cayley, who was also a lawyer, thought that he had discovered "an elegant algebraic system that could not possibly have any application." He could not have been more wrong. There are few areas of mathematics today that have more applications than matrices.

coefficient matrix. Since the augmented matrix is just a simple way to write the system of equations, we can perform operations on the rows of the matrix just as if they were equations.

Row Operations on an Augmented Matrix

Type 1: We can interchange any two rows.
Type 2: We can multiply any row by any nonzero real number.
Type 3: We can multiply any row by a real number and add the result to another row.

Note that a Type 1 row operation corresponds to interchanging any two equations of the system. A Type 2 row operation corresponds to multiplying both sides of an equation by a nonzero real number. A Type 3 row operation is equivalent to multiplying both sides of an equation by a real number and then adding the resulting equation to another equation. None of these operations will change the solution set of the original system of equations.

We shall use row operations to write the augmented matrix in the form

$$\left[\begin{array}{cc|c} 1 & a & b \\ 0 & 1 & c \end{array}\right]$$

This last matrix represents a system that is easy to solve. Note that it contains 1s down the main diagonal of the coefficient matrix, and a 0 below the first 1.

Example 1   Solve the following system by matrices:

$$\begin{aligned} 3x - y &= 17 \\ x + 2y &= -6 \end{aligned}$$

Solution: First, write the augmented matrix of the system.

$$\left[\begin{array}{cc|c} 3 & -1 & 17 \\ 1 & 2 & -6 \end{array}\right]$$

Interchange the two rows of the matrix (Type 1 row operation) to get a 1 in the first row, first column position.

$$\begin{array}{c} R_1 \\ R_2 \end{array} \left[\begin{array}{cc|c} 1 & 2 & -6 \\ 3 & -1 & 17 \end{array}\right]$$

To get a 0 under the 1, multiply the first row by -3 and add the result to the second row (Type 3 row operation).

$$-3R_1 + R_2 \rightarrow \left[\begin{array}{cc|c} 1 & 2 & -6 \\ 0 & -7 & 35 \end{array}\right]$$

To get a 1 in the second row, second column position, multiply the second row by $-\frac{1}{7}$ (Type 2 row operation).

$$-\tfrac{1}{7}R_2 \rightarrow \left[\begin{array}{cc|c} 1 & 2 & -6 \\ 0 & 1 & -5 \end{array}\right]$$

This last matrix represents the system

$$\begin{aligned} x + 2y &= -6 \\ y &= -5 \end{aligned}$$

To find x, substitute $y = -5$ into $x + 2y = -6$.

$$x + 2(-5) = -6$$
$$x - 10 = -6$$
$$x = 4$$

Try Exercise 27

Solve by matrices:

$$2x - y = 12$$
$$x + 4y = -3$$

The solution set is $\{(4, -5)\}$. Check in the original system.   ❑

Solving a Linear System of Three Equations by Matrices

We can solve a linear system of three equations in a similar fashion. Simply use row operations to write the augmented matrix in the form

$$\left[\begin{array}{ccc|c} 1 & a & b & c \\ 0 & 1 & d & e \\ 0 & 0 & 1 & f \end{array}\right]$$

Example 2   Solve the following system by matrices:

$$x + y + 5z = 12$$
$$2x - 3y - z = 7$$
$$4x - y + 4z = 21$$

Solution: Write the augmented matrix of the system.

$$\left[\begin{array}{ccc|c} 1 & 1 & 5 & 12 \\ 2 & -3 & -1 & 7 \\ 4 & -1 & 4 & 21 \end{array}\right]$$

There is already a 1 in the first row, first column position. To get 0s below this 1, multiply the first row by -2 and add the result to the second row. Then multiply the first row by -4 and add the result to the third row.

$$-2R_1 + R_2 \rightarrow \left[\begin{array}{ccc|c} 1 & 1 & 5 & 12 \\ 0 & -5 & -11 & -17 \\ 4 & -1 & 4 & 21 \end{array}\right]$$

$$-4R_1 + R_3 \rightarrow \left[\begin{array}{ccc|c} 1 & 1 & 5 & 12 \\ 0 & -5 & -11 & -17 \\ 0 & -5 & -16 & -27 \end{array}\right]$$

To get a 1 in the second row, second column position, multiply the second row by $-\frac{1}{5}$.

$$-\tfrac{1}{5}R_2 \rightarrow \left[\begin{array}{ccc|c} 1 & 1 & 5 & 12 \\ 0 & 1 & \frac{11}{5} & \frac{17}{5} \\ 0 & -5 & -16 & -27 \end{array}\right]$$

To get a 0 below the 1 in the second row, multiply the second row by 5 and add the result to the third row.

$$5R_2 + R_3 \rightarrow \left[\begin{array}{ccc|c} 1 & 1 & 5 & 12 \\ 0 & 1 & \frac{11}{5} & \frac{17}{5} \\ 0 & 0 & -5 & -10 \end{array}\right]$$

To get a 1 in the third row, third column position, multiply the third row by $-\frac{1}{5}$.

$$-\tfrac{1}{5}R_3 \rightarrow \left[\begin{array}{ccc|c} 1 & 1 & 5 & 12 \\ 0 & 1 & \frac{11}{5} & \frac{17}{5} \\ 0 & 0 & 1 & 2 \end{array}\right]$$

This last matrix represents the system

$$x + y + 5z = 12$$
$$y + \tfrac{11}{5}z = \tfrac{17}{5}$$
$$z = 2$$

Substitute $z = 2$ into $y + \tfrac{11}{5}z = \tfrac{17}{5}$.

$$y + \tfrac{11}{5}(2) = \tfrac{17}{5}$$
$$y + \tfrac{22}{5} = \tfrac{17}{5}$$
$$y = -\tfrac{5}{5}$$
$$y = -1$$

Substitute $y = -1$ and $z = 2$ into $x + y + 5z = 12$.

$$x + (-1) + 5(2) = 12$$
$$x + 9 = 12$$
$$x = 3$$

The solution set is $\{(3, -1, 2)\}$. Check in the original system. ❐

Try Exercise 31

Solve by matrices:

$$x - 5y + 2z = -6$$
$$-x + 6y - z = 1$$
$$3x - 8y + 9z = -33$$

This method of using matrices to solve a linear system of equations is called **Gaussian elimination.*** The sequence of steps in solving a linear system of three equations by Gaussian elimination is shown below. The asterisks represent arbitrary elements.

$$
\begin{bmatrix} * & * & * & | & * \\ * & * & * & | & * \\ * & * & * & | & * \end{bmatrix} \rightarrow
\begin{bmatrix} 1 & * & * & | & * \\ * & * & * & | & * \\ * & * & * & | & * \end{bmatrix} \rightarrow
\begin{bmatrix} 1 & * & * & | & * \\ 0 & * & * & | & * \\ 0 & * & * & | & * \end{bmatrix} \rightarrow
$$

$$
\rightarrow \begin{bmatrix} 1 & * & * & | & * \\ 0 & 1 & * & | & * \\ 0 & * & * & | & * \end{bmatrix} \rightarrow
\begin{bmatrix} 1 & * & * & | & * \\ 0 & 1 & * & | & * \\ 0 & 0 & * & | & * \end{bmatrix} \rightarrow
\begin{bmatrix} 1 & * & * & | & * \\ 0 & 1 & * & | & * \\ 0 & 0 & 1 & | & * \end{bmatrix}
$$

To Solve a Linear System by Gaussian Elimination

1. Write the augmented matrix of the system.
2. Use a row operation to get a 1 in the first row, first column position. Then use row operations to get 0s below that 1.
3. Use a row operation to get a 1 in the second row, second column position. Then use row operations to get 0s below that 1.
4. Continue until the coefficient matrix contains 1s down the main diagonal and 0s below each 1.
5. Write the system that is represented by the matrix of step 4. Then solve this system.
6. Check the solution in the original system.

The Gauss-Jordan Method

A mathematician named Camille Jordan (1838–1922) continued to apply row operations on the augmented matrix of a linear system of equations to produce a matrix that con-

---

*HISTORICAL NOTE: Named after the German mathematician Carl Friedrich Gauss (1777–1855). Gauss also enjoyed considerable fame as an astronomer. He was director of the Göttingen observatory for almost 50 years, and he developed a scheme for computing orbits that is still used to track satellites.

tained 0s above the main diagonal of the coefficient matrix as well as below it. This produced matrices like the ones shown below:

$$\begin{bmatrix} 1 & 0 & | & a \\ 0 & 1 & | & b \end{bmatrix} \qquad \begin{bmatrix} 1 & 0 & 0 & | & a \\ 0 & 1 & 0 & | & b \\ 0 & 0 & 1 & | & c \end{bmatrix}$$

The numbers to the right of the dashed lines in the augmented matrices above then represented the solution of the system.

Example 3 Solve the following system using the Gauss-Jordan method:

$$\begin{aligned} x - 2y &= -4 \\ -5y + z &= -9 \\ 4x - 3z &= -10 \end{aligned}$$

Solution: Write the augmented matrix.

$$\begin{bmatrix} 1 & -2 & 0 & | & -4 \\ 0 & -5 & 1 & | & -9 \\ 4 & 0 & -3 & | & -10 \end{bmatrix}$$

Multiply the first row by -4 and add the result to the third row.

$$-4R_1 + R_3 \to \begin{bmatrix} 1 & -2 & 0 & | & -4 \\ 0 & -5 & 1 & | & -9 \\ 0 & 8 & -3 & | & 6 \end{bmatrix}$$

Multiply the second row by $-\frac{1}{5}$.

$$-\frac{1}{5}R_2 \to \begin{bmatrix} 1 & -2 & 0 & | & -4 \\ 0 & 1 & -\frac{1}{5} & | & \frac{9}{5} \\ 0 & 8 & -3 & | & 6 \end{bmatrix}$$

Multiply the second row by 2 and add the result to the first row. Then multiply the second row by -8 and add the result to the third row.

$$\begin{aligned} 2R_2 + R_1 &\to \\ -8R_2 + R_3 &\to \end{aligned} \begin{bmatrix} 1 & 0 & -\frac{2}{5} & | & -\frac{2}{5} \\ 0 & 1 & -\frac{1}{5} & | & \frac{9}{5} \\ 0 & 0 & -\frac{7}{5} & | & -\frac{42}{5} \end{bmatrix}$$

Multiply row three by $-\frac{5}{7}$.

$$-\frac{5}{7}R_3 \to \begin{bmatrix} 1 & 0 & -\frac{2}{5} & | & -\frac{2}{5} \\ 0 & 1 & -\frac{1}{5} & | & \frac{9}{5} \\ 0 & 0 & 1 & | & 6 \end{bmatrix}$$

Multiply row three by $\frac{2}{5}$ and add the result to row one. Then multiply row three by $\frac{1}{5}$ and add the result to row two.

$$\begin{aligned} \frac{2}{5}R_3 + R_1 &\to \\ \frac{1}{5}R_3 + R_2 &\to \end{aligned} \begin{bmatrix} 1 & 0 & 0 & | & 2 \\ 0 & 1 & 0 & | & 3 \\ 0 & 0 & 1 & | & 6 \end{bmatrix}$$

This last matrix represents the system

$$\begin{aligned} x &= 2 \\ y &= 3 \\ z &= 6 \end{aligned}$$

Try Exercise 41

Solve using the Gauss-Jordan method:

$$\begin{aligned} x - 4y &= -10 \\ -3y + z &= -6 \\ 6x - 5z &= 15 \end{aligned}$$

Therefore, the solution set is $\{(2, 3, 6)\}$. Check in the original system. ❐

The sequence of steps in solving a linear system of three equations using the Gauss-Jordan method is shown below. The asterisks represent arbitrary elements.

$$\begin{bmatrix} * & * & * & | & * \\ * & * & * & | & * \\ * & * & * & | & * \end{bmatrix} \rightarrow \begin{bmatrix} 1 & * & * & | & * \\ * & * & * & | & * \\ * & * & * & | & * \end{bmatrix} \rightarrow \begin{bmatrix} 1 & * & * & | & * \\ 0 & * & * & | & * \\ 0 & * & * & | & * \end{bmatrix} \rightarrow$$

$$\rightarrow \begin{bmatrix} 1 & * & * & | & * \\ 0 & 1 & * & | & * \\ 0 & * & * & | & * \end{bmatrix} \rightarrow \begin{bmatrix} 1 & 0 & * & | & * \\ 0 & 1 & * & | & * \\ 0 & 0 & * & | & * \end{bmatrix} \rightarrow \begin{bmatrix} 1 & 0 & * & | & * \\ 0 & 1 & * & | & * \\ 0 & 0 & 1 & | & * \end{bmatrix} \rightarrow$$

Solution

$$\rightarrow \begin{bmatrix} 1 & 0 & 0 & | & * \\ 0 & 1 & 0 & | & * \\ 0 & 0 & 1 & | & * \end{bmatrix}$$

Example 4 Solve the following system using the Gauss-Jordan method:

$$\begin{aligned} -x + 3y &= 0 \\ 2x - 6y &= 5 \end{aligned}$$

Solution: Write the augmented matrix.

$$\begin{bmatrix} -1 & 3 & | & 0 \\ 2 & -6 & | & 5 \end{bmatrix}$$

Multiply the first row by -1.

$$-R_1 \rightarrow \begin{bmatrix} 1 & -3 & | & 0 \\ 2 & -6 & | & 5 \end{bmatrix}$$

Multiply the first row by -2 and add the result to the second row.

$$-2R_1 + R_2 \rightarrow \begin{bmatrix} 1 & -3 & | & 0 \\ 0 & 0 & | & 5 \end{bmatrix}$$

Because of the 0s in the second row, we cannot get a 1 in the second row, second column position. This last matrix represents the system

$$\begin{aligned} x - 3y &= 0 \\ 0x + 0y &= 5 \end{aligned}$$

The contradiction $0 = 5$ means the system is inconsistent and the solution set is $\varnothing$. ❐

Try Exercise 47

Solve using the Gauss-Jordan method:

$$\begin{aligned} -x + 2y &= 0 \\ 4x - 8y &= 11 \end{aligned}$$

CAUTION

If you are solving a linear system of two equations by matrices, and you obtain an augmented matrix like

$$\begin{bmatrix} 1 & 2 & | & -6 \\ 0 & 0 & | & -4 \end{bmatrix}$$

then the system is inconsistent and there is no solution. If you obtain an augmented matrix like

$$\begin{bmatrix} 1 & 2 & | & -6 \\ 0 & 0 & | & 0 \end{bmatrix}$$

then the system is dependent and there are an infinite number of solutions. In either case, there is no unique solution. ■

Exercises 11.7

Completion Exercises

1. A rectangular array of numbers is called a(n) _____.

2. The matrix

$$\begin{bmatrix} 1 & -3 & 6 \\ 7 & 0 & 4 \end{bmatrix}$$

is a _____ $(2 \times 3, 3 \times 2)$ matrix. The numbers 1, -3, 6, 7, 0, and 4 are called the _____ of the matrix.

3. For the system

$$x + 2y = 7$$
$$3x - 5y = 10$$

the matrix

$$\begin{bmatrix} 1 & 2 & \vdots & 7 \\ 3 & -5 & \vdots & 10 \end{bmatrix}$$

is called the _____ matrix, and the matrix

$$\begin{bmatrix} 1 & 2 \\ 3 & -5 \end{bmatrix}$$

is called the _____ matrix.

4. Solving a linear system of three equations by Gaussian elimination involves writing the augmented matrix in the form _____; to solve using the Gauss-Jordan method, we write the augmented matrix in the form _____.

Write the augmented matrix for each system. Do not solve the system.

5. $2x - 3y = 13$
   $6x - 4y = 4$

6. $3x - 2y = 0$
   $6x - 5y = 6$

7. $3x - 5y + 6z = 20$
   $-2x + 2y + 3z = -2$
   $4x - y - 9z = 4$

8. $5x + 3y - 4z = 14$
   $-2x - 4y - z = 5$
   $3x + 6y - 2z = 3$

Write a linear system of equations represented by each augmented matrix. Do not solve the system.

9. $\begin{bmatrix} 1 & 0 & \vdots & -3 \\ 0 & 1 & \vdots & 0 \end{bmatrix}$

10. $\begin{bmatrix} 1 & 4 & \vdots & 9 \\ 0 & 1 & \vdots & 1 \end{bmatrix}$

11. $\begin{bmatrix} 1 & 2 & -1 & \vdots & 10 \\ 0 & 1 & 6 & \vdots & -4 \\ 0 & 0 & 1 & \vdots & 1 \end{bmatrix}$

12. $\begin{bmatrix} 1 & 0 & 0 & \vdots & 7 \\ 0 & 1 & 0 & \vdots & -1 \\ 0 & 0 & 1 & \vdots & 0 \end{bmatrix}$

True-False Exercises

13. The main diagonal of the matrix

$$\begin{bmatrix} 4 & -1 & 7 \\ 2 & 8 & -5 \\ 0 & 3 & -6 \end{bmatrix}$$

consists of the numbers 0, 8, and 7.

14. A Type 1 row operation states that we can change any two rows of the augmented matrix.

15. A Type 2 row operation states that we can multiply any row of the augmented matrix by a nonzero number.

16. A Type 3 row operation states that we can multiply any row of the augmented matrix by a real number and add the result to another row.

Perform the indicated row operation on the given matrix.

17. $R_1 \leftrightarrow R_3$

$$\begin{bmatrix} 5 & -2 & 6 & \vdots & 4 \\ 3 & -1 & -8 & \vdots & 12 \\ 1 & 0 & 2 & \vdots & -3 \end{bmatrix}$$

18. $R_2 \leftrightarrow R_3$

$$\begin{bmatrix} 1 & 2 & -6 & \vdots & -2 \\ 3 & 3 & 5 & \vdots & 0 \\ 0 & 1 & 4 & \vdots & -1 \end{bmatrix}$$

19. $-\frac{1}{3} R_2$

$$\begin{bmatrix} 1 & 9 & -2 & \vdots & 15 \\ 0 & -3 & 6 & \vdots & -7 \\ 0 & 3 & -1 & \vdots & 10 \end{bmatrix}$$

20. $\frac{3}{4} R_2$

$$\begin{bmatrix} 1 & -4 & 5 & \vdots & -3 \\ 0 & \frac{4}{3} & -12 & \vdots & 6 \\ 0 & 8 & -2 & \vdots & 11 \end{bmatrix}$$

21. $2R_1 + R_3$

$$\begin{bmatrix} 1 & -3 & 0 & \vdots & 7 \\ 0 & 3 & -6 & \vdots & 0 \\ -2 & 1 & -5 & \vdots & -4 \end{bmatrix}$$

22. $-5R_2 + R_1$

$$\begin{bmatrix} 1 & 5 & -2 & | & 6 \\ 0 & 1 & -3 & | & 0 \\ 0 & -4 & 7 & | & -1 \end{bmatrix}$$

True-False Exercises

23. When solving a linear system of three equations by Gaussian elimination, we try to write the augmented matrix in the form

$$\begin{bmatrix} 1 & a & b & | & c \\ 0 & 1 & d & | & e \\ 0 & 0 & 1 & | & f \end{bmatrix}$$

24. When solving a linear system of three equations using the Gauss-Jordan method, we try to write the augmented matrix in the form

$$\begin{bmatrix} 1 & 0 & 0 & | & a \\ 0 & 1 & 0 & | & b \\ 0 & 0 & 1 & | & c \end{bmatrix}$$

Solve each system by Gaussian elimination.

25. $x + y = 6$
    $x + 2y = 8$

26. $x + 2y = 7$
    $x + 3y = 9$

27. $2x - y = 12$
    $x + 4y = -3$

28. $4x - y = 24$
    $x + 3y = -7$

29. $-3x + 10y = 6$
    $6x - 5y = 0$

30. $-4x + 8y = 4$
    $6x - 4y = 4$

31. $x - 5y + 2z = -6$
    $-x + 6y - z = 1$
    $3x - 8y + 9z = -33$

32. $x - 2y - 4z = 2$
    $-3x + 7y + 8z = 11$
    $x + 3y + 5z = 0$

33. $x - y + 3z = 12$
    $2x + 4y - z = -3$
    $5x + y - 4z = -3$

34. $x + y - 2z = 2$
    $2x - 4y + z = 12$
    $3x - 3y - 7z = 26$

Solve each system using the Gauss-Jordan method.

35. $x - 2y = -7$
    $3x + 4y = 9$

36. $x - 3y = -5$
    $2x + 5y = 1$

37. $3x - 2y = 3$
    $-5x + y = 9$

38. $5x - y = -7$
    $4x + 3y = 2$

39. $4x - 3y + 5z = 34$
    $2x - y - z = 6$
    $x + y + 4z = 15$

40. $4x - 3y + z = 10$
    $x - 2y - z = 5$
    $-x + 5y - 2z = 7$

41. $x - 4y = -10$
    $-3y + z = -6$
    $6x - 5z = 15$

42. $2y - 5z = -13$
    $x - 3y = -11$
    $4x + z = 33$

Discussion Exercises

43. Describe the steps in solving a linear system of three equations by Gaussian elimination.

44. Describe the steps in writing the augmented matrix for a linear system of three equations in the form

$$\begin{bmatrix} 1 & 0 & 0 & | & a \\ 0 & 1 & 0 & | & b \\ 0 & 0 & 1 & | & c \end{bmatrix}$$

45. Suppose you are solving a linear system of three equations and you obtain the augmented matrix

$$\begin{bmatrix} 1 & 3 & 2 & | & -2 \\ 0 & 1 & 5 & | & 1 \\ 0 & 0 & 0 & | & 7 \end{bmatrix}$$

What is your conclusion?

46. Suppose you are solving a linear system of three equations and you obtain the augmented matrix

$$\begin{bmatrix} 1 & -2 & 4 & | & 1 \\ 0 & 1 & 6 & | & -3 \\ 0 & 0 & 0 & | & 0 \end{bmatrix}$$

What is your conclusion?

Try to solve each system using matrices.

47. $-x + 2y = 0$
    $4x - 8y = 11$

48. $-x + 3y = 0$
    $2x - 6y = 5$

49. $2x - 10y = \frac{1}{3}$
    $-3x + 15y = -\frac{1}{2}$

50. $3x - 6y = \frac{1}{4}$
    $-4x + 8y = -\frac{1}{3}$

51. $x - 3y + 2z = 3$
    $x - y + 2z = 5$
    $-2x + 6y - 4z = 7$

52. $x + 2y - z = 8$
    $-x - y + 3z = 2$
    $2x + 3y - 4z = 6$

Solve each system of four equations by matrices.

53. $x + 2y - z + w = -4$
    $-x - 3y + 4z - w = 10$
    $x - 6z + 2w = -1$
    $2x + y + 3w = 5$

54. $x + 3y - z - w = -5$
    $y + 5z - 2w = -1$
    $-x + 2z - 3w = -8$
    $2x + y - 4z + w = 4$

Getting Acquainted with Your Graphing Calculator

Use your calculator to perform the given row operation on the given matrix.

55. $R_1 \leftrightarrow R_2$

$$\left[\begin{array}{cc|c} 1 & 2 & 3 \\ 4 & 5 & 6 \end{array}\right]$$

56. $\frac{1}{5}R_1$

$$\left[\begin{array}{cc|c} 5 & -15 & 8 \\ -4 & 10 & 0 \end{array}\right]$$

57. $-2R_1 + R_3$

$$\left[\begin{array}{ccc|c} 1 & 3 & 0 & -5 \\ 0 & 4 & -1 & 7 \\ 2 & 9 & -6 & 11 \end{array}\right]$$

Solve each system using matrices. Use your calculator to perform the necessary row operations.

58. $\begin{aligned} 2x + 4y &= 4 \\ 3x - y &= -15 \end{aligned}$

59. $\begin{aligned} x + y - z &= -3 \\ -2x - y + 3z &= 2 \\ 4x + 5y &= -16 \end{aligned}$

 ## 11.8 Nonlinear Systems of Equations

Earlier in this chapter, we learned how to solve a system of linear equations. In this section we learn how to solve a *nonlinear* system of equations.

A **nonlinear system** of equations is a system of equations that contains at least one nonlinear equation. For example, the system

$$x^2 + y^2 = 25$$
$$2x + y = 10$$

is a nonlinear system because the first equation is a nonlinear equation.

We solve nonlinear systems using the substitution method, the addition method, or a combination of the two methods.

Example 1 Solve the following system:

$$x^2 + y^2 = 25$$
$$2x + y = 10$$

Solution: Since one of the equations is linear (the second equation), we use the substitution method. Solve the linear equation for y.

$$2x + y = 10 \qquad \text{Original equation}$$
$$y = 10 - 2x \qquad \text{Subtract } 2x.$$

Substitute $10 - 2x$ for y in the nonlinear equation, $x^2 + y^2 = 25$.

$$\begin{aligned} x^2 + (10 - 2x)^2 &= 25 & \text{Let } y = 10 - 2x. \\ x^2 + 100 - 40x + 4x^2 &= 25 & \text{Multiply out } (10 - 2x)^2. \\ 5x^2 - 40x + 75 &= 0 & \text{Write in standard form.} \\ x^2 - 8x + 15 &= 0 & \text{Divide by 5.} \\ (x - 3)(x - 5) &= 0 & \text{Factor.} \\ x = 3 \quad \text{or} \quad x = 5 & & \text{Set each factor equal to 0 and solve.} \end{aligned}$$

Substitute the x-values into the equation $y = 10 - 2x$ to find the corresponding y-values.

When $x = 3$, $y = 10 - 2 \cdot 3$ When $x = 5$, $y = 10 - 2 \cdot 5$
          $y = 4$ $y = 0$

Therefore the solution set is $\{(3, 4), (5, 0)\}$. To check, each solution must satisfy *both* equations in the original system.

Check (3, 4):

$x^2 + y^2 = 25$ $2x + y = 10$

$3^2 + 4^2 = 25$ True $2 \cdot 3 + 4 = 10$ True

Check (5, 0):

$x^2 + y^2 = 25$ $2x + y = 10$

$5^2 + 0^2 = 25$ True $2 \cdot 5 + 0 = 10$ True   ❏

Try Exercise 13

Solve:

$x^2 + y^2 = 25$
$3x + y = 15$

The solutions to Example 1 correspond to the two intersection points of the circle $x^2 + y^2 = 25$ and the line $2x + y = 10$, as shown in Fig. 11.16.

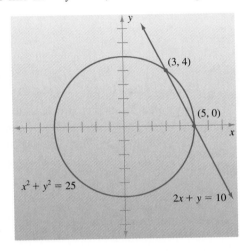

Figure 11.16

CAUTION

In Example 1, suppose we had substituted the solution $x = 3$ into $x^2 + y^2 = 25$ (instead of into $y = 10 - 2x$) to find the corresponding y-value. The result would have been *two* values for y.

$$3^2 + y^2 = 25$$
$$y^2 = 16$$
$$y = \pm 4$$

The ordered pair (3, 4) is a solution, but the ordered pair (3, −4) is not (it is an *extraneous* solution). **Always use the linear equation (if there is one) to find the value of the other variable.** In any case, it is always a good idea to check your solutions.   ■

Example 2   Solve the following system:

$$x^2 + y^2 = 16$$
$$y = x^2 - 4$$

Solution: Since both equations contain an x^2-term, and one of the equations (the second equation) contains a first-degree term, we use the substitution method. Solve the second equation for x^2.

$$y = x^2 - 4 \qquad \text{Second equation}$$
$$y + 4 = x^2 \qquad \text{Add 4.}$$

Substitute $y + 4$ for x^2 in the equation $x^2 + y^2 = 16$.

$$(y + 4) + y^2 = 16 \qquad \text{Let } x^2 = y + 4.$$
$$y^2 + y - 12 = 0 \qquad \text{Write in standard form.}$$
$$(y + 4)(y - 3) = 0 \qquad \text{Factor.}$$
$$y = -4 \qquad \text{or} \qquad y = 3 \qquad \text{Set each factor equal to 0 and solve.}$$

Use $y = x^2 - 4$ to find the corresonding x-values.

When $y = -4$, $-4 = x^2 - 4$ When $y = 3$, $3 = x^2 - 4$
$$0 = x^2 \qquad\qquad 7 = x^2$$
$$0 = x \qquad\qquad \pm\sqrt{7} = x$$

Try Exercise 19

The solution set is $\{(0, -4), (\sqrt{7}, 3), (-\sqrt{7}, 3)\}$. Check in the original system.   ❐

Solve:

$$x^2 + y^2 = 4$$
$$y = x^2 - 2$$

The solutions to Example 2 correspond to the three intersection points of the circle $x^2 + y^2 = 16$ and the parabola $y = x^2 - 4$, as shown in Fig. 11.17.

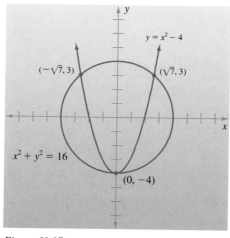

Figure 11.17

Figure 11.18

Example 3   Solve the following system:

$$3x^2 + 2y^2 = 116$$
$$x^2 - y^2 = 32$$

Solution: Since neither equation contains a first-degree term, the addition method is probably the best method. Multiply the second equation by 2 and add it to the first.

$3x^2 + 2y^2 = 116$	First equation
$2x^2 - 2y^2 = 64$	Second equation multiplied by 2.
$5x^2 = 180$	Add.
$x^2 = 36$	Divide by 5.
$x = \pm 6$	Square root property

Use $x^2 - y^2 = 32$ to find the corresponding y-values.

When $x = 6$, $6^2 - y^2 = 32$ When $x = -6$, $(-6)^2 - y^2 = 32$
$$-y^2 = -4 \qquad\qquad -y^2 = -4$$
$$y^2 = 4 \qquad\qquad y^2 = 4$$
$$y = \pm 2 \qquad\qquad y = \pm 2$$

Try Exercise 25

The solution set is $\{(6, 2), (6, -2), (-6, 2), (-6, -2)\}$. Check in the original system.   ❐

Solve:

$$3x^2 + 2y^2 = 140$$
$$x^2 - y^2 = 20$$

The solutions to Example 3 correspond to the four intersection points of the ellipse $3x^2 + 2y^2 = 116$ and the hyperbola $x^2 - y^2 = 32$, as shown in Fig. 11.18.

A combination of the addition method and the substitution method offers the best approach to solving Example 4.

Example 4   Solve the following system:

$$x^2 + 2xy + y^2 = 16$$
$$x^2 + y^2 = 10$$

Solution: Multiply the second equation by -1 and add it to the first.

$x^2 + 2xy + y^2 = 16$	First equation
$\underline{-x^2 - y^2 = -10}$	Second equation multiplied by -1
$2xy = 6$	Add.
$y = \dfrac{6}{2x}$	Divide by $2x$.
$y = \dfrac{3}{x}$	Simplify.

Substitute $y = \dfrac{3}{x}$ into $x^2 + y^2 = 10$.

$x^2 + \left(\dfrac{3}{x}\right)^2 = 10$	Let $y = \dfrac{3}{x}$.
$x^2 + \dfrac{9}{x^2} = 10$	Simplify.
$x^4 + 9 = 10x^2$	Multiply by x^2.
$x^4 - 10x^2 + 9 = 0$	Write in standard form.
$(x^2 - 1)(x^2 - 9) = 0$	Factor.

$x^2 - 1 = 0$	or	$x^2 - 9 = 0$	Set each factor equal to 0.
$x^2 = 1$		$x^2 = 9$	Solve for x^2.
$x = \pm 1$		$x = \pm 3$	Square root property

Use $y = \dfrac{3}{x}$ to find the corresponding y-values.

When $x = 1$, $y = \frac{3}{1} = 3$.   When $x = -1$, $y = \frac{3}{-1} = -3$.

When $x = 3$, $y = \frac{3}{3} = 1$.   When $x = -3$, $y = \frac{3}{-3} = -1$.

The solution set is $\{(1, 3), (-1, -3), (3, 1), (-3, -1)\}$. Check in the original system. ❏

Try Exercise 35

Solve:

$$x^2 + 3xy + y^2 = 29$$
$$x^2 + y^2 = 17$$

The solutions to Example 4 correspond to the intersection points of the graphs of $x^2 + 2xy + y^2 = 16$ and $x^2 + y^2 = 10$. These graphs are shown in Fig. 11.19. Note that $x^2 + 2xy + y^2 = 16$ represents a degenerate conic whose graph consists of two parallel lines.

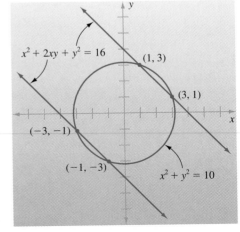

Figure 11.19

One application of a system of nonlinear equations involves locating the center of an earthquake (the *epicenter*). For example, suppose seismic station A determines that the epicenter is 130 mi away. Then the epicenter must lie on a circle whose center is at station A and whose radius is 130 mi (see Fig. 11.20). If similar circles are drawn centered at two other seismic stations, then the intersection point of the three circles gives the location of the epicenter (see Fig. 11.21).

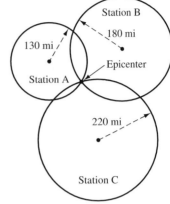

Figure 11.20

Figure 11.21

Exercises 11.8

True-False Exercises

1. A system of equations that contains at least one nonlinear equation is called a nonlinear system.

2. We solve nonlinear systems using the substitution method, the addition method, or a combination of the two methods.

Discussion Exercises

3. Can a line and a circle intersect at (a) two points? (b) one point? (c) zero points? Draw graphs to illustrate your answers.

4. Can a parabola and a circle intersect at (a) four points? (b) three points? (c) two points? (d) one point? (e) zero points? Draw graphs to illustrate your answers.

5. Suppose you are solving a system of two equations in two variables where one equation is linear and one is nonlinear. After you find a value for one of the variables, which equation should you use to find the corresponding value for the other variable? Why?

6. Discuss two different ways to check your answer to a nonlinear system.

Solve each system using the substitution method.

7. $y = x^2$
   $y = 2x - 1$

8. $y = x^2$
   $y = 4x - 4$

9. $y = x^2 + 2x + 1$
   $y = x + 3$

10. $y = x^2 + 3x + 1$
    $y = x + 4$

11. $y^2 - 8x^2 = 32$
    $y = 4x$

12. $y^2 - 4x^2 = 45$
    $y = 3x$

13. $x^2 + y^2 = 25$
    $3x + y = 15$

14. $x^2 + y^2 = 25$
    $2x + y = 5$

15. $x^2 + y^2 = 4$
    $x - 2y = 4$

16. $x^2 + y^2 = 16$
    $x - 2y = 8$

17. $x + y = 8$
    $xy = 16$

18. $x + y = 6$
    $xy = 9$

19. $x^2 + y^2 = 4$
    $y = x^2 - 2$

20. $x^2 + y^2 = 9$
    $y = x^2 - 3$

21. $x^2 + xy - y^2 = 19$
    $y - x = 3$

22. $x^2 + xy - y^2 = 31$
    $y - x = 2$

Solve each system using the addition method.

23. $x^2 + y^2 = 1$
    $x^2 - y^2 = 1$

24. $x^2 + y^2 = 16$
    $x^2 - y^2 = 16$

25. $3x^2 + 2y^2 = 140$
    $x^2 - y^2 = 20$

26. $3x^2 + 2y^2 = 200$
    $x^2 - y^2 = 60$

27. $x^2 + y^2 = 9$
    $9x^2 + 4y^2 = 36$

28. $x^2 + y^2 = 4$
    $4x^2 + 9y^2 = 36$

29. $4x^2 - 3y^2 = -11$
    $3x^2 + 2y^2 = 30$

30. $2x^2 - 3y^2 = 6$
    $3x^2 + 2y^2 = 35$

31. $xy - y^2 = 25$
    $2xy - y^2 = 75$

32. $xy - y^2 = 30$
    $3xy - y^2 = 162$

33. $4x^2 + 4y^2 = 25$
    $4x^2 + 2y = 13$

34. $9x^2 + 9y^2 = 13$
    $9x^2 + 3y = 7$

Use a combination of the addition method and the substitution method to solve each system.

35. $x^2 + 3xy + y^2 = 29$
    $x^2 \qquad + y^2 = 17$

36. $x^2 - 2xy + y^2 = 1$
    $x^2 \qquad + y^2 = 5$

37. $x^2 - xy + y^2 = 21$
    $x^2 + xy + y^2 = 31$

38. $x^2 + 2xy + y^2 = 25$
    $x^2 + \ xy + y^2 = 19$

Solve each system. Some of the solutions contain imaginary numbers.

39. $5x^2 - y^2 = 16$
    $y = 3x$

40. $11x^2 - y^2 = 45$
    $y = 4x$

41. $3x^2 + 2y^2 = 2$
    $x^2 + \ y^2 = 9$

42. $2x^2 + 3y^2 = 19$
    $x^2 + \ y^2 = 5$

43. $x^2 - 3xy - y^2 = 17$
    $x^2 \qquad - y^2 = 8$

44. $x^2 + 2xy - y^2 = 7$
    $x^2 \qquad - y^2 = 3$

Problem Solving

Write a nonlinear system of equations for each exercise. Then solve the system.

45. The sum of the squares of two positive numbers is 74, and the difference of the squares is 24. Find the numbers.

46. The product of two numbers is 96, and the sum of the two numbers is 20. Find the numbers.

47. Find x and y given that the area of the triangle in Fig. 11.22 is 3.

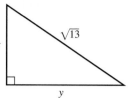

Figure 11.22

48. In Fig. 11.23, the area of the triangle is 104 and the area of the rectangle is 247. Find x and y.

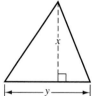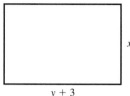

Figure 11.23

49. Determine the width and length of a rectangle whose perimeter is 20 ft and whose area is 24 ft².

50. A rectangle has an area of 12 ft² and a diagonal of length 5 ft. Find its width and length.

 Getting Acquainted with Your Graphing Calculator

Solve each nonlinear system by graphing the two equations in the same Viewing Rectangle and zooming in on the intersection points of the two graphs. Write your answers correct to two decimal places.

51. $\quad y = x^2 - 1$
    $x^2 + y^2 = 9$

52. $y = e^x + 1$
    $y = 2 - \ln x$

11.9 Second-Degree Inequalities; Systems of Inequalities

TAPE AU23

In Sec. 7.5, we graphed linear inequalities, such as $3x + 5y \leq 0$. In this section, we graph *second-degree inequalities,* such as $x^2 + y^2 < 25$. We also graph *systems of inequalities.*

Second-Degree Inequalities

A **second-degree inequality** is an inequality that contains at least one second-degree (and no higher-degree) term. We graph second-degree inequalities in much the same way that we graph linear inequalities.

Example 1   Graph: $x^2 + y^2 < 25$.

Solution: First graph the boundary curve $x^2 + y^2 = 25$. This is the circle with center at the origin and radius 5, shown in Fig. 11.24(a). The circle is dashed because the original inequality symbol is $<$, and not $\leq$. The boundary curve divides the plane into two regions—the region inside the circle and the region outside the circle. To determine which region represents the solution of $x^2 + y^2 < 25$, choose a test point that is not on the boundary itself, say $(0, 0)$. Substitute the test point into the original inequality.

$$x^2 + y^2 < 25 \qquad \text{Original inequality}$$
$$0^2 + 0^2 < 25 \qquad \text{Let } x = 0 \text{ and } y = 0.$$
$$0 < 25 \qquad \text{True}$$

Since $(0, 0)$ satisfies the inequality $x^2 + y^2 < 25$, shade the region that contains $(0, 0)$, as shown in Fig. 11.24(b). ◻

Try Exercise 5

Graph: $x^2 + y^2 < 9$.

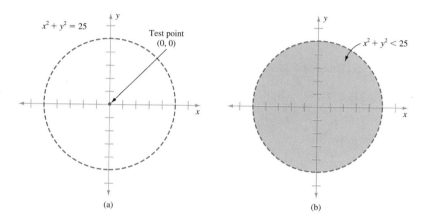

Figure 11.24   (a)   (b)

Example 2   Graph: $y \leq x^2 - 1$.

Solution: Graph the boundary $y = x^2 - 1$. This is the parabola with intercepts at $(0, -1)$, $(-1, 0)$, and $(1, 0)$, shown in Fig. 11.25(a). The parabola is solid (not dashed), because the original inequality symbol is $\leq$, and not $<$. Substitute the test point $(0, 0)$ into the original inequality.

$$y \leq x^2 - 1 \qquad \text{Original inequality}$$
$$0 \leq 0^2 - 1 \qquad \text{Let } (x, y) = (0, 0).$$
$$0 \leq -1 \qquad \text{False}$$

Since $(0, 0)$ does *not* satisfy $y \leq x^2 - 1$, shade the region that does *not* contain $(0, 0)$, as shown in Fig. 11.25(b). ◻

Try Exercise 13

Graph: $y \leq x^2 - 4$.

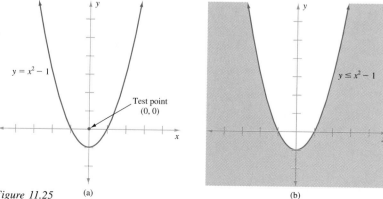

Figure 11.25   (a)   (b)

Systems of Inequalities

A **system of inequalities** consists of two or more inequalities that are true simultaneously. The **graph** of a system of inequalities, such as

$$x + y > 3$$
$$y \leq x$$

is the set of all points (x, y) that satisfy *both* inequalities.

Example 3 Graph the following system:

$$x + y > 3$$
$$y \leq x$$

Solution: First graph $x + y > 3$. This is the open half-plane that lies above the line $x + y = 3$, as shown in Fig. 11.26(a). Then graph $y \leq x$ on the same axes. This is the closed half-plane that lies on and below the line $y = x$, as shown in Fig. 11.26(b). The intersection of the two half-planes, shown as the shaded region in Fig. 11.26(c), represents the solution of the system. ❏

Try Exercise 27

Graph:

$$x + y > 1$$
$$y \leq x$$

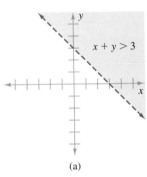

(a)

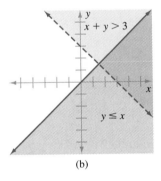

(b)

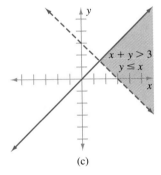

(c)

Figure 11.26

Example 4 Graph the following system:

$$x^2 + y^2 < 16$$
$$x - 2y < -4$$

Solution: Graph $x^2 + y^2 < 16$, as shown in Fig. 11.27(a). On the same axes, graph $x - 2y < -4$, as shown in Fig. 11.27(b). The intersection of the two shaded regions, shown in Fig. 11.27(c) is the graph of the system. ❏

Try Exercise 31

Graph:

$$x^2 + y^2 < 36$$
$$x - 2y < -6$$

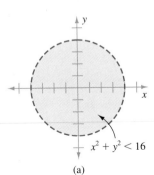

(a)

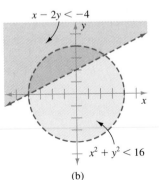

(b)

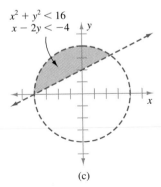

(c)

Figure 11.27

Example 5   Graph the following system:

$$x^2 - y^2 \geq 1$$
$$4x^2 + 9y^2 \leq 36$$

Solution: Graph $x^2 - y^2 \geq 1$, as shown in Fig. 11.28(a). On the same axes, graph $\dfrac{x^2}{9} + \dfrac{y^2}{4} \leq 1$, as shown in Fig. 11.28(b). The intersection of the two regions, shown in Fig. 11.28(c), is the graph of the system.   ❒

Try Exercise 39

Graph:

$$x^2 - y^2 \geq 4$$
$$x^2 + 4y^2 \leq 16$$

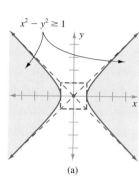
(a)

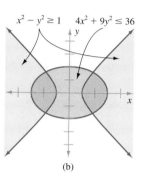
(b)

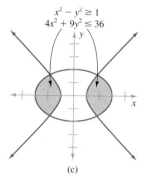
(c)

Figure 11.28

Exercises 11.9

True-False Exercises

1. A second-degree inequality is an inequality that contains at least one second-degree (and no higher-degree) term.

2. A system of inequalities consists of two or more inequalities that are true simultaneously.

3. To indicate that the boundary curve is not part of the graph, we draw the boundary curve as a dashed curve.

4. The graph of a system of inequalities is the set of all those points that satisfy at least one of the inequalities.

Matching Exercises

Match each inequality in Exercises 5 through 8 with its graph in letters A through D.

5. $x^2 + y^2 < 9$

6. $x^2 + y^2 \leq 9$

7. $x^2 + y^2 \geq 9$

8. $x^2 + y^2 > 9$

A.
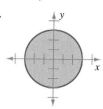

B.
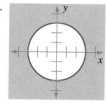

C.
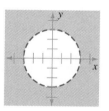

D.
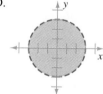

Graph each inequality.

9. $\dfrac{x^2}{9} + \dfrac{y^2}{4} \leq 1$

10. $\dfrac{x^2}{25} + \dfrac{y^2}{4} \geq 1$

11. $y > x^2$

12. $y < x^2$

13. $y \leq x^2 - 4$

14. $y \geq x^2 - 4$

15. $\dfrac{x^2}{9} - \dfrac{y^2}{25} \leq 1$

16. $\dfrac{x^2}{9} - \dfrac{y^2}{16} \geq 1$

17. $\dfrac{y^2}{9} - \dfrac{x^2}{9} \geq 1$

18. $\dfrac{y^2}{25} - \dfrac{x^2}{25} \leq 1$

19. $(x - 3)^2 + y^2 < 1$

20. $x^2 + (y - 4)^2 > 4$

Discussion Exercises

21. What determines whether the graph of an inequality contains its boundary curve or not?

22. Once we graph the boundary curve of an inequality, how do we determine which side of the boundary curve to shade?

23. Describe how to graph an inequality.

24. Describe how to graph a system of inequalities.

Graph each system of inequalities.

25. $y \leq x + 3$
    $y \geq 3$

26. $y \geq x + 2$
    $y \leq 2$

27. $x + y > 1$
    $y \leq x$

28. $x - y < 3$
    $y \leq -x$

29. $x - y \leq 4$
    $x - y \geq 0$

30. $x + y \leq 4$
    $x + y \geq 0$

31. $x^2 + y^2 < 36$
    $x - 2y < -6$

32. $x^2 + y^2 < 16$
    $x + 2y > 4$

33. $x^2 + y^2 \leq 16$
    $y < x^2$

34. $x^2 + y^2 \leq 25$
    $y > x^2$

35. $y \leq x^2 - 1$
    $y \geq -x^2 + 1$

36. $y \geq x^2 - 4$
    $y \leq -x^2 + 4$

37. $x^2 + y^2 \leq 9$
    $4x^2 + 9y^2 \geq 36$

38. $x^2 + y^2 \geq 4$
    $4x^2 + 9y^2 \leq 36$

39. $x^2 - y^2 \geq 4$
    $x^2 + 4y^2 \leq 16$

40. $x^2 - y^2 \leq 4$
    $4x^2 + y^2 \geq 16$

41. $x^2 + y^2 \geq 1$
    $-2 < x \leq 2$

42. $x^2 + y^2 \leq 25$
    $-3 \leq y < 3$

Problem Solving

For Exercises 43 and 44, (a) graph the solution of the system, (b) list three ordered pairs that lie in the graphed region, and (c) state the meaning of the three ordered pairs you listed in part (b).

43. A neighborhood-theater owner wants to be able to admit a family of two adults and one child for no more than $10. Also, the owner wants the price of a child's ticket, x, to be no more than half the price of an adult's ticket, y. A system of inequalities that describes this situation is

$$x + 2y \leq 10$$
$$x \leq \tfrac{1}{2}y$$
$$x \geq 0$$
$$y \geq 0$$

44. The storeroom of a shoe store will hold at most 300 pairs of shoes, and the owner wants to order at least twice as many women's shoes as men's shoes. If x and y are the number of pairs of women's shoes and men's shoes, respectively, then a system of inequalities that describes this situation is

$$x + y \leq 300$$
$$x \geq 2y$$
$$x \geq 0$$
$$y \geq 0$$

 Getting Acquainted with Your Graphing Calculator

Graph each inequality or system of inequalities.

45. $y \geq x^2 - 2x - 3$

46. $x^2 + y^2 \leq 16$

47. $x + y \leq 3$
    $y \geq x$

48. $y \geq x - 1$
    $y \leq x^3 - 4x$

 11.10 Linear Programming

During World War II, *linear programming* became an important method for tracking huge quantities of men and material as efficiently as possible. Since that time, the emergence of electronic computers has accelerated the development of linear programming as an invaluable tool for improving efficiency in a variety of fields.

To illustrate how linear programming works, consider Example 1.

Example 1   A company manufactures dolls and teddy bears. The profit from each doll is $10, and the profit from each bear is $8. Therefore, if P is the total profit, x is the number of dolls, and y is the number of bears, then

$$P = 10x + 8y$$

The company is able to manufacture no more than 60 dolls and bears combined. That is,

$$x + y \leq 60$$

Because of demand, the company wants to make at least twice as many bears as dolls. That is,

$$y \geq 2x$$

Neither the number of dolls nor the number of bears can be negative, so

$$x \geq 0 \quad \text{and} \quad y \geq 0$$

How many dolls and bears should be manufactured each day to maximize profit?

Solution: We want to find the values of x and y that give the greatest value of P in the equation

$$P = 10x + 8y$$

Note that P is a function of the two variables x and y. The equation $P = 10x + 8y$ is called the *objective function*. From the objective function $P = 10x + 8y$, we see that the greater the number of dolls, x, and bears, y, the greater the profit P. However, there are certain restrictions, or *constraints*, on x and y. These constraints form the system of inequalities below:

$$x + y \leq 60$$
$$y \geq 2x$$
$$x \geq 0$$
$$y \geq 0$$

The graph of this system, shown as the shaded region in Fig. 11.29, forms the *region of feasible solutions* (or simply, the *feasible region*). The feasible region includes the edges.

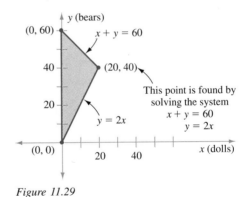

Figure 11.29

In other words, we can only choose points (x, y) from the shaded region in Fig. 11.29 to substitute into

$$P = 10x + 8y$$

But there are an infinite number of points in the shaded region; we cannot test them all! Fortunately, it can be shown that the maximum value of P (as well as the minimum value of P) will always occur at a **vertex** (corner point) of the feasible region. (It may also occur at every point along an entire edge of the feasible region.) Therefore, we simply test the vertices, looking for the greatest value of P.

Vertex	Profit: $P = 10x + 8y$
(0, 0)	$P = 10(0) + 8(0) = 0$
(0, 60)	$P = 10(0) + 8(60) = 480$
(20, 40)	$P = 10(20) + 8(40) = 520 \leftarrow$ Maximum profit

Therefore, the company should manufacture 20 dolls and 40 bears each day. This will produce a maximum daily profit of $520. ☐

We can summarize the ideas introduced in Example 1 by the following theorem.

Try Exercise 11

Find the maximum value of $P = 5x + 3y$ on the feasible region below:

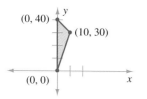

Fundamental Theorem of Linear Programming

If a function of the form $z = ax + by + c$ (**objective function**) is defined for the set of ordered pairs (x, y) that are solutions of a system of linear inequalities (**constraints**), then any **optimum** (maximum or minimum) value of z will occur at a vertex of the graph of the system (**feasible region**).

Example 2   Find both the maximum and the minimum value of the objective function $z = 2x + 9y$, subject to the constraints $y \leq x$, $y \geq 1$, $x \geq 2$, and $x \leq 5$.

Solution: Graph the feasible region, as shown in Fig. 11.30.

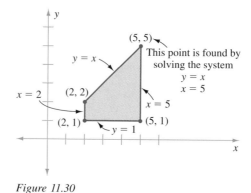

Figure 11.30

Test each vertex in the objective function.

Vertex	$z = 2x + 9y$
(2, 1)	$z = 2(2) + 9(1) = 13 \leftarrow$ Minimum value
(2, 2)	$z = 2(2) + 9(2) = 22$
(5, 5)	$z = 2(5) + 9(5) = 55 \leftarrow$ Maximum value
(5, 1)	$z = 2(5) + 9(1) = 19$

The maximum value is $z = 55$, and it occurs when $(x, y) = (5, 5)$. The minimum value is $z = 13$, and it occurs when $(x, y) = (2, 1)$. ☐

Try Exercise 13

Find the maximum and minimum value of $z = 3x + 10y$ subject to the constraints $y \leq x$, $y \geq 2$, $x \geq 3$, and $x \leq 6$.

Example 3   A patient is put on a liquid diet consisting of two different drinks. Drink X costs $3 per pint, and Drink Y costs $4 per pint. Each pint of Drink X contains 200 cal, 2 units of vitamin A, and 6 units of vitamin B. Each pint of Drink Y contains 200 cal, 5 units of vitamin A, and 3 units of vitamin B. The diet is to provide at least 1000 cal, at least 16 units of vitamin A, and at least 18 units of vitamin B per day. How many pints of each drink should the patient consume each day to minimize cost, while meeting the daily nutritional requirements?

Solution: Let C be the total cost of the two drinks, let x be the number of pints of Drink X, and let y be the number of pints of Drink Y. Then the objective function is

$$C = 3x + 4y$$

The constraints are given below:

Calories:

$$200x + 200y \geq 1000$$

Vitamin A:

$$2x + 5y \geq 16$$

Vitamin B:

$$6x + 3y \geq 18$$

Graph the feasible region, as shown in Fig. 11.31.

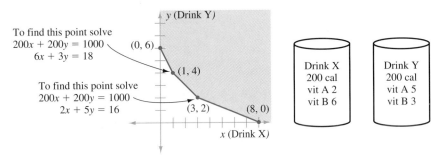

To find this point solve
$200x + 200y = 1000$
$6x + 3y = 18$

$(0, 6)$

$(1, 4)$

To find this point solve
$200x + 200y = 1000$
$2x + 5y = 16$

$(3, 2)$ $(8, 0)$

y (Drink Y)

x (Drink X)

Drink X
200 cal
vit A 2
vit B 6

Drink Y
200 cal
vit A 5
vit B 3

Figure 11.31

Test each vertex in the objective function.

Vertex	$C = 3x + 4y$
$(0, 6)$	$C = 3(0) + 4(6) = 24$
$(1, 4)$	$C = 3(1) + 4(4) = 19$
$(3, 2)$	$C = 3(3) + 4(2) = 17 \leftarrow$ Minimum cost
$(8, 0)$	$C = 3(8) + 4(0) = 24$

To meet the daily requirements at a minimum daily cost of \$17, the patient should consume 3 pints of Drink X and 2 pints of Drink Y.   ❏

Try Exercise 17

Repeat Example 3 using the following information:

	Cost	Calories	A	C
X	3	300	6	2
Y	2	300	3	8

The patient needs at least 1200 cal, at least 15 units of A, and at least 14 units of C.

CAUTION

In Example 3, the objective function $C = 3x + 4y$ has a minimum value at the point $(x, y) = (3, 2)$. However, C has no maximum value on the feasible region, since the farther the point (x, y) is from the origin, the greater the value of C. There is no maximum value of C because the feasible region is *unbounded.* A region in the plane is **unbounded** if it cannot be contained in some circle. The region of Example 2 is **bounded,** since it can be contained in some circle. If the feasible region is bounded, the objective function will have both a maximum and a minimum value on that region. If the feasible region is unbounded, the objective function may or may not have an optimum value (maximum or minimum value) on that region.   ■

Here is a summary of the steps in solving a linear programming problem.

> **To Solve a Linear Programming Problem**
> 1. Write the objective function.
> 2. Write the constraints to form a system of linear inequalities.
> 3. Graph the system of inequalities to obtain the feasible region.
> 4. Find the vertices of the feasible region.
> 5. Test all vertices in the objective function to find the optimum (maximum or minimum) value.
>
> Note:
>
> For a bounded feasible region, the objective function will attain both a maximum and a minimum value at a vertex of the feasible region. For an unbounded feasible region, *if* the objective function has an optimum value, it will occur at a vertex of the feasible region.

Exercises 11.10

Completion Exercises

A linear programming problem seeks to maximize the function $P = 7x + 12y$, using points (x, y) that lie in the region given by the system of linear inequalities

$$x + 2y \leq 6$$
$$y \leq x$$
$$y \geq 0$$

1. The function $P = 7x + 12y$ is called the _____ function. The inequalities are called the _____. The graph of the system of inequalities is called the _____ region.

2. Since the region determined by the system of inequalities is bounded, $P = 7x + 12y$ attains both a maximum and a minimum value on that region, and each optimum value occurs at _____ of that region.

3. A region in the plane is bounded if _____.

4. A region in the plane is unbounded if _____.

True-False Exercises

5. If the feasible region is bounded, the objective function will attain both a maximum and a minimum value on that

region, and both values will occur at a vertex of the region.

6. For an unbounded feasible region, if the objective function has an optimum value, it will occur at a vertex of the region.

Find the maximum and the minimum value of the given objective function on the given feasible region.

7. Objective function:
$z = 4x + 3y$
Feasible region:

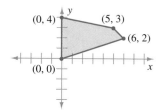

8. Objective function:
$z = 2x + 5y$
Feasible region:

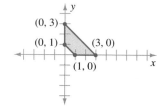

9. Objective function:
$z = x + 6y$
Feasible region:

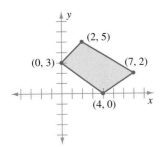

10. Objective function:
$z = 7x + y$
Feasible region:

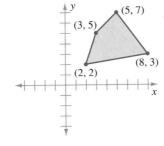

12. A specialty firm makes hats and belts. The profit from each hat is $8, and the profit from each belt is $4. Therefore, if P is the total profit, x is the number of hats, and y is the number of belts, then $P = 8x + 4y$. The company is able to manufacture no more than 50 hats and belts combined each day. That is, $x + y \leq 50$. Because of demand, the company wants to make at least as many belts as hats. That is, $y \geq x$. Neither the number of hats nor the number of belts can be negative, so $x \geq 0$ and $y \geq 0$. How many hats and belts should be manufactured each day to maximize profit? See Fig. 11.33.

Find the maximum and the minimum value of the given objective function subject to the given constraints.

Problem Solving

11. A company manufactures toy cars and toy trucks. The profit from each car is $5, and the profit from each truck is $3. Therefore, if P is the total profit, x is the number of cars, and y is the number of trucks, then $P = 5x + 3y$. The company is able to manufacture no more than 40 cars and trucks combined each day. That is, $x + y \leq 40$. Because of demand, the company wants to make at least 3 times as many trucks as cars. That is, $y \geq 3x$. Neither the number of cars nor the number of trucks is negative, so $x \geq 0$ and $y \geq 0$. How many cars and trucks should be manufactured each day to maximize profit? See Fig. 11.32.

13. Objective function:
$z = 3x + 10y$
Constraints:
$$y \leq x$$
$$y \geq 2$$
$$x \geq 3$$
$$x \leq 6$$

14. Objective function:
$z = 11x + 2y$
Constraints:
$$y \geq x$$
$$y \leq 5$$
$$x \geq 1$$
$$x \leq 4$$

15. Objective function:
$z = 8x + y$
Constraints:
$$x + y \leq 5$$
$$x + y \geq 2$$
$$x \geq 0$$
$$y \geq 0$$

16. Objective function:
$z = x + 9y$
Constraints:
$$x + y \leq 3$$
$$x + 2y \leq 4$$
$$x \geq 0$$
$$y \geq 0$$

Problem Solving

17. A patient is put on a liquid diet consisting of two different drinks. Drink X costs $3 per pint, and Drink Y costs $2 per pint. Each pint of Drink X contains 300 cal, 6 units of vitamin A, and 2 units of vitamin C. Each pint of Drink Y contains 300 cal, 3 units of vitamin A, and 8 units of vitamin C. The diet is to provide at least 1200 cal, at least 15 units of vitamin A, and at least 14 units of vitamin C per day. How many pints of each drink should the patient consume each day to minimize cost, while meeting the daily nutritional requirements?

18. A particular diet requires at least 16 units of vitamin A, at least 30 units of vitamin B, and at least 36 units of vitamin C per day. Each gram of food supplement X costs 50¢ and contains 2 units of vitamin A, 5 units of vitamin B, and 4 units of vitamin C. Each gram of food supplement Y costs 80¢ and contains 3 units of vitamin A, 5 units of vitamin B, and 8 units of vitamin C. How many grams of each supplement should be taken each day to minimize cost and yet fulfill the daily nutritional requirements?

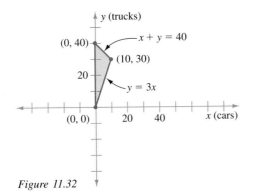

Figure 11.32

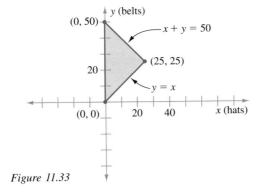

Figure 11.33

19. A pharmaceutical firm manufactures two different types of pills for headaches. Pink pills yield a profit of 30¢ per pill, and white pills yield a profit of 35¢ per pill. The demand for white pills is less than or equal to half the demand for pink pills. No more than 1200 pills of both types combined can be manufactured per minute. The number of white pills that can be manufactured each minute is at least 200 less than one-third of the number of pink pills. (a) Determine the production level that will maximize profit. (b) What should the production level be if the profit on each pink pill is 40¢?

20. A grain supplier has two warehouses, W_1 and W_2. There are 90 tons of grain stored at W_1 and 70 tons at W_2. Two farmers, Smith and Jones, order 45 tons and 60 tons, respectively. The shipping cost from each warehouse to each farmer is shown in the table below:

Warehouse	Farmer	Shipping cost per ton, $
W_1	Smith	5
W_1	Jones	3
W_2	Smith	7
W_2	Jones	8

(a) How should the two orders be filled to minimize total shipping costs: (*Hint:* Let x be the number of tons shipped from W_1 to Smith, and let y be the number of tons shipped from W_1 to Jones.)

(b) How should the two orders be filled if the shipping costs are, in order from top to bottom, $3, $5, $8, and $7?

Discussion Exercises

21. Sketch a bounded feasible region and test various points, both on the boundary and inside the region, to convince yourself that the objective function $z = 2x + 3y$ attains both a maximum and a minimum value on that region, and that those values occur at vertices. Then try the same exercise with an unbounded region.

22. Outline the steps in solving a linear programming problem.

Chapter 11 Key Terms

11.1 System of equations Two or more equations that are true simultaneously

Linear system (of equations) A system whose equations are all linear

Solution (of a system of two equations in two variables) An ordered pair that satisfies both equations

Independent system (of two linear equations in two variables) A system whose two lines are different

Dependent system (of two linear equations in two variables) A system whose two lines are the same line

Consistent system A system that has at least one solution

Inconsistent system A system that has no solution

11.4 Ordered triple (of numbers) Three numbers written in a specified order

Inconsistent system (of three linear equations in three variables) A system whose three planes do not contain a common point

Dependent system (of three linear equations in three variables) A system whose planes contain an infinite number of points in common

11.5 Determinant A square array of numbers enclosed between two vertical lines

Element (of a determinant) One of the numbers in a determinant

3 × 3 (third-order) determinant A determinant with three rows and three columns

Minor (of an element in a 3 × 3 determinant) The 2 × 2 determinant that results when both the row and the column that contain the element are deleted

Expansion by minors A method for evaluating a determinant

Sign array (for a 3 × 3 determinant) $\begin{vmatrix} + & - & + \\ - & + & - \\ + & - & + \end{vmatrix}$

11.6 Cramer's rule A method for solving a linear system of equations using determinants

11.7 Matrix A rectangular array of numbers enclosed in square brackets

3 × 4 matrix A matrix with three rows and four columns

Element (of a matrix) One of the numbers in a matrix

Square matrix A matrix that contains the same number of rows as columns

Main diagonal (of a square matrix) The elements from the upper left corner to the lower right corner of the matrix

Augmented matrix A matrix that represents the coefficients and the constant terms of a system of equations

Coefficient matrix The matrix formed by that part of the augmented matrix that represents the coefficients of a system of equations

Gaussian elimination A method for solving a linear system of equations that involves writing the augmented matrix in a form like the matrix below.

$$\begin{bmatrix} 1 & a & b & c \\ 0 & 1 & d & e \\ 0 & 0 & 1 & f \end{bmatrix}$$

Gauss-Jordan method A method for solving a linear system of equations that involves writing the augmented matrix in a form like the matrix below.

$$\begin{bmatrix} 1 & 0 & 0 & a \\ 0 & 1 & 0 & b \\ 0 & 0 & 1 & c \end{bmatrix}$$

11.8 **Nonlinear system** (of equations) A system of equations that contains at least one nonlinear equation

11.9 **Second-degree inequality** An inequality that contains at least one second-degree (and no higher-degree) term

System of inequalities Two or more inequalities that are true simultaneously

Graph (of a system of inequalities in two variables) The set of all those points (x, y) that satisfy all of the inequalities in the system

11.10 **Objective function** The function to be maximized or minimized in a linear programming problem

Constraints Restrictions that are placed on the domain of the objective function

Feasible region The graph of the system of inequalities that make up the constraints

Vertex (of a feasible region) A corner point of the feasible region

Optimum value A maximum value or a minimum value

Unbounded region (in the plane) A region that cannot be contained in some circle

Bounded region (in the plane) A region that can be contained in some circle

Chapter 11 Key Rules/Steps

11.1 The Graphing Method

Use the Graphing Method to Solve $x + y = 5$
$$x - y = 1$$

Graph each line. The solution is the intersection point $(3, 2)$. See Fig. 11.34.

Check: $\quad x + y = 5 \qquad\qquad x - y = 1$
$\qquad\qquad 3 + 2 = 5 \qquad\qquad 3 - 2 = 1$
$\qquad\qquad\quad 5 = 5 \quad$ True $\qquad\quad 1 = 1 \quad$ True

Independent and Consistent System

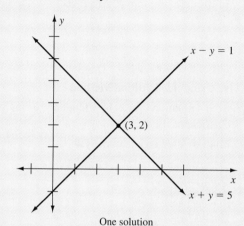

Figure 11.34

One solution

Independent and Inconsistent System

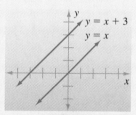

No solution

Dependent and Consistent System

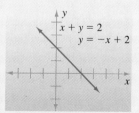

Infinite number of solutions

11.2 The Substitution Method and the Addition Method

Given the system $3x + y = 11$
$$x - 2y = 6$$

Solve Using the Substitution Method

1. Solve the first equation for y.

$$y = 11 - 3x$$

2. Substitute into the second equation.

$$x - 2(11 - 3x) = 6$$

3. Solve.

$$x - 22 + 6x = 6$$
$$7x = 28$$
$$x = 4$$

4. Substitute $x = 4$ into $y = 11 - 3x$.

$$y = 11 - 3(4)$$
$$y = -1$$

5. Check the solution $(4, -1)$ in both equations.

Solve Using the Addition Method

1. Both equations are in standard form with integer coefficients.

2. Multiply the first equation by 2.

$$6x + 2y = 22$$

3. Add and solve.

$$6x + 2y = 22$$
$$\underline{x - 2y = 6}$$
$$7x = 28$$
$$x = 4$$

4. Substitute $x = 4$ into $x - 2y = 6$.

$$4 - 2y = 6$$
$$-2y = 2$$
$$y = -1$$

5. Check the solution $(4, -1)$ in both equations.

Note:

If a contradiction (like $0 = 1$) arises in the solving process, the lines are parallel and there is no solution. If an identity (like $0 = 0$) arises, the lines coincide and any point on the line is a solution.

11.3 Problem Solving with Linear Systems

To Solve a Word Problem Using a System of Equations

Two numbers have a sum of 19 and a difference of 7. Find the numbers.

1. Represent each unknown by a different variable.

$$x = \text{first number} \qquad y = \text{second number}$$

2. Write a system of equations.

$$x + y = 19$$
$$x - y = 7$$

3. Solve the system.

$$2x = 26 \qquad \text{Add the equations.}$$
$$x = 13 \qquad \text{Divide by 2.}$$
$$y = 6 \qquad \text{Find } y.$$

4. Check in the original problem.

13 and 6 have a sum of 19 and a difference of 7

11.4 Linear Systems of Three Equations

To Solve
$$x + y - z = 2$$
$$x - 2y + z = 3$$
$$x + 3y - z = 4$$

1. Use any two equations to get an equation in two variables.

$$x + y - z = 2 \qquad \text{First equation}$$
$$\underline{x - 2y + z = 3} \qquad \text{Second equation}$$
$$2x - y = 5 \qquad \text{Add.}$$

2. Use a different pair of equations to get an equation in the same two variables as step 1.

$$x - 2y + z = 3 \qquad \text{Second equation}$$
$$\underline{x + 3y - z = 4} \qquad \text{Third equation}$$
$$2x + y = 7 \qquad \text{Add.}$$

3. Solve the system resulting from steps 1 and 2.

$$2x - y = 5 \qquad \text{Equation from step 1}$$
$$\underline{2x + y = 7} \qquad \text{Equation from step 2}$$
$$4x = 12 \qquad \text{Add.}$$
$$x = 3 \qquad y = 1 \qquad \text{and} \qquad z = 2$$

4. Check $(3, 1, 2)$ in all three equations.

Note:

If at any time a contradiction (like $-2 = 5$) or an identity (like $4 = 4$) is obtained, the system has no unique solution.

11.5 Determinants

Value of a 2 × 2 Determinant

$$\begin{vmatrix} a_1 & b_1 \\ a_2 & b_2 \end{vmatrix} = a_1 b_2 - a_2 b_1$$

Example: $\begin{vmatrix} 3 & 2 \\ 4 & 5 \end{vmatrix} = 3(5) - 4(2) = 7$

Value of a 3 × 3 Determinant

$$\begin{vmatrix} a_1 & b_1 & c_1 \\ a_2 & b_2 & c_2 \\ a_3 & b_3 & c_3 \end{vmatrix} = a_1 \begin{vmatrix} b_2 & c_2 \\ b_3 & c_3 \end{vmatrix} - a_2 \begin{vmatrix} b_1 & c_1 \\ b_3 & c_3 \end{vmatrix} + a_3 \begin{vmatrix} b_1 & c_1 \\ b_2 & c_2 \end{vmatrix}$$

Example:
$$\begin{vmatrix} 3 & 2 & -6 \\ -1 & 5 & 4 \\ 0 & 7 & 1 \end{vmatrix}$$
$$= 3\begin{vmatrix} 5 & 4 \\ 7 & 1 \end{vmatrix} - (-1)\begin{vmatrix} 2 & -6 \\ 7 & 1 \end{vmatrix} + 0\begin{vmatrix} 2 & -6 \\ 5 & 4 \end{vmatrix}$$
$$= 3(5 - 28) + 1(2 + 42) + 0(8 + 30)$$
$$= 3(-23) + 1(44) + 0(38)$$
$$= -25$$

11.6 Cramer's Rule

Cramer's Rule for a System of Three Equations

The solution of the system
$$a_1x + b_1y + c_1z = d_1$$
$$a_2x + b_2y + c_2z = d_2$$
$$a_3x + b_3y + c_3z = d_3$$

is given by
$$x = \frac{D_x}{D} \qquad y = \frac{D_y}{D} \qquad z = \frac{D_z}{D}$$

where
$$D = \begin{vmatrix} a_1 & b_1 & c_1 \\ a_2 & b_2 & c_2 \\ a_3 & b_3 & c_3 \end{vmatrix} \qquad D_x = \begin{vmatrix} d_1 & b_1 & c_1 \\ d_2 & b_2 & c_2 \\ d_3 & b_3 & c_3 \end{vmatrix}$$
$$D_y = \begin{vmatrix} a_1 & d_1 & c_1 \\ a_2 & d_2 & c_2 \\ a_3 & d_3 & c_3 \end{vmatrix} \qquad D_z = \begin{vmatrix} a_1 & b_1 & d_1 \\ a_2 & b_2 & d_2 \\ a_3 & b_3 & d_3 \end{vmatrix}$$

so long as $D \neq 0$.

Note:

If $D = 0$, Cramer's rule does not apply and the system does not have a unique solution. Use the addition method or the substitution method to determine whether there is no solution or an infinite number of solutions.

Example: For the system
$$x + 2y - z = -4$$
$$3x - y + z = 15$$
$$5x + 4y - 2z = 1$$
$$D = \begin{vmatrix} 1 & 2 & -1 \\ 3 & -1 & 1 \\ 5 & 4 & -2 \end{vmatrix} = 3$$
$$D_x = \begin{vmatrix} -4 & 2 & -1 \\ 15 & -1 & 1 \\ 1 & 4 & -2 \end{vmatrix} = 9$$

$$D_y = \begin{vmatrix} 1 & -4 & -1 \\ 3 & 15 & 1 \\ 5 & 1 & -2 \end{vmatrix} = -3$$
$$D_z = \begin{vmatrix} 1 & 2 & -4 \\ 3 & -1 & 15 \\ 5 & 4 & 1 \end{vmatrix} = 15$$

By Cramer's rule,
$$x = \frac{9}{3} = 3 \qquad y = \frac{-3}{3} = -1 \qquad z = \frac{15}{3} = 5$$

The solution set is $\{(3, -1, 5)\}$.

11.7 Solving Linear Systems by Matrices

Row Operations on an Augmented Matrix

Type 1: Interchange any two rows.

Example: $\begin{bmatrix} 1 & 2 & | & 3 \\ 4 & 5 & | & 6 \end{bmatrix} \overset{R_1 \leftrightarrow R_2}{\to} \begin{bmatrix} 4 & 5 & | & 6 \\ 1 & 2 & | & 3 \end{bmatrix}$

Type 2: Multiply any row by a nonzero number.

Example: $\begin{bmatrix} 1 & 2 & | & 3 \\ 4 & 5 & | & 6 \end{bmatrix} \overset{-2R_1}{\to} \begin{bmatrix} -2 & -4 & | & -6 \\ 4 & 5 & | & 6 \end{bmatrix}$

Type 3: Multiply any row by a number and add the result to another row.

Example: $\begin{bmatrix} 1 & 2 & | & 3 \\ 4 & 5 & | & 6 \end{bmatrix} \overset{2R_1 + R_2}{\to} \begin{bmatrix} 1 & 2 & | & 3 \\ 6 & 9 & | & 12 \end{bmatrix}$

Solving a Linear System of Two Equations by Gaussian Elimination

$$2x - y = 8$$
$$x + y = 7$$
$$\begin{bmatrix} 2 & -1 & | & 8 \\ 1 & 1 & | & 7 \end{bmatrix} \overset{R_1 \leftrightarrow R_2}{\to} \begin{bmatrix} 1 & 1 & | & 7 \\ 2 & -1 & | & 8 \end{bmatrix}$$
$$\overset{-2R_1 + R_2}{\to} \begin{bmatrix} 1 & 1 & | & 7 \\ 0 & -3 & | & -6 \end{bmatrix} \overset{-\frac{1}{3}R_2}{\to} \begin{bmatrix} 1 & 1 & | & 7 \\ 0 & 1 & | & 2 \end{bmatrix}$$
$$\left. \begin{matrix} x + y = 7 \\ y = 2 \end{matrix} \right\} \to x + 2 = 7 \to x = 5$$

The solution set is $\{(5, 2)\}$.

Solving a Linear System of Two Equations Using the Gauss-Jordan Method

$$2x - y = 8$$
$$x + y = 7$$

Use the steps shown above to write the augmented matrix in the form
$$\begin{bmatrix} 1 & 1 & | & 7 \\ 0 & 1 & | & 2 \end{bmatrix}$$

Then continue.

$$\begin{bmatrix} 1 & 1 & | & 7 \\ 0 & 1 & | & 2 \end{bmatrix} \begin{array}{c} -R_2 + R_1 \\ \rightarrow \end{array} \begin{bmatrix} 1 & 0 & | & 5 \\ 0 & 1 & | & 2 \end{bmatrix} \begin{array}{c} \rightarrow x = 5 \\ \rightarrow y = 2 \end{array}$$

The solution set is $\{(5, 2)\}$.

Note:

If an augmented matrix of the form

$$\begin{bmatrix} 1 & 4 & | & 6 \\ 0 & 0 & | & -3 \end{bmatrix}$$

is obtained, the system is inconsistent and there is no solution.
If an augmented matrix of the form

$$\begin{bmatrix} 1 & -5 & | & 7 \\ 0 & 0 & | & 0 \end{bmatrix}$$

is obtained, the system is dependent and there are an infinite number of solutions. In either case, there is no unique solution.

11.8 Nonlinear Systems of Equations

Solve $x^2 + 2xy + y^2 = 9$
$\qquad\quad x^2 + y^2 = 5$

Multiply the second equation by -1 and add the result to the first equation.

$$
\begin{array}{ll}
x^2 + 2xy + y^2 = 9 & \text{First equation} \\
\underline{-x^2 - y^2 = -5} & \text{Second equation multiplied by } -1 \\
 2xy = 4 & \text{Add.} \\
 y = \dfrac{2}{x} & \text{Divide by } 2x \text{ and simplify.}
\end{array}
$$

Substitute $y = \frac{2}{x}$ into $x^2 + y^2 = 5$.

$$
\begin{array}{ll}
x^2 + \dfrac{4}{x^2} = 5 & \\
x^4 + 4 = 5x^2 & \text{Multiply by } x^2. \\
x^4 - 5x^2 + 4 = 0 & \text{Subtract } 5x^2. \\
(x^2 - 1)(x^2 - 4) = 0 & \text{Factor.} \\
x^2 - 1 = 0 \text{ or } x^2 - 4 = 0 & \text{Set each factor equal to 0.}
\end{array}
$$

$$
\begin{array}{cccc}
x = 1 & x = -1 & x = 2 & x = -2 \\
y = \dfrac{2}{1} = 2 & y = \dfrac{2}{-1} = -2 & y = \dfrac{2}{2} = 1 & y = \dfrac{2}{-2} = -1
\end{array}
$$

The solution set is $\{(1, 2), (-1, -2), (2, 1), (-2, -1)\}$. Check in the original system.

11.9 Second-Degree Inequalities; Systems of Inequalities

Graph $y \geq x^2$

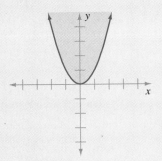

Graph $x^2 + y^2 < 9$
$\qquad\quad y < x + 1$

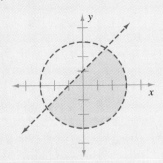

11.10 Linear Programming

To Solve a Linear Programming Problem

A pen yields a profit of $3, and a pencil yields a profit of $2. The number of pens, y, and of pencils, x, combined is no more than 60. The number of pens is less than or equal to half the number of pencils. Find the number of pencils and pens that maximize profit P.

1. Write the objective function.

$$P = 2x + 3y$$

2. Write the constraints.

$$
\begin{aligned}
x + y &\leq 60 \\
y &\leq \tfrac{1}{2}x \\
x &\geq 0 \\
y &\geq 0
\end{aligned}
$$

3 & 4. Graph the system of inequalities to obtain the feasible region. Find the vertices.

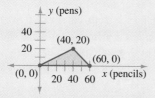

5. Test the vertices in the objective function.

Vertex	$P = 2x + 3y$
$(0, 0)$	$P = 2(0) + 3(0) = 0$
$(40, 20)$	$P = 2(40) + 3(20) = 140 \leftarrow$ Max
$(60, 0)$	$P = 2(60) + 3(0) = 120$

The maximum profit $P = \$140$ occurs with 40 pencils and 20 pens.

Note:

For a bounded feasible region, the objective function will attain both a maximum and a minimum value at a vertex of the feasible region. For an unbounded feasible region, *if* the objective function has an optimum value, it will occur at a vertex of the feasible region.

Chapter 11 Review Exercises

[11.1] Determine whether the given ordered pair is a solution of the given system.

1. $(-4, 9)$
   $$2x + y = 1$$
   $$y = 3x + 3$$

2. $(-3, -5)$
   $$6x - 4y = 2$$
   $$5x = 3y$$

Solve each system by the graphing method. Classify each system as dependent or independent. Also classify each system as consistent or inconsistent.

3. $x + y = 5$
   $x - y = 3$

4. $y = 2x$
   $5y - 10x = 18$

5. $2x - 3y = 6$
   $-4x + 6y = -12$

[11.2] Solve each system by the substitution method.

6. $3x + 4y = 1$
   $y = x + 2$

7. $x - 3y = 3$
   $3x - 4y = -1$

8. $4x + 5y = 2$
   $3x + 2y = 5$

9. $-8x + 4y = -56$
   $2x = y + 14$

Solve each system by the addition method.

10. $9x - 5y = 33$
    $-x + 5y = 7$

11. $6x - y = 24$
    $-7x + 3y = -28$

12. $2x + 3y = 25$
    $3x + 5y = 43$

13. $-4x + 6y = 11$
    $6x - 9y = 10$

Solve each system using any method.

14. $\frac{1}{2}x + \frac{5}{6}y = \frac{1}{4}$
    $\frac{1}{4}x - \frac{2}{3}y = \frac{2}{3}$

15. $-\frac{2}{x} + \frac{3}{y} = -4$
    $\frac{5}{x} - \frac{4}{y} = -11$

[11.3] Write a system of equations for each exercise. Then solve the system.

16. One placemat and two tablecloths cost \$17. Four placemats and one tablecloth cost \$19. Find the cost of each placemat and each tablecloth.

17. To manufacture one toy car requires 3 units of capital and 5 units of labor. One toy truck requires 5 units of capital and 6 units of labor. How many cars and how many trucks can be manufactured with 156 units of capital and 211 units of labor?

18. The total receipts for a talent show were \$600. Adult tickets were \$5 and student tickets were \$2. If a total of 165 tickets were sold, how many adults and how many students attended the talent show?

19. The sum of the digits of a two-digit number is 14. If the digits are reversed, the new number is 18 less than the original number. Find the original number.

[11.4] Solve each system.

20. $x - y + 2z = 3$
    $x + 2y - 2z = 12$
    $2x - 3y + z = 3$

21. $2x + y + 3z = 2$
    $3x + 2y + 4z = 0$
    $5x + 2y = 0$

22. $x + 2y + z = 7$
    $x + y - z = 7$
    $2x + 4y + 2z = 5$

23. $x + 2y - z = 1$
    $x - 2y + z = 1$
    $2x - 4y + 2z = 2$

Write a system of three equations for each exercise. Then solve the system.

24. The perimeter of a triangle is 20 m. The longest side is 2 m less than the sum of the other two sides. The shortest side is twice the difference of the longest side and the middle side. Find the length of each side.

25. A collection of 72 nickels, dimes, and quarters has a total value of \$8. The number of nickels plus the number of quarters equals twice the number of dimes. How many of each kind of coin are in the collection?

[11.5] Evaluate each 2 × 2 determinant.

26. $\begin{vmatrix} 5 & 4 \\ 3 & 2 \end{vmatrix}$

27. $\begin{vmatrix} 3 & -5 \\ 0 & 2 \end{vmatrix}$

28. $\begin{vmatrix} \frac{1}{2} & -\frac{1}{4} \\ \frac{1}{4} & \frac{3}{4} \end{vmatrix}$

Evaluate each 3 × 3 determinant. Expand about the row or column of your choice.

29. $\begin{vmatrix} 1 & 2 & -4 \\ -1 & 0 & 1 \\ 3 & 0 & 2 \end{vmatrix}$

30. $\begin{vmatrix} 3 & 4 & 5 \\ 2 & 1 & 0 \\ 0 & 4 & -1 \end{vmatrix}$

31. $\begin{vmatrix} 1 & 2 & 1 \\ 1 & 1 & -1 \\ 2 & 4 & 2 \end{vmatrix}$

Solve each determinant equation for x.

32. $\begin{vmatrix} x & 2 \\ 4x & 3 \end{vmatrix} = 5$

33. $\begin{vmatrix} x & \frac{1}{2} \\ 3x & 2 \end{vmatrix} = 3$

[11.6] Use Cramer's rule to solve each system.

34. $3x + 4y = 6$
$5x + 3y = -1$

35. $2x - y = 0$
$3x + 2y = 13$

36. $2x + 5y = -3$
$-3x + 10y = 8$

37. $x + 2y + z = 3$
$2x - y - z = 0$
$3x - 2y - z = 5$

38. $2x - y + z = 5$
$4x - 3y = 5$
$6x + 2y + 2z = 7$

39. $5x - 3y = 7$
$2x + 8y - 2z = 4$
$3x + 9z = 11$

[11.7] Solve each system by matrices.

40. $2x + 7y = 3$
$x + 5y = 6$

41. $4x + 3y = 10$
$-6x + 2y = 1$

42. $x - y + 4z = 8$
$3x - y + 6z = -10$
$-4x - 5z = 6$

43. $3x + 3y - z = 3$
$x + 6y + 5z = 0$
$x - 3y - 3z = 1$

[11.8] Solve each system.

44. $y = x^2$
$y - 2x = 3$

45. $x^2 + y^2 = 25$
$4x - 3y = 0$

46. $xy = -15$
$x + y = 2$

47. $x^2 + y^2 = 25$
$x^2 - y^2 = 7$

48. $x^2 + 7y^2 = 16$
$x^2 + y^2 = 10$

49. $4x^2 + y^2 = 17$
$x^2 + y = 5$

50. $x^2 - 4xy + y^2 = 6$
$x^2 + y^2 = 26$

51. $2x^2 + 3y^2 = 10$
$x^2 + y^2 = 3$

52. A nonlinear system consists of two equations, each of whose graph is a circle. Assuming the two circles are different, how many possible solutions does the system have? Draw graphs to illustrate your answer.

53. Two different squares have a combined perimeter totalling 68 in. One square contains 17 in.^{2} more than the other square. Find the length of a side of each square.

[11.9] Graph each inequality.

54. $(x - 3)^2 + y^2 < 9$

55. $y \geq x^2 - 2x$

56. $y^2 - x^2 \geq 4$

57. $\dfrac{x^2}{16} + y^2 < 1$

Graph each system of inequalities.

58. $y \geq -x$
$y \leq x + 3$

59. $x^2 + y^2 < 16$
$y > x^2$

60. $x^2 + y^2 < 25$
$x^2 + y^2 > 16$

61. $9x^2 + 16y^2 \leq 144$
$16x^2 + 9y^2 \leq 144$

[11.10] For Exercises 62 and 63, find the maximum and the minimum value of the objective function on the given feasible region.

62. Objective function:
$z = 7x + y$
Feasible region:

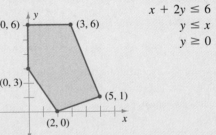

63. Objective function:
$z = 5x + 2y$
Constraints:
$x + 2y \leq 6$
$y \leq x$
$y \geq 0$

64. Green pills cost 20¢ each and contain 2 units of vitamin A, 6 units of vitamin B, and 3 units of vitamin C. Yellow pills cost 25¢ each and contain 3 units of vitamin A, 4 units of vitamin B, and 9 units of vitamin C. A patient needs at least 13 units of vitamin A, at least 24 units of vitamin B, and at least 24 units of vitamin C each day. How many of each type of pill should be taken to minimize cost, while meeting the daily nutritional requirements?

65. Explain the difference between a bounded and an unbounded region in the plane.

66. If the feasible region is bounded, will the objective function have a maximum value on the feasible region? Will it have a minimum value? What if the feasible region is unbounded?

Chapter 11 Test

1. Determine whether $(2, -3)$ is a solution of the system

$$3x + y = 3$$
$$x - 2y = 8$$

Evaluate each determinant.

2. $\begin{vmatrix} 6 & -3 \\ 5 & -2 \end{vmatrix}$

3. $\begin{vmatrix} 2 & -1 & 3 \\ 1 & 2 & -1 \\ 3 & -2 & 3 \end{vmatrix}$

Solve each system by the substitution method.

4. $\begin{aligned} 2x &= 5 + y \\ 5x + 6y &= 4 \end{aligned}$

5. $\begin{aligned} 4x + 3y &= 8 \\ 2x - 5y &= 30 \end{aligned}$

Solve each system by the addition method.

6. $\begin{aligned} 3x - 2y &= 3 \\ x - y &= -1 \end{aligned}$

7. $\begin{aligned} \frac{x}{3} - \frac{y}{4} &= \frac{5}{6} \\ \frac{1}{2}x + \frac{2}{5}y &= \frac{5}{2} \end{aligned}$

Solve each system by the graphing method. Classify each system as dependent or independent. Also classify each system as consistent or inconsistent.

8. $\begin{aligned} x + y &= 6 \\ x - 2y &= 0 \end{aligned}$

9. $\begin{aligned} 3x + y &= 3 \\ 6x + 2y &= -8 \end{aligned}$

Solve each system.

10. $\begin{aligned} x + y + z &= 3 \\ y + 2z &= 3 \\ 2x - z &= 2 \end{aligned}$

11. $\begin{aligned} x - y + z &= -4 \\ 2x + 3y + z &= 5 \\ 4x + y - z &= 9 \end{aligned}$

Use Cramer's rule to solve each system.

12. $\begin{aligned} 4x - 5y &= 11 \\ y &= 3x + 11 \end{aligned}$

13. $\begin{aligned} 2x + 3y + z &= 4 \\ x - z &= -5 \\ x + 6y &= 4 \end{aligned}$

Solve each system by matrices.

14. $\begin{aligned} 3x + 5y &= -9 \\ -2x - 4y &= 1 \end{aligned}$

15. $\begin{aligned} x - 2y + z &= -3 \\ 2x + 5y - 4z &= 9 \\ 3x - 6z &= -39 \end{aligned}$

Write a system of equations for each exercise. Then solve the system.

16. A number x added to 5 times a number y is 17. Find the numbers if x is 1 less than 4 times y.

17. A store owner wants to mix nuts worth $1.90 per pound with nuts worth $1.10 per pound to make 10 lb of nuts worth $1.40 per pound. How many pounds of each should she use?

18. A motorboat travels 33 mi upstream in 3 hr, then turns and travels 28 mi downstream in 2 hr. Find the speed of the boat and the speed of the current.

19. The sum of the angles of a triangle is 180°. The largest angle is 16° more than the sum of the other two. The middle angle is 5° less than twice the smallest angle. Find the angles.

Solve each system.

20. $\begin{aligned} x^2 + y^2 &= 25 \\ y &= 2x - 5 \end{aligned}$

21. $\begin{aligned} x - y &= 4 \\ xy &= 5 \end{aligned}$

22. $\begin{aligned} x^2 + 4y^2 &= 25 \\ x^2 - y^2 &= 5 \end{aligned}$

Graph each inequality.

23. $x^2 + y^2 < 9$

24. $y \le x^2 - 4$

Graph each system of inequalities.

25. $\begin{aligned} y &> x \\ x + y &\le 5 \end{aligned}$

26. $\begin{aligned} y^2 - x^2 &\le 9 \\ \frac{x^2}{4} + \frac{y^2}{9} &\ge 1 \end{aligned}$

27. The profit on a gold necklace is $15, and the profit on a gold bracelet is $25. A jeweler's showcase has room for at most 60 necklaces and bracelets combined. The demand for necklaces is at least 3 times the demand for bracelets. How many of each should the jeweler order to maximize profit?

Sequences, Series, and Probability

A ball is dropped on a hard surface from a height of 10 ft. On each rebound it bounces three-fifths as high as its previous height. In Sec. 12.4 we show how to find the total distance traveled by the ball by finding the sum of an infinite geometric series.

12.1 Defining Sequences and Series

Many real-world problems are best described using a particular type of function known as a *sequence*. In this section we define both a sequence and a close relative of a sequence, a *series*.

Sequences

Suppose a radio station announces that it will award $5 to the first person who correctly identifies a certain "mystery voice." Moreover, for each day the mystery voice remains unidentified, an additional $5 will be added to the award. If $a(n)$ denotes the size of the award on the nth day, then

$$a(n) = 5n$$

For example, the size of the award on the third day is

$$a(3) = 5 \cdot 3 = 15$$

The following numbers represent the size of the award on each successive day:

$$5, \ 10, \ 15, \ 20, \ 25, \ \ldots$$

The function $a(n) = 5n$ is valid only when n is a positive integer. That is, $a(2.5)$ and $a(-3)$ have no meaning. A function whose domain consists of only positive integers is called a **sequence function,** or simply a **sequence.** The range values of a sequence function are called the **terms** of the sequence.

> ### DEFINITION OF A SEQUENCE
>
> An **infinite sequence** is a function whose domain is the entire set of positive integers. A **finite sequence** is a function whose domain is the first k positive integers.

A finite sequence has a last term, whereas an infinite sequence does not. For example, the *finite* sequence $a(n) = 5n$ with domain $n = 1, 2, 3, 4$ is written

$$5, \ 10, \ 15, \ 20$$

The *infinite* sequence $a(n) = 5n$ with domain $n = 1, 2, 3, 4, \ldots$ is written

$$5, \ 10, \ 15, \ 20, \ \ldots$$

It is customary to name sequence functions with the letters a, b, and c, as opposed to the usual f, g, and h. It is also customary to use the letter n as the independent variable rather than x. Finally, we denote the value of a sequence function as a_n (read "a sub n") rather than $a(n)$. We can summarize this notation as shown below:

$$a_1 = \text{first term of the sequence}$$
$$a_2 = \text{second term of the sequence}$$
$$\vdots$$
$$a_n = n\text{th term of the sequence}$$

The nth term of the sequence is also called the **general term,** since from it we can obtain all the terms of the sequence. To conserve space, the terms of a sequence are usually written horizontally, as shown below:

$$a_1, \ a_2, \ a_3, \ \ldots, \ a_n, \ \ldots$$

Example 1 Find the first four terms of the sequence whose general term is given by $a_n = 2n - 1$.

Solution: Replace n with 1, 2, 3, and 4.

$$a_1 = 2(1) - 1 = 1$$
$$a_2 = 2(2) - 1 = 3$$
$$a_3 = 2(3) - 1 = 5$$
$$a_4 = 2(4) - 1 = 7$$

Try Exercise 5

Find the first four terms of the sequence whose general term is $a_n = 3n - 2$. Then find the tenth term.

Therefore the first four terms are 1, 3, 5, 7.   ❏

Example 2 Find the first four terms of the sequence whose general term is given by $a_n = \dfrac{(-1)^n}{n + 1}$.

Solution: Replace n with 1, 2, 3, and 4.

$$a_1 = \frac{(-1)^1}{1 + 1} = -\frac{1}{2}$$
$$a_2 = \frac{(-1)^2}{2 + 1} = \frac{1}{3}$$
$$a_3 = \frac{(-1)^3}{3 + 1} = -\frac{1}{4}$$
$$a_4 = \frac{(-1)^4}{4 + 1} = \frac{1}{5}$$

Try Exercise 15

Find the first four terms of the sequence whose general term is $a_n = \dfrac{(-1)^n}{n}$. Then find the tenth term.

The first four terms are $-\frac{1}{2}, \frac{1}{3}, -\frac{1}{4}, \frac{1}{5}$.   ❏

Sometimes we must reverse the procedure. That is, given some of the terms of a sequence, we must find a formula for the general term, a_n. There are no rules for finding the general term; we simply inspect the given terms and try to identify a pattern. Do not be afraid to guess at a formula for the general term. Often an incorrect guess will lead to the correct formula.

Example 3 Find the general term a_n for the sequence 1, 4, 9, 16, 25, ...

Solution: Observe that each term is the square of the term number. That is,

$$a_1 = 1 = 1^2$$
$$a_2 = 4 = 2^2$$
$$a_3 = 9 = 3^2$$

Try Exercise 25

Find the general term a_n for the sequence 1, 8, 27, 64, 125, ...

and so on. Therefore the general term is $a_n = n^2$.   ❏

Series

Consider the finite sequence below:

$$5, 10, 15, 20$$

The indicated sum of this sequence, written

$$5 + 10 + 15 + 20$$

is called a **series**.

> **DEFINITION OF A SERIES**
>
> The indicated sum of a finite sequence is called a **finite series.** The indicated sum of an infinite sequence is called an **infinite series.**

In Secs. 12.1, 12.2, and 12.3, we deal with finite series only. In Sec. 12.4 we will study a particular type of infinite series.

We can write a series in a more compact form, called **summation notation,** using the symbol Σ (the Greek letter *sigma*) along with the general term of the corresponding sequence. For example, since the general term of the sequence 5, 10, 15, 20 is $5n$, we can write the corresponding series $5 + 10 + 15 + 20$ as $\sum_{n=1}^{4} 5n$. To evaluate the expression $\sum_{n=1}^{4} 5n$, replace n by 1, 2, 3, and 4. Then add the resulting terms.

$$\sum_{n=1}^{4} 5n = 5(1) + 5(2) + 5(3) + 5(4)$$
$$= 5 + 10 + 15 + 20$$
$$= 50$$

The expression $\sum_{n=1}^{4} 5n$ is read "the summation of $5n$ as n goes from 1 to 4." The letter n is called the **index** of summation. The number 1 is the **lower limit** of summation, and 4 is the **upper limit** of summation.

Example 4   Expand $\sum_{n=1}^{3} (n^2 - n)$ and simplify.

Solution: Replace n with 1, 2, and 3. Then add.

$$\sum_{n=1}^{3} (n^2 - n) = (1^2 - 1) + (2^2 - 2) + (3^2 - 3)$$
$$= 0 + 2 + 6$$
$$= 8 \qquad \square$$

Try Exercise 47

Expand and simplify:

$$\sum_{n=1}^{3} (n^3 - n)$$

Example 5   Expand: $\sum_{n=1}^{5} \dfrac{x^n}{n}$.

Solution: Replace n (*not* x) with 1, 2, 3, 4, and 5.

$$\sum_{n=1}^{5} \frac{x^n}{n} = \frac{x^1}{1} + \frac{x^2}{2} + \frac{x^3}{3} + \frac{x^4}{4} + \frac{x^5}{5}$$
$$= x + \frac{x^2}{2} + \frac{x^3}{3} + \frac{x^4}{4} + \frac{x^5}{5} \qquad \square$$

Try Exercise 53

Expand:

$$\sum_{n=1}^{5} \frac{x^{n+1}}{n+1}$$

We now reverse the procedure and convert from expanded form to summation notation. Generally, there is more than one answer to this type of exercise.

Example 6   Write $4 + 7 + 10 + 13 + 16 + 19$ in summation notation.

Solution: The difference between any two successive terms is 3. This suggests writing $3n$. To obtain the first term 4 when $n = 1$, write $a_n = 3n + 1$. Therefore

$$4 + 7 + 10 + 13 + 16 + 19 = \sum_{n=1}^{6} (3n + 1)$$

Try Exercise 55

Write in summation notation:
$2 + 5 + 8 + 11 + 14 + 17$

You can verify that the answer may also be written as $\sum_{n=0}^{5} (3n + 4)$. $\square$

CAUTION

Make sure you use parentheses correctly when writing a series in summation notation.

$$\sum_{n=1}^{3} (n + 5) = (1 + 5) + (2 + 5) + (3 + 5) = 21$$

$$\sum_{n=1}^{3} n + 5 = (1 + 2 + 3) + 5 = 11$$

■

Exercises 12.1

Completion Exercises

1. A function whose domain consists of only positive integers is a(n) _____.

2. The range values of a sequence function are called the _____ of the sequence. The nth term of a sequence is also called the _____ term of the sequence.

3. The indicated sum of a sequence is a(n) _____.

4. The notation $\sum_{n=1}^{5} n^2$ is called _____ notation. The letter n is called the _____ of summation, the number 1 is called the _____ of summation, and the number 5 is called the _____ of summation.

Find the first four terms of the sequence whose general term is given. Then find the 10th term.

5. $a_n = 3n - 2$

6. $a_n = 4n + 5$

7. $a_n = n^2 + n$

8. $a_n = n^2 - n$

9. $a_n = \dfrac{n - 1}{n}$

10. $a_n = \dfrac{n + 1}{n}$

11. $a_n = n + \dfrac{1}{n}$

12. $a_n = n - \dfrac{1}{n}$

13. $a_n = 6$

14. $a_n = 8$

15. $a_n = \dfrac{(-1)^n}{n}$

16. $a_n = \dfrac{(-1)^n}{n + 2}$

17. $a_n = 2^n$

18. $a_n = 3^n$

19. $a_n = \dfrac{4n - 6}{5n + 2}$

20. $a_n = \dfrac{3n + 8}{2n - 7}$

Find the general term a_n for each sequence.

21. $2, 4, 6, 8, 10, \ldots$

22. $3, 6, 9, 12, 15, \ldots$

23. $1, 2, 3, 4, 5, \ldots$

24. $1, 3, 5, 7, 9, \ldots$

25. $1, 8, 27, 64, 125, \ldots$

26. $1, 16, 81, 256, 625, \ldots$

27. $1, \frac{1}{4}, \frac{1}{9}, \frac{1}{16}, \frac{1}{25}, \ldots$

28. $1, \frac{1}{8}, \frac{1}{27}, \frac{1}{64}, \frac{1}{125}, \ldots$

29. $-1, 1, -1, 1, -1, \ldots$

30. $1, -1, 1, -1, 1, \ldots$

31. $2, -4, 6, -8, 10, \ldots$

32. $-3, 6, -9, 12, -15, \ldots$

Discussion Exercises

33. Describe the difference between a finite sequence and an infinite sequence. Describe the difference between a finite series and an infinite series.

34. How would you read the expression $\sum_{n=2}^{10} n^3$?

35. Show that $\sum_{n=1}^{5} n^2$ and $\sum_{i=1}^{5} i^2$ represent the same number. What does this suggest?

36. Describe the difference between $\sum_{i=1}^{3} i^2 + 1$ and $\sum_{i=1}^{3} (i^2 + 1)$.

Expand and simplify.

37. $\sum_{n=1}^{6} 4n$

38. $\sum_{n=1}^{7} 10n$

39. $\sum_{n=1}^{6} 4$

40. $\sum_{n=1}^{7} 10$

41. $\sum_{n=2}^{5} (2n + 1)$

42. $\sum_{n=2}^{6} (2n - 1)$

43. $\sum_{n=2}^{5} 2n + 1$

44. $\sum_{n=2}^{6} 2n - 1$

45. $\sum_{n=0}^{4} (n^2 - 3n + 4)$

46. $\sum_{n=0}^{4} (n^2 - 4n + 5)$

47. $\sum_{n=1}^{3} (n^3 - n)$

48. $\sum_{n=1}^{3} (n^3 + n)$

49. $\displaystyle\sum_{n=1}^{4} \frac{1}{n}$

50. $\displaystyle\sum_{n=0}^{3} \frac{1}{n+1}$

51. $\displaystyle\sum_{n=1}^{3} (-n)^n$

52. $\displaystyle\sum_{n=0}^{3} (-n)^{n+1}$

53. $\displaystyle\sum_{n=1}^{5} \frac{x^{n+1}}{n+1}$

54. $\displaystyle\sum_{n=1}^{5} \frac{x^{2n}}{2n}$

Write each series in summation notation.

55. $2 + 5 + 8 + 11 + 14 + 17$

56. $1 + 5 + 9 + 13 + 17 + 21$

57. $5 + 8 + 11 + 14 + 17 + 20$

58. $5 + 9 + 13 + 17 + 21 + 25$

59. $3 + 6 + 9 + 12$

60. $2 + 4 + 6 + 8$

61. $3 + 9 + 27 + 81$

62. $2 + 4 + 8 + 16$

63. $\frac{1}{2} + \frac{2}{3} + \frac{3}{4} + \frac{4}{5}$

64. $\frac{3}{4} + \frac{4}{5} + \frac{5}{6} + \frac{6}{7}$

65. $2 + 5 + 10 + 17 + 26$

66. $0 + 3 + 8 + 15 + 24$

67. $1 + x + x^2 + x^3 + x^4$

68. $x + x^2 + x^3 + x^4 + x^5$

Problem Solving

69. An automobile costs $15,000 when new. If it depreciates 10 percent each year, write a sequence that describes its value for each of the first 5 years.

70. Your starting salary is $25,000. If you receive a 10 percent raise each year, write a sequence that describes your salary for each of the first 5 years.

71. Determine the fifth term of a sequence if $a_1 = 3$, and $a_n = 2a_{n-1}$ when $n \geq 2$.

72. Determine the fifth term of a sequence if $a_1 = 1$, and $a_n = 4a_{n-1}$ when $n \geq 2$.

73. Try to determine the pattern of the Fibonacci sequence

$$1, 1, 2, 3, 5, 8, 13, 21, 34, \ldots*$$

 Getting Acquainted with Your Scientific Calculator

74. An old legend has it that the King of Persia invited the inventor of the game of chess to name his own reward. What the inventor requested did not seem like much. He asked for one grain of wheat for the first square of the chessboard, two grains for the second square, four grains for the third square, eight for the fourth, and so on for all 64 squares. Write a series, first in expanded form and then

in summation notation, that describes the total amount of wheat requested. Then calculate the amount of wheat needed for (a) the first 10 squares and (b) the last (64th) square. (Incidentally, the total amount of wheat for all 64 squares would cover the entire state of California with a layer over a foot deep!)

 Getting Acquainted with Your Graphing Calculator

By experimenting with the range settings and the connected versus dot mode, try to duplicate the graphs of the given sequence functions.

75. $a_n = \dfrac{6}{n}$

76. $a_n = \dfrac{6n}{n+1}$

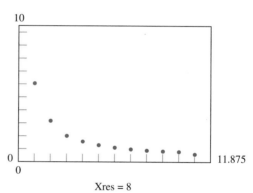

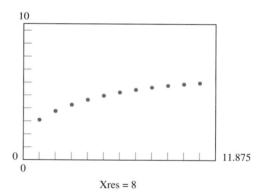

---

*HISTORICAL NOTE: The Fibonacci sequence is named after Leonardo Fibonacci (ca. 1175–1250), the most talented mathematician of the Middle Ages, who introduced it in 1202. Among other things, the sequence describes the reproductive behavior of rabbits.

12.2 Arithmetic Sequences and Series

The two most basic types of sequences are *arithmetic sequences* and *geometric sequences.* We study arithmetic sequences and series in this section and geometric sequences and series in Sec. 12.3.*

Arithmetic Sequences

The sequence

$$3, 7, 11, 15, 19, \ldots$$

is an arithmetic sequence, since each term after the first is obtained by adding 4 to the preceding term. The next two terms in the sequence are 23 and 27.

DEFINITION OF AN ARITHMETIC SEQUENCE

An **arithmetic sequence** is a sequence in which each term after the first is obtained by adding a fixed constant to the preceding term. This fixed constant is called the **common difference** and is denoted by the letter d.

Given an arithmetic sequence, we can determine the common difference by subtracting any term in the sequence from the term that follows it. That is,

$$d = a_{n+1} - a_n$$

Example 1   Find the common difference for the arithmetic sequence

$$10, 2, -6, -14, -22, \ldots$$

Then write the next two terms.

Solution: To find d, subtract 10 from 2 (or 2 from -6, etc.).

$$d = a_2 - a_1 = 2 - 10 = -8$$

The common difference is -8. Therefore the next two terms are

$$a_6 = -22 + (-8) = -30$$
$$a_7 = -30 + (-8) = -38$$ ❐

Try Exercise 11

Find the common difference for the arithmetic sequence

$$8, 3, -2, -7, -12, \ldots$$

Then write the next two terms.

An arithmetic sequence is completely defined when its first term a_1 and its common difference d are known. For then we can write the first few terms as shown below:

First term: a_1
Second term: $a_2 = a_1 + d$
Third term: $a_3 = a_2 + d = (a_1 + d) + d = a_1 + 2d$ Since $a_2 = a_1 + d$
Fourth term: $a_4 = a_3 + d = (a_1 + 2d) + d = a_1 + 3d$ Since $a_3 = a_1 + 2d$

This suggests the following formula.

General Term of an Arithmetic Sequence

$$a_n = a_1 + (n - 1)d$$

---

*HISTORICAL NOTE: Sequences are among the oldest topics of mathematical study. Arithmetic and geometric sequences appear in the Rhind papyrus, a document containing 85 mathematical problems copied around 1650 B.C. from an earlier work by the Egyptian scribe Ahmes.

This formula has four variables. They are a_n, a_1, n, and d. If any three of the variables are known, we can use the formula to find the fourth.

Example 2   Find the 11th term of the arithmetic sequence with first term 6 and common difference 7.

Solution: Substitute $n = 11$, $a_1 = 6$, and $d = 7$ into the formula for a_n.

$$a_n = a_1 + (n - 1)d \qquad \text{Formula for } a_n$$
$$a_{11} = 6 + (11 - 1)7 \qquad \text{Let } n = 11, a_1 = 6, \text{ and } d = 7.$$
$$= 6 + (10)7 \left.\begin{matrix} \\ \\ \end{matrix}\right\}$$
$$= 76 \qquad \text{Simplify.}$$

The 11th term is 76.   ❐

Try Exercise 17

Find a_{11} for the arithmetic sequence with $a_1 = 4$ and $d = 6$.

CAUTION

The variables n and a_n are *not* the same. The variable n represents the position of the term in the sequence, whereas a_n represents the *value* of the nth term.

Sequence: 4, 6, 8, 10, 12

Term position: n	*Value of term: a_n*
$n = 3$	$a_n = 8$
$n = 5$	$a_n = 12$

■

Example 3   Find the 40th term of the arithmetic sequence

$$\tfrac{1}{2}, \tfrac{7}{6}, \tfrac{11}{6}, \tfrac{5}{2}, \ldots$$

Solution: First find d.

$$d = \tfrac{11}{6} - \tfrac{7}{6} = \tfrac{4}{6} = \tfrac{2}{3}$$

Then substitute $n = 40$, $a_1 = \tfrac{1}{2}$, and $d = \tfrac{2}{3}$ into the formula for a_n.

$$a_n = a_1 + (n - 1)d$$
$$a_{40} = \tfrac{1}{2} + (40 - 1)\tfrac{2}{3}$$
$$= \tfrac{1}{2} + (39)\tfrac{2}{3}$$
$$= \tfrac{1}{2} + 26$$
$$= \tfrac{53}{2}$$

The 40th term is $\tfrac{53}{2}$.   ❐

Try Exercise 25

Find a_{101} for the arithmetic sequence
$$\tfrac{1}{5}, \tfrac{9}{5}, \tfrac{17}{5}, 5, \ldots$$

Example 4   Find the common difference for the arithmetic sequence with first term 3 and sixth term 28.

Solution: Write down the formula for a_n and substitute the known values.

$$a_n = a_1 + (n - 1)d \qquad \text{Formula for } a_n$$
$$28 = 3 + (6 - 1)d \qquad \text{Let } a_n = 28, a_1 = 3, \text{ and } n = 6.$$

Solve for d.

$$28 = 3 + 5d \qquad \text{Simplify.}$$
$$25 = 5d \qquad \text{Subtract 3.}$$
$$d = 5 \qquad \text{Divide by 5.}$$

Try Exercise 27

Find d for the arithmetic sequence with $a_1 = 9$ and $a_5 = 25$.

The common difference is 5.   ❐

Example 5   How many terms are in the arithmetic sequence 5, 11, 17, ... , 83?

Solution: Note that $d = 11 - 5 = 6$.

$$
\begin{aligned}
a_n &= a_1 + (n - 1)d && \text{Formula for } a_n \\
83 &= 5 + (n - 1)6 && \text{Let } a_n = 83, a_1 = 5, \text{ and } d = 6. \\
78 &= (n - 1)6 && \text{Subtract 5.} \\
13 &= n - 1 && \text{Divide by 6.} \\
n &= 14 && \text{Add 1.}
\end{aligned}
$$

Try Exercise 35

How many terms are in the arithmetic sequence below?

8, 13, 18, ... , 88

There are 14 terms in the sequence.   ❑

Arithmetic Series

The indicated sum of an arithmetic sequence is called an **arithmetic series.** If we denote the sum of the first n terms as S_n, then

First term   Second term   Third term   Next-to-last term   Last term

$$
S_n = a_1 + (a_1 + d) + (a_1 + 2d) + \cdots + (a_n - d) + a_n
$$

To develop a formula for S_n, write the same series in reverse order.

$$
S_n = a_n + (a_n - d) + (a_n - 2d) + \cdots + (a_1 + d) + a_1
$$

Then add the two equations as shown below:

$$
\begin{aligned}
S_n &= a_1 + (a_1 + d) + (a_1 + 2d) + \cdots + (a_n - d) + a_n \\
S_n &= a_n + (a_n - d) + (a_n - 2d) + \cdots + (a_1 + d) + a_1 \\
\hline
2S_n &= (a_1 + a_n) + (a_1 + a_n) + (a_1 + a_n) + \cdots + (a_1 + a_n) + (a_1 + a_n)
\end{aligned}
$$

Since the expression $(a_1 + a_n)$ appears n times, write this last equation as

$$
2S_n = n(a_1 + a_n)
$$

Divide each side by 2 to produce the following formula.

Sum of the First n Terms of an Arithmetic Series

$$
S_n = \frac{n}{2}(a_1 + a_n)
$$

Example 6   Find the sum of the first 10 terms of the arithmetic series with first term 4 and 10th term 31.

Solution: Substitute $n = 10$, $a_1 = 4$, and $a_{10} = 31$ into the formula for S_n.

$$
S_n = \frac{n}{2}(a_1 + a_n)
$$

$$
S_{10} = \frac{10}{2}(4 + 31) = 5(35) = 175
$$

Try Exercise 41

Find S_{10} for the arithmetic series with $a_1 = 5$ and $a_{10} = 32$.

The sum of the first 10 terms is 175.   ❑

Problem Solving

Example 7   A contract specifies that construction of an office building must be completed by June 1. If not, the general contractor will be fined $100 for the first late day, $150 for the second late day, $200 for the third late day, and so on. If the contractor is 2 weeks late, what is the total fine?

Solution: The total fine is the sum of the first 14 terms of the following arithmetic series:

$$100 + 150 + 200 + 250 + \cdots$$

First find a_{14}.

$$a_n = a_1 + (n - 1)d$$
$$a_{14} = 100 + (14 - 1)50 = 100 + (13)50 = 750$$

Then find S_{14}.

$$S_n = \frac{n}{2}(a_1 + a_n)$$
$$S_{14} = \frac{14}{2}(100 + 750) = 7(850) = 5950$$

The total fine is $5950.   ❑

Try Exercise 59

Heather saved $100 in January, $125 in February, $150 in March, and so on. How much did she save for the year?

Exercises 12.2

Completion Exercises

1. A sequence in which each term after the first is obtained by adding a fixed constant to the preceding term is called a(n) _____ sequence. The fixed constant is called the _____.

2. The common difference d of an arithmetic sequence can be found by _____.

3. The general term of an arithmetic sequence is given by the formula $a_n =$ _____.

4. The sum of the first n terms of an arithmetic series is given by the formula $S_n =$ _____.

Identify each sequence as arithmetic or nonarithmetic. For those that are arithmetic, find the common difference and write the next two terms.

5. $1, 2, 3, 4, 5, \ldots$

6. $1, 3, 5, 7, 9, \ldots$

7. $3, 9, 15, 21, 27, \ldots$

8. $4, 11, 18, 25, 32, \ldots$

9. $-6, -2, 2, 6, 10, \ldots$

10. $-12, -4, 4, 12, 20, \ldots$

11. $8, 3, -2, -7, -12, \ldots$

12. $9, 3, -3, -9, -15, \ldots$

13. $1, \frac{3}{2}, 2, \frac{5}{2}, 3, \ldots$

14. $1, \frac{4}{3}, \frac{5}{3}, 2, \frac{7}{3}, \ldots$

15. $1, \frac{1}{2}, \frac{1}{3}, \frac{1}{4}, \frac{1}{5}, \ldots$

16. $1, 4, 9, 16, 25, \ldots$

Find the indicated unknown for each arithmetic sequence.

17. $a_1 = 4, d = 6; a_{11} = ?$

18. $a_1 = 2, d = 5; a_{21} = ?$

19. $a_1 = 4, d = 6; a_{110} = ?$

20. $a_1 = 2, d = 5; a_{111} = ?$

21. $a_1 = 4, d = 6; a_n = ?$

22. $a_1 = 2, d = 5; a_n = ?$

23. $10, 7, 4, 1, \ldots; a_{15} = ?$

24. $15, 11, 7, 3, \ldots; a_{25} = ?$

25. $\frac{1}{5}, \frac{9}{5}, \frac{17}{5}, 5, \ldots; a_{101} = ?$

26. $\frac{1}{2}, 3, \frac{11}{2}, 8, \ldots; a_{99} = ?$

27. $a_1 = 9, a_5 = 25; d = ?$

28. $a_1 = 7, a_6 = 22; d = ?$

29. $a_1 = -1, a_7 = 11; d = ?$

30. $a_1 = -2, a_8 = 19; d = ?$

31. $1, 3, 5, 7, \ldots; a_n = ?$

32. $2, 4, 6, 8, \ldots; a_n = ?$

33. $a_5 = 6, a_4 = \frac{16}{3}; a_1 = ?$

34. $a_6 = 5, a_5 = \frac{17}{4}; a_1 = ?$

35. $8, 13, 18, \ldots, 88; n = ?$

36. $6, 10, 14, \ldots, 58; n = ?$

37. $5, 2, -1, \ldots, -43; n = ?$

38. $3, -2, -7, \ldots, -52; n = ?$

Discussion Exercises

39. Discuss the difference between a_n and n in the arithmetic sequence $a_1, a_2, a_3, \ldots$.

40. What is the difference between an arithmetic sequence and an arithmetic series?

Find the indicated unknown for each arithmetic series.

41. $a_1 = 5, a_{10} = 32; S_{10} = ?$
42. $a_1 = 6, a_{10} = 15; S_{10} = ?$
43. $a_n = 3n + 4; S_{20} = ?$ 44. $a_n = 2n + 3; S_{20} = ?$
45. $a_1 = \frac{1}{3}, d = \frac{2}{3}; S_{10} = ?$ 46. $a_1 = \frac{1}{2}, d = \frac{1}{3}; S_{10} = ?$
47. $3 + 11 + 19 + \cdots; S_{18} = ?$
48. $2 + 11 + 20 + \cdots; S_{19} = ?$
49. $S_9 = -36, a_9 = -12; a_1 = ?$
50. $S_{11} = 275, a_{11} = 60; a_1 = ?$
51. $S_{13} = 325, a_{13} = 61; a_8 = ?$
52. $S_{15} = -135, a_{15} = -30; a_{23} = ?$

Use the formula for S_n to find the sum of each arithmetic series.

53. $\displaystyle\sum_{i=1}^{86} 2i$

54. $\displaystyle\sum_{i=1}^{54} 3i$

55. $\displaystyle\sum_{k=1}^{2000} (4k + 1)$

56. $\displaystyle\sum_{k=1}^{2000} (6k - 1)$

Problem Solving

57. Find the sum of the first 100 positive integers.

58. Find the sum of the first 200 positive integers.

59. Heather saved $100 in January, $125 in February, $150 in March, and so on. How much did she save for the year?

60. A disgruntled baseball player is told to report to spring training or be fined $50 for the first late day, $60 for the second late day, $70 for the third late day, and so on. If the player reports 3 weeks late, what is the total fine?

61. An auditorium has 21 rows of seats. The back row contains 36 seats and each row has one less seat than the row behind it. (a) How many seats are in the first row? (b) How many seats are in the auditorium?

62. Bottles are stacked in 9 rows so that the bottom row contains 27 bottles, and each row contains three fewer bottles than the row below it. (a) How many bottles are in the top row? (b) How many bottles are there in all?

63. Find the sum of the odd integers from 7 through 859.

64. Find the sum of the even integers from 8 through 542.

65. A piece of machinery that cost $6425 loses $175 of its value each year. In how many years will the machine be worth $650?

66. Your starting salary is $18,500 and you receive a $750 raise each year. In how many years will your salary be $50,000?

67. Find the general term of the arithmetic sequence with 4th term 10 and 12th term 26.

68. Combine the formulas for S_n and a_n to show that the formula for the sum of the first n terms of an arithmetic series can also be written as

$$S_n = \frac{n}{2}[2a_1 + (n-1)d]$$

Getting Acquainted with Your Graphing Calculator

You can display the arithmetic sequence 4, 7, 10, 13, ... on your calculator by pressing

4 [Enter] [+] 3 [Enter] [Enter] [Enter] , etc.
  ↑   ↑   ↑   ↑
Execute   Execute Execute Execute

Display the first 25 terms of the given arithmetic sequence on your calculator.

69. 4, 7, 10, 13, ...   70. 11, 6, 1, −4, ...

12.3 Geometric Sequences and Series

In Sec. 12.2 we studied arithmetic sequences and series. In this section we turn our attention to *geometric* sequences and series.

Geometric Sequences

In an arithmetic sequence, each term after the first is obtained by adding a fixed constant to the preceding term. In a geometric sequence, we *multiply* each term by a fixed constant to obtain the next term. For example, the sequence

$$3, 6, 12, 24, \ldots$$

is a geometric sequence, since each term after the first is obtained by multiplying the preceding term by 2. The next two terms in the sequence are 48 and 96.

DEFINITION OF A GEOMETRIC SEQUENCE

A **geometric sequence** is a sequence in which each term after the first is obtained by multiplying the preceding term by a fixed constant. This fixed constant is called the **common ratio** and is denoted by the letter r.

Given a geometric sequence, we can determine the common ratio by dividing any term in the sequence by the term that precedes it. That is,

$$r = a_{n+1} \div a_n$$

Example 1 Find the common ratio for the geometric sequence

$$1, \tfrac{1}{2}, \tfrac{1}{4}, \tfrac{1}{8}, \ldots$$

Then write the next two terms.

Solution: To find r, divide $\tfrac{1}{2}$ by 1 (or $\tfrac{1}{4}$ by $\tfrac{1}{2}$, etc.).

$$r = a_2 \div a_1 = \tfrac{1}{2} \div 1 = \tfrac{1}{2}$$

The common ratio is $\tfrac{1}{2}$. Therefore the next two terms are

$$a_5 = \tfrac{1}{8} \cdot \tfrac{1}{2} = \tfrac{1}{16}$$

$$a_6 = \tfrac{1}{16} \cdot \tfrac{1}{2} = \tfrac{1}{32}$$

Try Exercise 7

Find the common ratio for the geometric sequence

$$1, \tfrac{1}{3}, \tfrac{1}{9}, \tfrac{1}{27}, \ldots$$

Then write the next two terms.

A geometric sequence is completely defined when its first term a_1 and its common ratio r are known. For then we can write the first few terms as shown below:

First term: a_1
Second term: $a_2 = a_1 r$
Third term: $a_3 = a_2 r = (a_1 r)r = a_1 r^2$ Since $a_2 = a_1 r$
Fourth term: $a_4 = a_3 r = (a_1 r^2)r = a_1 r^3$ Since $a_3 = a_1 r^2$

This suggests the following formula.

General Term of a Geometric Sequence
$$a_n = a_1 r^{n-1}$$

This formula contains the four variables a_n, a_1, r, and n. If any three of the variables are known, we can use the formula to find the fourth.

Example 2 Find the seventh term of the geometric sequence with first term 10 and common ratio 3.

Solution: Substitute $n = 7$, $a_1 = 10$, and $r = 3$ into the formula for a_n.

$$a_n = a_1 r^{n-1} \qquad \text{Formula for } a_n$$
$$a_7 = 10(3)^{7-1} \qquad \text{Let } n = 7, a_1 = 10, \text{ and}$$
$$= 10(3)^6 \qquad\qquad r = 3.$$
$$= 10(729) \qquad \text{Simplify.}$$
$$= 7290$$

The seventh term is 7290.   ❏

Try Exercise 19

Find a_4 for the geometric sequence with $a_1 = 2$ and $r = 3$.

Example 3   Find the fifth term of the geometric sequence

$$\tfrac{1}{4}, \; -\tfrac{1}{6}, \; \tfrac{1}{9}, \; \ldots$$

Solution: First find r.

$$r = -\tfrac{1}{6} \div \tfrac{1}{4} = -\tfrac{1}{6} \cdot \tfrac{4}{1} = -\tfrac{2}{3}$$

Then substitute $n = 5$, $a_1 = \tfrac{1}{4}$, and $r = -\tfrac{2}{3}$ into the formula for a_n.

$$a_n = a_1 r^{n-1}$$
$$a_5 = \tfrac{1}{4}\left(-\tfrac{2}{3}\right)^{5-1}$$
$$= \tfrac{1}{4}\left(-\tfrac{2}{3}\right)^4$$
$$= \tfrac{1}{4}\left(\tfrac{16}{81}\right)$$
$$= \tfrac{4}{81}$$

Try Exercise 25

Find a_4 for the geometric sequence

$$\tfrac{1}{3}, \; -\tfrac{1}{2}, \; \tfrac{3}{4}, \; \ldots$$

The fifth term is $\tfrac{4}{81}$.   ❏

Example 4   Find the common ratio for the geometric sequence with first term 5 and fourth term 0.135.

Solution: Write down the formula for a_n and substitute the known values.

$$a_n = a_1 r^{n-1} \qquad \text{Formula for } a_n$$
$$0.135 = 5r^{4-1} \qquad \text{Let } a_n = 0.135, a_1 = 5, \text{ and } n = 4.$$

Solve for r.

$$0.135 = 5r^3 \qquad \text{Simplify.}$$
$$0.027 = r^3 \qquad \text{Divide by 5.}$$
$$r = 0.3 \qquad \text{Take the cube root.}$$

Try Exercise 29

Find r for the geometric sequence with first term 5 and fifth term 0.008.

The common ratio is 0.3.   ❏

Geometric Series

The indicated sum of a geometric sequence is called a **geometric series.** If we denote the sum of the first n terms as S_n, then

First term	Second term	Third term	Fourth term		Next-to-last term	Last term
↓	↓	↓	↓		↓	↓

$$S_n = a_1 + a_1 r + a_1 r^2 + a_1 r^3 + \cdots + a_1 r^{n-2} + a_1 r^{n-1}$$

To develop a formula for S_n, multiply each side by $-r$.

$$-rS_n = -a_1r - a_1r^2 - a_1r^3 - \cdots - a_1r^{n-1} - a_1r^n$$

Then add the two equations as shown below:

$$S_n = a_1 + a_1r + a_1r^2 + a_1r^3 + \cdots + a_1r^{n-1}$$
$$- rS_n = \quad\quad - a_1r - a_1r^2 - a_1r^3 - \cdots - a_1r^{n-1} - a_1r^n$$
$$\overline{S_n - rS_n = a_1 \quad\quad\quad\quad\quad\quad\quad\quad\quad\quad\quad - a_1r^n}$$

Factor out S_n on the left side and a_1 on the right side.

$$(1 - r)S_n = a_1(1 - r^n)$$

Divide each side by $1 - r$ to produce the following formula.

Sum of the First n Terms of a Geometric Series
$$S_n = \frac{a_1(1 - r^n)}{1 - r} \quad\quad (r \neq 1)$$

Example 5 Find the sum of the first four terms of the geometric series with first term 6 and common ratio 5.

Solution: Substitute $n = 4$, $a_1 = 6$, and $r = 5$ into the formula for S_n.

$$S_n = \frac{a_1(1 - r^n)}{1 - r}$$
$$S_4 = \frac{6(1 - 5^4)}{1 - 5} = \frac{6(1 - 625)}{-4} = 936$$

The sum of the first four terms is 936.

Try Exercise 41

Find S_5 for the geometric series with $a_1 = 7$ and $r = 4$.

Problem Solving

Example 6 Suppose your first year's salary is $20,000, and you are guaranteed a 7 percent raise each year. Find your total income for the first 10 yr.

Solution: First year's salary = $20,000
Second year's salary = first year's salary + raise

$$= 20,000 + 20,000(0.07)$$
$$= 20,000(1 + 0.07) \quad \text{Factor out 20,000.}$$
$$= 20,000(1.07) \quad \text{Add in parentheses.}$$

Third year's salary = second year's salary + raise

$$= 20,000(1.07) + 20,000(1.07)(0.07)$$
$$= 20,000(1.07)(1 + 0.07) \quad \text{Factor out 20,000(1.07).}$$
$$= 20,000(1.07)(1.07) \quad \text{Add in parentheses.}$$
$$= 20,000(1.07)^2 \quad \text{Write with an exponent.}$$

Therefore the sum of your yearly salaries forms the following geometric series:

$$20{,}000 + 20{,}000(1.07) + 20{,}000(1.07)^2 + 20{,}000(1.07)^3 + \cdots$$

Your total income for the first ten years is the sum of the first 10 terms.

$$S_n = \frac{a_1(1 - r^n)}{1 - r}$$

$$S_{10} = \frac{20{,}000(1 - 1.07^{10})}{1 - 1.07}$$

$$\approx \frac{20{,}000(1 - 1.967151)}{-0.07} \qquad \text{From a calculator}$$

$$\approx 276{,}329$$

Your total income for the first 10 yr is approximately \$276,329.   ❒

Try Exercise 63

Suppose your first year's salary is \$10,000, and you are guaranteed an 8 percent raise each year. Find your salary for the 15th year. Find your total income for the first 15 yr.

Exercises 12.3

Completion Exercises

1. A sequence in which each term after the first is obtained by multiplying the preceding term by a fixed constant is called a(n) _____ sequence. The fixed constant is called the _____.

2. The common ratio r of a geometric sequence can be found by _____.

3. The general term of a geometric sequence is given by the formula $a_n =$ _____.

4. The sum of the first n terms of a geometric series is given by the formula $S_n =$ _____.

Identify each sequence as geometric or nongeometric. For those that are geometric, find the common ratio and write the next two terms.

5. $1, 2, 4, 8, 16, \ldots$

6. $1, 3, 9, 27, 81, \ldots$

7. $1, \frac{1}{3}, \frac{1}{9}, \frac{1}{27}, \ldots$

8. $1, \frac{1}{4}, \frac{1}{16}, \frac{1}{64}, \ldots$

9. $1, 5, 10, 15, 20, \ldots$

10. $1, 2, 4, 6, 8, \ldots$

11. $-5, -25, -125, \ldots$

12. $-2, -4, -8, \ldots$

13. $1, 4, 9, 16, 25, \ldots$

14. $1, \frac{1}{2}, \frac{1}{3}, \frac{1}{4}, \frac{1}{5}, \ldots$

15. $6, -6, 6, -6, 6, \ldots$

16. $9, -9, 9, -9, 9, \ldots$

17. $-2, 1, -\frac{1}{2}, \frac{1}{4}, \ldots$

18. $-3, 1, -\frac{1}{3}, \frac{1}{9}, \ldots$

Find the indicated unknown for each geometric sequence.

19. $a_1 = 2, r = 3; a_4 = ?$

20. $a_1 = 3, r = 2; a_7 = ?$

21. $a_1 = 2, r = 3; a_n = ?$

22. $a_1 = 3, r = 2; a_n = ?$

23. $4, 2, 1, \ldots; a_6 = ?$

24. $9, 3, 1, \ldots; a_5 = ?$

25. $\frac{1}{3}, -\frac{1}{2}, \frac{3}{4}, \ldots; a_4 = ?$

26. $\frac{1}{2}, -\frac{1}{3}, \frac{2}{9}, \ldots; a_5 = ?$

27. $-6, -12, -24, \ldots; a_5 = ?$

28. $-4, -8, -16, \ldots; a_4 = ?$

29. $a_1 = 5, a_5 = 0.008; r = ?$

30. $a_1 = 5, a_4 = 0.32; r = ?$

31. $2, 2, 2, 2, \ldots; a_{56} = ?$

32. $-7, -7, -7, -7, \ldots; a_{67} = ?$

33. $1, -1, 1, -1, \ldots; a_n = ?$

34. $-1, 1, -1, 1, \ldots; a_n = ?$

35. $a_4 = -12, a_3 = 9; a_{18} = ?$

36. $a_5 = 10, a_4 = -6; a_{21} = ?$

37. $16, 24, 36, \ldots, \frac{243}{2}; n = ?$

38. $81, 54, 36, \ldots, \frac{32}{3}; n = ?$

Discussion Exercises

39. Explain the difference between a geometric sequence and a geometric series.

40. What is the difference between an arithmetic sequence and a geometric sequence?

Find the indicated unknown for each geometric series.

41. $a_1 = 7, r = 4; S_5 = ?$

42. $a_1 = 9, r = 5; S_4 = ?$

43. $a_1 = 40, r = \frac{1}{2}; S_6 = ?$ 44. $a_1 = 80, r = \frac{1}{2}; S_7 = ?$

45. $2 + (-6) + 18 + \cdots ; S_7 = ?$

46. $3 + (-6) + 12 + \cdots ; S_9 = ?$

47. $S_8 = 1020, r = 2; a_1 = ?$

48. $S_6 = 1820, r = 3; a_1 = ?$

49. $S_5 = \frac{550}{9}, r = -\frac{2}{3}; a_9 = ?$ 50. $S_4 = \frac{75}{16}, r = -\frac{3}{4}; a_{10} = ?$

51. $1 + (-1) + 1 + (-1) + \cdots ; S_{50} = ? \; S_{51} = ?$

52. $-1 + 1 + (-1) + 1 + \cdots ; S_{100} = ? \; S_{101} = ?$

Use the formula for S_n to find the sum of each geometric series.

53. $\displaystyle\sum_{i=0}^{8} 2^i$ 54. $\displaystyle\sum_{i=0}^{9} 3^i$

55. $\displaystyle\sum_{j=1}^{6} \left(-\tfrac{4}{5}\right)^j$ 56. $\displaystyle\sum_{j=1}^{7} \left(-\tfrac{5}{2}\right)^j$

57. $\displaystyle\sum_{k=1}^{10} 81\left(\tfrac{2}{3}\right)^k$ 58. $\displaystyle\sum_{k=1}^{10} 16\left(\tfrac{3}{2}\right)^k$

Problem Solving

59. What will a decorative plate that costs $18 today cost in 6 yr if inflation doubles its value each year?

60. A car that costs $15,000 new loses one-fifth of its value each year. What is its value after 5 yr?

61. A tank contains 500 gal of water. With each stroke a pump removes one-fifth of the water in the tank. How much water remains after four strokes?

62. A rubber ball is dropped on a hard surface from a height of 80 ft. On each rebound it bounces three-quarters as high as on its previous bounce. How high does it bounce on its fourth bounce?

63. Suppose your first year's salary is $10,000, and you are guaranteed an 8 percent raise each year. Find your salary for the 15th year. Then find your total income for the first 15 yr.

64. Suppose a nation consumes 5,000,000 bbl of oil this year. If consumption increases by 4 percent each year, how many barrels will be consumed 10 yr from now? How many total barrels will be consumed over the next 10 yr?

65. Find the general term of the geometric sequence with third term 15 and sixth term 120.

66. Combine the formula for S_n and a_n to show that the formula for the sum of the first n terms of a geometric series can also be written as

$$S_n = \frac{a_1 - a_n r}{1 - r} \qquad (r \neq 1)$$

 Getting Acquainted with Your Graphing Calculator

You can display the geometric sequence $3, 15, 75, \ldots$ on your calculator by pressing

3 (Enter) (×) 5 (Enter)   (Enter)   (Enter) , etc.

      ↑   ↑   ↑   ↑

  Execute   Execute   Execute   Execute

Display the first 10 terms of the given geometric sequence on your calculator.

67. $3, 15, 75, \ldots$ 68. $1, -0.8, 0.64, \ldots$

You can find the sum of the first 10 terms of the geometric series $5 + 10 + 20 + 40 + \cdots$ by pressing

5 (Enter) (×) 2 (+) 5 (Enter) (9 times)

Find the sum of the first 10 terms of the given geometric series.

69. $5 + 10 + 20 + 40 + \cdots$ 70. $1 + \frac{1}{2} + \frac{1}{4} + \frac{1}{8} + \cdots$

 ## 12.4 Infinite Geometric Series

Is it possible to find the sum of an infinite geometric series? In some cases it is. For example, the series

$$1 + 2 + 4 + 8 + \cdots$$

does not have a sum. But the series

$$0.3 + 0.03 + 0.003 + \cdots$$

can be written as the repeating decimal $0.33\overline{3}$, which equals $\frac{1}{3}$.

     In this section we learn how to tell whether or not an infinite geometric series has a sum, and we learn how to find that sum if it exists.

Sum of an Infinite Geometric Series

If $|r| < 1$, the sum S of the infinite geometric series with first term a_1 and common ratio r is given by the formula

$$S = \frac{a_1}{1 - r}$$

If $|r| \geq 1$, the sum does not exist.

To see where the formula $S = a_1/(1 - r)$ comes from, consider the formula for the sum of the first n terms of a geometric series.

$$S_n = \frac{a_1(1 - r^n)}{1 - r}$$

If $|r| < 1$ (that is, if $-1 < r < 1$), then r^n approaches 0 as n grows larger. For example, the table below illustrates that if $r = 0.1$, then r^n approaches 0 as n grows larger:

n	$r^n = (0.1)^n$
1	$r^1 = (0.1)^1 = 0.1$
5	$r^5 = (0.1)^5 = 0.00001$
10	$r^{10} = (0.1)^{10} = 0.0000000001$

Therefore if n grows large without bound, r^n approaches 0, and the right side of the formula for S_n approaches the fraction below:

$$\frac{a_1(1 - 0)}{1 - r} = \frac{a_1}{1 - r}$$

If $|r| \geq 1$ (that is, if $r \geq 1$ or $r \leq -1$), then the sum S_n does not exist.

Note that the infinite geometric series $0.3 + 0.03 + 0.003 + \cdots$ has a sum because

$$|r| = \left| \frac{0.03}{0.3} \right| = |0.1| = 0.1$$

so $|r| < 1$. The infinite geometric series $1 + 2 + 4 + 8 + \cdots$ does not have a sum because $|r| = |\frac{2}{1}| = |2| = 2$, so $|r| \geq 1$.

Example 1 Find the sum of the series

$$0.3 + 0.03 + 0.003 + \cdots$$

Solution: This is an infinite geometric series with $r = 0.1$. Since $|r| < 1$, substitute $a_1 = 0.3$ and $r = 0.1$ into the formula for S.

$$S = \frac{a_1}{1 - r} = \frac{0.3}{1 - 0.1} = \frac{0.3}{0.9} = \frac{3}{9} = \frac{1}{3}$$

The sum is $\frac{1}{3}$. ❏

Try Exercise 11

Find the sum of
$$10 + 3 + 0.9 + \cdots$$

Example 2 Find the sum of the series

$$9 - 6 + 4 - \tfrac{8}{3} + \cdots$$

Solution: First write the series as a sum.

$$9 + (-6) + 4 + (-\tfrac{8}{3}) + \cdots$$

This is an infinite geometric series with

$$r = \frac{-6}{9} = -\frac{2}{3}$$

Since $|r| < 1$, substitute $a_1 = 9$ and $r = -\frac{2}{3}$ into the formula for S.

$$S = \frac{a_1}{1-r} = \frac{9}{1-(-\frac{2}{3})} = \frac{9}{1+\frac{2}{3}} = \frac{9}{\frac{5}{3}} = \frac{27}{5}$$

Try Exercise 13

Find the sum of

$$16 - 12 + 9 - \frac{27}{4} + \cdots$$

The sum is $\frac{27}{5}$.   ❐

Example 3   Write $0.\overline{45}$ as a ratio of two integers.

Solution: First write $0.\overline{45}$ as follows:

$$0.\overline{45} = 0.45 + 0.0045 + 0.000045 + \cdots$$

Note that the right side is an infinite geometric series with $a_1 = 0.45$ and $r = 0.01$. Since $|r| < 1$, we have

$$S = \frac{a_1}{1-r} = \frac{0.45}{1-0.01} = \frac{0.45}{0.99} = \frac{45}{99} = \frac{5}{11}$$

Try Exercise 31

Write $0.\overline{36}$ as a ratio of two integers.

Therefore $0.\overline{45} = \frac{5}{11}$. This answer can be checked by division.   ❐

Problem Solving

Try Exercise 43

On the first swing, the bob of a pendulum travels an arc of length 45 cm. The arc on each subsequent swing is four-fifths the length of the previous arc. What is the total distance the bob travels before coming to rest?

Example 4   On the first swing, the bob of a pendulum travels an arc of length 80 cm. The arc on each subsequent swing is seven-eighths the length of the previous arc. What is the total distance the bob travels before coming to rest?

Solution: Find the sum of the infinite geometric series with $a_1 = 80$ and $r = \frac{7}{8}$.

$$S = \frac{a_1}{1-r} = \frac{80}{1-\frac{7}{8}} = \frac{80}{\frac{1}{8}} = 640$$

The total distance traveled is 640 cm.   ❐

Exercises 12.4

◆◆

Completion Exercises

1. If $|r| < 1$, the sum of the infinite geometric series with first term a_1 and common ratio r is given by the formula
   $S =$ _____.

2. If $|r| \geq 1$, the sum of the infinite geometric series with first term a_1 and common ratio r is _____.

Find the sum, if it exists, of each infinite geometric series.

3. $a_1 = 16, r = \frac{1}{5}$

4. $a_1 = 12, r = \frac{2}{3}$

5. $a_1 = 16, r = 2$

6. $a_1 = 12, r = 3$

7. $a_1 = 4, r = -\frac{1}{3}$

8. $a_1 = 3, r = -\frac{1}{2}$

9. $1 + \frac{1}{2} + \frac{1}{4} + \frac{1}{8} + \cdots$

10. $1 + \frac{1}{3} + \frac{1}{9} + \frac{1}{27} + \cdots$

11. $10 + 3 + 0.9 + \cdots$

12. $10 + 7 + 4.9 + \cdots$

13. $16 - 12 + 9 - \frac{27}{4} + \cdots$

14. $25 - 20 + 16 - \frac{64}{5} + \cdots$

15. $\frac{1}{9} - \frac{1}{6} + \frac{1}{4} - \cdots$

16. $\frac{1}{16} - \frac{1}{12} + \frac{1}{9} - \cdots$

17. $1 + 1 + 1 + 1 + \cdots$

18. $1 - 1 + 1 - 1 + \cdots$

19. $-\frac{1}{2} - \frac{1}{3} - \frac{2}{9} - \cdots$

20. $-\frac{3}{4} - \frac{3}{5} - \frac{12}{25} - \cdots$

Use the formula for S to find the sum of each infinite geometric series.

21. $\displaystyle\sum_{n=1}^{\infty} 8(\tfrac{2}{3})^n$

22. $\displaystyle\sum_{n=1}^{\infty} 5(\tfrac{3}{4})^n$

23. $\displaystyle\sum_{i=1}^{\infty} \tfrac{5}{9}(-\tfrac{3}{7})^i$

24. $\displaystyle\sum_{i=1}^{\infty} \tfrac{3}{8}(-\tfrac{4}{5})^i$

Find the indicated unknown for each infinite geometric series.

25. $S = 8, r = \tfrac{1}{4}; a_1 = ?$

26. $S = 18, r = \tfrac{1}{6}; a_1 = ?$

27. $S = 12, a_1 = 2; r = ?$

28. $S = 21, a_1 = 6; r = ?$

Write each repeating decimal as a ratio of two integers.

29. $0.\overline{3}$

30. $0.\overline{6}$

31. $0.\overline{36}$

32. $0.\overline{18}$

33. $0.0\overline{7}$

34. $0.0\overline{8}$

35. $0.\overline{123}$

36. $0.\overline{471}$

37. $0.\overline{9}$

38. $9.\overline{9}$

39. $5.\overline{7}$

40. $7.\overline{8}$

Problem Solving

41. A charity is to receive $60,000 from an estate the first year, $45,000 the second year, $33,750 the third year, and so on. Determine the total amount the charity is to receive from the estate.

42. Seventy-two grams of an insecticide are sprayed on a redwood tree the first year, 60 g the second year, 50 g the third year, and so on. Determine the total amount of insecticide that is to be sprayed on the tree.

43. On the first swing, the bob of a pendulum travels an arc of length 45 cm. The arc on each subsequent swing is four-fifths the length of the previous arc. What is the total distance the bob travels before coming to rest?

44. A ball is dropped on a hard surface from a height of 10 ft. On each rebound it bounces three-fifths as high as its previous height. What is the total distance the ball travels before coming to rest?

45. Find the sum S of the series $1 + \tfrac{1}{2} + \tfrac{1}{4} + \tfrac{1}{8} + \cdots$. Then use the formula for S_n from Sec. 12.3 to calculate S_5, S_{10}, and S_{15}, and note that these sums approach S.

46. Find the sum S of the series $1 - \tfrac{1}{2} + \tfrac{1}{4} - \tfrac{1}{8} + \cdots$. Then use the formula for S_n from Sec. 12.3 to calculate S_5, S_{10}, and S_{15}, and note that these sums approach S.

Discussion Exercise

47. Why do you think it is possible to find the sum of an infinite number of terms like $0.3 + 0.03 + 0.003 + \cdots$, but it is not possible to find the sum of $1 + 2 + 4 + \cdots$?

12.5 The Binomial Theorem

In Sec. 3.5 we learned that the square of the binomial $a + b$ is

$$(a + b)^2 = a^2 + 2ab + b^2$$

The expression $a^2 + 2ab + b^2$ is called the *expansion* of $(a + b)^2$. In this section we state a formula for expanding $(a + b)^n$, where n is any positive integer.

Expanding $(a + b)^n$

When a binomial of the form $a + b$ is raised to a power, the result can be thought of as a series. To see the pattern of the series, consider the following expansions of $(a + b)^n$:

$$
\begin{aligned}
(a + b)^0 &= 1 \\
(a + b)^1 &= a + b \\
(a + b)^2 &= a^2 + 2ab + b^2 \\
(a + b)^3 &= a^3 + 3a^2b + 3ab^2 + b^3 \\
(a + b)^4 &= a^4 + 4a^3b + 6a^2b^2 + 4ab^3 + b^4 \\
(a + b)^5 &= a^5 + 5a^4b + 10a^3b^2 + 10a^2b^3 + 5ab^4 + b^5
\end{aligned}
$$

Based on the expansions above, we draw the following conclusions.

Expansion of $(a + b)^n$

1. There are $n + 1$ terms.
2. The exponents on a start with n and decrease to 0.
3. The exponents on b start with 0 and increase to n.
4. The sum of the exponents on a and b in every term is n.

All we need now is a method for determining the coefficients. One method involves constructing an array of numbers called **Pascal's triangle.** *

Pascal's Triangle

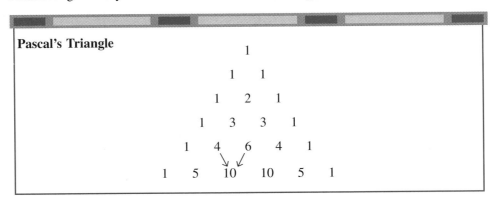

The sides of Pascal's triangle consist of 1s. To find a number in the interior of the triangle, add the two adjacent numbers in the row above. For example, $4 + 6 = 10$.

Note that the coefficient in the expansion of $(a + b)^0$ is given by the first row of Pascal's triangle, the coefficients of $(a + b)^1$ by the second row, the coefficients of $(a + b)^2$ by the third row, and so on.

Example 1   Expand $(a + b)^6$.

Solution: The coefficients in the expansion of $(a + b)^6$ are given by the seventh row of Pascal's triangle.

$$1 \quad 6 \quad 15 \quad 20 \quad 15 \quad 6 \quad 1$$

Use these coefficients, along with the four rules given earlier, to write the expansion.

$$(a + b)^6 = a^6 + 6a^5b + 15a^4b^2 + 20a^3b^3 + 15a^2b^4 + 6ab^5 + b^6 \; \square$$

Try Exercise 7

Expand: $(a + b)^7$.

Factorial Notation

Although Pascal's triangle provides a simple method for determining the coefficients of a binomial expansion, it is impractical for large values of n. More important, it does not give us a formula for the expansion of $(a + b)^n$ in terms of a, b, and n. Such a formula is essential in proving some of the theorems in higher mathematics.

Before we state such a formula, we introduce a notational shorthand called **factorial notation.**

---

*HISTORICAL NOTE: Pascal's triangle is named after the French mathematician and philosopher Blaise Pascal (1623–1662) because of the many applications Pascal made of the triangle, particularly in probability theory. Pascal was a mathematical prodigy, developing new theorems in projective geometry at age 16 and inventing the first calculating machine at age 18.

DEFINITION OF *n* FACTORIAL

If n is a positive integer, then **$n!$** (read "n factorial") is given by

$$n! = n(n - 1)(n - 2) \cdots 3 \cdot 2 \cdot 1$$

Examples of this definition are shown below:

$$4! = 4 \cdot 3 \cdot 2 \cdot 1 = 24$$
$$3! = 3 \cdot 2 \cdot 1 = 6$$
$$2! = 2 \cdot 1 = 2$$
$$1! = 1$$

In order to make certain formulas (including the binomial theorem) valid in all cases, we define 0! to be 1:   **$0! = 1$**

We now use factorial notation to define another shorthand notation.

DEFINITION OF A BINOMIAL COEFFICIENT

Let n and r be whole numbers with $n \geq r$. Then

$$\binom{n}{r} = \frac{n!}{r!(n - r)!}$$

The expression $\binom{n}{r}$ is read "the **binomial coefficient** of n over r," or simply "n choose r."

Example 2   Calculate: $\binom{5}{2}$.

Solution: Substitute $n = 5$ and $r = 2$ into the formula for $\binom{n}{r}$.

$$\binom{n}{r} = \frac{n!}{r!(n - r)!}$$
$$\binom{5}{2} = \frac{5!}{2!(5 - 2)!} = \frac{5!}{2!3!} = \frac{5 \cdot 4 \cdot 3 \cdot 2 \cdot 1}{(2 \cdot 1)(3 \cdot 2 \cdot 1)} = \frac{5 \cdot 4}{2 \cdot 1} = 10 \quad \square$$

Try Exercise 13

Calculate: $\binom{6}{2}$.

CAUTION

Note the difference in the expressions below:

$$(5 - 2)! = 3! = 6 \qquad\qquad (5 \cdot 2)! = 10! = 3{,}628{,}800$$
$$5! - 2! = 120 - 2 = 118 \qquad 5! \cdot 2! = 120 \cdot 2 = 240$$
$$5 - 2! = 5 - 2 = 3 \qquad\qquad 5 \cdot 2! = 5 \cdot 2 = 10 \quad \blacksquare$$

The binomial coefficient of Example 2 can be calculated more efficiently as follows:

$$\binom{5}{2} = \frac{5!}{2!(5 - 2)!} = \frac{5 \cdot 4 \cdot 3!}{(2 \cdot 1)3!} = 10$$

We use this technique in Example 3.

Example 3 Calculate each binomial coefficient.

(a) $\binom{4}{0}$ (b) $\binom{4}{1}$ (c) $\binom{4}{2}$ (d) $\binom{4}{3}$ (e) $\binom{4}{4}$

Solution:

(a) $\binom{4}{0} = \dfrac{4!}{0!(4-0)!} = \dfrac{4!}{1 \cdot 4!} = 1$

(b) $\binom{4}{1} = \dfrac{4!}{1!(4-1)!} = \dfrac{4 \cdot 3!}{1 \cdot 3!} = 4$

(c) $\binom{4}{2} = \dfrac{4!}{2!(4-2)!} = \dfrac{4 \cdot 3 \cdot 2!}{(2 \cdot 1)2!} = 6$

(d) $\binom{4}{3} = \dfrac{4!}{3!(4-3)!} = \dfrac{4 \cdot 3!}{3!1!} = 4$

(e) $\binom{4}{4} = \dfrac{4!}{4!(4-4)!} = \dfrac{4!}{4!0!} = 1$

❒

Try Exercise 23

Calculate: $\binom{15}{14}$.

The Binomial Theorem

Comparing the results of Example 3 with the fifth row of Pascal's triangle, we see that the values of $\binom{4}{0}$, $\binom{4}{1}$, $\binom{4}{2}$, $\binom{4}{3}$, and $\binom{4}{4}$ are the coefficients in the expansion of $(a+b)^4$. We use this fact to state the **binomial theorem.**[*]

The Binomial Theorem

If n is a positive integer, then

$$(a+b)^n = \binom{n}{0}a^n + \binom{n}{1}a^{n-1}b + \binom{n}{2}a^{n-2}b^2 + \cdots + \binom{n}{n}b^n$$

Example 4 Expand $(x+2)^4$ using the binomial theorem.

Solution: Apply the binomial theorem with $a = x$, $b = 2$, and $n = 4$.

$$(x+2)^4 = \binom{4}{0}x^4 + \binom{4}{1}x^3(2) + \binom{4}{2}x^2(2)^2 + \binom{4}{3}x(2)^3 + \binom{4}{4}2^4$$

$$= \frac{4!}{0!4!}x^4 + \frac{4!}{1!3!}x^3(2) + \frac{4!}{2!2!}x^2(4) + \frac{4!}{3!1!}x(8) + \frac{4!}{4!0!}(16)$$

$$= 1x^4 + 4x^3(2) + 6x^2(4) + 4x(8) + 1(16)$$

$$= x^4 + 8x^3 + 24x^2 + 32x + 16$$

❒

Try Exercise 29

Expand $(x+3)^4$ using the binomial theorem.

Example 5 Expand $(2m-3)^5$ using the binomial theorem.

Solution: Write $(2m-3)^5$ in the form $(a+b)^n$

$$(2m-3)^5 = (2m+(-3))^5$$

Apply the binomial theorem with $a = 2m$, $b = -3$, and $n = 5$.

---

[*]HISTORICAL NOTE: The Swiss mathematician Jacques Bernoulli (1654–1705) proved the binomial theorem in his treatise *Ars conjectandi*. He used a method of proof called *mathematical induction*, a method discussed in Sec. 12.6. No family in history has produced as many celebrated mathematicians (13) as did the Bernoulli family.

$$(2m + (-3))^5 = \binom{5}{0}(2m)^5 + \binom{5}{1}(2m)^4(-3) + \binom{5}{2}(2m)^3(-3)^2$$

$$+ \binom{5}{3}(2m)^2(-3)^3 + \binom{5}{4}(2m)(-3)^4 + \binom{5}{5}(-3)^5$$

$$= \frac{5!}{0!5!}(32m^5) + \frac{5!}{1!4!}(16m^4)(-3) + \frac{5!}{2!3!}(8m^3)(9)$$

$$+ \frac{5!}{3!2!}(4m^2)(-27) + \frac{5!}{4!1!}(2m)(81) + \frac{5!}{5!0!}(-243)$$

$$= 1(32m^5) + 5(16m^4)(-3) + 10(8m^3)(9)$$
$$+ 10(4m^2)(-27) + 5(2m)(81) + 1(-243)$$
$$= 32m^5 - 240m^4 + 720m^3 - 1080m^2 + 810m - 243$$

❐

Try Exercise 37

Expand $(3m - 2)^5$ using the binomial theorem.

Example 5 illustrates that **the expansion of the binomial $(a - b)^n$ produces terms with alternating signs.**

Example 6   Expand $\left(x - \dfrac{y}{5}\right)^3$ using the binomial theorem.

Solution: Apply the binomial theorem with $a = x$, $b = -\dfrac{y}{5}$, and $n = 3$.

$$\left(x - \frac{y}{5}\right)^3 = \binom{3}{0}x^3 + \binom{3}{1}x^2\left(-\frac{y}{5}\right) + \binom{3}{2}x\left(-\frac{y}{5}\right)^2 + \binom{3}{3}\left(-\frac{y}{5}\right)^3$$

$$= 1x^3 + 3x^2\left(-\frac{y}{5}\right) + 3x\left(\frac{y^2}{25}\right) + 1\left(-\frac{y^3}{125}\right)$$

$$= x^3 - \frac{3}{5}x^2y + \frac{3}{25}xy^2 - \frac{1}{125}y^3$$

❐

Try Exercise 39

Expand $\left(x - \dfrac{y}{2}\right)^4$ using the binomial theorem.

Finding a Single Term

If you look closely at the pattern of the terms in the binomial theorem, you will see that we can write down any single term without finding the other terms. For example, the third term in the expansion of $(a + b)^9$ is given by

$$\text{Third term} = \binom{9}{2}a^{9-2}b^2$$

This suggests the following rule.

The rth Term in the Expansion of $(a + b)^n$

Let n be a positive integer.

$$r\text{th term} = \binom{n}{r-1}a^{n-(r-1)}b^{r-1}$$

Example 7   Find the middle term in the expansion of $\left(x + \dfrac{y}{4}\right)^8$.

Solution: Since there are $8 + 1 = 9$ terms in the expansion, the middle term is the fifth term. Therefore substitute $r = 5$, $n = 8$, $a = x$, and $b = \frac{y}{4}$ into the formula for the rth term.

$$\text{Fifth term} = \binom{8}{4}x^{8-4}\left(\frac{y}{4}\right)^4$$

$$= \frac{8!}{4!4!}x^4\left(\frac{y^4}{4^4}\right)$$

$$= 70x^4\left(\frac{y^4}{256}\right)$$

$$= \frac{35}{128}x^4y^4$$ ❑

Try Exercise 53

Find the middle term in the expansion of $\left(x + \dfrac{y}{3}\right)^8$.

Exercises 12.5

True-False Exercises

1. The expansion of $(a + b)^7$ has eight terms.

2. The first term in the expansion of $(a + b)^7$ is a^7.

3. The last term in the expansion of $(a + b)^7$ is b^7.

4. The term $35a^4b^2$ could not be a term in the expansion of $(a + b)^7$.

5. The expansion of $(a - b)^7$ produces terms with alternating signs.

6. $0! = 0$.

Expand each binomial. Use Pascal's triangle to determine the coefficients.

7. $(a + b)^7$ 8. $(a + b)^8$

Completion Exercises

9. If n is a positive integer, then $n!$ is calculated as follows: $n! =$ _____.

10. If n and r are whole numbers with $n \geq r$, then $\binom{n}{r}$ is calculated as follows: $\binom{n}{r} =$ _____.

11. If n is a positive integer, the binomial theorem states that $(a + b)^n =$ _____.

12. If n is a positive integer, then the rth term in the expansion of $(a + b)^n$ is _____.

Calculate each binomial coefficient.

13. $\binom{6}{2}$ 14. $\binom{7}{2}$

15. $\binom{5}{3}$ 16. $\binom{7}{5}$

17. $\binom{11}{8}$ 18. $\binom{10}{7}$

19. $\binom{8}{1}$ 20. $\binom{9}{1}$

21. $\binom{8}{0}$ 22. $\binom{9}{0}$

23. $\binom{15}{14}$ 24. $\binom{20}{19}$

Discussion Exercises

25. Discuss the difference in $(7 - 4)!$ and $7! - 4!$ and $7 - 4!$

26. Discuss the difference in $(6 \cdot 3)!$ and $6! \cdot 3!$ and $6 \cdot 3!$

Expand each binomial using the binomial theorem.

27. $(x + y)^3$ 28. $(x + y)^5$

29. $(x + 3)^4$ 30. $(x + 5)^4$

31. $(x - y)^5$ 32. $(x - y)^3$

33. $(2x + y)^4$ 34. $(3x + y)^4$

35. $(a - 5)^3$ 36. $(a - 2)^5$

37. $(3m - 2)^5$ 38. $(3m - 4)^5$

39. $\left(x - \frac{y}{2}\right)^4$

40. $\left(x + \frac{y}{3}\right)^4$

41. $(t^2 + 1)^6$

42. $(t^2 - 1)^6$

Use the binomial theorem to write the first three terms in each expansion.

43. $(a + b)^{20}$

44. $(a + b)^{30}$

45. $(x - 4)^{15}$

46. $(x - 10)^{25}$

47. $(t^2 - 1)^{12}$

48. $(t^2 + 1)^{13}$

Find the indicated term in each expansion.

49. $(a + b)^9$; fourth term

50. $(a + b)^{11}$; fifth term

51. $(p - 2q)^{14}$; sixth term

52. $(p - 5q)^{16}$; fourth term

53. $\left(x + \frac{y}{3}\right)^8$; middle term

54. $\left(x + \frac{y}{2}\right)^{10}$; middle term

Show that each statement is true for all positive integers n.

55. $\binom{n}{0} = 1$

56. $\binom{n}{n} = 1$

57. $\binom{n}{1} = n$

58. $\binom{n}{n-1} = n$

Getting Acquainted with Your Scientific Calculator

To find 5! on your calculator, press

Clear 5 x!   120

Find each factorial on your calculator.

59. 0!

60. 7!

61. 11!

62. 69!

63. Why does your calculator register an error when you try to find 70!?

Your calculator may have a $\boxed{_nC_r}$ key. This is another notation for $\binom{n}{r}$. If your calculator has such a key, use it to do Exercises 13 through 24.

Getting Acquainted with Your Graphing Calculator

64. Verify that $(x - 2)^4 = x^4 - 8x^3 + 24x^2 - 32x + 16$ by showing that the graph of $y = (x - 2)^4$ is the same as the graph of $y = x^4 - 8x^3 + 24x^2 - 32x + 16$.

12.6 Mathematical Induction

Consider the expression $2^{2^n} + 1$, where n is a positive integer.

n	$2^{2^n} + 1$
1	$2^{2^1} + 1 = 2^2 + 1 = 5$
2	$2^{2^2} + 1 = 2^4 + 1 = 17$
3	$2^{2^3} + 1 = 2^8 + 1 = 257$
4	$2^{2^4} + 1 = 2^{16} + 1 = 65{,}537$

Since the numbers 5, 17, 257, and 65,537 are all prime numbers, the French mathematician Pierre de Fermat (1601–1655) conjectured that *all* numbers of the form $2^{2^n} + 1$, where n is a positive integer, were prime numbers. Later, the Swiss mathematician Leonhard Euler (1707–1783) showed that this was not the case. In fact, if $n = 5$, then

$$2^{2^n} + 1 = 2^{2^5} + 1 = 2^{32} + 1 = 4{,}294{,}967{,}297$$

which is not a prime since it is divisible by 641.

    Just because a statement involving n is true for several positive integers n does not mean that it is true for *every* positive integer n. Since we cannot test such statements for every positive integer n, we use a method of proof called **mathematical induction.***

---

*HISTORICAL NOTE: Mathematical induction was used extensively by Giuseppe Peano (1858–1932). Peano was among the first mathematicians to demonstrate that the great bulk of mathematics can be made to rest on a few simple assumptions.

The Principle of Mathematical Induction

A statement S_n involving n is true for all positive integers n if *both* of the following conditions are met:

1. The statement is true when $n = 1$. That is, S_1 is true.
2. If the statement is true when $n = k$, then it is true when $n = k + 1$. That is, if S_k is true, then S_{k+1} is true.

Example 1   Use mathematical induction to prove that the statement S_n (shown below) is true for all positive integers n.

$$S_n: 1 + 2 + 3 + \cdots + n = \frac{n(n + 1)}{2}$$

Solution: Step 1:   Show the statement is true when $n = 1$. Do this by replacing n with 1 in the given statement.

$$S_1: 1 = \frac{1(1 + 1)}{2} \qquad \text{Let } n = 1.$$

$$S_1: 1 = \frac{1(2)}{2} \qquad \text{Simplify.}$$

$$S_1: 1 = 1 \qquad \text{True}$$

Therefore, S_1 is true.

Step 2:   Show that if the statement is true when $n = k$, then it is true when $n = k + 1$. Do this by assuming the statement is true when n is replaced by k. That is, assume the statement below is true.

$$S_k: 1 + 2 + 3 + \cdots + k = \frac{k(k + 1)}{2}$$

Using the statement S_k, we must now show that the statement S_{k+1} is true. That is, we must show that the statement is true when n is replaced by $k + 1$ (see statement below).

$$S_{k+1}: 1 + 2 + 3 + \cdots + k + (k + 1) = \frac{(k + 1)((k + 1) + 1)}{2}$$

To show that S_{k+1} is true, add $k + 1$ to both sides of S_k (which we have assumed is true) to make the left side of S_k look like the left side of S_{k+1}.

$$1 + 2 + 3 + \cdots + k + (k + 1) = \frac{k(k + 1)}{2} + (k + 1)$$

The left side of this last statement is the same as the left side of S_{k+1}. Now show that the right side of this last statement is the same as the right side of S_{k+1}.

$$1 + 2 + 3 + \cdots + k + (k + 1) = \frac{k(k + 1)}{2} + \frac{2(k + 1)}{2}$$
$$= \frac{k(k + 1) + 2(k + 1)}{2}$$
$$= \frac{(k + 1)(k + 2)}{2}$$
$$= \frac{(k + 1)((k + 1) + 1)}{2}$$

Therefore, if S_k is true, then S_{k+1} is true.

By the principle of mathematical induction, we have shown that the statement

$$1 + 2 + 3 + \cdots + n = \frac{n(n + 1)}{2}$$

Try Exercise 19

Use mathematical induction to prove that the following statement is true for all positive integers n.

$$1^2 + 2^2 + 3^2 + \cdots + n^2$$
$$= \frac{n(n + 1)(2n + 1)}{6}$$

is true for all positive integers n.   ❏

Using the principle of mathematical induction to prove that a statement is true for all positive integers n is similar to getting a row of dominos to fall. When we show that "the statement is true when $n = 1$" we are knocking over the first domino. When we show that "if the statement is true when $n = k$, then it is true when $n = k + 1$," we are showing that if the kth domino falls, then the $(k + 1)$st domino will fall (see Fig. 12.1). Therefore, if the first domino falls, then the second domino will fall. But if the second domino falls, then the third domino will fall, and so on. In this way, we have shown that all of the dominos will fall—or that the statement is true for all positive integers n.

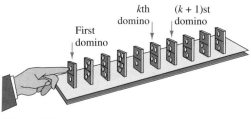

Figure 12.1

CAUTION

Both conditions in the principle of mathematical induction must be shown to complete the proof. Knocking over the first domino will not knock over the entire row, unless it is also true that any given domino will fall if the domino before it falls (see Fig. 12.2).

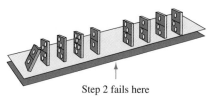

Step 2 fails here

Figure 12.2

Also, even if it is true that any given domino will fall if the domino before it falls, none of the dominos will fall unless you knock over the first domino (see Fig. 12.3).

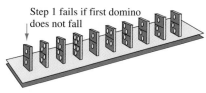

Step 1 fails if first domino does not fall

Figure 12.3   ■

Example 2   Use mathematical induction to prove that the sum of the first n odd positive integers is n^2. That is, prove that the statement S_n (shown below) is true for all positive integers n.

$$S_n: 1 + 3 + 5 + \cdots + (2n - 1) = n^2$$

Solution: Step 1:   Show that S_1 is true.

$$S_1: 1 = 1^2 \qquad \text{Let } n = 1.$$
$$S_1: 1 = 1 \qquad \text{True}$$

Therefore, S_1 is true.

Step 2:   Show that if S_k is true, then S_{k+1} is true.

$$S_k: 1 + 3 + 5 + \cdots + (2k - 1) = k^2$$
$$S_{k+1}: 1 + 3 + 5 + \cdots + (2k - 1) + (2(k + 1) - 1) = (k + 1)^2$$

To make the left side of S_k look like the left side of S_{k+1}, add $(2(k + 1) - 1)$ to both sides of S_k.

$$1 + 3 + 5 + \cdots + (2k - 1) + (2(k + 1) - 1) = k^2 + (2(k + 1) - 1)$$
$$= k^2 + 2k + 1$$
$$= (k + 1)^2$$

Therefore, if S_k is true, then S_{k+1} is true.
   By the principle of mathematical induction, we have shown that the statement

$$1 + 3 + 5 + \cdots + (2n - 1) = n^2$$

Try Exercise 21

is true for all positive integers n.   ❐

Use mathematical induction to prove that the following statement is true for all positive integers n.

$2 + 4 + 6 + \cdots + 2n =$
$\qquad n(n + 1)$

Example 3   Use mathematical induction to prove the statement S_n is true for all positive integers n.

$$S_n: 3^1 + 3^2 + 3^3 + \cdots + 3^n = \frac{3(3^n - 1)}{2}$$

Solution: Step 1:   Show that S_1 is true.

$$S_1: 3^1 = \frac{3(3^1 - 1)}{2} \qquad \text{Let } n = 1.$$

$$S_1: 3 = \frac{3(2)}{2} \qquad \text{Simplify.}$$

$$S_1: 3 = 3 \qquad \text{True}$$

Therefore, S_1 is true.

Step 2:   Show that if S_k is true, then S_{k+1} is true.

$$S_k: 3^1 + 3^2 + 3^3 + \cdots + 3^k = \frac{3(3^k - 1)}{2}$$

$$S_{k+1}: 3^1 + 3^2 + 3^3 + \cdots + 3^k + 3^{k+1} = \frac{3(3^{k+1} - 1)}{2}$$

To make the left side of S_k look like the left side of S_{k+1}, add 3^{k+1} to both sides of S_k.

$$3^1 + 3^2 + 3^3 + \cdots + 3^k + 3^{k+1} = \frac{3(3^k - 1)}{2} + 3^{k+1}$$

$$= \frac{3^{k+1} - 3}{2} + \frac{2 \cdot 3^{k+1}}{2}$$

But, since $3^{k+1} + 2 \cdot 3^{k+1} = (1 + 2)3^{k+1} = 3 \cdot 3^{k+1}$,

$$3^1 + 3^2 + 3^3 + \cdots + 3^k + 3^{k+1} = \frac{3 \cdot 3^{k+1} - 3}{2}$$

$$= \frac{3(3^{k+1} - 1)}{2}$$

Therefore, if S_k is true, then S_{k+1} is true.

By the principle of mathematical induction, we have shown that the statement

$$3^1 + 3^2 + 3^3 + \cdots + 3^n = \frac{3(3^n - 1)}{2}$$

is true for all positive integers n.   ❑

Try Exercise 25

Use mathematical induction to prove that the following statement is true for all positive integers n.

$$2^1 + 2^2 + 2^3 + \cdots + 2^n = 2(2^n - 1)$$

The principle of mathematical induction can also be used to show that a statement involving n is true for all integers greater than or equal to *any* given integer (not just 1). For example, the statement

$$S_n: 2^n > 2n + 1$$

is *not* true for all integers greater than or equal to 1. However, it *is* true for all integers greater than or equal to 3. To prove this, simply show that S_3 (rather than S_1) is true in step 1. Then verify step 2 as before.

Exercises 12.6

◆◆

True-False Exercise

1. If we can show that a statement S_n is true for $n = 1$, for $n = 2$, and for $n = 3$, then we can assume that it is true for all positive integers n.

Completion Exercise

2. To prove that a statement S_n is true for all positive integers n, we must show that _____ and that _____.

For each statement S_n, write both of the statements S_k and S_{k+1}. Then state what must be done to both sides of S_k to make the left side of S_k look like the left side of S_{k+1}.

3. $S_n: 1 + 2 + 3 + \cdots + n = \dfrac{n(n + 1)}{2}$

4. $S_n: 5 + 10 + 15 + \cdots + 5n = \dfrac{5n(n + 1)}{2}$

5. $S_n: 1^2 + 2^2 + 3^2 + \cdots + n^2 = \dfrac{n(n + 1)(2n + 1)}{6}$

6. $S_n: 1^3 + 2^3 + 3^3 + \cdots + n^3 = \dfrac{n^2(n + 1)^2}{4}$

7. $S_n: 2 + 4 + 6 + \cdots + 2n = n(n + 1)$

8. $S_n: 4 + 8 + 12 + \cdots + 4n = 2n(n + 1)$

9. $S_n: 3 + 7 + 11 + \cdots + (4n - 1) = n(2n + 1)$

10. $S_n: 1 + 3 + 5 + \cdots + (2n - 1) = n^2$

11. $S_n: 2^1 + 2^2 + 2^3 + \cdots + 2^n = 2(2^n - 1)$

12. $S_n: 3^1 + 3^2 + 3^3 + \cdots + 3^n = \dfrac{3(3^n - 1)}{2}$

13. S_n: $(ab)^n = a^n b^n$

14. S_n: $\left(\dfrac{a}{b}\right)^n = \dfrac{a^n}{b^n}$

15. S_n: $2^n > n$

16. S_n: $2^n > 2n + 1$ (for $n \geq 3$)

Use mathematical induction to prove each statement is true for all positive integers n.

17. $1 + 2 + 3 + \cdots + n = \dfrac{n(n + 1)}{2}$

18. $5 + 10 + 15 + \cdots + 5n = \dfrac{5n(n + 1)}{2}$

19. $1^2 + 2^2 + 3^2 + \cdots + n^2 = \dfrac{n(n + 1)(2n + 1)}{6}$

20. $1^3 + 2^3 + 3^3 + \cdots + n^3 = \dfrac{n^2(n + 1)^2}{4}$

21. $2 + 4 + 6 + \cdots + 2n = n(n + 1)$

22. $4 + 8 + 12 + \cdots + 4n = 2n(n + 1)$

23. $3 + 7 + 11 + \cdots + (4n - 1) = n(2n + 1)$

24. $1 + 3 + 5 + \cdots + (2n - 1) = n^2$

25. $2^1 + 2^2 + 2^3 + \cdots + 2^n = 2(2^n - 1)$

26. $3^1 + 3^2 + 3^3 + \cdots + 3^n = \dfrac{3(3^n - 1)}{2}$

27. $(ab)^n = a^n b^n$

28. $\left(\dfrac{a}{b}\right)^n = \dfrac{a^n}{b^n}$

29. $2^n > n$

30. $2^n > 2n + 1$ (for $n \geq 3$)

Discussion Exercises

31. Using a row of dominos, explain how the principle of mathematical induction works, and show why both conditions are necessary.

32. Consider the statement "every positive integer equals the next integer." That is, S_n: $n = n + 1$. What is wrong with the following "proof" of this statement?

$$S_k: k = k + 1 \qquad \text{Assume true.}$$
$$S_{k+1}: k + 1 = (k + 1) + 1 \qquad \text{Must show true.}$$

Add 1 to both sides of S_k.

$$k + 1 = k + 1 + 1$$

Therefore, if S_k is true, then S_{k+1} is true. By the principle of mathematical induction, the statement $n = n + 1$ is true for all positive integers n.

33. Does the expression $n^2 - n + 41$ represent a prime number for all positive integers n? Try $n = 1, 2, 3, 4,$ and 5. Then stop and think for a moment.

12.7 Permutations and Combinations*

To purchase a ticket in a particular state lottery, you must select six different numbers from the numbers $\{1, 2, 3, \ldots, 40\}$. In how many ways can you do this? In this section we will answer this question, as well as other related questions.

Fundamental Principle of Counting

A college student has a choice of taking math or chemistry at 9 a.m., and art, music, or history at 10 a.m. How many different two-course schedules can the student select?

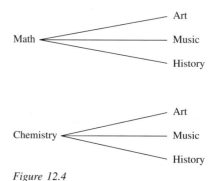

Figure 12.4

*HISTORICAL NOTE: The first substantial text on the theory of permutations, combinations, and probability was the *Ars conjectandi,* written by the Swiss mathematician Jacques Bernoulli (1654–1705).

From Fig. 12.4, we see that there are $2 \cdot 3 = 6$ different ways to make a schedule. This example illustrates the **fundamental principle of counting.**

Fundamental Principle of Counting

Let E_1 be an event that can occur in n_1 different ways. After E_1 has occurred, an event E_2 can occur in n_2 different ways. Then the number of ways that E_1 and E_2 can occur in succession is

$$n_1 \cdot n_2$$

The fundamental principle of counting can be extended to any number of events.

Example 1   Susan has seven blouses, five skirts, and two pairs of shoes. How many different outfits does she have? Assume all of the blouses, skirts, and shoes match.

Solution: Susan can perform the event of putting on a blouse in seven different ways. After that has occurred, she can put on a skirt in five different ways. After both of those events has occurred, she can put on a pair of shoes in two different ways. By the fundamental principle of counting, the three events can be performed in succession in

$$7 \cdot 5 \cdot 2 = 70 \text{ ways}$$

Try Exercise 11

Susan has 70 different outfits.   ❑

An evening college student has a choice of taking algebra, history, English, or philosophy at 6 p.m., and sociology or psychology at 7:30 p.m. How many different two-course schedules can she select?

Permutations

A **permutation** of r elements is an ordering of the r elements so that one of the elements is first, one is second, one is third, and so on.

Example 2   List all permutations of the three letters a, b, and c.

Solution: There are six permutations of the letters a, b, and c.

$$\begin{array}{ccc} abc & bac & cab \\ acb & bca & cba \end{array}$$
    ❑

Try Exercise 27

List all permutations of the numbers 1, 2, and 3.

In Example 2, we determined that there were six permutations of the letters a, b, and c by listing all of the permutations. However, this would be difficult to do if there were 10 letters (there would be 3,628,800 permutations!). Therefore, we shall use the fundamental principle of counting to develop a formula for finding the number of permutations of a given number of elements.

Consider one of the permutations of the three letters a, b, and c (Fig. 12.5). We can fill the first position in three ways (with a, b, or c). Having filled the first position, we can fill the second position in two ways (with one of the two letters left after filling the first position). Finally, we can fill the third position in just one way (with the last remaining letter). Therefore, we can fill the three positions in succession (that is, create a permutation) in

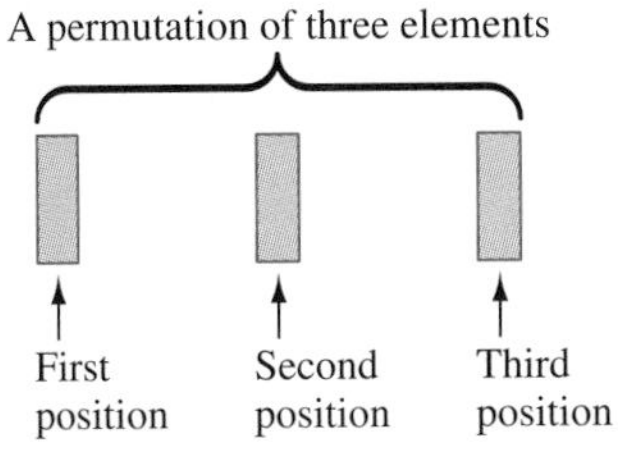

A permutation of three elements

First position   Second position   Third position

Figure 12.5

$$3 \cdot 2 \cdot 1 = 6 \text{ ways}$$

Example 3   In how many ways can the manager of a baseball team with 25 players make out a batting order?

Solution: Since there are only nine players in a batting order (Fig. 12.6), we want to know the number of permutations of 25 elements taken nine at a time.

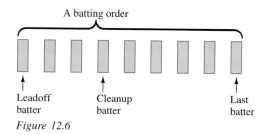

Figure 12.6

There are 25 ways to fill the first position (leadoff batter), 24 ways to fill the second position, 23 ways to fill the third position, and so on. Therefore, there are

$$25 \cdot 24 \cdot 23 \cdot 22 \cdot 21 \cdot 20 \cdot 19 \cdot 18 \cdot 17$$
$$= 741,354,768,000 \text{ different batting orders} \qquad \square$$

In order to develop a formula that will allow us to determine the number of permutations of a given number of elements, we need to review factorial notation.

Try Exercise 31

In how many ways can the manager of a girl's softball team with 10 players make out the batting order?

DEFINITION OF FACTORIAL NOTATION

Let n be a positive integer. Then $n!$ is read "n factorial," and

$$n! = n \cdot (n - 1) \cdot (n - 2) \cdots 3 \cdot 2 \cdot 1$$

Also, $0! = 1$.

By this definition, $5! = 5 \cdot 4 \cdot 3 \cdot 2 \cdot 1 = 120$, $3! = 3 \cdot 2 \cdot 1 = 6$, $1! = 1$, and $0! = 1$.

In Example 3, instead of determining the number of permutations of 25 elements taken nine at a time, suppose we wanted to determine the number of permutations of n elements taken r at a time. There are n ways to fill the first position, $n - 1$ ways to fill the second position, $n - 2$ ways to fill the third position, and so on (see Fig. 12.7).

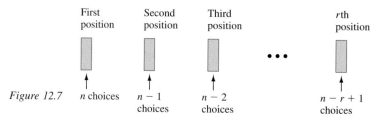

Figure 12.7

Therefore, the number of permutations of n elements taken r at a time is

$$n \cdot (n - 1) \cdot (n - 2) \cdots (n - r + 1)$$
$$= \frac{n \cdot (n - 1) \cdot (n - 2) \cdots (n - r + 1)}{1} \cdot \frac{(n - r) \cdot (n - r - 1) \cdots 3 \cdot 2 \cdot 1}{(n - r) \cdot (n - r - 1) \cdots 3 \cdot 2 \cdot 1}$$
$$= \frac{n!}{(n - r)!}$$

We have proved the following theorem.

Permutation Theorem

Let n and r be whole numbers with $n \geq r$. The number of permutations of n elements taken r at a time is denoted $P(n, r)$, and is given by

$$P(n, r) = \frac{n!}{(n - r)!}$$

Other notations for $P(n, r)$ are $_nP_r$ and P_r^n.

Example 4 Ten horses run a race and there are no dead heats (ties). (a) In how many ways can the first-, second-, and third-place prizes be awarded? (b) In how many ways can the 10 horses finish?

Solution: (a) This is the number of permutations of 10 elements taken three at a time.

$$P(10, 3) = \frac{10!}{(10 - 3)!} = \frac{10!}{7!} = \frac{10 \cdot 9 \cdot 8 \cdot 7!}{7!} = 10 \cdot 9 \cdot 8 = 720$$

There are 720 ways the first-, second-, and third-place prizes can be awarded.

(b) This is the number of permutations of 10 elements taken 10 at a time.

$$P(10, 10) = \frac{10!}{(10 - 10)!} = \frac{10!}{0!} = \frac{10!}{1} = 10! = 3,628,800$$

There are 3,628,800 ways that the 10 horses can finish. ❏

Example 4(b) illustrates the following special case of the permutation theorem:

$$P(n, n) = n!$$

Circular Permutations

Example 5 Suppose five people are seated at a round table. How many different seating arrangements are possible?

Solution: The number of ways to seat five people in a row is
$$P(5, 5) = 5! = 120$$

However, at a round table, moving each person to the right (or left) does not create a new seating arrangement. Similarly, moving each person two, three, four, or five places does not create a new seating arrangement. Therefore, we must divide 5! by 5 to eliminate these duplications.

$$\frac{5!}{5} = \frac{5 \cdot 4!}{5} = 4! = 24$$

There are 24 ways to seat five people at a round table. ❏

Generalizing on Example 5 gives the following formula.

Try Exercise 33

Seven Olympic swimmers compete in a 100-m freestyle race, and there are no ties. (a) In how many ways can the gold, silver, and bronze medals be awarded? (b) In how many ways can the seven swimmers finish?

Try Exercise 35

In how many ways can Bob, Sally, Jim, and Mary be seated (a) in a row? (b) in a circle?

Circular Permutations

Let n be a positive integer. There are $(n - 1)!$ ways to arrange n elements in a circle.

Distinguishable Permutations

There are six permutations of the letters in the word "mom."

$$m_1om_2 \qquad om_1m_2 \qquad m_2m_1o$$
$$m_2om_1 \qquad om_2m_1 \qquad m_1m_2o$$

However, only three of these permutations are distinguishable without the subscripts 1 and 2, because two of the letters are the same. To find the number of distinguishable permutations, we must divide the total number of permutations (namely, 3!) by the number of ways to permutate the two m's (namely, 2!).

$$\text{Number of distinguishable permutations of "mom"} = \frac{3!}{2!} = 3$$

Generalizing on this example gives the following theorem.

Distinguishable Permutations

Let n be a positive integer. The number of distinguishable permutations of n elements, where n_1 are of one kind, n_2 are of another kind, and so on, is

$$\frac{n!}{n_1!n_2!n_3! \cdots n_k!}$$

Example 6   How many distinguishable permutations of the letters in the word "Mississippi" are there?

Solution: The word "Mississippi" has 11 letters, of which four are i's, four are s's, two are p's, and one is an m. Therefore, the number of distinguishable permutations is

$$\frac{11!}{4!4!2!1!} = 34{,}650 \qquad \square$$

Try Exercise 37

How many distinguishable permutations of the letters in the word "horror" are there?

Combinations

When the 10 semifinalists in the Miss America Pageant are chosen from the original group of 50 contestants, the order in which they are chosen is not important. A group of r elements chosen from a group of n elements *without regard to order* is called a **combination** of n elements taken r at a time.

CAUTION

Changing the order of the elements produces a new permutation, but changing the order of the elements does *not* produce a new combination. ∎

Compare all *permutations* of the three letters a, b, and c taken two at a time with all *combinations* of a, b, and c taken two at a time.

Permutations		Combinations
ab	ba	ab
ac	ca	ac
bc	cb	bc

Since each pair of elements in the table above has 2! permutations, the number of combinations of three elements taken two at a time, denoted $C(3, 2)$, can be found by dividing the number of permutations by 2!.

$$C(3, 2) = \frac{P(3, 2)}{2!} = \frac{\frac{3!}{(3-2)!}}{2!} = \frac{3!}{(3-2)!2!} = 3$$

In general, we have the following theorem.

Combination Theorem

Let n and r be whole numbers with $n \geq r$. The number of combinations of n elements taken r at a time is denoted $C(n, r)$, and is given by

$$C(n, r) = \frac{n!}{(n-r)!r!}$$

Other notations for $C(n, r)$ are $_nC_r$, C_r^n, and $\binom{n}{r}$.

Example 7   There are 40 numbers in a particular state lottery. In how many ways can a person select six of the numbers? The numbers must be different, but the order of selection is not important.

Solution: This is the number of combinations of 40 elements taken six at a time.

$$C(40, 6) = \frac{40!}{(40-6)!6!} = \frac{40 \cdot 39 \cdot 38 \cdot 37 \cdot 36 \cdot 35 \cdot 34!}{34! \cdot 6 \cdot 5 \cdot 4 \cdot 3 \cdot 2 \cdot 1}$$
$$= 3,838,380$$

There are 3,838,380 ways to select 6 numbers from 40 numbers.   ❐

Try Exercise 47

There are 50 numbers in a particular state lottery. In how many ways can a person select five of the numbers? The numbers must be different, but the order of selection is not important.

Example 8   A committee of five is to be selected from a class of 20 students. One of the students has already been designated as the chairperson of the committee. In how many ways can the committee be selected?

Solution: Since one of the 20 students is already on the committee, the problem becomes one of selecting four students from a group of 19. Since the order of selection is not important, we have

$$C(19, 4) = \frac{19!}{15!4!} = \frac{19 \cdot 18 \cdot 17 \cdot 16 \cdot 15!}{15! \cdot 4 \cdot 3 \cdot 2 \cdot 1} = 3876$$

There are 3876 ways to select the committee.   ❐

Try Exercise 49

A committee of six is to be selected from a class of 25 students. One of the 25 students has already been designated as chairperson, and another as vice-chairperson. In how many ways can the committee be selected?

Example 9 A basketball team consists of three centers, four forwards, and five guards. In how many ways can the coach select a starting team consisting of one center, two forwards, and two guards?

Solution: The number of ways to select one center is

$$C(3, 1) = \frac{3!}{2!1!} = 3$$

The number of ways to select two forwards is

$$C(4, 2) = \frac{4!}{2!2!} = 6$$

The number of ways to select two guards is

$$C(5, 2) = \frac{5!}{3!2!} = 10$$

By the fundamental principle of counting, the number of ways to select, in succession, one center, two forwards, and two guards (that is, a starting team) is

$$3 \cdot 6 \cdot 10 = 180 \text{ ways}$$ ❐

Try Exercise 59

A basketball team consists of two centers, five forwards, and six guards. In how many ways can the coach select a starting team of one center, two forwards, and two guards?

Exercises 12.7

Completion Exercises

1. The fundamental principle of counting states that if event E_1 can occur in n_1 ways, and after E_1 has occurred, event E_2 can occur in n_2 ways, then E_1 and E_2 can occur in succession in _____ ways.

2. If n is a positive integer, then $n!$ is read _____, and $n! =$ _____.

3. An ordering of r elements so that one of the elements is first, one is second, and so on, is called a(n) _____ of the r elements.

4. The expression $P(n, r)$ is read _____, and $P(n, r) =$ _____.

5. There are _____ ways to arrange n elements in a circle.

6. The number of distinguishable permutations of n elements, where n_1 are one of a kind, n_2 are of another kind, and so on, is _____.

7. A group of r elements chosen from a group of n elements without regard to order is called a(n) _____ of n elements taken r at a time.

8. The expression $C(n, r)$ is read _____, and $C(n, r) =$ _____.

True-False Exercises

9. $0! = 0$.

10. Changing the order of the elements does not produce a new combination.

Problem Solving

11. An evening college student has a choice of taking algebra, history, English, or philosophy at 6 p.m., and sociology or psychology at 7:30 p.m. How many different two-course schedules can she select?

12. A banquet offers a choice of two appetizers, three main dishes, and one dessert. How many different three-course meals are possible?

13. A combination lock has 40 numbers on its dial. How many different right-left-right combinations are possible?

14. A test contains four multiple-choice questions. Each question has five choices. How many different ways are there to complete the test? Assume all questions are answered.

15. How many different license plates with two letters followed by a three-digit number are possible?

16. How many different seven-digit telephone numbers are possible if the first number cannot be 0 or 1?

Getting Acquainted with Your Scientific Calculator

To find 6! on your calculator, press

$$\boxed{\text{Clear}}\ 6\ \boxed{\text{x!}}\ \boxed{720}$$

Therefore, 6! = 720. Use your calculator to evaluate each expression.

17. 8!   18. 0!

19. 1!   20. 50!

Your calculator may have a key for permutations. Look for a key like one of the following:

$$\boxed{\text{P(n, r)}}\ \boxed{_n\text{P}_r}\ \boxed{\text{P}_r^n}$$

Your calculator may have a key for combinations. Look for a key like one of the following:

$$\boxed{\text{C(n, r)}}\ \boxed{_n\text{C}_r}\ \boxed{\text{C}_r^n}\ \boxed{\binom{n}{r}}$$

Evaluate each expression.

21. $P(8, 5)$ 22. $P(6, 3)$

23. $P(9, 2)$ 24. $P(10, 0)$

25. $P(20, 1)$ 26. $P(15, 15)$

Problem Solving

27. List all permutations of the numbers 1, 2, and 3.

28. List all permutations of the letters a, b, c, and d.

29. In how many ways can six books be arranged on a shelf?

30. In how many ways can seven people sit in a toboggan?

31. In how many ways can the manager of a girl's softball team with 10 players make out the batting order?

32. In how many ways can a president, a secretary, and a treasurer be selected from a class of 30 students?

33. Seven Olympic swimmers compete in a 100-m freestyle race, and there are no ties. (a) In how many ways can the gold, silver, and bronze medals be awarded? (b) In how many ways can the seven swimmers finish?

34. Eight runners compete in a 100-yd dash, and there are no ties. (a) In how many ways can the first-, second-, and third-place prizes be awarded? (b) In how many ways can the eight runners finish?

35. In how many ways can Bob, Sally, Jim, and Mary be seated (a) in a row? (b) in a circle?

36. In how many ways can Alice, Dan, Valerie, Ken, and Susan be seated (a) in a row? (b) in a circle?

37. How many distinguishable permutations of the letters in the word "horror" are there?

38. How many distinguishable permutations of the letters in the word "banana" are there?

Evaluate each expression.

39. $C(7, 3)$ 40. $C(11, 6)$

41. $C(14, 0)$ 42. $C(12, 9)$

43. $C(18, 18)$ 44. $C(25, 1)$

Problem Solving

45. List all combinations of the letters a, b, and c taken (a) one at a time, (b) two at a time, (c) three at a time.

46. List all combinations of the numbers 1, 2, 3, and 4 taken (a) one at a time, (b) two at a time, (c) three at a time, (d) four at a time.

47. There are 50 numbers in a particular state lottery. In how many ways can a person select five of the numbers? The numbers must be different, but the order of selection is not important.

48. How many different five-card hands can be dealt from a deck of playing cards that contains 52 different cards?

49. A committee of six is to be selected from a class of 25 students. One of the 25 students has already been designated as chairperson, and another has already been designated as vice-chairperson. In how many ways can the committee be selected?

50. A madrigal group of eight singers is to be selected from a choir of 40 singers. If three singers have already been chosen from the 40, how many different madrigal groups can be formed?

Discussion Exercises

51. Explain the difference between a permutation and a combination. Why would it be more accurate to call a combination lock a "permutation" lock?

52. Decide whether each of the following is a permutation or a combination: (a) your Social Security number, (b) your telephone number, (c) the five finalists for "Best Picture," (d) a 13-card bridge hand, (e) your license plate number.

53. Evaluate $C(n, n)$. Why does the answer make sense?

54. Evaluate $C(n, 1)$. Why does the answer make sense?

55. Show that $C(n, r) = C(n, n - r)$. Why does this make sense?

56. Compare and contrast the terms "permutation," "circular permutation," and "distinguishable permutation."

Problem Solving

57. The flip of a coin can produce a head or a tail. If a coin is flipped five times, how many different (ordered) outcomes are possible?

58. The roll of a die can produce a 1, 2, 3, 4, 5, or 6. If a die is rolled 3 times, how many different (ordered) outcomes are possible?

 59. A basketball team consists of two centers, five forwards, and six guards. In how many ways can the coach select a starting team of one center, two forwards, and two guards?

60. In how many ways can a grievance committee of three faculty members, one administrator, and two students be selected from a group of 10 faculty members, three administrators, and five students?

61. In how many ways can a true-false test consisting of 10 questions be completed? Assume all questions are answered.

62. How many three-letter license plates are possible if no letter is repeated?

63. In how many ways can four people sit in a canoe if only two of the four are willing to sit in the back?

64. In how many ways can five kids line up for a photograph if two particular kids will not stand next to each other?

65. In how many ways can six people sit at a round table if two of the six refuse to sit next to each other?

66. In how many ways can six people sit at a round table if two of the six insist on sitting next to each other?

67. In how many ways can three boys and two girls go through a haunted house single file (a) if there are no restrictions? (b) if the boys go before the girls? (c) if the girls go before the boys?

68. A bag contains eight Ping-Pong balls numbered 1 through 8. In how many ways can two balls with a sum of 12 be drawn from the bag (a) if the first ball is returned to the bag before the second ball is drawn? (b) if the first ball is not returned to the bag before the second ball is drawn?

12.8 Probability*

Probability influences the choices we make every day. If the weatherperson says, "There's a 90 percent chance of rain tomorrow," we might take an umbrella to work. If our chances of winning the lottery are 1 in a million, we might decide not to buy a ticket. In this section, we see how the probabilities of certain events can be measured in a precise way.

DEFINITION OF A SAMPLE SPACE

A **sample space** S of an experiment is the set of all outcomes of the experiment.

Example 1   List the elements in the sample space S of each experiment.

(a) Flipping a coin   (b) Flipping two coins   (c) Rolling a die

Solution: (a) The experiment of flipping a coin has two possible outcomes: landing heads up (denoted H), or landing tails up (denoted T). Therefore, the sample space is

$$S = \{H, T\}$$

---

*HISTORICAL NOTE: In 1654, the Chevalier de Méré, an experienced gambler, posed the following problem to the French mathematician Blaise Pascal: A player is to throw a 1 in eight throws of a die. After three unsuccessful tries, the game is interrupted. How should the player be compensated? Pascal in turn posed the problem to Pierre de Fermat, and the resulting correspondence between the two marked the beginning of the theory of probability.

(b) The possible outcomes are

$$HH = \text{head on both coins}$$
$$HT = \text{head on first coin, tail on second coin}$$
$$TH = \text{tail on first coin, head on second coin}$$
$$TT = \text{tail on both coins}$$

Therefore, the sample space is

$$S = \{HH, HT, TH, TT\}$$

(*Note:* To convince yourself that *HT* and *TH* are really different outcomes, think of one coin as a nickel and the other coin as a dime.)

(c) A die can land so that any of the numbers 1, 2, 3, 4, 5, or 6 is face up. Therefore, the sample space is

$$S = \{1, 2, 3, 4, 5, 6\}$$ ❐

Try Exercise 11

List the elements in the sample space for the experiment of answering every question of a true-false test that contains three questions.

> ### DEFINITION OF AN EVENT
>
> An **event** *E* of an experiment is any subset of the sample space *S* of the experiment.

Example 2 An experiment consists of rolling a single die. List the elements in each event, *E*.

(a) *E*: rolling a 5   (b) *E*: rolling an even number
(c) *E*: rolling a 7   (d) *E*: rolling a number between 0 and 7

Solution: (a) The event *E* is a subset of the sample space $S = \{1, 2, 3, 4, 5, 6\}$. In particular,

$$E = \{5\}$$

(b) $E = \{2, 4, 6\}$

(c) $E = \varnothing$ (the empty set)

Try Exercise 19

A card is drawn from a standard deck of 52 cards. List the elements in the event of drawing an ace.

(d) $E = \{1, 2, 3, 4, 5, 6\} = S$ ❐

> ### DEFINITION OF PROBABILITY
>
> Let *E* be an event of an experiment with a finite number of equally likely outcomes. The **probability** of event *E* is denoted *P(E)*, and is given by
>
> $$P(E) = \frac{n(E)}{n(S)}$$
>
> where $n(E)$ denotes the number of elements in *E*, and $n(S)$ denotes the number of elements in the sample space *S*.

Example 3 An experiment consists of rolling two dice. The first event (rolling the red die) can occur in six ways. The second event (rolling the green die) can also occur in six ways. By the fundamental principle of counting, the two events can occur in $6 \cdot 6 = 36$ ways. A convenient way to illustrate the 36 outcomes that make up the sample space is shown in Fig. 12.8.

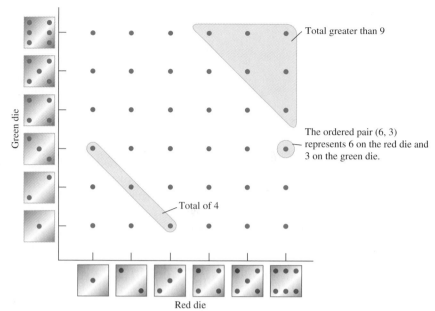

Figure 12.8

Find the probability of each event.

(a) E_1: a total of 4   (b) E_2: a total greater than 9

(c) E_3: a total of 16   (d) E_4: a total between 1 and 13

Solution: (a) The three ways to roll a 4 (see Fig. 12.8) are

$$(1, 3): 1 \text{ on the red die, 3 on the green die}$$
$$(2, 2): 2 \text{ on the red die, 2 on the green die}$$
$$(3, 1): 3 \text{ on the red die, 1 on the green die}$$

That is, event $E_1 = \{(1, 3), (2, 2), (3, 1)\}$. Therefore,

$$P(E_1) = \frac{n(E_1)}{n(S)} = \frac{3}{36} = \frac{1}{12}$$

(b) Since $E_2 = \{(4, 6), (5, 5), (6, 4), (5, 6), (6, 5), (6, 6)\}$,

$$P(E_2) = \frac{n(E_2)}{n(S)} = \frac{6}{36} = \frac{1}{6}$$

(c) The largest total that can be rolled with two dice is 12, represented by the ordered pair (6, 6). Therefore, $E_3 = \varnothing$, and

$$P(E_3) = \frac{n(E_3)}{n(S)} = \frac{0}{36} = 0$$

(d) Since each of the 36 ordered pairs in the sample space represents a total between 1 and 13,

$$P(E_4) = \frac{n(E_4)}{n(S)} = \frac{36}{36} = 1$$

Try Exercise 45

Two dice are rolled. Determine the probability of rolling a total of 5.

From Example 3, we observe that the probability of an impossible event (rolling a 16 with two dice) is 0, and the probability of a certain event (rolling a total between 1 and 13 with two dice) is 1.

Properties of Probability
1. P (an impossible event) $= 0$.
2. P (a certain event) $= 1$.
3. For any event E, $0 \leq P(E) \leq 1$.

Example 4   Three coins are tossed. What is the probability of getting at least two heads?

Solution: The sample space is

$$S = \{HHH, HHT, HTH, HTT, THH, THT, TTH, TTT\}$$

The event of getting at least two heads is

$$E = \{HHH, HHT, HTH, THH\}$$

Therefore, the probability of getting at least two heads is

$$P(E) = \frac{n(E)}{n(S)} = \frac{4}{8} = \frac{1}{2} \qquad \square$$

Try Exercise 53

Three coins are tossed. Determine the probability of getting at least one head.

When the sample space is large, we use the counting techniques discussed in Sec. 12.7 to determine $n(E)$ and $n(S)$.

Example 5   Three cards are drawn from a standard deck of 52 playing cards, like the one shown in Fig. 12.9. What is the probability the three cards are all hearts?

13 Hearts   13 Spades   13 Diamonds   13 Clubs

Figure 12.9 Standard deck of 52 playing cards.
Note: *Kings, queens, and jacks are called* face cards.

Solution: The sample space consists of all possible ways to draw (without regard to order) three cards from a deck of 52 cards. Therefore, $n(S)$ is simply the number of combinations of 52 elements taken three at a time. That is,

$$n(S) = C(52, 3) = \frac{52!}{49!3!} = 22,100$$

The event E consists of all possible ways to choose (without regard to order) three hearts from a group of 13 hearts. Therefore, $n(E)$ is the number of combinations of 13 elements taken three at a time. That is,

$$n(E) = C(13, 3) = \frac{13!}{10!3!} = 286$$

Try Exercise 61

Five cards are drawn from a standard deck of 52 cards. Find the probability of drawing five clubs.

Therefore,

$$P(E) = \frac{n(E)}{n(S)} = \frac{286}{22,100} = \frac{11}{850}$$

Exercises 12.8

Completion Exercises

1. The set of all outcomes of an experiment is called the _____ of the experiment.

2. A subset of a sample space of an experiment is called a(n) _____ of the experiment.

3. Let E be an event of an experiment with a finite number of equally likely outcomes. The probability of E is $P(E) = $ _____.

4. P (an impossible event) = _____.

5. P (a certain event) = _____.

6. For any event E, _____ $\leq P(E) \leq$ _____.

List the elements in the sample space S of each experiment.

7. Flipping a coin

8. Rolling a die

9. Rolling a die, then flipping a coin

10. Flipping three coins

11. Answering every question of a true-false test that contains three questions

12. Flipping a coin, then drawing a marble from a bag that contains a white marble, a red marble, and a blue marble

Determine the number of elements in the sample space S of each experiment.

13. Rolling two dice

14. Rolling three dice

15. Flipping 10 coins

16. Drawing a card from a deck of 52 cards

17. Drawing five cards (without regard to order) from a deck of 52 cards

18. Drawing five cards (in a particular order) from a deck of 52 cards

A card is drawn from a standard deck of 52 cards. List the elements in each event E.

19. E: drawing an ace

20. E: drawing an 8

21. E: drawing a face card

22. E: drawing a club

A bag contains five slips of paper numbered 1 through 5. Two slips are drawn at random. List the elements in each event E.

23. E: two even numbers

24. E: two odd numbers

25. E: a total of 4

26. E: an even total

Given a standard deck of 52 cards, determine the number of elements in each event E. The order in which the cards are drawn is not important.

27. E: drawing six spades in six draws

28. E: drawing five face cards in five draws

29. E: drawing three red cards in three draws

30. E: drawing two cards, both less than 5

31. E: drawing four aces and one 10 in five draws

32. E: drawing three queens and two jacks in five draws

A coin is flipped. Determine the probability of each event E.

33. E: a tail

34. E: a head

35. E: a head or a tail

36. E: a king of hearts

A single die is rolled. Find the probability of each event E.

37. E: a 3

38. E: a 2

39. E: an odd number

40. E: a prime number

41. E: a 9

42. E: a positive number

Discussion Exercises

43. The probability of getting a head on one flip of a coin is $\frac{1}{2}$. What does this mean? Does it mean that if we flip a coin two times it must come up heads exactly one of those

two times? Does it mean that if we flip a coin 10 times that it must come up heads exactly five of those times? Explain.

44. Suppose a coin is flipped nine times and it comes up heads all nine times. What is the probability that the coin will come up heads on the next toss? Explain.

Two dice are rolled. Find the probability of each event *E*.

45. *E*: a total of 5

46. *E*: a total of 8

47. *E*: a total less than 4

48. *E*: same number on both dice

49. What is the easiest total to roll with two dice? What is the probability of rolling it?

50. What is the hardest total to roll with two dice? What is the probability of rolling it?

Three coins are tossed. Determine the probability of each event *E*.

51. *E*: three heads

52. *E*: three tails

53. *E*: at least one head

54. *E*: at most one head

55. *E*: a tail on the first coin

56. *E*: a head on the second coin

A card is drawn from a standard deck of 52 cards. Find the probability of each event *E*.

57. *E*: a 9

58. *E*: a diamond

59. *E*: a face card

60. *E*: a number between 2 and 7

Five cards are drawn from a standard deck of 52 cards without regard to order. Find the probability of each event *E*.

61. *E*: five clubs

62. *E*: five face cards

63. *E*: five black cards

64. *E*: five even numbers

65. *E*: two aces and three jacks

66. *E*: four kings and one 6

A drawer contains three black socks and two white socks. Two socks are drawn at random. Determine the probability of each event *E*.

67. *E*: two black socks

68. *E*: two white socks

69. *E*: a sock of each color

70. *E*: a white sock on the first draw, and a black sock on the second draw

Problem Solving

71. Five different letters are randomly stuffed into five envelopes that have already been addressed. What is the probability that all of the letters are placed into the correct envelopes?

72. A box contains 12 light bulbs, two of which are defective. What is the probability that if three bulbs are drawn at random, they will all work?

73. A review sheet for an algebra exam contains 10 problems. The exam itself will consist of 5 of the 10 problems on the review sheet. If Alice can do 8 of the 10 review problems, what is the probability she will score 100 percent on the exam?

74. A student guesses at the answer to every question on a true-false quiz consisting of 10 questions. Find the probability the student scored 100 percent on the quiz.

75. Find the probability of getting three heads and five tails if eight coins are tossed.

76. Jim, Sally, Pam, Dave, and Nancy are seated in a row. What is the probability that Pam and Dave are seated so that one is at each end of the row?

77. Assuming a male birth and a female birth are equally likely, what is the probability that a couple will have (a) two girls and two boys in that order? (b) two girls and two boys in any order?

78. What is the probability of being dealt a poker hand of four of a kind? The hand would consist of four cards of the same face value (for example, four 5s or four kings) and one card that is different from the other four.

12.9 Compound Probabilities

Many probability problems involve finding the probability that both of two events will occur. For example, what is the probability that a baby born is both a redhead and a female? Other probability problems try to determine whether either of two events might occur. For example, what is the probability that it will either rain or snow tomorrow? A **compound probability** problem is a problem that attempts to determine the probability that all of two or more events will occur, or that any of two or more events will occur.

Probability of E_1 and E_2

Example 1   Suppose a bag contains two red marbles and one white marble. Two marbles are drawn from the bag. What is the probability that both marbles are red?

Solution: Let R_1 denote one red marble and R_2 the other. Then the sample space is $S = \{R_1R_2, R_2R_1, R_1W, R_2W, WR_1, WR_2\}$. The event of drawing two red marbles is $E = \{R_1R_2, R_2R_1\}$. Therefore, the probability of drawing two red marbles is

$$P(E) = \frac{n(E)}{n(S)} = \frac{2}{6} = \frac{1}{3}$$ ❏

Try Exercise 13

A bag contains two red marbles and two white marbles. Two marbles are drawn from the bag. Determine the probability that both marbles are red.

We can look at Example 1 in another way:

Probability of a red marble on the first draw $= \frac{2}{3}$

Probability of a red marble on the second draw, given that a red marble was taken on the first draw $= \frac{1}{2}$

Therefore, the probability of drawing two red marbles is

$$\tfrac{2}{3} \cdot \tfrac{1}{2} = \tfrac{1}{3}$$

This suggests the following theorem.

Probability of E_1 and E_2

Let $P(E_1 \cap E_2)$ denote the probability that events E_1 and E_2 both occur. Then

$$P(E_1 \cap E_2) = P(E_1) \cdot P(E_2/E_1)$$

where $P(E_2/E_1)$ denotes the probability that E_2 will occur given that E_1 has occurred.

This rule can be extended to any number of events.

Example 2   Three cards are drawn from a standard deck of 52 cards. Determine the probability that the three cards are hearts.

Solution: Let H_1 represent the event of a heart on the first draw, H_2 a heart on the second draw, and H_3 a heart on the third draw. Then the probability that H_1, H_2, and H_3 will all occur is

Probability of a heart on the second draw given that a heart was taken on the first draw
↓

$$P(H_1 \cap H_2 \cap H_3) = P(H_1) \cdot \overbrace{P(H_2/H_1)} \cdot \underbrace{P(H_3/H_1 \cap H_2)}$$
$$= \frac{13}{52} \cdot \frac{12}{51} \cdot \frac{11}{50}$$
$$= \frac{11}{850}$$

↑ Probability of a heart on the third draw given that a heart was taken on the first two draws

Try Exercise 17

Five cards are drawn from a standard deck of 52 cards. Determine the probability that the five cards are spades.

Try Exercise 21

Assume that 45 percent of babies born are male and 20 percent have blond hair (whether they are male or female). Find the probability that a baby born will be a blond male.

Note that this is the same answer obtained in Example 5 of Sec. 12.8. ☐

Example 3 Suppose that 55 percent of babies born are female, and that 15 percent of babies born have red hair (whether they are male or female). Find the probability that a baby born will be a redheaded girl.

Solution: Let F be a female birth, and R a redheaded birth. Then

$$P(F \cap R) = P(F) \cdot P(R/F) = \frac{55}{100} \cdot \frac{15}{100} = \frac{11}{20} \cdot \frac{3}{20} = \frac{33}{400}$$

The probability of a redheaded female birth can also be expressed as 0.0825, or 8.25 percent. ☐

In Example 3, note that the fact that the baby was female had no effect on whether the baby was a redhead. If the occurrence of one event has no effect on the occurrence of another event, we say that the two events are **independent** events. This suggests the following rule.

Independent Events

Events E_1 and E_2 are **independent** if and only if

$$P(E_1 \cap E_2) = P(E_1) \cdot P(E_2)$$

Probability of E_1 or E_2

In Example 3, we wanted to find the probability that a baby born would be both a female and a redhead. Suppose we wanted to find the probability that a baby born will be either a female or a redhead (or both). We denote this probability as $P(F \cup R)$.

Example 4 A card is drawn from a standard deck of 52 cards. Find the probability that the card is a diamond or a face card.

Solution: Let D be a diamond, F a face card, and $P(D \cup F)$ the probability of a diamond or a face card (see Fig. 12.10). Then

$$P(D \cup F) = \frac{n(D \cup F)}{n(S)} = \frac{22}{52} = \frac{11}{26}$$ ☐

Try Exercise 25(a)

A card is drawn from a standard deck of 52 cards. Find the probability that the card is a club or a face card.

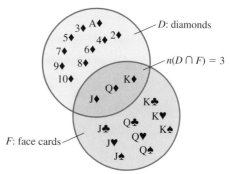

Figure 12.10

Here's another way to look at Example 4:

$$P(D \cup F) = P(D) + P(F) - P(D \cap F) = \frac{13}{52} + \frac{12}{52} - \frac{3}{52} = \frac{22}{52} = \frac{11}{26}$$

Note that $P(D \cap F)$, the probability of drawing a card that is *both* a diamond and a face card, is subtracted from $P(D) + P(F)$, so that $P(D \cap F)$ is not counted twice, once with $P(D)$ and once with $P(F)$. This suggests the following rule.

Probability of E_1 or E_2

Let $P(E_1 \cup E_2)$ denote the probability that E_1 or E_2 (or both) will occur. Then
$$P(E_1 \cup E_2) = P(E_1) + P(E_2) - P(E_1 \cap E_2)$$
where $P(E_1 \cap E_2)$ is the probability that both E_1 and E_2 will occur.

Example 5   Use the information from Example 3 to find the probability that a baby born is either a female or a redhead.

Solution: Using the formula for $P(E_1 \cup E_2)$ and the information from Example 3, the probability that a baby born is either a female or a redhead is

$$P(F \cup R) = P(F) + P(R) - P(F \cap R)$$
$$= \frac{55}{100} + \frac{15}{100} - \frac{33}{400}$$
$$= \frac{220 + 60 - 33}{400} = \frac{247}{400}$$

The probability of a female or a redhead can also be expressed as 0.6175, or 61.75 percent. ❏

Try Exercise 29

Use the information from Exercise 21 to find the probability that a baby born is either a blond or a male.

Example 6   A die is rolled. What is the probability that the number rolled is a 5 or an even number?

Solution: Let $F = \{5\}$ and $E = \{2, 4, 6\}$. Here are two different ways to look at this problem.
Method 1:   Use the definition of probability.

$$P(F \cup E) = \frac{n(F \cup E)}{n(S)} = \frac{n(\{5, 2, 4, 6\})}{n(\{1, 2, 3, 4, 5, 6\})} = \frac{4}{6} = \frac{2}{3}$$

Method 2:   Use the formula for $P(E_1 \cup E_2)$.
$$P(F \cup E) = P(F) + P(E) - P(F \cap E)$$
$$= \tfrac{1}{6} + \tfrac{3}{6} - \tfrac{0}{6} = \tfrac{4}{6} = \tfrac{2}{3}$$ ❏

Try Exercise 33

A die is rolled. What is the probability that the number rolled is a 6 or an even number?

In Example 6, note that $P(F \cap E) = 0$, since $F \cap E = \varnothing$. Two events E_1 and E_2 are said to be **mutually exclusive** if $E_1 \cap E_2 = \varnothing$. This suggests the following rule.

Mutually Exclusive Events

Events E_1 and E_2 are **mutually exclusive** if and only if
$$P(E_1 \cup E_2) = P(E_1) + P(E_2)$$

CAUTION

Independent events and mutually exclusive events are not the same.

Experiment: A baby is born (see Example 3).

Event *F*: A female   Event *R*: A redhead

F and *R* are independent.	*F* and *R* are not mutually exclusive.
$P(F \cap R) = P(F) \cdot P(R)$	$F \cap R \neq \varnothing$, since there are female redheads ∎

Complement of an Event

The **complement** of event *E*, denoted *E′*, is the set of all elements in the sample space *S* that do *not* belong to *E*. For example, if an experiment consists of rolling a single die, and *T* is the event of rolling a 2, then $T = \{2\}$ and $T' = \{1, 3, 4, 5, 6\}$, as shown in Fig. 12.11.

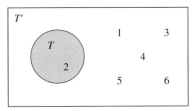

Figure 12.11 Experiment: A die is rolled. Sample space includes everything in rectangle.

Note that

$$T \cup T' = S \qquad \text{and} \qquad T \cap T' = \varnothing$$

Example 7   A die is rolled. Let *T* be the event of rolling a 2. Find $P(T)$ and $P(T')$.

Solution: From Fig. 12.11, we see that

$$P(T) = \frac{n(T)}{n(S)} = \frac{1}{6} \qquad \text{and} \qquad P(T') = \frac{n(T')}{n(S)} = \frac{5}{6} \qquad \qquad ❑$$

Try Exercise 41(a)

A die is rolled. Find the probability of rolling a 1.

Example 7 illustrates an important property of probability.

Probability of the Complement of *E*

For any event *E*,

$$P(E) + P(E') = 1 \qquad \text{or} \qquad P(E) = 1 - P(E')$$

Example 8   Find the probability that at least two people in a random group of 40 people have the same birthday. Assume there are 365 days in a year.

Solution: Let *B* be the event that at least two people in a random group of 40 people have the same birthday. Then *B′* is the event that everyone in the group has a different birthday. Rather than calculate $P(B)$ directly, we calculate $P(B')$ and subtract the result from 1. To calculate $P(B')$, let the first person have any of the 365 days in the year as a birthday. The probability that the second person will have a different birthday than the first person is $\frac{364}{365}$. If the first two people have different birthdays, the probability that the third person will have a different birthday than either of the first two people is $\frac{363}{365}$. Continuing to reason in this fashion, the probability that everyone in the group has a different birthday is

$$P(B') = \frac{365}{365} \cdot \frac{364}{365} \cdot \frac{363}{365} \cdot \frac{362}{365} \cdots \frac{326}{365} \approx 0.109$$

Therefore,

$$P(B) = 1 - P(B') \approx 1 - 0.109 = 0.891$$

Note that the probability that at least two people in a random group of 40 people have the same birthday is a surprisingly high 0.891, or 89.1 percent. ❑

Try Exercise 59

In a random class of 23 students, what is the probability that at least two students will have the same birthday?

Odds

Sometimes we express the probability of an event in terms of odds.

Converting Probability to Odds

The **odds** that event E *will* occur are $\dfrac{P(E)}{P(E')}$. The **odds** that event E *will not* occur are $\dfrac{P(E')}{P(E)}$.

Example 9   The probability of event E is $\frac{2}{5}$. Find the odds that E (a) will occur, (b) will not occur.

Solution: (a) The odds that E will occur are

$$\frac{P(E)}{P(E')} = \frac{2/5}{3/5} = \frac{2}{3} \quad \text{or} \quad 2 \text{ to } 3$$

(b) The odds that E will not occur are

$$\frac{P(E')}{P(E)} = \frac{3/5}{2/5} = \frac{3}{2} \quad \text{or} \quad 3 \text{ to } 2 \qquad ❑$$

Try Exercise 45

If $P(E) = \frac{3}{8}$, what are the odds that (a) E will occur? (B) E will not occur?

Note that when the odds that event E will occur are 2 to 3, the probability that E will occur is $\frac{2}{5}$. This suggests the following rule.

Converting Odds to Probability

If the odds that event E will occur are a to b, then

$$P(E) = \frac{a}{a + b} \quad \text{and} \quad P(E') = \frac{b}{a + b}$$

Example 10   The odds that a horse will win a race are 8 to 5. Find the probability the horse will win.

Solution: Let W be the event that the horse wins the race. Then

$$P(W) = \frac{8}{8+5} = \frac{8}{13}$$

Try Exercise 57

The odds that a horse will win a race are 9 to 5. What is the probability the horse (a) will win? (b) will not win?

The probability the horse will win is $\frac{8}{13}$.   □

Exercises 12.9

Completion Exercises

1. A probability problem that attempts to determine the probability that all of two or more events will occur, or the probability that any of two or more events will occur is called a(n) _____ problem.

2. If E_1 and E_2 are any two events, then $P(E_1 \cap E_2) =$ _____.

3. If E_1 and E_2 are independent events, then $P(E_1 \cap E_2) =$ _____.

4. If E_1 and E_2 are any two events, then $P(E_1 \cup E_2) =$ _____.

5. If E_1 and E_2 are mutually exclusive events, then $P(E_1 \cup E_2) =$ _____.

6. The set of all elements in the sample space that do not belong to event E is called the _____ of E and is denoted _____.

7. If E is an event, then $E \cup E' =$ _____ and $E \cap E' =$ _____.

8. If E is an event, then $P(E) + P(E') =$ _____.

9. The odds that event E will occur are _____. The odds that event E will not occur are _____.

10. If the odds that event E will occur are a to b, then $P(E) =$ _____ and $P(E') =$ _____.

True-False Exercises

11. Two events are independent if the occurrence of one has no effect on the occurrence of the other.

12. Two events are mutually exclusive if their intersection is the empty set.

A bag contains two red marbles and two white marbles. Two marbles are drawn from the bag. Determine the probability of each compound event using

(a) $P(E_1 \cap E_2) = \dfrac{n(E_1 \cap E_2)}{n(S)}$

(b) $P(E_1 \cap E_2) = P(E_1) \cdot P(E_2/E_1)$

13. Two red marbles   14. Two white marbles

15. A red marble, then a white marble

16. A white marble, then a red marble

Five cards are drawn from a standard deck of 52 cards. Determine the probability of each compound event.

17. Five spades   18. Five red cards

19. Five face cards

20. Five numbered cards (aces included)

Assume that 45 percent of babies born are male and 20 percent have blond hair (whether they are male or female). Find the probability of each compound event.

21. A blond male   22. A blond female

23. A nonblond female   24. A nonblond male

A card is drawn from a standard deck of 52 cards. Determine the probability of each compound event using

(a) $P(E_1 \cup E_2) = \dfrac{n(E_1 \cup E_2)}{n(S)}$

(b) $P(E_1 \cup E_2) = P(E_1) + P(E_2) - P(E_1 \cap E_2)$

25. A club or a face card   26. A diamond or a 5

27. A 3 or a black card   28. A red card or a face card

Use the results of Exercises 21 through 24 and the formula for $P(E_1 \cup E_2)$ to find the probability of each compound event.

29. A blond or a male

30. A blond or a female

31. A nonblond or a female

32. A nonblond or a male

A die is rolled. Determine the probability of each compound event using the formula for $P(E_1 \cup E_2)$.

33. A 6 or an even number

34. A 5 or a number less than 3

35. An even number or an odd number

36. An even number or a prime number

Discussion Exercises

37. Suppose that E_1 and E_2 are independent events. Then $P(E_2/E_1) = P(E_2)$. Why?

38. If E_1 and E_2 are mutually exclusive events, then $P(E_1 \cap E_2) = 0$. Why?

39. Why do we subtract $P(E_1 \cap E_2)$ from $P(E_1) + P(E_2)$ in the formula for $P(E_1 \cup E_2)$?

40. Explain why $P(E \cup E') = 1$ and $P(E \cap E') = 0$.

A die is rolled. Find the probability of E using

(a) $P(E) = \dfrac{n(E)}{n(S)}$ (b) $P(E) = 1 - P(E')$

41. E: A 1

42. E: An odd number

43. E: A prime number

44. E: A number greater than 2

In Exercises 45 and 46, the probability that event E will occur is given. What are the odds that

(a) E will occur?   (b) E will not occur?

45. $P(E) = \frac{3}{8}$

46. $P(E) = 0.3$

A bag contains five red marbles, three white marbles, two blue marbles, and one green marble. Two marbles are drawn from the bag. Find the probability of each compound event (a) if the first marble is returned to the bag before the second marble is drawn (this is called drawing *with replacement*), (b) if the first marble is not returned to the bag before the second marble is drawn (this is called drawing *without replacement*).

47. Two white marbles

48. Two red marbles

49. A blue marble, then a red marble

50. A green marble, then a white marble

A five-person committee is selected from Gary, Sal, Robin, Leanne, Holly, Amy, Tracy, and Jason. Determine the probability of each event.

51. Holly is not on the committee

52. Sal and Tracy are both on the committee

53. Neither Gary nor Leanne is on the committee

54. Either Amy or Jason, but not both, is on the committee

Problem Solving

55. If the probability that it will rain is 60 percent, what is the probability that it will not rain?

56. If the probability that the Bears will win the Super Bowl is $\frac{3}{10}$, what is the probability the Bears will not win the Super Bowl?

57. The odds that a horse will win a race are 9 to 5. What is the probability the horse (a) will win? (b) will not win?

58. The odds that it will snow are 6 to 5. What is the probability that (a) it will snow? (b) it will not snow?

59. In a random class of 23 students, what is the probability that at least two students will have the same birthday?

60. A random group of five people are on an elevator. What is the probability that at least two were born on the same day of the week?

61. Tari, Erin, and Jeremy are running for student body president. Tari is twice as likely to win as either of the other two, who have the same chance of winning. Find the probability of each person being elected.

62. If 20 percent of Americans have college degrees, and 75 percent of college graduates own their own homes, what percentage of the population are college graduates who own their own homes?

63. The batting average of a player is 0.300. Find the probability the player will get three hits in a row.

64. The probability that Jake will score a touchdown is $\frac{1}{3}$. The probability that Dale will score a touchdown is $\frac{3}{4}$. Find the probability that at least one of them will score a touchdown.

Chapter 12 Key Terms

12.1 Sequence A **finite sequence** is a function whose domain is the first k positive integers. An **infinite sequence** is a function whose domain is the entire set of positive integers.

Terms (of a sequence) The range values of a sequence
General term (of a sequence) An expression that describes any term of the sequence; the nth term of the sequence

Series A **finite series** is the indicated sum of a finite sequence. An **infinite series** is the indicated sum of an infinite sequence.

Summation notation An abbreviated way of writing a series

Index (of summation) The variable n in the summation notation $\sum_{n=1}^{5} n^2$

Lower limit (of summation) The number 1 in the summation notation $\sum_{n=1}^{5} n^2$

Upper limit (of summation) The number 5 in the summation notation $\sum_{n=1}^{5} n^2$

12.2 Arithmetic sequence A sequence in which each term after the first is obtained by adding a fixed constant, called the **common difference,** to the preceding term
Arithmetic series The indicated sum of an arithmetic sequence

12.3 Geometric sequence A sequence in which each term after the first is obtained by multiplying the preceding term by a fixed constant, called the **common ratio**
Geometric series The indicated sum of a geometric sequence

12.5 Pascal's triangle A triangular array of numbers that gives the coefficients in the expansion of $(a + b)^n$
Factorial notation An abbreviated way of writing products like $5 \cdot 4 \cdot 3 \cdot 2 \cdot 1$
Binomial coefficient A coefficient in the expansion of $(a + b)^n$

12.6 Mathematical induction A method of proving that certain types of statements are true for all positive integers

12.7 Permutation (of r elements) An ordering of the r elements so that one is first, one is second, and so on
Combination (of n elements taken r at a time) A group of r elements taken from the n elements without regard to order

12.8 Sample space (of an experiment) The set of all outcomes of the experiment

Event (of an experiment) A subset of the sample space of the experiment

12.9 Compound probability (problem) A problem that attempts to determine the probability that all of two or more events will occur, or that any of two or more events will occur
Independent events Events in which the occurrence of any one has no effect on the occurrence of any of the others
Mutually exclusive events Events whose intersection is the empty set
Complement (of event E) The set of all elements in the sample space that are not in E

Chapter 12 Key Rules/Steps

12.1 Defining Sequences and Series

Writing the Terms of $a_n = \dfrac{(-1)^n}{n}$

$$-1, \tfrac{1}{2}, -\tfrac{1}{3}, \tfrac{1}{4}, \ldots$$

Expanding $\displaystyle\sum_{n=1}^{5} x^n$

$$x + x^2 + x^3 + x^4 + x^5$$

12.2 Arithmetic Sequences and Series

General Term of an Arithmetic Sequence

The general term a_n of the arithmetic sequence with first term a_1 and common difference d is

$$a_n = a_1 + (n - 1)d$$

Example: The 20th term of the arithmetic sequence given by $3, 9, 15, 21, \ldots$ is

$$a_{20} = 3 + (20 - 1)6 = 117$$

Sum of an Arithmetic Series

The sum S_n of the first n terms of the arithmetic series with first term a_1 and nth term a_n is

$$S_n = \frac{n}{2}(a_1 + a_n)$$

Example: The sum of the first 20 terms of the arithmetic series $3 + 9 + 15 + 21 + \ldots$ is

$$S_{20} = \frac{20}{2}(3 + 117) = 1200$$

12.3 Geometric Sequences and Series
General Term of a Geometric Sequence

The general term a_n of the geometric sequence with first term a_1 and common ratio r is

$$a_n = a_1 r^{n-1}$$

Example: The sixth term of the geometric sequence given by $3, 6, 12, 24, \ldots$ is

$$a_6 = 3(2)^{6-1} = 96$$

Sum of a Geometric Series

The sum S_n of the first n terms of the geometric series with first term a_1 and common ratio r is

$$S_n = \frac{a_1(1 - r^n)}{1 - r} \qquad (r \neq 1)$$

Example: The sum of the first six terms of the geometric series $3 + 6 + 12 + 24 + \cdots$ is

$$S_6 = \frac{3(1 - 2^6)}{1 - 2} = 189$$

12.4 Infinite Geometric Series
Sum of an Infinite Geometric Series

If $|r| < 1$, the sum S of the infinite geometric series with first term a_1 and common ratio r is

$$S = \frac{a_1}{1 - r}$$

If $|r| \geq 1$, the sum does not exist.

Example: The sum of the infinite geometric series given by $2 + \frac{2}{3} + \frac{2}{9} + \ldots$ is

$$S = \frac{2}{1 - \frac{1}{3}} = 3$$

12.5 The Binomial Theorem
Pascal's Triangle

```
        1
      1   1
    1   2   1
  1   3   3   1
1   4   6   4   1
```

Factorial Notation

Let n be a positive integer.

$$n! = n(n - 1) \cdots 3 \cdot 2 \cdot 1$$
$$5! = 5 \cdot 4 \cdot 3 \cdot 2 \cdot 1 = 120$$
$$0! = 1$$

Binomial Coefficient

Let n and r be whole numbers with $n \geq r$. Then

$$\binom{n}{r} = \frac{n!}{r!(n - r)!}$$

Example: $\displaystyle\binom{10}{2} = \frac{10!}{2!(10 - 2)!} = \frac{10 \cdot 9 \cdot 8!}{2!8!} = 45$

Binomial Theorem

If n is a positive integer, then

$$(a + b)^n$$
$$= \binom{n}{0}a^n + \binom{n}{1}a^{n-1}b + \binom{n}{2}a^{n-2}b^2 + \cdots + \binom{n}{n}b^n$$

Example: $(x - 2y)^4$

$$= \binom{4}{0}x^4 + \binom{4}{1}x^3(-2y) + \binom{4}{2}x^2(-2y)^2$$
$$+ \binom{4}{3}x(-2y)^3 + \binom{4}{4}(-2y)^4$$
$$= 1x^4 + 4x^3(-2y) + 6x^2(4y^2) + 4x(-8y^3)$$
$$+ 1(16y^4)$$
$$= x^4 - 8x^3y + 24x^2y^2 - 32xy^3 + 16y^4$$

rth Term of $(a + b)^n$

$$r\text{th term} = \binom{n}{r - 1}a^{n-(r-1)}b^{r-1}$$

Example: The 4th term of $(3x + y)^5$ is

$$\binom{5}{3}(3x)^{5-3}y^3 = 10(9x^2)y^3 = 90x^2y^3$$

12.6 Mathematical Induction

Prove S_n: $2 + 4 + 6 + \cdots + 2n = n(n + 1)$ is true for all positive integers n.

Step 1: Show S_1 is true.

$$2(1) = 1(1 + 1) \quad \text{True}$$

Step 2: Show that if S_k is true, then S_{k+1} is true.

$$S_k: 2 + 4 + 6 + \cdots + 2k = k(k + 1)$$
$$2 + 4 + 6 + \cdots + 2k + 2(k + 1) = k(k + 1)$$
$$+ 2(k + 1)$$
$$S_{k+1}: 2 + 4 + 6 + \cdots + 2k + 2(k + 1) = (k + 1)(k + 2)$$

By the principle of mathematical induction, S_n is true for all positive integers n.

12.7 Permutations and Combinations

Fundamental Principle of Counting

If event E_1 can occur in n_1 ways and event E_2 can occur in n_2 ways, then E_1 and E_2 can occur in succession in $n_1 \cdot n_2$ ways.

Example: If a shirt can be selected in five ways and trousers can be selected in three ways, then a shirt-and-trousers outfit can be selected in $5 \cdot 3 = 15$ ways.

Permutation Theorem

Let n and r be whole numbers with $n \geq r$. The number of permutations of n elements taken r at a time is

$$P(n, r) = \frac{n!}{(n - r)!}$$

Example: The number of ways that the gold, silver, and bronze medals can be awarded among seven Olympic skaters is

$$P(7, 3) = \frac{7!}{(7 - 3)!} = 210 \text{ ways}$$

Circular Permutations

Let n be a positive integer. There are $(n - 1)!$ ways to arrange n elements in a circle.

Example: There are $(6 - 1)! = 120$ ways to arrange six people around a circular table.

Distinguishable Permutations

Let n be a positive integer. The number of distinguishable permutations of n elements, where n_1 are of one kind, n_2 are of another kind, and so on is

$$\frac{n!}{n_1! n_2! \cdots n_k!}$$

Example: The number of distinguishable permutations of the letters in the word "parallel" are

$$\frac{8!}{3! 2! 1! 1! 1!} = 3360$$

Combination Theorem

Let n and r be whole numbers with $n \geq r$. The number of combinations of n elements taken r at a time is

$$C(n, r) = \frac{n!}{(n - r)! r!}$$

Example: The number of different five-person committees that can be selected from a group of nine people is

$$C(9, 5) = \frac{9!}{(9 - 5)! 5!} = 126$$

12.8 Probability

Definition of Probability

Let E be an event of an experiment with a finite number of equally likely outcomes. The probability of E is

$$P(E) = \frac{n(E)}{n(S)}$$

where $n(E)$ is the number of elements in E and $n(S)$ is the number of elements in the sample space S.

Example: A die is rolled. If E is the event of rolling a 2 or a 5, then

$$P(E) = \frac{n(\{2, 5\})}{n(\{1, 2, 3, 4, 5, 6\})} = \frac{2}{6} = \frac{1}{3}$$

Properties of Probability

1. P (an impossible event) $= 0$.

2. P (a certain event) $= 1$.

3. For any event E, $0 \leq P(E) \leq 1$.

12.9 Compound Probabilities

Probability of $E_1 \cap E_2$

If $P(E_1 \cap E_2)$ is the probability that events E_1 and E_2 both occur, then

$$P(E_1 \cap E_2) = P(E_1) \cdot P(E_2/E_1)$$

where $P(E_2/E_1)$ is the probability that E_2 will occur given that E_1 has occurred.

Example: Find the probability of drawing two aces from a standard deck of 52 cards.

$$P(A_1 \cap A_2) = \frac{4}{52} \cdot \frac{3}{51} = \frac{1}{221}$$

Independent Events

Events E_1 and E_2 are independent if and only if

$$P(E_1 \cap E_2) = P(E_1) \cdot P(E_2)$$

Example: A die is rolled and a coin is flipped. If F is a 5 on the die and H is a head on the coin, then

$$P(F \cap H) = \tfrac{1}{6} \cdot \tfrac{1}{2} = \tfrac{1}{12}$$

Probability of $E_1 \cup E_2$

If $P(E_1 \cup E_2)$ is the probability that event E_1 or event E_2 (or both) will occur, then

$$P(E_1 \cup E_2) = P(E_1) + P(E_2) - P(E_1 \cap E_2)$$

Example: A card is drawn from a standard deck of 52 cards. If K is a king and H a heart, then

$$P(K \cup H) = \frac{4}{52} + \frac{13}{52} - \frac{1}{52} = \frac{16}{52} = \frac{4}{13}$$

Mutually Exclusive Events

Events E_1 and E_2 are mutually exclusive if and only if

$$P(E_1 \cup E_2) = P(E_1) + P(E_2)$$

Example: A coin is flipped. If H is a head and T is a tail, then

$$P(H \cup T) = \frac{1}{2} + \frac{1}{2} = 1$$

Probability of the Complement of E

For any event E,

$$P(E) + P(E') = 1$$
$$P(E) = 1 - P(E')$$

Example: A die is rolled. If $E = \{3\}$, then

$$E' = \{1, 2, 4, 5, 6\}$$

$$P(E) = 1 - P(E') = 1 - \frac{5}{6} = \frac{1}{6}$$

Converting Probability to Odds

The odds that event E will occur are $\dfrac{P(E)}{P(E')}$.

The odds that event E will not occur are $\dfrac{P(E')}{P(E)}$.

Example: If $P(E) = 30$ percent, the odds E will occur are

$$\frac{0.30}{0.70} = \frac{3}{7} \quad (3 \text{ to } 7)$$

and the odds E will not occur are 7 to 3.

Converting Odds to Probability

If the odds that event E will occur are a to b, then

$$P(E) = \frac{a}{a + b} \quad \text{and} \quad P(E') = \frac{b}{a + b}$$

Example: If the odds the Braves will win the pennant (event W) are 5 to 2, then

$$P(W) = \frac{5}{5 + 2} = \frac{5}{7} \quad \text{and} \quad P(W') = \frac{2}{5 + 2} = \frac{2}{7}$$

Chapter 12 Review Exercises

[12.1] Find the first four terms of the sequence whose general term is given. Then find the 10th term.

1. $a_n = 2n + 1$
2. $a_n = \dfrac{n - 2}{n + 2}$

3. $a_n = (-1)^n 2^n$

Find the general term a_n for each sequence.

4. 4, 8, 12, 16, ...
5. $\frac{2}{3}, \frac{4}{5}, \frac{6}{7}, \frac{8}{9}, \ldots$
6. $1, -\frac{1}{8}, \frac{1}{27}, -\frac{1}{64}, \ldots$

Expand and simplify.

7. $\sum_{n=1}^{6} 5n$
8. $\sum_{n=0}^{4} (n^2 - 2n + 5)$
9. $\sum_{n=1}^{5} \dfrac{1}{2^{n-1}}$
10. $\sum_{n=1}^{3} (-1)^n \dfrac{x^{2n+1}}{n}$

Write each series in summation notation.

11. $1 + 3 + 5 + 7 + 9 + 11 + 13$
12. $6 + 11 + 16 + 21$
13. $1 + \frac{1}{4} + \frac{1}{16} + \frac{1}{64} + \frac{1}{256}$
14. $1 + x^2 + x^4 + x^6 + x^8 + x^{10}$

[12.2] For each arithmetic sequence, find the common difference and write the next two terms.

15. 3, 7, 11, 15, 19, ...
16. $-6, -1, 4, 9, 14, \ldots$
17. $4, 3\frac{1}{2}, 3, 2\frac{1}{2}, 2, \ldots$

Find the indicated unknown for each arithmetic sequence or series.

18. $a_1 = 4, d = -3; a_{10} = ?$
19. $a_1 = 15, a_{11} = -25; d = ?$
20. $-4, 2, 8, \ldots; a_n = ?$ 21. $7, 11, 15, \ldots, 83; n = ?$
22. $a_1 = 7, a_{14} = 85; S_{14} = ?$
23. $a_1 = 6, d = 4; S_{12} = ?$
24. $4 + 7 + 10 + 13 + \cdots; S_{20} = ?$
25. $S_8 = -164, a_8 = -38; a_{12} = ?$

26. Boxes are stacked in 34 layers so that the bottom layer contains 289 boxes, and each of the other layers contains five fewer boxes than the layer below it. How many boxes are in the top layer?

27. A woman accepts a position at a salary of $30,000 for the first year with an increase in salary of $2000 for each year thereafter. Find the woman's total earnings if she works at the position for 15 yr.

[12.3] For each geometric sequence, find the common ratio and write the next two terms.

28. 2, 8, 32, 128, ...
29. $5, -10, 20, -40, \ldots$
30. $1, \frac{2}{3}, \frac{4}{9}, \frac{8}{27}, \frac{16}{81}, \ldots$

Find the indicated unknown for each geometric sequence or series.

31. $a_1 = 3, r = \frac{1}{3}; a_6 = ?$ 32. $a_1 = 5, r = -\frac{2}{3}; a_5 = ?$

33. $a_1 = 3, a_5 = 48; r = ?$

34. $a_3 = 125, a_4 = 100; a_6 = ?$

35. $\frac{1}{8}, \frac{1}{4}, \frac{1}{2}, \ldots, 128; n = ?$ 36. $a_1 = 3, r = 2; S_6 = ?$

37. $25 + 5 + 1 + \cdots; S_4 = ?$

38. $S_7 = -1094, r = -3; a_{10} = ?$

39. A ball is dropped from a height of 90 ft. On each rebound it bounces two-thirds of the height from which it last fell. How high does it bounce on its fifth bounce?

40. Suppose a country consumes 75 billion m³ of natural gas this year. If consumption increases by 10 percent each year, how many cubic meters will be consumed 12 yr from now? How many total cubic meters will be consumed over the next 12 yr?

[12.4] Find the sum, if it exists, of each infinite geometric series.

41. $a_1 = 9, r = \frac{1}{3}$ 42. $a_1 = 5, r = -0.2$

43. $16 + 4 + 1 + \cdots$ 44. $1 - \frac{1}{2} + \frac{1}{4} - \frac{1}{8} + \cdots$

45. $\sum_{n=1}^{\infty} \left(-\frac{1}{8}\right)^n$ 46. $\sum_{k=1}^{\infty} \frac{4}{9}\left(\frac{5}{6}\right)^k$

Find the indicated unknown for each infinite geometric series.

47. $S = 2, r = \frac{1}{2}; a_1 = ?$ 48. $S = 6, a_1 = 2; r = ?$

Write each repeating decimal as a ratio of two integers.

49. $0.\overline{12}$ 50. $2.\overline{7}$

51. Suppose a golf ball rebounds three-quarters of the distance it falls. Find the total distance the golf ball will travel before coming to rest if it is dropped from a height of 20 ft.

[12.5] Calculate each binomial coefficient.

52. $\binom{7}{4}$ 53. $\binom{11}{10}$

54. $\binom{13}{0}$ 55. $\binom{8}{2}$

Expand each binomial using the binomial theorem.

56. $(x + y)^6$ 57. $(3m + 2)^4$

58. $(t^2 - 2)^5$ 59. $(x - \frac{y}{2})^3$

60. Use the binomial theorem to write the first three terms in the expansion of $(x + 5)^{10}$.

61. Find the fifth term of $(a - 2b)^7$.

[12.6] For each statement S_n, write both of the statements S_k and S_{k+1}. Then state what must be done to both sides of S_k to make the left side of S_k look like the left side of S_{k+1}.

62. $S_n: 2 + 6 + 10 + \cdots + (4n - 2) = 2n^2$

63. $S_n: \frac{1}{2} + \frac{1}{4} + \frac{1}{8} + \cdots + \frac{1}{2^n} = 1 - \frac{1}{2^n}$

Use mathematical induction to prove each statement is true for all positive integers n.

64. $2 + 6 + 10 + \cdots + (4n - 2) = 2n^2$

65. $\frac{1}{2} + \frac{1}{4} + \frac{1}{8} + \cdots + \frac{1}{2^n} = 1 - \frac{1}{2^n}$

[12.7]

66. How many different four-letter radio station call letters are there if the first letter must be a W or a K?

67. There are five teams in the American League East. (a) In how many ways can the five teams finish first, second, and third? (b) In how many ways can the five teams finish?

68. In how many ways can 10 people sit (a) in a row? (b) in a circle?

69. How many distinguishable permutations of the letters in the word "daddy" are there?

70. How many different 13-card bridge hands can be dealt from a standard deck of 52 cards?

71. In how many ways can a committee of three Democrats and four Republicans be selected from a group of 10 Democrats and 12 Republicans?

[12.8]

72. Two dice are rolled. Find the probability of getting a total of 6.

73. Three coins are tossed. What is the probability of getting a head on the first and third coin?

74. A bag contains eight red marbles and three white marbles. If two marbles are drawn at random, find the probability that both marbles are red.

[12.9]

75. A card is drawn from a standard deck of 52 cards. What is the probability the card is (a) an even number and a club? (b) an even number or a club?

76. If 85 percent of the population are high school graduates, and 90 percent of high school graduates have jobs, what percentage of the population are employed high school graduates?

77. If 3 out of 1000 patients who have a certain type of elective surgery have complications, what is the probability that a patient who has this type of surgery will not have any complications?

78. If the probability of sunny skies is 75 percent, what are the odds the skies will not be sunny?

79. If the odds that a player will win in jai alai are 5 to 6, what is the probability the player will win?

Chapter 12 Test

1. Expand and simplify: $\sum_{n=1}^{4} n(n + 1)$.

2. Write in summation notation: $3 + 6 + 9 + 12 + 15$.

3. Find the general term a_n for the sequence whose terms are $1, -4, 9, -16, \ldots$.

4. Find the 10th term of the sequence whose general term is
$$a_n = \frac{2n}{n + 2}.$$

Find the indicated unknown for each arithmetic sequence or series.

5. $a_1 = 3, d = 2; a_8 = ?$ 6. $a_1 = 17, a_6 = -3; d = ?$

7. $a_{13} = 6, a_{12} = \frac{19}{4}; a_1 = ?$

8. $5 + 9 + 13 + \cdots; S_{10} = ?$

Find the indicated unknown for each geometric sequence or series.

9. $9, 3, 1, \ldots; a_7 = ?$ 10. $a_1 = 8, r = -\frac{3}{7}; a_n = ?$

11. $6 + 3 + \frac{3}{2} + \cdots; S_4 = ?$ 12. $S_5 = 217, r = 2; a_1 = ?$

Find the sum, if it exists, of each infinite geometric series.

13. $a_1 = 15, r = -\frac{4}{5}$ 14. $\sum_{n=1}^{\infty} 10\left(\frac{1}{12}\right)^n$

15. Write $0.\overline{024}$ as a ratio of two integers.

16. Expand $(5x - 2)^4$ using the binomial theorem.

17. Find the fourth term of $(a + 3b)^9$.

18. Use mathematical induction to prove the following statement is true for all positive integers n:

$$4 + 8 + 12 + \cdots + 4n = 2n(n + 1)$$

19. The number of bacteria in a culture doubles every hour. If the culture contains 12 bacteria after 1 hr, how many bacteria does it contain after 24 hr?

20. The first swing of a pendulum is 24 in., and the second swing is 20 in. The lengths of the successive swings form a geometric sequence. Find the total distance traveled by the pendulum before it comes to rest.

21. How many three-letter license plates are there if the middle letter cannot be a vowel?

22. In how many ways can a president, a vice-president, and a secretary be chosen from a class of 20 students?

23. In how many ways can a committee of five people be selected from a class of 15 students if one of the five students has already been selected?

24. Three coins are flipped. Find the probability of getting at least one tail.

25. A card is drawn from a standard deck of 52 cards. Find the probability the card is a club or a face card.

Cumulative Test for Chapters 10, 11, and 12

1. If $y = \log_3 \frac{1}{9}$, find y.   2. Evaluate: $\begin{vmatrix} 5 & 0 & -1 \\ 3 & -2 & -2 \\ 1 & 6 & 4 \end{vmatrix}$

3. Expand $\log \frac{\sqrt{x}}{y^3}$. Assume $x > 0$ and $y > 0$.

4. If $\ln x = 34.8$, find x to three significant figures.

5. Determine the value of \$3750 invested for 10 yr at 8 percent compounded continuously. Use $V = Pe^{rt}$.

6. Expand and simplify: $\sum_{n=1}^{6} (n^2 - 2)$.

7. Write in summation notation:
$4 + 9 + 14 + 19 + 24 + 29$.

8. Expand $(2x - y)^5$ using the binomial theorem.

9. Show that the function $f(x) = 2x - 10$ is one-to-one. Then find the inverse function.

10. Find the 17th term of the arithmetic sequence with second term 11 and common difference 4.

11. Find the seventh term of the geometric sequence with first term 5 and common ratio $\frac{2}{3}$.

12. Find the sum of the first 11 terms of the arithmetic series $3 + 7 + 11 + 15 + \cdots$.

13. Find the sum of the first eight terms of the geometric series $3 + 6 + 12 + 24 + \cdots$.

14. Use mathematical induction to prove the following statement is true for every positive integer n:

$$1 + 2 + 3 + \cdots + n = \frac{n(n + 1)}{2}$$

15. Solve by substitution.

$$x + 5y = 17$$
$$2x - 3y = -18$$

16. Solve by addition.

$$3x + 2y = 14$$
$$4x + 5y = 14$$

17. Use Cramer's rule to solve.

$$3x - 2y = -1$$
$$4x - 5y = -6$$

18. Solve by addition.

$$x + y + z = -1$$
$$x - y + 3z = -5$$
$$6x + 2y - z = 8$$

19. Solve by matrices.

$$x + 2y - z = -9$$
$$3x - y - z = 7$$
$$x - 4y - 5z = 3$$

20. Solve the following system of equations.

$$x^2 + y^2 = 16$$
$$2x + y = 4$$

21. Graph the following system of inequalities.

$$x^2 + y^2 < 16$$
$$2x + y > 4$$

Solve each equation. Round approximate answers to the nearest hundredth.

22. $5^x = 97$

23. $\log_2 (x + 2) - \log_2 (x - 5) = 3$

Graph each function.

24. $y = 3^{-x}$

25. $y = \log_2 x$

26. Two magazines and three books cost \$42. Six magazines and one book cost \$30. Find the cost of each magazine and each book.

27. A ball is dropped from a height of 16 ft. Each time it bounces, it rises to a height that is three-quarters of its previous height. Find the total distance traveled by the ball before it comes to rest.

28. In how many ways can a president, a vice-president, a secretary, and a treasurer be chosen from a class of 24 students?

29. A card is drawn from a standard deck of 52 cards. Find the probability that the card drawn is an ace or a red card.

30. Four coins are flipped. Find the probability of getting exactly two heads.

Answers to Odd-Numbered Exercises and Complete Chapter Review Exercises, Chapter Test Exercises, and Cumulative Test Exercises

Chapter 1

Exercises 1.1

1. Set; elements; members   3. ∈   5. ⊆; every element of A is also an element of B 7. Intersection; $A \cap B$ 9. True   11. True   13. E
15. E   17. C   19. D   21. Finite   23. Finite   25. Infinite   27. ∈
29. ∉   31. ∉   33. ∈   35. ∈   37. ∈   39. 2.236   41. 7.550
43. $\{2, 4, 6, 8\}$ 45. $\{b, a, n\}$ 47. $\{4, 5, 6, 7\}$ 49. ∅   51. $\{x \mid x$ is a whole number less than 40$\}$ 53. $\{y \mid y$ is a whole number greater than 4$\}$
55. $\{t \mid t$ is a month beginning with the letter J$\}$ 57. ⊆   59. ⊄   61. ⊆
63. ⊄   65. ⊄   67. ∅, $\{a\}$, $\{b\}$, $\{a, b\}$ 69. $\{1, 2, 3, 4, 5, 7\}$
71. $\{2\}$ 73. C   75. ∅   77. $\{1, 2, 3, 4, 5\}$ 79. D   81.
(a) $\{6, \sqrt{81}, \frac{18}{3}\}$, (b) $\{6, \sqrt{81}, 0, \frac{18}{3}\}$, (c) $\{6, \sqrt{81}, -24, 0, \frac{18}{3}\}$, (d) $\{6, 4\frac{2}{3}, \sqrt{81}, -24, 0.\overline{72}, 0, 9.83, \frac{18}{3}\}$, (e) $\{-\sqrt{15}, 0.7575575557 \ldots \}$,
(f) All of them   83. True   85. True   87. True   89. True   91. False
93. $\{ \ \}$ and ∅ both represent the empty set. $\{0\}$ is the set with the single element 0. $\{∅\}$ is the set with the single element ∅.   95. A finite set has a limited number of elements, while an infinite set has an unlimited number of elements.   97. The symbol ∈ is used between an element and a set to indicate the element is in the set. The symbol ⊆ is used between two sets when each element of the first set is an element of the second set.
99. If $\frac{8}{0} = x$, then $0 \cdot x = 8$. But $0 \cdot x = 0$ for every real number x. However, $\frac{0}{8} = 0$, since $8 \cdot 0 = 0$.   101. The only activity John, Bob, and Sue had in common was eating at Al's, so eating at Al's was probably the cause of their illness.   103. 43   105. 7.297   107. False   109. False $(\sqrt{2} \approx 1.414)$

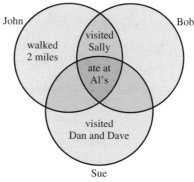

Exercises 1.2

1. True   3. Positive; negative   5. Absolute value; $|a|$ 7.

9. -5 11. $\frac{3}{8}$ 13. -6.99 15. -3.9 17. 0   19. Positive number. For example, if $x = -5$, then $-x = -(-5) = 5$.   21. 3   23. c
25. -25 27. 4   29. 1   31. 87.6   33. 8   35. 16   37. -6 39. 9
41. 9   43. 7, -7 45. False   47. False   49. $<$ 51. $>$ 53. $>$
55. $>$ 57. $<$ 59. $>$ 61. $x < 5$ 63. $y \geq -2$ 65. $z \leq 15$ 67. $m > 0$ 69. $p \geq 0$ 71. $0 < r < 7$ 73. $-5 \leq t < 13$ 75. $a \geq 18$ yr
77. 60 in. $\leq h \leq$ 80 in.   79. 299 mi $\leq d \leq$ 494 mi
81. $\{x \mid 4 > x\}$ 83. $\{x \mid -3 \leq x\}$

85. $\{x \mid 6 > x > 5\}$

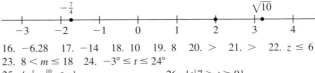

87. $\{x \mid 1 > x \geq -4\}$

89.

Symbol	Meaning	Example
$=$	is equal to	$5 = 5$
$\neq$	is not equal to	$4 \neq 1$
$<$	is less than	$-8 < 0$
$>$	is greater than	$10 > -3$
$\leq$	is less than or equal to	$-2 \leq -2$
$\geq$	is greater than or equal to	$9 \geq 7$

91. If a and b are real numbers, then $a < b$ if the graph of a lies to the left of the graph of b on the number line. 93. If a and b are real numbers, then $a < b$ and $a \not\geq b$ are equivalent statements. 95. -9 97. 17 99. True 101. False

Exercises 1.3

1. a 3. $a = c$ 5. $a < c$ 7. $a + b = b + a$; $ab = ba$ 9. $a + 0 = 0 + a = a$; $a \cdot 1 = 1 \cdot a = a$ 11. $a(b + c) = ab + ac$ 13. Inverse; inverse 15. B 17. F 19. D 21. $5k + 25$ 23. $2k = 7n$ 25. $x = y$, or $x < y$, or $x > y$ 27. No. $5 < 5$ is false. 29. $1(-3)$ 31. $y + (1 + 8)$ 33. t 35. 1 37. 0 39. Commutative of $+$ 41. Associative of $\cdot$ 43. Identity of $\cdot$ 45. Inverse of $+$ 47. Commutative of $+$ 49. Multiplication property of 0 51. Commutative of $\cdot$ 53. Distributive 55. Commutative of $\cdot$ 57. $\frac{1}{6}$ 59. $\frac{2}{3}$ 61. -1 63. $\frac{10}{47}$ 65. 0.5 67. -32 69. $4m + 4 \cdot 7$ 71. $x \cdot 2 + y \cdot 2 + 45 \cdot 2$ 73. $(3 + 7)z$ 75. $(10 + 1)p$ 77. $6(r + s + t)$ 79. $20x$ 81. $3z + 27$ 83. p 85. 47 87. $10k + 15$ 89. r 91. No. $3 - 2 \neq 2 - 3$; No. $3 - (2 - 1) \neq (3 - 2) - 1$ 93. A collection of constants, variables, and/or operations. For example, $3x + 7$. 95. 124

Exercises 1.4

1. Sum; difference 3. $a + (-b)$ 5. If the signs are the same, add the absolute values and prefix the common sign. If the signs are different, subtract the smaller absolute value from the larger and prefix the sign of the number having the larger absolute value. 7. $4 - 3$ means "4 subtract 3"; $4(-3)$ means "4 times -3". 9. -17 11. -7 13. 11 15. -5 17. 15 19. -10 21. 13 23. $-\frac{11}{24}$ 25. 6 27. 37 29. -24 31. -21 33. $-\frac{31}{36}$ 35. -21 37. 18 39. -21 41. $\frac{21}{50}$ 43. -66 45. 40 47. 3 49. -5 51. 3 53. -1 55. -19 57. Undefined 59. -12.24 61. -618.87 63. 0 65. $\frac{7}{9}$ 67. $-\frac{1}{4}$ 69. $-\frac{4}{5}$ 71. $-\frac{16}{9}$ 73. $\frac{3}{10}$ 75. $\frac{3}{5}$ 77. $x - 6$ 79. $y - 10$ 81. -14 83. $-20p$ 85. h 87. $-6r$ 89. $-t$ 91. k 93. $27x - 36$ 95. $-10y - 6$ 97. $-4m + 9$ 99. $a - 5b + 2c$ 101. $-13°$ 103. 30,314 ft 105. $-\$1$ 107. Yes. $a - b$ and $b - a$ are opposites, and opposites have the same absolute value. 109. $5 \boxed{-} 3$ subtracts 3 from 5 and gives the correct answer 2.

Exercises 1.5

1. Factors 3. Power 5. True 7. 243 9. 10 11. 1 13. 25 15. -25 17. $-\frac{27}{64}$ 19. 81 21. $-16,807$ 23. In $(-3)^2$, the base is -3, so $(-3)^2 = (-3)(-3) = 9$. In -3^2, the base is 3, so $-3^2 = -(3 \cdot 3) = -9$. 25. 76 27. -14 29. 26 31. 15 33. -99 35. 8 37. 1 39. 0 41. 24 43. 2 45. 65 47. $-\frac{12}{17}$ 49. $-\frac{7}{3}$ 51. 2 53. 6 55. 3 57. $\frac{11}{17}$ 59. $-\frac{2}{5}$ 61. 41 63. 13 65. 33 67. Undefined 69. -67 71. -1 73. 17 75. 5 77. $-12°$ 79. $\$923$ 81. To abbreviate the writing of repeated products, such as $x \cdot x \cdot x \cdot x \cdot x$. 83. (a) 9, (b) -9 85. 382 87. -6 89. 125 91. 47

Chapter 1 Review Exercises

1. Infinite 2. Finite 3. $\in$ 4. $\notin$ 5. $\{0, 1, 2, 3, 4, 5, 6\}$ 6. $\{5, 10, 15, 20, \ldots\}$ 7. $\{x \mid x \text{ is an even integer greater than 9}\}$ 8. $\{y \mid y \text{ is a vowel}\}$ 9. $\not\subseteq$ 10. $\subseteq$ 11. $\{1, 3, 5, 7\}$ 12. $\{2, 3, 5, 7, 9\}$ 13. (a) $\{12\}$, (b) $\{12, 0\}$, (c) $\{12, 0, -99\}$, (d) $\{12, -4.9, 0, 5\frac{3}{4}, -99, 8.\overline{7}\}$, (e) $\{\sqrt{27}\}$, (f) All of them 14. $\varnothing$, $\{a\}$, $\{b\}$, $\{c\}$, $\{d\}$, $\{a, b\}$, $\{a, c\}$, $\{a, d\}$, $\{b, c\}$, $\{b, d\}$, $\{c, d\}$, $\{a, b, c\}$, $\{a, b, d\}$, $\{a, c, d\}$, $\{b, c, d\}$, $\{a, b, c, d\}$ 15.

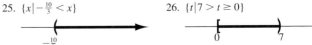

16. -6.28 17. -14 18. 10 19. 8 20. $>$ 21. $>$ 22. $z \leq 6$ 23. $8 < m \leq 18$ 24. $-3° \leq t \leq 24°$ 25. $\{x \mid -\frac{10}{3} < x\}$ 26. $\{t \mid 7 > t \geq 0\}$

27. Symmetric of equality 28. Transitive of inequality 29. Substitution of equality 30. Trichotomy of inequality 31. $z + 9$ 32. $x = y$ 33. Identity of $\cdot$ 34. Multiplication property of 0 35. Associative of $\cdot$ 36. Distributive 37. Inverse of $+$ 38. Commutative of $\cdot$ 39. $4 + 0$ 40. $(3 \cdot 9)x$ 41. y 42. 1 43. $-\frac{1}{8}$ 44. $\frac{5}{13}$ 45. $2p + 2 \cdot 14$ 46. $(5 + 7)y$ 47. x 48. $10y$ 49. z 50. $12p + 32$ 51. 0 52. 44 53. $-\frac{9}{8}$ 54. 42 55. $-\frac{88}{3}$ 56. $\frac{8}{45}$ 57. -12 58. $y + 11$ 59. $3z$ 60. $-8r - 28$ 61. -6 62. Undefined 63. $-4t$ 64. $8m$ 65. $\frac{5}{2}p - \frac{1}{8}$ 66. $7a + b - 13$ 67. $-\frac{125}{8}$ 68. -64 69. -7 70. -24 71. 22 72. -30 73. 0 74. $\frac{23}{4}$ 75. $-\frac{8}{15}$ 76. -4 77. 70 78. $\frac{3}{2}$ 79. $-1°$

Chapter 1 Test

1. $\{x \mid 1 \leq x \leq 88 \text{ and } x \text{ is an integer}\}$ 2. $\varnothing$, $\{0\}$, $\{9\}$, $\{0, 9\}$ 3. $\{6, 7\}$ 4. $\{x \mid -\frac{5}{2} \geq x\}$ 5. $\{y \mid 4 > y > 0\}$

6. $\{-\sqrt{36}\}$ 7. $\{\frac{18}{5}, 0.375, -\sqrt{36}\}$ 8. 0 9. $12x - 35$ 10. $-6y$ 11. 29 12. $z < 0$ 13. $2 \leq p \leq 15$ 14. Reflexive of equality 15. Commutative of $+$ 16. Distributive 17. Transitive of equality, or substitution of equality 18. True 19. True 20. False 21. 3 22. -26 23. 30 24. 22 25. $-\frac{1}{8}$ 26. $-\frac{17}{18}$ 27. -7 28. 21 29. 0 30. $-\frac{3}{5}$

Chapter 2

Exercises 2.1

1. False 3. True 5. $11y + 10$ 7. $8m - 7$ 9. $3y - 3$ 11. $-10r - 6$ 13. $-2p^2 - 3p - 11$ 15. $9x$ 17. 29 19. $3k - 6$ 21. $-x^2 - 18x + 15$ 23. Equation 25. Solution set 27. Coefficient 29. $a + c = b + c$ 31. $\{-6\}$ 33. $\{0\}$ 35. $\{-15\}$ 37. $\{-\frac{1}{2}\}$ 39. $\{3\}$ 41. $\{5\}$ 43. $\{1\}$ 45. $\{-\frac{5}{7}\}$ 47. $\{-3\}$ 49. $\{-4\}$ 51. $\{81\}$ 53. $\{-\frac{16}{7}\}$ 55. $\{\frac{1}{8}\}$ 57. $\{21\}$ 59. $\{15\}$ 61. $\{\frac{25}{9}\}$

63. $\{-6\}$ 65. $\{-14\}$ 67. Identity 69. Conditional equation
71. Contradiction 73. Contradiction 75. Identity 77. Terms are related by addition; for example, the terms of $x + 3$ are x and 3. Factors are related by multiplication; for example, the factors of $5y$ are 5 and y.
79. Like terms are terms that have the same variables with the same exponents. Like terms are combined by combining (adding or subtracting) their coefficients. 81. A linear equation in one variable is an equation that can be written in the form $ax + b = 0$ $(a \neq 0)$. 83. (a) 245, (b) 245, Yes 85. For example, store 8 in x and show $3(x - 2) + 7 = 3x + 1$ is true.

Exercises 2.2

1. Formula 3. π 5. 22 ft 7. 7 ft 9. 2.5 hr 11. \$685 13. 145 yd 15. 23° 17. 90 cm 19. 12.5 yr 21. 7058.405 cm² 23. The variable P appears on *both* sides of $P = A - Prt$. 25. $B = \dfrac{3A}{h}$

27. $r = \dfrac{A - P}{P}$ 29. $b^2 = c^2 - a^2$ 31. $l = \dfrac{T - \pi r^2}{\pi r}$ 33. $b_2 = \dfrac{2A - hb_1}{h}$ 35. $n = \dfrac{l - a + d}{d}$ 37. $T_2 = \dfrac{T_1 P_2 V_2}{P_1 V_1}$ 39. $r = \dfrac{S - a}{S}$

41. 73.12 cm² 43. 4396 ft² 45. 40 mi 47. π is *approximately* equal to 3.14.

Exercises 2.3

1. $5x - 12$ 3. $\frac{2}{3}y - 9$ 5. $3n + 7$ 7. $3(n + 7)$ 9. Addition is commutative, but subtraction is not commutative. 11. $24p$ 13. $35m$
15. $9.50x + 14.50y$ dollars 17. $0.09(x + 280)$ dollars 19. $\dfrac{1200}{k}$ mi
21. $x, x + 1, x + 2$ 23. $x, x + 2, x + 4, x + 6$ 25. x years, $x - 7$ years 27. y quarts, $15 - y$ quarts 29. 8 31. $-\frac{7}{2}$ 33. 38, 73
35. 116, 118 37. 11, 12, 13 39. \$875 41. \$595 43. \$4500 at 8 percent, \$3000 at 12 percent 45. \$3000 47. Width 75 yd, length 120 yd 49. 52°, 38° 51. 26°, 57°, 97° 53. Width 1.5 mi, length 4.5 mi 55. You may have written an incorrect equation.

Exercises 2.4

1. 16 nickels, 8 dimes, 11 quarters 3. 19 nickels, 21 dimes 5. 36 day hr, 10 night hr 7. 24 kg 9. 54 g of 20% alloy, 18 g of 28% alloy 11. 1 lb 13. 50 mi/hr, 55 mi/hr 15. 33 sec for P-wave, 60 sec for S-wave, 165 mi 17. 4.5 hr 19. 19.2 mi 21. 15 min 23. 650 mi 25. 97 27. 250 seats 29. Let x be the number. Then the given operations produce the following sequence of algebraic expressions: $x, x + 8, 3x + 24, 3x + 15, 6x + 30, x + 5, 5$.

Exercises 2.5

1. C 3. D 5. False 7. False 9. Solution 11. Solve
13. $(-3, \infty)$ 15. $(-\infty, -5)$

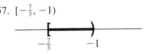

17. $(-\infty, 2)$ 19. $(-\infty, 1]$

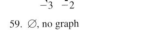

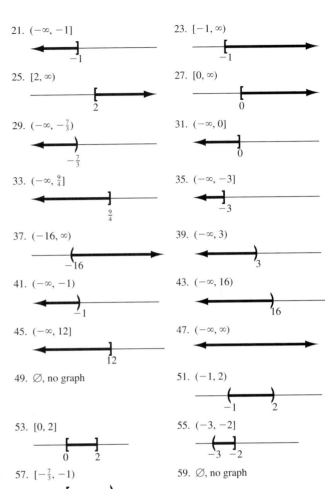

21. $(-\infty, -1]$ 23. $[-1, \infty)$
25. $[2, \infty)$ 27. $[0, \infty)$
29. $(-\infty, -\frac{7}{3})$ 31. $(-\infty, 0]$
33. $(-\infty, \frac{9}{4}]$ 35. $(-\infty, -3]$
37. $(-16, \infty)$ 39. $(-\infty, 3)$
41. $(-\infty, -1)$ 43. $(-\infty, 16)$
45. $(-\infty, 12]$ 47. $(-\infty, \infty)$
49. $\varnothing$, no graph 51. $(-1, 2)$
53. $[0, 2]$ 55. $(-3, -2]$
57. $[-\frac{7}{3}, -1)$ 59. $\varnothing$, no graph
61. $(-1, 4)$

63. $-2 < x < 2$ 65. 12 in. $<$ side $<$ 16 in. 67. More than 40 hr 69. More than 800 T-shirts 71. When you solve $7x - 5 > 3x + 3$, you get $x > 2$. Substitute 2 for x in the original inequality and the two sides should have the same value. Then substitute a number greater than 2 for x and the inequality should be a true statement. 73. We can add the same number to both sides of an inequality and the result is an equivalent inequality. 75. (1) Clear fractions. (2) Remove parentheses. (3) Simplify each side. (4) Collect variable terms on one side and constant terms on the other side. (5) Write the variable term with a coefficient of 1. 77. Provides a simple notation for writing intervals of real numbers 79. If $x = 1$, both $1 \leq 7 - 2x$ and $7 - 2x \leq 11$ are true. If $x = 4$, then $1 \leq 7 - 2x$ is false.

Exercises 2.6

1. C 3. B 5. F 7. And; or 9. Union 11. False 13. $(0, 3)$
15. $(-1, 8]$ 17. $(-\infty, -4)$ 19. $(-\infty, 1) \cup (2, \infty)$ 21. $\varnothing$
23. $(0, \infty)$

25. (1, 6)

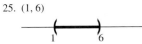

27. (−1, ∞)

41. (32, ∞)

42. [−2, ∞)

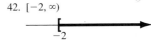

29. [4, 8]

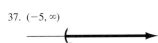

31. (−∞, −3)

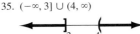

43. (−∞, 5)

44. (−∞, 3]

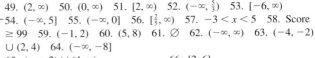

33. ∅, no graph

35. (−∞, 3] ∪ (4, ∞)

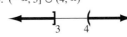

45. [−4, 3]

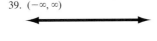

46. (0, 2]

37. (−5, ∞)

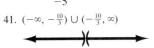

39. (−∞, ∞)

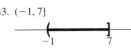

47. (−2, 5/2]

48. (−2, 6)

41. (−∞, −10/3) ∪ (−10/3, ∞)

43. (−1, 7]

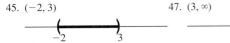

49. (2, ∞)   50. (0, ∞)   51. [2, ∞)   52. (−∞, 3/5)   53. [−6, ∞)
54. (−∞, 5]   55. (−∞, 0]   56. [2/5, ∞)   57. −3 < x < 5   58. Score
≥ 99   59. (−1, 2)   60. (5, 8)   61. ∅   62. (−∞, ∞)   63. (−4, −2)
∪ (2, 4)   64. (−∞, −8]

45. (−2, 3)

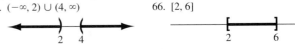

47. (3, ∞)

65. (−∞, 2) ∪ (4, ∞)

66. [2, 6]

49. (−∞, −15) ∪ (5, ∞)   51. (−5/2, 3]   53. [10/3, 12]   55. Only a single
connected interval can be described by a double inequality.

67. (18, ∞)

68. [−3, ∞)

Exercises 2.7

1. Absolute-value   3. x = 5 or x = −5   5. −7 < x < 7   7. F   9. G
11. H   13. D   15. {8, −8}   17. {7, −2}   19. {3, −3}   21. ∅
23. {20, −36}   25. {6, −4}   27. {1, −1/9}   29. {3/2}   31. {0, 5/2}
33. [−4, 4]   35. [3, 5]   37. ∅   39. {−1}   41. [−5, −1]
43. (−4, 7/2)   45. (−3, 3)   47. [−4, 3]   49. (−∞, −1) ∪ (1, ∞)
51. (−∞, −2) ∪ (8, ∞)   53. (−∞, −5/3] ∪ [2, ∞)   55. (−∞, ∞)
57. (−∞, −4) ∪ (−4, ∞)   59. (−∞, −2] ∪ [9/2, ∞)
61. (−∞, −8) ∪ (8, ∞)   63. (−∞, ∞)   65. {4, −2/3}   67. (−1, 4)
69. (−∞, −4) ∪ (−3, ∞)   71. {3, 1}   73. [−1/2, 2]   75. {5, −13}
77. [−18, 18]   79. (−∞, −6) ∪ (1, ∞)   81. x = y and −x = −y
are equivalent, as are x = −y and −x = y.   83. |any algebraic
expression| ≥ 0   85. If a > 0, the solution of |x| = a is x = a or
x = −a. The solution of |x| = |y| is x = y or x = −y. If a = 0, the
solution of |x| = a is x = 0. If a < 0, the solution of |x| = a is ∅.
87. (a) True, (b) True, (c) False. The numbers 4 and −1 are solutions.

69. (−∞, 7/3)

70. ∅, no graph

71. (3, 7)

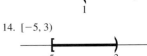

72. ∅, no graph

73. {2, −2}   74. (1, 4)   75. (−∞, −1) ∪ (3/2, ∞)   76. {6, −3}
77. {3, 1}   78. (−∞, 3] ∪ [5, ∞)   79. [−7, 8]   80. ∅   81. (−∞, ∞)
82. ∅   83. {−3, −11}   84. (−6, 1)

Chapter 2 Test

1. [−1, 0]   2. (2, ∞)   3. Identity   4. Conditional equation   5. {−2}
6. {16/7}   7. {24}   8. {2}   9. $b = \dfrac{2A}{h}$ 10. $w = \dfrac{P - 2l}{2}$
11. (−∞, −9]   12. (1, ∞)

13. (9, ∞)   14. [−5, 3)

15. [−4, 3)   16. (−∞, 5/3)

Chapter 2 Review Exercises

1. −5x − 5   2. 14y − 44   3. −2p² − 4p + 20   4. {−9}
5. {7}   6. {2/3}   7. {−2}   8. {3}   9. {−3}   10. {5}   11. {32/11}
12. Contradiction   13. Identity   14. $P = \dfrac{A}{1 + r}$ 15. $h = \dfrac{V}{\pi r^2}$
16. c = 2s − a − b   17. $R_1 = \dfrac{RR_2}{R_2 - R}$ 18. 54 mi/hr   19. 37 ft
20. 1650 ft   21. 19.625 ft³   22. $472.50   23. 2(x − 6)   24. 17 + 6y
25. 1440t   26. 0.35(40 − x) grams   27. x, x + 2, x + 4   28. x years,
x + 2 years, x − 6 years   29. 4   30. 17, 68   31. 19, 20, 21
32. $440   33. $6000 at 6 percent, $3600 at 10 percent   34. Width is
17 ft, length is 44 ft   35. Eight dimes, four quarters   36. 189 adult
tickets, 211 student tickets   37. 13 g of 22% alloy, 39 g of 30% alloy
38. 7.5 hr   39. 893.75 mi   40. Adam is 12, brother is 6

17. {3, −3}   18. {−4, −2}   19. (−∞, −7) ∪ (−1, ∞)   20. [−1, 4]
21. −13°   22. 8 percent   23. 16, 18   24. $2800   25. 24°, 39°, 117°
26. 36 day hr, 12 night hr   27. 2.5 kg   28. 45 mi/hr, 50 mi/hr

Chapter 3

Exercises 3.1

1. 1   3. m 5. 4   7. Keep; subtract   9. r^{10} 11. x^8 13. 3^8
15. 5^{12} 17. $(p + 3)^{15}$ 19. $-24r^{12}$ 21. $35a^7b^{11}$ 23. The base in
-5^0 is 5, not -5. Therefore, $-5^0 = -(5^0) = -1$.   25. 1   27. -1
29. 1   31. $\frac{1}{81}$ 33. $-\frac{1}{16}$ 35. $\frac{6}{x^2}$ 37. $\frac{1}{8}$ 39. 10   41. $-\frac{125}{8}$ 43. 25
45. $9p^6$ 47. $48r^4$ 49. $\frac{9}{20}$ 51. $\frac{3}{2}$ 53. 1   55. 0.001   57. y^3
59. $\frac{1}{z^5}$ 61. m^6 63. $3p - 1$ 65. $-5k^{45}$ 67. $\frac{2y^5}{x^{12}}$ 69. 7^9
71. $\frac{1}{x^4}$ 73. $\frac{72}{m^9}$ 75. 6^{21} 77. y^2 79. $-\frac{6}{z^{10}}$ 81. x^{2n+1} 83. x^{5r+2}
85. $\frac{1}{y^q}$ 87. (a) 100 mg, (b) 50 mg, (c) 25 mg   89. $3 \cdot 2^{18} = 786{,}432$
cells

91. (a)

	Question 1
Choice 1	True
Choice 2	False

(b)

	Question 1	Question 2
Choice 1	True	True
Choice 2	True	False
Choice 3	False	True
Choice 4	False	False

(c)

	Question 1	Question 2	Question 3
Choice 1	True	True	True
Choice 2	True	True	False
Choice 3	True	False	True
Choice 4	True	False	False
Choice 5	False	True	True
Choice 6	False	True	False
Choice 7	False	False	True
Choice 8	False	False	False

A test containing 10 questions can be completed in $2^{10} = 1024$ ways;
a test containing n questions can be completed in 2^n ways.
93. (a) \$1469.33, (b) \$2938.66, (c) \$2100.34, (d) \$2158.92
95. Because $2^3 \cdot 2^4 = (2 \cdot 2 \cdot 2) \cdot (2 \cdot 2 \cdot 2 \cdot 2) = 2^7$, not 4^7.
97. Sometimes. For example, $2^{-3} = \frac{1}{8}$, but $(-2)^{-3} = -\frac{1}{8}$.

Exercises 3.2

1. $8p^{12}$ 3. Numerator; denominator   5. x^8 7. 2^{21} 9. $\frac{1}{p^6}$
11. $81y^4$ 13. $49m^2n^2$ 15. $-32k^3$ 17. $\frac{9}{49}$ 19. $-\frac{1}{z^5}$ 21. $x^{12}y^{21}$
23. $\frac{p^{18}}{q^{30}}$ 25. r^{16} 27. $\frac{27}{k^{10}}$ 29. $\frac{v^4}{4u^6}$ 31. $m^{24}n^{20}$ 33. $\frac{1}{9x^2y^4}$
35. z^8 37. $-\frac{64y^{18}}{x^{36}}$ 39. $-243r^{46}s^{53}$ 41. $\frac{2p^{27}}{q^{11}}$ 43. $\frac{2b}{5a}$ 45. $\frac{ay^3}{bx^2}$
47. $-3a^4b^9$ 49. $\frac{128m^9}{5b^{32}}$ 51. 9   53. $\frac{c^{15}}{a^{27}b^{39}}$ 55. $\frac{5^{10}}{y^6}$
57. $\frac{25x^{10}y^8}{16z^6}$ 59. $\frac{5b^9c^{10}}{8a^{23}}$ 61. $\frac{x^{34}y^{16}}{16}$ 63. $\frac{4w^2}{81}$ 65. x^{10m}

67. x^{4m-4} 69. $\frac{1}{y^{10n}}$ 71. z^{3m+8} 73. 2^{30} 75. 5^{8m} 77. \$207.58,
\$2455.01   79. \$1,039,463.45   81. The base in $(2x)^3$ is $2x$. Therefore,
$(2x)^3 = (2x)(2x)(2x) = 8x^3$.

Exercises 3.3

1. D   3. A   5. False   7. 9; left   9. 3×10^4 11. 5×10^{-3}
13. 8.09×10^9 15. 10^7 17. 3.84×10^{-7} 19. 10^{-6} 21. 2.751×10^2 23. 5.59   25. 1.19×10^4 27. 7.43×10^{-3} 29. 5,500,000,000
31. 0.004   33. 0.000019   35. 990.7   37. 6.332   39. 0.101   41. To
write numbers that are very large or very small in a more compact form
43. $4.3 \times 10^{-4}, 6.5 \times 10^{-2}, 8.97, 2.7 \times 10^5$ 45. 6.3×10^7 47. 8×10
49. 4×10 51. 6×10^{-17} 53. 1.2×10^3 55. 6.5×10^{-11}
57. 2.4×10^9 59. 0.0000000000000000000000003   61. 10^{100}
63. 500 sec   65. 6.6×10^{21} 67. 2.4×10^{11} 69. 6.42×10^{-13}
71. 5.88×10^{24} 73. 9.5975   75. 1.25×10^{-2}

Exercises 3.4

1. Real number; whole number   3. Degree   5. Leading   7. Binomial
9. False   11. False   13. False

Descending Order	Degree	Leading Coefficient	Constant Term	Type
15. $x^2 + 4x - 13$	2	1	-13	Trinomial
17. z^3	3	1	0	Monomial
19. 8	0	8	8	Monomial
21. $-2m^5 + m^2 - 9m + 6$	5	-2	6	Polynomial
23. $3x^3t^4 - 6x^5t + 4x^2t^2$	7	3	0	Trinomial

25. (a) 34, (b) 49, (c) $5k^2 + 8k + 13$ 27. (a) 6, (b) 0, (c) $8a^3 - 2a$
29. (a) -6, (b) 0, (c) $-16a^4 - 4a^2 + 8a + 6$ 31. (a) 9, (b) $\frac{61}{9}$,
(c) $3z^6 - z^4 - 4z^2 + 9$ 33. -5.814 35. $13m - 6n$ 37. $-9x^2y + 5xy^2$ 39. $-2u^2 + 7u + 4$ 41. $-m + n$ 43. (a) $-6t^2 + 4t - 4$,
(b) $-6t^2 + 4t - 4$ 45. $2z^3 - 15z^2 + 7z - 23$ 47. $6r^5 + 5r^3 - 10r^2 + 6r - 3$ 49. (a) $3x + 3$, (b) $3x + 3$ 51. (a) $5p^2 - 14p + 16$,
(b) $5p^2 - 14p + 16$ 53. $2x^3 + 2xy^2 - 10y^3$ 55. $7a^2 + 2a - 17$
57. $-4m^5 - 2m^4 + 8m^2 + 2m$ 59. $-2x^2 + 7x + 11$ 61. $4m^3 + 5m^2 + 9m - 8$ 63. $6r^3 - r^2s + rs^2 + 11s^3$ 65. c 67. $4p^3 - 2pz^2 - 4p + 10z$ 69. $x^{3m} - x^{2m} + 3x^m - 3$ 71. $-4z^3 + 9z^2 - 21z - 3$ 73. $3n + 3$ 75. (a) 30 games, (b) 132 games   77. (a) 72.6 ft, (b) 224.4 ft
79. The degree of a polynomial is the same as the degree of the highest-
degree monomial in the polynomial.   81. Add their like terms
83. -1.6 85. 46.7

Exercises 3.5

1. $-15x^7$ 3. $-18a^{13}b^{15}c^{10}$ 5. $12r^2 + 10r$ 7. $-3m^3 - 27m^2 + 21m$ 9. $-20x^8y^4 - 12x^6y^6 + 36x^5y^7$ 11. First, Outer, Inner, Last. No,
FOIL can be used only when multiplying binomials.   13. $p^2 + 8p + 15$
15. $3x^2 + 2x - 16$ 17. $-7r^2 - 29r + 30$ 19. $12m^2 + 43mn + 35n^2$
21. $12p^2 + 8pq - 15q^2$ 23. $t^4 + 5t^2 - 36$ 25. $x^2 - xh + \frac{2}{9}h^2$
27. D   29. B   31. $(x + y)^2 = (x + y)(x + y)$, not $x^2 + y^2$. If $x = 2$ and
$y = 3$, then $(x + y)^2 = 25$, but $x^2 + y^2 = 13$.   33. The product of two
conjugates is the square of the first term minus the square of the second
term.   35. $k^2 - 36$ 37. $4z^2 - 25$ 39. $9p^2 - 64q^2$ 41. $25x^2 - y^6$
43. $r^2 - 18r + 81$ 45. $9t^2 + 24t + 16$ 47. $25a^2 - 20ab + 4b^2$
49. $36k^2 + 9km + \frac{9}{16}m^2$ 51. $6x^2 - 13x - 28$ 53. $32a^3x^2y - 80a^2xby^2 + 50ab^2y^3$ 55. $-m^5p^3 + 9m^3pq^2$ 57. $p^2 - 2pm + m^2 + 2p - 2m + 1$ 59. $k^2 + 2kh + h^2 - 25$ 61. $3m^3 - 21m^2 + 36m - 12$

63. $2x^3 - x^2y - 15xy^2 + 18xy + 45y^2$ 65. $12k^4 + 11k^3 - 23k^2 - 30k + 27$ 67. $10p^4 + 13p^3 - 45p^2 - 51p + 18$ 69. $r^3 - 12r^2 + 48r - 64$ 71. $8x^3 + 60x^2y + 150xy^2 + 125y^3$ 73. $24p^2 + 38p + 15$ 75. $14y^3 - 29y^2 + 30y - 27$ 77. $15m^4 + 20m^3 + 24m^2 - m - 44$ 79. $5x^5 - 30x^4 - 13x^3 + 53x^2 + 6x - 21$ 81. $11p^2q - 7pq - 6p$ 83. $-m^4 + 6m^2$ 85. $12k - 136$ 87. $10x^{2m+1} - 15x^{m+1} + 35$ 89. $18z^{2k} + 45p^{3k}z^k - 2p^{2k}z^k - 5p^{5k}$

Exercises 3.6

1. False   3. False   5. Factors (or divisors)   7. Prime   9. $2^2 \cdot 5$ 11. $3^3 \cdot 5^2$ 13. 59   15. $2^4 \cdot 3^2 \cdot 5 \cdot 7$ 17. 4   19. 12   21. $3p^2$ 23. $15x^3y^4$ 25. $2m + 5$ 27. $3z - 1$ 29. $9(x + 2)$ 31. $4x^2(2x + 3)$ 33. $a(a^4 + 1)$ 35. $7m(4m^2 - 2m + 5)$ 37. $-4p^2(p^3 - 3p + 4)$ 39. $15r^2t^3(3r^6t - 2t^5 - 4r^4)$ 41. $-8k^7m^8h^6(2k^4h^3 + 3m^2)$ 43. $\pi abc(a + b - c)$ 45. $18xz(3x^7z^3 - 6x^4z^2 - 2x^2z + 4)$ 47. $(m + n)(5x - 9y)$ 49. $(k - 3)(k - 2)$ 51. $(p + 2)(2p + 9)$ 53. $3(y - 2)^2(y + 8)^5(3y + 4)$ 55. $(y + 11)(x + 3)$ 57. $(p - 5)(p + 4)$ 59. $(m - 2)(m^2 + 8)$ 61. $(z + 1)(z^2 + 1)$ 63. $(x - y)(5x + 2)$ 65. $(a^2 - 3)(z^2 + 1)$ 67. $(8r + t^2)(r + 6t)$ 69. $(2k - 1)(k^2 - 5)$ 71. $(3p + 4)(4p - 3)$ 73. $2(m + 5)(3m^3 + 4)$ 75. $3y^2(y^{n+1} + 2y^n - 3)$ 77. $(z^m - 6)(z^{2m} + 2)$ 79. Multiply the factors.   81. When the polynomial has four or more terms   83. $\pi r(l + r)$ 85. $P(1 + r)^2$

Exercises 3.7

1. E   3. B   5. F   7. Descending   9. -1 11. True   13. $(x + 3)(x + 5)$ 15. $(y - 2)(y - 7)$ 17. $(m + 3)(m - 6)$ 19. $(p + 2)(p - 9)$ 21. $(k + 5)(k + 7)$ 23. Prime   25. $-(t - 2)(t + 8)$ 27. $(x + y)(x + 10y)$ 29. $(r + t)(r - 12t)$ 31. $(ab - 3)(ab + 7)$ 33. $5(k^2 + k + 1)$ 35. $12z(z + 2)(z - 4)$ 37. $(2x + 5)(x + 1)$ 39. $(3y - 2)(y - 1)$ 41. $(3x - 1)(2x + 3)$ 43. $(4r + 1)(3r - 5)$ 45. Prime   47. $(2y - 5)^2$ 49. $(5x + 2)(x + 3)$ 51. $-(2k - 3)(k + 4)$ 53. $(3p + 7)(p + 1)$ 55. $(5m + 3)(m - 3)$ 57. $(9p + 2q)(p - 4q)$ 59. $(3ab + 14)(ab - 4)$ 61. $(9x - 8)(4x + 5)$ 63. $4(2r - 9)(r + 8)$ 65. $2x^2y(5x + 2y)(x - 5y)$ 67. $9x^2 - 25$ 69. $3x - 6$ can be factored as $3(x - 2)$.   71. (1) List all pairs of integers whose product is ac. (2) Find the pair (say, m and n) from step 1 whose sum is b. (3) Write the trinomial as $ax^2 + mx + nx + c$. (4) Factor the polynomial from step 3 by grouping. (5) Check by multiplication.   73. $ax^2 + bx + c$, where a, b, and c are integers, will factor if and only if $b^2 - 4ac$ is a perfect square.   75. $(t^2 + 2)(t^2 + 3)$ 77. $(5z^2 - 3)(z^2 + 4)$ 79. $(w^3 + 2)(w^3 - 7)$ 81. $(x - y - 2)(x - y - 6)$ 83. $(2m + 2k - 3)(m + k + 10)$ 85. $(x^m + 2)^2$ 87. $(2y^{2n} - 5)(y^{2n} - 2)$

Exercises 3.8

1. H   3. G   5. D   7. E   9. $(a + b)(a - b)$ 11. $(a + b)(a^2 - ab + b^2)$ 13. True   15. $(x + 4)(x - 4)$ 17. $(3p + 5)(3p - 5)$ 19. $-(7h + 1)(7h - 1)$ 21. $(2z + 11k)(2z - 11k)$ 23. Prime   25. $(10ab + 9)(10ab - 9)$ 27. $(r^3 + 2t^2)(r^3 - 2t^2)$ 29. $3(p^2 + 12)(p^2 - 12)$ 31. $-8xy^2(x^2 + 1)(x + 1)(x - 1)$ 33. $85^2 - 15^2 = (85 + 15)(85 - 15) = (100)(70) = 7000$ 35. $\pi(R + r)(R - r)$ 37. $(x + 3)^2$ 39. $(x - 1)^2$ 41. $(2y - 5)^2$ 43. $(8k + m)^2$ 45. $-(r - 7)^2$ 47. $(9p + 11q)(4p + 11q)$ 49. $12m(m + 4n)^2$ 51. $(4t^2 + 5)^2$ 53. $(z + 2)(z^2 - 2z + 4)$ 55. $(y - 5)(y^2 + 5y + 25)$ 57. $(3p - 1)(9p^2 + 3p + 1)$ 59. $(4r + t)(16r^2 - 4rt + t^2)$

61. $(7m - 10n)(49m^2 + 70mn + 100n^2)$ 63. $(ab + c)(a^2b^2 - abc + c^2)$ 65. $-4k(k + 3)(k^2 - 3k + 9)$ 67. $16y(x^2 + 2y)(x^4 - 2x^2y + 4y^2)$ 69. The first and last terms are perfect squares, and the middle term is twice the product of the two terms of the squared binomial.   71. $x^6 - 64 = (x^3 + 8)(x^3 - 8) = (x + 2)(x^2 - 2x + 4)(x - 2)(x^2 + 2x + 4)$ is easier than
$x^6 - 64 = (x^2 - 4)(x^4 + 4x^2 + 16)$
$\qquad = (x + 2)(x - 2)(x^4 + 8x^2 + 16 - 4x^2)$
$\qquad = (x + 2)(x - 2)((x^2 + 4)^2 - (2x)^2)$
$\qquad = (x + 2)(x - 2)(x^2 + 4 + 2x)(x^2 + 4 - 2x)$
73. $(m + 2n + 5)(m - 2n - 5)$ 75. $(m - p + 2)^2$ 77. $(x - 1)^2$ 79. $(r - 6 + t)(r - 6 - t)$ 81. $(3a + 4b + c)(3a + 4b - c)$ 83. $(a + b - 3)(a^2 + 2ab + b^2 + 3a + 3b + 9)$ 85. $(2x^m + 5)(2x^m - 5)$ 87. $(3p^n + 4)^2$ 89. $(x^k - 5)(x^{2k} + 5x^k + 25)$

Exercises 3.9

1. E   3. F   5. C   7. True   9. True   11. $(x - 5)(x + 8)$ 13. $(3y + 7z)(3y - 7z)$ 15. $4x^4y(x^2 - 4x - 6)$ 17. $(k + 5)(k^2 + 1)$ 19. $(2t + 5)(4t^2 - 10t + 25)$ 21. $-(z - 3)(z + 12)$ 23. $(x^2y^2 + 4)(xy + 2)(xy - 2)$ 25. $(b + 1)(a + 4)(a - 4)$ 27. $-3p(p + 10)(p - 10)$ 29. $2(3r - 5t)(4r + 7t)$ 31. $4\pi(R + r)(R - r)$ 33. $2q^2(2p - 9q)(4p^2 + 18pq + 81q^2)$ 35. $(m + n)(m - n + 10)$ 37. $(5x - 1 + y)(5x - 1 - y)$ 39. $(5x^2 + 3y)(x^2 - 5y)$ 41. $(2z + 1)(4z^2 - 2z + 1)(2z - 1)(4z^2 + 2z + 1)$ 43. $12x(x - 3)^3$ $(x + 6)^7$ 45. $(3x^{2m} + 10y^n)(3x^{2n} - 10y^n)$ 47. $(p^n + 1)(p^{2n} - p^n + 1)(p^n - 1)(p^{2n} + p^n + 1)$ 49. $-(a + b - 10)(a^2 + 2ab + b^2 + 10a + 10b + 100)$ 51. When the leading coefficient is negative   53. Multiply the factors.

Exercises 3.10

1. Quadratic   3. $p = 0$; $q = 0$ 5. D   7. A   9. False   11. If $a = 0$, $ax^2 + bx + c = 0$ becomes $bx + c = 0$, which is not a quadratic equation.   13. $5 = 0$ is a contradiction and has no solution. The solution set of the original equation is $\{2, -7\}$.   15. $\{1, 4\}$ 17. $\{2, -5\}$ 19. $\{-3, -12\}$ 21. $\{\frac{1}{3}, -8\}$ 23. $\{\frac{5}{4}, -\frac{2}{3}\}$ 25. $\{0, 5\}$ 27. $\{\frac{4}{3}, -\frac{4}{3}\}$ 29. $\{-12\}$ 31. $\{-2, \frac{7}{9}\}$ 33. $\{0\}$ 35. $\{0, \frac{3}{2}\}$ 37. $\{10, -12\}$ 39. $\{0, \frac{2}{7}\}$ 41. $\{\frac{7}{4}, -\frac{7}{4}\}$ 43. $\{-1, 4, -\frac{9}{4}\}$ 45. $\{0, 3, -9\}$ 47. $\{0, -5, \frac{7}{2}\}$ 49. $\{1, -1, 2, -2\}$ 51. $\{1, -1, -2\}$ 53. 1 or -3 55. 9 and 10, or -9 and -10 57. Height $\frac{7}{3}$ m, base 6 m   59. 0 units, 440 units   61. 3.5 sec   63. 7 sec   65. -1.5 is a solution.

Chapter 3 Review Exercises

1. 1   2. -4 3. $\frac{1}{36}$ 4. $\frac{5}{y}$ 5. $-\frac{64}{27}$ 6. $\frac{2p^5}{7}$ 7. x^9 8. $\frac{1}{3^{14}}$ 9. $-28a^{10}b^8$ 10. $\frac{10p^3}{k^9}$ 11. t^8 12. 121   13. $-\frac{6h^2}{m^{11}}$ 14. y^{n-1} 15. $32p^5$ 16. $\frac{25b^8}{a^6}$ 17. $\frac{x^{28}}{y^{12}}$ 18. $\frac{-1}{z^3k^3}$ 19. $-\frac{3m^{17}}{16}$ 20. $\frac{2}{t^8}$ 21. $\frac{y^{26}}{729x^7}$ 22. x^{9m} 23. 3.4×10^7 24. 7×10^{-3} 25. 902,000   26. 0.0008911   27. 3.84×10^8 28. 1.65×10^{-15} 29. 5.859×10^{12}

	Descending Order	Degree	Leading Coefficient	Constant Term	Type
30.	$-6x^2 + 4x + 3$	2	-6	3	Trinomial
31.	$-x^3yz^2$	6	-1	0	Monomial
32.	$p - 7$	1	1	-7	Binomial

33. $-15; -9$ 34. $1.5; -4k^3 + 6k^2 + 3k + 1$ 35. $6x^2 - 3x + 4$
36. $3y^3 - 9y^2 + y + 10$ 37. $8m^2 + 3$ 38. $-4x^2 + 5xy - 7y^2$
39. $n - 4$ 40. $12y^4 - 18y^3 + 21y^2$ 41. $8z^2 - 10z - 63$ 42. $4x^2$
$- 20xy + 25y^2$ 43. $9a^2 - 64b^2$ 44. $6t^4 + 19t^2 - 20$ 45. $8r^4 - 6r^3$
$- 17r^2 + 39r - 30$ 46. $6x^4 - 34x^3 + 40x^2 + 3x - 5$ 47. $16t$
48. $-10x^{2k+1} + 20x^{k+1} - 10$ 49. $3 \cdot 5^2$ 50. $2^3 \cdot 3^2 \cdot 7 \cdot 11$
51. 28 52. $6x^9y^2$ 53. $9p^2(2p^3 - 4p + 3)$
54. $-4x^5y^2z^2(2x^2z - 5y)$ 55. $(5a + b)(3x - y)$
56. $x^{2n}(x^{3n} - 3x^n + 1)$ 57. $(s + 1)(r + 2)$ 58. $(x - 7)(x^2 + 2)$
59. $(5a + 4b)(a - 3b)$ 60. $(k - 1)(p - 1)$ 61. $(x + 2)(x - 4)$
62. $(p - 6)^2$ 63. $(y - 4z)(y + 7z)$ 64. $(2m - 3)(m + 4)$
65. $(9a + 4b)(2a - 5b)$ 66. $xy^2(2x - 3y)(3x - 4y)$
67. $(2r^2 - 5)(4r^2 + 3)$ 68. $(7y^{3k} + 4)(3y^{3k} - 1)$
69. $(8x + 3)(8x - 3)$ 70. $(2a + 11b)(2a - 11b)$
71. $10(p^2 + 4k^2)(p + 2k)(p - 2k)$
72. $xy(x^2y^2 + 25)(xy + 5)(xy - 5)$ 73. $-(z + 10)^2$ 74. $5(3a - 2b)^2$
75. $(p + 3)(p^2 - 3p + 9)$ 76. $(2x - 5y)(4x^2 + 10xy + 25y^2)$
77. $2(t - 4)(t^2 + 4t + 16)$ 78. $(x + y - 1)(x - y + 1)$
79. $(m - 4)(m^2 + 22m + 196)$ 80. $(5x^n + 9)(5x^n - 9)$
81. $6rt^2(2r^2 - 4rt + 3t^2)$ 82. $(x + 3)(x + 2)(x - 2)$ 83. $(2p - 7)^2$
84. $3(z - 1)(z^2 + z + 1)(z + 2)(z^2 - 2z + 4)$
85. $-(y^2 + 9)(y + 3)(y - 3)$ 86. $5a(ab + 6)(a^2b^2 - 6ab + 36)$
87. $\{-2, -5\}$ 88. $\{2, -3\}$ 89. $\{0, \frac{5}{4}\}$ 90. $\{3, -3\}$ 91. $\{1, -\frac{9}{5}\}$
92. Width 5 m, length 14 m   93. 4 sec

Chapter 3 Test

1. 1.6×10^5 2. 7.59×10^{-4} 3. $\frac{3}{4}$ 4. 1 5. $\frac{27}{m^4}$ 6. $-40x^6y^6$
7. 29 8. $\frac{8b^4}{3a^2}$ 9. $5x^3 - 2x^2 - 2x + 13$ 10. $2p^2 + 5p + 9$ 11. y^2
$- 8y - 28$ 12. $3r^4 + 2r^3 - 23r^2 + 41r - 28$ 13. -1 14. $6p^3 -$
$6p^2 + 4p + 18$ 15. 1.5×10^{20} 16. $(y + 2)(y - 6)$ 17. $(a + 3b)^2$
18. $(4p - 3)(16p^2 + 12p + 9)$ 19. $(s - 5)(r - 2)$
20. $6z^2(2z - 3)(2z + 5)$ 21. $(t^2 + 4)(t + 2)(t - 2)$ 22. $5(m + 2)(m^2$
$- 2m + 4)$ 23. $(x - 4 + y)(x - 4 - y)$ 24. $\{-4, \frac{3}{2}\}$ 25. $\{0, 3\}$
26. $\{\frac{2}{3}, -\frac{2}{3}\}$ 27. 10 or -1

Cumulative Test for Chapters 1, 2, and 3

1. $\varnothing, \{1\}, \{2\}, \{3\}, \{1, 2\}, \{1, 3\}, \{2, 3\}, \{1, 2, 3\}$ 2. $\{x \mid x$ is an integer$\}$ 3. $\{2, 3, 4, 5\}$ 4. $x \le 7$
5.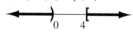
6. (a) Symmetric of equality,   (b) Commutative of addition,
(c) Distributive   7. $\frac{3}{2}$ 8. 110 9. $(3, \infty)$ 10. False   11. $9.13 \times$
10^{-5} 12. $2x^3 - 11x^2 + 29x - 35$ 13. -41 14. $P = \frac{A}{1 + rt}$
15. $(-\infty, 0) \cup [4, \infty)$

16. $6x^{13}$ 17. $\frac{y^7}{3x^5}$ 18. $(5x + 4)(5x - 4)$

19. $(m + 3)(m^2 - 3m + 9)$ 20. $(x - 7)(x^2 + 4)$ 21. $(4p - 5)(2p +$
$3)$ 22. $\{-3\}$ 23. $(-\infty, -6)$ 24. $\{6, 1\}$ 25. $(-4, \frac{20}{3})$ 26. $\{0, 2\}$
27. $\{-4, \frac{1}{3}\}$ 28. 15.6 hr   29. 23 nickels, 17 dimes   30. $\frac{1}{2}$ hr
31. 9 and 11, or -9 and -11

Chapter 4

Exercises 4.1

1. Rational expression   3. Factor   5. -1 7. False   9. False   11. $x = 0$ 13. $y = 6$ or $y = -6$ 15. $y = 2$ or $y = -5$ 17. $z = 0$ or $z = 1$
19. None   21. None   23. C   25. B   27. $\frac{5}{9}$ 29. $\frac{4b^2}{7}$ 31. $\frac{1}{3x^3y}$
33. $\frac{1}{6}$ 35. $\frac{1}{6}$ 37. $\frac{r - 10}{r + 10}$ 39. $\frac{1}{(w - 8)^4}$ 41. $\frac{x - y}{x + y}$ 43. -1
45. 1 47. $\frac{-3(t - 3)}{2t}$ 49. $\frac{3p + 8q}{p + 2q}$ 51. $\frac{-2(z - 7k)}{3(2z + k)}$
53. $-\frac{r^2 + rs + s^2}{r + s}$ 55. $\frac{y + z}{y - z}$ 57. $\frac{a^2 + 3ab + 9b^2}{a - 4}$ 59. $\frac{1}{p + 5}$
61. $2x + h$ 63. (a) $108°$, (b) $120°$ 65. (a) 506 vibrations/sec, (b) 462 vibrations/sec   67. You cannot cancel the 3's because they are terms, not factors. (a) If $x = 9$, then $\frac{x + 3}{3} = x$ becomes $4 = 9$, which is false. (b) If $x = 9$, then $\frac{x + 3}{3} = x + 1$ becomes $4 = 10$, which is false.
69. $3r^2 - 5$ 71. $\frac{5}{y^n - 9}$

Exercises 4.2

1. Reciprocals   3. $\frac{P \cdot R}{Q \cdot S}$ 5. $\frac{7a}{6}$ 7. $\frac{2y}{35x^3}$ 9. $\frac{7m^2}{20n}$ 11. $\frac{12}{7k^3}$
13. $-\frac{8b^5}{3a^8c^{15}}$ 15. $\frac{30z}{7}$ 17. $\frac{xy^3}{xy - 1}$ 19. $\frac{1}{k^2}$ 21. $-a(a + 1)$
23. $9p(p - 3)$ 25. 4 27. $\frac{1}{8s^2}$ 29. $-(x + 6y)$ 31. $\frac{m - 4}{m - 2}$
33. $\frac{c + d}{3c - d}$ 35. $\frac{2z}{z + 4}$ 37. $\frac{3w + 4}{4w + 3}$ 39. $\frac{k(k - m)}{5}$
41. $\frac{3(q - 4)^2}{7(q + 3)^2}$ 43. $x - y$ 45. $\frac{b - a}{x(x + y)}$ 47. $\frac{1}{s^2 - st + t^2}$
49. $\frac{5x + 7}{5x + 1}$ 51. $\frac{-y}{2y - 5}$ 53. (1) Factor numerators and denominators. (2) Divide out common factors. (3) Multiply the remaining factors.
55. It is usually easier. Also, if you multiply first, the numerator and/or the denominator of the resulting fraction may be too complicated for you to factor. Therefore, if there are common factors, you will not be able to find them, cancel them, and simplify your answer.

Exercises 4.3

1. False   3. True   5. False   7. $\frac{P - R}{Q}$ 9. Least common denominator   11. $\frac{3x + y}{5}$ 13. $\frac{2}{3r^2}$ 15. 1 17. $-q^2$ 19. $t + 5$
21. $\frac{3}{x - 3}$ 23. $\frac{x^2 + 9}{(x - 3)(x + 3)}$ 25. $\frac{a^2}{a - 6}$ 27. $\frac{1}{x - y}$

29. (a) $15a$, (b) $\dfrac{8}{15a}$ 31. (a) b^2, (b) $\dfrac{2b-1}{b^2}$ 33. (a) $9x^2$,

(b) $\dfrac{4x+21}{9x^2}$ 35.   (a) y,   (b) $\dfrac{6-3y^2}{y}$ 37.   (a) $360x^3y^2$,

(b) $\dfrac{20x^2+27y}{360x^3y^2}$ 39. (a) $z(z-4)$, (b) $\dfrac{-16}{z(z-4)}$ 41. (a) $p+9$,

(b) $\dfrac{9p}{p+9}$ 43. (a) $q(q+2)$, (b) $\dfrac{4}{q}$ 45. (a) $(h-10)(h-5)$,

(b) $\dfrac{5h}{(h-10)(h-5)}$ 47. (a) $(r-4)(r+1)$, (b) $\dfrac{4(2r-3)}{(r-4)(r+1)}$

49. (a) $5(t+10)$, (b) $-\frac{2}{5}$ 51. (a) $2(z+9)(z-9)$, (b) $\dfrac{-1}{2(z-9)}$

53. (a) $m(m-2)^2$, (b) $\dfrac{-12}{m(m-2)^2}$ 55. (a) $a(a+4b)(a-4b)$,

(b) $\dfrac{a^3+ab-4b^2}{a(a+4b)(a-4b)}$ 57. (a) $(k+4)(k-3)(k+2)$,

(b) $\dfrac{2(k+3)}{(k+4)(k-3)(k+2)}$ 59. (a) $m-1$, (b) $\dfrac{-m^2+3m+2}{m-1}$

61. (a) $(r+1)^2$, (b) $\dfrac{r^2}{(r+1)^2}$ 63. (a) $3y(2y+1)$, (b) $\dfrac{2(y+1)}{3y}$

65. (a) $4(x+2)(x-2)$, (b) $\dfrac{x+8}{4(x+2)}$ 67. (a) $m(m+4)(m+1)^2$,

(b) $\dfrac{-3m^2+2m-1}{m(m+4)(m+1)^2}$ 69. (a) $(x-1)(x-3)(x+2)$,

(b) $\dfrac{2}{x-1}$ 71. $\dfrac{a}{b}+\dfrac{c}{d}=\dfrac{ad}{bd}+\dfrac{cb}{db}=\dfrac{ad+bc}{bd}$ 73. $\dfrac{2(n+1)}{n(n+2)}$

75. (1) Add or subtract the numerators, keep the common denominator. (2) If possible, factor numerator and denominator and divide out common factors.   77. (1) Factor the denominators to find the LCD. (2) Write each expression with the LCD. (3) Multiply out the numerators, leave the denominators in factored form. (4) Add or subtract the numerators, keep the common denominator. (5) If possible, factor the numerator and divide out common factors.

Exercises 4.4

1. Complex fraction   3. Secondary fractions   5. True   7. $\dfrac{m+1}{3(m-1)}$

9. $\dfrac{2y^2}{x}$ 11. $\dfrac{3z}{2k}$ 13. $\frac{45}{4}$ 15. $\dfrac{t-r}{t+r}$ 17. $\dfrac{x}{y}$ 19. $\dfrac{4m-1}{m}$ 21. $-\dfrac{1}{x}$

23. $-7(a+b)$ 25. $\dfrac{c-3}{18}$ 27. $\dfrac{d+8}{d+1}$ 29. $\dfrac{p+1}{p+3}$

31. $\dfrac{k(k^2+4k+6)}{3k^2+16k-12}$ 33. $\dfrac{x^2-x+1}{x}$ 35. $\dfrac{22x+9}{12x+5}$

37. (a) Simplify numerator and denominator separately. Then invert the denominator and multiply. (b) Multiply both numerator and denominator by the LCD of all the secondary fractions.   39. $\dfrac{(y+1)(y-1)}{y^2+1}$

41. $\dfrac{1}{t(t+1)}$ 43. $\dfrac{9m^2n^2}{3n-m}$ 45. $\dfrac{xy}{y+x}$ 47. $\frac{4}{13}$ 49. 4.8 mi/hr

51. 3.75 mi/hr

Exercises 4.5

1. True   3. True   5. Proportion   7. Extraneous solution   9. $\{20\}$
11. $\{3\}$ 13. $\{1, -1\}$ 15. $\{-\frac{3}{2}\}$ 17. $\{1\}$ 19. $\{\frac{3}{5}\}$ 21. $\{-7\}$
23. $\{2, -\frac{1}{2}\}$ 25. $\{6\}$ 27. $\{0, -\frac{11}{2}\}$ 29. $\{0\}$ 31. $\{-18\}$

33. $\{1, 2\}$ 35. $\{-8\}$ 37. $\{2, -\frac{3}{5}\}$ 39. $\varnothing$ 41. $\{-24\}$
43. $\{1, -3\}$ 45. $\{2, -5\}$ 47. $\{-1\}$ 49. $\{-2\}$ 51. $\varnothing$ 53. $\{\frac{1}{2}\}$
55. $\{0\}$ 57. $\{10, -\frac{1}{2}\}$ 59. $\{-3\}$ 61. No. You cannot invert the *terms* of an equation (unless there is only one term on each side).   63. In Problem 1, you *solve* an equation and obtain a solution (a number). In Problem 2, you *add* two rational expressions and obtain a third rational expression.   65. $\dfrac{7x}{10}$ 67. $\{1\}$ 69. $\dfrac{(m+2)(m-2)}{(m-1)(m-4)}$ 71. $\{0, 9\}$

73. If $x=6$, then $\dfrac{x}{x+3}=\dfrac{2}{3}$ and $1-\dfrac{2}{x}=\dfrac{2}{3}$.

Exercises 4.6

1. Formula   3. $\frac{1}{3}$ 5. $\frac{d}{r}$ 7. 12   9. 3.75   11. 13.2   13. 224.70
15. Multiply by x; subtract $3x$; factor out x; divide by $y-3$.
17. $V=\dfrac{Q}{C}$ 19. $g=\dfrac{Fd^2}{m_1m_2}$ 21. $P_1=\dfrac{T_1P_2V_2}{V_1T_2}$ 23. $r=\dfrac{E-IR}{I}$
25. $R_2=\dfrac{RR_1}{R_1-R}$ 27. $y=\dfrac{1}{x+2}$ 29. $r=\dfrac{d}{1-dt}$ 31. $x=\dfrac{3(y+1)}{y-1}$ 33. 2   35. 5, 7   37. $1\frac{1}{5}$ hr   39. 15 days, 30 days   41. 6 hr, 12 hr   43. 9 min   45. 2 mi/hr   47. 50 mi/hr, 80 mi/hr   49. $4\frac{1}{3}$ mi/hr   51. (1) Write down the unknown quantities. Call one x and write the others in terms of x. (2) Write an equation involving x. (3) Solve the equation. (4) Check in the words of the original problem.

Exercises 4.7

1. True   3. True   5. $4x^2+2x-3$ 7. $3y-6+\dfrac{4}{y}$ 9. z^2+z-1
11. $-a^2+4a-\dfrac{5}{3}+\dfrac{3}{a^2}$ 13. $3n-\dfrac{2}{m}-\dfrac{1}{3mn}$
15. $\dfrac{8}{s^2}-\dfrac{6}{r^2}+\dfrac{5s}{4}-1$ 17. $3x+2$ 19. $2m+5$ 21. k^2+2k+3
23. $5y^2+y+3$ 25. $r^2-r+5-\dfrac{30}{4r+3}$ 27. $3p^2-p+4+\dfrac{6}{3p+1}$ 29. t^2+3 31. $2x+7+\dfrac{10x}{x^2-x-3}$ 33. $3a^2+a-3$
35. $4m^2+5m+3+\dfrac{9m-8}{5m^2-3m}$ 37. $4y-13-\dfrac{9}{2y^2+7}$ 39. $3x^2-8-\dfrac{40}{3x-5}$ 41. $5z+3-\dfrac{3z-11}{4z^2+z}$ 43. Multiply the divisor and the quotient and add the remainder. The result should be the dividend.
45. To ensure that like terms line up in the same vertical column
47. $4p^2+10p+25$ 49. $\frac{3}{2}p+\frac{1}{2}$ 51. $\dfrac{5}{4}r^3+\dfrac{1}{2}r+\dfrac{3}{2}+\dfrac{r+3}{2r^2-2}$
53. $1+\dfrac{1}{x-1}$ 55. $\dfrac{x+(x+1)+(x+2)}{3}=\dfrac{3x+3}{3}=x+1$
57. t^2+4 cm/sec

Chapter 4 Review Exercises

1. $x=0$ 2. $y=\frac{9}{4}$ 3. $p=5$ or $p=-3$ 4. None   5. $-\dfrac{5x^4}{4y^2}$
6. $6z$ 7. $\dfrac{5}{3t-4}$ 8. $\dfrac{b-a}{b+a}$ 9. $\dfrac{6p-5q}{3p-q}$ 10. $\dfrac{1}{m^2-2mn+4n^2}$

11. $\dfrac{x^3}{3}$ 12. $\dfrac{-b^6}{9a^5}$ 13. $\dfrac{25z}{3}$ 14. $\dfrac{p^4}{2(p-3)}$ 15. $\dfrac{x-4}{x-2}$

16. $c(c-d)$ 17. $(y-5)(y^2-5y+25)$ 18. $\dfrac{6r}{2t-5}$ 19. $\dfrac{1}{4x^2}$

20. 1 21. $\dfrac{5a+12}{8a^2}$ 22. $\dfrac{15y^2-4x}{80x^2y^3}$ 23. $\dfrac{-6p}{p+6}$ 24. $\dfrac{10}{h-8}$

25. $\dfrac{y^2+5y-4}{2(y-5)(y+5)}$ 26. $\dfrac{4b}{(a-b)(a+b)}$

27. $\dfrac{k^2+5k+3}{(k+3)(k-2)(k+4)}$ 28. $\dfrac{r^2}{r+1}$ 29. $\dfrac{z+6}{z-2}$

30. $\dfrac{5m+n}{6m(2m+n)}$ 31. $\dfrac{y+x}{y-x}$ 32. $\dfrac{b-1}{a+1}$ 33. $\dfrac{3m-1}{m}$

34. $\dfrac{4p+1}{2p+1}$ 35. $\frac{1}{2}$ 36. -1 37. $\dfrac{x^2y^2}{(y+x)(y-x)}$ 38. zk

39. $\{-\frac{2}{3}\}$ 40. $\{\frac{1}{2}\}$ 41. $\{-1,\frac{1}{3}\}$ 42. $\{0\}$ 43. $\{6,-1\}$ 44. $\varnothing$

45. $\{-1\}$ 46. $\{-2\}$ 47. $P_2=\dfrac{T_2P_1V_1}{T_1V_2}$ 48. $f_1=\dfrac{ff_2}{f_2-f}$ 49. $r=$

$\dfrac{a-S}{l-S}$ 50. $\frac{40}{3}$ 51. 7 52. 10, 12 53. New machine 20 hr, old machine 30 hr 54. 24 min 55. Motorcycle 50 mi/hr, plane 380 mi/hr

56. $\frac{3}{2}x^2-\frac{1}{2}x+\frac{4}{3}$ 57. $2a^7+3b^3c^3-9ab^2c^2$ 58. $x-5$ 59. $3y^2$

$-4y+2$ 60. $3p+2+\dfrac{9}{2p-3}$ 61. $2z^2-z-1+\dfrac{2}{2z+1}$

62. $6m^2+4m+5+\dfrac{25m-11}{4m^2-5m}$ 63. $3r+2+\dfrac{1}{2r^2-3r+2}$

Chapter 4 Test

1. $x=\frac{5}{2}$ or $x=-\frac{5}{2}$ 2. $\dfrac{2y}{y-1}$ 3. $\dfrac{b^2}{7a^3}$ 4. $p(p-2)$ 5. $\dfrac{1}{2t(t-2)}$

6. $\dfrac{3x+2}{x-1}$ 7. $\dfrac{12p^2+5}{6p}$ 8. $\dfrac{x^2+5}{(x-1)(x+2)}$ 9. $\dfrac{3(3m-7)}{5m(m-4)}$

10. $\dfrac{y-6}{y-3}$ 11. $\frac{1}{10}$ 12. $\dfrac{x}{y-1}$ 13. $\{-2\}$ 14. $\{4,-2\}$ 15. $\{-1\}$

16. $\{-5\}$ 17. $m_1=\dfrac{Fd^2}{gm_2}$ 18. $t=\dfrac{d-r}{dr}$ 19. $3y-1-\dfrac{2}{y}$

20. $2x^2-3x+4$ 21. $p^2-3p+3+\dfrac{2p+5}{p^2-2}$ 22. $\dfrac{t}{1+t^2}$

23. 5, 20 24. $3\frac{3}{4}$ hr 25. 15 mi/hr

Chapter 5

Exercises 5.1

1. 81 3. 2 5. 4, -4 7. 1.1, -1.1 9. Radical sign 11. $b^n=a$
13. Index 15. The square roots of 25 are 5 and -5. The symbol $\sqrt{25}$ represents only the *principal* square root of 25, which is 5. 17. 18
19. -4.872 21. 6 23. -2 25. ± 9 27. Not a real number
29. $\pm\frac{2}{3}$ 31. 4 33. -5 35. -2 37. 7 39. 2 41. -6 43. Not a real number 45. 3 47. 1 49. A 51. A 53. True 55. 13
57. 7 59. -3 61. $|x|$ 63. y 65. m^2 67. $a^4|b|$ 69. $|p-5|$
71. No real number squared is -9. 73. x^5 75. y^9 77. a^3b^7 79. $6z$

81. $-3p$ 83. $3t^2$ 85. $13x^2y^8$ 87. $-6a^3b^6$ 89. $10p^{22}q^{31}$
91. $\frac{2}{5}r^2h^2$ 93. $(5x+9)^4$ 95. $x+2y$ 97. 24 ft/sec 99. 18 cm
101. 12 m^2

Exercises 5.2

1. $\sqrt[n]{a}$ 3. True 5. True 7. 4 9. -6 11. Not a real number
13. -2 15. 3 17. 0 19. $-\frac{1}{5}$ 21. 9 23. $\frac{8}{125}$ 25. 4 27. $\frac{1}{12}$
29. $\frac{1}{125}$ 31. $-\frac{1}{3125}$ 33. $\frac{3}{2}$ 35. 81 37. 30.567 39. $\sqrt{3x}$
41. $3\sqrt{x}$ 43. $\sqrt{p^2+4}$ 45. $10\sqrt[4]{y^3}$ 47. $\dfrac{1}{\sqrt[5]{32a^5b^5}}$ 49. $\dfrac{1}{\sqrt[8]{(r+3)^3}}$

51. C 53. D 55. 5 57. $x^{3/2}$ 59. y^2 61. $y^{3/10}$ 63. $p^{1/2}$ 65. $\dfrac{1}{k}$

67. $\dfrac{1}{t^9}$ 69. $\dfrac{1}{z^{7/3}}$ 71. $\dfrac{r^{1/5}}{s^{1/2}}$ 73. $64x^9y^{10}$ 75. $4p^4q^6$ 77. $\dfrac{m^2}{n^3}$

79. $x^2y^3z^6$ 81. $\dfrac{-3a^2}{cb^5}$ 83. $\dfrac{p^3q^4}{2}$ 85. $\dfrac{1}{h^{8/5}}$ 87. $\dfrac{x^{29/2}}{y^9}$

89. $p+p^{3/2}$ 91. $z-1$ 93. $m^{1/2}(m^{1/2}+1)$ 95. x^{2r} 97. $x^{r/2+1}$
99. 89 earth days 101. 686 earth days 103. $R=15$ 105. In Sec. 3.2, we stated the laws of exponents for integer exponents. In this section, we stated the laws of exponents for rational number exponents (which include integer exponents). 107. The correct method is (a), because $a^{m/n}=(\sqrt[n]{a})^m$ only when m/n is in lowest terms.

Exercises 5.3

1. $\sqrt[n]{a}\cdot\sqrt[n]{b}$ 3. 16, 25, 36, 49, 64, 81, 100, 121, 144, 169 5. False
7. $2\sqrt{3}$ 9. $4\sqrt{5}$ 11. $40\sqrt{5}$ 13. Cannot be simplified 15. $2\sqrt[3]{9}$
17. $2\sqrt[4]{3}$ 19. $3x\sqrt{5}$ 21. $24y^2\sqrt{2y}$ 23. $4p^3q\sqrt[3]{q}$ 25. $-3m^2\sqrt[4]{7m^3}$

27. $\dfrac{8}{11}$ 29. $\dfrac{\sqrt{19}}{10}$ 31. $\dfrac{\sqrt[3]{z}}{6}$ 33. $\dfrac{2\sqrt{5}}{9}$ 35. $\dfrac{7\sqrt{m}}{12}$ 37. $\dfrac{3r\sqrt{3r}}{8}$

39. $\dfrac{3r\sqrt[3]{t^2}}{10}$ 41. $\dfrac{2b\sqrt[3]{b}}{5a^3}$ 43. True 45. B 47. C 49. $\sqrt[4]{3}$ 51. $\sqrt[6]{x}$
53. $\sqrt[8]{26}$ 55. $\sqrt[3]{z}$ 57. $\sqrt{5}$ 59. $\sqrt[3]{9}$ 61. $\sqrt[3]{6xy^2}$ 63. $\sqrt{2t+5}$
65. The radicand contains a factor, namely, 3, raised to a power equal to the index. Note that $\sqrt{9x}=\sqrt{3^2x}=\sqrt{3^2}\sqrt{x}=3\sqrt{x}$. 67. The exponent of the radicand and the index have a common factor, namely, 3. Note that
$\sqrt[6]{x^3}=x^{3/6}=x^{1/2}=\sqrt{x}$. 69. $3x^2\sqrt{11}$ 71. $\dfrac{2\sqrt{6}}{5}$ 73. $\sqrt[12]{z}$

75. $-5k^{12}\sqrt[4]{2}$ 77. $\dfrac{2p\sqrt{7p}}{q^3}$ 79. $-3k\sqrt[3]{k^2}$ 81. $\dfrac{3t^3\sqrt{3}}{4r}$ 83. $5\sqrt[4]{k}$
85. No. If $a=9$ and $b=16$, then $\sqrt{a+b}=\sqrt{9+16}=\sqrt{25}=5$, but $\sqrt{a}+\sqrt{b}=\sqrt{9}+\sqrt{16}=3+4=7$. 87. (1) The radicand contains no factor raised to a power greater than or equal to the index. ($\sqrt[3]{x^4}$ violates this rule.) (2) The radicand contains no fractions. $\left(\sqrt{\dfrac{3}{4}}\text{ violates this rule.}\right)$ (3) The exponent of the radicand and the index have no common factor. ($\sqrt[6]{x^3}$ violates this rule.) (4) No denominator contains a radical. $\left(\dfrac{1}{\sqrt{2}}\text{ violates this rule.}\right)$ 89. $\dfrac{\pi\sqrt{3}}{2}$ sec 91. $16\sqrt{2}$ ft

Exercises 5.4

1. $\sqrt[n]{\dfrac{a}{b}}$ 3. a 5. $\sqrt{15}$ 7. $\sqrt{10x}$ 9. 2 11. $19x$ 13. $3p\sqrt{2}$
15. $5y+9$ 17. $48m$ 19. $-8k\sqrt{k}$ 21. $z^2\sqrt{z}$ 23. $2\sqrt{4a+6}$
25. $\sqrt{b^2+9b+20}$ 27. $p^2+2p+12$ 29. $50t+25$ 31. $2p$
33. $6\sqrt{22z}$ 35. $60m\sqrt{2}$ 37. $10a\sqrt[3]{a^2}$ 39. $-400t^3\sqrt{6}$ 41. $36\sqrt[3]{4}$

43. $3r^2t\sqrt[3]{r^2t}$ 45. $2p^4\sqrt[4]{7}$ 47. $5k\sqrt[5]{k}$ 49. $20x^{10}y^9\sqrt{3y}$ 51. 3×10^5
53. $\sqrt{6}$ 55. 5   57. $2\sqrt{17}$ 59. $11\sqrt{x}$ 61. $\dfrac{1}{3y}$ 63. $\dfrac{4k^3z^2}{3}$
65. $rs^2t^3\sqrt{5}$ 67. $\dfrac{y\sqrt{11}}{7x^4}$ 69. $3b^2$ 71. $8m^2\sqrt{6m}$ 73. $\dfrac{2r^2\sqrt[3]{9r}}{t}$
75. $\dfrac{xy^2}{6}$ 77. $z^7k^{11}\sqrt[4]{3z^2}$ 79. $\sqrt{7pq}$ 81. 2×10^4 83. Yes. No.
(a) 2, (b) $\sqrt{6}$, (c) $\sqrt[6]{32}$, (d) $\sqrt[6]{72}$ 85. $x\sqrt[3]{x^5}$ 87. $\sqrt[3]{x}$ 89. $\sqrt[4]{a^2b}$
91. $\sqrt[15]{a^{12}b^{10}}$

Exercises 5.5

1. Index; radicand   3. False   5. $8\sqrt{7}$ 7. $7\sqrt{15}$ 9. $-4\sqrt{2y}$
11. Cannot be simplified   13. Cannot be simplified   15. $-3\sqrt{19}$
17. $6\sqrt{m} + 5m$ 19. $9a\sqrt{5}$ 21. $-7p\sqrt{6p}$ 23. $3\sqrt{3}$ 25. $5\sqrt{5}$
27. $9\sqrt{2}$ 29. $3\sqrt{6x}$ 31. $24\sqrt{15}$ 33. $13\sqrt{2}$ 35. $\sqrt{10y}$ 37. $15\sqrt{p}$
$+ 8p$ 39. $-z\sqrt{3}$ 41. $11\sqrt{5k} - 5\sqrt{5m}$ 43. $-\sqrt{7} - 21\sqrt{2}$
45. $7r\sqrt{11r}$ 47. $\dfrac{134\sqrt{6}}{5}$ 49. $7\sqrt[3]{6} + 4\sqrt{6}$ 51. $7\sqrt[5]{7} - 5\sqrt[4]{7}$
53. $4\sqrt[3]{3}$ 55. $2\sqrt[3]{y}$ 57. $5\sqrt[3]{3p}$ 59. $-7\sqrt[3]{2}$ 61. $-8z\sqrt[3]{5}$ 63. $5\sqrt[4]{t}$
65. $k\sqrt[3]{3k}$ 67. $70\sqrt{5}$ 69. $50\sqrt{3}$ volt

Exercises 5.6

1. F   3. C   5. D   7. $5\sqrt{3} + 20$ 9. $12\sqrt{10} - 2\sqrt{5}$ 11. $4\sqrt{6} - 2$
13. 168   15. $-x + x\sqrt{y}$ 17. $30p\sqrt{2} - 24\sqrt{p}$ 19. $2 - \sqrt[3]{14}$
21. $m + 5 + \sqrt{m^2 + 5m}$ 23. $72 + 12\sqrt[3]{12}$ 25. $17 + 8\sqrt{5}$
27. $\sqrt{22} - \sqrt{33} - \sqrt{10} + \sqrt{15}$ 29. $7\sqrt{2} + 12\sqrt{7} - 5\sqrt{14} - 60$
31. $79 + 16\sqrt{15}$ 33. $25 - 2\sqrt{46}$ 35. 4   37. -1 39. $6m -$
$13\sqrt{m} - 5$ 41. $16x - 8\sqrt{xy} + y$ 43. $2p + 3\sqrt{10p} - 10\sqrt{14p} -$
$30\sqrt{35}$ 45. $3x - 36$ 47. $z^2 - 56$ 49. $64 + 80\sqrt{2ab} + 50ab$
51. $15r + 7\sqrt{2rt} - 4t$ 53. $x + 8\sqrt{x + 8} + 24$ 55. -7 57. $9\sqrt[3]{25}$
$- 4\sqrt[3]{5} - 31$ 59. $49 + 28\sqrt[3]{9} + 12\sqrt[3]{3}$ 61. $1 - \sqrt[3]{13}$
63. $\dfrac{1 + 4\sqrt{22}}{3}$ 65. $-2 - \sqrt{2}$ 67. $\dfrac{1 \pm \sqrt{2}}{4}$ 69. $\dfrac{4 \pm \sqrt{11}}{3}$
71. $-5 + 6\sqrt{7x}$ 73. $5 - \sqrt{10}$

Exercises 5.7

1. False   3. Rationalizing the denominator   5. $\sqrt[3]{9}$ 7. $\dfrac{9\sqrt{2}}{2}$
9. $\dfrac{\sqrt{21}}{7}$ 11. $4\sqrt{5}$ 13. $\dfrac{2\sqrt{66}}{11}$ 15. $7\sqrt{2}$ 17. $\dfrac{\sqrt{10kz}}{5z}$
19. $-\dfrac{2m\sqrt{6m}}{15}$ 21. $\dfrac{6x^3\sqrt{7y}}{7y}$ 23. $\dfrac{\sqrt[3]{25}}{5}$ 25. $4\sqrt[3]{2}$ 27. $\dfrac{m\sqrt[3]{4mn}}{2n}$
29. $\dfrac{\sqrt[3]{6p^2}}{3p}$ 31. Conjugates   33. $2 - \sqrt{x}$ 35. $\sqrt{2} - 1$
37. $\dfrac{4 + \sqrt{6}}{2}$ 39. $\dfrac{5(3 - \sqrt{3})}{3}$ 41. $\dfrac{3\sqrt{5} + 1}{44}$ 43. $6(\sqrt{7} - \sqrt{5})$
45. $\dfrac{-9(\sqrt{26} + 2)}{11}$ 47. $-7\sqrt{2} - 6\sqrt{3}$ 49. $\sqrt{k} - 2$
51. $\dfrac{6x + 2\sqrt{6xy} + y}{6x - y}$ 53. $\dfrac{16a - 12\sqrt{ab}}{16a - 9b}$ 55. The indexes of $\sqrt[3]{2}$
and $\sqrt{2}$ are different. The product of $\sqrt[3]{2}$ and $\sqrt{2}$ is $\sqrt[3]{4}$, which cannot be
simplified.   57. $\dfrac{2\sqrt{5}}{15}$ 59. $\dfrac{7\sqrt{3}}{6}$ 61. $\dfrac{11\sqrt{2}}{2}$ 63. $\dfrac{5\sqrt{66}}{6}$ amperes

Exercises 5.8

1. Radicand   3. True   5. {5}   7. {42}   9. {49}   11. Ø   13. {0}
15. {2}   17. {1, $\frac{1}{4}$}   19. {64}   21. {7}   23. Ø   25. {5, -5}
27. {2}   29. {3, -15}   31. {2}   33. {5}   35. {9}   37. Ø
39. {-3}   41. {100}   43. {4}   45. {9}   47. {6400}   49. {4, 20}
51. {1}   53. The principal square root of a real number cannot be a
negative number.   55. 7   57. $\dfrac{128}{\pi^2}$ ft   59. 600 ft   61. If $x = 11$, then
$\sqrt{x + 25} = 6$ and $x - 5 = 6$.

Exercises 5.9

1. Complex number   3. $a + bi$ 5. 15; 0   7. B, C   9. B, C, D
11. False   13. $3i$ 15. $i\sqrt{2}$ 17. $2i\sqrt{3}$ 19. $\dfrac{i\sqrt{7}}{5}$ 21. $5 + 24i$
23. $14 - 3i\sqrt{2}$ 25. True   27. True   29. True   31. $10 + 9i$ 33. 2
35. $-8 + 3i$ 37. $25 + i$ 39. $20i$ 41. $7 + i$ 43. 0   45. $7 - 5i$
47. $-4 - 9i$ 49. In $\sqrt{2}i$, the i is *not* under the radical sign. In $\sqrt{2i}$, the
i *is* under the radical sign.   51. Method 2, since $\sqrt{a}\sqrt{b} = \sqrt{ab}$ is not true
when both a and b are negative numbers.   53. $-2 + 23i$
55. $-12 + 34i$ 57. $-21 + 57i$ 59. 41   61. $6 + 10i$
63. $-\sqrt{21}$ 65. $4i$ 67. $-3\sqrt{6}$ 69. $5 - 12i$ 71. $4 + 4i$
73. $3 + 4i$ 75. $\frac{10}{13} + \frac{24}{13}i$ 77. $\frac{3}{17} - \frac{5}{17}i$ 79. $2 - \frac{4}{5}i$ 81. -5
83. 2   85. B   87. A   89. 1   91. -1 93. $-i$ 95. $-i$
97. $(2i)^2 = 4i^2 = 4(-1) = -4$, and $(-2i)^2 = 4i^2 = 4(-1) = -4$

Chapter 5 Review Exercises

1. $6, -6$ 2. $\frac{7}{12}, -\frac{7}{12}$ 3. 5   4. Not a real number   5. -10 6. ±11
7. -3 8. 2   9. $3y^3$ 10. $-2x^2y^3$ 11. $(x + 7)^2$ 12. $5x^{5n}y^{4m}$
13. $\frac{2}{9}$ 14. -27 15. $\frac{1}{5}$ 16. $\frac{1}{216}$ 17. $4\sqrt[5]{x^2}$ 18. $\sqrt[3]{y} - 7$ 19. x
20. $\dfrac{p^6}{8q^{12}}$ 21. $\dfrac{1}{p^{5/4}}$ 22. $\dfrac{x^{3/2}}{y^2}$ 23. $z^3 - 1$ 24. $2\sqrt{5}$ 25. $15\sqrt[3]{2}$
26. $-5a^2b\sqrt[3]{b}$ 27. $2x\sqrt{2}$ 28. $\dfrac{3m^2\sqrt{2m}}{11}$ 29. $\dfrac{3y^2\sqrt{y}}{4}$ 30. $\sqrt[12]{x}$
31. $\sqrt[3]{7xy^2}$ 32. $\dfrac{k^6\sqrt{3}}{3z^5}$ 33. $4\sqrt[4]{p}$ 34. $5\sqrt{2}$ 35. $13r$ 36. $7x\sqrt{2}$
37. $5ht^2$ 38. $15p^3\sqrt{p}$ 39. $16(3z + 1)$ 40. $2a^2\sqrt[5]{a^2}$ 41. 7
42. $2y^2\sqrt{3}$ 43. $\dfrac{5}{14k}$ 44. $a^2b^4c\sqrt[3]{7}$ 45. $\dfrac{pq\sqrt{pq}}{3}$ 46. $\sqrt{2}$ 47. $\sqrt[6]{m}$
48. $-3\sqrt{3x}$ 49. $3\sqrt{6}$ 50. $-2\sqrt{5}$ 51. $9y\sqrt{xy}$ 52. $\sqrt[3]{2p^2}$
53. $-25\sqrt[3]{3}$ 54. $15\sqrt{7} - 15$ 55. $2 - \sqrt[3]{12}$ 56. $x - y$
57. $t^2 + 2t\sqrt{10} + 10$ 58. $18\sqrt{5} - 15\sqrt{3} + 4\sqrt{15} - 10$ 59. $6p -$
$8\sqrt{6p + 5} + 21$ 60. $3 + \sqrt{7}$ 61. $\dfrac{-3 + 2\sqrt{5}}{2}$ 62. $\dfrac{-2\sqrt{3}}{3}$
63. $\dfrac{4\sqrt{15}}{5}$ 64. $\dfrac{11x^3\sqrt{6y}}{6y}$ 65. $7\sqrt[3]{4}$ 66. $\dfrac{\sqrt[3]{30k^2}}{3k}$ 67. $\dfrac{3 + \sqrt{5}}{4}$
68. $6 - \sqrt{35}$ 69. $\dfrac{9(p\sqrt{m} + m\sqrt{2p})}{p - 2m}$ 70. $\dfrac{3\sqrt{2} - 2\sqrt{3}}{6}$
71. $\dfrac{(20z + 1)\sqrt{5z}}{5z}$ 72. {8}   73. Ø   74. {5, -5}   75. {2}
76. {1}   77. {6}   78. Ø   79. {-2}   80. $9 - 12i$, real part = 9,
imaginary part = -12 81. $\dfrac{5\sqrt{3}}{4}i$, real part = 0, imaginary part =
$\dfrac{5\sqrt{3}}{4}$ 82. $13 + 18i$ 83. $-2i$ 84. $-5 - i$ 85. $14 + 5i$ 86. $-3 +$
$12i$ 87. -18 88. $-\frac{3}{8}i$ 89. $\frac{10}{13} - \frac{15}{13}i$ 90. $\frac{1}{2} - \frac{1}{2}i$ 91. $-i$ 92. 1
93. -1

Chapter 5 Test

1. -4 2. -3 3. $12x^5y^7$ 4. $2p^3$ 5. 100 6. $\frac{1}{5}$ 7. $\frac{p^6}{7m^2}$

8. $\frac{1}{3^{9/4}}$ 9. $12x\sqrt{3}$ 10. $3a^4b^2\sqrt[3]{ab^2}$ 11. $\frac{z^3\sqrt{7z}}{8}$ 12. $t\sqrt{6t}$ 13. 30

14. $3r^3\sqrt[4]{5}$ 15. $\frac{h^4\sqrt{h}}{z}$ 16. $\sqrt{x}$ 17. $\sqrt{3}$ 18. $-2y\sqrt[3]{y}$ 19. $6 + \sqrt{6}$

20. $100p - 9$ 21. $2\sqrt{3} + 7\sqrt{42} - 5\sqrt{6} - 35\sqrt{21}$ 22. $\frac{-7\sqrt{2}}{2}$

23. $\frac{\sqrt[3]{25}}{5}$ 24. $3\sqrt{3} + 5$ 25. $\{8, -2\}$ 26. $\{\frac{1}{4}\}$ 27. $\{13\}$

28. $8 - 5i$ 29. $13 - 40i$ 30. $\frac{14}{13} - \frac{5}{13}i$

Chapter 6

Exercises 6.1

1. $x = k$; $x = -k$ 3. False 5. $\{8, -8\}$ 7. $\{\sqrt{13}, -\sqrt{13}\}$
9. $\{\frac{5}{2}, -\frac{5}{2}\}$ 11. $\{6i, -6i\}$ 13. $\{10i, -10i\}$ 15. $\{2i\sqrt{2}, -2i\sqrt{2}\}$
17. $\{11, -7\}$ 19. $\{9 + 2\sqrt{7}, 9 - 2\sqrt{7}\}$ 21. $\{-2 + 2\sqrt{5}, -2 - 2\sqrt{5}\}$ 23. $\{1 + 4i, 1 - 4i\}$ 25. $\{-11 + i\sqrt{3}, -11 - i\sqrt{3}\}$
27. $\{-\frac{7}{3} + \frac{2}{3}i, -\frac{7}{3} - \frac{2}{3}i\}$ 29. $x^2 + 8x + 16 = (x + 4)^2$ 31. $y^2 - 4y + 4 = (y - 2)^2$ 33. $p^2 + 3p + \frac{9}{4} = (p + \frac{3}{2})^2$ 35. $r^2 - \frac{4}{3}r + \frac{4}{9} = (r - \frac{2}{3})^2$ 37. $\{5, -1\}$ 39. $\{3 + \sqrt{6}, 3 - \sqrt{6}\}$ 41. $\{\frac{1}{2} + \frac{\sqrt{7}}{2},$
$\frac{1}{2} - \frac{\sqrt{7}}{2}\}$ 43. $\{1, -\frac{1}{5}\}$ 45. $\{-1 + \frac{\sqrt{15}}{3}, -1 - \frac{\sqrt{15}}{3}\}$
47. $\{1 + \frac{2\sqrt{6}}{3}, 1 - \frac{2\sqrt{6}}{3}\}$ 49. Factoring is faster, but it does not always work. Completing the square is slower, but it always works.
51. No significant difference. $\{-\frac{1}{4} + \frac{\sqrt{17}}{4}, -\frac{1}{4} - \frac{\sqrt{17}}{4}\}$
53. $\{3 + 5i, 3 - 5i\}$ 55. $\{-2 + \sqrt{11}, -2 - \sqrt{11}\}$ 57. $\{5 + 2\sqrt{3},$
$5 - 2\sqrt{3}\}$ 59. $\{-7 + i\sqrt{10}, -7 - i\sqrt{10}\}$ 61. $\{\frac{1}{2} + 4i, \frac{1}{2} - 4i\}$
63. $\{\frac{4}{3} + \frac{2\sqrt{5}}{3}i, \frac{4}{3} - \frac{2\sqrt{5}}{3}i\}$ 65. $x = \pm 2a$
67. $x = \pm \frac{\sqrt{a^2 + 81}}{2}$ 69. $x = a \pm 1$ 71. $12, -12$ 73. 20 percent
75. If $x = 3 + \sqrt{6}$ or $x = 3 - \sqrt{6}$, then $x^2 - 6x + 3 = 0$.

Exercises 6.2

1. $\frac{-b \pm \sqrt{b^2 - 4ac}}{2a}$ 3. True 5. False 7. True 9. $\{\frac{1}{3}, -1\}$
11. $\{\frac{-1 + \sqrt{13}}{2}, \frac{-1 - \sqrt{13}}{2}\}$ 13. $\{\frac{-7 + 3\sqrt{5}}{2}, \frac{-7 - 3\sqrt{5}}{2}\}$
15. $\{2 + \sqrt{3}, 2 - \sqrt{3}\}$ 17. $\{\frac{-9 + \sqrt{33}}{8}, \frac{-9 - \sqrt{33}}{8}\}$ 19. $\{4\}$
21. $\{\frac{2 + \sqrt{3}}{2}, \frac{2 - \sqrt{3}}{2}\}$ 23. $\{\frac{-6 + \sqrt{6}}{5}, \frac{-6 - \sqrt{6}}{5}\}$

25. $\{\frac{1 + \sqrt{3}}{3}, \frac{1 - \sqrt{3}}{3}\}$ 27. $\{\frac{2 + \sqrt{7}}{3}, \frac{2 - \sqrt{7}}{3}\}$
29. $\{\frac{2 + \sqrt{14}}{2}, \frac{2 - \sqrt{14}}{2}\}$ 31. $\{\frac{-6 + \sqrt{46}}{2}, \frac{-6 - \sqrt{46}}{2}\}$
33. The number 0 is isolated on one side and the other side is in descending order. 35. All of them. No, it is easier to solve them by the square root property or by factoring. 37. $\{-1 + i, -1 - i\}$
39. $\{3 + 2i, 3 - 2i\}$ 41. $\{\frac{-5 + i\sqrt{7}}{4}, \frac{-5 - i\sqrt{7}}{4}\}$
43. $\{1 + i\sqrt{6}, 1 - i\sqrt{6}\}$ 45. $\{\frac{5 + 2i\sqrt{5}}{5}, \frac{5 - 2i\sqrt{5}}{5}\}$ 47. $\{i, \frac{3}{2}i\}$
49. 3.2 51. Width 8.6 m and length 12.8 m, or width 6.4 m and length 17.2 m 53. $x = 7.2$ in., or $x = 0.8$ in. 55. $\{2.5, -3\}$ 57. No real solution

Exercises 6.3

1. $\{11, -11\}$ 3. $\{0, \frac{5}{2}\}$ 5. $\{\frac{2}{3}, -3\}$ 7. $\{9, -1\}$ 9. $\{3\sqrt{2}, -3\sqrt{2}\}$
11. $\{9i, -9i\}$ 13. $\{\frac{11 + \sqrt{37}}{6}, \frac{11 - \sqrt{37}}{6}\}$
15. $\{\frac{4 + \sqrt{11}}{5}, \frac{4 - \sqrt{11}}{5}\}$ 17. $\{-1 + \frac{\sqrt{5}}{2}i, -1 - \frac{\sqrt{5}}{2}i\}$
19. Discriminant

21. Method	Advantages	Disadvantages
Factoring	Easy and fast	Not all equations are factorable; some equations are difficult to factor because their coefficients are large.
Square root property	Best method for solving equations of the form $x^2 = k$ and $(ax + b)^2 = k$	Many equations are not of this form.
Completing the square	Will solve any quadratic equation; useful in other areas of mathematics	Sometimes tedious to use.
Quadratic formula	Will solve any quadratic equation; useful in proving properties involving the solutions of a quadratic equation	Not as easy to use as factoring or the square root property.

23. Two rational solutions 25. Two irrational solutions 27. Two imaginary solutions 29. One rational solution 31. True 33. False
35. $(4x + 5)(3x + 2)$ 37. Cannot be factored using integers 39. Cannot be factored using integers 41. $(2r - 7)^2$ 43. Sum $= -\frac{3}{4}$, product $= -\frac{5}{2}$ 45. Sum $= 5$, product $= 2$ 47. Sum $= \frac{3}{2}$, product $= 0$
49. Sum $= 0$, product $= \frac{81}{16}$ 51. Yes 53. Yes 55. No
57. $x^2 - 13x + 36 = 0$ 59. $7x^2 + 5x - 2 = 0$ 61. $6x^2 + 5x = 0$
63. $x^2 - 45 = 0$ 65. $x^2 - 6x + 25 = 0$ 67. $k = 9$
69. $k = 8$ or $k = -8$

Exercises 6.4

1. True   3. $\{-6, -9\}$ 5. $\{\frac{8}{3}, -\frac{5}{4}\}$ 7. $\left\{\dfrac{1 + \sqrt{33}}{4}, \dfrac{1 - \sqrt{33}}{4}\right\}$

9. $\left\{\dfrac{7 + \sqrt{17}}{4}, \dfrac{7 - \sqrt{17}}{4}\right\}$ 11. $\{-15 + 5\sqrt{3}, -15 - 5\sqrt{3}\}$

13. True   15. $\left\{\dfrac{1 + \sqrt{33}}{4}\right\}$ 17. $\left\{\dfrac{13 + \sqrt{101}}{2}\right\}$ 19. $\{8 + 2\sqrt{7}\}$

21. $\{6 + 2\sqrt{5}\}$ 23. $\{1, 2\}$ 25. x^2 27. $\{1, -1, 2, -2\}$
29. $\{3, -3, \sqrt{2}, -\sqrt{2}\}$ 31. $\{1, -1, \frac{3}{2}, -\frac{3}{2}\}$
33. $\{16, 81\}$ 35. $\{27, -1\}$ 37. $\{6, -\frac{1}{5}\}$ 39. $\{\sqrt{3} + \sqrt{6}, -\sqrt{3} + \sqrt{6}, \sqrt{3} - \sqrt{6}, -\sqrt{3} - \sqrt{6}\}$ 41. $\{1, -1, 6i, -6i\}$
43. $\{2i, -2i, 5i, -5i\}$ 45. $\{3, -5\}$ 47. It is an equation of the form $au^2 + bu + c = 0$ $(a \neq 0)$, where u is some algebraic expression.   49. If it takes x hours to do a job, then the portion of the job done in 1 hr is $\dfrac{1}{x}$.
51. Fast pipe 7.53 hr, slow pipe 8.53 hr   53. Fast copier 2.58 hr, slow copier 3.58 hr   55. Machine 5.65 hr, shovel 7.65 hr   57. If $x = 6 + 2\sqrt{5}$, then $\sqrt{2x - 3} - \sqrt{x} = 1$. The number $6 - 2\sqrt{5}$ does not check.

Exercises 6.5

1. $a^2 + b^2 = c^2$ 3. A triangle with a right angle (90° angle)   5. In any right triangle, the square of the hypotenuse equals the sum of the squares of the two legs.   7. $x = 10$ 9. $x = 3\sqrt{7}$ 11. $x = 3$ 13. 122 mi
15. $20\sqrt{2}$ ft   17. $x = 12$ 19. $a = \pm\sqrt{b^2 - c^2}$ 21. $W = \dfrac{V^2}{R}$

23. $d = \pm\sqrt{\dfrac{gm_1m_2}{F}}$ 25. $r = \pm\sqrt{\dfrac{V}{P} - 1}$

27. $r = \dfrac{-\pi h \pm \sqrt{\pi^2h^2 + 2\pi T}}{2\pi}$ 29. $n = \dfrac{-1 \pm \sqrt{1 + 8S}}{2}$

31. 2 ft   33. 15 in. by 15 in.   35. 6 in., 8 in., 10 in.   37. $8\sqrt{2}$ ft
39. Ship A 15 mi/hr, ship B 20 mi/hr   41. Parents 50 mi/hr, kids 60 mi/hr   43. 1 mi/hr   45. 5.17 mi/hr

Exercises 6.6

1. True   3. 9 and -6 5. B   7. A
9. $(-\infty, -1) \cup (5, \infty)$

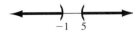

11. $[-4, -2]$

13. $(-\infty, -7] \cup [\frac{1}{2}, \infty)$

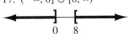

15. $[-4, 4]$

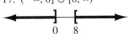

17. $(-\infty, 0] \cup [8, \infty)$

19. $(-\infty, 1 - \sqrt{3}) \cup (1 + \sqrt{3}, \infty)$

21. $\left[\dfrac{-2 - \sqrt{6}}{2}, \dfrac{-2 + \sqrt{6}}{2}\right]$

23. $(-\infty, -6) \cup (-2, 4)$

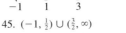

25. $[-1, \frac{3}{4}] \cup [9, \infty)$

27. $(0, 4) \cup (5, \infty)$

29. $(-\infty, 0] \cup \{2\}$

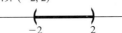

31. False   33. B   35. D
37. $(1, 6)$

39. $(-\infty, -5) \cup [\frac{1}{3}, \infty)$

41. $(-\infty, -6] \cup (-4, \infty)$

43. $(-1, 1] \cup [3, \infty)$

45. $(-1, \frac{1}{2}) \cup (\frac{3}{2}, \infty)$

47. $(-\infty, -1) \cup (0, 11)$

49. $(-2, 2)$

51. $\{-1\} \cup (7, \infty)$

53. (1) Replace the inequality symbol with an equals sign. Then solve the resulting equation to find the boundary values. (2) Graph the boundary values on a number line. This will divide the number line into regions. (3) Choose a number from each region and test it in the original inequality.

Using the results of these tests, label each region "True" or "False." (4) The numbers in those regions labeled "True" form the solution set. Include the boundary values in the solution set if the original inequality symbol is $\leq$ or $\geq$.

55. $ab > 0$ does not necessarily mean that $a > 0$ or $b > 0$. For example, $(-2)(-3) > 0$, but neither -2 nor -3 is greater than 0.   57. $1 \leq x \leq 4$

Chapter 6 Review Exercises

1. $\{10, -10\}$ 2. $\{\frac{4}{5}, -\frac{4}{5}\}$ 3. $\{9i, -9i\}$ 4. $\{2i\sqrt{3}, -2i\sqrt{3}\}$

5. $\left\{\dfrac{-10 + 3\sqrt{5}}{3}, \dfrac{-10 - 3\sqrt{5}}{3}\right\}$ 6. $\{\frac{2}{3} + \frac{1}{3}i, \frac{2}{3} - \frac{1}{3}i\}$

7. $x^2 - 12x + 36 = (x - 6)^2$ 8. $y^2 + \frac{2}{5}y + \frac{1}{25} = (y + \frac{1}{5})^2$

9. $\{1 + \sqrt{3}, 1 - \sqrt{3}\}$ 10. $\left\{3 + \dfrac{2\sqrt{21}}{3}, 3 - \dfrac{2\sqrt{21}}{3}\right\}$

11. $\left\{\dfrac{3}{4} + \dfrac{\sqrt{7}}{4}i, \dfrac{3}{4} - \dfrac{\sqrt{7}}{4}i\right\}$ 12. $x = \dfrac{-6a \pm 8b}{5}$ 13. $r = 0.3$

14. $\{\frac{1}{2}, -\frac{3}{4}\}$ 15. $\left\{\dfrac{-1 + \sqrt{17}}{4}, \dfrac{-1 - \sqrt{17}}{4}\right\}$

16. $\left\{\dfrac{-1 + \sqrt{145}}{4}, \dfrac{-1 - \sqrt{145}}{4}\right\}$ 17. $\left\{\dfrac{1 + 2\sqrt{3}}{2}, \dfrac{1 - 2\sqrt{3}}{2}\right\}$

18. $\left\{\dfrac{3 + \sqrt{57}}{2}, \dfrac{3 - \sqrt{57}}{2}\right\}$ 19. $\{\frac{1}{2}\}$

20. $\{6 + i, 6 - i\}$ 21. $\left\{\dfrac{3 + i\sqrt{6}}{3}, \dfrac{3 - i\sqrt{6}}{3}\right\}$ 22. $x = 7.85$ in. or $x = 1.15$ in.   23. $\{3, -3\}$

24. $\{\frac{7}{2}, -\frac{2}{3}\}$ 25. $\left\{\dfrac{3 + \sqrt{65}}{4}, \dfrac{3 - \sqrt{65}}{4}\right\}$ 26. $\{i, -i\}$ 27. Two imaginary solutions   28. One rational solution   29. $(2x + 1)(4x - 9)$
30. Cannot be factored using integers   31. Sum $= -\frac{1}{4}$, product $= -\frac{3}{2}$
32. Sum $= 6$, product $= 0$ 33. $9x^2 - 23x - 12 = 0$
34. $4x^2 - 3 = 0$ 35. $\left\{\dfrac{1 + \sqrt{133}}{6}, \dfrac{1 - \sqrt{133}}{6}\right\}$

36. $\left\{\dfrac{-7 + \sqrt{85}}{6}, \dfrac{-7 - \sqrt{85}}{6}\right\}$ 37. $\{2 + 2\sqrt{2}\}$ 38. $\left\{\dfrac{5 + \sqrt{5}}{2}\right\}$

39. $\{2, -2, 5, -5\}$ 40. $\{1, -1, 4i, -4i\}$ 41. $\{1, -27\}$
42. $\{-3 + 2\sqrt{3}, -3 - 2\sqrt{3}\}$ 43. New computer 7.12 hr,
old computer 9.12 hr   44. $x = 8$ 45. $x = 7$ 46. 58 mi   47. $g = \dfrac{v^2}{2r}$

48. $r = \pm\sqrt{R^2 - \dfrac{A}{\pi h}}$ 49. $x = y \pm 6$ 50. Width 11 m, length 29 m

51. 1 ft   52. 8 in., 15 in., 17 in.   53. Bike 20 mi/hr, moped 30 mi/hr
54. $(-\infty, -3) \cup (3, \infty)$

55. $[-2, \frac{5}{3}]$

56. $(-\infty, -\frac{3}{2}] \cup [0, \infty)$

57. $(4 - 2\sqrt{2}, 4 + 2\sqrt{2})$

58. $[-4, -1] \cup [6, \infty)$

59. $(-\infty, 0)$

60. $(-\infty, -2) \cup [5, \infty)$

61. $(\frac{3}{5}, \frac{3}{2})$

62. $(-\infty, -1) \cup (4, 5)$

63. $(-\infty, 0] \cup (3, 7]$

Chapter 6 Test

1. $x^2 - 8x + 16 = (x - 4)^2$ 2. Two irrational solutions   3. Sum $= \frac{1}{3}$, product $= \frac{3}{2}$ 4. $x^2 - 10x + 25 = 0$ 5. $\{2\sqrt{5}, -2\sqrt{5}\}$ 6. $\{3, -8\}$

7. $\{-3 + \sqrt{6}, -3 - \sqrt{6}\}$ 8. $\left\{-\dfrac{3}{2} + \dfrac{\sqrt{2}}{2}, -\dfrac{3}{2} - \dfrac{\sqrt{2}}{2}\right\}$ 9. $\{1, \frac{1}{3}\}$

10. $\left\{\dfrac{-1 + \sqrt{5}}{2}, \dfrac{-1 - \sqrt{5}}{2}\right\}$ 11. $\{-4 + \sqrt{17}, -4 - \sqrt{17}\}$

12. $\left\{\dfrac{5 + i\sqrt{23}}{4}, \dfrac{5 - i\sqrt{23}}{4}\right\}$ 13. $\{\frac{4}{3}, -\frac{15}{2}\}$ 14. $\left\{\dfrac{1 + \sqrt{61}}{6}\right\}$

15. $\{2, -2, 3, -3\}$ 16. $\{\sqrt{3}, -\sqrt{3}, i, -i\}$

17. $(-\infty, -1) \cup (\frac{7}{2}, \infty)$

18. $(-\infty, -3) \cup (3, 5)$

19. $(1, 2] \cup [4, \infty)$

20. $x = \pm 4a$ 21. $y = \pm\sqrt{\left(\dfrac{S}{\pi d}\right)^2 - 1}$ 22. $-1 + \sqrt{6}$ 23. 3 in., 4 in., 5 in.   24. Fast machine 7.12 hr, slow machine 9.12 hr   25. 16 in. by 16 in.

Cumulative Test For Chapters 4, 5, and 6

1. $x - 4$ 2. $x = \dfrac{1 + 2y}{1 - y}$ 3. $-\frac{3}{10} + \frac{11}{10}i$ 4. Two imaginary solutions

5. Sum $= -\frac{1}{3}$, product $= -\frac{7}{3}$ 6. $\dfrac{y + 1}{4y}$ 7. $\dfrac{p}{p + 3}$

8. $4m + 5 - \dfrac{1}{m}$ 9. $2x + 7 + \dfrac{10x}{x^2 - x - 3}$ 10. 9   11. $\dfrac{1}{6m^6}$

12. $20x^2y^2\sqrt{5y}$ 13. $2a^2\sqrt[3]{3a^2}$ 14. $3\sqrt{5m}$ 15. $6t - \sqrt{t} - 40$
16. $5\sqrt[3]{2}$ 17. $\sqrt{5} + 2$ 18. $\{-1, \frac{2}{5}\}$ 19. $\{9\}$ 20. $\{\frac{2}{3}, -4\}$

21. $\{2\sqrt{3}, -2\sqrt{3}\}$ 22. $\{4 + \sqrt{11}, 4 - \sqrt{11}\}$ 23. $\{-6 + \sqrt{11}, -6 - \sqrt{11}\}$

24. $[-5, 2]$

25. $(-\infty, -2) \cup (3, \infty)$

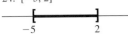

26. 10 in., 24 in., 26 in.   27. Anita 9 mi/hr, Tony 24 mi/hr   28. Fast machine 3.56 hr, slow machine 4.56 hr

Chapter 7

Exercises 7.1

1. Origin   3. Abscissa   5. $\sqrt{(x_2 - x_1)^2 + (y_2 - y_1)^2}$ 7. True
9. $A(5, 1)$, quadrant I; $B(-7, -3)$, quadrant III; $C(2, -2)$, quadrant IV; $D(0, 3)$, no quadrant (on the positive part of the y-axis); $E(-5, 4)$, quadrant II; $F(-6, 0)$, no quadrant (on the negative part of the x-axis); $G(0, 0)$, no quadrant

11, 13, 15, 17.

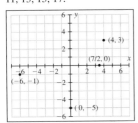

19.

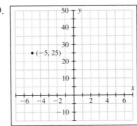

21.

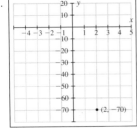

23. 5   25. $\sqrt{13}$ 27. $2\sqrt{34}$ 29. 1.3   31. $a^2 + b^2$ 33. 3; the points lie on the same horizontal line   35. 7; the points lie on the same vertical line   37. Collinear   39. Collinear   41. Not collinear   43. (4, 6)
45. $(3, -\frac{9}{2})$ 47. $(-\frac{3}{2}, -\frac{11}{2})$ 49. $(\frac{11}{15}, -\frac{5}{16})$ 51. Quadrant IV
53. Quadrant III   55. Quadrant II   57. $AC = BC = \sqrt{17}$ 59. $x = 1$ or $x = 5$ 61. $(-8, 7)$

63. *Ordered Pair*   *Interpretation*

(0, 0)   If taxable income is $0, Social Security tax is $0.
(20,000, 1530)   If taxable income is $20,000, Social Security tax is $1530.

(30,000, 2295)   If taxable income is $30,000, Social Security tax is $2295.

Exercises 7.2

1. $ax + by = c$ 3. False   5. True

7.
x	y
0	-4
1	-2
3	2
2	0

9.
x	y
0	-4
3	0
1	$-\frac{8}{3}$

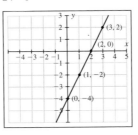

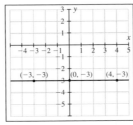

11.
x	y
-3	-3
0	-3
4	-3

13. An ordered pair of numbers that satisfies the equation; infinite number of solutions   15. Yes. If the x-intercept is 0, the point (0, 0) is on the line. Therefore, the y-intercept is 0 also.   17. We use the third point as a check. If the three points do not line up, we have made a mistake.

19.

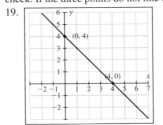

21.

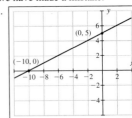

23.

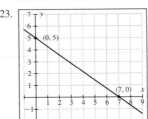

25.

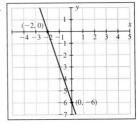

27.
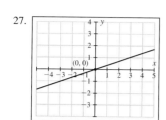

29.
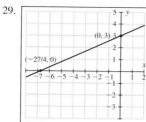

49. (a) −$400, (b) 80 pipes, (c) 280 pipes

51. (a)
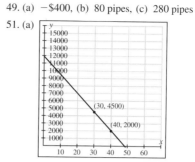

(b) 750 machines

31.
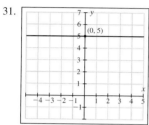

33.
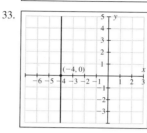

53.
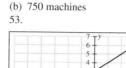

55.
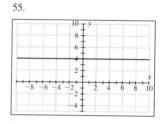

35.
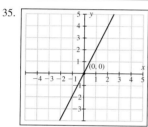

37.
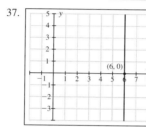

57.
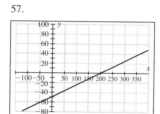

59.
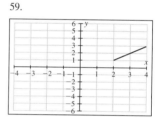

39.
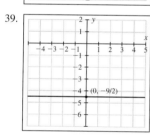

41.
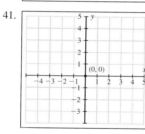

Exercises 7.3

1. $\dfrac{y_2 - y_1}{x_2 - x_1}$ 3. True   5. C   7. B   9. 2   11. −6   13. $\frac{3}{4}$ 15. $-\frac{7}{10}$
17. 0   19. Undefined   21. $\frac{4}{5}$ 23. $-\frac{3}{2}$ 25. 0   27. −3   29. 4
31. $-\frac{2}{7}$ 33. 0   35. Undefined   37. Collinear   39. Not collinear
41. Collinear

43. $y = 8 + 4x$
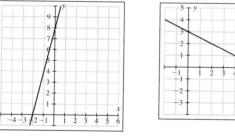

45. $x + 2y = 6$
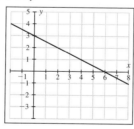

43.
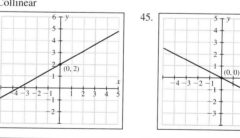

45.

47. $3x - 19 = 2$
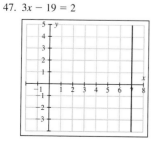

47.
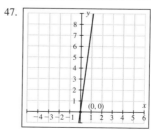

49.
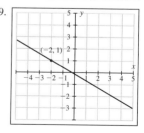

51.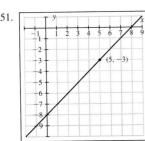

53.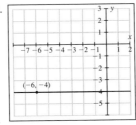

55.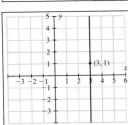

57. Parts (a) and (d) are correct. Part (b) is not correct because $m \neq \dfrac{y_2 - y_1}{x_1 - x_2}$. Part (c) is not correct because $m \neq \dfrac{x_2 - x_1}{y_2 - y_1}$. 59. (a) If the line is rising, and you travel from a point on the line with a smaller x-value to a point on the line with a larger x-value, both the rise and the run will be positive. (b) If the line is falling, and you travel from a point on the line with a smaller x-value to a point on the line with a larger x-value, the rise will be negative but the run will be positive.

61. $m_{AC} = m_{BD} = -\frac{4}{3}$, and $m_{AB} = m_{CD} = \frac{3}{2}$. 63. $t = 5$ or $t = -3$

65. 1300 ft^2 67.

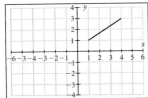

Exercises 7.4

1. $y - y_1 = m(x - x_1)$; point-slope form 3. $2x - y = 1$
5. $4x + 5y = -44$ 7. $7x - 2y = 0$ 9. $y = -10$ 11. $x = -4$
13. $x - y = 12$ 15. $7x - 4y = 21$ 17. $5x + 3y = -13$ 19. $x = 0$
21. $y = 2$ 23. $y = 6x - 3$ 25. $y = -x + 15$ 27. $y = \frac{1}{4}x$
29. $y = -\frac{5}{8}x - \frac{2}{5}$ 31. $y = 7$ 33. When you know one point on the line and the slope of the line 35. The slope is undefined. Yes, the slope is 0.
37. $m = 3, b = 2$ 39. $m = \frac{4}{5}, b = -1$

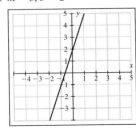 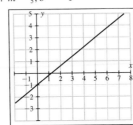

41. $m = -\frac{7}{2}, b = \frac{11}{2}$

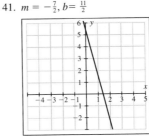

43. $m = 0, b = 8$

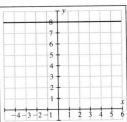

45. m is undefined, no y-intercept.

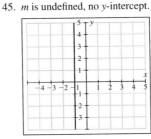

47. $m = \frac{1}{2}, b = 0$

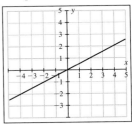

49. Slope; y-intercept 51. Parallel 53. Neither 55. Perpendicular
57. $3x + y = 6$ 59. $2x + 3y = -3$ 61. $x - 5y = 10$ 63. $x = 8$
65. (1) Solve for y. The coefficient of x is the slope. (2) Find any two ordered pairs that satisfy the equation and use $m = \dfrac{y_2 - y_1}{x_2 - x_1}$. (3) Graph the line and read the rise and run from the graph.
67. (a) $y = 3075x + 13{,}325$; (b) \$34,850 69. (a) $y = -15x + 750$;
(b) 225 members

71(a).

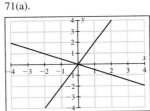

71(b).

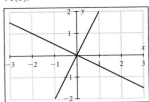

73.

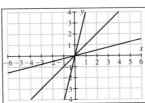

75. (a) \$360, (b) \$800, (c) \$640 77. 225 members

Exercises 7.5

1. True 3. D 5. A 7. The region of a plane that lies on one side of a line in the plane. A closed half-plane includes the boundary line, an open half-plane does not include the boundary line.

9.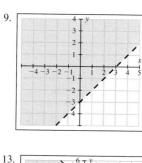

11.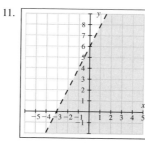

33.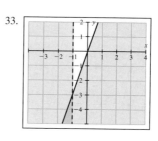

35.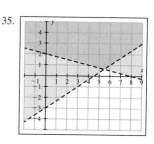

37. $-a < x < a$

13.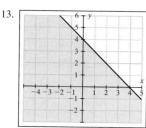

15.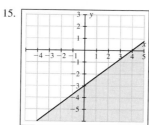

39.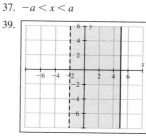

41.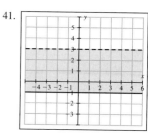

17.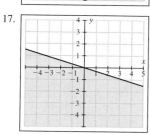

19.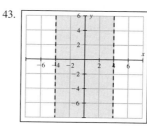

43.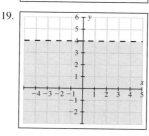

45.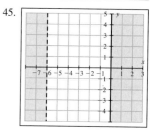

21. Replace the inequality symbol with an equals sign. If the inequality symbol is < or >, the boundary line is dashed; if it is ≤ or ≥, the boundary line is solid.   23. Both

25.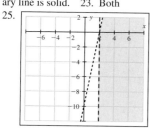

27.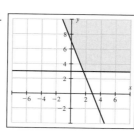

47.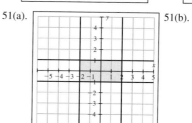

49.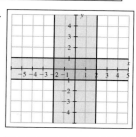

29.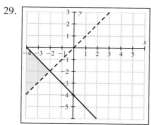

31.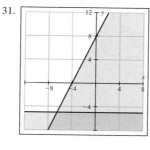

51(a). 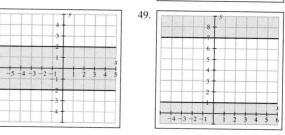 51(b).

53(a). 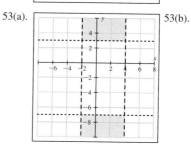 53(b).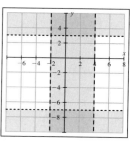

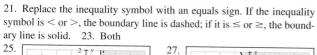

55. (200, 50) represents 200 g of protein and 50 g of fat.

57.

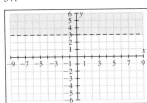

59.

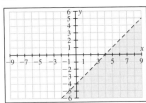

61.

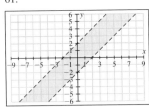

63.

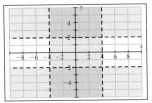

Exercises 7.6

1. General; specific   3. $p = kq$ 5. $r = ks^2$ 7. $w = \dfrac{k}{z^2}$ 9. $y = kxz^4$

11. $T = \dfrac{kr^2}{s^7}$ 13. $F = \dfrac{km_1m_2}{d^2}$ 15. (a) $k = 4$, (b) $y = 4x$, (c) $y = 10$

17. (a) $k = 3$, (b) $y = 3x^2$, (c) $y = \frac{75}{4}$ 19. (a) $k = 40$, (b) $y = \dfrac{40}{x}$,

(c) $y = 16$ 21. 30   23. 4   25. \$315 per oz   27. 256 ft   29. 18 hr
31. 9 Ω   33. 1536 lb   35. 12.6 in.   37. 4 times greater   39. A general variation equation, like $y = kx$, contains an arbitrary variation constant k. A specific variation equation, like $y = 3x$, contains the specific variation constant 3.   41. (1) Write the general variation equation. (2) Substitute the original values into the general variation equation. (3) Solve the equation from step 2 for k. (4) Substitute k into the general variation equation to get the specific variation equation. (5) Substitute the new values into the specific variation equation.   43. $y = 0.4x$ (a) 12 cm
(b) 62.5 kg

Chapter 7 Review Exercises

1. 13

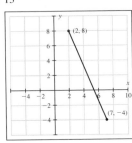

2. $2\sqrt{10}$

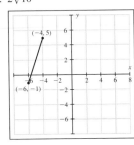

3. $(5, -\frac{3}{2})$ 4. $(\frac{1}{8}, -\frac{17}{16})$ 5. Collinear   6. Quadrant IV   7. $y = 5$ 8.
(a) No, (b) Yes   9. (a) Yes, (b) Yes

10.

x	y
0	0
−5	−2
5	2

11.

x	y
8	−3
8	0
8	4

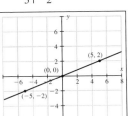

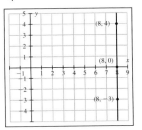

12.

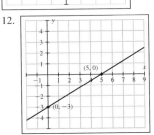

13.

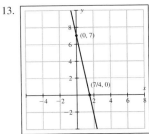

14.

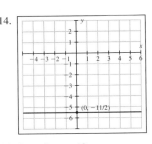

15.

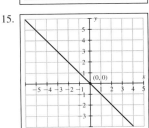

16. $2x + 6y = -10$

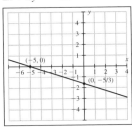

17. 6   18. $-\frac{2}{3}$ 19. $\frac{3}{2}$ 20. 3   21. $-\frac{4}{3}$ 22. −3   23. $\frac{4}{7}$ 24. 0
25. Not collinear   26. Collinear

27.

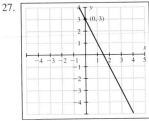

28.

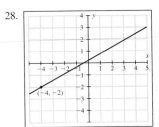

29.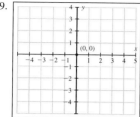

30. $t = 3$ 31. $2x - y = 5$ 32. $x + y = 2$ 33. $3x + 4y = -19$
34. $4x - 3y = -30$ 35. $x + 3y = 0$ 36. $x = 5$
37. $m = 2, b = -6$

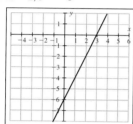

38. $m = -\frac{4}{7}, b = 3$

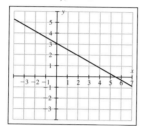

39. Perpendicular 40. Parallel 41. Neither

42. 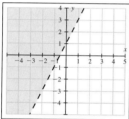 43.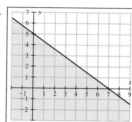

44. 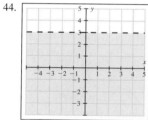 45.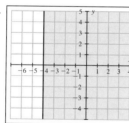

46. 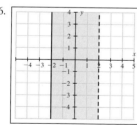 47.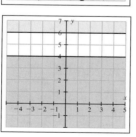

48(a). 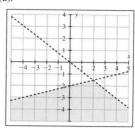 48(b).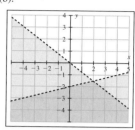

49(a). 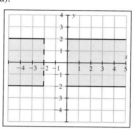 49(b).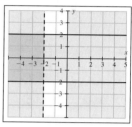

50. $s = kt$ 51. $A = \dfrac{k}{r^3}$ 52. $t = \dfrac{kxy}{z^2}$ 53. $y = \frac{27}{2}$ 54. $z = 4$
55. $V = 3\ \text{m}^3$ 56. $L = 2592\ \text{lb}$

Chapter 7 Test

1. $\frac{1}{2}$ 2. $(-4, \frac{13}{2})$ 3. $\sqrt{34}$ 4. Not collinear 5. Perpendicular
6.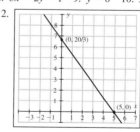

7. $m = \frac{3}{2}, b = -6$

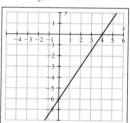

8. $3x - 2y = 1$ 9. $y = 8$ 10. $x = -7$ 11. $4x + 9y = -27$
12. 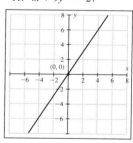 13.

14.

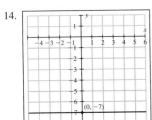

15.

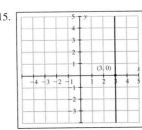

75. $R = 220 - a$; domain $= [18, 55]$; range $= [165, 202]$ 77. Domain $= (-\infty, \infty)$; range $= \{3\}$

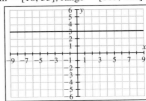

79. Domain $= (-\infty, \infty)$; range $= (-\infty, \infty)$

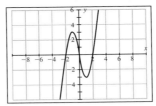

16.

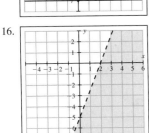

17.

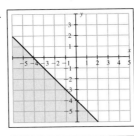

81. Domain $= (-\infty, \infty)$; range $= [-2, \infty)$

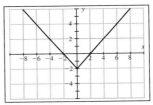

18.

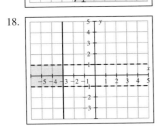

19.

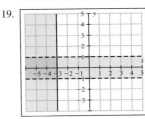

83. Domain $= [-4, \infty)$; range $= [0, \infty)$

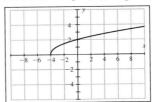

20. $y = 9$ 21. $z = 324$ 22. $R = 50\ \Omega$

Chapter 8

85. Domain $= [-3, 3]$; range $= [0, 3]$

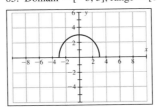

Exercises 8.1

1. Relation   3. Function   5. Domain $= \{2, -6, 0, 9\}$; range $= \{5, 4, 3\}$; is a function   7. Domain $= \{3, 1\}$; range $= \{9, 7, -8\}$; not a function   9. Domain $= \{1, 2, 3\}$; range $= \{1, 2, 3\}$; is a function   11. Domain $= \{-5, -6, -7\}$; range $= \{0\}$; is a function   13. Domain $= \{3, 7, -10\}$; range $= \{6, 7\}$; is a function   15. Domain $= \{-4, 5, 12\}$; range $= \{0, 1, -40\}$; not a function   17. Yes. No. A function is a relation in which each first coordinate corresponds to exactly one second coordinate. The set $\{(3, 5), (3, 7)\}$ is a relation since it is a set of ordered pairs, but this set is not a function.   19. Yes   21. Yes   23. Yes   25. Yes   27. No   29. No   31. No   33. D   35. A   37. $(-\infty, \infty)$ 39. $\{x | x \neq 3, -3\}$ 41. $(-\infty, \infty)$ 43. $\{x | x \neq 3, 4\}$ 45. $[-\frac{9}{2}, \infty)$ 47. $(-\infty, \infty)$ 49. $(-\infty, 5)$ 51. $(-\infty, \infty)$ 53. $[0, \infty)$ 55. $[-5, \infty)$ 57. Domain $= [-3, 3]$; range $= [-2, 2]$; not a function   59. Domain $= (-\infty, \infty)$; range $= [1, \infty)$; is a function   61. Domain $= (-\infty, \infty)$; range $= (-\infty, \infty)$; is a function   63. Domain $= (-\infty, \infty)$; range $= \{4\}$; is a function   65. Domain $= (-\infty, \infty)$; range $= (-\infty, -2] \cup [2, \infty)$; not a function   67. Domain $= (-\infty, \infty)$; range $= (-\infty, \infty)$; is a function   69. No. If a graph has more than one y-intercept, then the y-axis (a vertical line) intersects the graph at more than one point. Hence the graph would not be the graph of a function.   71. $A = x^2$; domain $= [0, \infty)$; range $= [0, \infty)$ 73. $y = 4.75x$; domain $= [0, 24]$; range $= [0, 114]$

Exercises 8.2

1. $f(x)$ 3. False   5. True   7. C   9. A, D   11. (a) 12, (b) 2, (c) $-4a$, (d) $4x + 4h$ 13. (a) 45, (b) $\frac{5}{2}$, (c) $6a^2 + 4a + 3$, (d) $6x^2 + 12xh + 6h^2 - 4x - 4h + 3$ 15. (a) 4, (b) 4, (c) 4, (d) 4   17. (a) 43, (b) -35, (c) -9, (d) $27b^3 - 18b^2 + 15b - 9$ 19. 3   21. $-x^2 + 3x + 11$; domain $= (-\infty, \infty)$ 23. 8   25. $3x^3 + 2x^2 - 27x - 18$; domain $= (-\infty, \infty)$ 27. $-\frac{2}{9}$ 29. $\dfrac{3x + 2}{x^2 - 9}$; domain $= \{x | x \neq 3, -3\}$ 31. (a) 25, (b) 54, (c) $10x + 5$; domain $= (-\infty, \infty)$, (d) $10x + 34$; domain $= (-\infty, \infty)$ 33. (a) 0, (b) 6, (c) $-x^2 + 4$; domain $= (-\infty, \infty)$, (d) $-x^2 + 4x + 2$; domain $= (-\infty, \infty)$ 35. (a) $\dfrac{x^2 + 5x - 5}{x - 1}$; domain $= \{x | x \neq 1\}$, (b) $\dfrac{x^2 + 5x}{x - 1}$;

domain = $\{x \mid x \neq 1\}$, (c) $\dfrac{x + 5}{x + 4}$; domain = $\{x \mid x \neq -4\}$,

(d) $\dfrac{6x - 5}{x - 1}$; domain = $\{x \mid x \neq 1\}$ 37. (a) $\dfrac{6}{x} + \sqrt{x + 2}$; domain

= $[-2, 0) \cup (0, \infty)$, (b) $\dfrac{6}{x}\sqrt{x + 2}$; domain = $[-2, 0) \cup (0, \infty)$,

(c) $\dfrac{6}{\sqrt{x + 2}}$; domain = $(-2, \infty)$, (d) $\sqrt{\dfrac{2x + 6}{x}}$;

domain = $(-\infty, -3] \cup (0, \infty)$ 39. $f(x) = x^{15}$, $g(x) = x^2 - 7$

41. $f(x) = \dfrac{1}{x}$, $g(x) = 4x + 5$ 43. 5   45. -27 47. 6   49. Undefined

51. 0   53. $t + a$ 55. $t + a - 7$ 57. 9   59. 10   61. 1

63. The notation $f(x) = 5x + 3$ gives the function a name (in this case f). Also, "$f(2) = 13$" is easier to write than "If $x = 2$, then $y = 13$."

65. 18   67. 100   69. 64   71. Yes   73. No   75. $f(a + b) = 3(a + b) + 5 = 3a + 3b + 5$, but $f(a) + f(b) = (3a + 5) + (3b + 5) = 3a + 3b + 10$ 77. $C(x) = 4x + 275$; $C(0) = 275$ means the weekly cost of producing no curtains is \$275; $C(235) = 1215$ means the weekly cost of producing 235 curtains is \$1215.   79. $A = 25\pi t^2$ 81. $D = 8t$ 83.

Exercises 8.3

1. Constant; $(0, k)$

3.
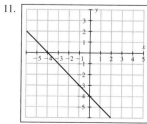

5.
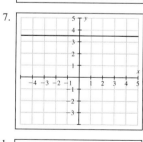

7.
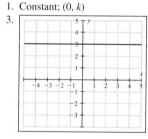

9.
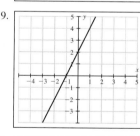

11.
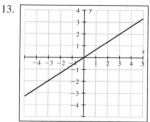

13.
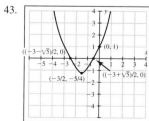

15. Quadratic; parabola; c; $\dfrac{-b}{2a}$; find $f\left(\dfrac{-b}{2a}\right)$ 17. Up; down

19. False   21. False   23. C   25. B

27.

x	$f(x)$
3	10
2	5
1	2
0	1
-1	2
-2	5
-3	10

29.

x	$g(x)$
2	-8
1	-2
$\frac{1}{2}$	$-\frac{1}{2}$
0	0
$-\frac{1}{2}$	$-\frac{1}{2}$
-1	-2
-2	-8

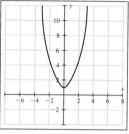

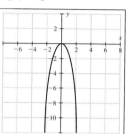

31.
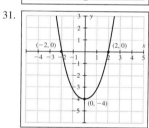

33.
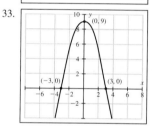

35.
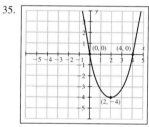

37.
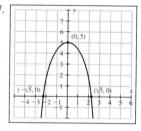

39.
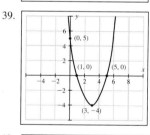

41.
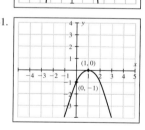

43.
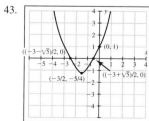

45.
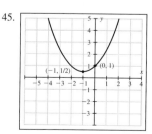

47. No *x*-intercepts, one *y*-intercept

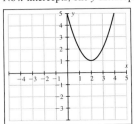

One *x*-intercept, one *y*-intercept

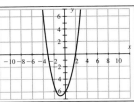

61.

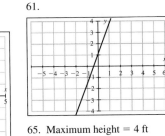

63.

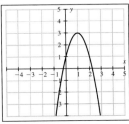

65. Maximum height = 4 ft

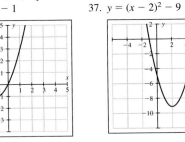

Two *x*-intercepts, one *y*-intercept

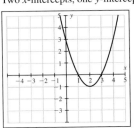

49. Linear function

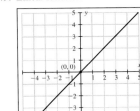

Exercises 8.4

1. 1.5 sec; 36 ft   3. Width 2 mi, length 4 mi; maximum area = 8 mi^2
5. $21 per day   7. F   9. A   11. E   13. B   15. C   17. Vertical; (h, k); $a > 0$; $a < 0$; $x = h$ 19. Down 5 units   21. Left 4 units
23. Right 1 unit, up 2 units   25. Thinner parabola—goes through $(1, 3)$ instead of $(1, 1)$ 27. Flip over *x*-axis and make fatter—goes through $(1, -\frac{1}{4})$ instead of $(1, -1)$.   29. Make fatter and move down 4 units.   31. Move left 1 unit, make thinner, and flip over *x*-axis.
33. Move right 3 units, flip over *x*-axis, and move up 1 unit.
35. $y = (x + 1)^2 - 1$

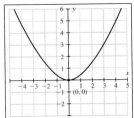

51. Quadratic function

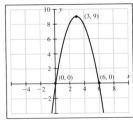

53. Constant function

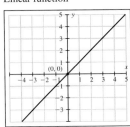

37. $y = (x - 2)^2 - 9$

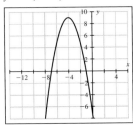

55. Quadratic function

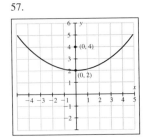

57.

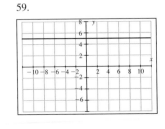

59.

39. $y = -(x + 4)^2 + 9$

41. $y = -2(x - 1)^2 + 3$

43. None. The vertex lies below the *x*-axis at $(4, -1)$, and the parabola opens downward (since $a < 0$).

45.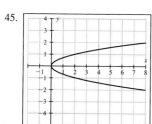

47.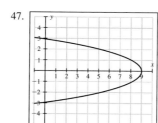

$(y + 2)^2 = 1$ 15. $x^2 + y^2 = 2$ 17. $C(0, 0), r = 2$ 19. $C(0, 0)$, $r = \frac{5}{2}$ 21. $C(0, 4), r = 4$ 23. $C(1, -6), r = 12$ 25. $C(-\frac{1}{4}, -\frac{3}{4})$, $r = \frac{1}{3}$ 27. $C(8, 8), r = 2\sqrt{2}$

29.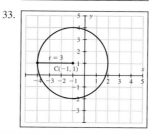

31.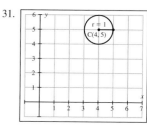

49.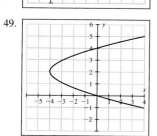

51.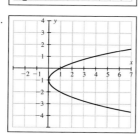

33.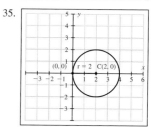

35.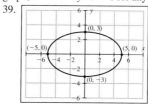

53.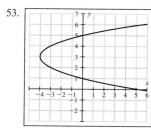

55. True

37. (a) A circle with center at $(0, 0)$ and radius 0; in other words, the graph is the origin. (b) The graph is the point $(4, -1)$. (c) There is no graph, since $x^2 + y^2 \neq -1$ for any point (x, y) with real coordinates.

39.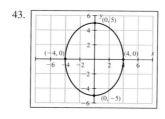

41.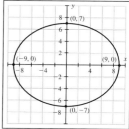

57.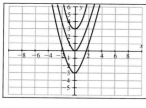

59.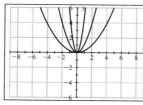

43.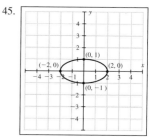

45.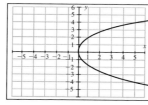

61.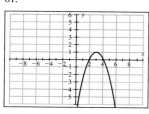

63.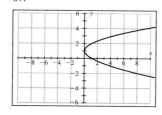

47.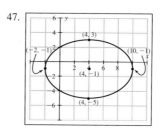

49.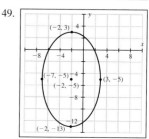

65.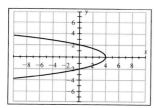

67.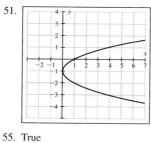

51. $x^2 + y^2 = 2.25$ 53. $(x + 1)^2 + (y + 6)^2 = 34$ 55. $\dfrac{x^2}{228^2} + \dfrac{y^2}{227^2} = 1$

Exercises 8.5

1. Circle; center; radius   3. Ellipse; foci   5. D   7. A   9. $(x - 3)^2 + (y - 5)^2 = 16$ 11. $(x - 4)^2 + (y + 10)^2 = 25$ 13. $(x + \frac{1}{2})^2 +$

57.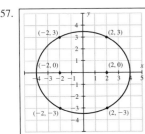

59.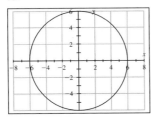

23. B 25. D 27. Circle

29. Hyperbola

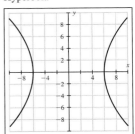

61.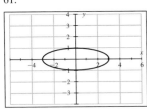

63.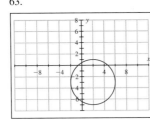

31. Parabola

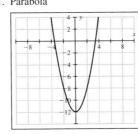

33. Ellipse

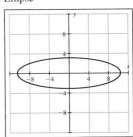

Exercises 8.6

1. Hyperbola; foci   3. a and $-a$; no y-intercepts   5. Asymptotes
7. D 9. C

11.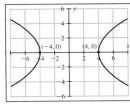

13.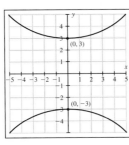

35. Circle

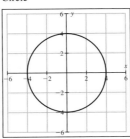

37. Parabola

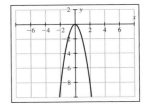

15.

17.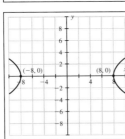

39.

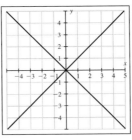

19.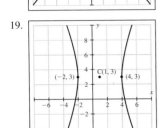

21.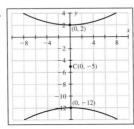

41. If the equation takes the form $\dfrac{x^2}{a^2} - \dfrac{y^2}{b^2} = 1$, the hyperbola opens right and left; if the equation takes the form $\dfrac{y^2}{b^2} - \dfrac{x^2}{a^2} = 1$, the hyperbola opens up and down.   43. If the face of the flashlight is parallel to the wall, the beam will produce a circle. If the face of the flashlight is perpendicular to the wall, the beam will produce one branch of a hyperbola. If the face of the flashlight is neither parallel to nor perpendicular to the wall, the beam will produce an ellipse or a parabola, depending on the angle between the face of the flashlight and the wall.

45.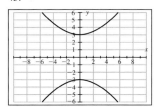

47.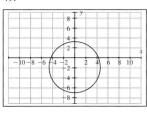

25.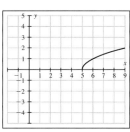

27.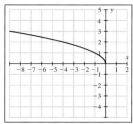

49.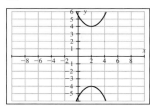

29. 7   31. 0   33. −4   35. −4   37. D   39. B

41. $C(x) = \begin{cases} 3 & \text{if } 0 < x \le 1 \\ 4 & \text{if } 1 < x \le 2 \\ 5 & \text{if } 2 < x \le 3 \\ 6 & \text{if } 3 < x \le 4 \end{cases}$

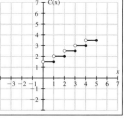

Exercises 8.7

1. Absolute value; $(-\infty, \infty)$; $[0, \infty)$ 3. Greatest integer; $(-\infty, \infty)$; $\{x \mid x \text{ is an integer}\}$ 5. B   7. A

9.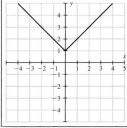

11.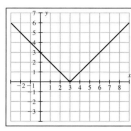

43. $C(x) = \begin{cases} 1.5 & \text{if } 0 < x \le 1 \\ 2 & \text{if } 1 < x \le 2 \\ 2.5 & \text{if } 2 < x \le 3 \\ 3 & \text{if } 3 < x \le 4 \\ 3.5 & \text{if } 4 < x \le 5 \end{cases}$

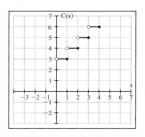

45. 2   47. 4   49. 25

51.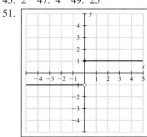

53.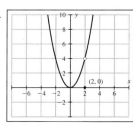

13.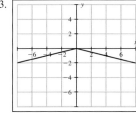

15.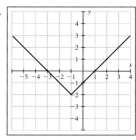

17. D   19. A

21.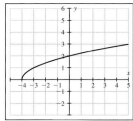

23.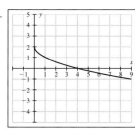

55.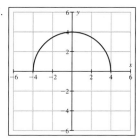

57.

59.

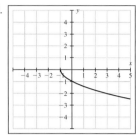

61. If $y = \sqrt{x}$, the values of y will be positive (or 0), since $\sqrt{x}$ represents the principal square root of x. If $y = -\sqrt{x}$, the values of y will be negative (or 0).

63.

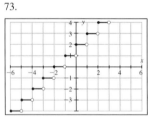

65.

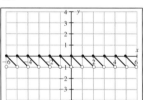

67.

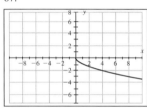

69.

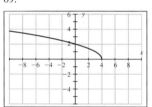

71. 5

73.

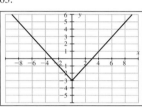

75.

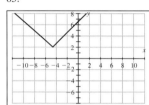

77.

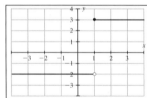

Exercises 8.8

1. $f(x_1) < f(x_2)$ 3. $f(x_1) = f(x_2)$ 5. $(x, -y)$ 7. False   9. Increasing on $(0, \infty)$; decreasing on $(-\infty, 0)$ 11. Decreasing on $(-\infty, \infty)$ 13. Increasing on $(-2, \infty)$; decreasing on $(-4, -2)$ 15. Decreasing on $(-\infty, -3) \cup (2, \infty)$; constant on $(-3, 2)$

17. Increasing on $(-\infty, \infty)$

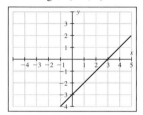

19. Increasing on $(-2, \infty)$; decreasing on $(-\infty, -2)$

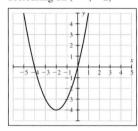

21. Increasing on $(1, \infty)$; constant on $(-\infty, 1)$

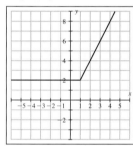

23.

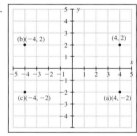

25.

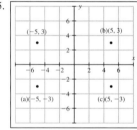

27.

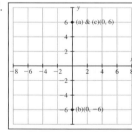

29. (a) No, (b) Yes, (c) No   31. (a) Yes, (b) Yes, (c) Yes   33. (a) No, (b) No, (c) Yes   35. Origin   37. y-axis   39. None   41. x-axis, y-axis, origin   43. x-axis   45. Origin

47.

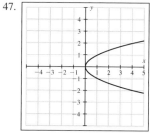

49. No. A graph cannot have exactly two of these types of symmetry.
51. Increasing on $(2, \infty)$;
decreasing on $(-\infty, 2)$

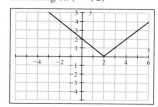

53. Increasing on $(-\infty, 0) \cup (2, \infty)$;
decreasing on $(0, 2)$

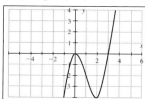

55. x-axis

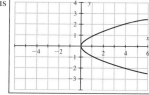

57. Origin

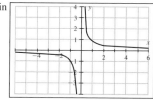

Chapter 8 Review Exercises

1. Domain = $\{1, 2, -6, 3\}$; range = $\{3, 1, 2\}$; is a function   2. Domain = $\{0, 4\}$; range = $\{4, 8, -10\}$; not a function   3. Yes   4. No
5. Yes   6. $(-\infty, \infty)$ 7. $(-\infty, 7]$ 8. $\{x \mid x \neq 3, -3\}$ 9. $[-5, \infty)$
10. $(-\infty, 13]$ 11. Domain = $[-3, 3]$; range = $[-2, 2]$; not a function
12. Domain = $(-\infty, \infty)$; range = $(-\infty, 0]$; is a function   13. Domain
= $\{x \mid x \neq 0\}$; range = $\{y \mid y \neq 0\}$; is a function   14. (a) 17, (b) -3,
(c) $4t + 9$, (d) $4x + 4h + 9$ 15. (a) 15, (b) 35, (c) $3t^2 - t + 5$,
(d) $3x^2 + 6xh + 3h^2 - x - h + 5$ 16. 10   17. $x^2 - x - 6$; domain
= $(-\infty, \infty)$ 18. 21   19. $\dfrac{x^2 - 2x}{6 - x}$; domain = $\{x \mid x \neq 6\}$ 20. -1
21. $-x^2 + 2x + 6$; domain = $(-\infty, \infty)$ 22. 2   23. 5   24. Undefined

25. Constant function

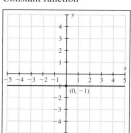

26. Linear function

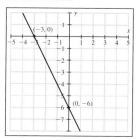

27. Quadratic function

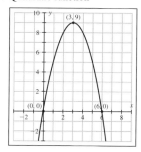

28. Linear function

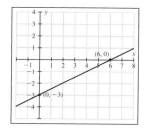

29. Constant function

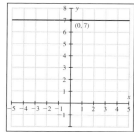

30. Quadratic function

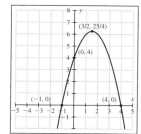

31.

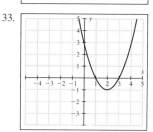

32.

33.

34.

35.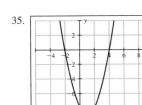

36.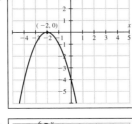

37.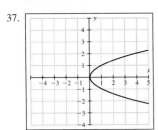

38.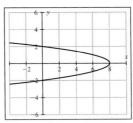

39.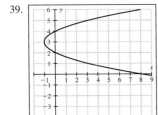

49. $(x + 2)^2 + (y - 1)^2 = 25$

50.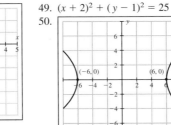

51.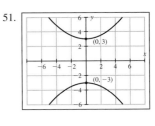

52.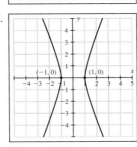

53.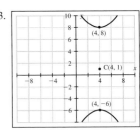

54. Circle 55. Parabola 56. Line 57. Hyperbola 58. Ellipse
59. Circle

60.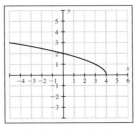

61.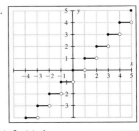

40. Width 65 ft, length 130 ft; maximum area = 8450 ft^2
41. $(x - 2)^2 + (y + 3)^2 = 64$ 42. $C(-4, 0), r = \sqrt{10}$

43.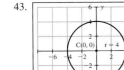
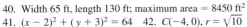

44.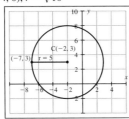

45.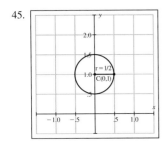

46.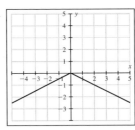

62.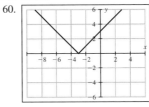

63.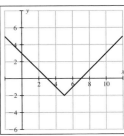

47.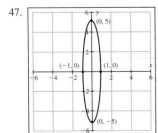

48.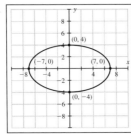

64.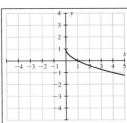

65.

66. 9 67. 25 68. −8 69. (a) 10, (b) 2, (c) 1

70.
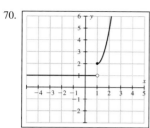

71. Increasing on $(-\infty, -2) \cup (2, \infty)$; decreasing on $(-2, 2)$ 72. Decreasing on $(-\infty, \infty)$ 73. Increasing on $(-\infty, 0)$; decreasing on $(3, 4)$; constant on $(0, 3)$ 74. None   75. Origin   76. x-axis, y-axis, origin   77. y-axis

Chapter 8 Test

1. Domain = $\{4, 2, 3\}$; range = $\{-6, 3, -9\}$; is a function   2. Domain = $(-\infty, \infty)$; range = $(-\infty, \infty)$; is a function   3. 7   4. 30   5. 51   6. $[\frac{13}{3}, \infty)$ 7. 8   8. Domain = $(-\infty, 5]$; range = $[0, \infty)$; is a function   9. Increasing on $(-\infty, -2)$; decreasing on $(-2, 0)$; constant on $(0, \infty)$ 10. y-axis   11. $(x + 3)^2 + (y - 2)^2 = 25$ 12. $t = 2.5$ sec; 135 ft

13.
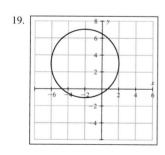

14.
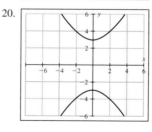

15.
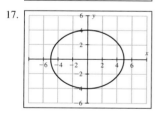

16.
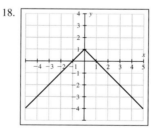

17.
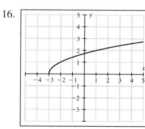

18.
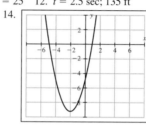

19.
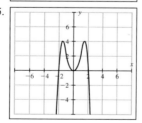

20.

Chapter 9

Exercises 9.1

1. Polynomial; n; a_n; a_0 3. Falls; rises; rises; rises   5. True   7. True   9. False   11. True

13.
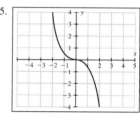

15.
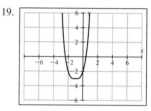

17.
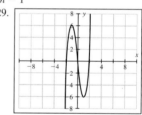

19.
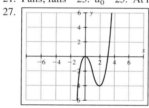

21. Falls; falls   23. a_0 25. At most $n - 1$

27.
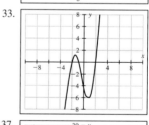

29.
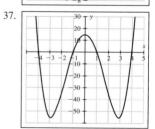

31.

33.

35.

37.

39.

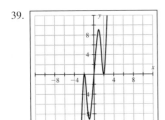

41.

x	1.5	1	0.5	0	−0.5	−1	−1.5
$f(x)$	7.59	1	0.03	0	−0.03	−1	−7.59
$g(x)$	3.38	1	0.13	0	−0.13	−1	−3.38

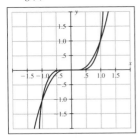

43. 1 hr $\le t \le$ 7.9 hr; at $t = 4.9$ hr   45. All real numbers. If you raise any real number to a whole number power, multiply the result by a real number, and add the result to a real number, the final result is a real number.   47. All real numbers. The graph rises on one end and falls on the other end.

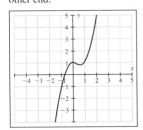

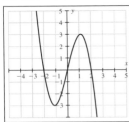

49. A polynomial equation in x is an equation that can be written so that a polynomial in x is on one side and 0 is on the other side (for example, $x^2 - 5x + 6 = 0$). A polynomial function is an equation that can be written so that a polynomial in x is on one side and y is on the other side (for example, $y = x^2 - 5x + 6$).   51. Even   53. Odd   55. Neither
57.   59.

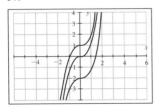

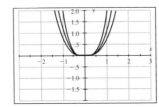

61. The highest-degree term

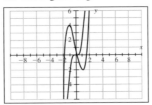

63. x-intercept $(-0.3, 0)$; turning points $(0.8, 4.1)$, $(2.5, 1.4)$

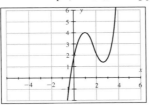

65. x-intercepts $(1.8, 0)$, $(-1.8, 0)$; turning points $(0, -1)$, $(-1.2, -3.25)$, $(1.2, -3.25)$

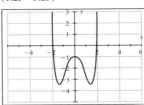

Exercises 9.2

1. False   3. False   5. Divisor is $x + 2$; dividend is $x^3 + 2x^2 - 6x - 3$; quotient is $x^2 - 6$; remainder is 9   7. $4x - 2 - \dfrac{1}{x + 3}$ 9. $x + 5$
11. $2x^2 + 3x + 6 + \dfrac{14}{x - 4}$ 13. $y^2 + 2y + 2 - \dfrac{2}{y - 1}$
15. $-4m^2 + 6m - 13 + \dfrac{13}{m + 1}$ 17. $x^3 - x^2 + 2x - 8 - \dfrac{10}{x + 4}$
19. $2k^3 - 5 + \dfrac{14}{k + 1}$ 21. $y^4 - 2y^3 + 4y^2 - 8y + 16$
23. $4p^2 + 2p + 6$ 25. 8   27. (a) 8, (b) 8. It is easier to multiply two numbers than raise a number to a power greater than 2.   29. 74
31. 50   33. −151   35. −315   37. Yes   39. No   41. No   43. Yes
45. $x - 6$ 47. 5   49. 5 (multiplicity 1)   51. 0 (multiplicity 3), −4 (multiplicity 1), 8 (multiplicity 6)   53. $2 + i$ (multiplicity 4), $2 - i$ (multiplicity 4)   55. $\{2, 1, -4\}$ 57. $\{-5, 3, -\frac{1}{2}\}$
59. $\{4, 0, -2, -7\}$ 61. $x^2 + 3x - 28 = 0$ 63. $x^3 - 6x^2 + 9x = 0$
65. $3x^3 - x^2 - 15x + 5 = 0$ 67. $2x^3 + x^2 + 18x + 9 = 0$
69. (a) 172, (b) 5185, (c) 111.8473

Exercises 9.3

1. a_0; a_n 3. $\sqrt{5}$ is not an integer   5. True   7. True   9. $\pm 1, \pm 2, \pm 4, \pm 8, \pm \frac{1}{2}$ 11. $\pm 1, \pm 2, \pm 3, \pm 4, \pm 6, \pm 12$ 13. $\pm 1, \pm 3, \pm \frac{1}{2}, \pm \frac{3}{2}, \pm \frac{1}{3}, \pm \frac{1}{6}$ 15. ± 1 17. $\pm 1, \pm 2, \pm 3, \pm 4, \pm 6, \pm 12$ 19. $\{2, 3, -3\}$
21. $\{1, -1, \sqrt{2}, -\sqrt{2}\}$ 23. $\{1, 4, -\frac{1}{2}\}$ 25. $\{5, -2, -3\}$
27. $\left\{\dfrac{2}{3}, \dfrac{5 + \sqrt{17}}{2}, \dfrac{5 - \sqrt{17}}{2}\right\}$ 29. $\{2, -2, -3\}$ 31. $\{0, \frac{3}{2}, 4i, -4i\}$

33. $\left\{5, -1, \dfrac{3 + \sqrt{5}}{4}, \dfrac{3 - \sqrt{5}}{4}\right\}$ 35. $\{1, 1 + 2i, 1 - 2i\}$

37. $\{\sqrt{7}, -\sqrt{7}, i, -i\}$ 39. $\{3, \frac{1}{2}\}$ 41. $\{1, 2, -1, -2, -3\}$

43. $\dfrac{p}{q} = \pm 1$ or ± 3. But none of these numbers are solutions.
45. Use synthetic division and the rational root test to find the quadratic factor. Then use the quadratic formula.   47. No, the coefficients are not real numbers.   49. 0.697 l or 4.303 l 51. $x = 26.619$ in. and $y = 1.524$ in., or $x = 3.381$ in. and $y = 94.476$ in.

Exercises 9.4

1. $f(k) = 0$ 3. $x - 5$ 5. False   7. 8   9. 3, -3 11. None   13. 0, $-4, \frac{1}{2}$ 15. $-2, -1, 1, 3$ 17. $(x + 2)(x + 1)(x - 1)(x - 3)$
19. $2(x - 3)(x + 3)(x - 3i)(x + 3i)$ 21. $5(x - \frac{2}{5})(x - 5i)(x + 5i)$
23. $(x - 4)(x + 4)(x - i\sqrt{3})(x + i\sqrt{3})$ 25. $x(x - 1)^3$
27. $(x - 2)(x + 1)(x + 5)$

29. $\left(x - \dfrac{9 + \sqrt{17}}{2}\right)\left(x - \dfrac{9 - \sqrt{17}}{2}\right)$

31. $4\left(x - \dfrac{2 + i}{2}\right)\left(x - \dfrac{2 - i}{2}\right)$

33. 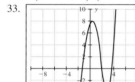   35.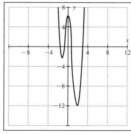

37. $\{2, 1 + \sqrt{7}, 1 - \sqrt{7}\}$ 39. $\left\{\dfrac{3}{2}, \dfrac{-3 + i\sqrt{7}}{2}, \dfrac{-3 - i\sqrt{7}}{2}\right\}$

41. (a) Yes,   (b) Yes,

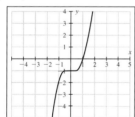

(c) Yes,

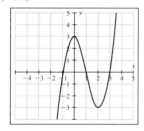

(d) No
43. Since f is a polynomial function, f is continuous. Therefore, if $f(2) = -1$ and $f(3) = 4$, then the graph of f must cross the x-axis between $x = 2$ and $x = 3$ (that is, f must have at least one real zero between 2 and 3). But if $f(2) = 1$ and $f(3) = 4$, then the graph of f may or may not cross the x-axis between $x = 2$ and $x = 3$.

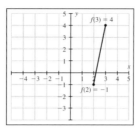   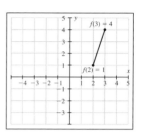

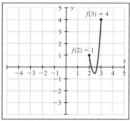

45. 2.36   47. 0.54   49. 0.54, 1.68, -2.21

Exercises 9.5

1. Polynomials   3. $y = 3$; horizontal   5. $x = k$ 7. True   9. True   11. True   13. False   15. C   17. A   19. D   21. B   23. V. A. is $x = 4$; H. A. is $y = 0$ 25. V. A. are $x = 5$ and $x = -5$; H. A. is $y = -2$ 27. V. A. is $x = -1$; S. A. is $y = 2x - 1$ 29. None
31. An asymptote of a graph is a line that a portion of the graph gets closer and closer to as that portion of the graph moves away from the origin.   33. A horizontal asymptote occurs when the degree of the numerator is less than or equal to the degree of the denominator. A slant asymptote occurs when the degree of the numerator is one more than the degree of the denominator.

35. 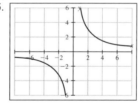   37.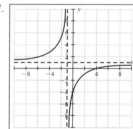

39. 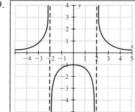   41.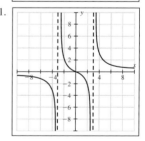

43.
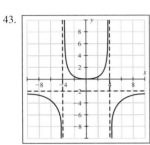

45.
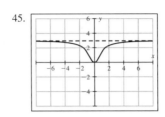

69.
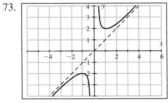

71.
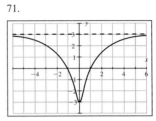

73.
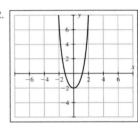

47.
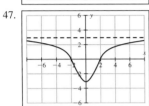

49.
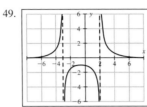

51.
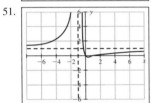

53.
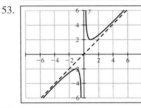

Chapter 9 Review Exercises

1.
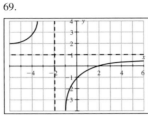

2.
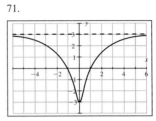

55.
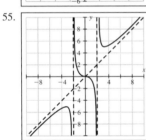

57.
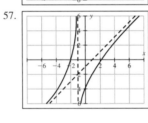

3.
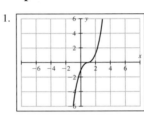

4.
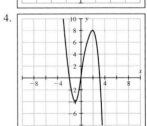

59.
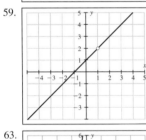

61.
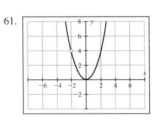

5.
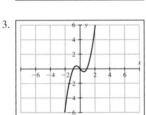

6.
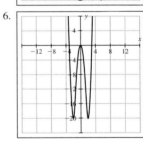

63.
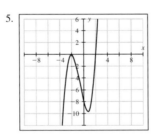

65. (a) 100,000 people, (b) At $t = 7$ days   67. 200 units

7. A point at which the graph changes from increasing to decreasing, or from decreasing to increasing   8. Yes. No, the range of the function graphed below is $[2, \infty)$.

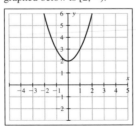

9. $f(-x) = 2(-x)^3 - 5(-x) = -2x^3 + 5x = -f(x)$ 10. $x + 2$

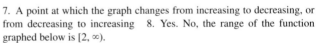

11. $2x^3 + 4x^2 + 8x + 29 + \dfrac{55}{x-2}$ 12. $t^2 - 4t + 3$

13. $p^2 + 5p + 7 - \dfrac{3}{p-3}$ 14. 5   15. -8 16. Yes   17. No   18. n

19. $\{4, 3, -6\}$ 20. $5x^3 + 28x^2 - 12x = 0$ 21. $\pm 1, \pm 2, \pm 3, \pm 4, \pm 6,$
$\pm 12, \pm\frac{1}{3}, \pm\frac{2}{3}, \pm\frac{4}{3}$ 22. $\{-\frac{5}{2}, 4, -4\}$ 23. $\{4, -4, 2i, -2i\}$
24. $\{-1, 2 + \sqrt{2}, 2 - \sqrt{2}\}$ 25. $\{-2, 1, -6\}$

26. $\dfrac{p}{q} = \pm 1$ or ± 3. But none of these numbers are solutions.
27. If the imaginary number $a + bi$ is a root of a polynomial equation
with real coefficients, then $a - bi$ must also be a root.   28. Height 2 ft,
width 3 ft, length 4 ft   29. $(x - 3i)(x + 3i)$
30. $(x + 3)(x - \sqrt{5})(x + \sqrt{5})$ 31. $(x - (4 + \sqrt{13}))(x - (4 - \sqrt{13}))$
32. $2(x - 3)(x + 4)(x - \frac{1}{2})$

33. 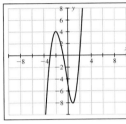   34. 0.45

35.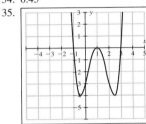

36. A polynomial function of positive degree with complex coefficients
has at least one complex zero.   37. V. A. is $x = 0$; H. A. is
$y = 0$ 38. V. A. are $x = 0$ and $x = 3$; H. A. is $y = \frac{1}{2}$
39. V. A. is $x = -2$; S. A. is $y = 3x - 6$

40. 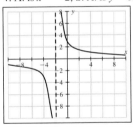   41.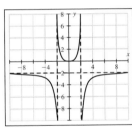

42. 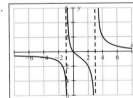   43.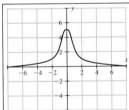

44.    45.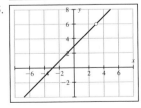

46. No. Yes. Yes

Chapter 9 Test

1. $5 + 3i$ 2. 2   3. The graph of a rational function may have any number of vertical asymptotes, but it will have at most one nonvertical asymp-

tote (horizontal or slant).   4. A graph is smooth if it has no corners. A
graph is continuous if it has no breaks.
Smooth and continuous

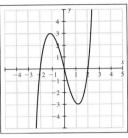

Smooth but not continuous

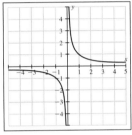

Continuous but not smooth

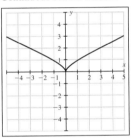

Not smooth, not continuous

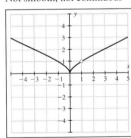

5. $f(-x) = (-x)^4 + 3(-x)^2 = x^4 + 3x^2 = f(x)$ 6. $x^3 + x^2 - 10x + 8$
$= 0$ 7. 0 (multiplicity 2), 6 (multiplicity 3)   8. $x^2 + x + 2 + \dfrac{8}{x-3}$

9. $\pm 1, \pm 3, \pm 5, \pm 15, \pm\frac{1}{2}, \pm\frac{3}{2}, \pm\frac{5}{2}, \pm\frac{15}{2}, \pm\frac{1}{4}, \pm\frac{3}{4}, \pm\frac{5}{4}, \pm\frac{15}{4}$
10. $(x - (2 + \sqrt{2}))(x - (2 - \sqrt{2}))$ 11. $\{-1, -2 + 2i, -2 - 2i\}$
12. 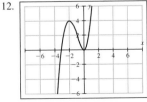   13.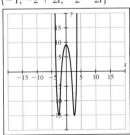

14.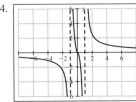

15.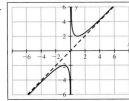

16. 1.5

Cumulative Test For Chapters 7, 8, and 9

1. 7   2. $\{x \mid x \neq 4, -4\}$ 3. $3\sqrt{2}$ 4. $(-4, \frac{5}{2})$

5. $x^3 + 3x^2 - 10x = 0$

6. Origin   7. $m = \frac{5}{3}, b = -2$

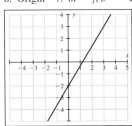

8. $\left(x - \dfrac{1 + \sqrt{21}}{2}\right)\left(x - \dfrac{1 - \sqrt{21}}{2}\right)$

9. $\pm 1, \pm 2, \pm 3, \pm 6, \pm \frac{1}{3}, \pm \frac{2}{3}$ 10. $2x^2 - 6x + 11 - \dfrac{20}{x + 3}$

11. $\left\{\dfrac{1}{2}, \dfrac{1 + \sqrt{5}}{2}, \dfrac{1 - \sqrt{5}}{2}\right\}$ 12. 12.5 cm   13. (a) 16, (b) 0,

(c) $x^2 + 6x$ 14. $9x + y = -29$ 15. $x + 4y = 24$

16.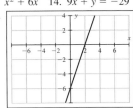

17.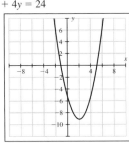

18.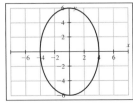

19.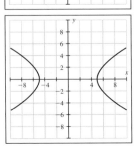

20.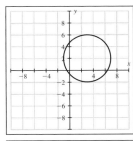

21.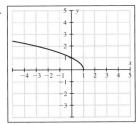

22.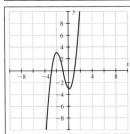

23.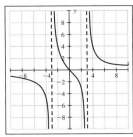

24.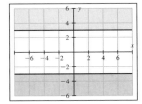

25.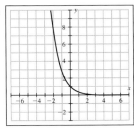

Chapter 10

Exercises 10.1

1. $f(x) = a^x$ 3. $a > 1$; $0 < a < 1$ 5. False   7. B   9. D   11. C

13. D

15.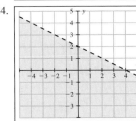

17.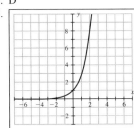

19.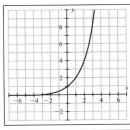

21.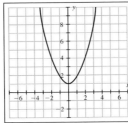

23.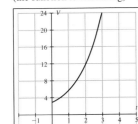

25. If $a = 0$, then for $x > 0$ $f(x) = a^x$ becomes $f(x) = 0$ (a constant function). If $a = 1$, then $f(x) = a^x$ becomes $f(x) = 1^x = 1$ (a constant function).

27. They are the same, since $y = 2^{-x} = \dfrac{1}{2^x} = \left(\dfrac{1}{2}\right)^x$.   29. $a > 1$
(the function is increasing)   31. (a) \$3, (b) \$6, (c) \$12, (d) \$24

33. $Q = 40(2)^{-t/8}$

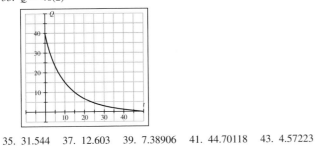

35. 31.544   37. 12.603   39. 7.38906   41. 44.70118   43. 4.57223
45. 0.74082   47.

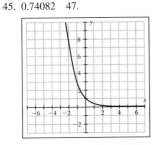

49. \$32,119.85   51. \$2983.65   53. 11 percent compounded continuously   55. 14.7 lb/in.^{2}; 1.8 lb/in.^{2}

57.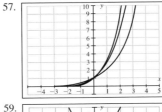

59.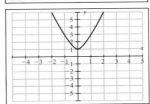

Exercises 10.2

1. (b, a) 3. The line $y = x$ 5. False   7. False   9. (a) Yes,
(b) $f^{-1} = \{(5, 2), (2, -1), (0, -3)\}$ 11. (a) Yes, (b) $f^{-1} = \{(0, 0),$
$(2, 2), (-4, 4)\}$ 13. (a) No, (b) No inverse function
15. $f^{-1}(x) = x - 2$

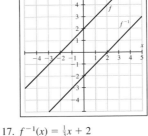

17. $f^{-1}(x) = \frac{1}{3}x + 2$ 19. $f^{-1}(x) = -x + 3$

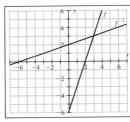

21. f is not one-to-one.
23. $f^{-1}(x) = \sqrt{x - 1}$

25. f is not one-to-one.   27. $f^{-1}(x) = \sqrt[3]{x+2}$

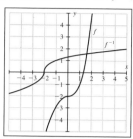

53. f is one-to-one.

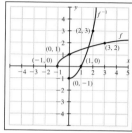

29. $f^{-1}(x) = \frac{3}{2}x + 3$

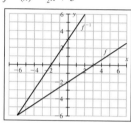

31. $f^{-1}(x) = x^2 \ (x \ge 0)$

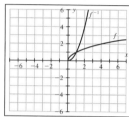

55. f is not one-to-one.
57. f is one-to-one.

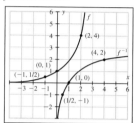

33. $f^{-1}(x) = \dfrac{1}{x}$

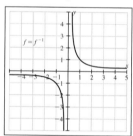

59.

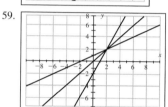

35. 2   37. 5   39. -1 41. 6   43. (a) 20, (b) 4, (c) a, (d) a
45. $f(f^{-1}(x)) = x$ and $f^{-1}(f(x)) = x$; these equations are true for any one-to-one function f and its inverse f^{-1}, so long as x is in the domain of the inside function; the inverse function "undoes" what the original function "does."   47. (1) Graph the function. If no horizontal line intersects the graph at more than one point, the function is one-to-one. (2) Assume x_1 and x_2 are arbitrary x-values and $x_1 \ne x_2$. If $f(x_1) \ne f(x_2)$, then f is one-to-one. But if you can find two different x-values, x_1 and x_2, such that $f(x_1) = f(x_2)$, then f is not one-to-one.
49. f is one-to-one.

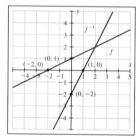

51. f is not one-to-one.

Exercises 10.3

1. $x = a^y$ 3. $1; 0$ 5. $w = \log_b z$ 7. True   9. True   11. 4   13. 2
15. 1   17. 0   19. -1 21. $\frac{1}{2}$ 23. $\log_3 81 = 4$
25. $\log_{10} 0.0001 = -4$ 27. $\log_4 2 = \frac{1}{2}$ 29. $\log_b t = s$
31. $5^3 = 125$ 33. $\left(\frac{1}{7}\right)^{-2} = 49$ 35. $27^{1/3} = 3$ 37. $(\sqrt{3})^2 = 3$
39. $x = 81$ 41. $x = 4$ 43. $a = 4$ 45. $y = 4$ 47. $y = 3$
49. $a = \frac{1}{5}$ 51. $y = 0$ 53. $x = \frac{1}{32}$ 55. $y = -3$ 57. $y = -3$
59. $a = 16$ 61. $y = 1$ 63. If $y = \log_2 0$, then $2^y = 0$.
If $y = \log_2(-4)$, then $2^y = -4$. But $2^y > 0$ for any real number y.
65. $a > 1$ 67. C   69. A   71. C   73. A
75.

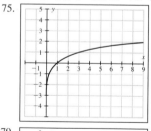

77.

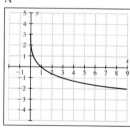

79.

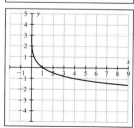

81.

83. (a) 0, (b) 6.7, (c) 8.3

85.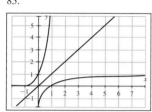

87.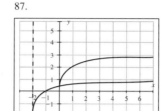

65.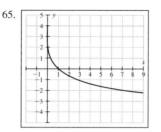

Exercises 10.4

1. $\log_a M + \log_a N$; $\log_a M - \log_a N$ 3. False   5. False
7. $\log_2 3 + \log_2 7$ 9. $\log_3 2 - \log_3 y$ 11. $2 \log_6 x$ 13. $-3 \log_8 z$
15. $\log_{10} x + \log_{10} (x + 1)$ 17. $\log_b (x - 4) - \log_b (x + 2)$ 19. 3
21. $\frac{4}{5}$ 23. $2x$ 25. 1   27. 13   29. $y - 1$ 31. The sum of the logarithms of the factors; the logarithm of the numerator minus the logarithm of the denominator; the power times the logarithm of the number
33. $\log_{10} 3 + 2$ 35. $\log_5 x - 2$ 37. $-2 \log_6 x$ 39. $1 + 4 \log_2 x$
41. $5 \log_3 x - 2$ 43. $\log_5 x + \frac{1}{2} \log_5 y - 2 \log_5 z$
45. $\frac{1}{4} \log_b x + \frac{3}{4} \log_b y$ 47. $\frac{2}{3} \log_2 x - \frac{1}{3} \log_2 y - \frac{5}{3}$

49. $4 \log_{10} (x + 8) + 6 \log_{10} (x - 5)$ 51. $\log_5 x^7$ 53. $\log_{10} \frac{1}{x^3}$

55. $\log_3 \frac{1}{8y}$ 57. $\log_2 rs$ 59. $\log_4 \frac{15}{s}$ 61. $\log_a \frac{\sqrt{x}}{yz}$ 63. $\log_a \frac{x}{y^2 z}$

65. $\log_{10} (x^2 - 1)$ 67. $\log_{10} (x + 3)$
69.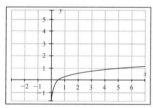

71.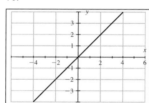

73. $\log_{10} x^2 = 2 \log_{10} x$ only when $x > 0$.

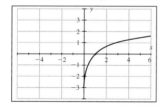 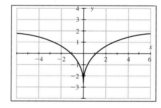

Exercises 10.5

1. $\log_{10} x$; $\log_e x$ 3. Scientific notation; product rule; evaluate each logarithm; add   5. t 7. Common; natural   9. 0.8762   11. 1.8129
13. 3.8129   15. -0.2890 17. 10.5740   19. -9.4260 21. 5.55
23. 555   25. 263   27. 3.98×10^{17} 29. 0.268   31. 4.13×10^{-16}
33. 10.5   35. 2.51×10^{-6} 37. 10 times   39. 7.1389   41. 4.6052
43. 1.6351   45. -2.6493 47. 6.75   49. 1.55   51. 8.00×10^8
53. 0.963   55. The numbers -5 and 0 are not in the domain of $y = \ln x$.   57. 1   59. 3   61. -1 63. $\frac{1}{2}$

67. 5.5 yr   69. $V = Pe^{rt}$; $3P = Pe^{rt}$; $3 = e^{rt}$; $rt = \ln 3$; $t = \dfrac{\ln 3}{r}$

71. 1.5850   73. 4.0069   75. -0.2775 77. -3.1859

79. $\log_2 15 = \dfrac{\log_9 15}{\log_9 2}$ 81. $\log_7 5 = \dfrac{1}{\log_5 7}$

83.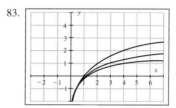

Exercises 10.6

1. Exponential   3. $x_1 = x_2$ 5. True   7. False   9. $\{2\}$ 11. $\{2.32\}$
13. $\{2\}$ 15. $\{1.17\}$ 17. $\{\frac{1}{2}\}$ 19. $\{0.95\}$ 21. $\{-3\}$
23. $\{-2.86\}$ 25. $\{\frac{1}{3}\}$ 27. $\{-1, -2\}$ 29. $\{3\}$ 31. $\{1\}$ 33. $\{5\}$
35. $\{32\}$ 37. $\{1001\}$ 39. $\{9, -1\}$ 41. $\{6\}$ 43. $\{20\}$ 45. $\{\frac{4}{3}\}$
47. $\{6\}$ 49. (1) Write both sides with the same base, then equate exponents. (2) If step 1 fails, take the common logarithm of both sides. Then use the power rule for logarithms to bring the exponent down as a multiplier. (3) Solve the equation resulting from step 1 or step 2. (4) Check in the original equation.   51. 5.3 yr   53. 35 yr   55. 4:35 a.m.
57. $\{2.10\}$ 59. $\{0.12, 1.60\}$

Chapter 10 Review Exercises

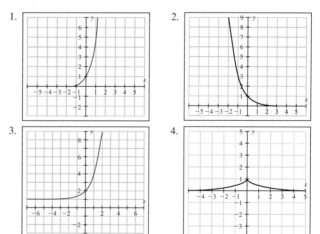

5. 11.02318   6. 0.60653   7. Increasing if $a > 1$; decreasing if $0 < a < 1$ 8. (a) \$9962.81, (b) \$10,190.52, (c) \$10,272.17

9. $f^{-1} = \{(2, 2), (4, 3), (3, -1)\}$ 10. f is not one-to-one.

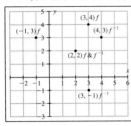

11. $f^{-1}(x) = -\dfrac{x}{3}$ 12. $f^{-1}(x) = \frac{1}{2}x - 2$

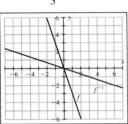   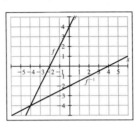

13. f is not one-to-one.
14. $f^{-1}(x) = \sqrt{x - 4}$ 15. (a) $\frac{4}{7}$, (b) 0, (c) a, (d) a

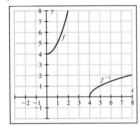

16. f is one-to-one.   17. f is one-to-one.

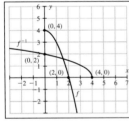   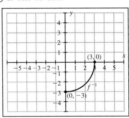

18. 5   19. -4 20. 0   21. $\log_3 243 = 5$ 22. $\log_8 4 = \frac{2}{3}$
23. $4^5 = 1024$ 24. $\left(\frac{1}{2}\right)^{-4} = 16$ 25. $y = 4$ 26. $x = 625$
27. $y = -6$ 28. $a = 12$ 29. $x = 27$
30. $a = \frac{1}{3}$
31.   32.

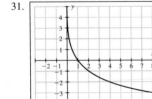   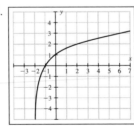

33. Increasing if $a > 1$; decreasing if $0 < a < 1$ 34. $\log_4 5 + \log_4 7$
35. $\log_2 x - \log_2 (x + 1)$ 36. $-\log_6 y$ 37. $\frac{1}{2} \log_a z$ 38. $\frac{2}{3}$ 39. t
40. $3t$ 41. $-4 \log_7 x$ 42. $\log_5 7 + \log_5 x - 2 \log_5 y - 2$

43. $\frac{1}{2} \log_3 x + \frac{3}{4} \log_3 y - \frac{1}{4} \log_3 z$ 44. $\log_9 \dfrac{1}{x}$ 45. $\log_3 2xy^4$

46. $\log_a \dfrac{\sqrt[3]{x}}{y^2 z^5}$ 47. $\log_{10} (x + 1)$ 48. Let $x = -3$.

Then $\log_3 (-3)^2 = \log_3 9 = 2$, but $2 \log_3 (-3)$ is undefined.   49. 4.5328
50. -3.3716 51. 27.6   52. 5.25×10^{15} 53. 0.402   54. 2.3026
55. -0.1567 56. 11.7   57. 0.403   58. -3 59. $\frac{1}{4}$ 60. 0.71

61. -2.63 62. $\log_5 93 = \dfrac{\log_8 93}{\log_8 5}$

63.

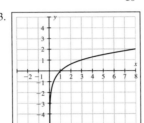

64. They are symmetric with respect to the line $y = x$.   65. 6.7 yr
66. Round off after the final calculation. Rounding off sooner may cause
the error to accumulate and produce a less accurate answer.   67. {3}
68. {2.86}   69. {5, -1}   70. {2.77}   71. {4}   72. {2}   73. {0}
74. {9}   75. 8.6 yr

Chapter 10 Test

1. $f^{-1} = \{(-6, 8), (5, 3), (9, 2)\}$ 2. $x = 9$ 3. $a = 2$ 4. $y = 0$ 5. $5t$
6. 12   7. $3 \log_2 x + \frac{1}{2} \log_2 y$ 8. $1 - \log_9 x - \log_9 y$ 9. 5.5094
10. -3.2478 11. 0.200   12. 8.33×10^{31} 13. \$12,719.69   14. No
horizontal line intersects the graph at more than one point.

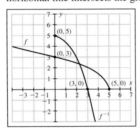

15. You must make sure each solution makes the argument of every logarithm positive.   16. Suppose $x_1 \neq x_2$. Then $5x_1 \neq 5x_2$ and
$5x_1 + 3 \neq 5x_2 + 3$. Therefore, $f(x_1) \neq f(x_2)$. $f^{-1}(x) = \frac{1}{5}x - \frac{3}{5}$ 17. {2}
18. {1.58}   19. {8}
20.   21.

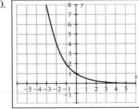   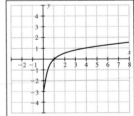

Chapter 11

Exercises 11.1

1. System of equations; linear system   3. Independent; dependent
5. A, B   7. B   9. C   11. A   13. Yes   15. No   17. No
19. $\{(3, -1)\}$; independent; consistent   21. $\{(-1, -4)\}$; independent;
consistent   23. $\{(-3, 0)\}$; independent; consistent   25. $\{(\frac{7}{2}, -\frac{3}{2})\}$; independent; consistent   27. $\varnothing$; independent; inconsistent
29. $\{(x, y) \mid y = -5x - 4\}$; dependent; consistent   31. It may be inaccurate.   33. The graphs may be two intersecting lines, two parallel
lines, or the same line.   35. (a) Independent, (b) consistent, (c) one
37. (a) Independent, (b) inconsistent, (c) none   39. (a) Dependent,
(b) consistent, (c) infinite number   41. (a) Independent,
(b) consistent, (c) one   43. (a) $45, (b) 350, (c) supply 250, demand
750, (d) 1500   45. $\{(4, 2)\}$ 47. $\varnothing$ 49. No

Exercises 11.2

1. $\{(3, 2)\}$ 3. $\{(-4, 5)\}$ 5. $\{(2, -1)\}$ 7. $\{(\frac{1}{2}, \frac{2}{3})\}$
9. $\{(x, y) \mid y = 4x\}$ 11. $\varnothing$ 13. $\{(-\frac{6}{7}, \frac{11}{7})\}$ 15. $\{(10, 6)\}$ 17. $\{(2,
1)\}$ 19. $\{(-1, -\frac{2}{3})\}$ 21. $\{(-4, 0)\}$ 23. $\{(1, 3)\}$ 25. $\{(-\frac{1}{4}, -\frac{3}{4})\}$
27. $\{(x, y) \mid y = \frac{1}{2}x + \frac{3}{2}\}$ 29. $\varnothing$ 31. $\{(5, 6)\}$ 33. The lines are parallel and the solution set is $\varnothing$; the lines coincide and any point on the
line is a solution.   35. (1) Solve one equation for one of its variables.
(2) Substitute the result of step 1 into the other equation. (3) Solve the
equation resulting from step 2. If a contradiction results, the lines are
parallel and there is no solution. If an identity results, the lines coincide
and any point on the line is a solution. (4) Substitute the value found in
step 3 into the equation resulting from step 1 to find the value of the
other variable. (5) Check in both of the original equations.
37. $\{(\frac{1}{5}, \frac{7}{15})\}$ 39. $\{(\frac{31}{23}, \frac{20}{23})\}$ 41. $\{(\frac{1}{2}, 1)\}$ 43. $\{(2, 4)\}$ 45. $x = \dfrac{2}{a}$,

$y = 4$ 47. $x = \dfrac{c}{a + b}$, $y = \dfrac{c}{a + b}$

Exercises 11.3

1. 37, 39   3. 6, 12   5. 10 kg of 5% solution, 30 kg of 25% solution
7. 16 g of pure gold, 56 g of 55% alloy   9. Marigold $3, geranium $4
11. 7 oz macaroni, 3 oz tuna   13. 90 adult tickets, 215 student tickets
15. Triangle 16 cm, square 12 cm   17. 33 nickels, 18 dimes   19. Boat
9 mi/hr, current 2 mi/hr   21. Jogger 9 mi/hr, cyclist 18 mi/hr   23. Boat
12 mi/hr, current 2.5 mi/hr   25. $7700 at 8 percent, $3850 at 12 percent
27. $26,250 at 6 percent, $15,750 at 10 percent   29. 83   31. Brett 16
yr, Kara 6 yr   33. (1) Write down the two unknown quantities and represent each by a different variable. (2) Write a system of two equations
involving the variables. (3) Solve the system. (4) Check in the words of
the original problem.

Exercises 11.4

1. Ordered triple   3. Is inconsistent and has no solution   5. No
7. Yes   9. $\{(3, 2, -2)\}$ 11. $\{(2, 1, 1)\}$ 13. $\{(-3, 5, 4)\}$ 15. $\{(-2,
0, \frac{1}{2})\}$ 17. $\{(-\frac{9}{5}, \frac{3}{5}, -2)\}$ 19. $\{(-4, -1, -2)\}$ 21. No unique so-
lution   23. No unique solution   25. $\{(0, 0, 0)\}$ 27. No unique solu-
tion   29. No unique solution   31. $\{(6, 3, 1)\}$ 33. Yes (see Fig. 11.14)
35. The system is either dependent, or inconsistent in the manner of Fig.
11.14(d). In either case, there is no unique solution.   37. First number
-4, second number 1, third number 7   39. 58 $1 tickets, 94 $2 tickets,
282 $5 tickets   41. Width of rectangle 10 cm, length of rectangle 65 cm,
side of triangle 25 cm   43. First test 71, second test 75, third test 82
45. Larry 12 bananas, Curly 10 bananas, Moe 19 bananas   47. $x = 1$,
$y = 5, z = -2, w = 3$ 49. Yes

Exercises 11.5

1. Determinant   3. 3×3; third   5. 7   7. -15 9. 0   11. 16
13. -25 15. -2 17. $\frac{1}{5}$ 19. 1   21. 0   23. $x^2 - y^2$
25. $\begin{vmatrix} 5 & 6 \\ 8 & 9 \end{vmatrix}$ 27. $\begin{vmatrix} 4 & 5 \\ 7 & 8 \end{vmatrix}$ 29. $\begin{vmatrix} 2 & 3 \\ 8 & 9 \end{vmatrix}$ 31. $\begin{vmatrix} 1 & 3 \\ 4 & 6 \end{vmatrix}$ 33. Multiply the element in the first row, first column by the element in the second
row, second column. Multiply the element in the second row, first column by the element in the first row, second column. Then subtract the
second product from the first product.   35. 33   37. 33   39. 40
41. -28 43. 55   45. 0   47. 1   49. $-3a + 3b - c$ 51. (1)
Choose a row or column to expand about. (2) Multiply each element in
the row or column chosen in step 1 by its minor. (3) Prefix the terms in
step 2 with the signs from the corresponding row or column in the sign
array.   53. $\{6\}$ 55. $\{-5\}$ 57. $\{-4, \frac{1}{2}\}$ 59. (a) 0, (b) 0, (c) 0. If
any row of a 3×3 determinant consists of zeros, the value of the determinant is 0.   61. 154   63. 26   65. $-63{,}362$

Exercises 11.6

1. $\begin{vmatrix} a_1 & b_1 \\ a_2 & b_2 \end{vmatrix}$; $\begin{vmatrix} c_2 & b_1 \\ c_2 & b_2 \end{vmatrix}$; $\begin{vmatrix} a_1 & c_1 \\ a_2 & c_2 \end{vmatrix}$ 3. $\{(1, 2)\}$ 5. $\{(6, -5)\}$
7. $\{(\frac{1}{3}, -\frac{7}{2})\}$ 9. $\{(3, 0)\}$ 11. $\varnothing$ 13. $\{(-3, 4)\}$ 15. $\{(3, -1, 2)\}$
17. $\{(-2, 0, 4)\}$ 19. $\{(5, -6, \frac{3}{4})\}$ 21. $\{(0, 0, 1)\}$ 23. No unique
solution   25. $\{(0, 0, 0)\}$ 27. Write the system in standard form. Then
D is the determinant whose elements are the coefficients of the variables
(in order). Find D_x from D by replacing the coefficients of x with the
constant terms. Find D_y from D by replacing the coefficients of y with the
constant terms. Find D_z from D by replacing the coefficients of z with the
constant terms.   29. If $D = 0$, there is no unique solution. You can solve
by addition or substitution to determine whether there is no solution or an
infinite number of solutions.
31. $x = \dfrac{a + b^2}{ab + 1}$, $y = \dfrac{a^2 - b}{ab + 1}$ 33. $x = 2, y = -1, z = 0, w = 5$
35. $\{(-1, 2)\}$ 37. $x = 6.5, y = 6.5, z = -5.5, w = -2$

Exercises 11.7

1. Matrix   3. Augmented; coefficient   5. $\begin{bmatrix} 2 & -3 & \vdots & 13 \\ 6 & -4 & \vdots & 4 \end{bmatrix}$

7. $\begin{bmatrix} 3 & -5 & 6 & \vdots & 20 \\ -2 & 2 & 3 & \vdots & -2 \\ 4 & -1 & -9 & \vdots & 4 \end{bmatrix}$ 9. $x = -3$
$y = 0$

11. $x + 2y - z = 10$ 13. False   15. True
$y + 6z = -4$
$z = 1$

17. $\begin{bmatrix} 1 & 0 & 2 & \vdots & -3 \\ 3 & -1 & -8 & \vdots & 12 \\ 5 & -2 & 6 & \vdots & 4 \end{bmatrix}$ 19. $\begin{bmatrix} 1 & 9 & -2 & \vdots & 15 \\ 0 & 1 & -2 & \vdots & \frac{7}{3} \\ 0 & 3 & -1 & \vdots & 10 \end{bmatrix}$

21. $\begin{bmatrix} 1 & -3 & 0 & \vdots & 7 \\ 0 & 3 & -6 & \vdots & 0 \\ 0 & -5 & -5 & \vdots & 10 \end{bmatrix}$ 23. True   25. $\{(4, 2)\}$

27. $\{(5, -2)\}$ 29. $\{(\frac{2}{3}, \frac{4}{5})\}$ 31. $\{(4, 0, -5)\}$ 33. $\{(2, -1, 3)\}$
35. $\{(-1, 3)\}$ 37. $\{(-3, -6)\}$ 39. $\{(4, -1, 3)\}$ 41. $\{(10, 5, 9)\}$
43. (1) Write the augmented matrix. (2) Use a row operation to get a 1 in the first row, first column. Then use row operations to get 0s below that 1. (3) Use a row operation to get a 1 in the second row, second column. Then use row operations to get 0s below that 1. (4) Continue until the coefficient matrix contains 1s down the main diagonal and 0s below each 1. (5) Write the system that is represented by the matrix of step 4. Then solve this system. (6) Check in the original system.   45. The system is inconsistent and the solution set is $\varnothing$.   47. $\varnothing$ 49. $\{(x, y) | y = \frac{1}{5}x - \frac{1}{30}\}$ 51. No unique solution   53. $x = 1, y = -3, z = 1, w = 2$

55. $\begin{bmatrix} 4 & 5 & \vdots & 6 \\ 1 & 2 & \vdots & 3 \end{bmatrix}$ 57. $\begin{bmatrix} 1 & 3 & 0 & \vdots & -5 \\ 0 & 4 & -1 & \vdots & 7 \\ 0 & 3 & -6 & \vdots & 21 \end{bmatrix}$ 59. $\{(1, -4, 0)\}$

Exercises 11.8

1. True   3. (a) Yes

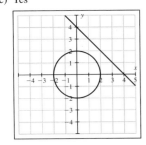

(b) Yes

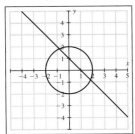

(c) Yes

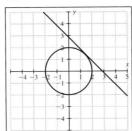

5. Use the linear equation. The nonlinear equation may produce extraneous solutions.   7. $\{(1, 1)\}$ 9. $\{(-2, 1), (1, 4)\}$ 11. $\{(2, 8),$
$(-2, -8)\}$ 13. $\{(4, 3), (5, 0)\}$ 15. $\{(0, -2), (\frac{8}{5}, -\frac{6}{5})\}$
17. $\{(4, 4)\}$ 19. $\{(0, -2), (\sqrt{3}, 1), (-\sqrt{3}, 1)\}$ 21. $\{(-4, -1), (7, 10)\}$ 23. $\{(1, 0), (-1, 0)\}$ 25. $\{(6, 4), (6, -4), (-6, 4), (-6, -4)\}$
27. $\{(0, 3), (0, -3)\}$ 29. $\{(2, 3), (2, -3), (-2, 3), (-2, -3)\}$
31. $\{(10, 5), (-10, -5)\}$ 33. $\{(\frac{3}{2}, 2), (-\frac{3}{2}, 2), (2, -\frac{3}{2}), (-2, -\frac{3}{2})\}$
35. $\{(1, 4), (-1, -4), (4, 1), (-4, -1)\}$ 37. $\{(1, 5), (-1, -5), (5, 1), (-5, -1)\}$ 39. $\{(2i, 6i), (-2i, -6i)\}$ 41. $\{(4i, 5), (4i, -5), (-4i, 5), (-4i, -5)\}$ 43. $\{(3, -1), (-3, 1), (i, 3i), (-i, -3i)\}$ 45. 7 and 5
47. $x = 3$ and $y = 2$, or $x = 2$ and $y = 3$ 49. Width 4 ft, length 6 ft
51. $\{(1.84, 2.37), (-1.84, 2.37)\}$

Exercises 11.9

1. True   3. True   5. D   7. B

9.
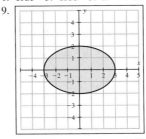

11.
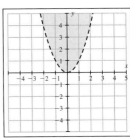

13.
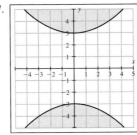

15.
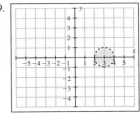

17.
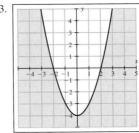

19.
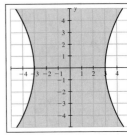

21. The graph contains the boundary curve if the inequality symbol is $\leq$ or $\geq$. The graph does not contain the boundary curve if the inequality symbol is $<$ or $>$.   23. Replace the inequality symbol with an equals sign and graph the resulting boundary curve. Then pick a test point not on the curve. If the test point satisfies the inequality, shade the region containing the test point. If the test point does not satisfy the inequality, shade the region not containing the test point.

25.
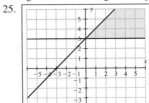

27.
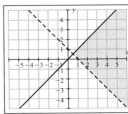

29.
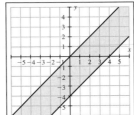

31.
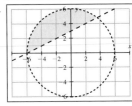

33.

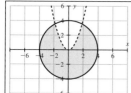

35.

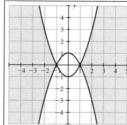

21.

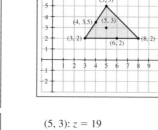 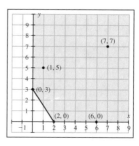

37.

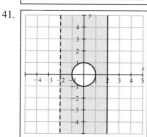

39.

(5, 3): $z = 19$ (0, 3): $z = 9$
(3, 2): $z = 12 \leftarrow$ Min   (2, 0): $z = 4 \leftarrow$ Min
(5, 5): $z = 25 \leftarrow$ Max   (6, 0): $z = 12$
(8, 2): $z = 22$ (1, 5): $z = 17$
(6, 2): $z = 18$ (7, 7): $z = 35$
(4, 3.5): $z = 18.5$ *Note:* No maximum value

Chapter 11 Review Exercises

1. No   2. Yes   3. $\{(4, 1)\}$; independent; consistent   4. $\varnothing$; independent; inconsistent   5. $\{(x, y) \mid y = \frac{2}{3}x - 2\}$; dependent; consistent
6. $\{(-1, 1)\}$ 7. $\{(-3, -2)\}$ 8. $\{(3, -2)\}$ 9. $\{(x, y) \mid y = 2x - 14\}$
10. $\{(5, \frac{12}{5})\}$ 11. $\{(4, 0)\}$ 12. $\{(-4, 11)\}$ 13. $\varnothing$ 14. $\{(\frac{4}{3}, -\frac{1}{2})\}$
15. $\{(-\frac{1}{7}, -\frac{1}{6})\}$ 16. Placement \$3, tablecloth \$7   17. 17 cars, 21 trucks   18. 90 adults, 75 students   19. 86   20. $\{(6, 3, 0)\}$
21. $\{(2, -5, 1)\}$ 22. No unique solution   23. No unique solution
24. 4 m, 7 m, 9 m   25. 32 nickels, 24 dimes, 16 quarters   26. -2
27. 6   28. $\frac{7}{16}$ 29. 10   30. 45   31. 0   32. $x = -1$ 33. $x = 6$
34. $\{(-2, 3)\}$ 35. $\{(\frac{13}{7}, \frac{26}{7})\}$ 36. $\{(-2, \frac{1}{5})\}$ 37. $\{(2, -3, 7)\}$
38. $\{(\frac{1}{2}, -1, 3)\}$ 39. $\{(\frac{113}{72}, \frac{61}{216}, \frac{151}{216})\}$ 40. $\{(-9, 3)\}$ 41. $\{(\frac{17}{26}, \frac{32}{13})\}$
42. $\{(1, 1, -2)\}$ 43. $\{(\frac{1}{2}, \frac{1}{3}, -\frac{1}{2})\}$ 44. $\{(3, 9), (-1, 1)\}$
45. $\{(3, 4), (-3, -4)\}$ 46. $\{(5, -3), (-3, 5)\}$ 47. $\{(4, 3), (4, -3),$
$(-4, 3), (-4, -3)\}$ 48. $\{(3, 1), (-3, 1), (3, -1), (-3, -1)\}$
49. $\{(2, 1), (-2, 1), (\sqrt{2}, 3), (-\sqrt{2}, 3)\}$ 50. $\{(1, 5), (-1, -5), (5, 1),$
$(-5, -1)\}$ 51. $\{(i, 2), (i, -2), (-i, 2), (-i, -2)\}$

41.

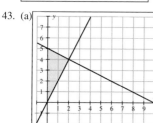

52. Two solutions

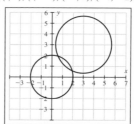

43. (a)

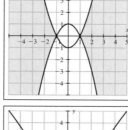

(b) and (c)   *Ordered Pair*   *Meaning*

(2, 4)   Child's ticket is \$2, adult ticket is \$4
(1, 4)   Child's ticket is \$1, adult ticket is \$4
(0, 5)   Children get in free, adults pay \$5

One solution

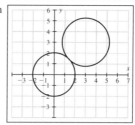

45.

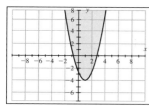

47.

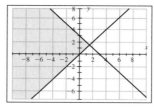

Exercises 11.10

1. Objective; constraints; feasible   3. It can be contained in some circle.
5. True   7. Max $= 30$, min $= 0$ 9. Max $= 32$, min $= 4$ 11. 10 cars, 30 trucks   13. Max $= 78$, min $= 29$ 15. Max $= 40$, min $= 2$ 17. 1 pt of Drink X, 3 pt of Drink Y   19. (a) 800 pink/min, 400 white/min
(b) 1050 pink/min, 150 white/min

No solutions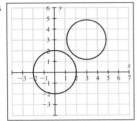

53. 8 in., 9 in.

54. 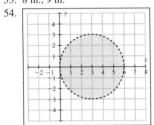   55.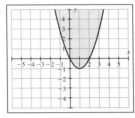

56. 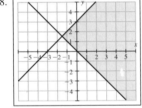   57.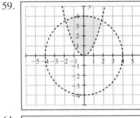

58. 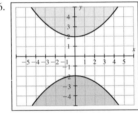   59.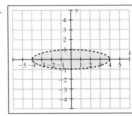

60. 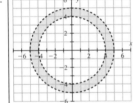   61.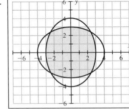

62. Max $= 36$, min $= 3$ 63. Max $= 30$, min $= 0$ 64. 2 green/day, 3 yellow/day   65. A bounded region can be contained in some circle; an unbounded region cannot be contained in any circle.   66. Yes; yes; the objective function may or may not have a maximum (or minimum) value on an unbounded region.

Chapter 11 Test

1. Yes   2. 3   3. -10 4. $\{(2, -1)\}$ 5. $\{(5, -4)\}$ 6. $\{(5, 6)\}$
7. $\{(\frac{115}{31}, \frac{50}{31})\}$ 8. $\{(4, 2)\}$; independent; consistent   9. $\varnothing$; independent; inconsistent   10. $\{(2, -1, 2)\}$ 11. $\{(1, 2, -3)\}$ 12. $\{(-6, -7)\}$
13. $\{(-\frac{6}{5}, \frac{13}{15}, \frac{19}{5})\}$ 14. $\{(-\frac{31}{2}, \frac{15}{2})\}$ 15. $\{(3, 7, 8)\}$ 16. $x = 7, y = 2$
17. 3.75 lb of \$1.90 nuts, 6.25 lb of \$1.10 nuts   18. Boat 12.5 mi/hr, current 1.5 mi/hr   19. $29°, 53°, 98°$ 20. $\{(0, -5), (4, 3)\}$ 21. $\{(5, 1),$

$(-1, -5)\}$ 22. $\{(3, 2), (-3, 2), (3, -2), (-3, -2)\}$

23. 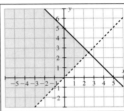   24.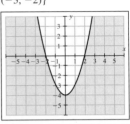

25. 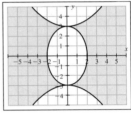   26.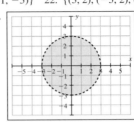

27. 45 necklaces, 15 bracelets

Chapter 12

Exercises 12.1

1. Sequence function, or sequence   3. Series   5. 1, 4, 7, 10; $a_{10} = 28$
7. 2, 6, 12, 20; $a_{10} = 110$ 9. $0, \frac{1}{2}, \frac{2}{3}, \frac{3}{4}; a_{10} = \frac{9}{10}$ 11. $2, \frac{5}{2}, \frac{10}{3}, \frac{17}{4}$;
$a_{10} = \frac{101}{10}$ 13. 6, 6, 6, 6; $a_{10} = 6$ 15. $-1, \frac{1}{2}, -\frac{1}{3}, \frac{1}{4}; a_{10} = \frac{1}{10}$
17. 2, 4, 8, 16; $a_{10} = 1024$ 19. $-\frac{2}{7}, \frac{1}{6}, \frac{6}{17}, \frac{5}{11}; a_{10} = \frac{17}{26}$ 21. $a_n = 2n$

23. $a_n = n$ 25. $a_n = n^3$ 27. $a_n = \dfrac{1}{n^2}$ 29. $a_n = (-1)^n$

31. $a_n = (-1)^{n+1}2n$ 33. A finite sequence is a function whose domain is the first k positive integers; an infinite sequence is a function whose domain is the entire set of positive integers. A finite series is the indicated sum of a finite sequence; an infinite series is the indicated sum of an infinite sequence.

35. $\displaystyle\sum_{n=1}^{5} n^2 = 1^2 + 2^2 + 3^2 + 4^2 + 5^2 = \sum_{i=1}^{5} i^2$

This suggests that the choice of the index of summation has no effect on the sum of the series.

37. $4 + 8 + 12 + 16 + 20 + 24 = 84$
39. $4 + 4 + 4 + 4 + 4 + 4 = 24$ 41. $5 + 7 + 9 + 11 = 32$
43. $(4 + 6 + 8 + 10) + 1 = 29$ 45. $4 + 2 + 2 + 4 + 8 = 20$
47. $0 + 6 + 24 = 30$

49. $1 + \frac{1}{2} + \frac{1}{3} + \frac{1}{4} = \frac{25}{12}$ 51. $-1 + 4 - 27 = -24$ 53. $\dfrac{x^2}{2} + \dfrac{x^3}{3} +$

$\dfrac{x^4}{4} + \dfrac{x^5}{5} + \dfrac{x^6}{6}$ 55. $\displaystyle\sum_{n=1}^{6} (3n - 1)$ 57. $\displaystyle\sum_{n=1}^{6} (3n + 2)$ 59. $\displaystyle\sum_{n=1}^{4} 3n$

61. $\displaystyle\sum_{n=1}^{4} 3^n$ 63. $\displaystyle\sum_{n=1}^{4} \dfrac{n}{n+1}$ 65. $\displaystyle\sum_{n=1}^{5} (n^2 + 1)$ 67. $\displaystyle\sum_{n=1}^{5} x^{n-1}$

69. \$15,000, \$13,500, \$12,150, \$10,935, \$9,841.50   71. 48   73. To obtain any term after the second, add the previous two terms. That is, $a_n = a_{n-1} + a_{n-2}$ for $n \geq 3$.

75.

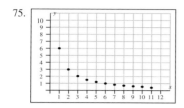

Exercises 12.2

1. Arithmetic; common difference   3. $a_1 + (n-1)d$ 5. Arithmetic; $d = 1; 6, 7$ 7. Arithmetic; $d = 6; 33, 39$ 9. Arithmetic; $d = 4; 14, 18$ 11. Arithmetic; $d = -5; -17, -22$ 13. Arithmetic; $d = \frac{1}{2}; \frac{7}{2}, 4$ 15. Nonarithmetic   17. 64   19. 658   21. $6n - 2$ 23. -32 25. $\frac{801}{5}$ 27. 4   29. 2   31. $2n - 1$ 33. $\frac{10}{3}$ 35. 17   37. 17   39. a_n represents the value of the nth term, whereas n represents the position of the term.   41. 185   43. 710   45. $\frac{100}{3}$ 47. 1278   49. 4   51. 31   53. 7482   55. 8,006,000   57. 5050   59. $2850   61. (a) 16, (b) 546   63. 184,891   65. 34 yr   67. $a_n = 2n + 2$ 69. 4, 7, 10, 13, 16, etc.

Exercises 12.3

1. Geometric; common ratio   3. $a_1 r^{n-1}$ 5. Geometric; $r = 2; 32, 64$ 7. Geometric; $r = \frac{1}{3}; \frac{1}{81}, \frac{1}{243}$ 9. Nongeometric   11. Geometric; $r = 5; -625, -3125$ 13. Nongeometric   15. Geometric; $r = -1; -6, 6$ 17. Geometric; $r = -\frac{1}{2}; -\frac{1}{8}, \frac{1}{16}$ 19. 54   21. $2(3)^{n-1}$ 23. $\frac{1}{8}$ 25. $-\frac{9}{8}$ 27. -96 29. 0.2   31. 2   33. $(-1)^{n-1}$ 35. $-\frac{4^{15}}{3^{13}}$ 37. 6   39. A geometric series is the indicated sum of a geometric sequence.   41. 2387   43. $\frac{315}{4}$ 45. 1094   47. 4   49. $\frac{2560}{729}$ 51. 0; 1   53. 511   55. $-\frac{5124}{15,625}$ 57. $\frac{116,050}{729}$ 59. $1152   61. 204.8 gal   63. $29,372; $271,521; etc.   65. $a_n = \frac{15}{4}(2)^{n-1}$ 67. 3, 15, 75, 375, 1875, etc.   69. 5115

Exercises 12.4

1. $\frac{a_1}{1-r}$ 3. 20   5. Does not exist   7. 3   9. 2   11. $\frac{100}{7}$ 13. $\frac{64}{7}$ 15. Does not exist   17. Does not exist   19. $-\frac{3}{2}$ 21. 16   23. $-\frac{1}{6}$ 25. 6   27. $\frac{5}{6}$ 29. $\frac{1}{3}$ 31. $\frac{4}{11}$ 33. $\frac{7}{99}$ 35. $\frac{41}{333}$ 37. 1   39. $\frac{52}{9}$ 41. $240,000   43. 225 cm   45. $S = 2; S_5 = \frac{31}{16} = 1.9375; S_{10} = \frac{1023}{512} \approx 1.9980; S_{15} = \frac{32,767}{16,384} \approx 1.9999$ 47. The terms of $0.3 + 0.03 + 0.003 + \cdots$ get smaller, but the terms of $1 + 2 + 4 + \cdots$ get larger.

Exercises 12.5

1. True   3. True   5. True   7. $a^7 + 7a^6b + 21a^5b^2 + 35a^4b^3 + 35a^3b^4 + 21a^2b^5 + 7ab^6 + b^7$ 9. $n(n-1)(n-2)\cdots 3 \cdot 2 \cdot 1$ 11. $\binom{n}{0}a^n + \binom{n}{1}a^{n-1}b + \binom{n}{2}a^{n-2}b^2 + \cdots + \binom{n}{n-1}ab^{n-1} + \binom{n}{n}b^n$ 13. 15   15. 10   17. 165   19. 8   21. 1   23. 15   25. For $(7-4)!$, subtract, then find the factorial of the result. For $7! - 4!$, find each factorial, then subtract the results. For $7 - 4!$, find $4!$, then subtract the result from 7.   27. $x^3 + 3x^2y + 3xy^2 + y^3$ 29. $x^4 + 12x^3 + 54x^2 + 108x + 81$ 31. $x^5 - 5x^4y + 10x^3y^2 - 10x^2y^3$

$+ 5xy^4 - y^5$ 33. $16x^4 + 32x^3y + 24x^2y^2 + 8xy^3 + y^4$ 35. $a^3 - 15a^2 + 75a - 125$ 37. $243m^5 - 810m^4 + 1080m^3 - 720m^2 + 240m - 32$ 39. $x^4 - 2x^3y + \frac{3}{2}x^2y^2 - \frac{1}{2}xy^3 + \frac{1}{16}y^4$ 41. $t^{12} + 6t^{10} + 15t^8 + 20t^6 + 15t^4 + 6t^2 + 1$ 43. $a^{20} + 20a^{19}b + 190a^{18}b^2 + \cdots$ 45. $x^{15} - 60x^{14} + 1680x^{13} - \cdots$ 47. $t^{24} - 12t^{22} + 66t^{20} - \cdots$ 49. $84a^6b^3$ 51. $-64,064p^9q^5$ 53. $\frac{70}{81}x^4y^4$ 55. $\binom{n}{0} = \frac{n!}{0!(n-0)!} = \frac{n!}{1 \cdot n!} = 1$ 57. $\binom{n}{1} = \frac{n!}{1!(n-1)!} = \frac{n \cdot (n-1)!}{1 \cdot (n-1)!} = \frac{n}{1} = n$ 59. 1   61. 39,916,800   63. The value of $70!$ is too large. (Your calculator probably will not display numbers that are larger than 10^{100}.)

Exercises 12.6

1. False

3. $S_k: 1 + 2 + 3 + \cdots + k = \frac{k(k+1)}{2}$

$S_{k+1}: 1 + 2 + \cdots + k + (k+1) = \frac{(k+1)(k+2)}{2}$; add $k+1$

5. $S_k: 1^2 + 2^2 + 3^2 + \cdots + k^2 = \frac{k(k+1)(2k+1)}{6}$

$S_{k+1}: 1^2 + 2^2 + 3^2 + \cdots + k^2 + (k+1)^2 = \frac{(k+1)(k+2)(2k+3)}{6}$; add $(k+1)^2$

7. $S_k: 2 + 4 + 6 + \cdots + 2k = k(k+1)$

$S_{k+1}: 2 + 4 + 6 + \cdots + 2k + 2(k+1) = (k+1)(k+2)$; add $2(k+1)$

9. $S_k: 3 + 7 + 11 + \cdots + (4k-1) = k(2k+1)$

$S_{k+1}: 3 + 7 + 11 + \cdots + (4k-1) + (4(k+1) - 1) = (k+1)(2k+3)$; add $4(k+1) - 1$

11. $S_k: 2^1 + 2^2 + 2^3 + \cdots + 2^k = 2(2^k - 1)$

$S_{k+1}: 2^1 + 2^2 + 2^3 + \cdots + 2^k + 2^{k+1} = 2(2^{k+1} - 1)$; add 2^{k+1}

13. $S_k: (ab)^k = a^k b^k; S_{k+1}: (ab)^{k+1} = a^{k+1}b^{k+1}$; multiply by ab 15. $S_k: 2^k > k; S_{k+1}: 2^{k+1} > k + 1$; multiply by 2   31. When we show S_1 is true, we are knocking over the first domino. When we show S_{k+1} is true if S_k is true, we are showing that the $(k+1)$st domino will fall if the kth domino falls. Both conditions are necessary because if we do not knock over the first domino, the entire row will not fall, and knocking over the first domino will not cause the entire row to fall unless any given domino will fall if the domino before it falls.   33. No. If $n = 41$, then $n^2 - n + 41 = 41^2 - 41 + 41$, which is divisible by 41.

Exercises 12.7

1. $n_1 \cdot n_2$ 3. Permutation   5. $(n-1)!$ 7. Combination   9. False   11. 8   13. 64,000   15. 676,000   17. 40,320   19. 1   21. 6720   23. 72   25. 20   27. 123, 132, 213, 231, 312, 321   29. 720   31. 3,628,800   33. (a) 210, (b) 5040   35. (a) 24, (b) 6   37. 60   39. 35   41. 1   43. 1   45. (a) a, b, c; (b) ab, ac, bc; (c) abc   47. 2,118,760   49. 8855   51. In a permutation of r elements, the order of the elements is important; in a combination of r elements, the order of the elements is not important. In a combination lock, the order in which you dial the numbers is important, so it would be more accurate to call it a permutation lock.   53. $C(n, n) = 1$. There is only one way to choose n elements (without regard to order) from a group of n elements.

55. $C(n, r) = \dfrac{n!}{(n-r)!r!} = \dfrac{n!}{r!(n-r)!} = \dfrac{n!}{(n-(n-r))!(n-r)!} = C(n, n-r)$. $C(n, r)$ denotes the number of ways to put r elements in one group (the chosen group) and the remaining $n - r$ elements in another group (the unchosen group). $C(n, n-r)$ denotes the number of ways to put $n - r$ elements in one group (the chosen group) and r elements in another group (the unchosen group). So both $C(n, r)$ and $C(n, n-r)$ denote the number of ways to put r elements in one group and $n - r$ elements in another group. 57. 32 59. 300 61. 1024 63. 12 65. 72 67. (a) 120, (b) 12, (c) 12

Exercises 12.8

1. Sample space 3. $\dfrac{n(E)}{n(S)}$ 5. 1 7. $\{H, T\}$ 9. $\{1H, 1T, 2H, 2T, 3H, 3T, 4H, 4T, 5H, 5T, 6H, 6T\}$ 11. $\{FFF, FFT, FTF, FTT, TFF, TFT, TTF, TTT\}$ 13. 36 15. 1024 17. 2,598,960 19. $\{AH, AD, AC, AS\}$ 21. $\{KH, KD, KC, KS, QH, QD, QC, QS, JH, JD, JC, JS\}$ 23. $\{(2, 4), (4, 2)\}$ 25. $\{(1, 3), (3, 1)\}$ 27. 1716 29. 2600 31. 4 33. $\frac{1}{2}$ 35. 1 37. $\frac{1}{6}$ 39. $\frac{1}{2}$ 41. 43. It means that when a coin is flipped, it is just as likely to come up heads as it is to come up tails. It also means that if a coin is flipped many times, approximately (perhaps exactly) one-half of the flips will come up heads. 45. $\frac{1}{9}$ 47. $\frac{1}{12}$ 49. 7; $\frac{1}{6}$ 51. $\frac{1}{8}$ 53. $\frac{7}{8}$ 55. $\frac{1}{2}$ 57. $\frac{1}{13}$ 59. $\frac{3}{13}$ 61. $\dfrac{33}{66,640}$ 63. $\dfrac{253}{9996}$ 65. $\dfrac{1}{108,290}$ 67. $\frac{3}{10}$ 69. $\frac{3}{5}$ 71. $\frac{1}{120}$ 73. $\frac{2}{9}$ 75. $\frac{7}{32}$ 77. (a) $\frac{1}{16}$, (b) $\frac{3}{8}$

Exercises 12.9

1. Compound probability 3. $P(E_1) \cdot P(E_2)$ 5. $P(E_1) + P(E_2)$ 7. S; $\varnothing$ 9. $\dfrac{P(E)}{P(E')}; \dfrac{P(E')}{P(E)}$ 11. True 13. (a) $\frac{1}{6}$, (b) $\frac{1}{6}$ 15. (a) $\frac{1}{3}$, (b) $\frac{1}{3}$ 17. $\dfrac{33}{66,640}$ 19. $\dfrac{33}{108,290}$ 21. 9 percent 23. 44 percent 25. (a) $\frac{11}{26}$, (b) $\frac{11}{26}$ 27. (a) $\frac{7}{13}$, (b) $\frac{7}{13}$ 29. 56 percent 31. 91 percent 33. $\frac{1}{2}$ 35. 1 37. If E_1 and E_2 are independent events, then the occurrence of E_1 has no effect on the occurrence of E_2. That is, $P(E_2/E_1) = P(E_2)$. 39. So that $P(E_1 \cap E_2)$ is not counted twice, once with $P(E_1)$ and once with $P(E_2)$. 41. (a) $\frac{1}{6}$, (b) $\frac{1}{6}$ 43. (a) $\frac{1}{2}$, (b) $\frac{1}{2}$ 45. (a) $\frac{3}{5}$, (b) $\frac{5}{3}$ 47. (a) $\frac{9}{121}$, (b) $\frac{3}{55}$ 49. (a) $\frac{10}{121}$, (b) $\frac{1}{11}$ 51. $\frac{3}{8}$ 53. $\frac{3}{28}$ 55. 40 percent 57. (a) $\frac{9}{14}$, (b) $\frac{5}{14}$ 59. Approx. 50.7 percent 61. Tari 50 percent, Erin 25 percent, Jeremy 25 percent 63. Approx. 2.7 percent

Chapter 12 Review Exercises

1. 3, 5, 7, 9; $a_{10} = 21$ 2. $-\frac{1}{3}, 0, \frac{1}{5}, \frac{1}{3}; a_{10} = \frac{2}{3}$ 3. $-2, 4, -8, 16$; $a_{10} = 1024$ 4. $a_n = 4n$ 5. $a_n = \dfrac{2n}{2n+1}$ 6. $a_n = \dfrac{(-1)^{n+1}}{n^3}$ 7. $5 + 10 + 15 + 20 + 25 + 30 = 105$ 8. $5 + 4 + 5 + 8 + 13 = 35$ 9. $1 + \frac{1}{2} + \frac{1}{4} + \frac{1}{8} + \frac{1}{16} = \frac{31}{16}$ 10. $-x^3 + \dfrac{x^5}{2} - \dfrac{x^7}{3}$ 11. $\displaystyle\sum_{n=1}^{7} (2n - 1)$ 12. $\displaystyle\sum_{n=1}^{4} (5n + 1)$ 13. $\displaystyle\sum_{n=1}^{5} (\tfrac{1}{4})^{n-1}$ 14. $\displaystyle\sum_{n=1}^{6} x^{2n-2}$ 15. $d = 4$; 23, 27 16. $d = 5$; 19, 24 17. $d = -\frac{1}{2}$; $1\frac{1}{2}$, 1 18. -23 19. -4 20. $6n - 10$ 21. 20 22. 644 23. 336 24. 650 25. -58 26. 124 27. \$660,000 28. $r = 4$; 512, 2048 29. $r = -2$; 80, -160 30. $r = \frac{2}{3}; \frac{32}{243}, \frac{64}{729}$ 31. $\frac{1}{81}$ 32. $\frac{80}{81}$ 33. 2 34. 64 35. 11 36. 189 37. $\frac{156}{5}$ 38. 39,366 39. $\frac{320}{27}$ ft 40. Approx. 235.4 billion

m³; approx. 1764.2 billion m³ 41. $\frac{27}{2}$ 42. $\frac{25}{6}$ 43. $\frac{64}{3}$ 44. $\frac{2}{3}$ 45. $-\frac{1}{9}$ 46. $\frac{20}{9}$ 47. 1 48. $\frac{2}{3}$ 49. $\frac{4}{33}$ 50. $\frac{25}{9}$ 51. 140 ft 52. 35 53. 11 54. 1 55. 28 56. $x^6 + 6x^5y + 15x^4y^2 + 20x^3y^3 + 15x^2y^4 + 6xy^5 + y^6$ 57. $81m^4 + 216m^3 + 216m^2 + 96m + 16$ 58. $t^{10} - 10t^8 + 40t^6 - 80t^4 + 80t^2 - 32$ 59. $x^3 - \frac{3}{2}x^2y + \frac{3}{4}xy^2 - \frac{1}{8}y^3$ 60. $x^{10} + 50x^9 + 1125x^8 + \cdots$ 61. $560a^3b^4$ 62. $S_k: 2 + 6 + 10 + \cdots + (4k - 2) = 2k^2$ $S_{k+1}: 2 + 6 + 10 + \cdots + (4k - 2) + (4(k + 1) - 2) = 2(k + 1)^2$; add $4(k + 1) - 2$ 63. $S_k: \dfrac{1}{2} + \dfrac{1}{4} + \dfrac{1}{8} + \cdots + \dfrac{1}{2^k} = 1 - \dfrac{1}{2^k}$ $S_{k+1}: \dfrac{1}{2} + \dfrac{1}{4} + \dfrac{1}{8} + \cdots + \dfrac{1}{2^k} + \dfrac{1}{2^{k+1}} = 1 - \dfrac{1}{2^{k+1}}$; add $\dfrac{1}{2^{k+1}}$ 66. 35,152 67. (a) 60, (b) 120 68. (a) 3,628,800; (b) 362,880 69. 20 70. 635,013,559,600 71. 59,400 72. $\frac{5}{36}$ 73. $\frac{1}{4}$ 74. $\frac{28}{55}$ 75. (a) $\frac{5}{52}$, (b) $\frac{7}{13}$ 76. 76.5 percent 77. 99.7 percent 78. $\frac{1}{3}$ 79. $\frac{5}{11}$

Chapter 12 Test

1. $2 + 6 + 12 + 20 = 40$ 2. $\displaystyle\sum_{n=1}^{5} 3n$ 3. $a_n = (-1)^{n+1}n^2$ 4. $\frac{5}{3}$ 5. 17 6. -4 7. -9 8. 230 9. $\frac{1}{81}$ 10. $8(-\frac{3}{7})^{n-1}$ 11. $\frac{45}{4}$ 12. 7 13. $\frac{75}{9}$ 14. $\frac{10}{11}$ 15. $\frac{8}{333}$ 16. $625x^4 - 1000x^3 + 600x^2 - 160x + 16$ 17. $2268a^6b^3$ 19. $3 \cdot 2^{25} = 100,663,296$ 20. 144 in. 21. 14,196 22. 6840 23. 1001 24. $\frac{7}{8}$ 25. $\frac{11}{26}$

Cumulative Test For Chapters 10, 11, and 12

1. -2 2. 0 3. $\frac{1}{2} \log x - 3 \log y$ 4. 1.30×10^{15} 5. \$8345.78 6. $-1 + 2 + 7 + 14 + 23 + 34 = 79$ 7. $\displaystyle\sum_{n=1}^{6} (5n - 1)$ 8. $32x^5 - 80x^4y + 80x^3y^2 - 40x^2y^3 + 10xy^4 - y^5$ 9. If $x_1 \neq x_2$, then $2x_1 \neq 2x_2$, and $2x_1 - 10 \neq 2x_2 - 10$. Therefore $f(x_1) \neq f(x_2)$. $f^{-1}(x) = \frac{1}{2}x + 5$ 10. 71 11. $\frac{320}{729}$ 12. 253 13. 765 15. $\{(-3, 4)\}$ 16. $\{(6, -2)\}$ 17. $\{(1, 2)\}$ 18. $\{(1, 0, -2)\}$ 19. $\{(2, -4, 3)\}$ 20. $\{(0, 4), (\frac{16}{5}, -\frac{12}{5})\}$ 21.

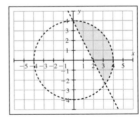

22. 2.84 23. 6 24.

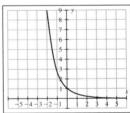

25.

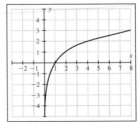

26. Magazine \$3, book \$12 27. 112 ft 28. 12,144 29. $\frac{7}{13}$ 30. $\frac{3}{8}$

APPENDIX

1

Abbreviations
and Symbols

Abbreviation	Meaning
a.m.	ante meridiem
amp	ampere(s)
B.C.	before Christ
bbl	barrel(s)
cal	calorie(s)
calc/sec	calculation(s) per second
cal/g	calorie(s) per gram
cm	centimeter(s)
¢	cent(s)
ca.	circa (around)
cm^3	cubic centimeter(s)
ft^3	cubic foot (feet)
$in.^3$	cubic inch(es)
m^3	cubic meter(s)
°	degree(s)
°C	degree(s) Celsius
°F	degree(s) Fahrenheit
$	dollar(s)
ft/hr	foot (feet) per hour
FOIL	first, outer, inner, last
ft	foot (feet)
ft/sec	foot (feet) per second
gal	gallon(s)
g	gram(s)
GCF	greatest common factor
hr	hour(s)
$[H^+]$	hydrogen ion concentration
in.	inch(es)
in./sec	inch(es) per second
kg	kilogram(s)
LCD	least common denominator
l	liter(s)
LORAN	long range navigation
m	meter(s)
mi	mile(s)
mi/gal	mile(s) per gallon
mi/hr	mile(s) per hour
mi/sec	mile(s) per second
mg	milligram(s)
min	minute(s)
mol/l	mole(s) per liter
Ω	ohm(s)
oz	ounce(s)
%	percent
pt	pint(s)
p.m.	post meridiem
lb	pound(s)
lb/ft^2	pound(s) per square foot (feet)
$lb/in.^2$	pound(s) per square inch

Abbreviation	Meaning
pH	power of hydrogen
qt	quart(s)
sec	second(s)
sec/yr	second(s) per year
cm^2	square centimeter(s)
ft^2	square foot (feet)
$in.^2$	square inch(es)
m^2	square meter(s)
mi^2	square mile(s)
t	ton(s)
vib/sec	vibration(s) per second
V	volt(s)
W	watt(s)
words/min	word(s) per minute
yd	yard(s)
yd/sec	yard(s) per second
yr	year(s)

Symbol	Meaning
$\{a, b\}$	set containing a and b
$\varnothing$	empty set
$\in$	is an element of
$\notin$	is not an element of
$\subseteq$	is a subset of
$\not\subseteq$	is not a subset of
$\cup$	union
$\cap$	intersection
$|a|$	absolute value of a
$=$	is equal to
$\neq$	is not equal to
$\approx$	is approximately equal to
$<$	is less than
$\leq$	is less than or equal to
$>$	is greater than
$\geq$	is greater than or equal to
$\pm$	plus or minus
π	Greek letter pi ($\pi \approx 3.14$)
(a, b)	open interval from a to b
$[a, b]$	closed interval from a to b
$[a, b), (a, b]$	half-open intervals from a to b
∞	infinity
$-\infty$	negative infinity
a^n	nth power of a
$\sqrt[n]{a}$	principal nth root of a
$\sqrt{a}$	principal square root of a
i	imaginary unit ($i = \sqrt{-1}$)
(a, b)	ordered pair
$f(x)$	value of function f at x
$[\![x]\!]$	greatest integer $\leq x$
$f \circ g$	composite of functions f and g

$x \to 5$	x approaches 5
e	$e \approx 2.718$
$\log_a x$	logarithm of x to the base a
$\log x$	common (base 10) logarithm
$\ln x$	natural (base e) logarithm
antilog x	antilogarithm of x
$\begin{vmatrix} a_1 & b_1 \\ a_2 & b_2 \end{vmatrix}$	2×2 determinant
$\begin{bmatrix} a_1 & b_1 & c_1 \\ a_2 & b_2 & c_2 \end{bmatrix}$	2×3 matrix
a_n	nth (general) term of a sequence
$\sum_{n=1}^{5}$	summation as n goes from 1 to 5
S_n	sum of first n terms of a series
$n!$	n factorial

$\binom{n}{r}$	binomial coefficient
$P(n, r)$	number of permutations of n elements taken r at a time
$C(n, r)$	number of combinations of n elements taken r at a time
$n(E)$	number of elements in event E
$n(S)$	number of elements in sample space S
$P(E)$	probability of event E
$P(E_1 \cap E_2)$	probability E_1 and E_2 will both occur
$P(E_1/E_2)$	probability of E_1 given E_2 has occurred
$P(E_1 \cup E_2)$	probability E_1 or E_2 (or both) will occur
E'	complement of E

2

Geometric Formulas

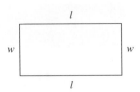

Rectangle

Area
$$A = lw$$
Perimeter
$$P = 2l + 2w$$

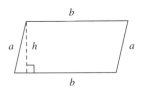

Parallelogram

Area
$$A = bh$$
Perimeter
$$P = 2a + 2b$$

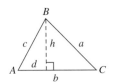

Triangle
$$c^2 = d^2 + h^2$$
$$\angle A + \angle B + \angle C = 180°$$
Area
$$A = \tfrac{1}{2}bh$$

Circle
$$d = 2r$$
Area
$$A = \pi r^2$$
Circumference
$$C = 2\pi r = \pi d$$

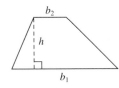

Trapezoid

Area
$$A = \tfrac{1}{2}h(b_1 + b_2)$$

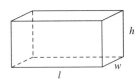

Parallelepiped

Volume
$$V = lwh$$
Surface area
$$S = 2(lw + lh + wh)$$

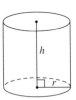

Right circular cylinder

Volume
$$V = \pi r^2 h$$
Lateral surface area
$$S = 2\pi rh$$
Total surface area
$$T = 2\pi r(h + r)$$

Right circular cone

Volume
$$V = \tfrac{1}{3}\pi r^2 h$$
Lateral surface area
$$S = \pi rl$$
Total surface area
$$T = \pi r(l + r)$$

Sphere

Volume
$$V = \tfrac{4}{3}\pi r^3$$
Surface area
$$S = 4\pi r^2$$

Powers of *e*

x	e^x	e^{-x}	x	e^x	e^{-x}	x	e^x	e^{-x}
0.00	1.00000	1.00000	0.40	1.49182	0.67032	0.80	2.22554	0.44933
0.01	1.01005	0.99005	0.41	1.50682	0.66365	0.81	2.24791	0.44486
0.02	1.02020	0.98020	0.42	1.52196	0.65705	0.82	2.27050	0.44043
0.03	1.03045	0.97045	0.43	1.53726	0.65051	0.83	2.29332	0.43605
0.04	1.04081	0.96079	0.44	1.55271	0.64404	0.84	2.31637	0.43171
0.05	1.05127	0.95123	0.45	1.56831	0.63763	0.85	2.33965	0.42741
0.06	1.06184	0.94176	0.46	1.58407	0.63128	0.86	2.36316	0.42316
0.07	1.07251	0.93239	0.47	1.59999	0.62500	0.87	2.38691	0.41895
0.08	1.08329	0.92312	0.48	1.61607	0.61878	0.88	2.41090	0.41478
0.09	1.09417	0.91393	0.49	1.63232	0.61263	0.89	2.43513	0.41066
0.10	1.10517	0.90484	0.50	1.64872	0.60653	0.90	2.45960	0.40657
0.11	1.11628	0.89583	0.51	1.66529	0.60050	0.91	0.48432	0.40252
0.12	1.12750	0.88692	0.52	1.68203	0.59452	0.92	2.50929	0.39852
0.13	1.13883	0.87810	0.53	1.69893	0.58860	0.93	2.53451	0.39455
0.14	1.15027	0.86936	0.54	1.71601	0.58275	0.94	2.55998	0.39063
0.15	1.16183	0.86071	0.55	1.73325	0.57695	0.95	2.58571	0.38674
0.16	1.17351	0.85214	0.56	1.75067	0.57121	0.96	2.61170	0.38289
0.17	1.18530	0.84366	0.57	1.76827	0.56553	0.97	2.63794	0.37908
0.18	1.19722	0.83527	0.58	1.78604	0.55990	0.98	2.66446	0.37531
0.19	1.20925	0.82696	0.59	1.80399	0.55433	0.99	2.69123	0.37158
0.20	1.22140	0.81873	0.60	1.82212	0.54881	1.00	2.71828	0.36788
0.21	1.23368	0.81058	0.61	1.84043	0.54335	1.01	2.74560	0.36422
0.22	1.24608	0.80252	0.62	1.85893	0.53794	1.02	2.77319	0.36059
0.23	1.25860	0.79453	0.63	1.87761	0.53259	1.03	2.80107	0.35701
0.24	1.27125	0.78663	0.64	1.89648	0.52729	1.04	2.82922	0.35345
0.25	1.28403	0.77880	0.65	1.91554	0.52205	1.05	2.85765	0.34994
0.26	1.29693	0.77105	0.66	1.93479	0.51685	1.06	2.88637	0.34646
0.27	1.30996	0.76338	0.67	1.95424	0.51171	1.07	2.91538	0.34301
0.28	1.32313	0.75578	0.68	1.97388	0.50662	1.08	2.94468	0.33960
0.29	1.33643	0.74826	0.69	1.99372	0.50158	1.09	2.97427	0.33622
0.30	1.34986	0.74082	0.70	2.01375	0.49659	1.10	3.00417	0.33287
0.31	1.36343	0.73345	0.71	2.03399	0.49164	1.11	3.03436	0.32956
0.32	1.37713	0.72615	0.72	2.05443	0.48675	1.12	3.06485	0.32628
0.33	1.39097	0.71892	0.73	2.07508	0.48191	1.13	3.09566	0.32303
0.34	1.40495	0.71177	0.74	2.09594	0.47711	1.14	3.12677	0.31982
0.35	1.41907	0.70469	0.75	2.11700	0.47237	1.15	3.15819	0.31664
0.36	1.43333	0.69768	0.76	2.13828	0.46767	1.16	3.18993	0.31349
0.37	1.44773	0.69073	0.77	2.15977	0.46301	1.17	3.22199	0.31037
0.38	1.46228	0.68386	0.78	2.18147	0.45841	1.18	3.25437	0.30728
0.39	1.47698	0.67706	0.79	2.20340	0.45384	1.19	3.28708	0.30422

x	e^x	e^{-x}	x	e^x	e^{-x}	x	e^x	e^{-x}
1.20	3.32012	0.30119	1.60	4.95303	0.20190	4.0	54.598	0.01832
1.21	3.35348	0.29820	1.61	5.00281	0.19989	4.1	60.340	0.01657
1.22	3.38719	0.29523	1.62	5.05309	0.19790	4.2	66.686	0.01500
1.23	3.42123	0.29229	1.63	5.10387	0.19593	4.3	73.700	0.01357
1.24	3.45561	0.28938	1.64	5.15517	0.19398	4.4	81.451	0.01228
1.25	3.49034	0.28650	1.65	5.20698	0.19205	4.5	90.017	0.01111
1.26	3.52542	0.28365	1.66	5.25931	0.19014	4.6	99.484	0.01005
1.27	3.56085	0.28083	1.67	5.31217	0.18825	4.7	109.947	0.00910
1.28	3.59664	0.27804	1.68	5.36556	0.18637	4.8	121.510	0.00823
1.29	3.63279	0.27527	1.69	5.41948	0.18452	4.9	134.290	0.00745
1.30	3.66930	0.27253	1.70	5.47395	0.18268	5.0	148.41	0.00674
1.31	3.70617	0.26982	1.71	5.52896	0.18087	5.1	164.02	0.00610
1.32	3.74342	0.26714	1.72	5.58453	0.17907	5.2	181.27	0.00552
1.33	3.78104	0.26448	1.73	5.64065	0.17728	5.3	200.34	0.00499
1.34	3.81904	0.26185	1.74	5.69734	0.17552	5.4	221.41	0.00452
1.35	3.85743	0.25924	1.75	5.75460	0.17377	5.5	244.69	0.00409
1.36	3.89619	0.25666	1.80	6.04965	0.16530	5.6	270.43	0.00370
1.37	3.93535	0.25411	1.85	6.35982	0.15724	5.7	298.87	0.00335
1.38	3.97490	0.25158	1.90	6.68589	0.14957	5.8	330.30	0.00303
1.39	4.01485	0.24908	1.95	7.02869	0.14227	5.9	365.04	0.00274
1.40	4.05520	0.24660	2.0	7.3891	0.13534	6.0	403.43	0.00248
1.41	4.09596	0.24414	2.1	8.1662	0.12246	6.1	455.86	0.00224
1.42	4.13712	0.24171	2.2	9.0250	0.11080	6.2	492.75	0.00203
1.43	4.17870	0.23931	2.3	9.9742	0.10026	6.3	544.57	0.00184
1.44	4.22070	0.23693	2.4	11.0232	0.09072	6.4	601.85	0.00166
1.45	4.26311	0.23457	2.5	12.1825	0.08208	6.5	665.14	0.00150
1.46	4.30596	0.23224	2.6	13.4637	0.07427	6.6	735.10	0.00136
1.47	4.34924	0.22993	2.7	14.8797	0.06721	6.7	812.41	0.00123
1.48	4.39295	0.22764	2.8	16.4446	0.06081	6.8	897.85	0.00111
1.49	4.43710	0.22537	2.9	18.1741	0.05502	6.9	992.27	0.00101
1.50	4.48169	0.22313	3.0	20.086	0.04979	7.0	1096.6	0.00091
1.51	4.52673	0.22091	3.1	22.198	0.04505	7.5	1808.0	0.00055
1.52	4.57223	0.21871	3.2	24.533	0.04076	8.0	2981.0	0.00034
1.53	4.61818	0.21654	3.3	27.113	0.03688	8.5	4914.8	0.00020
1.54	4.66459	0.21438	3.4	29.964	0.03337	9.0	8103.1	0.00012
1.55	4.71147	0.21225	3.5	33.115	0.03020	9.5	13360	0.00007
1.56	4.75882	0.21014	3.6	36.598	0.02732	10.0	22026	0.00005
1.57	4.80665	0.20805	3.7	40.447	0.02472	10.5	36316	0.00003
1.58	4.85496	0.20598	3.8	44.701	0.02237	11.0	59874	0.00002
1.59	4.90375	0.20393	3.9	49.402	0.02024	11.5	98716	0.00001

Common
Logarithms

4

x	0	1	2	3	4	5	6	7	8	9
1.0	0.0000	0.0043	0.0086	0.0128	0.0170	0.0212	0.0253	0.0294	0.0334	0.0374
1.1	0.0414	0.0453	0.0492	0.0531	0.0569	0.0607	0.0645	0.0682	0.0719	0.0755
1.2	0.0792	0.0828	0.0864	0.0899	0.0934	0.0969	0.1004	0.1038	0.1072	0.1106
1.3	0.1139	0.1173	0.1206	0.1239	0.1271	0.1303	0.1335	0.1367	0.1399	0.1430
1.4	0.1461	0.1492	0.1523	0.1553	0.1584	0.1614	0.1644	0.1673	0.1703	0.1732
1.5	0.1761	0.1790	0.1818	0.1847	0.1875	0.1903	0.1931	0.1959	0.1987	0.2014
1.6	0.2041	0.2068	0.2095	0.2122	0.2148	0.2175	0.2201	0.2227	0.2253	0.2279
1.7	0.2304	0.2330	0.2355	0.2380	0.2405	0.2430	0.2455	0.2480	0.2504	0.2529
1.8	0.2553	0.2577	0.2601	0.2625	0.2648	0.2672	0.2695	0.2718	0.2742	0.2765
1.9	0.2788	0.2810	0.2833	0.2856	0.2878	0.2900	0.2923	0.2945	0.2967	0.2989
2.0	0.3010	0.3032	0.3054	0.3075	0.3096	0.3118	0.3139	0.3160	0.3181	0.3201
2.1	0.3222	0.3243	0.3263	0.3284	0.3304	0.3324	0.3345	0.3365	0.3385	0.3404
2.2	0.3424	0.3444	0.3464	0.3483	0.3502	0.3522	0.3541	0.3560	0.3579	0.3598
2.3	0.3617	0.3636	0.3655	0.3674	0.3692	0.3711	0.3729	0.3747	0.3766	0.3784
2.4	0.3802	0.3820	0.3838	0.3856	0.3874	0.3892	0.3909	0.3927	0.3945	0.3962
2.5	0.3979	0.3997	0.4014	0.4031	0.4048	0.4065	0.4082	0.4099	0.4116	0.4133
2.6	0.4150	0.4166	0.4183	0.4200	0.4216	0.4232	0.4249	0.4265	0.4281	0.4298
2.7	0.4314	0.4330	0.4346	0.4362	0.4378	0.4393	0.4409	0.4425	0.4440	0.4456
2.8	0.4472	0.4487	0.4502	0.4518	0.4533	0.4548	0.4564	0.4579	0.4594	0.4609
2.9	0.4624	0.4639	0.4654	0.4669	0.4683	0.4698	0.4713	0.4728	0.4742	0.4757
3.0	0.4771	0.4786	0.4800	0.4814	0.4829	0.4843	0.4857	0.4871	0.4886	0.4900
3.1	0.4914	0.4928	0.4942	0.4955	0.4969	0.4983	0.4997	0.5011	0.5024	0.5038
3.2	0.5051	0.5065	0.5079	0.5092	0.5105	0.5119	0.5132	0.5145	0.5159	0.5172
3.3	0.5185	0.5198	0.5211	0.5224	0.5237	0.5250	0.5263	0.5276	0.5289	0.5302
3.4	0.5315	0.5328	0.5340	0.5353	0.5366	0.5378	0.5391	0.5403	0.5416	0.5428
3.5	0.5441	0.5453	0.5465	0.5478	0.5490	0.5502	0.5514	0.5527	0.5539	0.5551
3.6	0.5563	0.5575	0.5587	0.5599	0.5611	0.5623	0.5635	0.5647	0.5658	0.5670
3.7	0.5682	0.5694	0.5705	0.5717	0.5729	0.5740	0.5752	0.5763	0.5775	0.5786
3.8	0.5798	0.5809	0.5821	0.5832	0.5843	0.5855	0.5866	0.5877	0.5888	0.5899
3.9	0.5911	0.5922	0.5933	0.5944	0.5955	0.5966	0.5977	0.5988	0.5999	0.6010
4.0	0.6021	0.6031	0.6042	0.6053	0.6064	0.6075	0.6085	0.6096	0.6107	0.6117
4.1	0.6128	0.6138	0.6149	0.6160	0.6170	0.6180	0.6191	0.6201	0.6212	0.6222
4.2	0.6232	0.6243	0.6253	0.6263	0.6274	0.6284	0.6294	0.6304	0.6314	0.6325
4.3	0.6335	0.6345	0.6355	0.6365	0.6375	0.6385	0.6395	0.6405	0.6415	0.6425
4.4	0.6435	0.6444	0.6454	0.6464	0.6474	0.6484	0.6493	0.6503	0.6513	0.6522
4.5	0.6532	0.6542	0.6551	0.6561	0.6571	0.6580	0.6590	0.6599	0.6609	0.6618
4.6	0.6628	0.6637	0.6646	0.6656	0.6665	0.6675	0.6684	0.6693	0.6702	0.6712
4.7	0.6721	0.6730	0.6739	0.6749	0.6758	0.6767	0.6776	0.6785	0.6794	0.6803
4.8	0.6812	0.6821	0.6830	0.6839	0.6848	0.6857	0.6866	0.6875	0.6884	0.6893
4.9	0.6902	0.6911	0.6920	0.6928	0.6937	0.6946	0.6955	0.6964	0.6972	0.6981
5.0	0.6990	0.6998	0.7007	0.7016	0.7024	0.7033	0.7042	0.7050	0.7059	0.7067
5.1	0.7076	0.7084	0.7093	0.7101	0.7110	0.7118	0.7126	0.7135	0.7143	0.7152
5.2	0.7160	0.7168	0.7177	0.7185	0.7193	0.7202	0.7210	0.7218	0.7226	0.7235
5.3	0.7243	0.7251	0.7259	0.7267	0.7275	0.7284	0.7292	0.7300	0.7308	0.7316
5.4	0.7324	0.7332	0.7340	0.7348	0.7356	0.7364	0.7372	0.7380	0.7388	0.7396

x	0	1	2	3	4	5	6	7	8	9
5.5	0.7404	0.7412	0.7419	0.7427	0.7435	0.7443	0.7451	0.7459	0.7466	0.7474
5.6	0.7482	0.7490	0.7497	0.7505	0.7513	0.7520	0.7528	0.7536	0.7543	0.7551
5.7	0.7559	0.7566	0.7574	0.7582	0.7589	0.7597	0.7604	0.7612	0.7619	0.7627
5.8	0.7634	0.7642	0.7649	0.7657	0.7664	0.7672	0.7679	0.7686	0.7694	0.7701
5.9	0.7709	0.7716	0.7723	0.7731	0.7738	0.7745	0.7752	0.7760	0.7767	0.7774
6.0	0.7782	0.7789	0.7796	0.7803	0.7810	0.7818	0.7825	0.7832	0.7839	0.7846
6.1	0.7853	0.7860	0.7868	0.7875	0.7882	0.7889	0.7896	0.7903	0.7910	0.7917
6.2	0.7924	0.7931	0.7938	0.7945	0.7952	0.7959	0.7966	0.7973	0.7980	0.7987
6.3	0.7993	0.8000	0.8007	0.8014	0.8021	0.8028	0.8035	0.8041	0.8048	0.8055
6.4	0.8062	0.8069	0.8075	0.8082	0.8089	0.8096	0.8102	0.8109	0.8116	0.8122
6.5	0.8129	0.8136	0.8142	0.8149	0.8156	0.8162	0.8169	0.8176	0.8182	0.8189
6.6	0.8195	0.8202	0.8209	0.8215	0.8222	0.8228	0.8235	0.8241	0.8248	0.8254
6.7	0.8261	0.8267	0.8274	0.8280	0.8287	0.8293	0.8299	0.8306	0.8312	0.8319
6.8	0.8325	0.8331	0.8338	0.8344	0.8351	0.8357	0.8363	0.8370	0.8376	0.8382
6.9	0.8388	0.8395	0.8401	0.8407	0.8414	0.8420	0.8426	0.8432	0.8439	0.8445
7.0	0.8451	0.8457	0.8463	0.8470	0.8476	0.8482	0.8488	0.8494	0.8500	0.8506
7.1	0.8513	0.8519	0.8525	0.8531	0.8537	0.8543	0.8549	0.8555	0.8561	0.8567
7.2	0.8573	0.8579	0.8585	0.8591	0.8597	0.8603	0.8609	0.8615	0.8621	0.8627
7.3	0.8633	0.8639	0.8645	0.8651	0.8657	0.8663	0.8669	0.8675	0.8681	0.8686
7.4	0.8692	0.8698	0.8704	0.8710	0.8716	0.8722	0.8727	0.8733	0.8739	0.8745
7.5	0.8751	0.8756	0.8762	0.8768	0.8774	0.8779	0.8785	0.8791	0.8797	0.8802
7.6	0.8808	0.8814	0.8820	0.8825	0.8831	0.8837	0.8842	0.8848	0.8854	0.8859
7.7	0.8865	0.8871	0.8876	0.8882	0.8887	0.8893	0.8899	0.8904	0.8910	0.8915
7.8	0.8921	0.8927	0.8932	0.8938	0.8943	0.8949	0.8954	0.8960	0.8965	0.8971
7.9	0.8976	0.8982	0.8987	0.8993	0.8998	0.9004	0.9009	0.9015	0.9020	0.9025
8.0	0.9031	0.9036	0.9042	0.9047	0.9053	0.9058	0.9063	0.9069	0.9074	0.9079
8.1	0.9085	0.9090	0.9096	0.9101	0.9106	0.9112	0.9117	0.9122	0.9128	0.9133
8.2	0.9138	0.9143	0.9149	0.9154	0.9159	0.9165	0.9170	0.9175	0.9180	0.9186
8.3	0.9191	0.9196	0.9201	0.9206	0.9212	0.9217	0.9222	0.9227	0.9232	0.9238
8.4	0.9243	0.9248	0.9253	0.9258	0.9263	0.9269	0.9274	0.9279	0.9284	0.9289
8.5	0.9294	0.9299	0.9304	0.9309	0.9315	0.9320	0.9325	0.9330	0.9335	0.9340
8.6	0.9345	0.9350	0.9355	0.9360	0.9365	0.9370	0.9375	0.9380	0.9385	0.9390
8.7	0.9395	0.9400	0.9405	0.9410	0.9415	0.9420	0.9425	0.9430	0.9435	0.9440
8.8	0.9445	0.9450	0.9455	0.9460	0.9465	0.9469	0.9474	0.9479	0.9484	0.9489
8.9	0.9494	0.9499	0.9504	0.9509	0.9513	0.9518	0.9523	0.9528	0.9533	0.9538
9.0	0.9542	0.9547	0.9552	0.9557	0.9562	0.9566	0.9571	0.9576	0.9581	0.9586
9.1	0.9590	0.9595	0.9600	0.9605	0.9609	0.9614	0.9619	0.9624	0.9628	0.9633
9.2	0.9638	0.9643	0.9647	0.9652	0.9657	0.9661	0.9666	0.9671	0.9675	0.9680
9.3	0.9685	0.9689	0.9694	0.9699	0.9703	0.9708	0.9713	0.9717	0.9722	0.9727
9.4	0.9731	0.9736	0.9741	0.9745	0.9750	0.9754	0.9759	0.9763	0.9768	0.9773
9.5	0.9777	0.9782	0.9786	0.9791	0.9795	0.9800	0.9805	0.9809	0.9814	0.9818
9.6	0.9823	0.9827	0.9832	0.9836	0.9841	0.9845	0.9850	0.9854	0.9859	0.9863
9.7	0.9868	0.9872	0.9877	0.9881	0.9886	0.9890	0.9894	0.9899	0.9903	0.9908
9.8	0.9912	0.9917	0.9921	0.9926	0.9930	0.9934	0.9939	0.9943	0.9948	0.9952
9.9	0.9956	0.9961	0.9965	0.9969	0.9974	0.9978	0.9983	0.9987	0.9991	0.9996

Example 1: $\log 3.17 = 0.5011$

Example 2: $\log 3170 = \log (3.17 \times 10^3)$
$= \log 3.17 + \log 10^3$
$= 0.5011 + 3$
$= 3.5011$

Example 3: $\log 0.0317 = \log (3.17 \times 10^{-2})$
$= \log 3.17 + \log 10^{-2}$
$= 0.5011 + (-2)$
$= -1.4989$

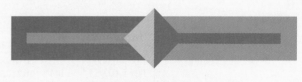

Natural Logarithms

x	0	1	2	3	4	5	6	7	8	9
1.0	0.0000	0.0100	0.0198	0.0296	0.0392	0.0488	0.0583	0.0677	0.0770	0.0862
1.1	0.0953	0.1044	0.1133	0.1222	0.1310	0.1398	0.1484	0.1570	0.1655	0.1740
1.2	0.1823	0.1906	0.1989	0.2070	0.2151	0.2231	0.2311	0.2390	0.2469	0.2546
1.3	0.2624	0.2700	0.2776	0.2852	0.2927	0.3001	0.3075	0.3148	0.3221	0.3293
1.4	0.3365	0.3436	0.3507	0.3577	0.3646	0.3716	0.3784	0.3853	0.3920	0.3988
1.5	0.4055	0.4121	0.4187	0.4253	0.4318	0.4383	0.4447	0.4511	0.4574	0.4637
1.6	0.4700	0.4762	0.4824	0.4886	0.4947	0.5008	0.5068	0.5128	0.5188	0.5247
1.7	0.5306	0.5365	0.5423	0.5481	0.5539	0.5596	0.5653	0.5710	0.5766	0.5822
1.8	0.5878	0.5933	0.5988	0.6043	0.6098	0.6152	0.6206	0.6259	0.6313	0.6366
1.9	0.6419	0.6471	0.6523	0.6575	0.6627	0.6678	0.6729	0.6780	0.6831	0.6881
2.0	0.6931	0.6981	0.7031	0.7080	0.7129	0.7178	0.7227	0.7275	0.7324	0.7372
2.1	0.7419	0.7467	0.7514	0.7561	0.7608	0.7655	0.7701	0.7747	0.7793	0.7839
2.2	0.7885	0.7930	0.7975	0.8020	0.8065	0.8109	0.8154	0.8198	0.8242	0.8286
2.3	0.8329	0.8372	0.8416	0.8459	0.8502	0.8544	0.8587	0.8629	0.8671	0.8713
2.4	0.8755	0.8796	0.8838	0.8879	0.8920	0.8961	0.9002	0.9042	0.9083	0.9123
2.5	0.9163	0.9203	0.9243	0.9282	0.9322	0.9361	0.9400	0.9439	0.9478	0.9517
2.6	0.9555	0.9594	0.9632	0.9670	0.9708	0.9746	0.9783	0.9821	0.9858	0.9895
2.7	0.9933	0.9969	1.0006	1.0043	1.0080	1.0116	1.0152	1.0188	1.0225	1.0260
2.8	1.0296	1.0332	1.0367	1.0403	1.0438	1.0473	1.0508	1.0543	1.0578	1.0613
2.9	1.0647	1.0682	1.0716	1.0750	1.0784	1.0818	1.0852	1.0886	1.0919	1.0953
3.0	1.0986	1.1019	1.1053	1.1086	1.1119	1.1151	1.1184	1.1217	1.1249	1.1282
3.1	1.1314	1.1346	1.1378	1.1410	1.1442	1.1474	1.1506	1.1537	1.1569	1.1600
3.2	1.1632	1.1663	1.1694	1.1725	1.1756	1.1787	1.1817	1.1848	1.1878	1.1909
3.3	1.1939	1.1969	1.2000	1.2030	1.2060	1.2090	1.2119	1.2149	1.2179	1.2208
3.4	1.2238	1.2267	1.2296	1.2326	1.2355	1.2384	1.2413	1.2442	1.2470	1.2499
3.5	1.2528	1.2556	1.2585	1.2613	1.2641	1.2669	1.2698	1.2726	1.2754	1.2782
3.6	1.2809	1.2837	1.2865	1.2892	1.2920	1.2947	1.2975	1.3002	1.3029	1.3056
3.7	1.3083	1.3110	1.3137	1.3164	1.3191	1.3218	1.3244	1.3271	1.3297	1.3324
3.8	1.3350	1.3376	1.3403	1.3429	1.3455	1.3481	1.3507	1.3533	1.3558	1.3584
3.9	1.3610	1.3635	1.3661	1.3686	1.3712	1.3737	1.3762	1.3788	1.3813	1.3838
4.0	1.3863	1.3888	1.3913	1.3938	1.3962	1.3987	1.4012	1.4036	1.4061	1.4085
4.1	1.4110	1.4134	1.4159	1.4183	1.4207	1.4231	1.4255	1.4279	1.4303	1.4327
4.2	1.4351	1.4375	1.4398	1.4422	1.4446	1.4469	1.4493	1.4516	1.4540	1.4563
4.3	1.4586	1.4609	1.4633	1.4656	1.4679	1.4702	1.4725	1.4748	1.4770	1.4793
4.4	1.4816	1.4839	1.4861	1.4884	1.4907	1.4929	1.4951	1.4974	1.4996	1.5019
4.5	1.5041	1.5063	1.5085	1.5107	1.5129	1.5151	1.5173	1.5195	1.5217	1.5239
4.6	1.5261	1.5282	1.5304	1.5326	1.5347	1.5369	1.5390	1.5412	1.5433	1.5454
4.7	1.5476	1.5497	1.5518	1.5539	1.5560	1.5581	1.5602	1.5623	1.5644	1.5665
4.8	1.5686	1.5707	1.5728	1.5748	1.5769	1.5790	1.5810	1.5831	1.5851	1.5872
4.9	1.5892	1.5913	1.5933	1.5953	1.5974	1.5994	1.6014	1.6034	1.6054	1.6074
5.0	1.6094	1.6114	1.6134	1.6154	1.6174	1.6194	1.6214	1.6233	1.6253	1.6273
5.1	1.6292	1.6312	1.6332	1.6351	1.6371	1.6390	1.6409	1.6429	1.6448	1.6467
5.2	1.6487	1.6506	1.6525	1.6544	1.6563	1.6582	1.6601	1.6620	1.6639	1.6658
5.3	1.6677	1.6696	1.6715	1.6734	1.6752	1.6771	1.6790	1.6808	1.6827	1.6845
5.4	1.6864	1.6882	1.6901	1.6919	1.6938	1.6956	1.6974	1.6993	1.7011	1.7029

$\ln 10 = 2.3026$	$7 \ln 10 = 16.1181$
$2 \ln 10 = 4.6052$	$8 \ln 10 = 18.4207$
$3 \ln 10 = 6.9078$	$9 \ln 10 = 20.7233$
$4 \ln 10 = 9.2103$	$10 \ln 10 = 23.0259$
$5 \ln 10 = 11.5129$	$11 \ln 10 = 25.3284$
$6 \ln 10 = 13.8155$	$12 \ln 10 = 27.6310$

x	0	1	2	3	4	5	6	7	8	9
5.5	1.7047	1.7066	1.7084	1.7102	1.7120	1.7138	1.7156	1.7174	1.7192	1.7210
5.6	1.7228	1.7246	1.7263	1.7281	1.7299	1.7317	1.7334	1.7352	1.7370	1.7387
5.7	1.7405	1.7422	1.7440	1.7457	1.7475	1.7492	1.7509	1.7527	1.7544	1.7561
5.8	1.7579	1.7596	1.7613	1.7630	1.7647	1.7664	1.7681	1.7699	1.7716	1.7733
5.9	1.7750	1.7766	1.7783	1.7800	1.7817	1.7834	1.7851	1.7867	1.7884	1.7901
6.0	1.7918	1.7934	1.7951	1.7967	1.7984	1.8001	1.8017	1.8034	1.8050	1.8066
6.1	1.8083	1.8099	1.8116	1.8132	1.8148	1.8165	1.8181	1.8197	1.8213	1.8229
6.2	1.8245	1.8262	1.8278	1.8294	1.8310	1.8326	1.8342	1.8358	1.8374	1.8390
6.3	1.8405	1.8421	1.8437	1.8453	1.8469	1.8485	1.8500	1.8516	1.8532	1.8547
6.4	1.8563	1.8579	1.8594	1.8610	1.8625	1.8641	1.8656	1.8672	1.8687	1.8703
6.5	1.8718	1.8733	1.8749	1.8764	1.8779	1.8795	1.8810	1.8825	1.8840	1.8856
6.6	1.8871	1.8886	1.8901	1.8916	1.8931	1.8946	1.8961	1.8976	1.8991	1.9006
6.7	1.9021	1.9036	1.9051	1.9066	1.9081	1.9095	1.9110	1.9125	1.9140	1.9155
6.8	1.9169	1.9184	1.9199	1.9213	1.9228	1.9242	1.9257	1.9272	1.9286	1.9301
6.9	1.9315	1.9330	1.9344	1.9359	1.9373	1.9387	1.9402	1.9416	1.9430	1.9445
7.0	1.9459	1.9473	1.9488	1.9502	1.9516	1.9530	1.9544	1.9559	1.9573	1.9587
7.1	1.9601	1.9615	1.9629	1.9643	1.9657	1.9671	1.9685	1.9699	1.9713	1.9727
7.2	1.9741	1.9755	1.9769	1.9782	1.9796	1.9810	1.9824	1.9838	1.9851	1.9865
7.3	1.9879	1.9892	1.9906	1.9920	1.9933	1.9947	1.9961	1.9974	1.9988	2.0001
7.4	2.0015	2.0028	2.0042	2.0055	2.0069	2.0082	2.0096	2.0109	2.0122	2.0136
7.5	2.0149	2.0162	2.0176	2.0189	2.0202	2.0215	2.0229	2.0242	2.0255	2.0268
7.6	2.0281	2.0295	2.0308	2.0321	2.0334	2.0347	2.0360	2.0373	2.0386	2.0399
7.7	2.0412	2.0425	2.0438	2.0451	2.0464	2.0477	2.0490	2.0503	2.0516	2.0528
7.8	2.0541	2.0554	2.0567	2.0580	2.0592	2.0605	2.0618	2.0631	2.0643	2.0656
7.9	2.0669	2.0681	2.0694	2.0707	2.0719	2.0732	2.0744	2.0757	2.0769	2.0782
8.0	2.0794	2.0807	2.0819	2.0832	2.0844	2.0857	2.0869	2.0882	2.0894	2.0906
8.1	2.0919	2.0931	2.0943	2.0956	2.0968	2.0980	2.0992	2.1005	2.1017	2.1029
8.2	2.1041	2.1054	2.1066	2.1078	2.1090	2.1102	2.1114	2.1126	2.1138	2.1150
8.3	2.1163	2.1175	2.1187	2.1199	2.1211	2.1223	2.1235	2.1247	2.1258	2.1270
8.4	2.1282	2.1294	2.1306	2.1318	2.1330	2.1342	2.1353	2.1365	2.1377	2.1389
8.5	2.1401	2.1412	2.1424	2.1436	2.1448	2.1459	2.1471	2.1483	2.1494	2.1506
8.6	2.1518	2.1529	2.1541	2.1552	2.1564	2.1576	2.1587	2.1599	2.1610	2.1622
8.7	2.1633	2.1645	2.1656	2.1668	2.1679	2.1691	2.1702	2.1713	2.1725	2.1736
8.8	2.1748	2.1759	2.1770	2.1782	2.1793	2.1804	2.1815	2.1827	2.1838	2.1849
8.9	2.1861	2.1872	2.1883	2.1894	2.1905	2.1917	2.1928	2.1939	2.1950	2.1961
9.0	2.1972	2.1983	2.1994	2.2006	2.2017	2.2028	2.2039	2.2050	2.2061	2.2072
9.1	2.2083	2.2094	2.2105	2.2116	2.2127	2.2138	2.2148	2.2159	2.2170	2.2181
9.2	2.2192	2.2203	2.2214	2.2225	2.2235	2.2246	2.2257	2.2268	2.2279	2.2289
9.3	2.2300	2.2311	2.2322	2.2332	2.2343	2.2354	2.2364	2.2375	2.2386	2.2396
9.4	2.2407	2.2418	2.2428	2.2439	2.2450	2.2460	2.2471	2.2481	2.2492	2.2502
9.5	2.2513	2.2523	2.2534	2.2544	2.2555	2.2565	2.2576	2.2586	2.2597	2.2607
9.6	2.2618	2.2628	2.2638	2.2649	2.2659	2.2670	2.2680	2.2690	2.2701	2.2711
9.7	2.2721	2.2732	2.2742	2.2752	2.2762	2.2773	2.2783	2.2793	2.2803	2.2814
9.8	2.2824	2.2834	2.2844	2.2854	2.2865	2.2875	2.2885	2.2895	2.2905	2.2915
9.9	2.2925	2.2935	2.2946	2.2956	2.2966	2.2976	2.2986	2.2996	2.3006	2.3016

Example 1: $\ln 6.06 = 1.8017$

Example 2: $\ln 606 = \ln 6.06 \times 10^2$
$= \ln 6.06 + 2 \ln 10$
$= 1.8017 + 4.6052$
$= 6.4069$

Example 3: $\ln 0.000606 = \ln 6.06 \times 10^{-4}$
$= \ln 6.06 - 4 \ln 10$
$= 1.8017 - 9.2103$
$= -7.4086$

INDEX

Abbreviations, table of, 685

Abel, Niels Henrik, 453n

Abscissa, 306

Absolute value, 12

Absolute-value equation, 87

Absolute-value function, 406

Absolute-value inequality, 89, 338

Addition:

   of complex numbers, 251

   of functions, 362

   of polynomials, 118

   properties of 21, 48, 74

   of radical expressions, 234

   of rational expressions, 176

Addition property of equality, 48

Addition property of inequality, 74

Additive identity, 21

Additive inverse, 11, 21

Ahmes papyrus, 47

   (*See also* Rhind papyrus)

Algebraic expressions, 19

   evaluating, 37, 118

   simplifying, 23, 32, 46

al-Karkhî, 282n

And, in compound inequalities, 82

Antilogarithms, 501

Appollonius of Perga, 402n

Approximating real roots, 453

Archimedes, 56n, 156n, 402n

Argument of a function, 501

Arithmetic sequence, 589

   common difference, 589

   general term, 589

Arithmetic series, 591

   sum, 591

Associative properties, 21

Asymptotes, 458

   horizontal, 458, 477

   of a hyperbola, 398

   slant, 460

   vertical, 458

Augmented matrix, 551

Axis of a parabola, 372

Babylonians, 106n, 155n, 215n, 269n, 285n

Base:

   of an exponential function, 476

   of a logarithmic function, 489

   of a power, 35, 100

Bell, Alexander Graham, 495n

Bernoulli, Jacques, 604n, 612n

Binomial coefficient, 603

Binomial theorem, 604

   rth term, 605

Binomials, 117

   expanding, 601

   factoring, 144, 146

   multiplying, 124

   special products, 126

   squaring, 126

Bounded region, 571

Boyle, Robert, 346n

Brahe, Tycho, 222n

Branches of a hyperbola, 398

Briggs, Henry, 500n

Cantor, Georg, 2n

Cardan, Geronimo, 453n

Cartesian coordinate system, 307

Cauchy, Augustin-Louis, 551n

Cavanaugh, Joseph, 345

Cayley, Arthur, 551n

Celsius, Anders, 59n

Center:

   of a circle, 389

   of an ellipse, 392

   of a hyperbola, 398

Change-of-base formula, 505

Chinese, 11n

Christina, Queen of Sweden, 307n

Chudnovsky, David, 56n

Chudnovsky, Gregory, 56n

Circle, 389

   center, 389

   general form, 390

   graph, 389

   radius, 389

   standard form, 389

Circular permutation, 615

Clark, A. J., 170

Closed interval, 76

Closed half-plane, 337

Coefficient, 46, 117

   binomial, 603

   numerical, 46

Coefficient matrix, 552

Collinear points, 308

Combination, 616

   theorem, 617

Combining like terms, 46

Common difference, 589

Common logarithms, 500

   table of, 695

Common ratio, 594

Commutative properties, 21

Compass, 391

Composition of functions, 364

Compound inequality, 82

Compound probability, 625

Complement of an event, 629

Completing the square, 263

Complex conjugates, 253

Complex fractions, 184

   secondary fractions of, 184

   simplifying, 184

Complex numbers, 250

   adding, 251

   conjugates, 253

   dividing, 253

   imaginary part of, 250

   multiplying, 252

   standard form of, 250

   subtracting, 251

   real part of, 250

Conditional equation, 51

Conic section, 402

   degenerate, 402

Conjugate pairs theorem, 446

Conjugates, 126, 241, 253

   multiplying, 126

Connective word "and," 82

Connective word "or," 83

Consecutive integers, 61

Consistent system, 522

Constant, 3

Constant function, 369, 415

Constant of proportionality, 341

Constant of variation, 341

Constraints, 570

Continuous curve, 430

Continuously compounded interest, 479

Contradiction, 52

Coordinate system, 11, 306, 535

Coordinates of a point, 11, 306, 535

Counting number, 6

Cramer, Gabriel, 547n

Cramer's rule, 546

Cross-product rule, 190

Cube of a number, 35

Cube root, 214

De Morgan, Augustus, 21n
Decreasing function, 415, 477
Dedekind, Richard, 6n
Degenerate conic, 402
Degree:
  of a monomial, 117, 118
  of a polynomial, 117, 118
  of a polynomial function, 428
Dependent system, 522, 538
Dependent variable, 355
Descartes, René, 35n, 60n, 249n, 307n
Descending order, 118
Determinant, 542
  elements, 542
  expansion by minors, 543
  minor, 543
  second order, 542
  sign array, 544, 545
  solving a system by, 546
  third order, 543
  3×3, 543
  2×2, 542
  value, 542, 543
Difference, 28
Difference quotient, 366
Difference of two cubes, 146
Difference of two squares, 144
Diophantus, 520n
Direct variation, 341
Directed number, 11
Directrix of a parabola, 371
Dirichlet, Lejeune, 361n
Discriminant, 275
Distance formula, 307
Distinguishable permutation, 616
Distributive property, 21

Division:
  of complex numbers, 253
  of functions, 362
  by a monomial, 200
  of polynomials, 200
  of radical expressions, 230
  of rational expressions, 173
  of real numbers, 30
  synthetic, 438
  by zero, 6
Divisor, 130
Domain, 354, 484
Doppler, Christian Johann, 171n
Double inequality, 15
  graph, 16
  solving, 78
Double negative rule, 12

e, 478
  powers of, 691
Element of a determinant, 542
Element of a set, 2
Ellipse, 392
  foci, 392
  graph, 392
  standard form, 392
Empty set, 2
Equal sets, 2
Equality, properties of, 20, 48
Equality symbol, 11
Equals sign, 11
Equations, 47
  absolute-value, 87
  conditional, 51
  contradiction, 52
  equivalent, 48
  exponential, 507
  first-degree, 47

Equations *(cont'd)*

  higher-degree, 154

  identity, 52

  leading to quadratic equations, 280

  logarithmic, 509

  linear, 47, 313

  polynomial, 430

  quadratic, 152

  quadratic in form, 282

  radical, 244, 281

  with rational expressions, 189, 280

  second-degree, 151

  solution of, 48, 313, 535

  solution set, 48

  solving for a specified variable, 54, 195, 286

  system, 520, 535, 559

Equivalent equations, 48

Equivalent inequalities, 74

Euclid, $31n$, $190n$, $402n$

Euler, Leonhard, $249n$, $361n$, $478n$, $551n$, $607n$

Evaluating an expression, 37, 118

Even function, 436

Events, 621

  complement of, 629

  independent, 627

  mutually exclusive, 628

  probability of, 621

  odds of, 630

Expanding a binomial, 601

Expanding a logarithm, 497

Expansion by minors, 543

Exponential equation, 507

Exponential expression, 100

Exponential functions, 476

  base of, 476

  graphing, 476

  one-to-one property of, 491, 507

Exponential notation, 35

Exponents, 35, 100

  base, 35, 100

  negative, 101

  rational, 219

  rules of, 109, 221, 476

  zero, 101

Expression:

  algebraic, 19

  exponential, 100

  radical, 224

  rational, 166

Extraneous solution, 192, 244

Factor, 35, 46, 130, 450

Factor theorem, 440

Factorial notation, 602, 614

Factoring, 130

  difference of two cubes, 146

  difference of two squares, 144

  greatest common factor, 131

  by grouping, 133

  perfect-square trinomials, 145

  trinomials, 136

  strategy, 149

  sum of two cubes, 146

  test for factorability, 138, 276

Fahrenheit, Gabriel, $59n$

Feasible region, 569

  vertex of, 569

Fermat, Pierre de, $607n$, $620n$

Ferrari, Ludovico, $453n$

Ferro, Scipione del, $453n$

Fibonacci, Leonardo, $588n$

Finite sequence, 584

Finite series, 586

Finite set, 2

First-degree equation, 47

First-degree inequality, 73

Flipping a graph, 382, 429

Foci of an ellipse, 392

Foci of a hyperbola, 397

Focus of a parabola, 371

FOIL, 125

Fontana, Niccolo, 453n

Forms of a linear equation, 330

Formulas, 54, 194

  list of geometric, 689

  solving for a variable, 54, 195, 286

Fraction:

  complex, 184

  secondary, 184

  simple, 184

Fractional exponent, 219

Functions, 354

  absolute-value, 406

  composition of, 364

  constant, 369, 415

  decreasing, 415, 477

  domain of, 354

  even, 436

  exponential, 476

  greatest-integer, 410

  increasing, 415, 477

  inverse, 485

  linear, 370

  logarithmic, 489

  notation, 360

  odd, 436

  one-to-one, 483

  operations on, 362

  piecewise-defined, 411

  polynomial, 428

  quadratic, 370

  range of, 354

  rational, 456

  sequence, 584

Functions *(cont'd)*

  square-root, 408

  step, 410

  value of, 350

  vertical line test for, 356

  zeros of, 449, 450

Fundamental principle of counting, 612

Fundamental principle of rational expressions, 167

Fundamental rectangle, 398

Fundamental theorem of algebra, 450

Galilei, Galileo, 156n, 267n

Gauss, Carl Friedrich, 156n, 450n, 554n

Gauss-Jordan method, 554

Gaussian elimination, 554

General form of a circle, 390

General term of a sequence, 584

General variation equation, 341

Geometric formulas, table of, 689

Geometric sequence, 594

  common ratio, 594

  general term, 594

Geometric series, 595

  infinite, 598

  sum, 596, 599

Girard, Albert, 49n

Graphing:

  absolute-value functions, 406

  circles, 389

  constant functions, 369

  double inequalities, 16

  ellipses, 392

  exponential functions, 476

  by flipping, 382, 429

  greatest-integer functions, 410

  horizontal lines, 316

  hyperbolas, 398

  inequalities, 16, 336, 564

  inverse functions, 486

Graphing *(cont'd)*

   linear equations, 313

   linear functions, 370

   lines using a point and the slope, 323

   logarithmic functions, 491, 504

   ordered pairs, 307

   numbers, 11

   parabolas, 370, 379

   piecewise-defined functions, 411

   polynomial functions, 428

   quadratic functions, 370

   rational functions, 456

   by shifting, 381, 429

   by stretching, 382, 429

   square-root functions, 408

   systems of equations, 521, 559

   systems of inequalities, 566

   vertical lines, 316

Graphing calculator exercises:

   absolute value, 19

   changing the scale, 313

   checking:

      absolute-value equations, 93

      absolute-value inequalities, 93

      contradictions, 54

      cubic equations, 158

      equations with rational expressions, 194, 285

      linear equations, 54

      identities, 54

      inequalities, 82, 93

      irrational solutions, 267, 285

      quadratic equations, 158

      quadratic form equations, 285

      radical equations, 249, 285

      systems of equations, 524, 542

   combining functions, 369

   connected mode, 414

   displaying a sequence, 593, 598

Graphing calculator exercises *(cont'd)*

   displaying a series, 598

   dot mode, 414

   drawing a line segment, 320

   drawing a line using the cursor, 325

   drawing a point, 312

   evaluating a determinant, 546

   evaluating expressions, 54, 60, 123, 443

   graphing:

      circles, 396

      conic sections, 406

      constant functions, 379

      ellipses, 396

      exponential functions, 482

      inequalities, 341, 568

      inverse functions, 489

      linear functions, 379

      lines, 320

      logarithmic functions, 495, 507

      miscellaneous equations, 419

      miscellaneous functions, 360

      parabolas, 388

      piecewise-defined functions, 414

      polynomial functions, 436

      quadratic functions, 379

      rational functions, 467

      relations, 388, 396

      sequences, 588

      special functions, 414

      systems of inequalities, 568

      variation equations, 347

      verifying equations by, 607

   panning, 335

   parentheses, 39

   pixels, 312

   programming, 274

   range values, 312

   reading the coordinates of a point, 312

Graphing calculator exercises *(cont'd)*

   right and left behavior of a graph, 437

   rounding a number, 10

   row operations, 559

   scientific notation mode, 116

   shading a graph, 341

   solving an equation by graphing, 456, 513

   solving a system by graphing, 524, 564

   solving a system by determinants, 551

   solving a system by matrices, 559

   squaring a graph, 334, 397

   storing a value, 26

   subtraction versus opposite of, 34

   testing a statement, 10, 60

   tracing a graph, 335

   turning points of a graph, 437

   viewing rectangle, 312

   x-intercepts of a graph, 437

   zooming in, 335

   zooming out, 436

Greater than, 15

Greatest common factor, 131

Greatest-integer function, 410

Gregory, Duncan, $21n$

Grouping, factoring by, 133

Half-open interval, 76

Half-plane, 336

   closed, 337

   open, 337

Harriot, Thomas, $13n$

Heron of Alexandria, $219n$

Higher-degree equations, 154

Higher-degree inequalities, 296

Hooke, Robert, $346n$

Horizontal asymptote, 458, 477

Horizontal line, 316, 369

Horizontal line test, 483

Horizontal parabola, 384

Hyperbola, 397

   asymptotes, 398

   branches, 398

   foci, 397

   fundamental rectangle, 398

   graph, 398

   standard form, 398

Hypotenuse, 285

i, 249

   powers of, 254

Identity, 52

Identity element of addition, 21

Identity element of multiplication, 21

Identity properties, 21

Imaginary number, 249

Imaginary part of a complex number, 250

Imaginary unit, 249

Inconsistent system, 522, 538

Increasing function, 415, 477

Independent events, 627

Independent system, 522

Independent variable, 355

Index of a radical, 214

Index of summation, 586

Inequalities, 13, 73

   absolute-value, 89, 338

   compound, 82

   double, 15

   equivalent, 74

   first-degree, 73

   graphing, 16, 336, 564

   higher-degree, 296

   intersection of, 82, 337, 566

   linear, 73, 335

   nonlinear, 292

   quadratic, 292

Inequalities *(cont'd)*

  rational, 296

  second-degree, 564

  solution of, 74

  solution set, 74

  symbols for, 15

  systems of, 566

  union of, 83, 338

Inequality, properties of, 20, 74

Infinite geometric series, 599

Infinite sequence, 584

Infinte series, 586

Infinite set, 2

Integer, 6

Intercepts of a line, 314

Intersection of sets, 5, 82, 337, 566

Interval, 75

  closed, 76

  half-open, 76

  open, 76

Interval notation, 75

Inverse:

  additive, 21

  of a function, 485

    domain, 484

    graph, 486

    range, 484

  multiplicative, 21

  properties, 21

  variation, 343

Irrational number, 6

  decimal form, 7

Joint variation, 344

Jordan, Camille, 554

Kepler, Johann, 222*n*

Lambert, Johann, 56*n*

Leading coefficient, 117, 428

Least common denominator, 50, 178

  steps in finding, 178

Legs of a right triangle, 285

Leibniz, Gottfried, 542*n*

Less than, 14

Like radicals, 234

Like terms, 46, 119

  combining, 46

Line, 314

  horizontal, 316, 369

  vertical, 316

Linear equations, 47, 313

  forms of, 330

  graphing, 313

  solving, 47

  standard form of, 313

  in three variables, 535

  in two variables, 313

Linear factorization theorem, 450

Linear function, 370

Linear inequalities, 73

  graphing, 336

  intersection of, 337

  solving, 73

  in two variables, 335

  union of, 338

Linear programming, 568

  constraints, 570

  feasible region, 569

  objective function, 570

  optimum value, 570

Linear systems of equations, 520

  choosing a method for solving, 528

  consistent, 522

Linear systems of equations *(cont'd)*
  dependent, 522, 538
  inconsistent, 522, 538
  independent, 522
  solution of, 520
  solving by addition, 526
  solving by determinants, 546
  solving by graphing, 521
  solving by matrices, 551
  solving by substitution, 525
  in three variables, 535
  in two variables, 520
Logarithmic equation, 509
Logarithmic functions, 489
  antilogarithm, 501
  base of, 489
  change-of-base formula for, 505
  common, 500
  expanding, 497
  graphing, 491, 504
  natural, 503
  one-to-one property of, 510
  power rule for, 497
  product rule for, 496
  properties of, 490, 495, 497
  quotient rule for, 496
  tables for, 695, 699
Long division of polynomials, 201

Maclaurin, Colin, 547*n*
Magnitude of a number, 12
Mathematical induction, 607
Matrix, 551
  augmented, 551
  coefficient, 552
  main diagonal of, 551
  row operations on, 552
  solving a system by, 551
  square, 551

Member of a set, 2
Méré, Chevalier de, 620*n*
Midpoint formula, 309
Minor of a determinant, 543
Monomial, 117
  coefficient, 117
  degree, 117, 118
Multiplication:
  of binomials, 124
  of complex numbers, 252
  of functions, 362
  of polynomials, 123
  properties of, 21
  of radical expressions, 230, 237
  of rational expressions, 171
  of real numbers, 29
Multiplication property of equality, 48
Multiplication property of inequality, 74
Multiplication property of zero, 24
Multiplicative identity, 21
Multiplicative inverse, 21
Multiplicity of a solution, 441
Mutually exclusive events, 628

Napier, John, 489*n*, 503*n*
Natural logarithms, 503
  table of, 699
Natural number, 6
Negative exponent, 101
Negative exponent shortcuts, 102
Negative number, 11
Newton, Isaac, 156*n*, 345, 455*n*
Nonlinear inequality, 292
Nonlinear system of equations, 559
Notation:
  exponential, 35
  factorial, 602, 614

Notation *(cont'd)*
  function, 360
  interval, 75
  polynomial, 118
  scientific, 112
  set-builder, 3
  summation, 586
nth root, 214
nth term of a sequence, 584
Null set, 2
Number line, 11
Numbers:
  complex, 250
  counting, 6
  directed, 11
  graphing, 11
  imaginary, 249
  integer, 6
  irrational, 6
  natural, 6
  negative, 11
  positive, 11
  prime, 131
  pure imaginary, 250
  rational, 6
  real, 6
  signed, 11
  whole, 2, 6
Numerical coefficient, 46

Objective function, 570
  optimum value of, 570
Odd function, 436
Odds, 630
  converting to probability, 630
One-to-one function, 483
  horizontal line test for, 483
One-to-one property of the exponential function, 491, 507

One-to-one property of the logarithmic function, 510
Open half-plane, 337
Open interval, 76
Opposite of a number, 11
Or, in compound inequalities, 83
Order of operations, 36
Ordered pair, 306
Ordered triple, 535
Ordinate, 306
Oresme, Nicole, 219n
Origin, 306

Parabola, 371
  axis of symmetry, 372
  directrix, 371
  focus, 371
  graph, 370, 379
  horizontal, 384
  standard form, 383, 386
  vertex, 372
  vertical, 383
Parallel lines, 330
Pascal, Blaise, 602n, 620n
Pascal's triangle, 602
Peacock, George, 21n
Peano, Giuseppe, 607n
Perfect square, 7
Perfect-square trinomial, 145
Permutation, 613
  circular, 615
  distinguishable, 616
  theorem, 615
Perpendicular lines, 330
pH of a solution, 502
π, 7, 56
Piecewise-defined function, 411
Point-slope form, 326

Polynomial, 117

  adding, 118

  binomial, 117

  constant term, 117

  degree, 117, 118

  descending order, 118

  dividing, 200

  evaluating, 118

  leading coefficient, 117

  monomial, 117

  multiplying, 123

  notation, 118

  prime, 133

  prime factored form, 131

  subtracting, 120

  trinomial, 117

Polynomial equation, 430

  number of solutions, 441

Polynomial functions, 428

  constant term, 428

  degree, 428

  domain, 430

  graph, 428

  leading coefficient, 428

  left behavior, 431

  right behavior, 431

  turning points, 434

Positive integer, 6

Positive number, 11

Power, 35

Power rule for equality, 244

Power rule for logarithms, 497

Power rules for exponents, 107

Power-to-a-power rule, 107

Powers of e, 691

Powers of i, 254

Prime factored form, 131

Prime number, 131

Prime polynomial, 133

Principal nth root, 214

Principal square root, 214

Probability, 621

  of the complement of an event, 629

  compound, 625

  converting to odds, 630

  of independent events, 627

  of mutually exclusive events, 628

  properties of, 623

Product, 29

Product rule for exponents, 100

Product rule for logarithms, 496

Product rule for radicals, 225, 231, 250

Product of solutions of a quadratic equation, 278

Product of two conjugates, 126

Product-to-a-power rule, 108

Properties of addition and multiplication, 21

Properties of equality, 20, 48

Properties of exponents, 109, 221, 476

Properties of inequality, 20, 74

Properties of logarithms, 490, 495, 497

Properties of radicals, 225, 226, 231

Proportional to, 341

Proportions, 190

  cross-product rule for, 190

Pure imaginary number, 250

Pythagoras, 7n, 285n

Pythagorean theorem, 285

Quadrant, 306

Quadratic equation, 152

  choosing a method for solving, 274

  determining from the solutions, 278

  discriminant, 275

  number of solutions, 154

  product of solutions, 277

  solving by completing the square, 262, 274

Quadratic equation *(cont'd)*
  solving by factoring, 151, 274
  solving by the quadratic formula, 268, 274
  solving by the square root property, 262, 274
  standard form, 152
  sum of solutions, 277
Quadratic in form, 282
Quadratic formula, 269
  discriminant, 275
Quadratic function, 370
Quadratic inequality, 292
Quotient, 29
Quotient rule for exponents, 103
Quotient rule for logarithms, 496
Quotient rule for radicals, 226, 231
Quotient-to-a-power rule, 108

Radical, index of, 214
Radical equations, 244, 281
Radical expressions, 224
  adding, 234
  dividing, 230
  multiplying, 230, 237
  rationalizing the denominator, 239
  simplifying, 224, 237
  squaring, 231, 237
  subtracting, 234
Radical sign, 214
Radicand, 214
Radius of a circle, 389
Range, 354, 484
Rational exponents, 219
Rational expressions, 166
  adding, 176
  dividing, 173
  equations with, 189, 280
  fundamental principle of, 167
  multiplying, 171

Rational expressions *(cont'd)*
  simplifying, 166
  subtracting, 176
Rational functions, 456
  asymptotes, 458
  domain, 457
  graphing, 456
  horizontal asymptotes, 457
  not in lowest terms, 464
  slant asymptotes, 460, 463
  vertical asymptotes, 457
Rational inequalities, 296
Rational number, 6
  decimal form, 7
Rational root test, 443
Rationalizing the denominator, 237
Real numbers, 6
  operations on, 26
Real part of a complex number, 250
Reciprocal, 21, 173
Recorde, Robert, 19n
Rectangular coordinate system, 306
Reflexive property, 20
Region of feasible solutions, 569
Relation, 354
  domain, 354
  range, 354
Remainder theorem, 439
Rhind papyrus, 47n, 56n, 62n, 190n, 589n
Richter, Charles, 495n
Right angle, 285
Right triangle, 285
Root:
  cube, 214
  nth, 214
  square, 214
Root of an equation, 443, 450
  approximating, 453

Row operations, 552

*r*th term of a binomial expansion, 605

Ruffin, Paolo, 453*n*

Rule for solving work problems, 196

Rules for exponents, 109, 221, 476

Rules for logarithms, 490, 495, 497

Rules for radicals, 225, 226, 231

Sample space, 620

Scientific notation, 112

Secondary fraction, 184

Second-degree equation, 151

Second-degree inequality, 564

Sequence function, 584

Sequences, 584

    arithmetic, 589

    finite, 584

    general term, 584

    geometric, 593

    infinite, 584

    *n*th term, 584

    terms, 584

Series, 586

    arithmetic, 591

    finite, 586

    geometric, 595

    infinite, 586

Set, 2

    element, 2

    empty, 2

    equal, 2

    finite, 2

    infinite, 2

    intersection, 5

    member, 2

    null, 2

    subset, 3

    union, 4

Set-builder notation, 3

Set operation, 4

Shifting a graph, 381, 429

Sigma notation, 586

    (*See also* Summation notation)

Signed number, 11

Signs of a fraction, 31

Simple fraction, 184

Simplified form of a radical expression, 228

Slant asymptote, 460, 463

Slope-intercept form, 328

Slope of a line, 320

Smooth curve, 430

Solution of an equation, 48, 313

    extraneous, 192, 244

    multiplicity of, 441

    number, 154, 441

Solution of an inequality, 74

Solution set, 48, 74

Solve, 48, 74

Solving for a specified variable, 54, 195, 286

Special products, 126

Specific variation equation, 341

Square of a number, 35, 214

Square matrix, 551

Square root, 214

Square root function, 408

Square root property, 262

Squaring a binomial, 126

Standard form:

    of a circle, 389

    of a complex number, 250

    of an ellipse, 392

    of a hyperbola, 398

    of a linear equation, 313

    of a parabola, 383, 386

    of a quadratic equation, 152

Steinmetz, Charles, 249*n*

Step function, 410

Steps in solving word problems, 62

Stretching a graph, 382, 429

Subscripts, 57

Subset, 3

Substitution property, 20

Subtraction:

   of complex numbers, 251

   of functions, 362

   of polynomials, 120

   of radical expressions, 234

   of rational expressions, 176

   of real numbers, 28

Sum, 26

   of an arithmetic series, 591

   of a geometric series, 596

   of an infinite geometric series, 599

Sum of solutions of a quadratic equation, 278

Sum of two cubes, 146

Sum of two squares, 145

Summation notation, 586

   index of, 586

   lower limit of, 586

   upper limit of, 586

Symbol for equality, 11

Symbols, table of, 685

Symbols for inequality, 15

Symmetric property, 20

Symmetry:

   origin, 417

   test for, 417

   x-axis, 416

   y-axis, 416

Synthetic division, 438

System of equations, 520

   consistent, 522

   dependent, 522, 538

System of equations (cont'd)

   inconsistent, 522, 538

   independent, 522

   linear, 520

   nonlinear, 559

   solution of, 520, 535

   in three variables, 535

   in two variables, 520

System of inequalitites, 566

Table:

   of abbreviations, 685

   of common logarithms, 695

   of geometric formulas, 689

   of natural logarithms, 699

   of powers of e, 691

   of powers of i, 254

   of symbols, 685

Tartaglia, Niccolo, 453n

Term, 46, 117

   like, 46, 119

   of a sequence, 584

Test for factorability, 138, 276

Theorem, 24

Three-dimensional coordinate system, 535

Torricelli, Evangelista, 267n

Transitive property of equality, 20

Transitive property of inequality, 20

Translating from English to math, 60

Trichotomy property, 20

Trinomials, 117

   factoring, 136

   perfect-square, 145

Truth diagram, 292

Turning point of a graph, 434

Unbounded region, 571

Union of sets, 4, 83, 338

Variables, 3
  dependent, 355
  independent, 355
  solving for, 54, 195, 286
Variation, 341
  constant of, 341
  direct, 341
  inverse, 343
  joint, 344
Venn, John 4n
Venn diagram, 4
Vertex of a feasible region, 569
Vertex of a parabola, 372
Vertical asymptote, 458
Vertical line, 316
Vertical line test, 356
Vertical parabola, 383
Vertices of an ellipse, 393
Viète, François, 3n

Wallis, John, 101n, 219n
Weierstrass, Karl, 12n
Whole number, 2, 6
Word problems, steps in solving, 62
Work problems, rule for solving, 196

x-axis, 306
x-coordinate, 306
x-intercept:
  of an ellipse, 393
  of a graph, 450
  of a hyperbola, 400
  of a line, 314
  of a parabola, 375, 385
  of a polynomial function, 432
  of a rational function, 462

y-axis, 306
y-coordinate, 306
y-intercept:
  of an ellipse, 393
  of a hyperbola, 400
  of a line, 314
  of a parabola, 375, 385
  of a polynomial function, 432
  of a rational function, 462

Zero, multiplication property of, 24
Zero exponent, 101
Zero factor property, 152
Zeros of a function, 449, 450